The Sun: A Laboratory for Astrophysics

NATO ASI Series

Advanced Science Institutes Series

A Series presenting the results of activities sponsored by the NATO Science Committee, which aims at the dissemination of advanced scientific and technological knowledge, with a view to strengthening links between scientific communities.

The Series is published by an international board of publishers in conjunction with the NATO Scientific Affairs Division

A Life Sciences	Plenum Publishing Corporation
B Physics	London and New York
C Mathematical	Kluwer Academic Publishers
and Physical Sciences	Dordrecht, Boston and London
D Behavioural and Social Sciences	
E Applied Sciences	
F Computer and Systems Sciences	Springer-Verlag
G Ecological Sciences	Berlin, Heidelberg, New York, London,
H Cell Biology	Paris and Tokyo
I Global Environmental Change	

NATO-PCO-DATA BASE

The electronic index to the NATO ASI Series provides full bibliographical references (with keywords and/or abstracts) to more than 30000 contributions from international scientists published in all sections of the NATO ASI Series.
Access to the NATO-PCO-DATA BASE is possible in two ways:

– via online FILE 128 (NATO-PCO-DATA BASE) hosted by ESRIN,
Via Galileo Galilei, I-00044 Frascati, Italy.

– via CD-ROM "NATO-PCO-DATA BASE" with user-friendly retrieval software in English, French and German (© WTV GmbH and DATAWARE Technologies Inc. 1989).

The CD-ROM can be ordered through any member of the Board of Publishers or through NATO-PCO, Overijse, Belgium.

Series C: Mathematical and Physical Sciences - Vol. 373

The Sun: A Laboratory for Astrophysics

edited by

Joan T. Schmelz

Applied Research Corporation/Lockheed,
Landover, MD, U.S.A.

and

John C. Brown

Department of Physics and Astronomy,
University of Glasgow, Glasgow, Scotland

Kluwer Academic Publishers

Dordrecht / Boston / London

Published in cooperation with NATO Scientific Affairs Division

Proceedings of the NATO Advanced Study Institute on
The Sun: A Laboratory for Astrophysics
Crieff, Scotland
June 16–29, 1991

Library of Congress Cataloging-in-Publication Data

NATO Advanced Study Institute "The Sun: a Laboratory for Astrophysics"
(1991 : Crieff, Scotland)
    The sun : a laboratory for astrophysics : proceedings of the NATO
Advanced Study Institute "The Sun: a Laboratory for Astrophysics
held in Crieff, Scotland on 16-29 June 1991 / edited by Joan T.
Schmelz and John C. Brown.
    p.   cm. -- (NATO ASI series. Series C, Mathematical and
physical sciences ; v. 373.)
    ISBN 0-7923-1811-0
    1. Sun--Congresses. 2. Stars--Congresses. 3. Astrophysics-
-Congresses.   I. Schmelz, Joan T.   II. Brown, John C., 1947-   .
III. North Atlantic Treaty Organization.   IV. Title.   V. Series.
QB520.N38   1991
523.7--dc20   92-16845

ISBN 0-7923-1811-0

Published by Kluwer Academic Publishers,
P.O. Box 17, 3300 AA Dordrecht, The Netherlands.

Kluwer Academic Publishers incorporates the publishing programmes of
D. Reidel, Martinus Nijhoff, Dr W. Junk and MTP Press.

Sold and distributed in the U.S.A. and Canada
by Kluwer Academic Publishers,
101 Philip Drive, Norwell, MA 02061, U.S.A.

In all other countries, sold and distributed
by Kluwer Academic Publishers Group,
P.O. Box 322, 3300 AH Dordrecht, The Netherlands.

Printed on acid-free paper

TABLE OF CONTENTS

PART III
SOLAR INSTRUMENTATION

PART IV
SOLAR AND STELLAR ACTIVITY

PREFACE

As in the days following *Skylab*, solar physics came to the end of an era when the *Solar Maximum Mission* re-entered the earth's atmosphere in December 1989. The 1980s had been a pioneering decade not only in space- and ground-based studies of the solar atmosphere (*Solar Maximum Mission, Hinotori*, VLA, Big Bear, Nançay, *etc.*) but also in solar-terrestrial relations (*ISEE*, AMPTE), and solar interior neutrino and helioseismology studies. The pace of development in related areas of theory (nuclear, atomic, MHD, beam-plasma) has been equally impressive. All of these raised tantalizing further questions about the structure and dynamics of the Sun as the prototypical and best observed star. This Advanced Study Institute was timed at a pivotal point between that decade and the realisation of *Yohkoh, Ulysses, SOHO, GRANAT, Coronas*, and new ground-based optical facilities such as LEST and GONG, so as to teach and inspire the up and coming young solar researchers of the 1990s. The topics, lecturers, and students were all chosen with this goal in mind, and the result seems to have been highly successful by all reports.

The great success of the meeting would have been impossible without many interwoven factors. First, its generous funding by NATO, NASA, and NSF enabled the adoption of the Crieff Hydro Hotel as venue, the location, facilities, food, and staff of which contributed greatly to the fine social and interactive atmosphere of the gathering. Second, the Local Organizing Committee (David Alexander, Declan Diver, Lyndsay Fletcher, Graeme Stewart, and Alan Thompson) did a marvellous job in attending unstintingly to the innumerable needs to which such meetings give rise at all hours of the day, making these even less distinguishable than did the northerly latitude. Third, no matter how good the organisation, such an event will flop without the right spirit among the participants and we have never seen better in this regard! Fourth, all of these factors are ones which are promoted by the wisdom of NATO's guidelines in respect of such matters as locality, duration, group size, and national mix. Thanks are also due to many for the success of the social events, most notably to Cruachan Folk Group, Lyndsay Fletcher for leading the dancing, and Morrisons Academy for their Piping and Dance display. Among the hotel staff, special thanks are due to Irene and the Reception crew, Isobel and the other bartenders, to the Chef and dining room staff for tending the inner man/woman, and especially to Norman Murray, the manager of the Crieff Hydro, for all his help with the local organization. Neither the meeting nor these Proceedings would have been possible without the administrative support of Lillias Williamson, Daphne Davidson, and Alastair McLachlan, nor the photographic work of Ian MacVicar.

The Lecture material in this volume is organised under four broad headings following an opening address by one of us (JCB), the chair of the Local Organizing Committee: Interiors (Part I), Atmospheres (Part II), Instrumentation (Part III), and Activity (Part IV). In general, Chapters on both solar and stellar material are included, in the spirit of seeing the Sun in an astrophysical context and so helping further break down past mental barriers which inhibited more rapid cultivation of the very fruitful "solar-stellar" connection. Theory and observation have, for similar reasons, been mixed within and between Chapters.

The exception is Part III on Instrumentation where the special observational opportunities and problems offered by the Sun's proximity require a distinctive approach, although stellar parallels and relevant theory are cited where appropriate. Part I covers the Solar Interior from various interwoven viewpoints, reflecting the diversity both of the astrophysics involved and of the individual Lecturer's approaches. After details of the solar structure models and their helioseismic exploration, the processes of convection and dynamo action are presented alongside the basics of MHD theory. Part II describes solar and stellar photospheres, chromospheres, and coronae, radio and X-ray diagnostics for these, plus sunspot phenomena and the modelling of winds and outflows. Part III deals with solar instrumentation with details of methodology for all frequencies from radio to gamma-ray, giving both a historical perspective and the future outlook for space- and ground-based solar instruments. Part IV addresses solar and stellar activity, covering observational and theoretical aspects of flares on the Sun and other stars, and related magnetised plasma phenomena in accretion disks.

Finally, we would like to thank the lecturers, all of whom took their jobs of teaching, writing, refereeing, and, especially, socializing very seriously! One lecturer, in particular, A.G. Emslie, spent extra time and effort helping us with the editing and even started the index, our most daunting editorial job. With hindsight, we both realize that neither directing an ASI nor editing its Proceedings is an easy task, even with the support of an excellent Local Organizing Committee and lecturing team and an appreciative crowd of students. If, however, the meeting and the book help speed newcomers down the fascinating path of solar physics, then our efforts and tribulations will have been vindicated.

J.T. Schmelz, Applied Research Corporation/Lockheed, USA.

J.C. Brown, University of Glasgow, Scotland.

March 1992

LIST OF PARTICIPANTS

Al'Malki, M.B. – University of Glasgow, U.K.

Alevizos, A. – University of Athens, Greece

Alexander, D. – University of Glasgow, U.K.

Atakan, K.A. – Marmara University, Turkey

Babcock, T. – University of Cambridge, U.K.

Basurah, H.M. – University of Glasgow, U.K.

Baudin, F. – Inst. d'Astrophysique Spatiale, Verrieres le Buisson, France

Belkora, L. – Caltech, U.S.A.

Brasja, R. – Hvar Observatory, University of Zagreb, Yugoslavia

Bravo, S. – Instituto de Geofisica, UNAM, Mexico

Brekke, P – Institute of Theoretical Astrophysics, Oslo, Norway

Brown, J.C. – University of Glasgow, U.K.

Burke, N. – University of Calgary, Canada

Campbell, J. – NASA MSFC, AL, U.S.A.

Chae, J.-C. – Seoul National University, Republic of Korea

Chan, K.L. – NASA GSFC, MD, U.S.A.

Choudhuri, A.R. – Indian Institute of Science, Bangalore, India

Christensen-Dalsgaard, J. – University of Aarhus, Denmark

Cornille, M. – Observatoire de Paris (Meudon), France

Correia, E. – Escola Politecnica, Universidade Sao Paolo, Brazil

Costa, J.E.R. – Caltech, U.S.A.

Crosby, N.B. – NASA GSFC, MD, U.S.A.

Diver, D.A. – University of Glasgow, U.K.

Drillia, G.A. – University of Athens, Greece

Dubau, J. – Observatoire de Paris (Meudon), France

Emslie, A.G. – University of Alabama in Huntsville, U.S.A.

Erdelyi, R. – Katholieke Universiteit Leuven, Belgium

Fletcher, L – University of Glasgow, U.K.

Fullerton, S. – University of Glasgow, U.K.

Gabriel, A.H. – Inst. d'Astrophys. Spatiale, Verrieres le Buisson, France

Galsgaard, K. – Astronomisk Observatory, Copenhagen, Denmark

Gibson, S. – University of Colorado, U.S.A.

Gierlinski, M. – University of Krakow, Poland

Golovko, A. – USSR Academy Sibizmir, Irkutsk, U.S.S.R.

Grandpierre, A. – Konkoly Observatory, Budapest, Hungary

Guhathakurta, M.– University of Colorado, U.S.A.

Gurman, J. – NASA GSFC, MD, U.S.A.

Harrah, L. – Queens University, Belfast, U.K.

Hick, P. – NASA GSFC, MD, U.S.A.

Howe, R. – University of Birmingham, U.K.

Hoyng, P. – Space Research Laboratory, Utrecht, The Netherlands

Hurford, G. – Caltech, U.S.A.

Jackson, M. – Clwyd, U.K.

Johnston, A.T.F. – University of St. Andrews, U.K.

Kenny, H. – University of Calgary, Canada

Kile, J. – Tufts University, MA, U.S.A.

Kucera, A. – Astron. Inst. Slovak Acad., Tatranska Lomnica, Czechoslovakia

Kucuk, I. – Middle East Technical University, Ankara, Turkey

Kuijpers, J. – Astronomical Institute, Utrecht, The Netherlands

Kurths, J.– Institute of Astrophysics, Potsdam, Germany

Lanzafame, A. – Armagh Observatory, U.K.

Lefevre, E. – Observatoire de Paris (Meudon), France

Leifsen, T. – Institute of Theoretical Astrophysics, Oslo, Norway

Lima, J.– University of St. Andrews, U.K.

Lopes, I.P. – Centre d'Etudes Nucleaires de Saclay, Gif-sur-Yvette, France

MacDonald, A.K. – University of St. Andrews, U.K.

Marmolino, C. – Univerzite de Napoli, Italy

Marti, A.T. – University of Berne, Switzerland

Mathioudakis, M. – Armagh Observatory, U.K.

Monteiro, M.J.P.F.G. – Centro de Astrofisico, Porto, Portugal

Moon, Y.-J. – Seoul National University, Republic of Korea

Moorthy, S. – Mullard Space Science Lab., Dorking, U.K.

Moussas, X. – University of Athens, Greece

Nash, A.G. – NRL, Washington D.C., U.S.A.

O'Neal, R. – Stanford University, CA, U.S.A.

Ozguc, A. – Kandilli Observatory, Istanbul, Turkey

Pagano, I. – Catania University, Italy

Pallavicini, R. – Osservatorio Astrofisico di Arcetri, Italy

Petersen, S.T. – University of Aarhus, Denmark

Raulin, J.-P. – Observatoire de Paris (Meudon), France

Reale, F. – Osservatorio Astronomico, Palermo, Italy

Rolli, E.J. – University of Berne, Switzerland

Roulias, D. – University of Athens, Greece

Sabbah, I. – Moharum Bak, Alexandria, Egypt

Sauty, C. – Observatoire de Paris (Meudon), France

Schmelz, J.T. – NASA GSFC, MD, U.S.A.

Schuessler, M. – Kiepenheuer Institute, Freiburg, Germany

Singh, H. – University of Kiel, Germany

Stewart, G. – University of Glasgow, U.K.

Surlantzis, G. – University of Crete, Greece

Thompson, A.M. – University of Glasgow, U.K.

Tomczak, M. – Wroclaw University, Poland

Tsinganos, K. – University of Crete, Greece

Tsiropoula, G. – University of Athens, Greece

van Driel Gesztelyi, L. – Astronomical Institute, Utrecht, The Netherlands

Waljewski, K. – NRL, Washington DC, U.S.A.

Wang, J.-X. – Beijing Astron. Observatory, Chinese Acad. Sciences, China

Wang, Y. – NRL, Washington DC, U.S.A.

Wilkinson, L.K. – University of Alabama in Huntsville, U.S.A.

Zirin, H. – Caltech, U.S.A.

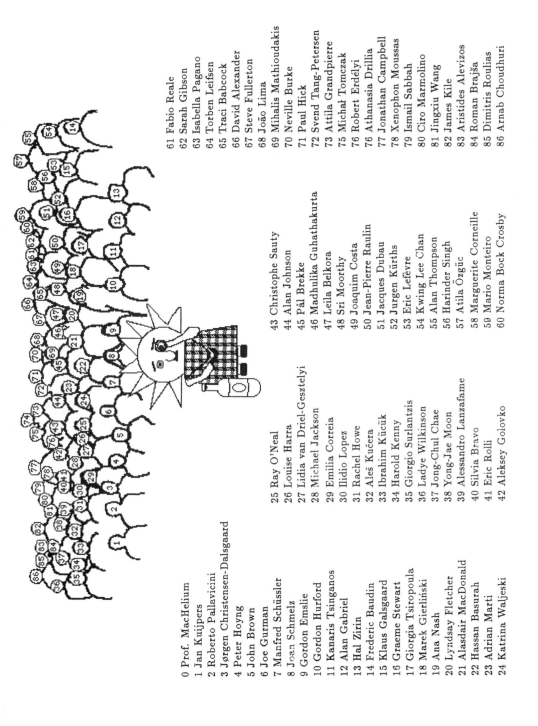

0 Prof. MacHelium
1 Jan Kuijpers
2 Roberto Pallavicini
3 Jørgen Christensen-Dalsgaard
4 Peter Hoyng
5 John Brown
6 Joe Gurman
7 Manfred Schüssler
8 Joan Schmelz
9 Gordon Emslie
10 Gordon Hurford
11 Kanaris Tsinganos
12 Alan Gabriel
13 Hal Zirin
14 Frederic Baudin
15 Klaus Galsgaard
16 Graeme Stewart
17 Giorgia Tsiropoula
18 Marek Gierliński
19 Ana Nash
20 Lyndsay Fletcher
21 Alasdair MacDonald
22 Hassan Basurah
23 Adrian Marti
24 Katrina Waljeski

25 Ray O'Neal
26 Louise Harra
27 Lidia van Driel-Gesztelyi
28 Michael Jackson
29 Emilia Correia
30 Ilidio Lopez
31 Rachel Howe
32 Aleš Kučera
33 Ibrahim Küçük
34 Harold Kenny
35 Giorgio Surlantzis
36 Ladye Wilkinson
37 Jong-Chul Chae
38 Yong-Jae Moon
39 Alessandro Lanzafame
40 Silvia Bravo
41 Eric Rolli
42 Aleksey Golovko

43 Christophe Sauty
44 Alan Johnson
45 Pál Brekke
46 Madhulika Guhathakurta
47 Leila Belkora
48 Sri Moorthy
49 Joaquim Costa
50 Jean-Pierre Raulin
51 Jacques Dubau
52 Jurgen Kürths
53 Eric Lefèvre
54 Kwing Lee Chan
55 Alan Thompson
56 Harinder Singh
57 Atila Özgüc
58 Marguerite Corneille
59 Mario Monteiro
60 Norma Bock Crosby

61 Fabio Reale
62 Sarah Gibson
63 Isabella Pagano
64 Torben Leifsen
65 Traci Babcock
66 David Alexander
67 Steve Fullerton
68 João Lima
69 Mihalis Mathioudakis
70 Neville Burke
71 Paul Hick
72 Svend Tang-Petersen
73 Attila Grandpierre
75 Michal Tomczak
76 Robert Erdélyi
76 Athanasia Drillia
77 Jonathan Campbell
78 Xenophon Moussas
79 Ismail Sabbah
80 Ciro Marmolino
81 Jingxiu Wang
82 James Kile
83 Aristides Alevizos
84 Roman Brajša
85 Dimitris Roulias
86 Arnab Choudhuri

1 OPENING ADDRESS

J.C.BROWN
[Codirector and LOC Chairman]
Department of Physics and Astronomy
University of Glasgow
Glasgow G12 8QW
Scotland, U.K.

ABSTRACT. After personalthanks to all who made this ASI a success, a brief introduction is given to Scotland and the Scots, covering both the miserly, hard drinking, tough stereotype and the reality, including practical matters of local colour like the weather and the "English" used. This is followed by a short historical overview of solar science in Scotland from megailithic times through Alexander Wilson of the "Wilson Effect," with particular reference to the lessons to be learnt from Lord Kelvin. These include topics of publication mania and the importance of making mistakes. The present healthy state and institutional distribution of astronomical research in Scotland is then summarised. Praises are sung of the special and prominent position occupied by the Sun as the fundamental laboratory for astrophysics.

1. Retrospective Remarks

Though I had previously participated in NATO ASIs and ARWs, this was my first in an organisational capacity, but I hope not my last. Despite numerous unforeseen problems, I found enormous satisfaction in helping, and welcoming to my homeland, about 100 of the next generation of solar astrophysicists from around the globe for two weeks. The fact that this ASI was, by all accounts and thanks received, a resounding success, can be attributed to many factors, and I would like to add some personal comments and thanks to those already given in the Preface to these Proceedings. One of the major chores performed by the LOC was the transporting of many Lecturers and Students from the urban centres of Scotland to the tranquillity of Crieff. In addition to preplanning (mainly by Declan Diver, Graeme Stuart, and David Alexander) on the scale of the Norman invasion, this involved an extraordinary amount of LOC chauffeuring, from which neither the Crieff-Perth road nor, alas, one young deer will ever recover but it also established the astonishing sociability and carrying capacity of a Renault 21, best not revealed to the rental company. The great success of many of the social events can be traced to the efforts of many specific individuals and my personal gratitude is due to Steve for bringing Cruachan, Bill Hall for the loan of a kilt, Isobel for a friendly non-ASI ear in the bar, and Ben Lawers for kind weather. I am also indebted to Alan Thompson for his computer wizardry, without which all of the LOC's work on manuscripts and schedules would have reached maximum entropy in no time, as would lecture slides without Lyndsay's team. Least enjoyable of all the Organisers' jobs is the preparatory paperwork and retrospective accounting and reporting, which I only survived with the support of David and Declan plus Daphne Davidson and Lillias Williamson. Most

1

J. T. Schmelz and J. C. Brown (ed.), The Sun, A Laboratory for Astrophysics, 1–7.

of all, I would like to thank my family for all their support: Stuart as creator of Professor MacHelium, who pervaded the meeting as mascot and cartoon character; Lorna for graphics assistance, and the Danish sacrificial midsummer witch; and Margaret for all sorts of help, not least putting up with it all.

In thinking how to open this ASI, I had contemplated giving an overview of hot topics in solar physics. However, with two full weeks of such lectures looming, I decided it was better and kinder to try to set the mood of the meeting by giving some insight into the Scottish setting, to the Scottish dimension in the subject of the ASI, and to the broad nature of that subject.

2. Scotland and the Scots

One of the joys of travel is to see places and people as they really are, rather than as dots on maps or as jaded Hollywood stereotypes, of which Scotland has suffered an ample share. By the time the ASI WAs over, each participant no doubt had formed his/her own impression as to the truth of some of the mist-swaithed, kilt-adorned, mythology of the land and people. For certain they will have found some of the language hard to recognise as English and the weather distinctly changeable

PROBLEMS WITH SCOTLAND

The Accent

The Weather

Both the Hotel and local bars seemed, in retrospect, well enough frequented for realistic assessement of the "hard drinking northerner" idea, though the Glenturret Distillery visit suggested this trait is not confined to polar peoples, and some of the Lecturers even seemed to find the local water of life of scientific value also

SCOTLAND – THE PEOPLE – I

DRINK A LOT

THE REAL THING : SCOTCH FLUX TUBE MODELS

EXPANDING   CONTRACTING

M. Schuessler

4

I do hope, however, that no-one encountered any evidence for the aggressive or stingy Celt ideas

As regards the country, we were mostly fortunate on tours in seeing Scotland at its best, the castles and Highlands being at their most impressive in good but not too fine weather - even if not everyone saw all they wanted, Nessie making only one brief appearance in the guise of a fluxtube

3. Early Solar Physics in Scotland

In its broadest sense, solar science in Scotland dates back five millenia to the megalithic stone circles which abound here, especially on the west coast. These have been shown by the work of Professor A.E. Thom, Professor A.E.Roy, and Dr. E. McKie of Glasgow University, among others, to be elaborate calendrical sites. They are based on a length unit ("Megalithic Yard") shared with continental European counterparts which reveals a level of organisation and communication which archaeologists were slow to accept. Modern studies of the Sun, in an astrophysical sense, were first conducted by Professor Alexander Wilson, first Chair of Practical Astronomy (1760) in Glasgow University and discoverer of the Wilson Effect in sunspots. While it is not conventional for Glasgow residents (Glaswegians) to pay much heed to Edinburgh activities, or vice versa, out of scientific deference mention should be made here of the the the work of Piazzi Smyth, Astronomer Royal for Scotland in the 1860s, as a pioneer of mountaintop observing and an early student of the solar spectrum. Most notable, however, of historical Scottish figures on the solar physics, and almost every scientific, scene was Lord Kelvin, Chair of Natural Philosophy in Glasgow University 1846 -1899 (age 22-75). A whole meeting could easily be based on the works of Kelvin, which resulted in 686 publications (about one per month) and 70 patents but here, without disrespect to his truly great achievements, I would rather draw some lessons from aspects of his work, less often emphasised, which I hope may be of value to those starting their scientific careers. First, our "publish or perish" society which forces production of fairly trivial papers, is not new - among Kelvin's publications is one on the efficacy of wearing loose clothing in winter which treats the human body as a sphere with a low conductivity shell! The lesson here is not to commend this practice but slightly to reduce your guilt since you are pursuing it in eminent company. Second, and more important, is to learn early not to be afraid of sticking your neck out at the risk of being wrong in your conjectures or questions. Remember that it is in what your ASI Lecturers cannot teach you that the future innovations lie, not in mimicking them, and to venture in that territory demands risk-taking. Dr. Johnson, a famous writer and early arrival on the growing Scottish tourism scene said "No-one ever become wise by imitation" (even if he also said "The only good thing that ever came out of Scotland is the road to England"). Among Kelvin's greatest bloomers were his statement concerning magnetic storms that "This [theoretical] result is absolutely conclusive evidence against [these being] due to the magnetic action of the Sun" and his kinetic theory of radioactivity. Though the latter was admittedly formulated late in his declining years, I cannot resist to let you hear the great man's pronouncement on this topic. Unfortunately ASI Proceedings cannot include a CD in the sleeve so those readers wishing to hear this for themselves will need to send me a blank cassette or visit Glasgow University's Hunterian Museum. Here I can only quote an extract - "In the equilibrium of kinetic averages in any solid or liquid body every electrion [sic] must occasionally have so high a velocity that it is shot out of the body."

4. Contemporary Solar- and Astro-Physics in Scotland

Scotland currently has a thriving astrophysics community, despite the ravages wrought on funding by a decade of Thatcherism. This community is mainly concentrated on the sites of the three ancient universities, founded ca. 1400, and each having an established Chair of Astronomy or Astrophysics. Edinburgh is the largest focus of astronomical activity because of its being the site of the Government Royal Observatory of Edinburgh which, together with the University Department of Astronomy, musters some 200 personnel including those at remote sites. Interests there are diverse, but with the largest emphasis on extragalactic studies and on instrumentation. The Department of Physics and Astronomy in Glasgow is home to about 40 personnel with astronomical interests, namely about 12 in the Gravitational Wave Group, and 28 in the Astronomy and Astrophysics Group (which now incorporates Plasma Theory) with interests in Solar Physics, Astrophysics, Plasma Theory and Diagnostics, Observational Spectropolarimetry, Dynamics, Cosmology, and Statistical and Inverse Problems. The Department of Mathematics also has interests in Planetary Magnetism. St. Andrews, oldest of the Scottish Universities, covers a range of stellar, galactic and extragalactic topics in the Department of Physics and Astronomy, and Solar Magnetohydrodynamics in the School of Mathematics, with a total personnel similar to Glasgow.

5. The Solar Laboratory for Astrophysics

Aside from its intrinsic interest both scientifically and as the source of life on earth and of the planetary environment, the proximity of the Sun makes it the fundamental testing ground for virtually all astrophysical techniques. The signal to noise associated with the collection in one second of more solar photons than arrive from a similar source at even 1 parsec in 1000 years, permits resolution in solar data far transcending anything possible elsewhere, whether in the polarimetric, spectral, temporal, or spatial domain - the last rendered all the more dramatic by the Sun's large angular extent seen from earth. It is these factors which, for example, have resulted in solar physics giving birth to atomic and nuclear spectroscopy in astrophysics, to cosmic magnetometry, to neutrino astrophysics, and to asteroseismology. In addition, since we are embedded in the outer layers of the Sun, and since spacecraft permit a three-dimensional stereoscopic perspective, the Sun is not so strictly "remotely-sensed" as most cosmic objects though it is not immune from the problems of remotely-sensed data inversion which permeate astrophysics and stretch our ingenuity to decipher what tricks the Sun is performing.

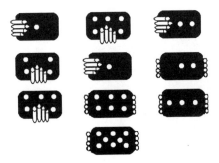

In the next two weeks we will all learn, from our assembled body of eminent Lecturers, a great deal about our fascinating star and its relationship to its distant cousins. As rough guidelines to making the most of the meeting, I would encourage Lecturers to keep to the essential basics, while indicating the doubts and frontiers, and Students to ask lots of questions. I hope the setting and organisation of this ASI will help inspire your efforts in these directions - the LOC are here to ensure that it does, and on behalf of all of them, especially our Superdirector Professor MacHelium, I wish you a very warm welcome to Scotland and to this ASI.

PART I
SOLAR INTERIOR

2 THE STRUCTURE AND EVOLUTION OF THE SUN

J. CHRISTENSEN-DALSGAARD
Institut for Fysik og Astronomi
Aarhus Universitet
DK-8000 Aarhus C
Denmark

ABSTRACT. Computations of "standard" solar models are based on assumptions of equilibrium between the forces acting on different parts of the Sun and between the amount of energy produced in the core and radiated from the surface. The structure of the resulting models depends on the assumed physics of the solar interior (equation of state, opacity, energy generation). Although accurate calculations are required to obtain detailed information about solar structure, simple estimates can give a feel for the general properties of the solar interior. A firm prediction of the calculation is that the solar luminosity has increased somewhat since the formation of the Sun. All standard models predict a neutrino flux which is substantially larger than the observed value.

1. Introduction

It is normally assumed that the Sun is very nearly in equilibrium: the forces of gravity and pressure balance, maintaining *hydrostatic equilibrium*; and there is a near balance between the energy produced by nuclear reactions in the solar core and the energy radiated from the solar surface, implying overall *thermal equilibrium* where matter in the Sun is neither heated nor cooled. These conditions essentially determine the structure of the Sun; the evolution in time is determined by the slow conversion of hydrogen into helium which takes place in the solar core and leads to the generation of energy. Given the assumptions of hydrostatic equilibrium, assumptions about the form of the energy transport, and an assumption that no material motion takes place except in the convection zone in the outer parts of the Sun, models of solar evolution can be computed. The computation furthermore requires knowledge about the physical properties of matter in the Sun: the equation of state, the opacity, and the rate of nuclear energy generation and transformation of hydrogen into helium. The result of such calculations is often called a (or even, sometimes, the) "standard" solar model. This term is somewhat misleading, however, in that the model depends sensitively on the assumed physics; in particular, there remain considerable uncertainties in the opacity. Therefore, there exist many somewhat different "standard" models; the more neutral term "normal" solar models is probably preferable.

No justice can be given to the field of stellar evolution theory in this Chapter. Reference may be made to the many books on this topic, for example the recent work by Kippenhahn and Weigert (1990); specific details about the computation of solar models are addressed in several chapters in Cox, Livingston, and Matthews (1991). Here, I discuss very briefly the basic equations controlling the structure and evolution of a star and some simple approximations to the physics. This is used to derive estimates for the properties of a star which are

11

J. T. Schmelz and J. C. Brown (ed.), The Sun, A Laboratory for Astrophysics, 11–28.
© 1992 *Kluwer Academic Publishers. Printed in the Netherlands.*

compared with the results of calculations of solar structure. Finally, I discuss two possible problems with the computed models: the discrepancy between the predicted and observed flux of neutrinos and the so-called "faint early Sun problem". It may be noted that further detail about computed models and additional references are given in Chapter 3.

2. The Basic Equilibrium of a Star

2.1. HYDROSTATIC AND THERMAL EQUILIBRIUM

Equilibrium of a star demands that the forces acting on each part of the star cancel. Assuming spherical symmetry, the gravitational acceleration at a distance r from the centre is $-Gm/r^2$, where G is the gravitational constant and m is the mass of the sphere inside r. Balancing gravity requires a gradient of pressure p given by

$$\frac{dp}{dr} = -\frac{Gm\rho}{r^2} \, , \tag{1}$$

where ρ is the density (recall that pressure is force per unit area). Also, obviously,

$$\frac{dm}{dr} = 4\pi r^2 \rho \, . \tag{2}$$

To maintain equilibrium, there must be a balance between the energy generated by nuclear reactions and the energy radiated by the star. Locally, this balance relates the rate of energy release to the change in the local luminosity $L(r)$, *i.e.*, the rate of flow of energy through a spherical surface of radius r. This condition may be obtained from the first law of thermodynamics as

$$\frac{dL}{dr} = 4\pi r^2 \left[\rho\epsilon - \rho\frac{du}{dt} + \frac{p}{\rho}\frac{d\rho}{dt} \right] \, , \tag{3}$$

where ϵ is the rate of energy production and u the internal energy per unit mass. On the right-hand side, the first term corresponds to the energy produced by the nuclear reactions, whereas the second and third terms give the contributions resulting from changes in the internal energy of the gas and from the work done on the gas.

2.2. ENERGY TRANSPORT

The condition that energy must be transported from the region where it is generated to the surface where it is radiated determines the temperature gradient in the Sun. In much of the Sun, the transport is by radiation. The energy density of radiation per unit volume is $u_R = aT^4$, where T is temperature and a is the radiation density constant. When T, and hence u_R, decreases with increasing distance r from the centre of the star, photons moving away from the centre carry a slightly higher energy on average than photons that move towards the centre; this gives rise to a net radiative energy flux

$$F_R = -\frac{4a\tilde{c}T^3}{3\kappa\rho}\frac{dT}{dr} \, . \tag{4}$$

Here $\tilde{c}$ is the speed of light and κ is the opacity, defined such that $\lambda = (\kappa\rho)^{-1}$ is the mean free path of a photon. If the energy transport occurs only through radiation, the total amount of energy transported through a sphere of radius r is $L(r) = 4\pi r^2 F_R$. Hence

$$\frac{dT}{dr} = -\frac{3\kappa\rho L(r)}{16\pi a\tilde{c}r^2T^3} , \tag{5}$$

which determines the temperature gradient.

If the temperature gradient as given by Equation (5) is too steep, the layer becomes unstable towards *convection* (see Chapter 4); the instability condition may be written as $\nabla > \nabla_{ad}$; here $\nabla = d\ln T/d\ln p$ and $\nabla_{ad} = (\partial\ln T/\partial\ln p)_s$, where the former derivative is taken along conditions in the model and the latter at constant specific entropy. This instability sets in in the outer parts of the Sun, due to the increasing opacity as the temperature decreases. Where the instability occurs, energy is predominantly transported by convective motions. In most of the convection zone, this process is so efficient that it requires only a very small superadiabatic gradient $\nabla - \nabla_{ad}$; as a result, the temperature gradient may be approximated by

$$\frac{dT}{dr} \simeq -\nabla_{ad}\frac{T}{p}\frac{Gm\rho}{r^2} , \tag{6}$$

(where Equation (1) was used).

This approximation is inadequate near the solar surface; since the density is low, vigorous motion is needed to transport the energy, and hence a substantial superadiabatic gradient is required. Here, a more detailed description of convection is required, introducing considerable uncertainty into the calculation. This region is quite thin, however: only in roughly the outermost 1000 km is $\nabla - \nabla_{ad}$ greater than 0.01. The principal effect of the superadiabatic region on the structure of the model is to determine the adiabat in the adiabatic part of the convection zone. In practice, convection is usually treated by means of the so-called mixing-length formalism, which provides a very rough physical picture of the convective processes (see Chapter 4) It is characterized by a parameter α_C which determines the scale of the convective elements in units of the local pressure scale height. As discussed in section 5.1 below, this parameter is determined by requiring that the model have the observed radius.

2.3. THE CHANGE IN COMPOSITION

The evolution of the Sun is controlled by the conversion of hydrogen into helium. Apart from a possible brief period at the start of solar evolution, the cores of solar models are generally found to be stable to convection. It is an assumption of standard solar model calculations that there is no other motion in the solar core which might have caused mixing. Hence the change in composition is a local process, which may be described by the equation

$$\frac{dX}{dt} = r_X ; \tag{7}$$

here the time derivative is assumed to be at fixed mass, and the local rate of change r_X of the hydrogen abundance is determined by the rate of nuclear reactions.

3. The Physics of Stellar Interiors

As discussed in section 2, the basic equations of stellar structure and evolution are quite simple; but they are incomplete. In addition, we need expressions for the thermodynamic state of the gas to relate the density ρ and the internal energy u to pressure, temperature, and chemical composition. Similarly, we must be able to evaluate the opacity κ, the rate of energy generation ϵ, and the rate of change r_X in the hydrogen abundance. In calculations of solar models the evaluation of these quantities constitutes by far the major effort. However, it is useful to consider extremely simple approximations which nevertheless capture some of the essential features of the physics.

3.1. EQUATION OF STATE

In most of the Sun, matter is essentially completely ionized. Thus the simplest approximation to the equation of state is obtained by regarding the gas as a completely ionized ideal gas; in particular, we neglect the interactions between the constituents of the gas and the effects of degeneracy. Then, the pressure is given by

$$p = \frac{\rho k_B T}{\mu m_u} \,, \tag{8}$$

where μ is the mean molecular weight, k_B is Boltzmann's constant, and m_u is the atomic mass unit. Also, we assume that the internal energy comes solely from the kinetic energy of the particles in the gas; the mean internal energy per particle is $3/2\, k_B T$, and hence the internal energy per unit mass is $u = 3/2\, p/\rho$. It may be shown that $\nabla_{ad} = 2/5$. Also, the adiabatic relation between pressure and density, which is essential for adiabatic stellar pulsations, is determined by $\Gamma_1 = (\partial \ln p/\partial \ln \rho)_s = 5/3$.

The mean molecular weight depends on the chemical composition. It is conventional to denote the mass fractions of H and He by X and Y, respectively, and the mass fraction of the remaining, so-called heavy, elements by Z. This separation is useful because, in most stellar compositions, $Z \ll X, Y$. Clearly, the normalization $X + Y + Z = 1$ must hold. Then, counting the number of particles (nuclei and electrons) contributed by each of the species in the gas shows that

$$\mu \simeq \frac{4}{3 + 5X - Z} \,. \tag{9}$$

This simple equation of state fails in the outer parts of the Sun; here hydrogen and helium changes from being unionized in the atmosphere to essentially complete ionization at temperatures exceeding 2×10^5 K. In addition to changing the expression for pressure at fixed temperature and density, ionization locally decreases ∇_{ad} and Γ_1. This is always taken into account when computing solar models, although the detailed description of the ionization balance is complicated and still somewhat uncertain. Additional complications which are included at varying levels of detail are the effects of degeneracy in the solar core and the Coulomb interaction between the charged particles of the gas (for a review of the treatment of the equation of state and further references, see Däppen, Keady, and Rogers 1991). Such effects change the details of the computed structure; in particular, they have significant effects on the computed oscillation frequencies, and hence on helioseismic investigations of the solar interior.

3.2. OPACITY

The mean free path of a photon, and hence the opacity, depends on the microscopic interaction between radiation and matter. This is a complex problem, where account must be taken of the detailed interaction between the radiation and the different atoms in the gas (see Däppen *et al.* 1991). Hence, it is common in computations of stellar models to use tables of opacity as a function of ρ, T, and the chemical composition. However, simple approximations can give a feel for the dependence of the opacity on the thermodynamical state. In most of the Sun, the opacity can be expressed approximately as

$$\kappa = \kappa_0(1 + X)\rho T^{-3.5} \qquad (10)$$

(the so-called *Kramers approximation*) where κ_0 is a constant which depends on the abundance of heavy elements. In the solar atmosphere, the opacity is dominated by absorption by H^-. The abundance of H^- is determined by the density of free electrons, which derive predominantly from ionization of metals; hence the electron density, the abundance of H^-, and consequently the opacity increase steeply with temperature.

3.3. NUCLEAR REACTIONS

The energy generation in the solar core occurs through the fusion of hydrogen into helium, according to the following net reaction

$$4\,{}^1\mathrm{H} \rightarrow {}^4\mathrm{He} + 2\mathrm{e}^+ + 2\nu_e \ . \qquad (11)$$

In this reaction, charge conservation requires that two of the four protons on the left-hand side are converted into neutrons and positrons; the positrons are immediately annihilated by two electrons so that the reaction can be thought of as the fusion of four hydrogen atoms into one helium atom. The reaction, furthermore, must conserve the number of *leptons*; since two anti-leptons (the positrons) are created, this must be balanced by the creation of two leptons, the neutrinos. Thus, the fusion of four hydrogen atoms into one helium atom always leads to the production of two neutrinos.

Hydrogen fusion may proceed through a number of different paths. The total amount of energy Q_{tot} liberated in the reaction leading to the formation of a single ${}^4\mathrm{He}$ is, in any case, given by the mass difference between the four hydrogen atoms on the left-hand side of Equation (11) and the helium atom on the right-hand side, *i.e.*, $Q_{\mathrm{tot}} = 26.73\,\mathrm{MeV}$. However, the neutrinos escape directly, and hence do not contribute to the energy release within the Sun. Therefore, to compute the energy generation rate ϵ we must subtract the neutrino energy. This depends on the detailed reactions in which the neutrinos are produced.

The detection of solar neutrinos provides a potential diagnostic of conditions in the solar core. To interpret the results (*cf.* section 6), we need to consider in more detail the actual reactions taking place. It is convenient to introduce a compact notation for nuclear reactions, such that the reaction $A + a \rightarrow Y + y$ is written as $A(a,y)Y$. Then the basic sequences of reactions may be written as

$$^1\mathrm{H}(^1\mathrm{H},\mathrm{e}^+\nu_e)\,^2\mathrm{D}(^1\mathrm{H},\gamma)\,^3\mathrm{He}(^3\mathrm{He},2\,^1\mathrm{H})\,^4\mathrm{He} \qquad\qquad\text{(PP-I)}$$

$$\Downarrow$$

$$^3\mathrm{He}(^4\mathrm{He},\gamma)\,^7\mathrm{Be}(\mathrm{e}^-,\nu_e)\,^7\mathrm{Li}(^1\mathrm{H},\,^4\mathrm{He})\,^4\mathrm{He} \qquad\qquad\text{(PP-II)}$$

$$\Downarrow$$

$$^7\mathrm{Be}(^1\mathrm{H},\gamma)\,^8\mathrm{B}(,\mathrm{e}^+\nu_e)\,^8\mathrm{Be}(,\,^4\mathrm{He})\,^4\mathrm{He}\ . \qquad\text{(PP-III)}$$

$$\text{(12)}$$

As indicated, these reactions are known as the PP-I, PP-II and PP-III chains. The average neutrino losses are 0.26 MeV, 1.06 MeV, and 7.46 MeV, respectively. The high-energy neutrinos emitted in the $^8\mathrm{B}(,\mathrm{e}^+\nu_e)\,^8\mathrm{Be}$ reaction dominate current experiments to detect solar neutrinos.

To react, two nuclei must overcome their mutual Coulomb repulsion. Since the average energy of the nuclei in the gas is far lower than typical nuclear energies, the reaction requires that the nuclei tunnel through an extensive Coulomb barrier. As a result, the temperature sensitivity of the reactions increases with the charges of the reacting nuclei. A convenient way of parameterizing this is to approximate the reaction rate by a power law so that the total number of reactions per unit volume between two nuclei, i and j, is expressed as

$$r_{ij} = r_{ij,0}\left(\frac{\rho}{\rho_0}\right)^2\left(\frac{T}{T_0}\right)^n . \qquad\qquad\text{(13)}$$

Here $r_{ij,0}$ is the reaction rate at the density and temperature ρ_0 and T_0, and the exponent n characterizes the temperature sensitivity (ρ enters with the power two because the number of reactions is proportional to the number of *pairs* of particles (i,j)).

Despite having the lowest possible Coulomb barrier, the first reaction in the PP chains is by far the slowest because it involves the conversion of a proton into a neutron through the effect of the weak interaction. The remaining reactions are generally nearly *in equilibrium*, the abundances of the nuclei involved being adjusted such that equal amounts are produced and destroyed. Hence, the combined rate of energy generation is controlled by the $^1\mathrm{H}(^1\mathrm{H},\mathrm{e}^+\nu_e)\,^2\mathrm{D}$ reaction. According to Equation (13), we may approximate it as

$$\epsilon = \epsilon_0 X^2 \rho\, T^n , \qquad\qquad\text{(14)}$$

where ϵ_0 is a constant; in the solar core $n \simeq 4$. The branching ratios between the different parts of the PP chains depend on the balance between the competing reactions, in such a way that the number of reactions going through the PP-II and PP-III chains increases rapidly with increasing temperature (since electron capture in $^7\mathrm{Be}$ depends weakly on temperature, the branching ratio between the PP-II and the PP-III chains, and hence the flux of PP-III neutrinos, is highly temperature-sensitive). In the Sun, the PP-I chain dominates and the PP-III chain makes a very small contribution to the energy generation.

A recent detailed treatment of nuclear energy generation in the Sun was given by Parker and Rolfs (1991).

4. Simple Estimates of Stellar Internal Properties

By applying very simple approximations to the equations of stellar structure, one may obtain rough estimates of the conditions in the solar interior. Such estimates serve a very useful purpose in giving an order-of-magnitude idea about characteristic properties and their dependence on the overall parameters of the star.

4.1. THE PRESSURE AND TEMPERATURE

From Equation (1), we may obtain an estimate of the central pressure p_c of a star with mass M and radius R. We replace dp/dr by $-p_c/R$, m by M, r by R, and approximate ρ by the mean density, which in turn is approximated as M/R^3. Then Equation (1) gives

$$p_c \simeq \frac{GM^2}{R^4} .$$

(15)

If we assume the ideal gas law, Equation (8), we may estimate the central temperature as

$$T_c = \frac{\mu_c m_u p_c}{k_B \rho_c} \simeq \frac{G\mu_c m_u M}{k_B R} ,$$

(16)

where μ_c is the central mean molecular weight. In terms of solar mass $M_\odot$ and radius $R_\odot$, we obtain

$$p_c \simeq 1.1 \times 10^{16} \left(\frac{M}{M_\odot}\right)^2 \left(\frac{R}{R_\odot}\right)^{-4} \text{ dyn cm}^{-2} ,$$

(17)

$$T_c \simeq 1.9 \times 10^7 \left(\frac{M}{M_\odot}\right) \left(\frac{R}{R_\odot}\right)^{-1} \left(\frac{\mu_c}{0.85}\right) \text{ K} ,$$

(18)

where the reference value of μ_c was obtained from Equation (9), with the values $X = 0.35$ and $Z = 0.02$ appropriate for the core of the present Sun.

These are obviously not accurate estimates of the central pressure or temperature of a star, but they provide an order of magnitude. In fact, the estimates are reasonably accurate: realistic computations of solar models show that the solar central pressure is $p_c = 2.4 \times 10^{17} \text{ dyn cm}^{-2}$ and the solar central temperature is $1.5 \times 10^7 \text{ K}$. Furthermore, they indicate how pressure and temperature scale with the stellar mass and radius. This is very helpful for the interpretation of detailed numerical results.

4.2. THE LUMINOSITY

The luminosity of stars can be estimated from Equation (5) for radiative transport, combined with the estimates of the temperature and density given above. We assume the ideal gas law, estimate the temperature by Equation (16), and approximate the density by the mean density. The opacity is approximated by Equation (10). Finally, we replace r by R, and approximate $-dT^4/dr$ by T^4/R to obtain

$$L \simeq \frac{a\tilde{c} R T^{7.5}}{\kappa_0 (1+X)\rho^2} \simeq \frac{a\tilde{c}}{\kappa_0 (1+X)} \left(\frac{Gm_u \mu}{k_B}\right)^{7.5} R^{-0.5} M^{5.5} .$$

(19)

Using $\kappa_0 \simeq 8 \times 10^{23}(1 + X)$ (in cgs units), Equation (19) gives

$$
L \simeq \frac{a\tilde{c}}{\kappa_0(1 + X)} \left(\frac{G m_u \mu}{k_B}\right)^{7.5} R^{-0.5} M^{5.5}
$$

$$
\simeq 10^{35} \left(\frac{\mu}{0.62}\right)^{7.5} \left(\frac{1.7}{1 + X}\right) \left(\frac{M}{M_\odot}\right)^{5.5} \left(\frac{R}{R_\odot}\right)^{-0.5} \mathrm{erg\ sec^{-1}} .
$$
(20)

where we assumed $X = 0.7$ as reference composition. This estimate is rather high compared with the solar luminosity of $3.846 \times 10^{33}\,\mathrm{erg\,sec^{-1}}$; in view of the approximations made, this is hardly surprising. In Equation (20), L depends on both M and R. However, the exponent in the dependence on R is much smaller than the exponent in the M-dependence. Numerical calculations show that for main-sequence stars R is approximately proportional to M; hence, the variation of L is dominated by the M-dependence. One might also note the remarkably strong dependence on the mean molecular weight.

A more precise justification of the relations derived here can be obtained from *homology analysis* (see, for example, Kippenhahn, and Weigert 1990; Gough 1990).

4.3. CHARACTERISTIC TIMESCALES OF A STAR

Estimates of the timescales of change in a star that is not in equilibrium may indicate whether or not departures from equilibrium might give rise to observable consequences.

4.3.1. The dynamical timescale. If the star is not in hydrostatic equilibrium (*i.e.*, the balance between pressure and gravity has been disturbed), changes occur on the *dynamical timescale* t_{dyn}. To estimate it, we assume that the pressure gradient can be neglected; then motion occurs on the free-fall time scale. The gravitational acceleration at the surface of the star is $g_s = GM/R^2$. Hence, the time required for a particle to fall a distance of order R in the gravitational field of the star is of order

$$
t_{\mathrm{dyn}} = \left(\frac{R}{g_s}\right)^{1/2} = \left(\frac{R^3}{GM}\right)^{1/2} \simeq 30\ \mathrm{min} \left(\frac{R}{R_\odot}\right)^{3/2} \left(\frac{M}{M_\odot}\right)^{-1/2} .
$$
(21)

This provides an estimate of the typical timescale for dynamical changes to a star. In particular, it gives an estimate of a characteristic pulsation period.

4.3.2. The timescale for release of gravitational energy. If a star has no internal sources of energy, it can still radiate energy by contracting. In this way, its gravitational potential energy decreases (*i.e.*, becomes of larger negative magnitude). It follows from the virial theorem that half of the energy released goes to heat up the star and the other half is radiated away. An estimate for the timescale of this process can be obtained by calculating the time a star could radiate at a given rate on the energy released through gravitational contraction to a given radius. The gravitational binding energy of the star may be estimated as $-GM^2/R$. Hence, the relevant timescale, known as the *Kelvin-Helmholtz timescale*, is

$$
t_{\mathrm{KH}} = \frac{GM^2}{RL} \simeq 30\ \mathrm{million\ years} \left(\frac{M}{M_\odot}\right)^2 \left(\frac{R}{R_\odot}\right)^{-1} \left(\frac{L}{L_\odot}\right)^{-1} .
$$
(22)

This value gave rise to some controversy in the 19th century when gravitational contraction was considered a possible source of the solar energy (*cf.* Chapter 1); in this case, the age of the Sun would clearly be at most of order t_{KH}. On the other hand, it was becoming clear from geological evidence that the Earth had to be much older. The problem was resolved when it was realized that solar energy derives from nuclear reactions.

4.3.3. The nuclear timescale. The duration of the phase of stellar evolution where the energy comes from the fusion of hydrogen into helium may be estimated by noting that in the process about 0.7 per cent of the mass is lost. The reaction occurs only in the inner 10 per cent of the mass of the star. Hence, the total amount of energy available is approximately $7 \times 10^{-4} M \tilde{c}^2$ and the corresponding timescale is

$$ t_{\mathrm{nuc}} = 7 \times 10^{-4} \frac{M \tilde{c}^2}{L} \simeq 10^{10} \, \text{years} \quad \left(\frac{M}{M_\odot} \right) \left(\frac{L}{L_\odot} \right)^{-1} . \qquad (23) $$

The present age of the Sun is about 4.5×10^9 years; hence, the Sun is approximately halfway through its hydrogen-burning phase.

5. Computation of Stellar Models

Given the basic equations of stellar structure and evolution and a description of the physics, models of the Sun may be computed. The computation requires the adoption of a suitable numerical procedure. Care is required to ensure that adequate numerical precision is achieved. This condition is particularly stringent because of the high accuracy of the observed solar oscillation frequencies (see Chapter 3); it is evidently desirable that the computed frequencies and, therefore, the computed models should at least in a numerical sense match the precision of the observations. Figure 1 illustrates schematically the structure of the Sun obtained from the computations.

5.1. FITTING THE RADIUS AND LUMINOSITY

The computation of a model of the present Sun requires the specification of certain parameters. The *solar mass* $M_\odot$ is known with great precision from the dynamics of the solar system. Estimates of the *present solar age* $t_\odot$ can be obtained from ages of meteorites, leading to a value of about 4.55×10^9 years (*e.g.* Guenther 1989); here, I shall use $t_\odot = 4.75 \times 10^9$ years. In addition, the *initial solar composition* must be specified; although it has been discussed here in terms of X, Y, and Z, the detailed distribution amongst the heavy elements has a substantial effect on the opacity. Spectroscopic observations of the solar photosphere yield measurements of the ratio between the abundances of the heavy elements and the abundance of hydrogen. However, since the photospheric spectrum of the Sun contains no lines of helium, a direct determination of the helium abundance is not possible; hence, the initial helium abundance Y_0 is an undetermined parameter. A second undetermined quantity is the mixing-length parameter α_C (*cf.* section 2.2).

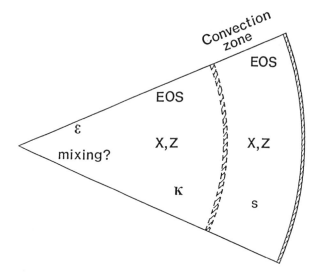

Figure 1. Schematic representation of solar structure. The thin hatched area near the surface indicates the region where the physics is uncertain, because of the effects of convection, nonadiabaticity, *etc.* At the base of the convection zone, overshoot introduces additional uncertainty. The structure of the adiabatic part of the convection zone is determined by the equation of state, the constant values of specific entropy s, and the composition (given by the abundances X and Z of hydrogen and heavy elements). Beneath the convection zone, the structure depends also on opacity κ and the energy generation rate ϵ.

The radius and surface luminosity of the computed model must agree with the observed values $R_\odot$ and $L_\odot$. This can be achieved by choosing the values of α_C and Y_0 suitably. Indeed, Equation (20) shows that the luminosity is quite sensitive to the composition, whereas the properties of the convection zone, which are determined by α_C, have a substantial effect on the radius of the model. A model that has been adjusted in this manner, as have the models discussed in the following, is called a *calibrated solar model*. Needless to say, the accuracy of the calibration has to match the desired accuracy of the model (*cf.* Christensen-Dalsgaard and Thompson 1991).

The calibration is essentially insensitive to the details of the description adopted for convection; all that matters is that the adiabatic part of the convection zone is on the correct adiabat (Gough and Weiss 1976). On the other hand, the value obtained for α_C evidently depends on the details of the mixing-length formalism which is used. Here, I have used the version of Böhm-Vitense (1958).

5.2. SOME RESULTS

It is of interest to consider the properties of computed solar models in a little more detail, and in particular to illustrate the effects of assuming different physics. Here, I consider the effect of changing the opacity; the calculation was described in more detail by Christensen-Dalsgaard (1991), who also discussed a substantial number of additional models.

The computation used the Eggleton, Faulkner, and Flannery (1973) equation of state and the Parker (1986) and Bahcall and Ulrich (1988) nuclear reaction rates. Model 1 was computed with the Cox and Tabor (1976) opacity tables (in the following CT) with $Z = 0.02$. Models 2 and 3 used opacities computed with the Los Alamos Opacity Library (LAOL) in the implementation used at Saclay (see Courtaud *et al.* 1990), and based on the abundances given by Anders and Grevesse (1989); the two models differ in the assumed iron abundance: for Model 2 an abundance by mass, relative to the total heavy element abundance Z, of 0.095 was used, corresponding to the value observed in the solar photosphere; for Model 3 the relative abundance was 0.067, corresponding to the meteoritic value. The values of Z for Model 2 and Model 3, 0.01934 and 0.01917 respectively, were obtained from the value of Z/X which has been determined spectroscopically and the value of X which results from the calibration of the models.

Some properties of the models are summarized in Table 1.

Table 1

No.	Y_0	α_C	T_c (10^6 K)	ρ_c (g cm^{-3})	p_c	d_b/R	T_b (10^6 K)	ν rate (SNU)
1	0.2496	1.6425	15.052	153.25	2.3632	0.2767	2.084	4.90
2	0.2776	2.5149	15.638	151.38	2.3303	0.2696	2.050	8.46
3	0.2709	2.5154	15.493	152.11	2.3454	0.2671	2.014	7.36

Summary of solar model calculations for models distinguished by the opacity tables used; the models have been calibrated to have solar radius and luminosity. The first column gives an identification number. Y_0 and α_C are the initial helium abundance and mixing-length parameter required to calibrate the model to have the correct luminosity and radius. T_c, ρ_c and p_c are the central temperature, density and pressure (the latter in 10^{17} dyn cm^{-2}); d_b/R is the depth of the convection zone in units of the surface radius, and T_b is the temperature at the base of the convection zone. The last column gives the neutrino capture rate in the ^{37}Cl experiment (see section 6).

There are several features distinguishing the two opacity tables which have direct effects on the computed models:

- The LAOL opacities are approximately a factor two higher than the CT opacities at conditions corresponding to the solar atmosphere. This changes the structure of the atmosphere, in particular the specific entropy, and hence the value of α_C required to reach the correct adiabat in the bulk of the convection zone. This feature, taken in isolation, has essentially no effect on the interior of the model, however.

- The LAOL opacity used for Model 2 is, on average, about 10 per cent higher than the CT opacity in the radiative interior. According to Equation (19), this would lead to roughly a 10 per cent decrease in luminosity, if nothing else were changed. Thus, to maintain the calibration of the model, the mean molecular weight must be increased by increasing the initial helium abundance; this accounts for the difference of about 0.03 in Y_0 between the two models. In accordance with Equation (18), the increase in μ in Model 2 is accompanied by an increase in the central temperature. Similarly, the reduction in iron abundance between Models 2 and 3 leads to smaller opacities in the interior, and hence to a slightly lower value of Y_0 in Model 3.

- At conditions near the base of the convection zone, the LAOL opacities are slightly smaller than the CT opacities. As a result, the convection zone extends to a slightly smaller depth and slightly lower temperature in Model 2 than in Model 1. The same effect can be noticed in going from Model 2 to Model 3.

A more detailed comparison of Models 2 and 3 and of their oscillation frequencies is given in section 4.1.1 of Chapter 3.

6. The Solar Neutrino Problem

The measurement of the flux of solar neutrinos might provide a direct measure of the rate of nuclear reactions in the core of the Sun. In practice, the very small cross section of the neutrino makes it extremely hard to detect. The results obtained over the past two decades have been consistently lower than the theoretically predicted values. This *solar neutrino problem* has led to extensive efforts to find modifications of the solar models that would bring the predicted neutrino flux into agreement with observations. A comprehensive discussion of the observations and model computations and of the possible solutions to the problem was given by Bahcall (1989).

It follows from the analysis in section 3.3 that the neutrinos may be emitted in a number of different reactions, with very different energies. The distribution of the neutrinos depends on the branching ratios between the PP-I, PP-II, and PP-III chains; an additional contribution comes from the so-called CNO cycle, which constitutes another (and, in the solar case, minor) route whereby hydrogen burning can take place. Since the sensitivity of the neutrino detectors depends strongly on the neutrino energy, the observed number of neutrinos depends on the details of the neutrino spectrum.

The longest-running experiment to detect solar neutrinos uses the reaction

$$\nu_e + {}^{37}\text{Cl} \rightarrow \text{e}^- + {}^{37}\text{Ar} . \tag{24}$$

It was set up by R. Davies in the Homestake Mine, South Dakota, around 1967. The detector consists of a tank containing about 380 000 liters of C_2Cl_4 (which is a common cleaning fluid). Even with this large amount of detector material, a solar neutrino event is recorded on average only every second day. Predicted capture rates for this experiment from the different sources of neutrinos are listed in Table 2, in units of the *Solar Neutrino Unit (SNU)* which is defined as 10^{-36} captures per target atom in the detector per second. The reaction is sensitive to neutrinos with energies exceeding about 0.8 MeV, but the predicted capture rate is dominated by the ^{8}B neutrinos, with the ^{7}Be neutrinos also making a significant contribution. On the other hand, the experiment is insensitive to the neutrinos emitted in the PP-I chain which dominates the energy generation in the Sun.

The measured capture rate is shown in Figure 2. An average over all the data gives

$$\text{Observed capture rate} = (2.05 \pm 0.2)\,\text{SNU} , \tag{25}$$

which is clearly inconsistent with the predicted rates.

Table 2

Neutrino source	Capture rate (SNU)		
	Model 1	Model 2	Model 3
pp	0.00	0.00	0.00
pep	0.24	0.22	0.23
^{7}Be	0.90	1.16	1.09
^{8}B	3.53	6.70	5.71
CNO	0.24	0.39	0.34
Total	4.90	8.46	7.36

Predicted capture rates for neutrinos in the ^{37}Cl detector, from the models in Table 1.

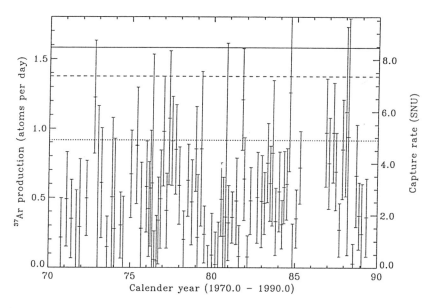

Figure 2. Observational results from the chlorine solar neutrino experiment, adapted from Bahcall (1989) and Bahcall and Press (1991). The lines across the top of the figure show the predictions of Models 1 (··············), 2 (————), and 3 (------------), in Tables 1 and 2.

It is evident from Table 2 that changes in the opacity introduce substantial differences in the computed capture rate, due to the differences in central temperature. This is related to the great temperature sensitivity of the branching to the PP-III chain. Particularly striking is the fact that even the relatively modest change in iron abundance and central temperature between Models 2 and 3 leads to a change in the capture rate by more than 1 SNU, almost entirely as a result of a change in the ^{8}B neutrinos. However, even for Model 1, the computed rate exceeds the observed rate by more than a factor two. Furthermore, it is likely that the older CT opacities are in fact somewhat too low in the solar interior.

It appears that, regardless of the assumed physics, "standard" solar models yield neutrino capture rates that are substantially higher than the observed values.

There may be some indication of a systematic, almost periodic, variation with time in the observed capture rate. This appears to be anti-correlated with the solar cycle: the observed capture rate tends to be large at solar minimum and small at solar maximum. The reality of this correlation is still hotly debated, however (see Bahcall and Press 1991).

A second experiment, the Kamiokande II experiment, has measured the scattering of neutrinos on electrons; the experiment is located in the Kamioka mine in Japan and consists a tank containing 3000 tons of water, of which 680 tons are used for neutrino detection. This experiment is exclusively sensitive to the ^{8}B neutrinos. It also records the direction of the neutrinos, confirming that they do indeed come from the Sun. As in the case of the Cl experiment, the observed rate of neutrino events is about half the value obtained from the theoretical predictions.

The high temperature sensitivity of the observed rates means that to reduce the predicted capture rate to the observed value requires a reduction in the core temperature of the Sun by only about 5 per cent while, of course, maintaining the correct total integrated energy generation rate and surface luminosity. It is possible to construct such models although only by making fairly drastic modifications to the normal assumptions or parameters of stellar evolution calculations. For example, Schatzman et al. (1981) considered a model with partial mixing of the core; since this brings additional hydrogen to the centre, the energy production can occur at a slightly lower temperature, leading to a lower neutrino capture rate. Alteratively, it was noted that if part of the energy transport in the solar core were to take place through the motion of hypothetical, so-called "weakly interacting massive particles" (WIMPs), the temperature gradient required for radiative transport in the core would be reduced; this would again lead to a a lower central temperature and a lower neutrino capture rate (see Steigman et al. 1978; Spergel and Press 1985; Faulkner and Gilliland 1985). However, as discussed in Chapter 3, such models have oscillation frequencies which are inconsistent with the observed values; in contrast, the frequencies of normal models generally agree reasonably well with the observations.

It is of great interest to carry out measurements that are sensitive to the neutrinos from the basic ^{1}H + ^{1}H reaction. Such measurements may be based on the reaction

$$\nu_e + {}^{71}\text{Ga} \rightarrow e^- + {}^{71}\text{Ge}\,, \tag{6}$$

which is sensitive to neutrinos of energy exceeding 0.23 MeV. Two such experiments are now being set up. Preliminary results from one of them (Abazov et al. 1991) indicate a counting rate of less than 79 SNU (with a 90 per cent confidence limit), substantially below the predicted value of about 132 SNU for standard solar models. However, the significance of this result is difficult to evaluate until further observations are available.

Given the apparent success of normal solar models in reproducing the observed solar oscillation frequencies, it is natural to seek the solution to the neutrino problem elsewhere. A possibility is the Mikheyev-Smirnov-Wolfenstein (MSW) effect (Wolfenstein 1978; Mikheyev and Smirnov 1985; see also Bahcall and Bethe 1990). It is based on the existence of two additional types of neutrinos: the *muon neutrino* ν_μ and the *tau neutrino* ν_τ. If neutrinos have mass, the electron neutrinos generated by hydrogen fusion may be transformed, through interaction with matter in the Sun, into neutrinos of different types which are not detected in current experiments. It is possible to choose parameters of the neutrinos such that the fraction of ν_e remaining is consistent with observations. Unfortunately, the effect is so small that it will probably never be within range of laboratory experiments. Thus, the only possibility for studying it experimentally is through observations of solar neutrinos. This requires that the neutrino spectrum generated by nuclear reactions is known with sufficient accuracy. Hence, by providing information about conditions in the solar core, helioseismology may contribute to the use of the Sun as a laboratory for neutrino physics.

7. Is There a "Faint Early Sun" Problem?

The luminosity of solar models increases as the model evolves. As hydrogen is converted into helium, the mean molecular weight μ increases. Since the pressure must be sufficient to balance the weight of the overlying material, it follows from Equation (8) that ρT must increase to compensate for the increase in μ. This is accomplished by contraction of the core which increases ρ. The contraction releases gravitational potential energy which, according to the virial theorem, goes partly towards increasing the internal energy of the gas; as a result, the temperature increases. This tends to increase the nuclear energy generation rate, as does the increase in density; furthermore, the increase in temperature increases the radiative flux of energy, partly through the resulting reduction in opacity (*cf.* Equations (4) and (10)). The result is that the luminosity of the star increases (see also Gough 1977, 1990).

The same trend is predicted by the simple expression for the luminosity given in Equation (20), if μ is taken to be an average mean molecular weight. It should be noted that in the derivation of that equation, the effect of μ comes exclusively from the increase in radiative transport caused by the increase in T which follows from an increase in μ.

To illustrate this effect in more detail, Figure 3 shows the changes in a number of quantities during the evolution of a solar model from its formation to the present age. These are largely consistent with the simplified discussion given above. It is interesting, however, that the change in the central energy generation rate ϵ_c is fairly modest; the effect of the increase in T_c is largely compensated by the decrease in the hydrogen abundance. Also, there is a substantial reduction in the central opacity κ_c. The combined effect is an increase in the surface luminosity L, such that L was approximately 30 per cent lower than at present at the beginning of hydrogen burning in the Sun.

Could this have affected the climate of the Earth? A simple estimate suggests that the answer might be affirmative. Assuming balance between the energy received and radiated by the Earth, and approximating the Earth by a black body, the equilibrium terrestrial temperature is approximately proportional to $L^{1/4}$. Hence, the initial average temperature of the Earth should have been 20 K lower than the present temperature which is about 290 K. This suggests that the Earth may initially have been frozen which conflicts with indications of the presence of liquid water at least 3.5×10^9 years ago. This so-called "faint early Sun" problem has caused attempts to produce solar models that did not display an increase in luminosity during main sequence evolution. In fact, this increase is one of the most robust predictions of stellar evolution theory. Even if the Sun were assumed to be completely mixed, a luminosity increase of about 20 per cent is still predicted. Models with a constant solar luminosity can be constructed, but only by involving drastic modifications

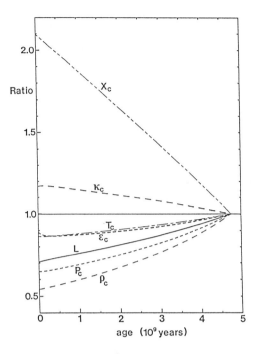

Figure 3. Variation in the surface luminosity and in properties at the centre during the evolution of a solar model. All variables have been normalized with their value in the present Sun. The heavy lines show the surface luminosity L (————), the central energy generation rate ϵ_c (------------), and the central opacity κ_c (——————). The thin lines show the central density ρ_c (——————), the central pressure p_c (------------), the central temperature T_c (——·——·——), and the central hydrogen abundance X_c (——·——··——).

in the basic physics such as a change with time in the gravitational constant G or possibly the solar mass.

The estimate given above of the change in the Earth's temperature is extremely naive, however, in that it ignores terrestrial climatology, particularly the so-called *greenhouse effect*: the atmosphere of the Earth is largely transparent to solar radiation; but because the temperature of the Earth is much lower, it radiates predominantly at much longer wavelengths where there is strong absorption in the atmosphere due to certain molecules such as carbon dioxide (CO_2). Hence, the energy is partially trapped in the Earth's atmosphere and the degree of trapping depends on the abundance of CO_2. Thus, with a suitable reduction in the abundance of CO_2, the decrease in the greenhouse effect might have compensated for the increase in solar luminosity, keeping the temperature at the Earth's surface approximately constant during solar evolution. In fact, there are geological mechanisms which may link the abundance of CO_2 in the Earth's atmosphere to the surface temperature of the Earth in such a way as to decrease the efficiency of the greenhouse effect if the temperature increases (see Kastings, Toon, and Pollack 1988).

8. Discussion

The computation of normal solar and stellar models follows well-defined and well-tested procedures. In addition to their use in understanding the properties of the Sun, the results are of importance to broad areas of astrophysics including models of element synthesis in the Universe and investigations of the ages and distribution of stars in the Galaxy. There are remaining uncertainties in the physics used in the calculations, particularly the opacity, which must be resolved. Nonetheless, there is general agreement concerning the main aspects of the results. On the other hand, the discrepancy between the predicted and observed neutrino capture rates must cause concern.

Of course, the predictions of normal stellar models depend on the validity of the underlying assumptions, and hence the possibility of failures of these assumptions must be kept in mind; indeed, the solar neutrino problem has led to numerous investigations of models that depart from the standard assumptions (for a review, see Maeder 1990). For example, departures from thermal equilibrium would have observable effects only over timescales of thousands of years. Thus, although on average the Sun must be close to thermal equilibrium during its evolution, we cannot totally exclude a temporary departure from equilibrium at the present time; Dilke and Gough (1972) suggested that such a departure, initiated by mixing caused by the onset of instability towards g-mode oscillations, might account for the scarcity of detected neutrinos at present. Another important departure from the standard assumptions would be mixing of the solar interior, modifying the variation with position in the hydrogen abundance resulting from nuclear burning. Additional potentially significant features that are not usually taken into account in stellar evolution calculations are settling of helium or heavy elements (*e.g.* Proffit and Michaud 1991) or convective overshoot (see Chapter 4).

Such departures from standard stellar evolution theory could have serious consequences for general computations of stellar models. To test for them, detailed observations of the solar interior would obviously be very helpful. As discussed in Chapter 3, helioseismic analyses of observed frequencies of solar oscillations provide such data. It is striking, and perhaps surprising, that the results obtained so far have generally been in accordance with the "standard" models; also, they have excluded many of the non-standard effects that have been proposed to solve the neutrino problem. New observations of solar oscillations over the coming decade are expected to lead to a much more sensitive probing of the solar interior, and hence should allow the study of more subtle features of the physics of stellar interiors.

References.

Abazov, A. I. *et al.*, 1991. *Phys. Rev. Lett.*, **67**, 3332 – 3335.

Anders, E. and Grevesse, N., 1989. *Geochim. Cosmochim. Acta*, **53**, 197 – 214.

Bahcall, J. N., 1989. *Neutrino astrophysics*, Cambridge University Press, Cambridge.

Bahcall, J. N. and Bethe, H. A., 1990. *Phys. Rev. Lett.*, **65**, 2233 – 2235.

Bahcall, J. N. and Press, W. H., 1991. *Astrophys. J.*, **370**, 730 – 742.

Bahcall, J. N. and Ulrich, R. K., 1988. *Rev. Mod. Phys.*, **60**, 297 – 372.

Böhm-Vitense, E., 1958. *Z. Astrophys.*, **46**, 108 – 143.

28

Christensen-Dalsgaard, J., 1991. In *Challenges to theories of the structure of moderate-mass stars, Lecture Notes in Physics*, vol. **388**, p. 11 – 36, eds Gough, D. O. and Toomre, J., Springer, Heidelberg.

Christensen-Dalsgaard, J. and Thompson, M. J., 1991. *Astrophys. J.*, **367**, 666 – 670.

Courtaud, D., Damamme, G., Genot, E., Vuillemin, M. and Turck-Chièze, S., 1990. *Solar Phys.*, **128**, 49 – 60.

Cox, A. N. and Tabor, J. E., 1976. *Astrophys. J. Suppl.*, **31**, 271 – 312.

Cox, A. N., Livingston, W. C. and Matthews, M., (eds), 1991. *Solar interior and atmosphere*, Space Science Series, University of Arizona Press, in the press.

Däppen, W., Keady, J. and Rogers, F., 1991. In Cox *et al.* (1991).

Dilke, F. W. W. and Gough, D. O., 1972. *Nature*, **240**, 262 – 264 and 293 – 294.

Eggleton, P. P., Faulkner, J. and Flannery, B. P., 1973. *Astron. Astrophys.*, **23**, 325 – 330.

Faulkner, J. and Gilliland, R. L., 1985. *Astrophys. J.*, **299**, 994 – 1000.

Gough, D. O., 1977. In *The solar output and its variation*, p. 451 – 474, ed. White, O. R., Colorado Ass. Univ. Press, Boulder.

Gough, D. O., 1990. In *Astrophysics. Recent progress and future possibilities, Mat.-fys. Meddel. Kgl. Danske Vidensk. Selsk.*, **vol. 42, No. 4**, 13 – 50.

Gough, D. O. and Weiss, N. O., 1976. *Mon. Not. R. astr. Soc.*, **176**, 589 – 607.

Guenther, D. B., 1989. *Astrophys. J.*, **339**, 1156 – 1159.

Kastings, J. F., Toon, O. B. and Pollack, J. B., 1988. *Scientific American*, **258** (February), p. 46 – 53 (US p. 90 – 97).

Kippenhahn, R. and Weigert, A., 1990. *Stellar structure and evolution*, Springer-Verlag, Berlin.

Maeder, 1990. In *Inside the Sun, Proc. IAU Colloquium No. 121*, eds Berthomieu G. and Cribier M., Kluwer, Dordrecht, p. 133 – 144.

Mikheyev, S. P. and Smirnov, A. Yu., 1985. *Yad. Fiz.*, **42**, 1441 – 1448 (English translation: *Sov. J. Nucl. Phys.*, **42**, 913 – 917).

Parker, P. D., 1986. *Physics of the Sun*, Vol. 1, p. 15 – 32, eds Sturrock, P. A., Holzer, T. E., Mihalas, D. and Ulrich, R. K., Reidel, Dordrecht, Holland.

Parker, P. D. and Rolfs, C. E., 1991. In Cox *et al.* (1991).

Proffit, C. R. and Michaud, G., 1991. *Astrophys. J.*, **380**, 238 – 250.

Schatzman, E., Maeder, A., Angrand, F. and Glowinski, R., 1981. *Astron. Astrophys.*, **96**, 1 – 16.

Spergel, D. N. and Press, W. H., 1985. *Astrophys. J.*, **294**, 663 – 673.

Steigman, G., Sarazin, C. L., Quintana, H. and Faulkner, J., 1978. *Astron. J.*, **83**, 1050 – 1061.

Wolfenstein, L., 1978. *Phys. Rev. D*, **17**, 2369 – 2374.

3 SEISMIC INVESTIGATION OF THE SOLAR INTERIOR

J. CHRISTENSEN-DALSGAARD
Institut for Fysik og Astronomi
Aarhus Universitet
DK-8000 Aarhus C
Denmark

ABSTRACT. Observation of a large number of modes of solar oscillation has permitted detailed investigation of the solar interior. To illustrate the diagnostic potential of the frequencies, some properties of the observed p-modes are discussed in terms of a simple ray picture of the oscillations. Solar models and their frequencies are used to illustrate how the frequencies depend on the physics of the solar interior. From inverse analyses of the frequencies, one may determine, *e.g.*, the variation of sound speed with position. Solar rotation causes fine structure in the frequencies. By inverting the observations, it is possible to infer the angular velocity as a function of depth and latitude in much of the Sun. Recent observations have given detailed information about frequency changes during the solar cycle. In future, greatly expanded sets of observations will result from networks of observing stations around the Earth and from space. Also, it is hoped that investigations of this nature can be extended to other stars.

1. Introduction

The Sun is observed to oscillate simultaneously in a very large number of individual modes. These oscillations typically have periods between about 15 and 3 minutes; they range in spatial scale from radially symmetric modes to modes whose wavelength on the solar surface is only a few thousand kilometers. The amplitudes for each mode in Doppler velocity is generally well below $1 \, \mathrm{m \, sec^{-1}}$, corresponding to a relative variation in intensity of order 10^{-6}. The frequencies of thousands of modes have been determined, with relative precisions that, in some cases, are better than 10^{-5}. These frequencies provide detailed and accurate information about the properties of the solar interior. *Helioseismology* is concerned with the utilization of these data.

Modes of solar oscillation are characterized by their degree l, azimuthal order m, and radial order n. The Sun can support three essentially different types of oscillations: standing acoustic waves (or p-modes), standing internal gravity waves (or g-modes), and surface gravity waves (or f-modes). At the lowest degrees, the observed periods correspond to high-order p-modes, whereas at moderate or high degrees both p-modes of low and moderate order and f-modes are observed.

Observation of solar oscillations is complicated by their very small amplitudes and by the complexity of the oscillation spectrum. On the other hand, the very long phase stability of the modes means that data can be combined coherently over long timeseries, bringing out the oscillations amongst the incoherent solar "noise." The modes have been observed by means of a variety of techniques, the most extensive sets of data having been obtained

J. T. Schmelz and J. C. Brown (ed.), The Sun, A Laboratory for Astrophysics, 29–80.

by means of observations of Doppler velocity. From spatially resolved observations, one can achieve considerable separation of the modes by means of suitable spatial transforms. Observations in light integrated over the solar disk, corresponding to observations of stars, are primarily sensitive to modes of very low degree. Problems are introduced by gaps in the timeseries, which give rise to sidebands surrounding the real peaks in the power spectra. These complicate the mode identification and may introduce systematic errors in the frequencies or contribute to the noise. To get long continuous time strings, projects are under way to combine observations from several sites.

For a given solar model, it is relatively straightforward to compute the frequencies numerically. However, to understand the behaviour of the modes and the relation between the structure of the model and the frequencies, simplified asymptotic descriptions are very useful. Through its effect on the structure of the model, the physics used in the computation plays an important role in determining the oscillation frequencies. This can be illustrated by comparing frequencies of models computed with different physics. It is found that there is a substantial sensitivity to the assumed equation of state and opacity. Indeed, one may hope from helioseismic data to obtain information about the properties of matter in the solar interior, although this requires that the effects of the physics can be separated from other uncertainties in the models, including departures from the assumptions underlying "standard" solar model computations.

In a non-rotating star, the frequencies are independent of the azimuthal order m. The rotation of the Sun introduces a splitting of the frequencies according to m, analogous to the Zeeman splitting of the energy levels of an atom in a magnetic field, with different modes sampling the rotation in different parts of the Sun. Frequency splitting, although with a different dependence on m, would also be introduced by a large-scale magnetic field, or by departures from spherical symmetry of the structure of the Sun.

The frequencies can be regarded as integrals over the solar interior structure or angular velocity. In this sense, the observed frequencies provide a set of integral equations for the properties of the solar interior. The solution of integral equations of this nature is usually known as *inverse analysis*. In the case of the observed solar acoustic oscillations, approximate inverse techniques can be based on the asymptotic properties of the modes. These have been used to determine the sound speed in much of the Sun. More general inversions for solar structure that do not rely on the asymptotic approximation have also been performed, leading to fairly accurate inferences about the density and sound speed in most of the Sun, including much of the energy-generating core. In addition, extensive inverse analyses have enabled inferences about the angular velocity in the solar interior.

Dramatic improvements in the helioseismic data may be expected over the next decade. These include the Global Oscillation Network Group (GONG) project to observe solar oscillations from a network of six stations, and helioseismic instruments on the *SOHO* satellite (for SOlar and Heliospheric Observatory), to be launched in 1995. Finally, several efforts are under way or being planned to search for solar-like oscillations of other stars. Altogether, the field of seismological investigations of the Sun and other stars promises to be very active and dynamic in the coming years.

2. Properties of Solar Oscillations

To get an impression of the possibilities of helioseismology and evaluate the results, some understanding of the properties of solar oscillations is required. Since the oscillations have small amplitudes, they can be described in terms of linear perturbation theory. Then, for a single mode, the dependence on colatitude θ and longitude ϕ is given by a *spherical harmonic* $Y_l^m(\theta, \phi)$. Here, the *degree l* determines the overall complexity of the perturbation over spherical surfaces. More precisely, the horizontal wavelength λ_h and the magnitude k_h of the horizontal component of the wavenumber at a distance r from the centre of the Sun are

$$k_h = \frac{2\pi}{\lambda_h} = \frac{L}{r} \tag{1}$$

(*cf.* section 2.2), where $L \equiv \sqrt{l(l+1)}$. The *azimuthal order m* determines the number of nodes around the equator. For a spherically symmetric star, the orientation of the coordinate system used to describe the oscillation can have no effect on the properties of the modes, including their frequencies. Since the definition of m is related to the location of the equator, it follows that frequencies of a spherical star must be independent of m. However, as discussed in section 2.4, the presence of rotation (or any other symmetry-breaking agent) lifts this degeneracy and causes a splitting according to m.

For each l and m, there is a spectrum of modes, distinguished by the dependence of the perturbation on r and characterized by the *radial order n*, which measures the number of zeros in the radial direction in, say, the radial component of the displacement associated with the perturbation.

As discussed in section 2.1, the excitation and damping of the modes is neglected in this Chapter; then, as a function of time t, the modes are purely harmonic functions $\cos \omega t$, where ω is the *angular frequency* of oscillation. For a theoretical description, the frequency is most simply characterized by ω; on the other hand, observed frequencies are normally given in terms of the *cyclic frequency*

$$\nu = \frac{\omega}{2\pi} = \frac{1}{P}, \tag{2}$$

where P is the oscillation period.

To illustrate the basic properties of the spectrum of possible oscillations, Figure 1 shows computed frequencies for a typical solar model. It is immediately obvious that there are two distinct classes of modes, which are conventionally distinguished by the sign of n. Those with $n > 0$ are normally called *p-modes*; they have frequencies that, for large degree, increase with l roughly as $l^{1/2}$. For modes with $n < 0$, normally called *g-modes*, the frequencies tend to a limit as l increases. The modes with $n = 0$ form an intermediate case, despite the similarity of the behaviour of their frequencies to that of the p-modes; these modes are called *f-modes*.

It should be noted that Figure 1 shows the modes that are *possible* in a solar model. What is of interest, from the point of view of helioseismology, are the modes that are actually excited in the Sun. The only modes for which definite data are available are the so-called *five-minute oscillations*, which occupy the range in period between about 15 and about 3 minutes, at all degrees between 0 and an upper limit of $l \simeq 1500$ which is mainly set by

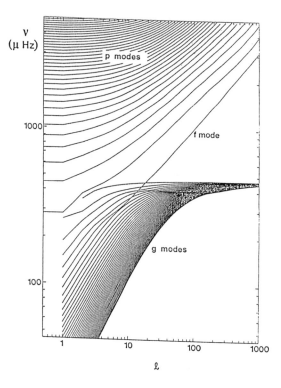

Figure 1. Adiabatic oscillation frequencies for a normal model of the present Sun, as functions of the degree l. For clarity, points corresponding to modes with a given radial order have been connected by straight lines. Only g-modes with radial order less than 40 have been included.

observational constraints. Figure 1 shows that they correspond to p-modes and, at higher degree, to f-modes. In the following, I am concerned only with these modes.

2.1. THE ADIABATIC APPROXIMATION

In most of the Sun, the typical thermal timescale $\tau_{\rm th}$ is much longer than typical pulsation periods (in particular, $\tau_{\rm th}$ for the entire Sun is roughly the Kelvin-Helmholz time $t_{\rm KH}$, of order 3×10^7 years; *cf.* Chapter 2). The only exception is very near the solar surface, where $\tau_{\rm th}$ is comparable with the oscillation periods. As a result, it is generally a good approximation to neglect loss or gain of energy during a pulsation period. This corresponds to assuming that the oscillations occur *adiabatically*, so that

$$\frac{\delta p}{p} = \Gamma_1 \frac{\delta \rho}{\rho} , \tag{3}$$

where p is pressure, ρ is density, $\Gamma_1 = (\partial \ln p / \partial \ln \rho)_s$ (the derivative being at constant specific entropy s), and δ indicates the *Lagrangian* perturbation, *i.e.*, the perturbation following the motion.

The adiabatic approximation is evidently highly restrictive. In particular, the motion is conservative, and hence the total energy is constant. Thus, when considering a single mode of oscillation in the adiabatic approximation, the amplitude of the mode is constant and, therefore, it is not possible to treat its excitation or damping. Also, close to the surface and in the atmosphere, the thermal time scale is comparable with or smaller than the periods and there are large departures from adiabaticity. Nonadiabatic effects modify the dynamical properties of the motion, and hence affect the oscillation frequencies; however, by far the largest fraction of the mass involved in the oscillations is in regions where τ_{th} is large and the dynamics is barely affected by nonadiabaticity. Thus, although the nonadiabatic effects on the frequencies are not negligible, they are generally small.

The adiabatic approximation simplifies considerably the discussion of the properties of the oscillations. Thus, in the following, I shall assume that the oscillations are adiabatic, keeping in mind that I thereby introduce errors in the computed frequencies.

Additional complications are introduced by the instability of the outer layers to convection, which gives rise to convective motion. An obvious effect of this is the convective flux, which in most of the convection zone dominates the energy transport; however, convection also gives rise to a "turbulent pressure" which affects the hydrostatic structure of the model and the dynamics of the oscillations. No satisfactory procedure exists for dealing with the convection in a static star, let alone in the presence of pulsations. Balmforth and Gough (1990a) and Balmforth (1992) used a time-dependent generalization of mixing-length theory to treat the effect of convection on the oscillations, whereas Stein, Nordlund, and Kuhn (1988) and Stein and Nordlund (1991) studied such effects by means of detailed hydrodynamic models of convection; these calculations showed that convection may introduce significant, and still somewhat uncertain, shifts in the frequencies. Here, I shall ignore effects of convection; the resulting errors in the properties of the oscillations, in particular the frequencies, must be kept in mind when computed frequencies are compared with the observations.

2.2. BASIC OSCILLATION EQUATIONS

A complete derivation of the equations of linear stellar oscillation is beyond the scope of this Chapter; however, for the the subsequent discussion, it is convenient to include a brief summary. (More detailed discussions are given by, for example, Unno *et al.* 1989 and Christensen-Dalsgaard and Berthomieu 1991). The derivation starts from the general hydrodynamical equations of motion and continuity and Poisson's equation; viscosity is neglected. The equations are linearized; this is done most conveniently in terms of the *Eulerian* perturbations, *i.e.*, the perturbations at a given point in space, which are denoted by a prime. Also, the adiabatic relation, Equation (3), is applied. Because of the spherical symmetry and time independence of the equilibrium model, one can separate out the dependence of the perturbation quantities on co-latitude θ, longitude ϕ, and time, using spherical polar coordinates (r, θ, ϕ). The relevant functions of θ and ϕ are eigenfunctions of the tangential Laplace operator

$$r^2 \nabla_t^2 \equiv \frac{1}{\sin \theta} \frac{\partial}{\partial \theta} \left(\sin \theta \frac{\partial}{\partial \theta} \right) + \frac{1}{\sin^2 \theta} \frac{\partial^2}{\partial \phi^2} .$$

These may be chosen as spherical harmonics

$$Y_l^m(\theta, \phi) = (-1)^m c_{lm} P_l^m(\cos \theta) e^{im\phi} , \tag{4}$$

where P_l^m is a Legendre function, and the normalization constant c_{lm} is determined such that the integral of $|Y_l^m|^2$ over the unit sphere is unity. Similarly, the time dependence can be expressed in terms of a harmonic function. Thus, *e.g.*, the pressure perturbation is written as

$$p'(r, \theta, \phi, t) = \sqrt{4\pi} Re \left[p'(r) Y_l^m(\theta, \phi) e^{-i\omega t} \right] , \tag{5}$$

where, for simplicity, I use p' to denote both the full perturbation and the amplitude function. Also, it follows from the momentum equation that the displacement vector can be written as

$$\delta \mathbf{r}(r, \theta, \phi, t) = \sqrt{4\pi} Re \left\{ \left[\xi_r(r) Y_l^m \mathbf{a}_r + \xi_h(r) \left(\frac{\partial Y_l^m}{\partial \theta} \mathbf{a}_\theta + \frac{1}{\sin \theta} \frac{\partial Y_l^m}{\partial \phi} \mathbf{a}_\phi \right) \right] e^{-i\omega t} \right\} , \tag{6}$$

where $\mathbf{a}_r$, $\mathbf{a}_\theta$, and $\mathbf{a}_\phi$ are unit vectors in the r, θ, and ϕ directions. It may be noted that, with this separation of variables, the effect of ∇_t^2 on any scalar variable corresponds to multiplication by $-k_h^2$, where k_h is given by Equation (1). The root mean squares, over a spherical surface at radius r and time, of the vertical and horizontal components of the displacement are given by

$$\langle \delta r \rangle_{rms} = \frac{1}{\sqrt{2}} \xi_r(r), \qquad \langle \delta h \rangle_{rms} = \frac{L}{\sqrt{2}} \xi_h(r) . \tag{7}$$

The amplitude ξ_h of the horizontal component of displacement is related to p' and the perturbation Φ' in the gravitational potential by

$$-\omega^2 \rho \xi_h = -\frac{1}{r}(p' - \rho \Phi') . \tag{8}$$

The equations may now be arranged into the following fourth-order set of ordinary differential equations for the amplitude functions $(\xi_r(r), \xi_h(r), \Phi'(r))$:

$$\frac{d\xi_r}{dr} = -\left(\frac{2}{r} + \frac{1}{\Gamma_1} \frac{d \ln p}{dr} \right) \xi_r + \frac{r\omega^2}{c^2} \left(\frac{S_l^2}{\omega^2} - 1 \right) \xi_h - \frac{1}{c^2} \Phi' \tag{9}$$

$$\frac{d\xi_h}{dr} = \frac{1}{r} \left(1 - \frac{N^2}{\omega^2} \right) \xi_r + \left(\frac{1}{\Gamma_1} \frac{d \ln p}{dr} - \frac{d \ln \rho}{dr} - \frac{1}{r} \right) \xi_h + \frac{N^2}{rg\omega^2} \Phi' \tag{10}$$

$$\frac{d^2 \Phi'}{dr^2} = -\frac{2}{r} \frac{d\Phi'}{dr} - 4\pi G \rho \left(\frac{N^2}{g} \xi_r + \frac{r\omega^2}{c^2} \xi_h \right) + \left(\frac{L^2}{r^2} - \frac{4\pi G \rho}{c^2} \right) \Phi' . \tag{11}$$

Here c is the adiabatic sound speed,

$$c^2 = \frac{\Gamma_1 p}{\rho} \simeq \frac{\Gamma_1 k_B T}{\mu m_u} , \tag{12}$$

where the last approximation is valid for an ideal gas; here, T is temperature, k_B is Boltzmann's constant, m_u is the atomic mass unit, and μ is the mean molecular weight. Also, I have introduced the Lamb frequency

$$S_l = \frac{Lc}{r} , \tag{13}$$

and the buoyancy frequency N defined by

$$N^2 = g \left(\frac{1}{\Gamma_1} \frac{d \ln p}{dr} - \frac{d \ln \rho}{dr} \right) , \tag{14}$$

where g is the gravitational acceleration.

The amplitude functions must satisfy boundary conditions at the centre and surface of the Sun. The centre is a regular singular point of Equations (9) – (11). The behaviour of the solution near $r = 0$ can be analyzed by expansion in r, using the expansion of the equilibrium quantities; in this way, boundary conditions that isolate the regular solutions can be established. At the surface, one boundary condition is obtained by requiring that Φ' and its first derivative match continuously onto the decreasing solution to Laplace's equation outside the Sun. The dynamical surface boundary condition should, in principle, take into account the detailed behaviour of the oscillation in the solar atmosphere. In the idealized case, where the model is assumed to have a free outer surface, the absence of forces on the surface corresponds to demanding that the pressure perturbation vanish on the perturbed surface, *i.e.*,

$$\delta p = p' + \frac{dp}{dr} \xi_r = 0, \quad \text{at} \quad r = R . \tag{15}$$

In practice, a somewhat more realistic condition is applied at a suitable point within the solar atmosphere (*e.g.* Unno *et al.* 1989). This condition is qualitatively similar to Equation (15), particularly at fairly low frequencies; furthermore, the solution is relatively insensitive to the details of the boundary condition, as long as it is applied sufficiently high in the atmosphere (see also Ulrich and Rhodes 1983; Noels, Scuflaire, and Gabriel 1984).

By using Equation (8) and the equation of hydrostatic equilibrium, Equation (15) can also be written as

$$\xi_h = \frac{GM}{R^3 \omega^2} \xi_r - \frac{1}{R \omega^2} \Phi' , \tag{16}$$

where M is the total mass of the Sun. As argued below, Φ' is often small. Hence, Equations (16) and (7) show that the ratio between the root-mean-squares of the horizontal and vertical components of the surface velocity is given approximately by

$$\frac{\langle V_h \rangle_{rms}}{\langle V_r \rangle_{rms}} \simeq \frac{GM}{R^3 \omega^2} L , \tag{17}$$

and hence is a function of the frequency and the degree, but not of the detailed nature of the mode. Also, it is evident that for high-frequency modes of low degree the surface velocity is predominantly in the radial direction.

It should be noted that Equations (9) – (11), as well as the boundary conditions, are independent of the azimuthal order m; it follows that the frequency and the eigenfunctions do not depend on m. This is a consequence of the assumed spherical symmetry: m is related to the choice of polar axis in the coordinate system used to describe the oscillations, and that choice can have no physical effects in a spherically symmetric system. As discussed in section 2.4, rotation introduces a preferred axis, and hence a dependence of the frequencies on m.

The differential equations and boundary conditions constitute a boundary value problem. This is in principle straightforward to solve numerically; in practice, the high radial order of the observed modes means that some care is required in the calculations to match the precision of the observed frequencies. A summary of computational techniques and remarks about their precision was given by Christensen-Dalsgaard and Berthomieu (1991).

When using observed frequencies to probe the structure of the solar interior, it is crucially important how the frequencies depend on the properties of the solar model. In fact, the coefficients in Equations (9) – (11) are essentially completely determined if density ρ and the adiabatic exponent Γ_1 are given as functions of r, assuming that the model is in hydrostatic equilibrium. Given ρ, the interior mass $m(r)$ is obtained by integration, with the obvious boundary condition $m(0) = 0$. Then $p(r)$ can be obtained from Equation (2.1) of Chapter 2 by integrating from the surface; this requires an assumption about the surface pressure which, for example, can be obtained from semi-empirical models of the solar atmosphere. Given $p(r)$, $\rho(r)$, and $\Gamma_1(r)$, the coefficients can be evaluated. Instead of ρ and Γ_1, one may choose any other pair of variables from which ρ and Γ_1 can be derived; given that acoustic modes are predominantly influenced by the sound speed, a natural choice is to use $c(r)$ and $\Gamma_1(r)$.

2.3. Asymptotic theory of p-modes

Numerical solution of the equations of adiabatic oscillations is straightforward. However, the understanding of the numerical results and the interpretation of the observed oscillations have been greatly assisted by asymptotic analyses of the oscillation equations. The usefulness of asymptotics is, to a large extent, due to the fact that the observed acoustic modes in the five-minute region have high radial order or high degree. Hence, asymptotics can provide relations of acceptable accuracy between the properties of the Sun and the properties of the oscillation frequencies.

The asymptotic analysis is simplified considerably by noting that at high degree or radial order the perturbation Φ' in the gravitational acceleration can be neglected. This approximation, known as the *Cowling approximation*, was first suggested by Cowling (1941). It may be justified by noting that at high degree or order a mode gives rise to many regions of alternating sign in the density perturbation, the effects of which largely cancel in the perturbation of the gravitational potential.

When Φ' is neglected, the system (9) – (11) of equations is reduced to a second-order system, which is amenable to JWKB analysis (for Jeffreys, Wentzel, Kramers, Brillouin; see, for example, Gough 1986a, Unno *et al.* 1989). However, the asymptotic behaviour of p-modes can, in fact, be understood very simply in terms of the propagation of sound waves (*e.g.*, Christensen-Dalsgaard *et al.* 1985). The wavelength of high-order modes is small compared with the typical scale over which the equilibrium structure changes; furthermore, the frequencies of the modes are high compared with N and, therefore, effects of buoyancy can be neglected. Hence, the modes can be approximated locally by plane sound waves with the dispersion relation

$$k^2 \equiv k_r^2 + k_h^2 = \frac{\omega^2}{c^2} , \tag{18}$$

where k_r and k_h are the radial and horizontal components of the wave vector. For a wave corresponding to a mode of oscillation, k_h is given by Equation (1). From Equation (18), one then obtains

$$k_r^2 = \frac{\omega^2}{c^2} - \frac{L^2}{r^2} . \tag{19}$$

Close to the surface, c is small, and hence k_r is large. Here, the wave propagates almost vertically. With increasing depth, c increases and k_r decreases (see Figure 2) until $k_r = 0$ and the wave propagates horizontally. This happens at the so-called *lower turning point* $r = r_t$ where

$$\frac{r_t}{c(r_t)} = \frac{L}{\omega} . \tag{20}$$

Thus, the location of the lower turning point is a function of ω/L alone. The reflection of the wave at the surface is not immediately contained in this simple description; however, it is at least plausible that it results from the steep density gradient. A more detailed asymptotic analysis shows that the wave is reflected when its frequency is less than the acoustical cutoff frequency, ω_{ac} which, for the approximately isothermal solar atmosphere, is given by

$$\omega_{\mathrm{ac}} = \frac{c}{2H_p} \tag{21}$$

(Lamb 1909), where H_p is the pressure scale height (in the solar atmosphere, $\omega_{\mathrm{ac}}/2\pi$ is around $5300\,\mu\mathrm{Hz}$). Thus, at frequencies below ω_{ac} the wave propagates in a series of "bounces" between the surface and the turning point (*cf.* Figure 2). A mode of oscillation is formed as the interference pattern between such bouncing waves.

A more direct derivation of this behaviour, which avoids explicit separation of the spherical harmonics, can be carried out on the basis of ray theory (Gough 1984, 1986a).

The behaviour of r_t determines the regions over which the modes propagate and which, therefore, affect their frequencies. Figure 3 shows how r_t varies with l and the frequency for a normal solar model. Low-degree modes extend almost to the centre of the Sun, whereas modes at the highest degrees observed are confined to the outer fraction of a per cent of the solar radius.

This description of the p-modes also yields the asymptotic dispersion relation for their frequencies. To get a standing wave, the change in radial phase between the lower turning point and the surface must be an integral multiple of π, apart from contributions coming from the turning points. It follows from Equation (19) that this condition may be expressed as

$$\omega \int_{r_t}^{R} \left(1 - \frac{L^2 c^2}{\omega^2 r^2}\right)^{1/2} \frac{dr}{c} \simeq \pi(n + \alpha) . \tag{22}$$

Here, α describes the phase change at the turning points; a more careful analysis shows that $\alpha = \alpha(\omega)$ is, in general, a function of frequency.

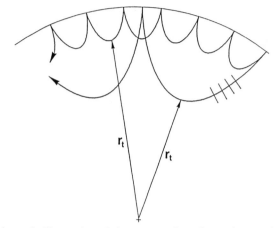

Figure 2. Schematic illustration of the propagation of sound waves in a star. Due to the increase of the sound speed with depth, the deeper parts of the wave fronts move faster. This causes the refraction of the wave described by Equation (19). Notice that waves with a smaller wavelength, corresponding to a higher value of the degree l, penetrate less deeply.

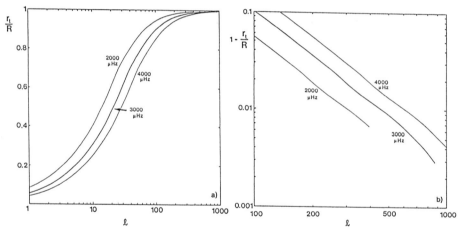

Figure 3. The turning point radius r_t (a) and the penetration depth $R - r_t$ (b), in units of the solar radius R, as a function of degree l for three values of the frequency ν. The curves have been calculated from Equation (20) for a normal model of the present Sun.

As r_t is a function of ω/L, Equation (22) may be written as

$$\frac{\pi[n + \alpha(\omega)]}{\omega} = F\left(\frac{\omega}{L}\right) , \qquad (23)$$

where

$$F(w) = \int_{r_t}^{R} \left[1 - \left(\frac{c}{rw}\right)^2\right]^{1/2} \frac{dr}{c} . \qquad (24)$$

A relation like Equation (23) was first found by Duvall (1982) for observed frequencies.

It is of interest to use Equations (23) and (24) to estimate the effects on the frequencies of *changes* in the equilibrium model. If δc denotes the difference in c between two solar models at fixed r, and $\delta\alpha$ denotes the difference in α at fixed frequency, it follows (by linearization), assuming δc and $\delta\alpha$ to be small, that

$$S\frac{\delta\omega}{\omega} \simeq \int_{r_t}^{R}\left(1 - \frac{L^2 c^2}{r^2\omega^2}\right)^{-1/2}\frac{\delta c}{c}\frac{dr}{c} + \pi\frac{\delta\alpha}{\omega} \,, \tag{25}$$

where

$$S = \int_{r_t}^{R}\left(1 - \frac{L^2 c^2}{r^2\omega^2}\right)^{-1/2}\frac{dr}{c} - \pi\frac{d\alpha}{d\omega} \tag{26}$$

(Christensen-Dalsgaard, Gough, and Pérez Hernández 1988). Since α is assumed to be a function of ω alone, and r_t is determined by ω/L, the scaled frequency difference $S\,\delta\omega/\omega$ is predicted to be of the form

$$S\frac{\delta\omega}{\omega} \simeq H_1\left(\frac{\omega}{L}\right) + H_2(\omega) \,, \tag{27}$$

where the two functions H_1 and H_2 are determined by Equation (25). Furthermore, the last term on the right-hand side of Equation (26) is, in general, relatively small compared with the first, and hence S is approximately a function of ω/L. Equation (27) evidently describes a very special dependence of $\delta\omega$ on ω and L.

Since c/r decreases quite rapidly with increasing r, $(Lc/r\omega)^2 \ll 1$ except near the turning point r_t; hence, as a rough approximation, $1 - L^2 c^2/r^2\omega^2$ may be replaced by 1 in the integrals in Equations (25) and (26). If, furthermore, the term in $\delta\alpha$ can be neglected, the result is the very simple relation between the changes in sound speed and frequency:

$$\frac{\delta\omega}{\omega} \simeq \frac{\int_{r_t}^{R}\dfrac{\delta c}{c}\dfrac{dr}{c}}{\int_{r_t}^{R}\dfrac{dr}{c}} \,. \tag{28}$$

This shows that the change in sound speed in a region of the Sun affects the frequency with a weight determined by the time the mode, regarded as a superposition of traveling waves, spends in that region. Thus changes near the surface, where the sound speed is low, have relatively large effects on the frequencies.

As discussed in section 5.3.1, the asymptotic relations (22) and (25) permit an almost direct, and surprisingly precise, inversion for the solar sound speed or the sound-speed difference between the Sun and a model.

When l is small, $(Lc/\omega r)^2 \ll 1$ except close to the centre of the Sun. Also, it is found that $L = \sqrt{l(l+1)}$ should be replaced by $l + 1/2$ (*e.g.* Brodsky and Vorontsov 1987). By expanding Equation (22) to take out the dependence on l, one then obtains

$$\nu_{nl} = \frac{\omega_{nl}}{2\pi} \simeq \left(n + \frac{l}{2} + \frac{1}{4} + \alpha\right)\Delta\nu \,, \tag{29}$$

where

$$\Delta\nu = \left[2\int_{0}^{R}\frac{dr}{c}\right]^{-1} \tag{30}$$

is the inverse of twice the sound travel time between the centre and the surface (see also Vandakurov 1967, Tassoul 1980). Thus, to this asymptotic order, there is a uniform spacing $\Delta\nu$ between modes of same degree but different order. In addition, Equation (29) predicts the approximate equality

$$\nu_{nl} \simeq \nu_{n-1\,l+2} \; . \tag{31}$$

This frequency pattern has been observed for the solar five-minute modes of low degree (*cf.* section 3.3) and may be used in the search for stellar oscillations of solar type.

The *deviations* from the simple relation (29) have considerable diagnostic potential. By taking the expansion of Equation (22) to the next order (Gough 1986a), or from direct JWKB analysis of the oscillation equations (Tassoul 1980), one obtains

$$\delta\nu_{nl} \equiv \nu_{nl} - \nu_{n-1\,l+2} \simeq -(4l+6)\frac{\Delta\nu}{4\pi^2\nu_{nl}}\int_0^R \frac{dc}{dr}\frac{dr}{r} \; . \tag{32}$$

Thus $\delta\nu_{nl}$ is determined predominantly by conditions in the solar core. This may be understood physically by noting that only near the centre is k_h comparable with k_r. Elsewhere, the wave vector is almost vertical and the dynamics of the oscillations are largely independent of their horizontal structure, *i.e.*, of l; therefore, at given frequency the contributions of these layers to the frequency are nearly the same, and hence almost cancel in the difference $\delta\nu_{nl}$.

It should be noted that the accuracy of Equation (32) is questionable: the derivation ignores effects of the perturbation in the gravitational potential, which are not negligible for low-degree modes; in addition, the asymptotic description assumes that the equilibrium model varies on a scale that is large compared with the wavelength of the modes, and this condition is not satisfied in the core of the model. Largely fortuitously, these two errors almost cancel in models of the present Sun where, consequently, Equation (32) agrees with computed frequencies; it is less successful for models of different ages or masses (Christensen-Dalsgaard 1988, 1992a). However, the general form of the dependence of $\langle\delta\nu_{nl}\rangle$ on l shown in Equation (32), as well as the argument that this quantity is most sensitive to conditions in stellar cores, has a broader range of validity.

In comparisons between computed and observed values of $\delta\nu_{nl}$, it is convenient to use a parametrization in terms of a small number of parameters. Equation (32) suggests that one consider the scaled separation

$$d_{nl} = \frac{3}{2l+3}\delta\nu_{nl} \; , \tag{33}$$

thus effectively reducing the separation to the value for $l=0$; the dependence of d_{nl} on l can then be used as an indication of departures from the asymptotic behaviour. Elsworth *et al.* (1990a) proposed that d_{nl} be analyzed in terms of a linear fit of the form

$$d_{nl} = \overline{d_l} + s_l(n - n_1) \; , \tag{34}$$

where n_1 is a suitable reference value of the order n. Examples of such fits to observed and computed frequencies are discussed in section 4.2.

Asymptotic analysis also shows the existence of modes whose frequencies are given by

$$\omega^2 = \frac{g_s L}{R} \; , \tag{35}$$

where g_s is the surface gravity. For these modes, $\text{div}\delta\mathbf{r}$ vanishes and ξ_r decreases with increasing depth as $\exp[-k_h(R-r)]$; they are entirely equivalent to surface gravity waves at a free surface (for example on a Scottish loch). They correspond to the f-modes at moderate or high degree in Figure 1.

2.4. ROTATIONAL SPLITTING

I assume that rotation is sufficiently slow to allow neglect of the rotational distortion of the star or, more generally, of terms of order Ω^2, where Ω is the angular velocity. Then, the structure of the star is unchanged and the effects of rotation can be treated as a small perturbation in the equations describing the oscillations.

That rotation causes a splitting of the frequencies can be seen from a purely geometrical argument. Assume the angular velocity Ω to be uniform, and consider an oscillation with a frequency ω_0, independent of m, in the frame rotating with the star. I introduce a coordinate system in this frame, with coordinates (r', θ', ϕ') which are related to the coordinates (r, θ, ϕ) in an inertial frame through

$$(r', \theta', \phi') = (r, \theta, \phi - \Omega t) \,. \tag{36}$$

It follows from Equation (6) that, in the rotating frame, the perturbations depend on ϕ' and t as $\cos(m\phi' - \omega_0 t)$; hence, the dependence in the inertial frame is $\cos(m\phi - \omega_m t)$, where

$$\omega_m = \omega_0 + m\Omega \,. \tag{37}$$

Thus an observer in the inertial frame finds that the frequency is split uniformly according to m.

This description is obviously incomplete. Even in the case of uniform rotation, the effects of the Coriolis force must be taken into account in the rotating frame, causing a contribution to the frequency splitting (Cowling and Newing 1949; Ledoux 1949). Furthermore, in general the angular velocity is a function $\Omega(r, \theta)$ of position. Nevertheless, as shown below, the effect of the Coriolis force is often small and Equation (37) is approximately correct if Ω is replaced by a suitable average of the position-dependent angular velocity.

The general case was considered by Hansen, Cox, and van Horn (1977), Cuypers (1980), and Gough (1981). The result is that the *rotational splitting*, *i.e.*, the perturbation in the frequencies caused by rotation, can be written as

$$\omega_{nlm} - \omega_{nl0} = \delta\omega_{nlm} = m\frac{R_{nlm}}{I_{nlm}} \,, \tag{38}$$

where

$$R_{nlm} = \int_0^\pi \sin\theta d\theta \int_0^R \left\{ \xi_r^2 P_l^m(\cos\theta)^2 + \xi_h^2 \left[\left(\frac{dP_l^m}{d\theta}\right)^2 + \frac{m^2}{\sin^2\theta} P_l^m(\cos\theta)^2 \right] \right. \tag{39}$$

$$\left. -2P_l^m(\cos\theta)^2\xi_r\xi_h - 2P_l^m(\cos\theta) \frac{dP_l^m}{d\theta}\frac{\cos\theta}{\sin\theta} \xi_h^2 \right\} \Omega(r, \theta)\rho(r)r^2 dr \,,$$

and

$$I_{nlm} = \frac{2}{2l+1}\frac{(l+|m|)!}{(l-|m|)!}\int_0^R [\xi_r^2 + l(l+1)\xi_h^2]\,\rho(r)r^2dr \ . \tag{40}$$

It should be noted from Equations (38) – (40) and the properties of the P_l^m that $\delta\omega_{nlm}$ is an odd function of m, $\delta\omega_{nl-m} = -\delta\omega_{nlm}$; this is a consequence of the fact that rotation imparts orientation, and hence distinguishes between modes travelling East and West. In this, rotation differs from other effects, such as asphericity of solar structure or magnetic fields, that might cause frequency splitting according to m; such splittings are even functions of m.

2.4.1. Splitting for spherically symmetric rotation. To proceed, an explicit assumption about the variation of Ω with θ is required. For simplicity, I shall assume first that Ω is independent of θ. In this case, the integrals over θ in Equation (39) only involve Legendre functions and may be evaluated analytically. The result is

$$\delta\omega_{nlm} = m\,\frac{\int_0^R \Omega(r)\,[\xi_r^2 + (L^2-1)\xi_h^2 - 2\xi_r\xi_h]\,r^2\rho dr}{\int_0^R (\xi_r^2 + L^2\xi_h^2)\,r^2\rho dr}\ . \tag{41}$$

It should be noticed that the integrands in Equation (41) are given solely in terms of ξ_r, ξ_h, and l, and hence are independent of m. Therefore, in the case of spherically symmetric rotation the splitting is proportional to m.

It is convenient to write Equation (41) as

$$\delta\omega_{nlm} = m\beta_{nl}\int_0^R K_{nl}(r)\Omega(r)dr\ , \tag{42}$$

where

$$K_{nl} = \frac{[\xi_r^2 + (L^2-1)\xi_h^2 - 2\xi_r\xi_h]\,r^2\rho}{\int_0^R [\xi_r^2 + (L^2-1)\xi_h^2 - 2\xi_r\xi_h]\,r^2\rho dr}\ , \tag{43}$$

and

$$\beta_{nl} = \frac{\int_0^R [\xi_r^2 + (L^2-1)\xi_h^2 - 2\xi_r\xi_h]\,r^2\rho dr}{\int_0^R (\xi_r^2 + L^2\xi_h^2)\,r^2\rho dr}\ . \tag{44}$$

This definition ensures that the *rotational kernel* K_{nl} is unimodular, *i.e.*,

$$\int_0^R K_{nl}(r)dr = 1\ . \tag{45}$$

Hence for uniform rotation, where $\Omega = \Omega_s$ is constant,

$$\delta\omega_{nlm} = m\beta_{nl}\Omega_s\ . \tag{46}$$

Comparison with Equation (37) shows that, in this case, the effects of the detailed dynamics of the oscillations are contained in β_{nl}. For high-order or high-degree p-modes, the terms in ξ_r^2 and $L^2\xi_h^2$ dominate and $\beta_{nl} \simeq 1$. Thus, the rotational splitting between adjacent m-values is given simply by the angular velocity, as in Equation (37). Physically, the neglected terms in Equation (44) do indeed arise from the Coriolis force.

When Ω depends on r, the integral in Equation (42) provides a weighted average

$$\langle\Omega\rangle = \int_0^R K_{nl}(r)\Omega(r)dr \qquad (47)$$

of $\Omega(r)$. For high-order p-modes, one obtains from the asymptotic behaviour of the eigenfunctions that

$$\delta\omega_{nlm} \simeq m\langle\Omega\rangle \simeq \frac{\int_{r_t}^R \left(1 - \frac{L^2c^2}{r^2\omega^2}\right)^{-1/2} \Omega(r)\frac{dr}{c}}{\int_{r_t}^R \left(1 - \frac{L^2c^2}{r^2\omega^2}\right)^{-1/2}\frac{dr}{c}} \simeq m\frac{\int_{r_t}^R \Omega(r)\frac{dr}{c}}{\int_{r_t}^R \frac{dr}{c}}, \qquad (48)$$

where in the last equality I crudely approximated $(1 - L^2c^2/r^2\omega^2)$ by 1. Thus, in this case one obtains the intuitively appealing result that the splitting between adjacent values of m is an average of the angular velocity, weighted by the sound-travel time. It should be noted that Equation (48) is entirely analogous to Equations (25), (26), and (28) for the frequency change arising from a perturbation to the sound speed.

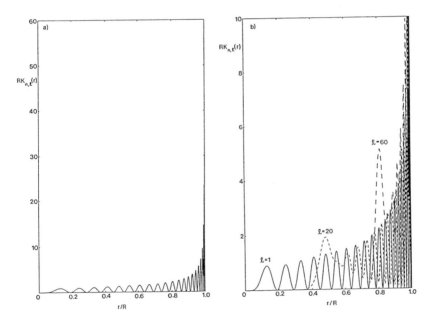

Figure 4. Kernels K_{nl} for the frequency splitting in p-modes caused by spherically symmetric rotation (*cf.* Equation (43)). In a) is plotted $RK_{nl}(r)$ for a mode with $l = 1$, $n = 22$ and $\nu = 3233\,\mu$Hz. The maximum value of $RK_{nl}(r)$ is 57. In b) is shown the same mode, on an expanded vertical scale, (————) together with the modes $l = 20$, $n = 17$, $\nu = 3367\,\mu$Hz (·············), and $l = 60$, $n = 10$, $\nu = 3231\,\mu$Hz (– – – – –). Notice that the kernels almost vanish inside the turning point radius r_t, and that there is an accumulation just outside the turning point.

Figure 4 shows a few examples of kernels for high-order p-modes. They are clearly large near the solar surface, as is also implicit in Equation (48). Beneath the turning point, the kernels get very small but they are enhanced locally just above it. This effect arises from the term in ξ_h in Equation (43); physically, it corresponds to the fact that the waves travel approximately horizontally in this region, and hence spend a relatively long time there.

2.4.2. General rotation laws. In the general case, where Ω depends on both r and θ, the rotational splitting may be computed from Equations (38) – (40) by evaluating the two-dimensional integral in Equation (39). This integral is, in general, m-dependent and the splitting is no longer a linear function of m. To illustrate the properties of the splitting, it is convenient to rewrite Equation (39) for R_{nlm} using integration by parts:

$$R_{nlm} = \int_0^\pi \sin\theta d\theta \int_0^R P_l^m(\cos\theta)^2 \left\{ [\xi_r^2 + (L^2 - 1)\xi_h^2 - 2\xi_r\xi_h] \sin\theta\, \Omega(r,\theta) \right.$$
$$\left. + \xi_h^2 \left(\frac{3}{2}\cos\theta \frac{\partial\Omega}{\partial\theta} + \frac{1}{2}\sin\theta \frac{\partial^2\Omega}{\partial\theta^2} \right) \right\} d\theta \tag{49}$$

(Cuypers 1980). I consider again the case of high-order p-modes; here the terms in ξ_r^2 and $L^2\xi_h^2$ dominate, and consequently

$$\delta\omega_{nlm} \simeq m \frac{\int_0^\pi \sin\theta \, [P_l^m(\cos\theta)]^2 \int_{r_t}^R \Omega(r,\theta)[\xi_r(r)^2 + L^2\xi_h(r)^2] dr d\theta}{\int_0^\pi \sin\theta \, [P_l^m(\cos\theta)]^2 \, d\theta \int_{r_t}^R [\xi_r(r)^2 + L^2\xi_h(r)^2] dr}. \tag{50}$$

Hence, the splitting is simply an average of the angular velocity $\Omega(r,\theta)$, weighted by $[\xi_r(r)^2 + L^2\xi_r(r)^2]P_l^m(\cos\theta)^2$. Approximating the eigenfunction as in the derivation of Equation (48) and using, furthermore, an asymptotic approximation to P_l^m, this may be written as

$$\delta\omega_{nlm} \simeq m \frac{\int_{-\cos\Theta}^{\cos\Theta}(\cos^2\Theta - \cos^2\theta)^{-1/2}\int_{r_t}^R \left(1 - \frac{L^2 c^2}{r^2\omega^2}\right)^{-1/2}\Omega(r,\theta)\frac{dr}{c}\, d(\cos\theta)}{\frac{1}{\pi}\int_{r_t}^R \left(1 - \frac{L^2 c^2}{r^2\omega^2}\right)^{-1/2}\frac{dr}{c}}, \tag{51}$$

where $\Theta = \sin^{-1}(m/L)$ (Gough and Thompson 1990, 1991; Gough 1991). It should be noted that a given spherical harmonic is confined essentially to the latitude band between $\pm\Theta$. The variation of the extent of the P_l^m with m/L allows resolution of the latitudinal variation of the angular velocity, in much the same way as the variation of the depth of penetration with ω/L allows resolution of the variation with radius. In particular, with increasing l the sectoral modes (with $l = |m|$) get increasingly confined towards the equator. Thus, the rotational splitting of sectoral modes provides a measure of the solar equatorial angular velocity.

To study the splitting without making the asymptotic approximation, it is convenient to consider a parameterized representation of $\Omega(r,\theta)$, of the form

$$\Omega(r,\theta) = \sum_{s=0}^{s_{max}} \Omega_s(r)\psi_s(\theta), \tag{52}$$

where the ψ_s are suitable expansion functions (*e.g.* Brown *et al.* 1989, Ritzwoller and Lavely 1991). Then, the integrals over θ can be evaluated and the rotational splitting can be written as

$$\delta\omega_{nlm} = m \sum_{s=0}^{s_{\max}} \int_0^R K_{nlms}(r)\Omega_s(r)dr \,, \tag{53}$$

where the kernels K_{nlms} obviously depend on the expansion functions. In the common case where the ψ_s are polynomials in $\cos^2\theta$, the splitting is a polynomial in odd powers of m. As discussed in section 3.3, it is conventional from the point of view of data analysis to write this in terms of Legendre polynomials:

$$\delta\omega_{nlm} = L \sum_{j=0}^{s_{\max}} \Delta^j \omega_{nl} P_{2j+1}\left(\frac{m}{L}\right) \,. \tag{54}$$

Here, the expansion coefficients $\Delta^j \omega_{nl}$ of the splitting are related to the expansion coefficients Ω_s of the angular velocity through expressions of the form

$$\Delta^j \omega_{nl} = \sum_{s=j}^{s_{\max}} \int_0^R K_{nls}^j(r)\Omega_s(r)dr \,. \tag{55}$$

3. Observation of Solar Oscillations

The detectability of solar oscillations with velocity amplitudes less than $1\,\mathrm{cm\,sec^{-1}}$ or relative irradiance amplitudes below 10^{-6} may *a priori* seem incredible. One might have expected that such small effects would have been completely masked by the far larger fluctuations in velocity and intensity which occur on the solar surface as a result of granulation and the magnetic activity. That the oscillations can nevertheless be seen is a result of the fact that they maintain phase over long periods of time, unlike the other fluctuations. Observations that average over N independent random fluctuations, each with an *rms* amplitude of A, produce a signal of $N^{-1/2}A$. Hence, the spatial average over 10^6 granules, each with an *rms* velocity of order $10^5\,\mathrm{cm\,sec^{-1}}$, results in a velocity noise of order $10^2\,\mathrm{cm\,sec^{-1}}$ per measurement. When, in addition, data over several weeks are combined, the solar "noise" is reduced even further, to a level of a few $\mathrm{mm\,sec^{-1}}$ per frequency bin (*cf.* Jiménez *et al.* 1988).

Needless to say, to utilize this potentially very low noise level, the observing techniques must be such that the least possible noise is introduced in the data by instrumental effects, or effects in the Earth's atmosphere. It is certainly a triumph of observational ingenuity that it has been possible, in some cases, to achieve data that are largely limited by the solar noise. A detailed summary of the observing techniques that have been utilized to observe solar oscillations is beyond the scope of this Chapter. Instead, I sketch the principles in the observations and discuss some of the problems that are encountered in the analysis of the data. More extensive treatments of observations and data analysis can be found, for example, in Brown (1988).

3.1. Some observing techniques

The oscillations affect a number of properties of the solar atmosphere, such as its velocity and the continuum intensity and line spectrum that it emits. Hence, there is a corresponding range of potential techniques for observing the oscillations. The choice of techniques depends on noise properties, particularly the ratio between the solar oscillation signal and the solar noise and the effect of the Earth's atmosphere, and on technical convenience. It should be noted also that, in general, the oscillation amplitude increases more rapidly with altitude in the solar atmosphere as the frequency increases. Hence, techniques that probe the upper parts of the solar atmosphere are relatively more sensitive to high-frequency modes.

3.1.1. Doppler observations. Broadly speaking, it appears that the ratio of solar signal to noise is highest for observations of the Doppler velocity. This technique has, in fact, been used for the most extensive studies of solar oscillations. The principle is illustrated schematically in Figure 5: intensities I_r and I_b are measured in narrow passbands on the red and the blue sides of a spectral line; if the line shifts towards the red, I_r decreases and I_b increases. Hence the ratio

$$\frac{I_r - I_b}{I_r + I_b} \tag{56}$$

is a measure of the line shift, and hence of the Doppler velocity (the normalization with the sum of intensities eliminates effects of intensity variations).

The different Doppler techniques are distinguished by the ways in which the passbands are determined and the intensities are measured. Figure 5 refers specifically to the *resonant scattering technique*, where the passbands are defined by scattering off the Zeeman-split components of a K or Na vapour in a permanent magnetic field (Grec, Fossat, and Vernin 1976; Brookes, Isaak, and van der Raay 1978). Their locations, which are determined by the field-strength, are extremely stable; on the other hand, it is not possible with this technique to achieve spatial resolution and, therefore, it has been used mainly to observe oscillations in light integrated over the solar surface. Alternatively, the passbands can be defined by a "filter" which is tuned alternately to the red and the blue wing of the line. If an imaging detector (such as a CCD camera) is placed behind the filter, the intensities I_r and I_b, and hence the velocity, can be determined as a function of position on the solar disk. Such filters can be either based on atomic resonance applied in transmission (Cacciani and Fofi 1978), or one can use a birefringent filter (*e.g.* Libbrecht and Zirin 1986) or a Fabry-Perot (*e.g.* Rust and Appourchaux 1988). A somewhat more elaborate technique is used in the Fourier tachometer (Brown 1984) where the wavelength resolution is achieved by means of a Michelson interferometer. Unlike the simple filter techniques, the signal from the Fourier tachometer is essentially linear in the velocity over a substantial range.

3.1.2. Observations in continuum intensity. The solar five-minute oscillations of low degree were detected in solar irradiance by the Active Cavity Radiometer Irradiance Monitor (ACRIM) instrument on the *Solar Maximum Mission* satellite (Woodard and Hudson 1983); the relative amplitude was, at most, a few parts per million. Detailed results on the five-minute oscillations were obtained from the Soviet satellite *Phobos II* (Fröhlich *et al.* 1990, Toutain and Fröhlich 1992). Observations of the intensity oscillations, both integrated and with spatial resolution, have been made also from the ground (Nishikawa *et al.* 1986,

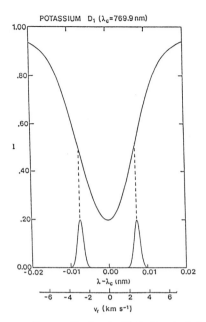

POTASSIUM D₁ (λ_c=769.9 nm)

Figure 5. The average line profile for the K D_1 line in light integrated over the solar disk, normalized to unity in the continuum. For convenience, the abscissa is given both in terms of wavelength shift from the line centre and equivalent Doppler velocity. Also shown are the scattering profiles for K vapour in a cell in a magnetic field of 2000 Gauss. The measurements consist of switching between the two scattering profiles, by changing polarization, to measure alternatively the intensities I_r and I_b in the red and the blue wings of the line. From the ratio in Equation (56), the line-of-sight velocity may then be determined.

Jiménez et al. 1987), although fluctuations in the Earth's atmosphere make substantial contributions to the noise. More generally, it appears that the ratio of solar noise to oscillation signal is less favourable for the continuum intensity observations than for the case of Doppler velocity (Harvey 1988). On the other hand, continuum intensity observations are attractive for studying solar-like oscillations in other stars. For stellar observations, the limiting source of noise is photon statistics; hence, Doppler velocity measurements, which only utilize a very small part of the light from the star, require large telescopes or very sophisticated techniques, whereas broad-band intensity observations, in principle, can be made with fairly small telescopes. The main difficulty is again the effect of the Earth's atmosphere, particularly scintillation, which essentially excludes terrestrial observations of this nature. However, from space such observations should be possible even with fairly modest instruments (Baglin 1991).

3.1.3. Observations in spectral line intensity. Problems with fluctuations in the Earth's atmosphere can be reduced substantially by making differential intensity observations, where the ratio between the intensity in the core of a spectral line and in the neighbouring continuum is measured. The amplitude of the oscillations increases with height in the atmosphere.

Hence, the intensity amplitude in the line core, which may be formed high in the atmosphere, is larger than the amplitude in the continuum which is formed at the photosphere. On the other hand, the atmospheric effects are largely the same in the line core and the continuum and cancel in the ratio. Extensive observations of this nature have been made from the South Pole (*e.g.* Jefferies *et al.* 1988, Duvall *et al.* 1991), utilizing the core of the Ca II K line, which is formed roughly at the temperature minimum. An interesting feature of these results is that the observations are more sensitive to oscillations at high frequency than are Doppler observations using spectral lines formed lower in the atmosphere. In fact, mode-like structure was detected at frequencies as high as 6.5 mHz, well above the normally assumed acoustical cut-off frequency $\omega_{ac}/2\pi$ of 5.3 mHz (*cf.* section 2.3). The interpretation of this result is still debated (*e.g.* Balmforth and Gough 1990b, Kumar *et al.* 1990, Kumar and Lu 1991). Leifsen and Maltby (1990) detected oscillations with an unexpectedly large amplitude in an infrared band around 2.23 μm. Very recent observations (Leifsen, private communication) will allow a detailed investigation of the oscillation amplitude and phase as a function of (radiative) wavelength and oscillation frequency. If their solar origin is confirmed observation of such oscillations will, in addition to supplementing other measurements of the oscillation frequencies, enable investigation of the behaviour of the oscillations in the solar atmosphere; this is crucial for understanding the damping and excitation of the modes, and may eventually provide an independent means for probing the structure of the solar atmosphere (Staiger 1987, Deubner *et al.* 1990, Frandsen 1988).

3.2. THE SEPARATION OF INDIVIDUAL MODES

In the Sun, a very large number of modes are excited simultaneously. As a result, observation at a single point on the solar disk shows an interference between many individual harmonic signals, the result of which has an almost chaotic appearance. It was this fact which for a long time led to the five-minute oscillations being regarded as local phenomena, possibly excited in the solar atmosphere by granulation (see, for example, Noyes and Leighton 1963). The true large-scale nature of the oscillations only came to be realized with the introduction of observations which were more extensive in space and time, hence allowing some separation between the modes (Deubner 1975, Rhodes, Ulrich, and Simon 1977). An interpretation in terms of standing waves in the solar interior had previously been proposed by Ulrich (1970) and Leibacher and Stein (1971).

To illustrate the principles in the mode separation, I note that, according to Equation (6), the observed Doppler velocity on the solar surface is of the form

$$V_D(\theta, \phi, t) = \sin\theta \cos\phi \sum_{n,l,m} A_{nlm}(t) c_{lm} P_l^m(\cos\theta) \cos[m\phi - \omega_{nlm}t - \delta_{nlm}(t)]. \qquad (57)$$

Here, the axis of the coordinate system was taken to be in the plane of the sky; longitude ϕ is measured from the central meridian. For simplicity, I assumed that the velocity is predominantly in the radial direction, as is the case for five-minute oscillations of low or moderate degree (*cf.* Equation (17)); the factor $\sin\theta \cos\phi$ results from the projection of the velocity vector onto the line of sight. The amplitudes A_{nlm} and phases δ_{nlm} may vary with time, as a result of the excitation and damping of the modes.

As discussed above, it may be assumed that V_D has been observed as a function of position (θ, ϕ) on the solar surface. The first step in the analysis is then to perform a spatial transform, to isolate a limited number of spherical harmonics. This may be thought of as an integration of the observations multiplied by a weight function $W_{l_0 m_0}(\theta, \phi)$ designed to give greatest weight to modes in the vicinity of $l = l_0, m = m_0$. The result is the filtered time string

$$
\begin{aligned}
V_{l_0 m_0}(t) &= \int_A V_D(\theta, \phi, t) W_{l_0 m_0}(\theta, \phi) dA \\
&= \sum_{n,l,m} S_{l_0 m_0 lm} A_{nlm} \cos[\omega_{nlm} t + \hat{\delta}_{nlm, l_0 m_0}] .
\end{aligned}
\tag{58}
$$

Here, the integral is over area on the solar disk, and $dA = \sin^2 \theta \cos \phi \, d\theta d\phi$; also, I introduced the *spatial response function* $S_{l_0 m_0 lm}$, defined by

$$
(S_{l_0 m_0 lm})^2 = \left(S_{l_0 m_0 lm}^{(+)}\right)^2 + \left(S_{l_0 m_0 lm}^{(-)}\right)^2 ,
\tag{59}
$$

where

$$
S_{l_0 m_0 lm}^{(+)} = c_{lm} \int_A W_{l_0 m_0}(\theta, \phi) P_l^m(\cos \theta) \cos(m\phi) \sin \theta \cos \phi \, dA ,
\tag{60}
$$

and

$$
S_{l_0 m_0 lm}^{(-)} = c_{lm} \int_A W_{l_0 m_0}(\theta, \phi) P_l^m(\cos \theta) \sin(m\phi) \sin \theta \cos \phi \, dA .
\tag{61}
$$

The new phases $\hat{\delta}_{nlm, l_0 m_0}$ in Equation (58) depend on the original phases δ_{nlm} and on $S_{l_0 m_0 lm}^{(+)}$ and $S_{l_0 m_0 lm}^{(-)}$.

It is evident that to simplify the subsequent analysis of the time string $V_{l_0 m_0}(t)$, it is desirable that it contain contributions from a limited number of spherical harmonics (l, m). This is to be accomplished through a suitable choice of the weight function $W_{l_0 m_0}(\theta, \phi)$ such that $S_{l_0 m_0 lm}$ is large for $l = l_0$, $m = m_0$ and "small" otherwise. Indeed, it follows from the orthogonality of the spherical harmonics that, if $W_{l_0 m_0}$ is taken to be the spherical harmonic $Y_{l_0}^{m_0}$, if the integrals in Equations (60) and (61) are extended to the full sphere, and if, in the integrals, $\sin \theta \cos \phi \, dA$ is replaced by $\sin \theta \, d\theta d\phi$, then essentially $S_{l_0 m_0 lm} \propto \delta_{l_0 l} \delta_{m_0 m}$. It is obvious that, with realistic observations restricted to one hemisphere of the Sun, this optimal level of concentration cannot be achieved. However, the result suggests that suitable weights can be obtained from spherical harmonics. Weights of this nature are almost always used in the analysis. The resulting response functions are typically of order unity for $|l - l_0| \lesssim 2$, $|m - m_0| \lesssim 2$ and relatively small elsewhere. That the timestring contains modes over a range in l and m is analogous to the quantum-mechanical uncertainty principle between localization in space and momentum (here represented by wavenumber). If the area being analyzed is reduced, the spread in l and m is increased; conversely, intensity observations, which do not include the projection factor $\sin \theta \cos \phi$, effectively sample a larger area of the Sun and therefore, in general, lead to somewhat greater concentration in l and m.

Whole-disk velocity observations, analyzing light integrated over the solar surface, correspond roughly to unit weight, *i.e.*, essentially to $l_0 = 0$, $m_0 = 0$. The resulting response function is very small except for $l \leq 3$ (*e.g.* Dziembowski 1977, Hill 1978, Christensen-Dalsgaard and Gough 1982); for whole-disk observations in intensity, the sensitivity is restricted to $l \leq 2$. Hence, such observations are sensitive only to low-degree modes. The same is evidently true of observations of stellar oscillations.

Having isolated a relatively small number of modes through the spatial analysis, the remaining mode identification is carried out through Fourier analysis in the temporal domain by computing

$$\tilde{V}_{l_0 m_0}(\omega) = \int_0^T V_{l_0 m_0}(t) \exp(-i\omega t) dt , \qquad (62)$$

and the power spectrum

$$P_{l_0 m_0}(\omega) = |\tilde{V}_{l_0 m_0}(\omega)|^2 . \qquad (63)$$

Here, I have arbitrarily taken the start of the observations as the zero-point in time, and T is the total duration of the observations. For a single undamped oscillation, $V_D(t) = \cos(\omega_0 t)$, the power spectrum is

$$P(\omega) = \frac{1}{4} T^2 \mathrm{sinc} \left[\frac{1}{2} T(\omega - \omega_0) \right]^2 , \qquad (64)$$

where $\mathrm{sinc}\, x = (\sin x)/x$ (I have neglected a corresponding contribution on the negative ω-axis). As shown in Figure 6a and 6b, this gives rise to a single dominant peak in the power spectrum, centred at ω_0 and of width approximately $2\pi/T$; hence, the frequency resolution is improved by extending the duration of the observation (another analogy to the uncertainty principle).

Figure 6 also illustrates another important aspect of the observations: the effects of gaps in the data. Except for the observations from the South Pole, data from a single site typically consist of stretches of 8 – 12 hours, separated by night-time gaps. Fourier analysis of such data results in power spectra such as those shown in Figures 6c and 6f. The central peak is surrounded by *sidebands* separated from it by $1/(1\,\mathrm{day}) = 11.56\,\mu\mathrm{Hz}$. In reality, the data contain several closely spaced peaks, arising partly from rotational splitting, partly from the presence of several values of l and m in a single time string; the resulting spectrum can get quite complicated (*cf.* Figure 6f). The presence of noise obviously adds to the confusion. Such effects of gaps motivate setting up networks of observing stations to ensure largely uninterrupted data, or observing from space (see section 7).

3.3. THE OBSERVED SPECTRUM OF OSCILLATIONS

Figure 7 shows an example of an observed power spectrum for the five-minute oscillations. This was obtained by means of Doppler velocity measurements in light integrated over the solar disk and hence, according to section 3.2, is dominated by modes of degrees $0 - 3$. The data were obtained from two stations widely separated in longitude, to suppress the daily side-bands, and span 53 days. Thus, the intrinsic frequency resolution, as determined by Equation (64), is smaller than the thickness of the lines. It should be noticed that the peaks occur in pairs, corresponding to the expression for the frequencies given in Equation (29). Also, there is a visible increase in the line-width when going from low to high frequency (see also Grec, Fossat, and Pomerantz 1980). The broadening of the peaks at high frequency is probably caused by the damping and excitation processes; thus, the observations indicate that the damping rate increases with increasing frequency. More careful analyses (see, for example, Libbrecht and Zirin 1986; Libbrecht 1988; Elsworth *et al.* 1990b; Jefferies *et al.* 1991) have shown that the mode lifetimes vary from about a day at high frequency

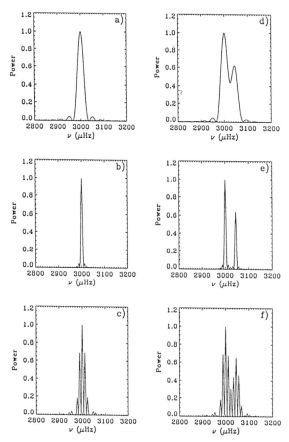

Figure 6. Illustration of power spectrum analysis. Panels a) – c) show spectra of a single oscillation with frequency $3000\,\mu$Hz, observed for 8 hours (panel a), 32 hours (panel b), and two 8 hour segments, separated by a gap of 16 hours (panel c). Panels d) – f) similarly show spectra of a timeseries consisting of the superposition of two modes; one with frequency $3000\,\mu$Hz and relative amplitude 1, the second with frequency $3045\,\mu$Hz and relative amplitude 0.8. The power is on an arbitrary scale.

(as directly visible in the width of the peaks in Figure 7) to several months at the lowest frequencies observed. Another striking aspect of Figure 7 is the well-defined distribution of amplitudes, with a maximum around 3000 μHz and very small values below 2000 μHz and above 4500 μHz. The maximum velocity amplitude for a single mode is about $15\,\mathrm{cm\,sec^{-1}}$. Libbrecht *et al.* (1986) analyzed the observed dependence of mode amplitudes on degree, azimuthal order, and frequency. They found that the distribution of amplitude is largely independent of degree, and that the mode amplitudes depend only weakly on degree (see also Christensen-Dalsgaard and Gough 1982).

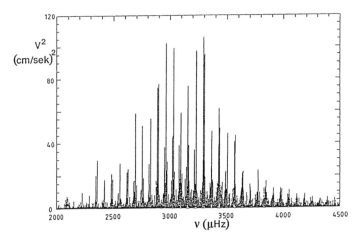

Figure 7. Power spectrum of solar oscillations, obtained from Doppler observations in light integrated over the disk of the Sun. The ordinate is normalized to show velocity power per frequency bin. The data were obtained from two observing stations, on Hawaii and Tenerife, and span 53 days. (See Claverie *et al.* 1984).

Extensive tables of five-minute oscillation frequencies, compiled from several sources, were given by Duvall *et al.* (1988) and Libbrecht, Woodard, and Kaufman (1990). To illustrate the quality of current frequency determinations, Figure 8 shows observed frequencies at low and moderate degree from the latter compilation, with error bars magnified by a factor 1000 over the usual 1σ error bars. For the most accurate measurements, the relative standard deviation is less than 10^{-5}, thus substantially exceeding the precision with which the solar mass is known. Precise measurements of frequencies and frequency separations for low-degree modes were recently published by Elsworth *et al.* (1990a, 1991) and Toutain and Fröhlich (1992); such measurements are of great diagnostic importance for the properties of the solar core (*cf.* section 4.2).

From spatially resolved observations, individual frequencies ω_{nlm} can in principle be determined. Because of observational errors and the large amount of data resulting from such determination, it has been common to present the results in terms of coefficients in fits to the m-dependence of the frequencies, either averaged over n at given l (Brown and Morrow 1987) or for individual n and l (*e.g.*, Libbrecht 1989). A common form of the fit is in terms of Legendre polynomials,

$$\omega_{nlm} = \omega_{nl0} + L \sum_{j=1} a_j(n,l) P_j\left(\frac{m}{L}\right) . \tag{65}$$

As discussed in section 2.4.2 (*cf.* Equations (54) and (55)) the coefficients a_j with odd j arise from rotational splitting; the coefficients with even j are caused by departures from spherical symmetry in solar structure, or from effects of magnetic fields. It was pointed out by Ritzwoller and Lavely (1991) that a more suitable expansion of the rotational splitting could be obtained in terms of Clebsch-Gordon coefficients; with a proper choice of expansion functions ψ_s for the angular velocity $\Omega(r,\theta)$ (*cf.* Equation (52)), the relations corresponding to Equation (55) decouple such that each expansion coefficient for the splitting is related to a single expansion function for the angular velocity. It should be noted that, in general, averaging or expansion of the observed frequencies may involve loss of information; for the purpose of inversion it is, in principle, preferable to work directly in terms of the observed frequencies. On the other hand, by suitably combining the frequencies before inversion, the computational effort required may be greatly reduced.

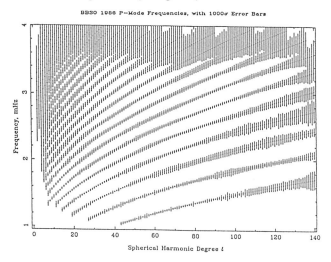

Figure 8. Plot of observed solar p-mode oscillation frequencies, as a function of the degree l. The vertical lines show the 1000σ error bars. Each ridge corresponds to a given value of the radial order n, the lowest ridge having $n = 1$. (From Libbrecht and Woodard 1990).

4. Computed Oscillation Frequencies

The most immediate use of the observed frequencies for investigating the solar interior is to compare the observations with computed frequencies of solar models. In this way, one obtains a test of the models. Furthermore, by analysing frequencies of models computed with different assumptions, one can get some impression of the sensitivity of the frequencies to the physics of the solar interior and the other ingredients of the solar model computations. This illustrates the diagnostic potential of the frequencies and may provide an indication of the changes required to improve the agreement between computations and observations.

In this section, the relations between physics, model, and frequencies are illustrated by considering three different changes to the physics: a minor change to the opacity, a substantial change to the equation of state, and the inclusion of overshoot below the convection zone. The analysis is carried out by considering differences between the models and the corresponding differences between the frequencies; the dependence of the frequency differences on the model differences can be understood qualitatively from the asymptotic behaviour of

54

the oscillations discussed in section 2.3. Furthermore, I compare observed frequencies with two different sets of computed frequencies.

A particular feature of the comparison relates to the dependence of the depth of penetration of the modes on the degree at fixed frequency: since modes of high degree penetrate less deeply than modes of low degree, they involve less of the solar mass, and hence their frequencies are easier to perturb. To compensate for this, frequency differences are shown after scaling by the ratio

$$Q_{nl} = \frac{E_{nl}}{\overline{E}_0(\omega_{nl})} \ . \tag{66}$$

Here, E_{nl} is a measure of the energy of the mode, at fixed surface amplitude, and $E_0(\omega_{nl})$ is the corresponding energy for radial modes, interpolated to the frequency ω_{nl} of the mode being considered. This compensates for the effect of the difference in penetration depth; it may be shown that changes in the model that are localized very near the solar surface cause frequency changes $\delta\omega_{nl}$ such that $Q_{nl}\delta\omega_{nl}$ is essentially independent of l at given frequency (*e.g.* Christensen-Dalsgaard and Berthomieu 1991). In particular, this property may be expected to hold true for the errors introduced into the frequencies by the uncertain physics of the model and the oscillations (effects of convection, nonadiabaticity *etc.*) concentrated near the solar surface.

4.1. SENSITIVITY TO CHANGES IN THE PHYSICS OF THE MODEL

4.1.1. Change in the opacity. To illustrate the effects of a subtle change in the opacity, I consider two models differing essentially only in the abundance of iron used in the opacity tables, namely Models 2 and 3 in Table 1 of Chapter 2.

The smaller iron abundance in the tables used for Model 3 leads to a general decrease in opacity in the radiative interior of the model. As noted in Chapter 2, this causes a decrease in the value of Y_0 required to calibrate the model. The detailed effects of the change are illustrated in Figure 9 which shows structure and frequency differences between Models 3 and 2. The decrease in the opacity near the base of the convection zone causes the convection zone to be shallower in Model 3 than in Model 2. In the region that is adiabatically stratified in Model 2 and sub-adiabatically stratified in Model 3, temperature increases more slowly with depth in Model 3. This causes the sharp decrease in the sound speed beneath the convection zone in Model 3 relative to Model 2. In the convection zone, the differences are mainly a result of the difference in Y_0; in particular, this causes a significant difference in sound speed (barely visible in the present figure) in the ionization zones of hydrogen and helium. The negative sound-speed difference at the base of the convection zone dominates the frequency differences shown in Figure 9b: the frequencies of modes penetrating into or beyond this region of negative δc are somewhat reduced in Model 3 relative to Model 2; for modes of high degree that are entirely trapped within the convection zone, the frequency change is small and dominated by δc in the near-surface ionization zones. It is interesting that this effect is of considerable magnitude compared with the precision of the observed frequencies. This suggests that helioseismic investigations may be sensitive to even quite subtle effects in the opacity. Indeed, analyses of the observed frequencies (*e.g.* Christensen-Dalsgaard *et al.* 1985, Korzennik and Ulrich 1988) have indicated that the opacity obtained from the CT and LAOL tables might be too low in the temperature region corresponding to the outer parts of the radiative interior; this has been confirmed by recent independent opacity calculations (Iglesias and Rogers 1991; Yan, Seaton, and Mihalas 1992). Dziembowski, Pamyatnykh, and Sienkiewicz (1992) demonstrated that the use of the new Iglesias and Rogers opacities led to a substantial improvement in the agreement between the computed model and the Sun. However, it remains to be seen whether effects of opacity differences of the magnitude considered in Figure 9 can be distinguished among the other potential uncertainties in the model.

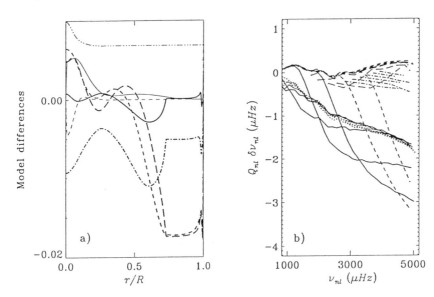

Figure 9. Structure differences, at fixed fractional radius r/R (a), and scaled frequency differences (b), between Model 3 computed with the opacity tables based on the meteoritic Fe abundance and Model 2 computed with tables based on the photospheric Fe abundance, in the sense (Model 3) – (Model 2). In panel a) the lines have the following meaning (note that ln denotes the natural logarithm):
Thin lines: $\delta \ln q$, where $q \equiv m/M$ is the mass fraction (————); $\delta \ln L$, where L is the luminosity at r (------------); δX, where X is the hydrogen abundance (—···—···—···). *Heavy lines:* $\delta \ln c$ (————); $\delta \ln p$ (------------); $\delta \ln \rho$ (— — — — —); $\delta \ln T$ (—·—·—·—).

For the frequency differences shown in panel b) points corresponding to a given value of l have been connected, according to the following convention: $l = 0 - 3$ (················); $l = 4, 5, 10, 20, 30$ (————); $l = 40, 50, 70, 100$ (------------); $l = 150, 200, 300, 400$ (— — — — —); and $l = 500, 600, 700, 800, 900, 1000$ (—·—·—·—).

4.1.2. Change in the equation of state. The Eggleton, Faulkner, and Flannery (1973) equation of state (in the following EFF; see Chapter 2) which was used in the models considered so far is extremely simple. In particular, it assumes all ions to be in the ground state, neglects the Coulomb interaction between the particles in the gas, and treats "pressure ionization" in a very crude, although thermodynamically consistent, way. Christensen-Dalsgaard, Däppen, and Lebreton (1988) found that agreement between observed and theoretically computed frequencies could be markedly improved by using the so-called MHD equation of state (Hummer and Mihalas 1988, Mihalas, Däppen, and Hummer 1988, Däppen *et al.* 1988); this is determined from minimization of an approximation to the free energy containing a substantial number of physical effects. The consequences of the change in the equation of state are illustrated in Figure 10, which shows differences between structure and frequencies of models computed with the EFF and MHD equations of state. The EFF model is Model 1 in Table 1 of Chapter 2, whereas the MHD model differs from that model only in the equation of state. There are clearly substantial differences between the EFF and the MHD models; in particular, the sound speed differs by up to two per cent in the hydrogen and helium ionization zones. This leads to frequency differences of up to 10 μHz. For most modes, the frequencies are decreased by the predominantly negative δc. However, very near the surface there is a region where δc is positive, leading to positive frequency differences for high-degree modes which are predominantly trapped in that region. The general effect of the frequency change caused by the introduction of the MHD equation of state is to reduce the difference between the computed and the observed frequencies.

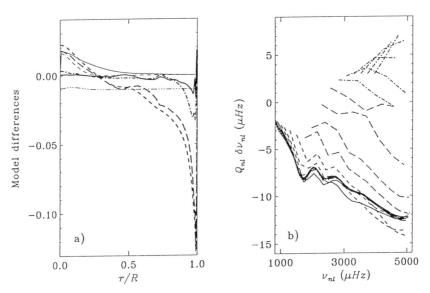

Figure 10. Structure differences, at fixed fractional radius r/R, (panel a) and scaled frequency differences (panel b) between Model 1 computed with the EFF equation of state and an otherwise similar model computed with the MHD equation of state, in the sense (EFF model) – (MHD model). See caption to Figure 9 for details.

It was pointed out by Däppen (1990) that the dominant departures of the MHD equation of state from the simple EFF treatment arise from the Coulomb terms. Indeed, Christensen-Dalsgaard (1991) found that by including the Coulomb effects in the EFF formulation most of the differences with MHD were removed, leaving frequency differences of less than 2 μHz relative to the MHD models. The consequences of improvements in the equation of state, such as the inclusion of Coulomb effects, were also considered by Stix and Skaley (1990). Baturin, Kononovich, and Mironova (1991) and Pamyatnykh, Vorontsov, and Däppen (1991) analysed the effects of various treatments of the equation of state on the frequencies, in particular as expressed in terms of a phase function $\beta(\omega)$ which is closely related to the Duvall phase $\alpha(\omega)$ introduced in Equation (22). It was found again that in general the more complex, and hence presumably more accurate, treatments of the equation of state improved the agreement between theory and observation. A detailed analysis of the effects of the equation of state on the solar oscillation frequencies was given by Christensen-Dalsgaard and Däppen (1992).

4.1.3. Convective overshoot. As discussed in Chapter 4, convective motion is likely to extend beyond the unstable region; the result is a region where the temperature gradient is slightly subadiabatic, followed by a sharp transition to the purely radiative gradient. The detailed dynamics of this overshoot region, and it extent, is quite uncertain, however. Hence it would be of great interest to be able to study it by means of helioseismology.

To investigate the effect of overshoot on the solar model and its oscillation frequencies, I have considered a simplified model simulating the behaviour discussed in Chapter 4: the temperature gradient was forced to be nearly adiabatic for a specified distance beneath the point of transition from convective instability to convective stability, and this was followed by a discontinuous transition to the radiative gradient. As usual, the model was calibrated to have the observed radius and luminosity; in particular, the inclusion of overshoot requires a change of the mixing-length parameter. The changes in structure resulting from overshoot by $0.029R$, corresponding to 0.36 pressure scale heights at the base of the convection zone, are shown in panel a) of Figure 11. Since the temperature gradient is steeper just beneath the convection zone in the overshoot model, there is a sharp increase in the temperature difference at this point. However, at slightly greater depth the temperature gradient is more shallow in the overshoot model, and the temperature difference decreases again. Hence, there is a sharp peak in $\delta \ln T$ localized to a small region just beneath the convection zone, with a corresponding peak in the sound-speed difference. The change in the mixing-length parameter causes changes of pressure and density within the convection zone, whereas in the bulk of the radiative interior the changes are very small.

The effects on the oscillations are seen most clearly by plotting scaled relative frequency differences against ν/L which, according to Equation (20), determines the depth of the lower turning point; this has been done in panel b) of Figure 11. The results may be understood from the asymptotic expressions (25) and (26) for the frequency differences. For $\nu/L \lesssim 60\,\mu\text{Hz}$ the modes are trapped entirely within the convection zone where the sound-speed change is small; so, therefore, are the frequency differences. Modes that penetrate beyond the convection zone experience the sharp peak in $\delta \ln c$; this leads to the rapid increase in $\delta\nu$ as ν/L increases beyond $80\,\mu\text{Hz}$. The effect is particularly strong for those modes whose turning points are in the vicinity of the peak in $\delta \ln c$. Mathematically this follows from the fact that the integrand in Equation (25) has an integrable singularity at $r = r_t$; physically, the modes propagate almost horizontally at the lower turning point, spending relatively more time in this region and hence being more sensitive to the change in c. Modes that penetrate substantially beyond the peak in $\delta \ln c$ feel the integrated effect of the change in sound speed, and hence the frequency change is roughly constant. Overlying this general trend the frequency changes exhibit rapid oscillations, barely visible on the scale of the figure. These are caused by the shift with changing frequency of the location of the extrema in the eigenfunctions relative to the peak in $\delta \ln c$. Such a behaviour is characteristic of the frequency response to sharp features in the model (see, for example,

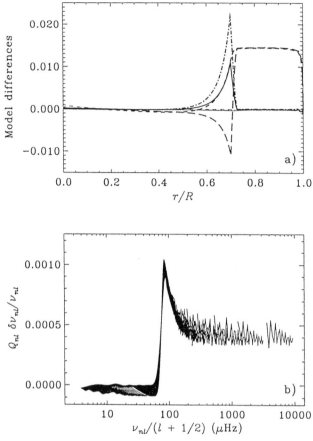

Figure 11. The effects of convective overshoot by a distance of $0.029R$, corresponding to about 0.36 pressure scale heights. Panel a) shows structure differences, in the sense (overshoot model) – (normal model), using the line styles of Figure 9. Panel b) shows the corresponding scaled relative frequency differences; they have been plotted against ν/L which determines the location of the lower turning point (*cf.* equation (20)).

Gough 1990); the "wavelength", measured in frequency units, of the oscillation in the frequency difference is determined by the depth of the sharp feature.

From these results it is clear that convective overshoot has characteristic effects on the structure and frequencies of the model. The magnitude of the effects evidently depends on the extent of the overshoot region; indeed, it is straightforward to show that the changes in structure and frequencies scale roughly as the square of the extent of overshoot. Whether or not such effects can be studied observationally depends on the ability to extract them from the noise in the observed frequencies, and to distinguish them from the consequences of, for example, errors in the opacities. These questions require further study.

4.2. Comparison with observed frequencies

A detailed analysis of the observed frequencies is outside the scope of this Chapter. However, to indicate the typical behaviour found in comparisons of the computed frequencies with observations, Figure 12 shows scaled differences between observed frequencies from the compilation by Libbrecht *et al.* (1990) and the frequencies of two models computed with the MHD equation of state. In panel a) the MHD model considered in Figure 10 was used; this was computed with the CT opacities. The model considered in panel b) was computed with a version of the LAOL tables similar to the one used for Model 2 in Table 1 of Chapter 2, although with a heavy element abundance of 0.02. There are striking differences between the two sets of results: the variation of the differences with frequency is considerably larger in the CT than in the LAOL case, whereas the scatter with the degree of the mode is slightly larger in the LAOL case. These properties are a result of the differences between the two opacities and the resulting differences between the models, which are similar to those between Models 1 and 2. In particular, inversion of the observed frequencies indicates that the depth of the solar convection zone is $0.287 \pm 0.003R$ (Christensen-Dalsgaard, Gough, and Thompson 1991), somewhat larger than the values of $0.279R$ and $0.274R$, respectively, for the CT and the LAOL models. The effect on the frequencies is similar to the differences illustrated in Figure 9, with a transition between modes trapped within the convection zone and modes penetrating beyond it. This leads to the l-dependence of the frequency differences; since the CT model is closer to the Sun in this respect, the variation with degree is slightly smaller. The smaller variation of the differences with frequency for the LAOL model is a result of the substantially larger opacity at low temperature in the LAOL tables; since the resulting changes in the model are concentrated near the surface, they lead to scaled frequency differences that depend predominantly on frequency. These results might suggest that the atmospheric behaviour of the LAOL opacities should be preferred. It is important to remember, however, that the calculation neglects a number of complications near the solar surface, such as nonadiabaticity and dynamical effects of convection; such effects would be expected to lead to scaled frequency shifts relative to the observations that depend on frequency but not on degree, much as is obtained for the CT model, and of a comparable magnitude. Thus, the apparently better agreement between theory and observation in Figure 12b is not immediately significant.

There has been extensive discussion about the observed and computed values of the frequency separation $\delta\nu_{nl}$ (see Equation (32)). As mentioned in section 2.3, the importance of this quantity lies in the fact that it is sensitive to the core structure of the model, and hence may help elucidating those properties of the model which are involved in the discrepancy between the predicted and observed flux of solar neutrinos (*cf.* Chapter 2). Faulkner, Gough, and Vahia (1986) and Däppen, Gilliland, and Christensen-Dalsgaard (1986) noted that $\delta\nu_{nl}$ was decreased by the inclusion in the model of energy transport by "weakly interacting massive particles" (WIMPs) as had been suggested previously as a way of reducing the solar neutrino flux. Indeed, these early results suggested that $\delta\nu_{nl}$ was somewhat higher than observed for normal solar models, so that the agreement between theory and observation could be improved by considering models with WIMPs. In contrast, partial mixing of the solar core, which had also been proposed as a means of reducing the predicted neutrino flux, led to a substantial increase in $\delta\nu_{nl}$. The current situation can be judged from Figure 13, in a form first proposed by Elsworth *et al.* (1990a): this shows

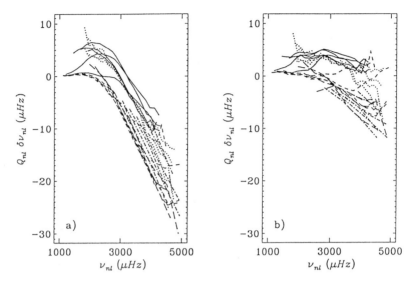

Figure 12. Differences between observed frequencies of solar oscillations, from the compilation by Libbrecht *et al.* (1990), and adiabatic frequencies for two solar models, in the sense (observation) – (model). The models were computed with the MHD equation of state and had a heavy element abundance of 0.02. In panel a) the CT opacities were used, whereas the model in panel b) used LAOL opacities computed with the photospheric iron abundance. The mode sets and line styles are the same as in Figure 9b.

the coefficients $\overline{d_l}$ and s_l of the fit (34), for $l = 0$ and 1. The figure includes results for a substantial number of "normal" solar models using different physics, for models with reduction in the core opacity simulating the effects of WIMPs, and for a model with a partially mixed core. In addition, observed values obtained by Gelly *et al.* (1988) and Elsworth *et al.* (1990a) are indicated. Details about the models were given by Christensen-Dalsgaard (1991, 1992ab). The results for the normal models fall close to the observations, although with a slight but systematic tendency for $\overline{d_l}$ to be larger than observed. In contrast, $\overline{d_l}$ for WIMP-like models with the observed neutrino flux is far lower than observed (see also Elsworth *et al.* 1990a and Cox, Guzik, and Raby 1990) whereas, for the partially mixed model, $\overline{d_l}$ is much larger than observed, even though the neutrino flux is still in excess of the observed value (see, for example, Cox and Kidman 1984, Provost 1984). These results indicate that it will be very difficult to find a model that is consistent with both the observed neutrino flux and the observed oscillation frequencies. As discussed in Chapter 2, this strengthens the case for solutions to the neutrino problem in terms of the properties of the neutrinos.

5. Inverse Analyses

The observed solar oscillation frequencies are integral measures of conditions in the solar interior. A simple example is provided by the rotational splitting, which is a weighted

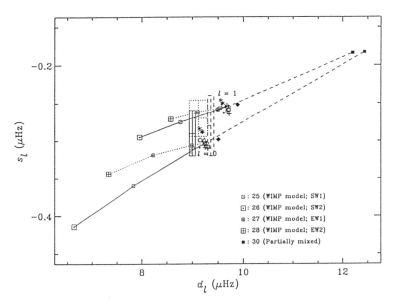

Figure 13. Observed and computed coefficients $\overline{d_l}$ and s_l in the fit in Equation (34) to the scaled frequency separation d_{nl}. The solid (for $l = 0$) and dashed (for $l = 1$) error boxes show the observed results of Elsworth *et al.* (1990a), whereas the dotted error box gives what are essentially averages of $\overline{d_0}$, $\overline{d_1}$ and s_0, s_1, based on observations by Gelly *et al.* (1988). The points clustering close to the error boxes are for normal solar models with a variety of physics. The remaining points are identified in the figure by the symbol type. Those to the left are for different simulations of the effects of WIMPs, including two models (SW2 and EW2) with approximately the observed flux of ^{37}Cl neutrinos. The points to the far right are for a model with a partially mixed core. Adapted from Christensen-Dalsgaard (1992a) where further details about the models are given.

integral of the angular velocity over the region where the mode is trapped. Because of the variation of the turning point radius r_t with degree and frequency, different modes sample different parts of the Sun. Hence, the variation of the splitting with degree, say, provides an indication of the variation of Ω with r. This also suggests that it may be possible to obtain localized information about Ω: roughly speaking, the difference between splittings for modes with different r_t should be a measure of the rotation in the region between the turning points of the modes.

The expression (42) for the splitting caused by spherically symmetric rotation is a particularly simple example of a relation between observable properties of oscillation frequencies and properties of the solar interior. The determination of $\Omega(r)$ from the $\delta\omega_{nlm}$ constitutes an *inverse problem*. Such problems have a vast literature, covering their application in, for example, geophysics and radiation theory (*e.g.* Deepak 1977, Parker 1977, Craig and Brown 1986, and Tarantola 1987). The application to the solar inverse problem was discussed by, for example, Gough (1978, 1985), Thompson (1991), and Gough and Thompson (1991). Christensen-Dalsgaard, Schou, and Thompson (1990) made a systematic comparison of different inversion techniques, as applied to the problem of spherically symmetric rotation.

5.1. BASIC PRINCIPLES OF INVERSE ANALYSIS

As an illustration of some general properties of inverse analyses, it is instructive to consider briefly the technique of *optimally localized averages*, developed by Backus and Gilbert (1970), as applied to inversion for a spherically symmetric angular velocity $\Omega(r)$. The inverse problem may be formulated as

$$\Delta_i = \int_0^R K_i(r)\Omega(r)dr \; , \tag{67}$$

where, for notational simplicity, I represent the pair (n,l) by the single index i. Δ_i is the scaled rotational splitting $m^{-1}\beta_{nl}^{-1}\delta\omega_{nlm}$, so that the kernels K_i are normalized as in Equation (45). The principle of the method is to construct a linear combination

$$\tilde{\Omega}(r_0) = \sum_i c_i(r_0)\Delta_i = \int_0^R K(r; r_0)\Omega(r)dr \tag{68}$$

of the observed data, where

$$K(r; r_0) \equiv \sum_i c_i(r_0)K_i(r) \; . \tag{69}$$

The goal is to choose coefficients $c_i(r_0)$ such as to make $K(r; r_0)$ approximate as far as possible a delta function $\delta(r - r_0)$ centred on r_0. Then, $\tilde{\Omega}(r_0)$ provides an approximation to $\Omega(r_0)$. If this can be done for all r_0, an estimate of $\Omega(r)$ is obtained.

The coefficients $c(r_0)$ are determined by minimizing

$$\cos\eta \int_0^R (r - r_0)^2 K(r; r_0)^2 dr + \sin\eta \sum_{ij} E_{ij} c_i c_j \; , \tag{70}$$

subject to the constraint

$$\int_0^R K(r; r_0)dr = 1 \; ; \tag{71}$$

here E_{ij} is the covariance matrix of the data. The effect of the minimization is most easily understood for $\eta = 0$. Minimizing Equation (70) subject to Equation (71), ensures that $K(r; r_0)$ is large close to r_0, where the weight function $(r - r_0)^2$ is small, and small elsewhere. This is precisely the required "delta-ness" of the combined kernel. However, with no further constraints, the optimization of the combined kernel may result in numerically large coefficients of opposite sign. Hence, the variance in $\tilde{\Omega}$, which can be estimated as

$$\sigma^2(\tilde{\Omega}) = \sum_{ij} E_{ij} c_i c_j \; , \tag{72}$$

would be large. The effect of the second term in Equation (70), when $\eta > 0$, is to restrict $\sigma^2(\tilde{\Omega})$. The size of η determines the relative importance of the localization and the size of the variance in the result. Hence, η must be determined to ensure a trade-off between the localization and the error, and η is generally known as *the trade-off parameter* .

The minimization problem defined by Equations (70) and (71) leads to a set of linear equations for the c_i. To characterize the properties of the inversion, it is convenient to consider a measure of the width of the averaging kernels $K(r; r_0)$ and the *error magnification*

$$\Lambda(r_0) = \left[\sum_i c_i(r_0)^2 \right]^{1/2} ; \qquad (73)$$

Λ is defined such that if the standard error $\sigma(\Delta_i)$ is the same for all the modes the standard error in the result of the inversion is

$$\sigma[\tilde{\Omega}(r_0)] = \Lambda(r_0)\sigma(\Delta_i) . \qquad (74)$$

By considering the *trade-off curve*, where Λ is plotted against the width for varying η, one may choose an appropriate value of η. It is evident that this value will depend on the properties of the data, particularly the level of errors, and on the desired properties of the solution.

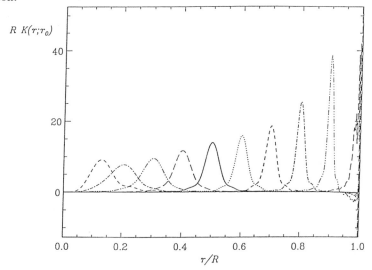

Figure 14. Averaging kernels $K(r; r_0)$ at selected radii ($r_0/R = 0.1, 0.2, \ldots, 1.0$) for inversion by means of optimally localized averages. The kernel at $r_0/R = 0.5$ is shown as a bolder curve. From Christensen-Dalsgaard *et al.* (1990).

An illustration of the use of this method is provided by the results obtained by Christensen-Dalsgaard *et al.* (1990). They considered a set consisting of about 830 modes at selected degrees between 1 and 200, and frequencies between 2000 and 4000 μHz. Examples of averaging kernels $K(r; r_0)$ are shown in Figure 14. The trade-off parameter was chosen such that the error magnification at $r_0 = 0.5R$ was close to 1. It should be realized that the kernels entering into the combination are of the form shown in Figure 4. Thus, a very large degree of cancellation has been achieved of the dominant contribution from near the surface. Nevertheless, it is obvious that the averaging kernels are only approximate realizations of delta functions; structure on a scale smaller than roughly 0.05 R is not resolved.

This limitation is inherent in any inversion method. Indeed, it is evident that from a finite set of data one can never completely resolve the function $\Omega(r)$. Thus, the solution must be constrained. The constraint that is invoked in the present method, and in most other inversion methods, is that the solution be smooth. This is ensured by the representation of the solution by the averaging kernels whose shape is determined by the minimization in Equation (70).

A second commonly used technique is the regularized least-squares method (see, for example, Craig and Brown 1986). Here $\Omega(r)$ is parameterized, often as a piecewise constant function, and the parameters are determined through a least-squares fit to the data by minimizing the sum of the squared differences between the observed splittings and the splittings computed from the parameterized representation of Ω. In general, this least-squares solution needs to be regularized to obtain a smooth solution. This is achieved in the minimization by adding to the sum of squared differences a multiple of the average of the square of Ω, or the square of its first or second derivative; the weight given to this term serves as a trade-off parameter, determining the balance between resolution and error for this method. Another inversion technique is spectral expansion, where Ω is approximated as a linear combination of the kernels (see, for example, Backus and Gilbert 1967; Gough 1985). Finally, from the asymptotic properties of p-modes it follows that the inverse problem can be formulated approximately as an integral equation of the Abel type, the solution to which can be written down analytically; this leads to a very simple, although approximate, technique for inversion of data consisting only of such modes (e.g. Gough 1984). In these methods, also, there are parameters which determine the trade-off between resolution and error.

These methods are all *linear*, in the sense that the result of the inversion depends linearly on the data. For any such method, there exist coefficients $c_i(r_0)$ such that the solution $\tilde{\Omega}(r_0)$ may be written in terms of the observed data Δ_i as in Equation (68) and hence, by using Equation (67), may be expressed from the original angular velocity $\Omega(r)$ through an averaging kernel $K(r; r_0)$. It should be noticed that, once the parameters of the inversion have been determined, the averaging kernels are independent of the data (however, they obviously depend on the weights given to each data point, and hence on the assumed errors in the data). As discussed extensively by Christensen-Dalsgaard *et al.* (1990), a quantitative comparison of different inversion methods can be carried out in terms of the averaging kernels and the coefficients $c_i(r_0)$.

So far, I have considered inversion for a function that depends on r alone. It is evidently desirable, however, to carry out inversion for more general properties, for example, the angular velocity $\Omega(r, \theta)$, which are functions both of r and θ. This may be carried out by expansion in terms of suitable functions of θ, with coefficients that are functions of r (see section 2.4.2). The inverse problem then reduces to inversions for the expansion functions (see, for example, Korzennik *et al.* 1988, Brown *et al.* 1989, Thompson 1990). Alternatively, one may perform a direct two-dimensional inversion by means of a regularized least-squares technique (*e.g.* Sekii 1990, 1991 and Schou 1991a). Finally, a two-dimensional asymptotic inversion technique may be developed by noting that, in the asymptotic Equation (51), the dependence on θ leads to an integral equation of the Abel type in latitude (Kosovichev and Parchevskii 1988, Gough 1991, Gough and Thompson 1990, 1991).

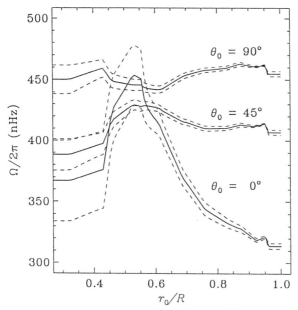

Figure 15. The inferred solar rotation rate obtained from inversion of rotational splitting observations, given on the form of coefficients in an expansion like Equation (54). Results are displayed at three target co-latitudes: $\theta_0 = 0°$ (the pole), $\theta_0 = 45°$, and $\theta_0 = 90°$ (the equator). Dashed lines indicate 1-σ error bars based on the observers' error estimates. (From Schou *et al.* 1992).

5.2. RESULTS ON SOLAR INTERNAL ROTATION

A substantial number of inversions have been carried out to investigate properties of the solar internal rotation (for a review, see, for example, Christensen-Dalsgaard 1990). Early analyses in general concentrated on the splitting for sectoral modes with $l = m$ which, according to section 2.4, predominantly give information about the angular velocity at the equator. Duvall *et al.* (1984) found that the result was generally at, or possibly somewhat below, the surface equatorial angular velocity, although there were slight indications of substantially faster rotation of a small core. Additional evidence for fast core rotation was obtained by Toutain and Fröhlich (1992); nevertheless, the rotation of the core is still uncertain. More recently, observations of the dependence of the splitting on m has enabled investigation of the latitude-dependence of the internal angular velocity (*e.g.* Brown *et al.* 1989, Christensen-Dalsgaard and Schou 1988, Korzennik *et al.* 1988, Dziembowski, Goode, and Libbrecht 1989, Rhodes *et al.* 1990, and Thompson 1990). Goode *et al.* (1991) recently carried out a comprehensive analysis of several different sets of data.

As an example, I present results obtained by Schou, Christensen-Dalsgaard, and Thompson (1992) from rotational splitting observations by Libbrecht (1989). The observations were given as coefficients $\Delta^j \omega_{nl}$ in the expansion (54), with $s_{max} = 2$. Consequently, the inversion was carried out by determining expansion functions $\Omega_s(r)$ as in Equation (52). The inversion for each of the Ω_s was performed by means of the technique of optimally localized averages, discussed in section 5.1. The results are presented in Figure 15, in the form of inferred angular velocities at the equator, the latitude 45°, and the pole. It is striking that in much of the convection zone the angular velocity is quite similar to the behaviour on the surface; it should be noted that the inversion does not impose continuity with the surface angular velocity. In the lower part of the convection zone there appears to be a transition such that the angular velocity in the radiative interior is roughly independent of latitude, at a value intermediate between the surface equatorial and polar values, but substantially closer to the former.

In interpreting results such as these, it is important to keep in mind the limited resolution of the inversion. Thus, with the data that were used, it is not possible to distinguish between the gradual transition to latitude-independent rotation in Figure 15 and a discontinuous transition at, for example, the base of the convection zone. Because of the simple three-term representation (54) of the splitting, the resolution in latitude is even poorer. To illustrate the resolution in both r and θ, Schou *et al.* (1992) used generalized averaging kernels $K(r, \theta; r_0, \theta_0)$ defined such that the inferred angular velocity $\tilde{\Omega}(r_0, \theta_0)$ is related to the true angular velocity $\Omega(r, \theta)$ through

$$\tilde{\Omega}(r_0, \theta_0) = \int_0^\pi \int_0^R K(r, \theta; r_0, \theta_0)\Omega(r, \theta)\, r dr d\theta \, . \tag{75}$$

Examples of such kernels are shown in Figure 16. It is evident that the kernels have a substantial extent in latitude. Also, the figure shows that what was inferred to be the polar angular velocity in fact corresponds to extrapolation from lower latitudes; indeed, it is obvious that the rotation of the region very near the pole has little effect on the frequency splittings, and hence cannot be determined from the inversion.

5.3. RESULTS ON SOLAR INTERNAL STRUCTURE

It follows from the discussion in section 2.2 that the frequencies of adiabatic oscillation of a solar model can be written as functionals of the dependence on r of density $\rho(r)$ and adiabatic exponent $\Gamma_1(r)$:

$$\omega_{nl} = \mathcal{F}_{nl}^{(\mathrm{ad})}[\rho(r), \Gamma_1(r)] \, . \tag{76}$$

Here, $\mathcal{F}_{nl}^{(\mathrm{ad})}$ is defined through the way in which $\rho(r)$ and $\Gamma_1(r)$ enter into the coefficients in Equations (9) – (11). Hence, the dependence is quite complicated and non-linear. To get information about solar structure from a given set of observed frequencies $\omega_{nl}^{(\mathrm{obs})}$, one must, therefore, "solve" the non-linear equations

$$\omega_{nl}^{(\mathrm{obs})} = \mathcal{F}_{nl}^{(\mathrm{ad})}[\rho(r), \Gamma_1(r)] \, , \tag{77}$$

to determine $\rho(r)$ and $\Gamma_1(r)$; this, in general, requires iterative techniques. A further difficulty arises from nonadiabaticity and other effects that were neglected in the simple dependence expressed formally in Equation (76). Hence, the inverse problem for solar structure is considerably more complicated than the simple linear problems considered in section 5.1 and 5.2.

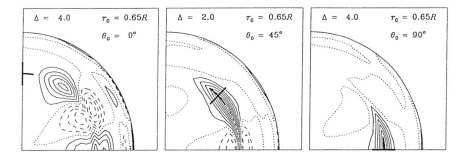

Figure 16. Contour plots of two-dimensional averaging kernels $R^{-2}K(r,\theta;r_0,\theta_0)$ (*cf.* Equation (75)) for the inversion shown in Figure 15, at a target radius $r_0 = 0.65R$ and target co-latitudes $\theta_0 = 0°, 45°$, and $90°$. The plots are in the (r,θ) plane, with the polar axis towards the top of the page. Positive contours are indicated by solid lines, negative contours by dashed lines; Δ is the value of the separation between contour levels. From Schou *et al.* (1992).

5.3.1. Asymptotic sound-speed inversion. Simple inversion procedures can be obtained from the asymptotic properties of the oscillations which were discussed in section 2.3. In the asymptotic limit, the frequencies are determined by the sound speed and by the function $\alpha(\omega)$ which describes the behaviour of the modes very near the solar surface. The dependence of frequencies on sound speed is given by Equations (23) and (24). Here, the functions $F(\omega/L)$ and $\alpha(\omega)$ can be determined by fitting the observed frequencies to a relation of the form given in Equation (23). Given F, Equation (24) provides an integral equation for the sound speed c as a function of r. This can be inverted analytically (Gough 1984) to yield

$$r = r\left(\frac{c}{r}\right) = R\exp\left[-\frac{2}{\pi}\int_{c_0/R}^{c/r}\left(w^{-2}-\frac{r^2}{c^2}\right)^{-1/2}\frac{dF}{dw}dw\right], \qquad (78)$$

where c_0 is the surface value of c; hence, $c(r)$ can be determined. This procedure was described in more detail by Gough (1986b). It was applied to observed frequencies by Christensen-Dalsgaard *et al.* (1985) who were able to determine the sound speed in much of the Sun with a precision of considerably better than 1 per cent. Similar inversion techniques based on the asymptotic expression (22) have been developed (*e.g.* Brodsky and Vorontsov 1987, Kosovichev 1988, Sekii and Shibahashi 1989, and Vorontsov and Shibahashi 1990). They are mainly distinguished by the methods of fitting the data to the asymptotic expression, particularly the separation into the parts depending on ω/L and ω. It should be noted that, in these procedures, the uncertain influence of nonadiabaticity and other effects near the solar surface is eliminated through the function $\alpha(\omega)$.

A very attractive feature of these inversion methods is that they are *absolute* : the sound speed $c(r)$ is obtained directly from the data, without any use of a solar model. However, they suffer from systematic errors arising from inaccuracies in the asymptotic Equation (22). It was shown by Christensen-Dalsgaard, Gough, and Thompson (1989) that these errors, to a large extent, cancel if one considers instead differences between frequencies of pairs of models; this suggests that a *differential asymptotic inversion* of the solar data may be more accurate. This can be carried out by fitting Equation (27) to differences between observed solar frequencies and those of a suitable reference model; the resulting function $H_1(\omega/L)$ is linearly related to the sound-speed difference $\delta c/c$ between the Sun and the model, and this relation can be inverted analytically to obtain an estimate of δc. The effects of near-surface uncertainties are eliminated in the fit through the term $H_2(\omega)$ in Equation (27). Christensen-Dalsgaard, Gough, and Thompson (1991) used this method to infer the solar sound speed and hence the depth of the solar convection zone.

5.3.2. Linearized structure inversion. To move beyond the asymptotic treatment, one must consider the general relation (76) for the frequencies, possibly also taking into account departures from the adiabatic approximation. As is common for non-linear equations, Equation (77) is "solved" through linearisation around an initial reference model. Let $(\rho_0(r), \Gamma_{1,0}(r))$ correspond to the reference model, which has oscillation frequencies $\omega_{nl}^{(0)}$. We seek to determine corrections $\delta\rho(r) = \rho(r) - \rho_0(r)$ and $\delta\Gamma_1(r) = \Gamma_1(r) - \Gamma_{1,0}(r)$ to match the differences $\omega_{nl}^{(\mathrm{obs})} - \omega_{nl}^{(0)}$ between the observed frequencies and those of the reference model. By linearizing Equation (77), assuming $\delta\rho$ and $\delta\Gamma_1$ to be small, one obtains

$$\omega_{nl}^{(\mathrm{obs})} - \omega_{nl}^{(0)} = \int_0^R K_{nl}^{(\rho)}(r)\delta\rho(r)dr + \int_0^R K_{nl}^{(\Gamma_1)}(r)\delta\Gamma_1(r)dr \; , \tag{79}$$

where the kernels $K_{nl}^{(\rho)}$ and $K_{nl}^{(\Gamma_1)}$ are determined from the eigenfunctions in the reference model. An additional constraint is that the mass of the Sun and the reference model be the same, *i.e.*,

$$\delta M = 4\pi \int_0^R \delta\rho(r)r^2 dr = 0 \; . \tag{80}$$

In Equation (79), a term may be included which takes into account the uncertainties introduced by the surface layers. Indeed, it was argued in section 4 that such uncertainties introduce frequency changes $\delta\omega_{nl}$ such that the scaled frequency difference $Q_{nl}\delta\omega_{nl}$ is a function of frequency alone. This suggests adding to the right-hand side of Equation (79) a term $Q_{nl}^{-1}\mathcal{G}(\omega)$, where the function $\mathcal{G}(\omega)$ is determined as part of the inversion (see Dziembowski, Pamyatnykh, and Sienkiewicz 1990, and Däppen *et al.* 1991). Note that this is closely analogous to the determination of the function $\alpha(\omega)$ or $H_2(\omega)$ in the absolute and differential asymptotic inversions.

After linearization, the inverse problem has been reduced to a form similar to that considered in section 5.1, and the techniques discussed there can be used, with comparatively little modification, to infer the corrections $\delta\rho$ and $\delta\Gamma_1$ to the reference model. The process could then, in principle, be iterated by adding the corrections to the reference model, computing a new reference model by imposing again the constraint of hydrostatic equilibrium, and repeating the inversion. So far, there is little experience with the properties of such iteration, however.

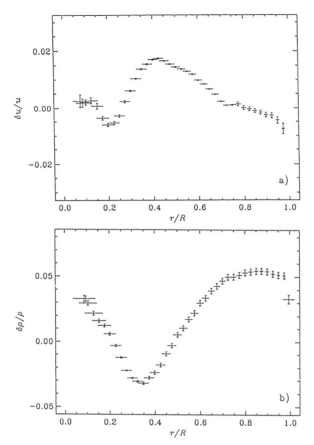

Figure 17. Corrections to a reference solar model, obtained by inverting differences between the observed frequencies and the frequencies of the model. Panel a) shows corrections to $u \equiv p/\rho$, and panel b) shows corrections to ρ. The vertical bars indicate the errors in the results, based on the errors in the observed frequencies, whereas the horizontal bars provide a measure of the resolution in the inversion (from Däppen *et al.* 1991).

As an example of recent inversions, one may consider the analysis by Däppen *et al.* (1991), who carried out the inversion for the corrections to the model by means of the method of optimally localized averages, discussed in some detail in section 5.1. Typical results are illustrated in Figure 17, for the corrections $\delta\rho$ and δu ($u = p/\rho$ being closely related to the sound speed) which should be applied to the model to approximate the Sun. Here, the reference model was Model 13 of Table 2 in Christensen-Dalsgaard *et al.* (1991); this used rather similar physics to the model with MHD equation of state and LAOL opacities considered in Figure 12; however, the opacity was increased somewhat in the vicinity of the base of the convection zone, to increase the convection zone depth to $0.286R$. The figure shows that there are systematic differences between the Sun and the model but that these are generally fairly small. It is probable that the positive values of

$\delta u/u$ around $r = 0.5R$ result from errors in the opacities used to compute the reference model. Further work is required to test whether all the discrepancies between the solar data and the model can be accounted for in terms of such simple changes to the physics used in the model calculation.

6. Frequency Variation with Time

It is of obvious interest to study the temporal variation of the solar oscillation frequencies, to look for effects of possible changes in solar structure associated with the solar cycle. Such effects might help elucidate the mechanisms responsible for the variations of solar magnetic activity.

Evidence for time variation of the frequencies was obtained by Woodard and Noyes (1985), who found an average frequency decrease of about $0.4\,\mu$Hz from 1980 (near solar maximum) to 1984 (approaching minimum) in low-degree modes. More detailed results, although still restricted to low-degree modes, were obtained by Elsworth *et al.* (1990c), who followed the frequency change through an entire solar cycle; the amplitude of the change was roughly consistent with the variation found by Woodard and Noyes (1985); Elsworth *et al.* (1990c) also presented evidence that the frequency variation was correlated with the smoothed sunspot number.

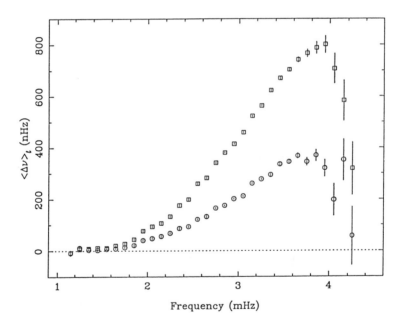

Figure 18. Frequency shift as a function of frequency, using frequencies from 1986 as a reference. The frequency dependence was obtained by averaging over modes in the range $4 \leq l \leq 140$ in degree. Data from 1988 are denoted by circles, data from 1989 by squares. From Woodard and Libbrecht (1991).

Much more detailed results were obtained by Libbrecht and Woodard (1990); they determined the frequency change between 1986 and 1988 for a large number of modes, and hence were able to study the dependence of the frequency shift on the frequency and degree of the modes. Typical results, which also include data from 1989, are shown in Figure 18. As 1986 was near solar minimum, 1988 was on the ascending branch of the solar cycle, and 1989 near solar maximum, the results again show a close correlation between the magnetic activity and the frequencies. Furthermore, the fact that the frequency change is a rapidly increasing function of frequency, being very small at low frequency, strongly suggests that it is caused by effects localized very near the solar surface. This is confirmed by analysis of the l-dependence of the change. An even closer association with the surface magnetic field was found by Woodard et al. (1991), who followed the frequency change on a month-by-month basis. Figure 19 shows the average frequency change relative to 1986. Also shown is a magnetic field index, defined as the average of the absolute value of the magnetic field as obtained from Kitt Peak magnetograms and assumed to correspond to the mean square field (see Chapter 9). The correlation between the frequency change and the field index is striking.

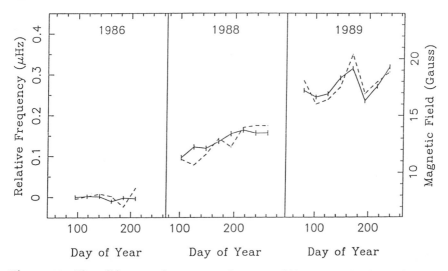

Figure 19. The solid curves show average frequency shifts over 23-day intervals, taking the average frequency for 1986 as a reference. The dashed curves show a properly scaled magnetic field index, obtained as an average of the absolute value of the magnetic field. From Woodard et al. (1991).

Libbrecht and Woodard (1990) also noticed another close connection between the oscillation frequencies and surface activity. As discussed in section 3.3, departures from spherical symmetry induce even terms in the expansion given in Equation (65) of the frequency splitting. Libbrecht and Woodard found that the even coefficients a_i, and their change with time, displayed a behaviour qualitatively similar to the frequency change shown in Figure 18. Furthermore, the latitude variation inferred for the asphericity responsible for the even coefficients was very similar to the measured variation with latitude of the solar-limb brightness; as the limb brightness is correlated with solar activity, this gives further support for the direct relation between frequency changes and activity.

The physical cause of the frequency changes has not been definitely established; however, it seems likely that direct magnetic effects dominate. Gough and Thompson (1988) showed that effects of a fibril magnetic field could account for the closely related even component of the frequency splitting. This description was further elaborated by Goldreich et al. (1991) and Dziembowski and Goode (1991) who showed that it could account for some, if not all, of the features in the observed frequency change; in particular, the frequency change was found to be proportional to the mean square field, in accordance with the results shown in Figure 19. An alternative explanation, in terms of variations in the chromospheric magnetic field which might change the atmospheric boundary conditions for the oscillations, was proposed by Campbell and Roberts (1989) and Evans and Roberts (1990) (see Chapter 9).

Goode and Dziembowski (1991) and Goode et al. (1991) speculated that there might be evidence for change during the solar cycle in the helioseismically inferred equatorial angular velocity near $r \simeq 0.4R$; the statistical significance of such variation was questioned by Schou (1991b), however. It might also be noted that Dziembowski and Goode (1989, 1991) found evidence for a strong (megagauss) field near the base of the solar convection zone, which apparently does not change during the solar cycle. It is evident that such results, if confirmed by improved observations and a more careful analysis, are of great importance for our understanding of the solar cycle.

7. Prospects

Although substantial efforts are under way to increase our theoretical understanding of solar oscillations and to develop new analysis tools, it is probably a fair assessment that the major developments in helioseismology in the coming years will result from improvements in the observations. The principal problems in current data are the presence of gaps, leading to sidebands in the power spectra, and the effects of atmospheric noise. The problem with gaps will be overcome through observations from global networks; nearly continuous observations, which are furthermore free of effects of the Earth's atmosphere, will be obtained from space. The result, ten years from now, should be greatly improved sets of frequencies, extending to high degree, from which one may expect a vast improvement in our knowledge about the solar interior.

7.1. NETWORK OBSERVATIONS OF SOLAR OSCILLATIONS

As shown in Figure 6, gaps in the timeseries introduce sidebands in the spectrum; these add confusion to the mode identification and contribute to the background of noise in the spectra. Largely uninterrupted timeseries of a few days' duration have been obtained from the South Pole (e.g. Grec et al. 1980 and Duvall et al. 1991); however, to utilize fully the phase stability of the modes at relatively low frequency requires continuous observations over far longer periods, and these cannot be obtained from a single terrestrial site.

Nearly continuous observations can be achieved from a network of observing stations, suitably placed around the Earth (*e.g.* Hill and Newkirk 1985). An overview of current network projects was given by Hill (1990). A group from the University of Birmingham has operated such a network for several years, to perform whole-disk observations using the resonant scattering technique (*e.g.* Aindow *et al.* 1988). Among the significant results already obtained from this effort are the frequency separations (Elsworth *et al.* 1990a), which were discussed in section 4.2 and illustrated in Figure 13. A similar network (the IRIS network) is under construction by a group at the University of Nice (Fossat 1991).

An even more ambitious network is being established in the GONG project, organized by the National Solar Observatory of the United States (for an introduction to the project, see Harvey, Kennedy, and Leibacher 1987). This project involves the setting up at carefully selected locations of six identical observing stations. They use the Michelson interferometer technique, to observe solar oscillations of degrees up to around 250. In addition to the design and construction of the observing equipment, a great deal of effort is going into preparing for the merging and analysis of the very large amounts of data expected, and into establishing the necessary theoretical tools. The network is expected to become operational in 1994.

7.2. HELIOSEISMIC INSTRUMENTS ON SOHO

Major efforts are going into the development of helioseismic instruments for the *SOHO* satellite, which has a planned launch in 1995. *SOHO* will be located near the L_1 point between the Earth and the Sun, and hence will be in continuous sunlight. This permits nearly unbroken observations of solar oscillations. A further advantage is the absence of effects from the Earth's atmosphere. These are particularly troublesome for observations of high-degree modes, where seeing is a serious limitation (*e.g.* Hill *et al.* 1991), and for intensity observations of low-degree modes, which suffer from transparency fluctuations.

SOHO will carry three instrument packages for helioseismic observations:

- The GOLF instrument (for Global Oscillations at Low Frequency; see Gabriel *et al.* 1991). This uses the resonant scattering technique in integrated light. Because of the great stability of this technique, it is hoped to measure oscillations at comparatively low frequency, possibly even g-modes. Unlike the p-modes, which have formed the basis for helioseismology so far, the g-modes have their largest amplitude near the solar centre; hence, detection of these modes would greatly aid the study of the structure and rotation of the core. Also, since the lifetime of p-modes increases rapidly with decreasing frequency, very great precision is possible for low-frequency p-modes.
- The SOI-MDI experiment (for Solar Oscillations Investigation – Michelson Doppler Imager; see Scherrer, Hoeksema, and Bush 1991) will use the Michelson interferometer technique. By observing the entire solar disk with a resolution of 4 arcseconds, and parts of the disk with a resolution of 1.2 arcseconds, it will be possible to measure oscillations of degree as high as a few thousand; furthermore, very precise data should be obtained on modes of degree up to about 1000, including those modes for which ground-based observation is severely limited by seeing. As a result, it will be possible to study the structure and dynamics of the solar convection zone, and of the radiative interior, in great detail.

- The VIRGO experiment (for Variability of solar IRradiance and Gravity Oscillations; see Andersen 1991). This contains radiometers and Sun photometers to measure oscillations in solar irradiance and broad-band intensity. It is hoped that this will allow the detection of g-modes; furthermore, the observations will supplement those obtained in Doppler velocity, particularly with regards to investigating the phase relations for the oscillations in the solar atmosphere.

7.3. ASTEROSEISMOLOGY

It is of obvious interest to extend seismic studies to stars other than the Sun. Here the definition of "seismic studies" is somewhat imprecise, since even inferences based on the periods of "classical" variable stars could be included. However, it is probably more reasonable to restrict the term to those cases where rich spectra of oscillations can be observed, and where, therefore, there is a considerable amount of information about the properties of the star. Very interesting examples are provided by the white dwarfs (for a review, see Winget 1988), where extensive spectra of g-mode frequencies have been obtained from intensity observations. Spectra of high-order p-modes have been observed in the rapidly oscillating Ap-stars (see Kurtz 1990); these spectra show some similarities to the observed spectrum of solar oscillations, although the amplitudes are larger by two to three orders of magnitude.

Detection of solar-like oscillations in other stars has proved more difficult, as a result of their expected very low amplitude. Brown *et al.* (1991) obtained power spectra from Doppler observations of Procyon, which strongly suggested the presence of p-mode oscillations, although no definite frequency determinations were possible because of the effects of sidebands caused by the gaps in the data. Possible evidence for oscillations has also been obtained for α Cen A (Butcher, Christensen-Dalsgaard, and Frandsen 1990; Pottasch, Butcher, and van Hoesel 1992); these included a slight suggestion of equally-spaced frequencies, as might have been anticipated from the asymptotic Equation (29), corresponding to a separation $\Delta\nu$ of about $110\,\mu$Hz.

As discussed in section 3.1.2, Doppler observations of stellar oscillations suffer from the fact that only a small part of the spectrum, and hence of the available photons, is used; as a result, photon statistics limit the accuracy that can be achieved, restricting the observations to fairly bright stars and demanding the use of large telescopes. No such limitation affects intensity measurements, where a sufficient number of photons can be obtained with a modest telescope; however, fluctuations in the Earth's atmosphere, particularly scintillation, make the detection of oscillations at the solar amplitude very difficult. This evidently suggests that the observations be made from space. Indeed, observations of solar-like intensity oscillations in a few selected stars, using a 6 cm telescope, will be made in the EVRIS experiment (for Etude de la Variabilité, de la Rotation et des Intérieurs Stellaires) on the Soviet Mars probe *MARS 94*. Substantially more ambitious observations are planned in the *PRISMA* project (for Probing Rotation and Interior of Stars: Microvariability and Activity), which is currently undergoing Phase A studies in ESA (see Lemaire *et al.* 1991). Here, it is expected to observe of order 100 stars with a precision sufficient to detect solar-like oscillations.

7.4. CONCLUDING REMARKS

I hope that this Chapter has given some impression of the exciting results and tremendous possibilities of seismic investigations of the Sun. Twenty years ago, the concept of getting even rough observational information about the properties of the solar interior might have seemed utterly unrealistic. Now, we can imagine studying the Sun to a level of detail exceeding even what is possible for the interior of the Earth. It seems possible that subtle effects in the physics of matter in the Sun can be investigated, potentially yielding new information about the properties of dense plasmas. Finally, observations of solar-like oscillations in other stars now seem to be at the point where observations of solar oscillations were about 15 years ago. Although the data for other stars will never be as detailed as what is available for the Sun, this is partly compensated for by the possibility of studying stars of different masses and in different stages of evolution. By testing the basic assumptions of stellar evolution calculations, the results of such work will be of great importance to large areas of astrophysics.

References.

Aindow, A., Elsworth, Y. P., Isaak, G. R., McLeod, C. P., New, R. and van der Raay, H. B., 1988. *Seismology of the Sun and Sun-like Stars*, p. 157 – 160, eds Domingo, V. and Rolfe, E. J., ESA SP-286.

Andersen, B. N., 1991. *Adv. Space Res.*, **vol. 11, No. 4**, 93.

Backus, G. and Gilbert, F., 1967. *Geophys. J. R. astr. Soc.*, **13**, 247.

Backus, G. and Gilbert, F., 1970. *Phil. Trans. R. Soc. London, Ser. A*, **266**, 123.

Baglin, A., 1991. *Solar Phys.*, **133**, 155.

Balmforth, N. J., 1992. *Mon. Not. R. astr. Soc.*, in the press.

Balmforth, N. J. and Gough, D. O., 1990a. *Solar Phys.*, **128**, 161.

Balmforth, N. J. and Gough, D. O., 1990b. *Astrophys. J.*, **362**, 256.

Baturin, V. A., Kononovich, E. V. and Mironova, I. V., 1991. *Solar Phys.*, **133**, 141.

Brodsky, M. A. and Vorontsov, S. V., 1987. *Pis'ma Astron. Zh.*, **13**, 438 (English translation: *Sov. Astron. Lett.*, **13**, 179).

Brookes, J. R., Isaak, G. R. and van der Raay, H. B., 1978. *Mon. Not. R. astr. Soc.*, **185**, 1.

Brown, T. M., 1984. *Proc. Conf. on Solar Seismology from Space,*, p. 157, eds Ulrich, R. K., Harvey, J., Rhodes, E. J. and Toomre, J., NASA, JPL Publ. 84 – 84.

Brown, T. M., 1988. *Proc. IAU Symposium No 123, Advances in helio- and asteroseismology*, p. 453, eds Christensen-Dalsgaard, J. and Frandsen, S., Reidel, Dordrecht.

Brown, T. M. and Morrow, C. A. 1987. *Astrophys. J.*, **314**, L21 - L26.

Brown, T. M., Christensen-Dalsgaard, J., Dziembowski, W. A., Goode, P., Gough, D. O., Morrow, C. A., 1989. *Astrophys. J.*, **343**, 526.

Brown, T. M., Gilliland, R. L., Noyes, R. W. and Ramsey, L. W., 1991. *Astrophys. J.*, **368**, 599.

Butcher, H. R., Christensen-Dalsgaard, J. and Frandsen, S., 1990. Unpublished report to the commission of the EEC.

Cacciani, A. and Fofi, M., 1978. *Solar Phys.*, **59**, 179.

Campbell, W. R. and Roberts, B., 1989. *Astrophys. J.*, **338**, 538.

Christensen-Dalsgaard, J., 1988. *Proc. IAU Symposium No 123, Advances in helio- and asteroseismology*, p. 295, eds Christensen-Dalsgaard, J. and Frandsen, S., Reidel, Dordrecht.

Christensen-Dalsgaard, J., 1990. *Reviews in Modern Astronomy*, vol. **3**, p. 313, *(Proceedings of the Spring Meeting of the Deutsche Astronomische Gesellschaft, Berlin, 1990)*, ed. Klare, G., Springer, Berlin.

Christensen-Dalsgaard, J., 1991. In *Challenges to theories of the structure of moderate-mass stars, Lecture Notes in Physics*, vol. **388**, 11, eds Gough, D. O. and Toomre, J., Springer, Heidelberg.

Christensen-Dalsgaard, J., 1992a. *Geophys. Astrophys. Fluid Dynamics*, in the press.

Christensen-Dalsgaard, J., 1992b. *Astrophys. J.*, in the press.

Christensen-Dalsgaard, J. and Berthomieu, G., 1991. In *Solar interior and atmosphere*, eds Cox, A. N., Livingston, W. C. and Matthews, M., Space Science Series, University of Arizona Press, in the press.

Christensen-Dalsgaard, J. and Däppen, W., 1992. *Astron. Astrophys. Rev.*, submitted.

Christensen-Dalsgaard, J. and Gough, D. O., 1982. *Mon. Not. R. astr. Soc.*, **198**, 141.

Christensen-Dalsgaard, J. and Schou, J., 1988. *Seismology of the Sun and Sun-like Stars*, p. 149, eds Domingo, V. and Rolfe, E. J., ESA SP-286.

Christensen-Dalsgaard, J., Däppen, W. and Lebreton, Y., 1988. *Nature*, **336**, 634.

Christensen-Dalsgaard, J., Duvall, T. L., Gough, D. O., Harvey, J. W. and Rhodes, E. J., 1985. *Nature*, **315**, 378.

Christensen-Dalsgaard, J., Gough, D. O. and Thompson, M. J., 1989. *Mon. Not. R. astr. Soc.*, **238**, 481.

Christensen-Dalsgaard, J., Gough, D. O. and Thompson, M. J., 1991. *Astrophys. J.*, **378**, 413.

Christensen-Dalsgaard, J., Gough, D. O. and Pérez Hernández, F., 1988. *Mon. Not. R. astr. Soc.*, **235**, 875.

Christensen-Dalsgaard, J., Schou, J. and Thompson, M. J., 1990. *Mon. Not. R. astr. Soc.*, **242**, 353.

Claverie, A., Isaak, G. R., McLeod, C. P., van der Raay, H. B., Palle, P. L. and Roca Cortes, T., 1984. *Mem. Soc. Astron. Ital.*, **55**, 63.

Cowling, T. G., 1941. *Mon. Not. R. astr. Soc.*, **101**, 367.

Cowling, T. G. and Newing, R. A., 1949. *Astrophys. J.*, **109**, 149.

Cox, A. N. and Kidman, R. B., 1984. *Theoretical problems in stellar stability and oscillations*, p. 259, (Institut d'Astrophysique, Liège)

Cox, A. N., Guzik, J. A. and Raby, S., 1990. *Astrophys. J.*, **353**, 698.

Craig, I. J. D. and Brown, J. C., 1986. *Inverse problems in astronomy: a guide to inversion strategies for remotely sensed data*, Adam Hilger, Bristol.

Cuypers, J. 1980. *Astron. Astrophys.*, **89**, 207.

Däppen, W., 1990. *Progress of seismology of the sun and stars, Lecture Notes in Physics*, vol. **367**, 33, eds Osaki, Y. and Shibahashi, H., Springer, Berlin.

Däppen, W., Gilliland, R. L. and Christensen-Dalsgaard, J., 1986. *Nature*, **321**, 229.

Däppen, W., Gough, D. O., Kosovichev, A. G. and Thompson, M. J., 1991. In *Challenges to theories of the structure of moderate-mass stars, Lecture Notes in Physics*, vol. **388**, 111, eds Gough, D. O. and Toomre, J., Springer, Heidelberg.

Däppen, W., Mihalas, D., Hummer, D. G. and Mihalas, B. W., 1988. *Astrophys. J.*, **332**, 261.

Deepak, A., 1977. *Inversion Methods in Atmospheric Remote Sounding*, Academic, New York.

Deubner, F.-L., 1975. *Astron. Astrophys.*, **44**, 371.

Deubner, F.-L., Fleck, B., Marmolino, C. and Severino, G., 1990. *Astron. Astrophys.*, **236**, 509.

Duvall, T. L., 1982. *Nature*, **300**, 242.

Duvall, T. L., Dziembowski, W. A., Goode, P. R., Gough, D. O., Harvey, J. W. and Leibacher, J. W., 1984. *Nature*, **310**, 22.

Duvall, T. L., Harvey, J. W., Jefferies, S. M. and Pomerantz, M. A., 1991. *Astrophys. J.*, **373**, 308.

Duvall, T. L., Harvey, J. W., Libbrecht, K. G., Popp, B. D. and Pomerantz, M. A., 1988. *Astrophys. J.*, **324**, 1158.

Dziembowski, W. A., 1977. *Acta Astron.*, **27**, 203.

Dziembowski, W. A. and Goode, P. R., 1989. *Astrophys. J.*, **347**, 540.

Dziembowski, W. A. and Goode, P. R., 1991. *Astrophys. J.*, **376**, 782.

Dziembowski, W. A., Goode, P. R. and Libbrecht, K. G., 1989. *Astrophys. J.*, **337**, L53 - L57.

Dziembowski, W. A., Pamyatnykh, A. A. and Sienkiewicz, R., 1990. *Mon. Not. R. astr. Soc.*, **244**, 542.

Dziembowski, W. A., Pamyatnykh, A. A. and Sienkiewicz, R., 1992. *Acta Astron.*, submitted.

Eggleton, P. P., Faulkner, J. and Flannery, B. P., 1973. *Astron. Astrophys.*, **23**, 325.

Elsworth, Y., Howe, R., Isaak, G. R., McLeod, C. P. and New, R., 1990a. *Nature*, **347**, 536.

Elsworth, Y., Howe, R., Isaak, G. R., McLeod, C. P. and New, R., 1990c. *Nature*, **345**, 322.

Elsworth, Y., Howe, R., Isaak, G. R., McLeod, C. P. and New, R., 1991. *Mon. Not. R. astr. Soc.*, **251**, 7p.

Elsworth, Y., Isaak, G. R., Jefferies, S. M., McLeod, C. P., New, R., Pallé, P. L., Régulo, C. and Roca Cortés, T., 1990b. *Mon. Not. R. astr. Soc.*, **242**, 135.

Evans, D. J. and Roberts, B., 1990. *Astrophys. J.*, **356**, 704.

Faulkner, J., Gough, D. O. and Vahia, M. N., 1986. *Nature*, **321**, 226.

Fossat, E., 1991. *Solar Phys.*, **133**, 1.

Fröhlich, C., Toutain, T., Bonnet, R. M., Bruns, A. V., Delaboudinière, J. P., Domingo, V., Kotov, V. A., Kollath, Z., Rashkovsky, D. N., Vial, J. C. and Wehrli, C., 1990. *Proc. IAU Colloquium No. 121, Inside the Sun*, eds Berthomieu G. and Cribier M., Kluwer, Dordrecht, p. 279.

Frandsen, S., 1988. *Proc. IAU Symposium No 123, Advances in helio- and asteroseismology*, p. 405, eds Christensen-Dalsgaard, J. and Frandsen, S., Reidel, Dordrecht.

Gabriel, A. H. and the GOLF team, 1991. *Adv. Space Res.*, **vol. 11, No. 4**, 103.

Gelly, B., Fossat, E., Grec, G. and Schmider, F.-X., 1988. *Astron. Astrophys.*, **200**, 207.

Goldreich, P., Murray, N., Willette, G. and Kumar, P., 1991. *Astrophys. J.*, **370**, 752.

Goode, P. R. and Dziembowski, W. A., 1991. *Nature*, **349**, 223.

78

Goode, P. R., Dziembowski, W. A., Korzennik, S. G. and Rhodes, E. J., 1991. *Astrophys. J.*, **367**, 649.

Gough, D. O., 1978. *Proc. Workshop on solar rotation*, p. 255, eds Belvedere, G. and Paterno, L., University of Catania Press.

Gough, D. O., 1981. *Mon. Not. R. astr. Soc.*, **196**, 731.

Gough, D. O., 1984. *Phil. Trans. R. Soc. London, Ser. A*, **313**, 27.

Gough, D. O., 1985. *Solar Phys.*, **100**, 65.

Gough, D. O., 1986a. *Hydrodynamic and magnetohydrodynamic problems in the Sun and stars*, ed. Osaki, Y., University of Tokyo Press, p. 117.

Gough, D. O., 1986b. *Seismology of the Sun and the distant Stars*, p. 125, ed. Gough, D. O., Reidel, Dordrecht.

Gough, D. O., 1990. *Progress of seismology of the sun and stars, Lecture Notes in Physics*, vol. **367**, 283, eds Osaki, Y. and Shibahashi, H., Springer, Berlin.

Gough, D. O., 1991. In *Astrophysical Fluid Dynamics*, eds Zahn, J.-P. and Zinn-Justin, J., Elsevier Science Publisher B.V.

Gough, D. O. and Thompson, M. J., 1988. *Proc. IAU Symposium No 123, Advances in helio- and asteroseismology*, p. 175, eds Christensen-Dalsgaard, J. and Frandsen, S., Reidel, Dordrecht.

Gough, D. O. and Thompson, M. J., 1990. *Mon. Not. R. astr. Soc.*, **242**, 25.

Gough, D. O. and Thompson, M. J., 1991. In *Solar interior and atmosphere*, eds Cox, A. N., Livingston, W. C. and Matthews, M., Space Science Series, University of Arizona Press, in the press.

Grec, G., Fossat, E. and Pomerantz, M., 1980. *Nature*, **288**, 541.

Grec, G., Fossat, E. and Vernin, J., 1976. *Astron. Astrophys.*, **50**, 221.

Hansen, C. J., Cox, J. P. and van Horn, H. M., 1977. *Astrophys. J.*, **217**, 151.

Harvey, J. W., 1988. *Proc. IAU Symposium No 123, Advances in helio- and asteroseismology*, p. 497, eds Christensen-Dalsgaard, J. and Frandsen, S., Reidel, Dordrecht, Holland.

Harvey, J. W., Kennedy, J. R. and Leibacher, J. W., 1987. *Sky and Telescope*, **74**, 470.

Hill, F., 1990. *Proc. IAU Colloquium No. 121, Inside the Sun*, p. 265, eds Berthomieu G. and Cribier M., Kluwer, Dordrecht.

Hill, F. and Newkirk, G., 1985. *Solar Phys.*, **95**, 201.

Hill, F., Gough, D. O., Merryfield, W. J. and Toomre, J., 1991. *Astrophys. J.*, **369**, 237.

Hill, H. A., 1978. *The New Solar Physics*, chapter 5, pp. 135, ed. Eddy, J. A., Westview Press, Boulder CO.

Hummer, D. G. and Mihalas, D., 1988. *Astrophys. J.*, **331**, 794.

Iglesias, C. A., and Rogers, F. J., 1991. *Astrophys. J.*, **371**, 408 - 417.

Jefferies, S. M., Duvall, T. L., Harvey, J. W., Osaki, Y. and Pomerantz, M. A., 1991. *Astrophys. J.*, **377**, 330.

Jefferies, S. M., Pomerantz, M. A., Duvall, T. L., Harvey, J. W. and Jaksha, D. B., 1988. *Seismology of the Sun and Sun-like Stars*, p. 279, eds Domingo, V. and Rolfe, E. J., ESA SP-286.

Jiménez, A., Pallé, P. L., Roca Cortés, T., Domingo, V. and Korzennik, S., 1987. *Astron. Astrophys.*, **172**, 323.

Jiménez, A., Pallé, P. L., Pérez Hernández, F., Régulo, C. and Roca Cortés, T., 1988. *Astron. Astrophys.*, **192**, L7.

Korzennik, S. G. and Ulrich, R. K., 1989. *Astrophys. J.*, **339**, 1144.

Korzennik, S. G., Cacciani, A., Rhodes, E. J., Tomczyk, S. and Ulrich, R. K., 1988. *Seismology of the Sun and Sun-like Stars*, p. 117, eds Domingo, V. and Rolfe, E. J., ESA SP-286.

Kosovichev, A. G. and Parchevskii, K. V., 1988. *Pis'ma Astron. Zh.*, **14**, 473 (English translation: *Sov. Astron. Lett.*, **14**, 201).

Kosovichev, A. G., 1988. *Seismology of the Sun and Sun-like Stars*, p. 533, eds Domingo, V. and Rolfe, E. J., ESA SP-286.

Kumar, P. and Lu, E., 1991. *Astrophys. J.*, **375**, L35.

Kumar, P., Duvall, T. L., Harvey, J. W., Jefferies, S. M., Pomerantz, M. A. and Thompson, M. J., 1990. *Progress of seismology of the sun and stars, Lecture Notes in Physics*, vol. **367**, 87, eds Osaki, Y. and Shibahashi, H., Springer, Berlin.

Kurtz, D. W., 1990. *Ann. Rev. Astron. Astrophys.*, **28**, 607.

Lamb, H., 1909. *Proc. London Math. Soc.*, **7**, 122.

Ledoux, P., 1949. *Mem. Soc. R. Sci. Liège, 4th ser.*, **9**, 263.

Leibacher, J. and Stein, R. F., 1971. *Astrophys. Lett.*, **7**, 191.

Leifsen, T. and Maltby, P., 1990. *Solar Phys.*, **125**, 241.

Lemaire, P., Appourchaux, T., Catala, C., Catalano, S., Frandsen, S., Jones, A. and Weiss, W., 1991. *Adv. Space. Res.*, **Vol. 11, No. 4**, 141.

Libbrecht, K. G., 1988. *Astrophys. J.*, **334**, 510.

Libbrecht, K. G., 1989. *Astrophys. J.*, **336**, 1092.

Libbrecht, K. G. and Woodard, M. F., 1990. *Nature*, **345**, 779.

Libbrecht, K. G. and Zirin, H., 1986. *Astrophys. J.*, **308**, 413.

Libbrecht, K. G., Popp, B. D., Kaufman, J. M. and Penn, M. J., 1986. *Nature*, **323**, 235.

Libbrecht, K. G., Woodard, M. F. and Kaufman, J. M., 1990. *Astrophys. J. Suppl.*, **74**, 1129.

Mihalas, D., Däppen, W. and Hummer, D. G., 1988. *Astrophys. J.*, **331**, 815.

Nishikawa, J., Hamana, S., Mizugaki, K. and Hirayama, T., 1986. *Publ. Astron. Soc. Japan*, **38**, 277.

Noels, A., Scuflaire, R. and Gabriel, M., 1984. *Astron. Astrophys.*, **130**, 389.

Noyes, R. W. and Leighton, R. B., 1963. *Astrophys. J.*, **138**, 631.

Pamyatnykh, A. A., Vorontsov, S. V. and Däppen, W., 1991. *Astron. Astrophys.*, **248**, 263.

Parker, R. L., 1977. *Ann. Rev. Earth Planet. Sci.*, **5**, 35.

Pottasch, E. M., Butcher, H. R. and van Hoesel, F. H. J., 1992. *Astron. Astrophys.*, in preparation.

Provost, J., 1984. *Proc. IAU Symposium No 105, Observational Tests of the Stellar Evolution Theory*, p. 47, eds Maeder, A. and Renzini, A., Reidel, Dordrecht.

Rhodes, E. J., Cacciani, A., Korzennik, S. G., Tomczyk, S., Ulrich, R. K. and Woodard, M. F., 1990. *Astrophys. J.*, **351**, 687.

Rhodes, E. J., Ulrich, R. K. and Simon, G. W., 1977. *Astrophys. J.*, **218**, 901.

Ritzwoller, M. H. and Lavely, E. M., 1991. *Astrophys. J.*, **369**, 557.

Rust, D. M. and Appourchaux, T., 1988. *Seismology of the Sun and Sun-like Stars*, p. 227, eds Domingo, V. and Rolfe, E. J., ESA SP-286.

Scherrer, P. H., Hoeksema, J. T. and Bush, R. I., 1991. *Adv. Space Res.*, **vol. 11, No. 4**, 113.

Schou, J., 1991a. In *Challenges to theories of the structure of moderate-mass stars, Lecture Notes in Physics*, vol. **388**, 93, eds Gough, D. O. and Toomre, J., Springer, Heidelberg.

Schou, J., 1991b. In *Challenges to theories of the structure of moderate-mass stars, Lecture Notes in Physics*, vol. **388**, 81, eds Gough, D. O. and Toomre, J., Springer, Heidelberg.

Schou, J., Christensen-Dalsgaard, J. and Thompson, M. J., 1992. *Astrophys. J.*, **385**, L59.

Sekii, T., 1990. *Progress of seismology of the sun and stars, Lecture Notes in Physics*, vol. **367**, 337, eds Osaki, Y. and Shibahashi, H., Springer, Berlin.

Sekii, T., 1991. *Publ. Astron. Soc. Japan*, **43**, 381.

Sekii, T. and Shibahashi, H., 1989. *Publ. Astron. Soc. Japan*, **41**, 311.

Staiger, J., 1987. *Astron. Astrophys.*, **175**, 263.

Stein, R. F. and Nordlund, Å., 1991. In *Challenges to theories of the structure of moderate-mass stars, Lecture Notes in Physics*, vol. **388**, 195, eds Gough, D. O. and Toomre, J., Springer, Heidelberg.

Stein, R. F., Nordlund, Å. and Kuhn, J. R., 1988. *Seismology of the Sun and Sun-like Stars*, 529, eds Domingo, V. and Rolfe, E. J., ESA SP-286.

Stix, M. and Skaley, D., 1990. *Astron. Astrophys.*, **232**, 234.

Tarantola, A., 1987. *Inverse Problem Theory*, Elsevier, Amsterdam.

Tassoul, M., 1980. *Astrophys. J. Suppl.*, **43**, 469.

Thompson, M. J., 1990. *Solar Phys.*, **125**, 1.

Thompson, M. J., 1991. In *Challenges to theories of the structure of moderate-mass stars, Lecture Notes in Physics*, vol. **388**, 61, eds Gough, D. O. and Toomre, J., Springer, Heidelberg.

Toutain, T. & Frölich, C., 1992. *Astron. Astrophys.*, in the press.

Ulrich, R. K., 1970. *Astrophys. J.*, **162**, 993.

Ulrich, R. K. and Rhodes, E. J., 1983. *Astrophys. J.*, **265**, 551.

Unno, W., Osaki, Y., Ando, H., Saio, H. and Shibahashi, H., 1989. *Nonradial Oscillations of Stars, 2nd Edition* (University of Tokyo Press).

Vandakurov, Yu. V., 1967. *Astron. Zh.*, **44**, 786 (English translation: *Soviet Astronomy AJ*, **11**, 630).

Vorontsov, S. V. and Shibahashi, H., 1990. *Progress of seismology of the sun and stars, Lecture Notes in Physics*, vol. **367**, 326, eds Osaki, Y. and Shibahashi, H., Springer, Berlin.

Winget, D. E., 1988. *Proc. IAU Symposium No 123, Advances in helio- and asteroseismology*, p. 305, eds Christensen-Dalsgaard, J. and Frandsen, S., Reidel, Dordrecht.

Woodard, M. F. and Hudson, H. S., 1983. *Nature*, **305**, 589.

Woodard, M. F. and Libbrecht, K. G., 1991. *Astrophys. J.*, **374**, L61.

Woodard, M. F. and Noyes, R. W., 1985. *Nature*, **318**, 449.

Woodard, M. F., Kuhn, J. R., Murray, N. and Libbrecht, K. G., 1991. *Astrophys. J.*, **373**, L81.

Yan, Yu, Seaton, M. J., & Mihalas, D., 1992. *Proc. Workshop on Astrophysical Opacities, Revista Mexicana de Astronomia y Astrofisica*, eds Mendoza, C. & Zeippen, C.

4 CONVECTION

M. SCHÜSSLER
Kiepenheuer-Institut für Sonnenphysik
Schöneckstr. 6
D-7800 Freiburg
Germany

ABSTRACT. Convection in stars is discussed on a basic level, with emphasis on the convection zone of the Sun. The treatment starts with the classical criteria for convective instability and the role of temperature and molecular weight gradients. This gives a basis for introducing the mixing length formalism as a means for describing the large-scale properties of a convective region. It follows a discussion of convective overshoot into the stably stratified layers adjacent to a convection zone. With aid of a non-local extension of the mixing length formalism, the depth and structure of an overshoot layer below the solar convection zone can be calculated. Observed solar convective flow patterns are characterized briefly. Finally, numerical simulations of convection are discussed in some detail. After presenting the basic equations and some general results from simulations, we describe the new picture of stellar convection which emerges from these calculations. The lecture concludes with a comparison of these new insights with the assumptions which form the basis of the mixing length formalism.

1. Introduction

The Sun without convection would probably be a rather dull star. Most of the lectures contained in this volume deal with processes and phenomena which, directly or indirectly, depend on the existence of the Sun's outer convection zone and its supply of mechanical energy. Consider the fascinating phenomena of magnetic activity and the solar cycle, the complicated velocity, thermal, and magnetic structure of the photosphere, the global oscillations, the chromosphere, corona, and the solar wind – all most probably would be absent were it not for the convective motions in the outer 30% of the solar radius.

For our purposes we define convection as the transport of energy (internal, kinetic, latent heat, ...) by macroscopic motions (flows) in a fluid. These flows are driven by the dynamical instability of a hydrostatic equilibrium due to an entropy gradient. Convection in a star sets in when the temperature gradient necessary to transport the stellar energy flux by radiation exceeds the temperature gradient of an adiabatically stratified hydrostatic equilibrium. The criteria for the onset of convection are discussed in more detail in section 2.

Convection occurs in many places in the universe, *e.g.* in the envelopes of cool stars and the cores of hot stars, in accretion disks, and in planetary interiors and atmospheres. Besides being a basic energy transport mechanism, convection has a number of further important consequences for stellar physics and astrophysics in general:

J. T. Schmelz and J. C. Brown (ed.), The Sun, A Laboratory for Astrophysics, 81–98.

- Convection leads to strong *structuring* of late type star surface regions (which renders plane-parallel models inadequate for many purposes) and thus affects the determination of atmospheric structure and abundances of elements.

- Convective flows in a rotating star generate large-scale *magnetic fields* by a self-excited dynamo mechanism and create small-scale magnetic structures (see Chapter 5).

- Mechanical energy required for *heating* of chromospheres/coronae and driving stellar winds is provided by convection.

- Convection drives global *oscillations*.

- Convection generates *differential rotation*.

- Convective *mixing* of stellar material, especially in the energy-generating regions, has profound consequences for stellar evolution.

The *solar* atmosphere is a unique testing ground for understanding stellar convection: Only there can the basic processes and flow patterns be observed on their natural time scales and length scales. Unfortunately, the large amount of theoretical and experimental results obtained for *laboratory* convection cannot easily be applied to a description of the solar convection zone: The Reynolds number $Re = uL/\nu$ (where u is the characteristic velocity, L is the length scale, and ν is the kinematic viscosity) is huge ($> 10^{12}$) due to the large dimensions of the system, the Prandtl number $\sigma = \nu/\kappa$ (κ is the thermal diffusivity) is very small (because radiation is much more efficient in transporting energy than thermal conduction), the system is strongly stratified (*i.e.* comprises many pressure scale heights), and, since in the upper layers the velocities are of the order of the sound speed, pressure fluctuations are important there. These conditions are drastically different from those relevant for laboratory convection.

This lecture cannot cover the topic of solar/stellar convection in an exhaustive way. We shall restrict ourselves to discussing some basic concepts, describe briefly the convective flow patterns observed in the solar photosphere, and comment on attempts to simulate convection numerically. Among the topics we do *not* touch upon but which are nevertheless very important are: the interaction of convection with (differential) rotation (*e.g.* Part 4 of Durney and Sofia 1987, Rüdiger 1989) and magnetic fields (*e.g.* Proctor and Weiss 1982), the excitation of acoustic waves (Musielak 1991) and of global oscillations (Goldreich and Kumar 1988, Press 1981) by convective flows, and convection on other stars (*e.g.* Dravins 1990). More complete overviews of solar convection have been given by Stix (1989, Chapter 6) and Nordlund (1985a), and by Spruit *et al.* (1990). A number of recent conference proceedings deal with various aspects of solar convection: Schmidt (1985), Rutten and Severino (1989), Stenflo (1990). Somewhat less recent, but still very valuable, are the papers in Spiegel and Zahn (1977).

The plan of the lecture is as follows: After a discussion of the convective instability in section 2 we proceed in section 3 to describe the mixing length formalism (MLF). For lack of an accepted theory of convective energy transport, the MLF and similar approaches containing adjustable parameters are still the only means to describe the effects of fully developed convection on stellar structure and evolution. A short discussion of convective overshoot on the basis of a modified (non-local) MLF follows in section 4. Observable convective flows in the solar photosphere are described very briefly in section 5; for more details the reader is referred to Chapter 8. In section 6 we discuss numerical simulations and their implications for our understanding of solar and stellar convection.

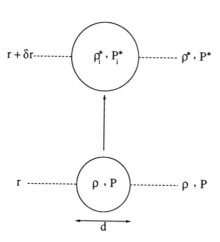 83

2. Convective Instability

In this section, we discuss the criteria for the onset of convective instability of a hydro-statically stratified fluid in a simplified, heuristic way. The formal mathematical approach (linear stability analysis) is described, for example, in the classical book of Chandrasekhar (1961), in the paper by Lebovitz (1966), and in reviews by Spiegel (1971, 1972) and Ledoux (1977).

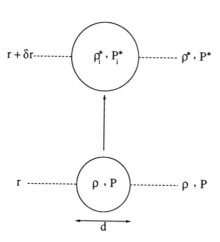

Fig. 1: Perturbation of a hydrostatic equilibrium by displacement of a small parcel of material from radius $r \rightarrow r + \delta r$. In its equilibrium position (r), internal pressure and density of the parcel are the same as the values P, ρ of the surrounding fluid. The difference between parcel density (ρ_i^*) and external density (ρ^*) in the displaced position $(r + \delta r)$ determines whether the parcel is driven back towards its original position (stability) or its displacement increases (instability).

Consider a star in hydrostatic equilibrium described by the thermodynamical quantities pressure (P), density (ρ), temperature (T), mean molecular weight (μ), and entropy (S) as functions of the radial coordinate (r). Assume a parcel of fluid which is small enough that all quantities may be taken uniform within, *i.e.* the parcel diameter d should be small compared to the local pressure scale height $H_P = \mathcal{R}T/\mu g$ (where $\mathcal{R}$ is the ideal gas constant and g is the gravitational acceleration). As sketched in Figure 1, we now assume a small perturbation of the system such that the parcel is slightly displaced from its equilibrium position at radius r to radius $r + \delta r$ where $\delta r > 0$ is of the order of the parcel diameter, d. While at radius r interior and exterior of the parcel are identical, at $r + \delta r$ the interior quantities (*e.g.* pressure P_i^* and density ρ_i^*) generally differ from those in the exterior (P^*, ρ^*). Whether the system is stable or not with respect to this perturbation is decided by the *density difference:* If the internal density is smaller than the external density, we have a positive buoyancy force and the parcel will continue to rise while, in the opposite case, buoyancy acts as a restoring force which drives the parcel back in the direction of its equilibrium position and, without dissipation, leads to an oscillation around the equilibrium position.

84

Consequently, we have as *criterion for instability:*

$$\rho_i^* < \rho^* . \tag{1}$$

In order to determine the density difference, we have to specify the perturbation in somewhat more detail. We choose the displacement velocity in such a way that the perturbation timescale, $\tau = d/v$, is much smaller than the timescale of thermal exchange, $\tau_{th} = d^2/\eta_R$ (where η_R is the radiative thermal diffusivity), and, at the same time, much larger than the dynamical timescale, $\tau_{dyn} = d/v_S$ (where v_S is the velocity of sound). Such a choice is always possible in a stellar interior since the radiative timescale (Kelvin-Helmholtz time) is much larger than the dynamical timescale. This choice of timescale has the following consequences:

$$\tau \ll \tau_{th} \quad \Rightarrow \quad S_i^* = S \quad \text{adiabatic displacement,} \tag{2a}$$
$$\tau \gg \tau_{dyn} \quad \Rightarrow \quad P_i^* = P^* \quad \text{no pressure fluctuations.} \tag{2b}$$

We can use the properties expressed in Equation (2) to rewrite the condition for instability. For $\delta r \ll H_P$, we express Equation (1) in terms of the difference between the density gradient in the external fluid $(d\rho/dr)$ and the density gradient $(d\rho/dr)_{ad}$ of an adiabatic stratification (stratification with constant entropy equal to the local value at the equilibrium position, r):

$$\rho_i^* - \rho^* = \left[\left(\frac{d\rho}{dr} \right)_{ad} - \frac{d\rho}{dr} \right] \delta r < 0 . \tag{3}$$

Using the equation of state $P = \mathcal{R}T/\mu g$ and $P_i^* = P^*$, we obtain the criterion for instability as a condition on the gradients of temperature and mean molecular weight:

$$\frac{dT}{dr} < \left(\frac{dT}{dr} \right)_{ad} + \frac{T}{\mu} \left[\frac{d\mu}{dr} - \left(\frac{d\mu}{dr} \right)_{ad} \right] . \tag{4}$$

There are two possible reasons for a molecular weight gradient in the interior of a star: (a) spatial variations of chemical composition due to nuclear processes in stellar cores, and (b) ionization of abundant elements (H, He) in stellar envelopes. We consider both cases separately. In case (a), *i.e.* compositional variations and no ionization, the molecular weight within the parcel does not change during the displacement,

$$\left(\frac{d\mu}{dr} \right)_{ad} = 0 , \tag{5}$$

and from Equation (4) we obtain the *Ledoux criterion* for convective instability

$$\frac{dT}{dr} < \left(\frac{dT}{dr} \right)_{ad} + \frac{T}{\mu} \frac{d\mu}{dr} . \tag{6}$$

If, due to nuclear burning in a stellar core, He and other heavier elements are enriched with respect to the outer regions, we have $d\mu/dr < 0$, a stabilizing μ-gradient. For the solar convection zone, case (b) is relevant, *i.e.* ionization in a chemically homogeneous medium. Under these circumstances, the molecular weight is a given function $\mu(T, P)$, both in the parcel and in the external fluid. Equation (4) may then be written as

$$\left[\frac{dT}{dr} - \left(\frac{dT}{dr} \right)_{ad} \right] \cdot \left[1 - \left(\frac{d\ln\mu}{d\ln T} \right)_{p=const.} \right] \equiv \left[\frac{dT}{dr} - \left(\frac{dT}{dr} \right)_{ad} \right] \cdot \chi_P < 0 . \tag{7}$$

Since χ_P is always positive, instability ensues if the *Schwarzschild criterion*

$$\frac{dT}{dr} < \left(\frac{dT}{dr}\right)_{ad} \qquad (8)$$

is fulfilled, which is equivalent to the condition $dS/dr < 0$ on the entropy gradient. In a stellar interior, the Schwarzschild criterion means that convection sets in if the temperature decreases outward more rapidly than that of a locally adiabatic stratification.

Near the cores of hot stars, the interesting situation may evolve that the Schwarzschild criterion, Equation (8), is fulfilled but (due to a μ-gradient) the Ledoux criterion, Equation (6), is not:

$$\left(\frac{dT}{dr}\right)_{ad} + \frac{T}{\mu}\frac{d\mu}{dr} < \frac{dT}{dr} < \left(\frac{dT}{dr}\right)_{ad}. \qquad (9)$$

Although the stratification is dynamically stable, there is free energy available due to the superadiabatic temperature gradient (negative entropy gradient). If thermal energy exchange (by radiation) between a parcel and its environment is allowed, growing oscillations are driven. Consider again Figure 1 in this situation: Since the inequality in Equation (6) is not fulfilled, the parcel is driven back towards its equilibrium position. However, since Equation (8) is satisfied, the displaced parcel is hotter than its surroundings and, therefore, cools by radiation. Because of pressure equilibrium, the internal density increases above its adiabatic value and the restoring force is enhanced. Similarly, for a downward displacement the upward restoring force is increased. As a consequence, the parcel performs oscillations with growing amplitude, a phenomenon called *overstability*. In the present context, the process is also known as *semiconvection*. Energy transport by this kind of oscillatory convection is not very efficient since the whole process is driven by the small amount of radiative energy exchange between the parcel and its surroundings. The importance of semiconvection lies in its ability to mix "fresh" hydrogen into the regions of nuclear burning; in this way, it can significantly influence the evolution of hot stars.

It is technically convenient to replace the radius r by $\ln P$ as the independent variable and rewrite the stability criteria in a non-dimensionalized form. For example, the temperature gradient may be written as

$$\frac{dT}{dr} = \frac{T}{P}\frac{d\ln T}{d\ln P}\frac{dP}{dr} = -\frac{T\rho g}{P}\nabla = -\frac{T}{H_P}\nabla. \qquad (10)$$

In this equation, we have defined $\nabla = d\ln T/d\ln P$ and have used the condition of hydrostatic equilibrium, $dP/dr = -\rho g$. In a similar way, we define the adiabatic temperature gradient, $\nabla_{ad} = (d\ln T/d\ln P)_{ad}$, and the non-dimensional μ-gradient, $\nabla_\mu = d\ln\mu/d\ln p$. The stability criteria Equations (6) and (8) can then be written

$$\text{Ledoux criterion}: \quad \nabla \;>\; \nabla_{ad} + \nabla_\mu. \qquad (11a)$$

$$\text{Schwarzschild criterion}: \quad \nabla \;>\; \nabla_{ad}. \qquad (11b)$$

It is clear from the special choice of the perturbations that these criteria as "derived" here are only sufficient. However, it has been shown by Lebovitz (1966) by way of a complete (linear) stability analysis that they are both sufficient *and* necessary. Finally, in fully developed stellar convective region, we can define and distinguish four temperature gradients (in "∇ notation"):

$$\nabla_{rad} \quad > \quad \nabla \quad > \quad \nabla_i \quad > \quad \nabla_{ad} \qquad (12)$$

∇_{rad} is the temperature gradient necessary to transport the given stellar energy flux purely by radiation. Since convection is much more efficient in transporting energy, the actual average temperature gradient, ∇, is much smaller than ∇_{rad} and nearly equal to ∇_{ad} (*cf.* Figure 2). In order to drive the convection, however, ∇ must be larger than ∇_i, the temperature gradient of a displaced parcel. If energy exchange between parcel and surrounding medium is taken into account, ∇_i becomes slightly larger than ∇_{ad}. A more detailed treatment of these relations and of convective instability in general can be found in Cox and Giuli (1968, Chapters 13 and 14).

3. Mixing Length Formalism

An accepted nonlinear theory of fully developed stellar convection does not exist. This is caused by the lack of a reliable theory of turbulence (the Reynolds number in a stellar convective region is very large) and the complications arising from the effects of stratification, rotation, ionization and radiation. On the other hand, for many purposes (*e.g.* stellar structure and evolution), detailed information about the properties of convective flows may not be needed and only average or global quantities matter. For these kind of objectives, the Mixing Length Formalism (MLF) developed, among others, by Prandtl (1925), Biermann (1948), Öpik (1950), and Böhm-Vitense (Vitense 1953) has been successfully used. The idea is to consider the convecting fluid as a mixture of parcels (in the sense of the preceding section) which move vertically over a finite distance l (the *mixing length,* which is a free parameter) and then dissolve. The mixing length is often taken proportional to the pressure scale height,

$$l = \alpha \cdot H_P, \qquad (13)$$

where α is a constant of order unity. Other prescriptions for the mixing length can be found in the literature; they do not alter the basic concept described below, assuming the relation given by Equation (13).

The aim of the MLF is to determine the convective energy flux as function of the actual temperature gradient, ∇, or superadiabatic gradient, $\nabla - \nabla_{ad}$. For the sake of clarity in what follows, we suppress numerical factors of order unity which appear in the classical formulations of MLF; they can always be absorbed in the free parameter, α. The temperature difference, ΔT, between a parcel and its surroundings after having moved upward over a distance $\delta r = l$ is given by

$$\Delta T = \left[\left(\frac{dT}{dr} \right)_i - \frac{dT}{dr} \right] \delta r = (\nabla - \nabla_i) T \frac{\delta r}{H_P} = (\nabla - \nabla_i) T \alpha . \qquad (14)$$

The index i indicates parcel quantities which may differ from their adiabatic values due to radiative energy exchange between parcel and surrounding medium. However, since radiation in the bulk of the solar convection zone is very inefficient, except for the outermost layers, the rise of the parcel is practically adiabatic. If the average speed v of a parcel during its rise is much smaller than the sound velocity, the energy flux (energy per unit area and unit time) carried by the parcels (*i.e.* the convective flux, F_c) is given by the enthalpy flux

$$F_c = \Delta T \rho c_P v = \rho c_P v T (\nabla - \nabla_i) \alpha. \tag{15}$$

c_P is the specific heat at constant pressure. The velocity is determined by considering the acceleration of the parcel by the buoyancy force, *viz.*

$$\ddot{\delta r} = -g \frac{\Delta \rho}{\rho} = g \chi_P \frac{\Delta T}{T} = g \chi_P (\nabla - \nabla_i) \frac{\delta r}{H_P} \tag{16}$$

where Equation (14) has been used and χ_P is defined in Equation (7). Since the parcel is assumed small enough to be taken homogeneous, Equation (16) may easily be integrated and we obtain

$$(\dot{\delta r})^2 = \frac{g \chi_P}{H_P} (\nabla - \nabla_i) (\delta r)^2 \tag{17}$$

which expresses the fact that the kinetic energy is equal to the work done by the buoyancy force. After having moved over a length $\delta r = l$, the parcel has reached the velocity $(v = \dot{\delta r})$

$$v = \left[\frac{g \chi_P}{H_P} (\nabla - \nabla_i) \right]^{1/2} \cdot l \tag{18}$$

which can be inserted into Equation (15) for the convective flux:

$$F_c = \rho c_P T (g \chi_P H_P)^{1/2} \alpha^2 (\nabla - \nabla_i)^{3/2}. \tag{19}$$

The total energy flux (sum of radiative and convective flux) is fixed by the given stellar luminosity. Except for the upper and lower boundary layers, radiative flux can be neglected in the solar convection zone and Equation (19) is used to determine the temperature gradient, $\nabla(r)$. In the case $\nabla_i = \nabla_{ad}$ this can be done immediately while inclusion of radiative exchange between parcels and environment ($\to \nabla_i > \nabla_{ad}$) requires some additional considerations which, in the end, lead to a cubic equation for ∇ (see, for example, Stix 1989, p. 198f).

When a model of the Sun including a mixing length treatment of its convection zone is constructed, the free parameter α is fixed by requiring that the model matches the correct solar radius. Consequently, the one free parameter of the MLF is fixed by the one observable quantity – the theory has no predictive capability. Typically, α is assigned a value between 1 and 2. An often-used MLF model of the solar convection zone has been provided by Spruit (1974, 1977). In the deep parts of this model convection zone we have

$$\nabla - \nabla_{ad} \approx 10^{-5} \ll 1$$
$$\Delta T \approx 2\,\mathrm{K} \ll T$$
$$v \approx 100\ \mathrm{m\ sec^{-1}} \ll v_S.$$

Consequently, convection is very efficient in the sense that only a small amount of superadiabaticity is necessary to drive it. Temperature and velocity fluctuations are small and satisfy *a posteriori* the assumptions on which the MLF is based. This situation changes drastically if we consider the region near the solar surface; here we have

$$\nabla - \nabla_{ad} \approx 0.6$$
$$\Delta T \approx 2000\,\mathrm{K}$$
$$v \approx 2\ \mathrm{km\ sec^{-1}}.$$

The deviation from adiabatic stratification and the fluctuations become large near the surface. Figure 2 illustrates this difference between surface and bulk of the convection zone; it shows the temperature gradients ∇_{rad}, ∇_{ad}, and ∇ for Spruit's model of 1977 as functions of logarithmic temperature.

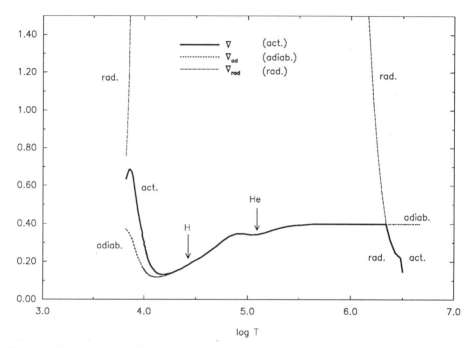

Figure 2: Temperature gradients ∇_{rad}, ∇_{ad}, and ∇ as functions of $\log T$ for a mixing length model of the solar convection zone (Spruit 1977). At the bottom of the convection zone ($\log T \approx 6.3$, depth $\approx 2 \cdot 10^5$ km) the actual temperature gradient changes from $\nabla \approx \nabla_{rad}$ to $\nabla \approx \nabla_{ad}$. The superadiabaticity becomes significant only in the surface layers ($\log T < 4.$) where the whole flow is driven. The ionization regions of hydrogen and helium are indicated; the adiabatic temperature gradient decreases there due to the effect of latent heat.

We see that $\nabla \approx \nabla_{ad}$ over most of the superadiabatic region. Near the surface, convection must be driven by a larger superadiabaticity since the radiative losses of rising parcels become non-negligible. It is apparent from Figure 2 and Equations (11b) and (12) that two factors contribute to the driving of convection: On the one hand, ∇_{rad} grows strongly (due to increasing opacity and decreasing temperature) near the bottom of the convection zone (approx. 200,000 km below the surface) and comes down rapidly (due to the drastically decreasing opacity) in the visible photosphere which becomes convectively stable. On the other hand, ionization of hydrogen and helium reduces ∇_{ad} and thus provides additional driving for the convective instability. The ionization regions are indicated in Figure 2.

For many years, the MLF has been used as a tool for studies of stellar structure and evolution. Results from helioseismology (see Chapter 3) have demonstrated that MLF models of the solar convection zone predict the average structure remarkably well. MLF has its drawbacks, however: First of all, it contains an adjustable parameter and is not a predictive theory. Its local character excludes overshooting of convective motions into stable layers, a process which evidently takes place in the solar photosphere. The neglect of pressure fluctuations (Equation 2b) implies that the formalism assumes a weakly stratified, essentially incompressible fluid (Boussinesq approximation). Although this assumption may be valid in the deep parts of the convection zone it is clear that the MLF is inapplicable to the convective flow patterns observed in a stellar photosphere.

4. Overshoot

The classical MLF formulation implies that convection stops abruptly at the boundary of a convective region when the Schwarzschild criterion is no longer fulfilled (see Eqs. 18,19). This is unrealistic, however, since the convective motions approach the boundary with a finite velocity and thus "overshoot," i.e. intrude into the stable layers above and below a superadiabatically stratified region. In fact, the photospheric layers of the Sun are stably stratified and the observed convective flow patterns (e.g. granulation), strictly speaking, are motions overshooting from the underlying convection zone proper.

The photospheric overshoot is driven by pressure gradients and radiative heating; it cannot be described by simple models but requires numerical hydrodynamical simulations (see section 6). In the remainder of this section, we shall restrict ourselves to overshoot into the radiative layers below the solar convection zone ("undershoot"). In principle, this process should be somewhat simpler to describe theoretically, since all fluctuations are small and pressure fluctuations can be neglected. On the other hand, theory is not directly guided by observation * as in the case of the photosphere. An extended layer of overshooting convection at the bottom of the solar convection zone is of considerable interest. For example, if such an overshoot mixes the convection zone material down to depths with temperatures in excess of $2.5 \cdot 10^6$ K, Lithium would be consumed by nuclear processes and its low abundance in the solar photosphere (and similarly in other cool main sequence stars) could be understood (e.g. Sackmann et al. 1990). An overshoot layer below the convection zone proper has been proposed as a possible site for magnetic flux storage and for the operation of the solar dynamo (e.g. Spiegel and Weiss 1980, van Ballegooijen 1982).

In Figure 3, we again use the "blob picture" of section 2 to illustrate and discuss the overshoot process. Consider a cool ($\Delta T = T_i - T < 0$) convective parcel which is sinking adiabatically due to its positive density contrast ($\rho_i > \rho$). At the formal boundary of the unstable region ($r = r_S, \nabla = \nabla_{ad}$), density and temperature contrast start to decrease (the stratification becomes subadiabatic) but the parcel is accelerated further by the downward buoyancy force. The convective energy flux is still upward (positive) since cool material is transported downward. At $r = r_F$, the contrasts of temperature and density vanish, acceleration and convective flux become zero. By virtue of its inertia, however, the parcel continues to sink. But from now on the buoyancy force is directed upward and decelerates the parcel. The temperature contrast becomes positive (hot material moves downward) and the direction of the convective energy flux changes to downward. Eventually, at $r = r_C$, the parcel is stopped and it starts to be accelerated upwards by the positive buoyancy force.

---

* Helioseismology promises to supply an observational handle on the overshoot region at the bottom of the convection zone; see Chapter 3.

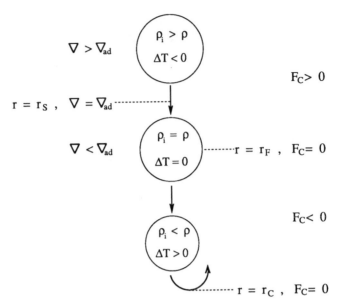

Figure 3: Convective overshoot in the "blob" picture. A heavy ($\rho_i > \rho$), sinking convective parcel penetrates into the subadiabatically ($\nabla < \nabla_{ad}$) stratified region. When the temperature fluctuation ($\Delta T = T_i - T$) changes sign ($r = r_F$), the parcel becomes neutrally buoyant but continues to sink by virtue of its inertia until the increasing upward buoyancy force reverses the direction of its motion. The direction of the convective energy flux, F_c, changes from upward (positive) to downward (negative) in the overshoot region.

If we determine the distance $r_S - r_C$ using a solar model with a standard MLF treatment of the convection zone, the amount of this "ballistic" overshoot is very small, almost negligible. Such a point of view, however, omits an important effect: Each overshooting parcel mixes high entropy material from the convection zone into the low-entropy, stable layers. Consequently, the stabilizing entropy gradient is decreased, the next parcel is somewhat less decelerated and able to move deeper down than its predecessor. This *penetrative* character of the overshooting convective motions must be accounted for, *i.e.* the thermodynamic consequences of the overshoot have to be treated simultaneously with its dynamics, if a determination of realistic values for the depth of an overshoot region is attempted.

One possibility for such a consistent treatment is a non-local generalization of the MLF as proposed by Shaviv and Salpeter (1973) which formalizes the heuristic arguments given above. Instead of using the local value given by Equation (18), the velocity of a parcel at $r = r_2$ which has started from rest at $r = r_1$ satisfying

$$\frac{1}{2} v^2(r_2, r_1) = \int_{r_1}^{r_2} \frac{g \chi_P \, \Delta T(r_1, r)}{T(r)} \, dr \qquad (20)$$

is calculated using the *integrated* temperature difference

$$\Delta T(r_1, r_2) = - \int_{r_1}^{r_2} \left[\frac{dT}{dr} - \left(\frac{dT}{dr} \right)_{ad} \right] dr . \qquad (21)$$

Of course, v is set to zero if $v^2 < 0$ in Equation (20). We now may have $v \neq 0$ also in stable layers since the accumulated ΔT enters into the determination of the velocity. The convective flux at radius r is determined by considering rising and sinking parcels which have started at $r - l$ and $r + l$, respectively.

Pidatella and Stix (1986) and Skaley and Stix (1991) have applied this recipe to the overshoot layer at the bottom of the solar convection zone. An example of the results is shown in Figure 4, taken from Stix (1989).

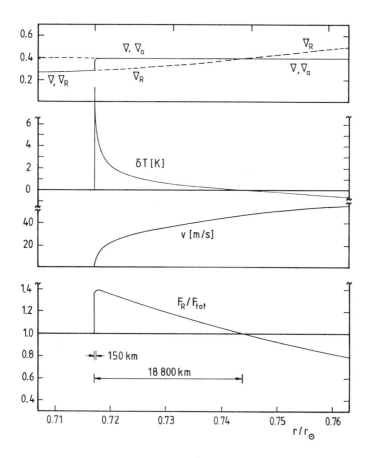

Figure 4 (from Stix 1989): Adiabatic (∇_a), radiative (∇_R), and actual (∇) temperature gradient, temperature excess of sinking parcels (δT), convection velocity (v), and the ratio of radiative (F_R) to total energy flux (F_{tot}) in a nonlocal mixing length model of the overshoot layer at the bottom of the solar convection zone.

The extension of the overshoot layer (defined as the distance $r_F - r_C$) in this model is about 2×10^4 km, a third of the pressure scale height in this depth. A layer thickness of a few tenths of a scale height is a common result for most models of overshooting convection. Like the convection zone proper, the overshoot layer is almost adiabatically stratified, but it is slightly *sub*adiabatic. Thus the convective motions proceed almost undisturbed in the overshoot layer until they are stopped rather abruptly in a narrow thermal boundary layer of about 150 km thickness which separates the overshoot region from the radiative interior. Here the temperature gradient changes from almost adiabatic to radiative. The temperature excess of the sinking parcels is positive in the overshoot region (*cf.* Figure 3) and increases sharply in the thermal boundary layer. Since the convective energy flux in the overshoot layer is directed downward (negative), the upward (positive) radiative flux has to be larger than the positive total flux in order to keep it constant with depth.

Alternative approaches to the structure of the overshoot layer at the bottom of the solar convection zone have been proposed, among others, by van Ballegooijen (1982) and by Schmitt *et al.* (1984). The results are similar to those described here. Numerical simulations of convective overshoot have been performed by Toomre *et al.* (1984).

5. Observed Solar Convective Flow Patterns

The inhomogeneity of the solar photosphere was recognized at the beginning of the 19th century when W. Herschel (1801) noticed structures which he described as similar to rice grains. W. R. Dawes (1864) coined the term "granules" for these structures which were photographed for the first time with outstanding quality by Janssen (1896). Spectroscopic measurements revealed the convective nature of granulation: The bright granules represent hot, rising material while the the interconnected network of dark intergranular lanes consists of cool, descending gas. Granules have a typical spatial scale $L \approx 1,000$ km, vertical rms velocities $v_{rms} \approx 1$ km sec-1, and a lifetime $\tau \approx 10$ minutes. Numerous studies of the physical properties of granulation have been carried out in the last decades; they include observations from the ground, balloon flights and observations from space (see Bray *et al.* 1984, Muller 1989, and Spruit *et al.* 1990 for overviews).

Apart from granulation which carries the majority of the convective energy flux in the uppermost layers of the convection zone, other velocity patterns which very probably are of convective origin have been discovered: *Supergranulation* with $L \approx 30,000$ km, $v_{rms} \approx 50$ m sec-1 and $\tau \approx 1$ day (Leighton *et al.* 1962), and *mesogranulation* with $L \approx 7,000$ km, $v_{rms} \approx 60$ m sec-1 and $\tau \approx 1$ hour (November *et al.* 1981). It is not clear, however, whether mesogranulation is really a distinct pattern or just part of a continuum of convective flow scales (*cf.* Nordlund and Stein 1990). So-called *giant cells, i.e.* convection on scales $L > 100,000$ km with a cell lifetime $\tau \approx 1$ month, which have been predicted by a number of theoretical investigations have not been detected.

6. Numerical Simulation of (Solar) Convection

Apart from the possibilities which may be opened in future by helioseismology, observational studies of solar convection are restricted to the photospheric surface layers. Unfortunately, a theoretical description is most complicated in this region and simplified concepts like the mixing length formalism, linear superposition of eigenmodes, and Boussinesq approximation fail. The most important complications are

• non-stationarity and dependence on three spatial coordinates,

• strong stratification, small scale heights,

- non-linearity of the flows,
- compressibility effects, shocks,
- non-local radiative energy transport.

Under these circumstances, numerical simulations provide a powerful tool. They are based on the (magneto)hydrodynamic equations which we briefly present below. Although we do not discuss magnetic fields here, for reasons of completeness (and to be used in Chapter 9) we include the effects of a magnetic field. The purely hydrodynamic equations are obtained by setting $\mathbf{B} = 0$ in the equations below.

Mass conservation and compressibility effects are expressed in the *equation of continuity*:

$$\frac{\partial \rho}{\partial t} = -\nabla \cdot (\rho \mathbf{v}) \tag{22}$$

(where ρ is the density and $\mathbf{v}$ is the velocity vector). The *equation of motion* determines the velocity:

$$\rho \left(\frac{\partial \mathbf{v}}{\partial t} + \mathbf{v} \cdot \nabla \mathbf{v} \right) = -\nabla P + \rho \mathbf{g} + \frac{1}{4\pi} (\nabla \times \mathbf{B}) \times \mathbf{B} + \mathbf{f}_{visc} \tag{23}$$

(where P is the pressure, $\mathbf{g}$ is the gravitational acceleration vector, and $\mathbf{f}_{visc}$ is the viscous force). Since the Reynolds number $Re = uL/\nu$ (where u is the characteristic velocity, L is the length scale, and ν is the kinematic viscosity) is huge ($> 10^{12}$), steep gradients and strongly concentrated flows are likely to form; the flow becomes turbulent and very small spatial scales appear which cannot be resolved by the spatial grid used in numerical simulations. In order to account (crudely) for the effect of these unresolved velocities, "sub-grid" transport coefficients (for viscosity, thermal, and magnetic diffusion) are introduced.

The *energy equation* takes care of energy conservation. It is written here as an equation for temperature, but a number of other forms (in terms of internal energy, total energy, entropy, ...) is possible:

$$\frac{\partial T}{\partial t} + \mathbf{v} \cdot \nabla T = -\frac{T}{\chi_P}(\gamma - 1)\nabla \cdot \mathbf{v} - \frac{\gamma}{\rho c_P} \nabla \cdot (\mathbf{F}_{rad} + \mathbf{F}_{visc} + \mathbf{F}_{mag}) \tag{24}$$

(where γ is the ratio of specific heats). The first term on the right hand side of Equation (24) describes compressional heating while the other terms express the effects of radiation ($\mathbf{F}_{rad}$), viscous dissipation ($\mathbf{F}_{visc}$), and Joule heating ($\mathbf{F}_{mag}$). The thermodynamical quantities $\chi_P, \gamma, c_P, ...$ must be determined self-consistently taking into account partial ionization and its dependence on temperature and density. For a realistic simulation of convective flows in the surface layers, radiation has to be treated accurately. This means that instead of the simple diffusion approximation (adequate in the deep convection zone and in the radiative interior) the non-local radiative transfer equation has to be solved, preferably for non-grey opacities (opacity distribution functions, *e.g.* Kurucz 1979).

Finally, the time evolution of the magnetic field is governed by the *magnetic induction equation*:

$$\frac{\partial \mathbf{B}}{\partial t} = \nabla \times (\mathbf{v} \times \mathbf{B}) - \nabla \times (\eta_m \nabla \times \mathbf{B}) \tag{25}$$

The small value of the magnetic diffusivity, η_m, leads to steep magnetic field gradients and supports the formation of concentrated magnetic fields.

In addition to the basic Equations (22-25), a number of further equations has to be treated:

- the equations of state and of ionization equilibrium,
- determination of thermodynamic quantities ($\chi_P, \gamma, c_P, ...$) and opacities,
- integration of radiative transfer equation along many lines of sight.

Furthermore, the numerical results have to be analyzed and presented in way which provides the maximum of useful information. This requires the determination of diagnostic information (mean quantities, correlations, single and average spectral line profiles, continuum intensity maps...) and a proper visualization of flow properties and time evolution.

Boundary and initial conditions deserve particular attention. Since no numerical calculation can encompass the whole Sun, from its core out to the solar wind, with sufficient spatial resolution, the domain of integration (the numerical "box") has to be artificially bounded. Similar to the introduction of "turbulent" or "sub-grid" diffusivities which depend on assumed properties of an unknown velocity field, the specification of boundary conditions therefore inevitably introduces some degree of arbitrariness to the simulation. Ideally, the boundary conditions should be "passive," *i.e.* allow free in- and outflow of material and permit transmittance of all kinds of waves without introducing artificial reflection. Apart from special cases (like a boundary with supersonic outflow), such conditions cannot be generally determined in a rigorous way; several kinds of "recipes" are used instead. Examples are the introduction of an extra boundary layer whose properties are tailored in a way to passively absorb disturbances from the proper domain of integration (Stein and Nordlund, 1989) or the exclusion of inwardly directed characteristics (Hedstrøm 1979).

The duration of the simulation must be sufficiently large that the initial conditions are "forgotten" by the system and their influence on the results be neglected. This means that the simulation must run at least for a few thermal relaxation times.

No simulation exists which fulfills all the requirements described above. Limitations of computer resources inevitably enforce compromises. Furthermore, in order to study particular problems in detail and under more "clean" conditions, it may even be advantageous to simplify the physics and study general properties of convection under different conditions without attempting to model the Sun. This approach may be called *numerical experiment* and has been pursued, among others, by Graham (1975, 1977), Hurlburt *et al.* (1984, 1986), Chan and Sofia (1986, 1987), Hossain and Mullan (1990), Malagoli *et al.* (1990), and Cattaneo *et al.* (1991). Another approach, simulation of *global convection*, has been taken by Gilman and Miller (1986) and by Glatzmayer (1985, 1987). They trade spatial resolution for total size of the numerical box in order to simulate the large-scale convective flows of the full solar convection zone, neglecting the strongly stratified surface regions. Finally, numerical simulations of solar surface convection on a granular to mesogranular scale including a realistic radiative transfer have been performed by Nordlund (1985a) and Stein and Nordlund (1989) in 3D, and by Steffen *et al.* (1989) in 2D (cylindrical symmetry).

Let us discuss some key results from the simulations of compressible convection relevant for the Sun. A characteristic property of convection in a strongly stratified medium is the *asymmetry of up- and downflows:* The downflows are strong and narrow while the upflows are slow and broad. The reason for this difference lies in the flow topology and the role of pressure fluctuations: In the upper layers, upflowing material is accelerated sideways, moves horizontally, until it is decelerated above a downdraft and flows downwards again. Consequently, above both upflows and downflows, the pressure is enhanced; the pressure perturbation $P' = P - \langle P \rangle$ is positive, where $\langle P \rangle$ is the horizontally averaged pressure at the same height. From the equation of state for an ideal gas we can determine the relation between the relative perturbations of density, pressure, and temperature, *viz.*

$$\frac{\rho'}{\langle\rho\rangle} = -\frac{T'}{\langle T\rangle} + \frac{P'}{\langle P\rangle}.$$ (26)

In a downflow, both terms on the right side are positive and the flow is driven by the resulting strongly positive density perturbation and negative buoyancy force. For an upflow, on the other hand, the first term, $-T'/\langle T\rangle$, is negative ($T' > 0$ is the driver of the upflow), but, as explained above, over most of the upflow the second term is positive (horizontal acceleration in a strongly stratified medium). This reduces the buoyancy of the upflowing material and may even lead to $\rho' > 0$ in the upper part of the upflow, a phenomenon called "buoyancy braking." Consequently, the pressure fluctuations brake the upflows and accelerate the downflows; mass conservation then requires that the upflows become much broader than the downflows. Under these circumstances, convection is mainly driven by the buoyancy force in the downflows, an effect which becomes even stronger when radiative surface cooling is taken into account. This mechanism has been demonstrated and discussed lucidly by Hurlburt *et al.* (1984; see their Figure 8) in the case of stationary cellular convection in 2D.

Buoyancy braking also provides an explanation for the observed sizes of granules. Assume, for simplicity, a stationary, cylindrically symmetric convection cell (the assumption of stationarity is unnecessary if the flow is subsonic and the "anelastic" approximation applies, *cf.* Nordlund 1982). The equation of continuity, Equation (22), in cylindrical coordinates is then given by

$$0 = \nabla\cdot(\rho\mathbf{v}) = \frac{\partial}{\partial z}(\rho v_z) + \frac{1}{r}\frac{\partial}{\partial r}(r\rho v_r)$$ (27)

where v_z and v_r are the vertical and radial velocity components, respectively. Taking density only as a function of height, z, and v_z approximately constant with height, and $v_r \propto r$ (first term of a Taylor expansion about the axis) we may rewrite Equation (27) in the form

$$\frac{\rho v_z}{H_\rho} = \frac{2\rho v_r}{r}$$ (28)

(where H_ρ is the density scale height) and obtain a relation between the radius of the cell, R, and the radial velocity at its periphery, $v_R = v_r(R)$:

$$R = 2H_\rho\left(\frac{v_R}{v_z}\right).$$ (29)

Consequently, the ratio of radial to vertical velocity increases for growing cell size; the radial outflow becomes faster and faster as the cell grows in order to transport the vertical mass flux. However, the pressure fluctuation, $P'/\langle P\rangle$, which drives this outflow cannot grow indefinitely since it retards the upflow at the same time and eventually chokes it. Such a destruction of growing granules by buoyancy braking of their upflows is indeed observed as the "exploding granule" phenomenon (*e.g.* Muller 1989): For many rapidly expanding granules, a growing dark spot appears in the center which signifies the choking of the upflow by buoyancy braking, supported by strong radiative cooling. After a short time the granule fades away and disappears. The same process is shown by numerical simulations (*e.g.* Nordlund, 1985a).

Equation (29) allows us to determine a maximum size for granules. With a typical upflow velocity of $v_z = 2$ km sec-1, and taking the sound speed as an upper limit for the radial outflow velocity, $v_r \leq v_S \approx 10$, km sec-1 we find

$$R \leq 10H_\rho \approx 2000 \text{ km} \tag{30}$$

for solar granulation which agrees well with the observed value. An interesting conclusion from this argument is that granules on other stars should scale like the density scale height on their surfaces if no other effects come into play. This means that stars with lower surface gravity (*e.g.* giants or F-type main sequence stars) should have larger granules than the Sun and lower main sequence stars. This is indeed qualitatively borne out by simulations of stellar granulation (see Dravins and Nordlund 1990, and references therein).

Three-dimensional simulations of non-stationary, compressible convection (Stein and Nordlund 1989, Cattaneo *et al.* 1991) show that the observed granulation pattern, *i.e.* the quasi-regular structure with isolated upflows and a connected network of downflows, is a shallow surface phenomenon. In deeper layers, the downflows become filamentary and with growing depth the individual filaments merge into fewer, more widely separated, downdrafts. Although this picture of an hierarchical "inverse cascade" of downflows throughout the whole convection zone (see the discussion in Spruit *et al.* 1990) is still speculative, it provides an attractive interpretation for the observed flows on larger spatial scales (meso- and supergranulation): Merging of the downdraft filaments is accomplished by horizontal flows which are driven by the entropy deficit of the downdrafts themselves and, therefore, ultimately, by the radiative cooling of the gas at the solar surface. In deeper layers, the spatial scales of horizontal flows permitted by Equation (30) are larger since the density scale height increases with depth. Since pressure fluctuations penetrate over heights which are comparable to their horizontal spatial scale (Nordlund 1985b), these large-scale flows become visible as the horizontal flows of meso- and supergranulation at the surface. On the basis of this scenario, there should be a continuum of scales (with an upper cutoff) for horizontal flows in the solar photosphere reflecting the continuous variation of the density scale height with depth.

The most advanced simulations provide a wealth of diagnostic information which can be compared with observations. This ranges from the sizes of granules (Wöhl and Nordlund 1985) over continuum intensity distributions and contrasts (Nordlund 1984, Stein and Nordlund 1989), to synthetic spectra (Dravins *et al.* 1981, Lites et al. 1989, Steffen 1989, Steffen and Freytag 1991).

How do the simplified models of convection, in particular the MLF, fit into the picture emerging from numerical simulations ? As expected, the surface flows are not well described by the MLF: Simulations demonstrate the significance of pressure fluctuations, supersonic velocities and shocks appear (Cattaneo *et al.* 1989, Malagoli *et al.* 1990, Steffen and Freytag 1991), strong inhomogeneities and steep gradients form. The dominant role of the downdrafts as the drivers of the whole convective flow and carriers of the major parts of the kinetic energy flux (downward) and enthalpy flux (upward) introduces a non-locality in the process which is in conflict with the basic assumptions of the MLF. On the other hand, Cattaneo *et al.* (1991) have shown that if 3D convection becomes turbulent, enthalpy and kinetic energy flux in the downdrafts nearly cancel and 98% of the *net energy flux* is carried by the weak, disorganized upflows in a local, "turbulent" manner!

Consequently, although the transport of any other quantity (enthalpy, kinetic energy, angular momentum, magnetic fields, ...) is strongly non-local, for the total energy the local approach on which the MLF is based may be valid, at least if the surface regions are excluded. Furthermore, Chan and Sofia (1987) found in their simulations that the correlation between the vertical velocity components at two height levels decreases proportional to the pressure scale height, giving substance even to Equation (13), the classical scaling of the mixing length. We may conclude that MLF models, after all, gain justification from the numerical simulations as far as the total convective energy transport is concerned. However, in view of the emerging picture of the subsurface topology of solar convective flows, application of the MLF to the transport of other quantities may lead to severe problems.

Acknowledgement: Thanks are due to Peter Caligari for helping me with the figures.

References

Biermann, L.: 1948, *Zs. f. Astrophys.*, **25**, 135

Bray, R. J., Loughhead, R. E., Durrant, C. J.: 1984, *The solar granulation* (2nd ed.), Cambridge University Press

Cattaneo, F., Brummell, N. H., Toomre, J., Malagoli, A., Hurlburt, N. E.: 1991, *Astrophys. J.* **370**, 282

Cattaneo, F., Hurlburt, N. E., Toomre, J.: 1989, in Rutten and Severino (1989), p. 415

Chan, K. L., Sofia, S.: 1986, *Astrophys. J.* **263**, 935

Chan, K. L., Sofia, S.: 1987, *Science* **235**, 465

Chandrasekhar, S.: 1961, *Hydrodynamic and Hydromagnetic Stability*, Cambridge University Press, England

Cox, J. P., Giuli, R. T.: 1968, *Stellar Structure* (2 Vol.), Gordon and Breach, New York

Dawes, W. R.: 1864, *Mon. Not. Roy. Astron. Soc.* **24**, 161

Dravins, D.: 1990, in Stenflo (1990), p. 397

Dravins, D., Lindegren, L., Nordlund, Å: 1981, *Astron. Astrophys.* **96**, 345

Dravins, D., Nordlund, Å: 1990, *Astron. Astrophys.* **228**, 203

Durney, B. R., Sofia, S. (eds): 1987, *The Internal Solar Angular Velocity*, Reidel/Kluwer, Dordrecht

Gilman, P. A., Miller, J.: 1986, *Astrophys. J. Suppl.* **46**, 211

Glatzmayer, G. A.: 1985, *Astrophys. J.* **291**, 300

Glatzmayer, G. A.: 1987, in Durney and Sofia (1987), p. 263

Goldreich, P., Kumar, P.: 1988, *Astrophys. J.* **326**, 462

Graham, E.: 1975, *J. Fluid Mech.* **70**, 689

Graham, E.: 1977, in Spiegel and Zahn (1977), p. 151

Hedstrøm, G. W.: 1979, *J. Comp/ Phys.* **30**, 222

Herschel, W.: 1801, Phil. Trans. Roy. Soc. Lon. Part 1, 265

Hossain, M., Mullan, D. J.: 1990, *Astrophys. J.* **354**, L33

Hurlburt, N. E., Toomre, J., Massaguer, J. M.: 1984, *Astrophys. J.* **282**, 557

Hurlburt, N. E., Toomre, J., Massaguer, J. M.: 1986, *Astrophys. J.* **311**, 563

Janssen, J.: 1896, *Ann. Obs. Astrophys. Paris Meudon*, **1**, 191

Kurucz, R.L.: 1979, *Astrophys. J. Suppl. Ser.* **40**, 1

Lebovitz, N. R.: 1966, *Astrophys. J.* **146**, 946

Ledoux, P.: 1977, in Spiegel and Zahn (1977), p. 87

Leighton, R. B., Noyes, R. W., Simon, G. W.: 1962, *Astrophys. J.* **135**, 474

98

Lites, B. W., Nordlund, Å, Scharmer, G. B.: 1989, in Rutten and Severino (1989), p. 349

Malagoli, A., Cattaneo, F., Brummell, N. H.: 1990, *Astrophys. J.* **361**, L33

Muller, R.: 1989, in Rutten and Severino (1989), p. 101

Musielak, Z. E.: 1991, in Ulmschneider *et al.* (1991), p. 369

Nordlund, Å: 1982, *Astron. Astrophys.* **107**,1

Nordlund, Å: 1984, in in S. L. Keil (ed.): *Small-scale Dynamical Processes in Quiet Stellar Atmospheres*, Sacramento Peak Observatory, Sunspot, USA, p. 181

Nordlund, Å: 1985a, *Solar Phys.* **100**, 209

Nordlund, Å: 1985b, in Schmidt (1985), p. 1

Nordlund, Å, Stein, R. F.: 1990, in Stenflo (1990), p. 191

November, L. J., Toomre, J., Gebbie, J., Simon, G. W.: 1981, *Astrophys. J.* **245**, L123

Öpik, E. J.: 1950, *Mon. Not. Roy. Astron. Soc.* **110**, 559

Pidatella, R. M., Stix, M.: 1986, *Astron. Astrophys.* **157**, 338

Prandtl, L.: 1925, *Zeitschr. f. angew. Math. u. Mech.*, **5**, 136

Press, W. H.: 1981, *Astrophys. J.* **245**, 286

Proctor, M. R. E., Weiss, N. O.: 1982, *Rep. Progr. Phys.* **45**, 1317

Rüdiger, G.: 1989, *Differential Rotation and Stellar Convection*, Akademie-Verlag, Berlin

Rutten, R. J., Severino, G. (eds.): 1989, *Solar and Stellar Granulation*, Kluwer, Dordrecht

Sackmann, I.-J., Bothroyd, A. I., Fowler, W. A.: 1990, *Astrophys. J.* **360**, 727

Schmidt, H. U. (ed.): 1985, *Theoretical Problems in High Resolution Solar Physics,* Proc. MPA/LPARL Workshop, MPA 212, Max-Planck-Institut für Astrophysik, Garching

Schmitt, J. H. M. M., Rosner, R., Bohn, H. U.:, *Astrophys. J.* **282**, 316

Shaviv, G., Salpeter, E. E.: 1973, *Astrophys. J.* **184**, 191

Skaley, D., Stix, M.: 1991, *Astron. Astrophys.* **241**, 227

Spiegel, E. A.: 1971, *Ann. Rev. Astron. Astrophys.*, **9**, 323

Spiegel, E. A.: 1972, *Ann. Rev. Astron. Astrophys.*, **10**, 261

Spiegel, E. A., Weiss, N. O.: 1980, *Nature* **287**, 616

Spiegel, E. A., Zahn, J. P.: 1977, *Problems of Stellar Convection*, Lecture Notes in Physics No. 71, Springer, Berlin

Spruit, H. C.: 1974, *Solar Phys.* **34**, 277

Spruit, H. C.: 1977, thesis, University of Utrecht, The Netherlands

Spruit, H. C., Nordlund, Å, Title, A. M.: 1990, *Annu. Rev. Astron. Astrophys.* **28**, 263

Steffen, M., Freytag, B.: 1991, *Reviews in Modern Astronomy*, **4**, Springer, Berlin, in press

Steffen, M., Ludwig, H.-G., Krüss, A.: 1989, *Astron. Astrophys.* **213**, 371

Stein, R. F., Nordlund, Å: 1989, *Astrophys. J.* **342**, L95

Stenflo, J. O. (ed.): 1990, *Solar Photosphere: Structure, Convection, and Magnetic Fields*, IAU-Symp. No. 138, Kluwer, Dordrecht

Stix, M.: 1989, *The Sun: An Introduction*, Springer, Berlin

Toomre, J, Hurlburt, N. E., Massaguer, J. M.: 1984, in S. L. Keil (ed.): *Small-scale Dynamical Processes in Quiet Stellar Atmospheres*, Sacramento Peak Observatory, Sunspot, USA, p. 222

Ulmschneider, P., Priest, E. R., Rosner, R. (eds.): 1991, *Mechanisms of Chromospheric and Coronal Heating*, Springer, Berlin

van Ballegooijen, A. A.: 1982, *Astron. Astrophys.* **113**, 99

Vitense, E.: 1953, *Zs. f. Astrophys.*, **32**, 135

Wöhl, H., Nordlund, Å: 1985, *Solar Phys.* **97**, 213

5 MEAN FIELD DYNAMO THEORY

P. HOYNG
Laboratory for Space Research
Sorbonnelaan 2, 3584 CA Utrecht
The Netherlands

ABSTRACT. Mean field dynamo theory has witnessed a very rapid development during the last 25 years, resulting in many models which reproduce most of the basic features of the large scale field of the Sun and the Earth. In this review the application to the Sun features prominently. After a brief discussion of the relevant observations and some basic results from laminar dynamo theory, the traditional picture of linear mean field theory is outlined, including an analysis of the properties of the plane wave solutions of the dynamo equation. The emphasis is on explaining the physical mechanisms and theoretical concepts in an elementary fashion, and not on a detailed comparison of various solar mean field models. A rigorous derivation of the dynamo equation under various conditions is also presented (isotropic/anisotropic turbulence, short/long correlation time and the two-scale approximation). Finally, I discuss some of the major problems and recent developments such as the structure of the magnetic field, nonlinear dynamos, boundary layer dynamos, internal and external forcing, and subcritical dynamo operation.

1. Introduction

Dynamo theory is the collective name for various theoretical schemes that try to explain the sustained presence of magnetic fields in planets (Mercury, Earth, Jupiter, Saturn), the Sun, stars, *etc.* In some objects the magnetic field may be 'primordial'. Such fields are maintained passively, simply by being frozen in. This could be case in Ap stars and white dwarfs, for example. However, we know on paleomagnetic evidence that the magnetic field of the Earth has existed for several billion years, although the resistive decay time of the core is only $10^4 - 10^5$ years. In the Sun, the decay time is of the order of 10^9 years, but its large scale field exhibits a periodic behaviour. This is the well-known solar cycle, which has a period of 22 years. Therefore, in the Earth and the Sun something else must happen which regenerates the field. This 'something else' is what dynamo theory is about. The motion u of the conducting fluid inside the body across the magnetic field B somehow manages to induce an electric field $u \times B/c$ that drives just the right the current J to generate B. This idea of a feedback, first suggested by Larmor (1919), still stands today. It was this dynamo problem that led to the development of what we now call Magnetohydrodynamics (*MHD*). Simple dynamos without feedback also exist, for example the Jupiter-Io system. Its dynamo action is trivial in the sense that a conductor (Io) moves through an *externally imposed* magnetic field. We may distinguish four different types of approach in dynamo theory:

99

J. T. Schmelz and J. C. Brown (ed.), The Sun, A Laboratory for Astrophysics, 99–138.
© 1992 *Kluwer Academic Publishers. Printed in the Netherlands.*

(1). *Laminar theory* is the oldest approach. Mathematically simple model flows are studied in view of their capability to generate magnetic fields. All 'toy-dynamos' as well as pro- and anti-dynamo theorems belong to this class.

(2). *Mean field theory* started with the work of Parker (1955). To handle turbulent convection an average is made, and the result is a simple equation for the 'mean field' (the dynamo equation).

(3). *Spectral theories*. The flow field u and the magnetic field B are Fourier transformed, resulting in a set of equations for the correlation functions of the Fourier coefficients. These theories yield *spectra* (distribution of magnetic helicity and energy density over spatial scales), but not the field B since the phase of the Fourier components is averaged out (see *e.g.* Moffatt 1978, Chapter 11).

(4). *Numerical simulations*. The MHD equations are solved in a box or a spherical shell. No average is made; the convective motion is actually followed numerically (see *e.g.* Galloway 1986; Krause 1991 – and other references therein).

I shall briefly discuss some aspects of the laminar theory (section 3), and then concentrate on mean field theory. Spectral theories and numerical simulations are large topics in their own right for which the reader may consult the references given. Mean field dynamo theory has been a very rapidly growing field since the mid-1960s, because models based on mean field theory appear to reproduce the most important features of the large scale magnetic field of the Sun and the Earth. This Chapter is mainly concerned with the solar dynamo. The relevant observations are discussed in section 2. The main part (sections 4 – 6) deals with traditional linear mean field theory, with an emphasis on a clear exposition of its theoretical basis and the physical mechanisms involved. In section 7 some outstanding problems and recent developments are discussed. Mean field theory has also been invoked to explain the magnetic field in other bodies, such as accretion discs (*e.g.* Stepinski and Levy 1990) and the galaxy (*e.g.* Ruzmaikin *et al.* 1988). The existence of dynamos in these objects, incidentally, though likely, has not been demonstrated beyond doubt.

2. Observations

The outer layers of the Sun are convectively unstable. This convection zone, about 2×10^5 km deep, is the seat of the dynamo. At the solar surface two convective patterns are conspicuous: the granulation (typical length scale $\lambda_c \approx 10^3$ km and time scale $\tau_c \approx 400$ s) and the supergranulation ($\lambda_c \approx 3 \times 10^4$ km and $\tau_c \approx 10^5$ s). Deep in the convection zone giant convection cells may occur ($\lambda_c \approx 10^5$ km and $\tau_c \approx 3 \times 10^6$ s). These are seen in global convection models, but no observational evidence has yet been found (for more details see Spruit *et al.* 1990). The bottom of the convection zone is not sharp; there is a thin transition layer, a few times 10^4 km thick, where convective overshoot takes place (see Skaley and Stix 1991). The convection is superposed on the solar rotation $\Omega_r(r, \theta)$. The message from helioseismology is that Ω_r is roughly independent of r in the convection zone, and that Ω_r has large radial gradients close to the transition layer (Libbrecht 1988). Below the transition layer Ω_r may be constant, *i.e.* solid body rotation.

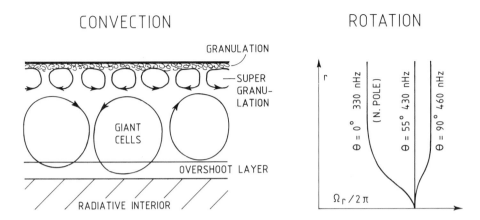

FIGURE 1. Convective patterns and mean rotation Ω_r in the solar convection zone (schematically).

2.1. SURFACE MAGNETIC FIELDS

The photospheric magnetic field is organised in patches of strong field which are embedded in regions with relatively weak field (the 'intranetwork field'). The strong field patches in turn consist of a hierarchy of magnetic elements ('flux tubes' and 'bundles') which all have roughly the same magnetic field but differ mainly in their total flux, *i.e.* in cross section. At the top of the scale are the dark sunspots, a large one having typically $\Phi \approx 10^{22}$ Mx and $B \approx 3000$ G, while the ubiquitous small bright network elements with $\Phi \approx 10^{18}$ Mx and $B \approx 1500$ G figure at the bottom of the scale (see Zwaan 1981, 1987). Due to strong buoyancy forces the orientation of magnetic elements in the photosphere is vertical.

During solar maximum, a large fraction of all flux is concentrated in active regions, *i.e.* groups of two (sometimes more) sunspots of opposite polarity with their immediate surroundings ('faculae'). Active regions have their bipolar axis nearly East-West oriented and they occur in two zones or belts parallel to the equator, one in each hemisphere. Their magnetic polarities obey Hale's polarity law, Figure 2, from which there are very few exceptions. Active regions emerge from deep in the convection zone and decay by turbulent diffusion in one or a few months (see Wang *et al.* 1989 and Zwaan 1992). In this way the small network elements originate, as decay products of active regions. At the beginning of each cycle, new bipolar active regions appear at high latitudes ($|\theta - 90°| \simeq 35°\text{-}40°$). As the cycle advances, the zones where active regions emerge moves closer to the equator. This leads to the famous *butterfly diagram*, Figure 3. The solar cycle can be regarded as a wave, featuring successive activity belts of alternating leading polarity moving to the equator. The period of the *magnetic* cycle is 22 years. The distribution of active regions in an activity belt is not uniform in longitude: new active regions have a tendency to

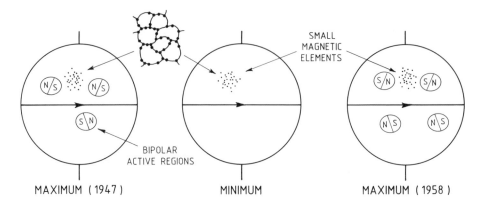

MAXIMUM (1947)　　　　　　MINIMUM　　　　　　MAXIMUM (1958)

FIGURE 2. All active regions in an activity belt have the same leading polarity, and these are opposite in the Northern and Southern hemisphere. In the next cycle, 11 years later, all polarities have reversed. These two statements comprise Hale's polarity law. The small magnetic elements are located at the boundaries of the supergranulation cells. They are distributed almost uniformly over the solar surface.

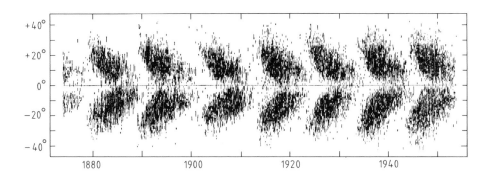

FIGURE 3. The butterfly diagram: a plot of the latitudes of occurrence of active regions *vs.* time (adapted from Kiepenheuer 1959).

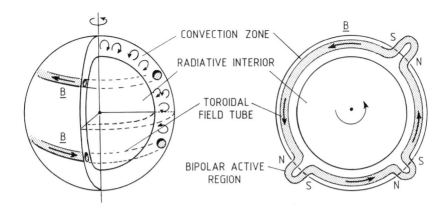

FIGURE 4. Apparent structure of the large-scale field in the convection zone.

emerge near the site of existing or previous active regions (see Kiepenheuer 1953; Zwaan 1987, 1992). The spatial distribution of smaller magnetic elements is progressively more homogeneous. For example, the small network elements ($\Phi \approx 10^{18}$ Mx) are located at the boundaries of supergranulation cells, Figure 2, but are otherwise distributed almost uniformly over the solar surface.

2.2. APPARENT BEHAVIOUR OF THE LARGE-SCALE FIELD

From the observations we arrive at the following hypothetical picture. The large scale field *inside* the convection zone behaves as if it consists of two 'tubes' of toroidal field, oppositely oriented in each hemisphere. Buoyancy causes the field to break occasionally through the surface, thus forming bipolar active regions, Figure 4. In this way, Hale's polarity law is observed. These tubes apparently move to the equator in about 11 years and disappear, after which, somehow, new tubes are generated at high latitudes in which the field direction is reversed. This behaviour can be explained almost by means of the differential rotation, as shown in Figure 5. Starting from an arbitrary field configuration, the field will be wound up to form two tubes of toroidal field of opposite polarity. However, such a dynamo would not be periodic as there is no mechanism to reverse the direction of the field. As we shall see, this is where the turbulent convection comes into play.

2.3. STABILITY OF THE SUNSPOT CYCLE; OTHER PERIODS

The cycle shows fairly large variations in period and amplitude. From the observed sunspot numbers, one obtains $\delta P_{rms}/P \simeq 0.1$ (P = mean half cycle period). Subsequent period variations δP appear to be statistically independent, implying that the magnetic oscillator would have a quality factor $Q \simeq 10$. The butterfly diagram is more regular

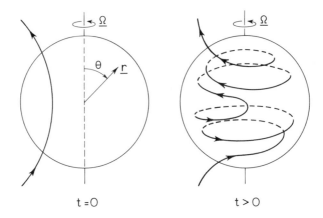

FIGURE 5. Differential rotation generates two toroidal flux tubes of opposite polarity in the convection zone.

than the sunspot numbers and leads to $Q \simeq 30$ (see Kiepenheuer 1953). However, the dataset (1610 - present) is too short to be able to exclude with certainty the possibility that the cycle is a noisy surface effect of a high-Q oscillator deep in the Sun (see Gough 1988). The sunspot data show a modulation of about 90 years (the 'Wolf-Gleissberg cycle'). Other proxy records such as tree ring ^{14}C data indicate irregular intensity modulations in the cycle on a time scale of several hundred years, including so-called *Grand Minima* (the latest being the Maunder Minimum, 1645 - 1715). It is not clear if these longer periods are true periods in the sense of frequency spectra showing distinct peaks, and this is again because the dataset is not sufficiently long. Very illustrative in this connection is a simple numerical experiment of Barnes *et al.* (1980), who simulated sunspot numbers from narrowband Gaussian noise ($\Delta\nu/\nu_0 = 0.05$ with $\nu_0 = 1/(22\,\text{year})$), and obtained a remarkable similarity with the observed sunspot record.

The global properties of the radial surface magnetic field have been analysed by spherical harmonic decomposition (Stenflo and Vogel 1986; Stenflo 1991). The power spectrum of the axisymmetric ($m = 0$) components shows the 22-year cycle at odd ℓ, as expected. However, there is a weak power ridge along which the periods decrease for increasing ℓ, down to about 1.5 year at $\ell = 14$, reminiscent of the k-ω diagram of solar oscillations. This is an interesting result which needs to be confirmed; it indicates that *overtones* of the 22-year magnetic oscillation may also be weakly excited.

3. Elementary results

The starting point of dynamo theory are the *MHD* equations:

$$\nabla \times \boldsymbol{E} = -\frac{1}{c}\frac{\partial \boldsymbol{B}}{\partial t} \quad (a) \qquad \boldsymbol{J} = \sigma\left\{\boldsymbol{E} + \frac{1}{c}\boldsymbol{u} \times \boldsymbol{B}\right\} \quad (c)$$

$$\nabla \times \boldsymbol{B} = \frac{4\pi}{c}\boldsymbol{J} \quad (b) \qquad \nabla \cdot \boldsymbol{B} = 0 \quad (d)$$

(3.1)

It is customary to take the conductivity σ constant. The electric field $\boldsymbol{E}$ and the current density $\boldsymbol{J}$ can be eliminated and the result is the induction equation for the magnetic field $\boldsymbol{B}$:

$$\frac{\partial \boldsymbol{B}}{\partial t} = \nabla \times (\boldsymbol{u} \times \boldsymbol{B}) + \eta \nabla^2 \boldsymbol{B} ; \tag{3.2a}$$

$$\eta = \frac{c^2}{4\pi\sigma} ; \qquad \nabla \cdot \boldsymbol{B} = 0 \tag{3.2b}$$

Here, η is the resistivity. The timescales associated with the two terms in Equation (3.2a) are the evolution timescale τ_e and the diffusion timescale τ_d:

$$\tau_e = \frac{L}{u} ; \qquad \tau_d = \frac{L^2}{\eta} = \frac{4\pi L^2 \sigma}{c^2} , \tag{3.3}$$

where L is a typical size of the system. The magnetic Reynolds number is defined by

$$R_m = \frac{\tau_d}{\tau_e} = \frac{uL}{\eta} = \frac{4\pi\sigma}{c^2}uL \tag{3.4}$$

If $R_m \gg 1$ the field lines are to a good approximation *frozen into* the plasma, while if $R_m \ll 1$ the magnetic field decays by ohmic dissipation.

If the velocity field $\boldsymbol{u}$ is considered as given, then dynamo action is entirely described by (3.2a), which is linear in $\boldsymbol{B}$, whence the name *linear* or *kinematic dynamo theory*. The kinematic approach, though rather powerful, is not complete. For example, it cannot determine the magnitude of $\boldsymbol{B}$. To this end we must consider the effect of the Lorentz force on $\boldsymbol{u}$, and supplement (3.2) with the momentum equation

$$\rho\left\{\frac{\partial \boldsymbol{u}}{\partial t} + \boldsymbol{u} \cdot \nabla \boldsymbol{u}\right\} = -\nabla p + \rho \boldsymbol{g} + \frac{1}{c}\boldsymbol{J} \times \boldsymbol{B} \tag{3.5}$$

The full problem posed by Equations (3.2) and (3.5) is referred to as *nonlinear dynamo theory*.

The existence of dynamo action implies that there are velocity fields $\boldsymbol{u}$ for which the solution $\boldsymbol{B} = 0$ of (3.2a) is unstable. That this is possible is illustrated by the *homopolar disc dynamo*. A conducting disc rotates with angular velocity $\boldsymbol{\Omega}$. A coil centered on the (conducting) axis is electrically connected with sliding contacts, see Figure 6. The coil generates a magnetic field $\boldsymbol{B}$, and the required current I is maintained by the potential

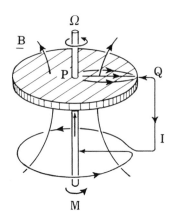

FIGURE 6. The homopolar disc dynamo (Bullard 1955).

drop which exists between P and Q because the disc moves through the magnetic field. The circuit equation is $V_{disc} + V_{coil} = RI$, where R is the total resistance. Now V_{disc} is proportional to Ω and I: $V_{disc} = a\Omega I$. The reader may verify that $a = \Phi/2\pi cI$, where Φ is the total magnetic flux through the disc (assuming $\boldsymbol{B}$ to be homogeneous and vertical). On the other hand, $V_{coil} = -L\dot{I}/c^2$, where L is the self-inductance of the coil. The result is

$$L\dot{I} = c^2(a\Omega - R)I \tag{3.6}$$

If $\Omega > R/a$, a spontaneous fluctuation δI will grow exponentially. The direction of δI determines that of I. The Lorentz force provides a breaking torque. Angular momentum balance along the vertical axis $\boldsymbol{e}_z$ requires $T\dot{\Omega} = M + M_L$ (T = moment of inertia; M = driving torque). Passing by details, $M_L = \boldsymbol{e}_z \cdot \int \boldsymbol{r} \times (\boldsymbol{J} \times \boldsymbol{B}/c)\, d^3r = -aI^2$ (the integral is over the current path in the disc). Hence

$$T\dot{\Omega} = M - aI^2 \tag{3.7}$$

In the stationary state we have that $\dot{I} = \dot{\Omega} = 0$, whence

$$\Omega = R/a \ ; \qquad I = \sqrt{M/a} \tag{3.8}$$

Increasing the driving torque enhances I, but not Ω! The power $M\Omega$ delivered by M equals the electrical power I^2R dissipated in R (friction is neglected).

The disc dynamo is a very simple nonlinear dynamo. It has two properties which are also essential for astrophysical dynamos: (1) *differential motion*, here concentrated in the sliding contacts, and (2) *reflectionally asymmetric motion*. Although I can flow either way, the sense of rotation cannot be reversed. If $\Omega < 0$ then $I \downarrow 0$ according to (3.6). The correct sense of rotation is determined by the sense of the winding of the coil. This feature

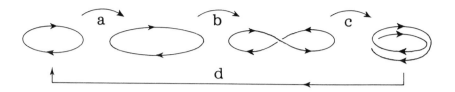

FIGURE 7. The *STFW* dynamo, a variation on an idea originally due to Alfvén (1950). (a) stretch, (b) twist, (c) fold, and (d) wait. Depending on the duration of the waiting phase, the field may be amplified (by a constant factor each cycle), or a quasi-stationary field may result as shown.

is inherent in eq. (3.2a): if a flow u can generate a field B, then it can also generate $-B$ since that is a solution, too, but the flow $-u$ is usually not able to excite a field. There are also fundamental differences: (1) in astrophysical dynamos the magnetic Reynolds number R_m is large; the field lines are frozen in. In the disc dynamo we tacitly assume that the field lines slip through the rotating disc unaffected, hence $R_m \ll 1$; (2) the disc dynamo is *multiply-connected*; it has the topology of a torus, and there are only two possible ways for the current to flow. Astrophysical dynamos are simply-connected (a sphere or spherical shell) and there is no preferred current path. If we embed the whole device of Figure 6 in a conducting fluid to make the system simply-connected, then dynamo action ceases due to short-circuiting.

A consequence of the last point is that a dynamo in a simply-connected body can only exist if the velocity field is sufficiently asymmetric. An example is Herzenberg's dynamo, another 'toy-dynamo', consisting of a spherical conductor, in which two small rotating spheres are embedded whose rotation axes are *not* parallel. Electrical contact is secured by a mercury film (Herzenberg 1958). Herzenberg's dynamo provided the first example that it is possible to maintain a magnetic field against ohmic decay by (asymmetric) motions inside a simply-connected spherical conductor, such that the field is nonzero outside the sphere (important existence proof for the dynamo of the Earth). There are many other toy-dynamos such as the coupled-discs dynamo of Rikitake (1958), the ring dynamo of Lortz (1972), and the screw dynamo of Ponomarenko (1973), see Moffatt (1978) and Parker (1979, section 18.1) for a discussion. Another example is the stretch-twist-fold-wait dynamo shown in Figure 7, which operates at $R_m \gg 1$ (at least for a fraction of the time).

The suggestion that absence of symmetry is a prerequisite for dynamo action is reinforced by Cowling's famous theorem, which says that it is impossible to maintain a stationary, axisymmetric magnetic field by dynamo action (Cowling 1934). To prove this, we split B in a meridional component B_m (in the plane through the z-axis) and an azimuthal component B_φ: $B = B_m + B_\varphi$. Figure 8 shows only B_m. Now B_m must have at least one line, for example $\mathcal{L}$, along which $B_m = 0$ (there may be more than one line; the field need

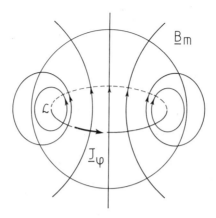

FIGURE 8. A stationary axisymmetric dynamo is impossible (Cowling's theorem).

not be dipolar). The current density J possesses a nonzero component J_φ along $\mathcal{L}$ since $\nabla \times B_m \neq 0$ there. With the help of (3.1c) we find that $J_\varphi = \sigma(E + u \times B/c)_\varphi = \sigma E_\varphi$, since $(u \times B)_\varphi = u_m \times B_m = 0$ on $\mathcal{L}$. However, $J_\varphi \neq 0$ is impossible since

$$\oint_{\mathcal{L}} J_\varphi \, d\ell = \oint_{\mathcal{L}} J_\varphi \cdot d\ell = \sigma \oint_{\mathcal{L}} E \cdot d\ell$$

$$= \sigma \int_S \nabla \times E \cdot d\sigma = -\frac{\sigma}{c} \int_S \frac{\partial B}{\partial t} \cdot d\sigma = 0 , \qquad (3.9)$$

for a stationary dynamo. Use has been made first of $E_\varphi \cdot d\ell = E \cdot d\ell$, then of Stokes' theorem with $\partial S = \mathcal{L}$, and finally of Equation (3.1a). No assumption has been made about u, nor about the location of $\mathcal{L}$.

Dynamos are therefore either non-axisymmetric or nonstationary (or both). The dynamos of the Sun and the Earth are indeed neither exactly axisymmetric nor stationary. The stationary field of the disc dynamo in Figure 6 is not *exactly* axisymmetric, and this is now seen to be an essential feature rather than an irrelevant detail. Anti-dynamo theorems such as that of Cowling have had a large impact on the development of dynamo theory because they say, loosely speaking, that simple constructions will not work ('simple' in the sense of mathematically simple flows u interacting with mathematically simple fields B). In the next section we shall see how one has succeeded in escaping from their grasp by introducing the concept of *mean fields*. This development took place only much later, from 1955 to 1965.

4. Mean field theory for pedestrians

Attempts to analyse dynamo action of prescribed velocity fields culminated in the efforts of Bullard and Gellman (1954) and Backus (1958), but the mathematical complications are enormous. The velocity fields in the Sun and the Earth are extremely complex due to the turbulent convection, which shreds the magnetic field lines into knots and filaments. This poses a totally unmanageable mathematical problem, and the idea emerges to take an average. To this end we put

$$\left. \begin{array}{ll} \boldsymbol{B} = \boldsymbol{B}_0 + \boldsymbol{B}_1 & \boldsymbol{E} = \boldsymbol{E}_0 + \boldsymbol{E}_1 \\ \boldsymbol{J} = \boldsymbol{J}_0 + \boldsymbol{J}_1 & \boldsymbol{u} = \boldsymbol{u}_0 + \boldsymbol{u}_1 \end{array} \right\} \tag{4.1}$$

Each physical quantity f is split in a 'large scale' component f_0 which remains after averaging, superposed on which is a 'fluctuating' component f_1. The average will be indicated by $<\cdot>$. The fluctuating components are zero on average: $<\boldsymbol{B}_1> = <\boldsymbol{J}_1> = <\boldsymbol{E}_1> = <\boldsymbol{u}_1> = 0$, so that $<\boldsymbol{B}> = \boldsymbol{B}_0$; $<\boldsymbol{u}> = \boldsymbol{u}_0$, etc. $\boldsymbol{B}_0$ is called the *mean field* and $\boldsymbol{u}_0$ is the *mean flow* (in the Sun, it is the differential rotation). Performing the average in (3.1) we get

$$\nabla \times \boldsymbol{E}_0 = -\frac{1}{c}\frac{\partial \boldsymbol{B}_0}{\partial t} \quad (a) \qquad \boldsymbol{J}_0 = \sigma\left\{\boldsymbol{E}_0 + \frac{1}{c}\boldsymbol{u}_0 \times \boldsymbol{B}_0 + \frac{1}{c}<\boldsymbol{u}_1 \times \boldsymbol{B}_1>\right\} \quad (c)$$

$$\tag{4.2}$$

$$\nabla \times \boldsymbol{B}_0 = \frac{4\pi}{c}\boldsymbol{J}_0 \quad (b) \qquad \nabla \cdot \boldsymbol{B}_0 = 0 \quad (d)$$

It is assumed that $<\cdot>$ commutes with ∇ and $\partial/\partial t$. Nothing exciting has happened in (4.2a,b) and (4.2d). However, the quadratic term $\boldsymbol{u} \times \boldsymbol{B}$ in Ohm's law gives rise to a new term $<\boldsymbol{u}_1 \times \boldsymbol{B}_1> \neq 0$ in (4.2c), because $\boldsymbol{u}_1$ and $\boldsymbol{B}_1$ are statistically correlated. The cross terms $<\boldsymbol{u}_0 \times \boldsymbol{B}_1>$ and $<\boldsymbol{u}_1 \times \boldsymbol{B}_0>$ vanish since we assume that $<\boldsymbol{u}_0 \times \boldsymbol{B}_1> = \boldsymbol{u}_0 \times <\boldsymbol{B}_1> = 0$, implying that $<f<g\gg$ should be equal to $<f><g>$. The properties which $<\cdot>$ must satisfy are called the *Reynolds rules*:

$$<f + g> = <f> + <g> \tag{4.3a}$$

$$<f<g\gg = <f><g> \tag{4.3b}$$

$$<c> = c \tag{4.3c}$$

$$<\cdot> \text{ commutes with } \nabla, \partial/\partial t, \text{ and } \int dt \tag{4.3d}$$

Here f and g are arbitrary functions of $\boldsymbol{r}$ and t; c is an arbitrary constant. That $<\cdot>$ should commute with $\int dt$ will only be used in section 6. On writing $Af = <f>$ we find with (4.3b,c) that $A^2f = A(1 \cdot Af) = (A1)(Af) = Af$. It follows that $<f_0> = f_0$ (e.g. $<\boldsymbol{u}_0> = \boldsymbol{u}_0$). Since $A^2 = A$ we conclude that A is a projection operator. This implies that information is lost; it is not possible to reconstruct f from $Af = f_0$ since A has at least one zero eigenvalue. The average may be regarded as an average over a large ensemble of copy systems, each having the same $\boldsymbol{B}_0, \boldsymbol{J}_0, \boldsymbol{E}_0$ and $\boldsymbol{u}_0$, but different realisations of $\boldsymbol{u}_1$ (and hence also of $\boldsymbol{B}_1, \boldsymbol{J}_1$ and $\boldsymbol{E}_1$). This average obeys the Reynolds rules, but there is no immediate connection between $\boldsymbol{B}_0$ and the field $\boldsymbol{B}$ of the dynamo (i.e. one ensemble member picked at random). However, an average over longitude in one system also satisfies the Reynolds rules (without proof); B_0 is then the large-scale, axisymmetric component of B.

In section 6 it will be shown that for isotropic turbulence u_1:

$$<u_1 \times B_1> = \alpha B_0 - \beta \nabla \times B_0 = \alpha B_0 - \frac{4\pi\beta}{c} J_0 \qquad (4.4)$$

The parameters α and β are determined by the statistical properties of u_1:

$$\alpha \simeq -\tfrac{1}{3} <u_1 \cdot \nabla \times u_1> \tau_c \quad ; \quad \beta \simeq \tfrac{1}{3} <u_1^2> \tau_c \qquad (4.5)$$

τ_c is the *correlation time* of the turbulence. In this section, I shall restrict myself solely to illustrating the consequences of (4.4) and to an explanation of the physics hidden in the coefficients α and β. Substitute (4.4) in (4.2c) and solve for J_0:

$$J_0 = \sigma_e \left\{ E_0 + \frac{1}{c} u_0 \times B_0 + \frac{\alpha}{c} B_0 \right\} \qquad (4.6)$$

with

$$\frac{1}{\sigma_e} = \frac{1}{\sigma} + \frac{4\pi\beta}{c^2} \quad \overset{\sigma=\infty}{\longrightarrow} \quad \sigma_e = \frac{c^2}{4\pi\beta} \propto <u_1^2>^{-1} \qquad (4.7)$$

Equations (4.2a,b,d) and (4.6) are the new equations for the mean field. Comparing with (3.1), we see that only Ohm's law has changed. Just as in the case of the induction equation it is again possible to eliminate J_0 and E_0. From (4.2a): $\partial_t B_0 = -c\nabla \times E_0$. Eliminate E_0 with (4.6) and next J_0 with (4.2b). The result is

$$\frac{\partial B_0}{\partial t} = \nabla \times \{ u_0 \times B_0 + \alpha B_0 - (\eta + \beta)\nabla \times B_0 \} \qquad (4.8)$$

This is a closed equation for B_0, called the *dynamo equation*. An equivalent form of it was first derived by Parker (1955) by heuristic arguments. A formal derivation was given ten years later by Steenbeck, Krause and Rädler (see Roberts and Stix 1971, and also Braginskii 1965a,b).

The terms in the dynamo equation have the following physical meaning:

(1). *Advection*: $\partial_t B_0 = \nabla \times (u_0 \times B_0)$. This says that the mean field is advected by the mean flow (just as the actual field B is advected by the actual flow u).

(2). *Turbulent diffusion*: $\partial_t B_0 = -\nabla \times (\eta + \beta)\nabla \times B_0 \simeq \beta \nabla^2 B_0$, assuming that $\beta + \eta \simeq \beta$ is constant. By comparing with (3.2a), it follows that the mean field B_0 diffuses much more rapidly than the field itself since $\beta \gg \eta$, as we shall see. The diffusion time of the *mean field* of the Sun has been reduced from $R_\odot^2/\eta \approx 4 \times 10^9$ years to $R_\odot^2/\beta \approx 10$ years, which is of the order of the period of the solar dynamo (this is not a coincidence - section 5). The explanation is given in Figure 9. The turbulence causes the field lines to get entangled (left). After averaging the result is that the *mean field* has spread over a considerable volume (right). Hence, even if σ is infinite, the turbulence causes B_0 to behave as if it were subject to a finite effective conductivity. That is why $\sigma_e \propto <u_1^2>^{-1}$ appears in (4.6) and (4.8).

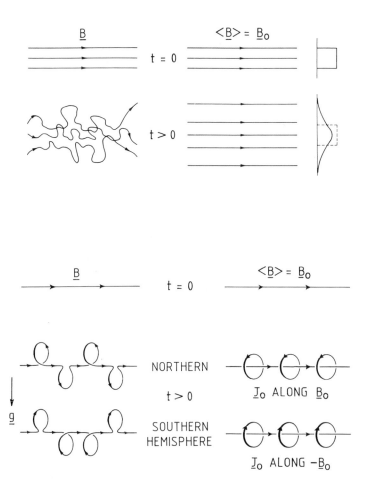

FIGURE 9. The physics of turbulent diffusion (top) and the α-effect (bottom); see text for details.

(3). The α-effect: $\partial_t \boldsymbol{B}_0 = \alpha \nabla \times \boldsymbol{B}_0$ (assuming α constant). Consider a flux tube of mean field $\boldsymbol{B}_0$, see Figure 9 (only one field line of $\boldsymbol{B}_0$ is drawn). Apparently, new $\boldsymbol{B}_0$ will now grow along $\nabla \times \boldsymbol{B}_0$, which is in circles along the mantle of the tube. Alternatively, we may say that there is, according to (4.6), a mean current $\alpha\sigma_e \boldsymbol{B}_0/c$ along the flux tube which generates new $\boldsymbol{B}_0$ around the tube according to (4.2b). This is called the 'α-current'. The explanation is given in the left figure. Rising (sinking) gas bubbles in the Northern hemisphere expand (shrink) laterally as they adapt themselves to the ambient density. The Coriolis force makes the gas and the field it carries rotate clockwise (anticlockwise). In the Southern hemisphere, the rotations are in the opposite sense. After averaging we get

112

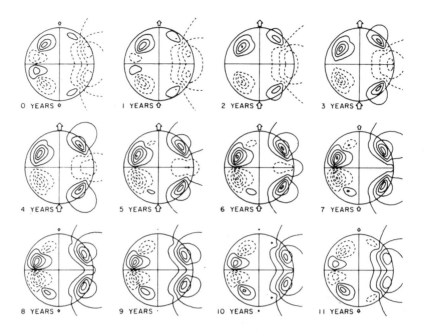

FIGURE 10. Numerical solution of an axisymmetric solar $\alpha\Omega$-dynamo (Deinzer and Stix 1971; Stix 1976a). The shear is radial with $\partial\Omega_r/\partial r < 0$, and $\alpha \propto \cos\theta$. Each frame is a meridional cross section through the Sun. To the left are contours of constant toroidal mean field; on the right are the field lines of the poloidal mean field. Solid curves indicate toroidal field pointing out of the figure and clockwise poloidal field lines. The field at the poles is indicated by vertical arrows.

the figure to the right. Hence, $\alpha > 0$ in the Northern hemisphere, while in the Southern hemisphere $\alpha < 0$. The effect is due to a coupling of rotation and convection, which induces a preferred sense of rotation for vertically moving gas. In this way we can understand why $\alpha \propto\, <\boldsymbol{u}_1 \cdot \nabla \times \boldsymbol{u}_1>$, the *mean helicity* of the turbulence, see (4.5). Contrary to β, the coefficient α is zero if there is no rotation. The existence of the α-effect has been demonstated in laboratory experiments (Krause and Rädler 1980, Chapter 3). Since the Coriolis force is involved one expects that

$$\alpha \propto\, <\boldsymbol{u}_1 \cdot \nabla \times \boldsymbol{u}_1> \propto\ \Omega_r \cos\theta \tag{4.9}$$

(Krause 1967; Durney 1981; Stix 1983). However, if the turbulence has a long correlation time τ_c, then rotations can persist and the blobs in Figure 9 may rotate over (much) more than 90° on average. Hence, there is no simple way to predict the sign of α if the correlation time is long. This could be one of the reasons why numerical simulations find contradictory results for α (see sections 6 and 7).

The dynamo equation (4.8) appears to describe the major properties of the global magnetic field of the Sun and the planets. A numerical solution for the Sun is shown in Figure 10. The solution is axisymmetric and features very clearly the qualitative behaviour outlined in section 2.2: tubes or waves of toroidal mean field of alternating polarity move to the equatorial plane. The functions α and β have been adapted to get a marginally stable oscillation with a period of 22 year. The toroidal component of B_0 is much larger than the poloidal component. This provides a basis for explaining the butterfly diagram: the idea is that this toroidal field occasionally breaks through the surface by buoyancy. There are many more numerical models; for brevity I refer to the reviews by Stix (1976a, 1981) and Parker (1987a) where many references are given. Here I shall restrict myself to an intuitive explanation of Figure 10, based on the three processes of advection, turbulent diffusion and α-effect, to which the mean field B_0 is subjected. This will be elaborated more quantitatively in section 5 with the help of a plane wave analysis.

Figure 11, at (a), shows two toroidal flux tubes of opposite polarity in each hemisphere. (b). The α-effect generates new loops around these tubes. Since α as well as B_0 change sign between hemispheres, this new (poloidal) field has the same orientation in both hemispheres.[†] At (c) we observe the system along the direction of the arrow. The differential rotation makes the loops tilt as in (d).[†] In this way we get cross section (e). Turbulent diffusion causes the tubes indicated by an } to 'annihilate', but some mean field of opposite polarity is left over. The result is (f): the original flux tubes have moved to the equator, leaving a small amount of opposite mean field in their wake. This process continues, (g): the flux tube reproduces itself at the equator side while destroying itself at the poleward side. The latter process is incomplete, leading to a growing wake field of opposite polarity.[††] Eventually the two leading flux tubes annihilate at the equator, (h), and we return to (a) with all polarities reversed. Note that only the mean field B_0 is subject to the α-effect and turbulent diffusion; it leads to inconsistencies to think that the actual field B is, too.

The dynamo equation has stationary solutions when differential rotation can be ignored. This is believed to be the case for the dynamo in the fluid outer core of the Earth. Numerical solutions of Equation (4.8) with $u_0 = 0$ and $\alpha \propto \cos\theta$ are given for example in Moffatt (1978, section 9.5), and Krause and Rädler (1980, section 16.3). A physical explanation of such a dynamo is given in Figure 12. The four panels (a)-(d) are snapshots taken at the same instant of time. We start at (a) with two tubes of toroidal mean field, oppositely oriented in both hemispheres. (b). The α-effect drives a mean α-current J_0 along these tubes, which has the same direction in both hemispheres since $\alpha \propto \cos\theta$. (c). The toroidal

---

[†] The loops would have the opposite orientation if α had the opposite sign, and the end result is that the flux tubes would then migrate to the pole! The reader may verify that for equatorial migration $\alpha(\partial\Omega_r/\partial r)$ must be negative in the Northern hemisphere. The model of Figure 10 assumes that the angular velocity Ω_r increases for smaller r which we know from helioseismology to be incorrect – see section 7.
[††] It is possible that the wake field produces yet another field of opposite polarity in its own wake. This corresponds to an *overtone* solution of Equation (4.8); Figure 10 shows the fundamental eigenmode of (4.8).

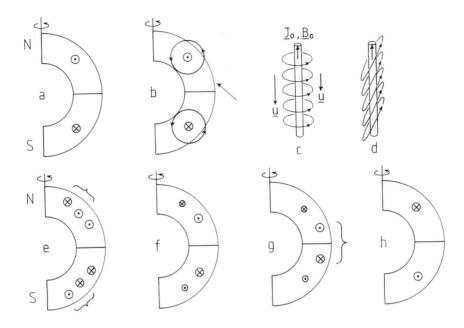

FIGURE 11. Physical explanation of the oscillatory $\alpha\Omega$-dynamo shown in Figure 10 (after Stix 1976a). See text for details.

α-current generates its own poloidal mean field, which extends partly outside the sphere. This is the field that we observe at the Earth's surface. (d). The α-effect now drives a mean α-current along those poloidal field lines which are closed inside the sphere. And this poloidal mean current, finally, is the source of the toroidal mean field at (a). According to this picture, the field is in a state of self-amplification, but turbulent diffusion provides a damping, allowing the field to become constant.

Apparently, the dynamo equation possesses stationary, axisymmetric solutions B_0. This is not in conflict with Cowling's theorem since that applies to B, but not to the mean field B_0. The actual field B need not be axisymmetric at all, even if B_0 is, so that there is no contradiction. This shows the strength of the mean field concept: averaging leads at first sight to an enormous simplification. However, the information contained in the mean field is far less than in the field itself.

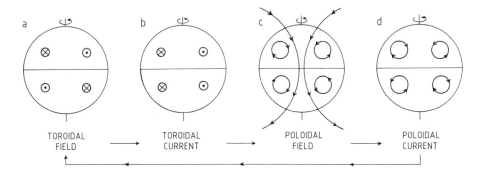

FIGURE 12. Physical explanation of a stationary α^2-dynamo with $\alpha \propto \cos\theta$, believed to be located in the Earth's core. See text for details.

5. Plane dynamo waves

It is instructive to analyse the plane wave solutions of the dynamo equation (4.8), as was first done by Parker (1955). In this way, it is possible to elucidate most of the properties of the numerical mean field models quantitatively, while avoiding the complications of a finite spherical geometry. Consider an infinite space filled with turbulent gas, such that α and β are constant. The mean flow $\boldsymbol{u}_0$ is in the y-direction; $\boldsymbol{u}_0$ is a linear function of x and z so that $\boldsymbol{a} \equiv \nabla u_0$ is a constant vector, see Figure 13. Axisymmetry as in Figure 10 implies here invariance for translations along the y-axis, that is, $\partial/\partial y = 0$. For $\boldsymbol{B}_0$ we choose the gauge

$$\boldsymbol{B}_0 = \nabla \times (P\boldsymbol{e}_y) + T\boldsymbol{e}_y = -\boldsymbol{e}_y \times \nabla P + T\boldsymbol{e}_y , \qquad (5.1)$$

with P and T functions of x, z and t. In this way $\nabla \cdot \boldsymbol{B}_0 = 0$ and $\boldsymbol{B}_0$ is split into its toroidal ($\parallel \boldsymbol{e}_y$) and poloidal ($\perp \boldsymbol{e}_y$) components. We now insert (5.1) into Equation (4.8). Since $\nabla \cdot \boldsymbol{u}_0 = \nabla \cdot \boldsymbol{B}_0 = 0$ and $\boldsymbol{u}_0 \cdot \nabla \propto \partial/\partial y = 0$, we have

$$\begin{aligned}
\nabla \times (\boldsymbol{u}_0 \times \boldsymbol{B}_0) &= (\boldsymbol{B}_0 \cdot \nabla)\boldsymbol{u}_0 - (\boldsymbol{u}_0 \cdot \nabla)\boldsymbol{B}_0 = \boldsymbol{e}_y(\boldsymbol{B}_0 \cdot \nabla)u_0 \\
&= \boldsymbol{e}_y(\boldsymbol{B}_0 \cdot \boldsymbol{a}) = -\boldsymbol{e}_y(\boldsymbol{a} \cdot \boldsymbol{e}_y \times \nabla P) \\
&= \boldsymbol{e}_y(\boldsymbol{e}_y \times \boldsymbol{a}) \cdot \nabla P
\end{aligned} \qquad (5.2)$$

The other terms can be dealt with in a similar way. Inserting these in (4.8) leads, after some rearranging, to an expression of the type $\nabla \times f\boldsymbol{e}_y + g\boldsymbol{e}_y = 0$, which we shall not write down here. The reader may verify that also $\nabla \cdot f\boldsymbol{e}_y = \partial f/\partial y = 0$. It follows that $f = g = 0$. Explicitly (we absorb η in β):

116

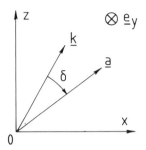

e_y = unit vector along y-axis
$u_0 = u_0(x, z)e_y$
$a = \nabla u_0$
$n = k/k$
k = wavevector
δ = angle between k and a

FIGURE 13. Co-ordinate system for analysing the plane wave solutions of the dynamo equation (4.8).

$$\frac{\partial P}{\partial t} \quad = \quad \alpha T \quad + \quad \beta\nabla^2 P \tag{5.3}$$

new P from T diffusion
field   by α-effect   term

$$\frac{\partial T}{\partial t} \quad = \quad (e_y \times a)\cdot\nabla P \quad - \quad \alpha\nabla^2 P \quad + \quad \beta\nabla^2 T \tag{5.4}$$

new T from P by   from P diffusion
field   shear flow   by α-effect   term
            ('Ω-term')

We see that new toroidal field can be created from poloidal field in two ways: by differential rotation and by the α-effect. But new poloidal field can *only* be created from toroidal fields by the α-effect. Toroidal and poloidal fields are both subject to turbulent diffusion. The crucial term is αT in (5.3). If it were absent, then Equation (5.3) predicts that $P \downarrow 0$ by turbulent diffusion, and, subsequently, also $T \downarrow 0$ according to Equation (5.4). However this does not mean that there is no generation of magnetic field: the *mean field* is zero, but, in general, B_1 is not.

The terms $(e_y \times a)\cdot\nabla P$ and $\alpha\nabla^2 P$ in (5.4) may have widely different relative magnitude, and this leads to qualitatively different dynamos. The jargon is as follows:

(1). $\alpha\Omega$-*dynamo*: the α-term is much smaller than the Ω-term in (5.4) and can be ignored. One α-term and the Ω-term remain in (5.3) and (5.4). These dynamos tend to have a periodic behaviour. The solar dynamo is believed to be of the $\alpha\Omega$-type.

(2). α^2-*dynamo*: the Ω-term is much smaller than the α-term in (5.4) and can be ignored, so that the two α-terms are left over. This comes down to ignoring the term $u_0 \times B_0$ in

Equation (4.8) altogether. These dynamos often have stationary solutions. The dynamo of the Earth is probably an α^2-dynamo.

(3). $\alpha^2\Omega$-dynamo: the shear term and the α-term in (5.4) are of comparable magnitude.

5.1. $\alpha\Omega$-DYNAMO WAVES

To find plane wave solutions, we substitute

$$(P,T) = (P_0, T_0) \exp\{i(\boldsymbol{k} \cdot \boldsymbol{r} - \omega t)\} \tag{5.5}$$

in (5.3) and (5.4), assuming that the term $\alpha\nabla^2 P$ can be ignored ($\alpha\Omega$-approximation). In matrix notation, the result is:

$$\begin{pmatrix} i\omega - \beta k^2 & \alpha \\ -iks & i\omega - \beta k^2 \end{pmatrix} \begin{pmatrix} P_0 \\ T_0 \end{pmatrix} = 0 \tag{5.6}$$

with

$$s = \boldsymbol{e}_y \cdot \boldsymbol{n} \times \boldsymbol{a} = a \sin \delta \tag{5.7}$$

A nontrivial solution exists if the 2×2 determinant vanishes, leading to the dispersion relation:

$$\omega^2 + 2i\beta k^2 \omega - (\beta k^2)^2 - i\alpha k s = 0 \tag{5.8}$$

whence

$$\omega = -i\beta k^2 \pm \sqrt{i\alpha k s} = -i\beta k^2 \pm (1 + i)\sqrt{\alpha k s/2} \tag{5.9}$$

Consequently we have found that

$$\text{frequency } \Omega = \text{Re}\,\omega = \pm\sqrt{\alpha k s/2} \tag{5.10}$$

$$\text{growth rate } \Gamma = \text{Im}\,\omega = -\beta k^2 \pm \sqrt{\alpha k s/2} \tag{5.11}$$

(αs under the square root should really be $|\alpha s|$). The lower sign gives a wave which is always damped. This solution can be ignored, unless one is interested in an initial value problem. If we repeat the analysis for the α^2-limit, $i.e.$ ignoring $\boldsymbol{e}_y \times \boldsymbol{a} \cdot \nabla P$ instead of $\alpha\nabla^2 P$ in (5.4), we obtain

$$\Omega = \text{Re}\,\omega = 0 \; ; \qquad \Gamma = \text{Im}\,\omega = -\beta k^2 \pm \alpha k \tag{5.12}$$

It follows that the solution is purely growing or decaying, and this illustrates the point that α^2-dynamos are usually non-oscillatory.

5.2. PROPERTIES OF SOLAR MEAN FIELD MODELS

We are now in a position to apply these results to the solar dynamo, in an attempt to explain the properties of mean field models such as the one shown in Figure 10. The period of the waves is determined by (5.10): $P_d = 2\pi(\alpha k s/2)^{-1/2}$. To translate this to the Sun, we take $s = a \sin \delta \approx |\nabla u_0| \approx \Delta u_0/R_\odot \approx \Delta\Omega_r$ = magnitude of the differential rotation ($e.g.$ the difference between equatorial and polar rotation). Furthermore, we take $k \approx 1/R_\odot$,

Table 1. Typical parameters of the solar $\alpha\Omega$-dynamo

$\Omega = 2\pi/P_d$	9×10^{-9}	s^{-1}
$k \approx R_\odot^{-1}$	1.4×10^{-11}	cm^{-1}
$\Delta\Omega_r \approx a \approx s$	6×10^{-7}	s^{-1}
α	10	$cm\,s^{-1}$
β	10^{13}	$cm^2\,s^{-1}$

since, as we shall see, it is usually the fundamental mode with the largest wavelength which is excited. The period of the solar dynamo is therefore given by

$$P_d \approx 2\pi \left\{ \frac{R_\odot}{\alpha\Delta\Omega_r} \right\}^{1/2} \tag{5.13}$$

Taking $P_d = 22$ years and $\Delta\Omega_r \simeq 6 \times 10^{-7}\,s^{-1}$, it follows that $\alpha \approx 10\,cm\,s^{-1}$, which is of the same order of magnitude as found in numerical models – see further section 7. If we also require that the wave is marginally stable ($\Gamma = 0$) we find $\beta k^2 = (\alpha ks/2)^{1/2}$, or $R_\odot^2/\beta \approx P_d/2\pi$. The turbulent diffusion timescale is thus of the order of the period of the dynamo. With the numbers in Table 1, we obtain $\beta \approx 4 \times 10^{13}\,cm^2\,s^{-1}$. Equation (4.5) leads to a similar value: with $\tau_c \approx 10^5\,s$ and $u_1 \approx 400\,m\,s^{-1}$ (supergranulation cell), we get $\beta \approx (4 \cdot 10^4)^2 10^5/3 \approx 5 \times 10^{13}\,cm^2\,s^{-1}$. Numerical models of the solar dynamo based on Equation (4.8) usually find $\beta \approx 10^{13}\,cm^2\,s^{-1}$. A slightly smaller value is inferred from the observed surface diffusion of solar magnetic fields ($\beta \approx 6 \times 10^{12}\,cm^2\,s^{-1}$; Wang $et\ al.$ 1989). Finally, $\eta \approx 2 \times 10^4\,cm^2\,s^{-1}$ in the convection zone near $T = 10^6\,K$, confirming that $\beta \gg \eta$.

The consistency of the $\alpha\Omega$-approximation requires $|e_y \times a \cdot k| \gg \alpha k^2$, see (5.4), that is $|s| \gg \alpha k$, or $R_\odot \Delta\Omega_r \gg \alpha$. This inequality is satisfied by several orders of magnitude. Note that the frequency of the wave is given by $\Omega = (\alpha ks/2)^{1/2}$, so that $\Omega \propto k^{1/2}$. It is confusing to say that $\Omega = \beta k^2$, as that merely expresses marginal stability. The difference is very important, for example, when we determine the group velocity of the dynamo wave.

Inserting (5.5) in (5.1) we find

$$B_0 = (-iP_0\,e_y \times k + T_0 e_y)\,\exp(i\psi)\,, \tag{5.14}$$

where $\psi = k \cdot r - \omega t$ is the phase. Hence, the wave is transverse, $k \cdot B_0 = 0$. From the first line of Equation (5.6) we obtain

$$\frac{P_0}{T_0} = \frac{\alpha}{\beta k^2 - i\omega} = \frac{\alpha}{2\Omega}\,(1+i) \tag{5.15}$$

At the second $=$ sign, the wave is assumed to be marginally stable. Inserting this in Equation (5.14) and taking the real part we get

$$B_0 \propto \frac{\alpha k}{\Omega\sqrt{2}}\,\cos(\psi + 3\pi/4)\,n \times e_y + \cos\psi\,e_y \tag{5.16}$$

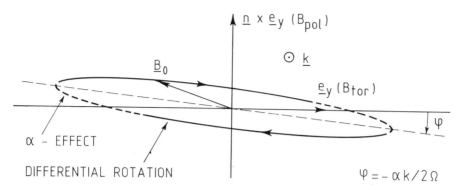

FIGURE 14. An $\alpha\Omega$-dynamo wave. The vector $\boldsymbol{B}_0$ runs over a narrow ellipse $\perp \boldsymbol{k}$, clockwise (anticlockwise) for $\alpha > 0$ ($\alpha < 0$). The wave mechanism can be explained as follows (*cf.* section 4, and Parker 1979, Figure 19.1). Differential rotation sweeps $\boldsymbol{B}_0$ rapidly from one edge to the other (solid line) and does work, so that $\boldsymbol{B}_0$ decreases and then increases again. The α-effect moves $\boldsymbol{B}_0$ very slowly around the edge (broken line), after which differential rotation takes over again. During most of the time $\boldsymbol{B}_0$ is (anti)parallel to $\boldsymbol{e}_y$. For a mechanical model consisting of needles advected by the flow, see Hoyng 1987a.

By eliminating ψ we find that $\boldsymbol{B}_0$ rotates over a tilted ellipse, Figure 14. The tilt angle $\varphi = -\alpha k/2\Omega$ is very small as $\alpha k/2\Omega \approx \alpha P_d/4\pi R_\odot \approx 10^{-2}$. The relative magnitude of the poloidal field B_{pol} and the toroidal field B_{tor} follows from (5.16):

$$\frac{B_{pol}}{B_{tor}} = \frac{\alpha k}{\Omega\sqrt{2}} \approx 10^{-2} , \tag{5.17}$$

for the numbers in Table 1. The ellipse in Figure 14 is therefore very flat; the toroidal field is about a factor 100 stronger than the poloidal field, a property which is also borne out by numerical models.

Finally we compute the group velocity of the wave:

$$\boldsymbol{v}_g = \frac{\partial \Omega}{\partial \boldsymbol{k}} = \frac{\partial}{\partial \boldsymbol{k}} \left(\alpha\, \boldsymbol{e}_y \cdot \boldsymbol{k} \times \boldsymbol{a}/2\right)^{1/2} = \frac{\alpha}{4\Omega}\, \boldsymbol{a} \times \boldsymbol{e}_y \tag{5.18}$$

A wave packet propagates perpendicular to $\boldsymbol{e}_y$ and $\boldsymbol{a}$, along the isoplanes of the 'rotation' u_0, as was first proved by Yoshimura (1975). In Figure 10, $\boldsymbol{a}$ points radially inwards, and since $\alpha \propto \cos\theta$, $\boldsymbol{v}_g$ is directed to the equator in both hemispheres, thus explaining the migration of the mean field toward the equator in that model. With the numbers in Table 1 we get $v_g = \alpha a/4\Omega \approx \alpha\Delta\Omega_r P_d/8\pi \approx 30\,\text{cm s}^{-1}$, or $10^5\,\text{km}$ in 11 year, about $10°$ in latitude, which is of the right order of magnitude.

5.3. MODE STRUCTURE AND CRITICAL DYNAMO ACTION

In an unbounded medium the wave vector can have any value, but k becomes quantised in a bounded medium such as the solar convection zone. There is a discrete spectrum of eigenmodes which can only be found by solving the full eigenvalue problem of Equation (4.8). Nevertheless, it is possible to understand some of the properties with the help of the plane wave modes. In Figure 15 the eigenmodes are indicated schematically by the dots on the k-axis. The largest possible wavelength is of the order of $R_\odot$, whence there is a lower bound on k, say $k \gtrsim 1/R_\odot$. We see in Figure 15 that larger k (smaller wavelength) means smaller growth rate Γ. The reason is that, for smaller wavelengths, the turbulent diffusion terms $\beta\nabla^2 P$ and $\beta\nabla^2 T$ in (5.3) and (5.4) become progressively more effective. From (5.11), we find that the growth rate Γ is zero if $(kR_\odot)^3 = |\alpha s| R_\odot^3/2\beta^2 \simeq |\alpha| \Delta\Omega_r R_\odot^3/2\beta^2$. This is usually expressed in terms of the (dimensionless) *dynamo number* D:

$$D = \frac{\alpha \Delta\Omega_r R_\odot^3}{2\beta^2} \qquad (5.19)$$

(for $\alpha\Omega$-dynamos, D is often written as the product of two dynamo numbers: $D = (\alpha R_\odot/\beta) \cdot (\Delta\Omega_r R_\odot^2/2\beta) \equiv D_\alpha \cdot D_\omega$). Hence, $\Gamma = 0$ if $(kR_\odot)^3 \simeq |D|$. Modes with $(kR_\odot)^3 > |D|$ are damped, while those with $(kR_\odot)^3 < |D|$ grow exponentially. For example, in the supercritical case drawn in Figure 15, the first five modes would grow, while the higher modes are damped. Since $k \gtrsim 1/R_\odot$, the possibility of growing modes requires that $|D| \geq D_c$ with $D_c \simeq 1$. The quantity D_c is called the *critical dynamo number*. When $|D| = D_c$, the fundamental mode is marginally stable and overtones are damped. The dynamo is *subcritical* if $|D| < D_c$. All modes are damped and there is no dynamo action – that is, there is no *mean* field. The fact that $D_c \simeq 1$ is artificial, because we took $k_{min} \simeq 1/R_\odot$, but k_{min} varies from model to model and is only known to be of the order of $1/R_\odot$. Since the excited modes have a characteristic scale of the order of $R_\odot$, we may conclude that dynamo action is a global phenomenon.

Inhomogeneous dynamos usually have $D_c \gg 1$, typically $D_c \approx 10^3 - 10^4$. To explain this, let us suppose that only α depends on position. Common practice is to write $\alpha = \alpha_0 f(r,\theta)$ where α_0 is a constant and f a given profile with a maximum value of 1. The dynamo number is now no longer uniquely defined but the obvious choice is to use α_0 in the definition (5.19) of D. Consider now a *homogeneous* dynamo with $\alpha \equiv \alpha_0$ such that its D is just critical. Its inhomogeneous counterpart has the same value of D, but everywhere a smaller α, *i.e.* a dynamo number which is effectively subcritical. Therefore, its D_c must be larger, but that is not the whole story. The eigenfunction B_0 adapts its structure and counteracts steep gradients and large values of α, so as to reduce the magnitude of the term $\nabla \times \alpha B_0$ in Equation (4.8), and this renders D_c even larger, see *e.g.* Meinel and Brandenburg (1990). However, there always exists a critical number D_c such that dynamo action requires $|D| \geq D_c$. Physically, it means that the amplifying effects of differential rotation and α-effect should be able to overcome the damping by turbulent diffusion.

A linear dynamo cannot operate permanently supercritically, as it implies $B_0 \uparrow \infty$. Nonlinear effects, *i.e.* the Lorentz force, will come into play and then things get very

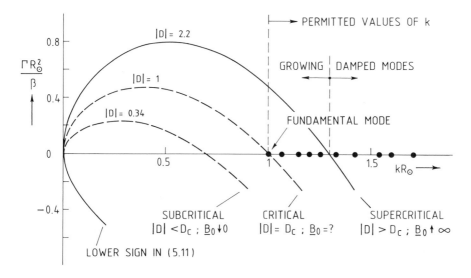

FIGURE 15. The growth rate Γ from relation (5.11) plotted as a function of k, for three values of $|D|$, illustrating subcritical, critical and supercritical dynamo action.

complicated, see section 7 for more details. They may for example reduce the value of $|D|$ (by changing α, $\Delta\Omega_r$ and/or β) but, at the same time, D_c changes because the magnetic field alters the profiles of α, β, etc. Usually this means that D_c is increased, as explained above, and, on average, $|D|$ will again be equal to D_c.[†] The details of the fine-tuning mechanism(s) at work in the solar dynamo, and in any other dynamo for that matter, are presently unknown.

[†] The notion of a *changing* D_c is not a very useful one and it has become customary to speak of 'highly supercritical dynamos' when $|D|$ is very large, *cf.* Meinel and Brandenburg (1990).

6. Rigorous mean field theory

We shall now present a rigorous derivation of Equation (4.4) and the dynamo equation (4.8). Unfortunately, I know of no simple and yet convincing method. This section is therefore somewhat technical, and more directed towards specialists. The starting point is the induction equation. It is useful to introduce a shorthand operator notation and to write (3.2a) as

$$\partial_t B = A_0 B + A_1 B \tag{6.1}$$

with

$$A_0 = \nabla \times (u_0 \times \cdots) + \eta \nabla^2 \tag{6.2}$$

$$A_1 = \nabla \times (u_1 \times \cdots) \tag{6.3}$$

The flow field u has been written as the sum of the differential rotation u_0 and the turbulent convection u_1. The operator A_1 depends explicitly on time, while A_0 is time independent. All quantities f are split in a similar way: $f = f_0 + f_1$, $<f_1> = 0$ so that $<f> = f_0$. Note that this is *not* an expansion for small A_1 and f_1; f_1 and A_1 may be much larger than f_0 and A_0, respectively. We now take the average of Equation (6.1), assuming that $<\cdot>$ obeys (4.3):

$$\partial_t B_0 = A_0 B_0 + <A_1 B_1> , \tag{6.4}$$

since $<A_0 B> = <A_0 B_0 + A_0 B_1> = A_0 B_0$, *etc.* Subtract (6.4) from (6.1):

$$\partial_t B_1 = A_0 B_1 + A_1 B_0 + \underbrace{(A_1 B_1 - <A_1 B_1>)}_{\text{ignore } (FOSA)} \tag{6.5}$$

We must now derive an equation for $<A_1 B_1>$ from (6.5). The last term in (6.5) can be ignored if the correlation time τ_c of A_1 is sufficiently short:

$$\tau_c |A_1| \ll 1 , \tag{6.6}$$

where $|A_1|$ is the typical magnitude of A_1. This approximation is called the First Order Smoothing Approximation (*FOSA*), or quasi-linear approximation. Since $|A_1| \approx u_1/\lambda_c$, where λ_c is the correlation length of the turbulence, (6.6) implies

$$u_1 \tau_c \ll \lambda_c \tag{6.7}$$

In other words, the correlation time τ_c should be much smaller than the eddy turnover time λ_c/u_1. The validity of *FOSA* when (6.7) holds is not in dispute, but the problem is that (6.7) is probably violated since it is likely that $u_1 \tau_c \approx \lambda_c$ ('a convection cell loses its identity after making about one turnover'). We leave this point to section 6.4 and shall now assume *FOSA*. Equation (6.5) is then a simple linear first order differential equation; $A_0 B_1$ describes the effect of differential rotation and ohmic dissipation on B_1, and $A_1 B_0$ is a source term. The solution is:

$$B_1 = \exp\{(t - t_0)A_0\} B_1^{t_0} + \int_{t_0}^{t} \exp\{(t - \tau)A_0\}(A_1 B_0)^{\tau} d\tau \tag{6.8}$$

Time arguments t_0, τ, *etc.* are sometimes written as an upper index for brevity, and if none is specified, it is understood to be t; t_0 is the initial time. The position argument is always suppressed as it is r everywhere. To compute $<A_1 B_1>$ we multiply B_1 with $A_1 = A_1^t$. If $t - t_0 \gg \tau_c$, then $<A_1^t \exp\{(t - t_0)A_0\} B_1^{t_0}> = 0$ since there is no correlation between A_1^t and $B_1^{t_0}$. The initial condition therefore drops out, and we are left with (take $t_0 = -\infty$ and substitute $\tau \to t - s$):

$$<A_1 B_1> = \int_0^\infty ds \ <A_1^t \, e^{s A_0} A_1^{t-s}> \, B_0^{t-s} \tag{6.9}$$

All operators work on everything to their right; *e.g.* $\exp(s A_0)$ in (6.9) operates on the divergence-free vector $\nabla \times (u_1 \times B_0)^{t-s}$, apart from the averaging. The exponential operator is really a Green's function: $\exp(s A_0) a^{t-s}$ is the solution of $\partial_t b = A_0 b$, advanced s seconds from the initial condition $b(r, 0) = a(r, t - s)$.

The final step is to re-express B_0^{t-s} in terms of $B_0 = B_0^t$. The correlation function in (6.9) is effectively zero for $s > \tau_c$. Hence we need only reconstruct B_0^{t-s} from B_0^t over a time interval of the order of τ_c, which is much shorter than the timescale for a big change in B_0 (the period of the dynamo). We might put $B_0^{t-s} \simeq B_0^t$. This is the *two-scale approximation* which we shall meet again below; it does not allow for the possibility that in about τ_c seconds a totally different vector B_0 may be advected to the (Eulerian) position r, possibly involving a considerable shear due to differential rotation. This would certainly happen in a rapidly rotating dynamo. A better approximation is to assume that on timescales of the order of τ_c, B_0 evolves only due to A_0, *i.e.* $\partial_t B_0 \simeq A_0 B_0$, so that $B_0^t \simeq \exp(s A_0) B_0^{t-s}$, or $B_0^{t-s} \simeq \exp(-s A_0) B_0^t$ (Hoyng 1985). We substitute this in (6.9) and then (6.9) in (6.4), with the following result:

$$\partial_t B_0 = \left\{ A_0 + \int_0^\infty ds \ <A_1(t) e^{s A_0} A_1(t - s)> \, e^{-s A_0} \right\} B_0 \tag{6.10}$$

This is a closed equation for the mean field B_0. In fact, (6.10) is a general result of the theory of stochastic differential equations since we have not used the definition of A_0 and A_1 (Van Kampen 1976).

6.1. VERY SHORT CORRELATION TIME

Equation (6.10) is the dynamo equation in disguise. The reason that we are able to find a closed equation for B_0 at all, is the fact that the induction equation is linear; it contains no terms which are *explicitly* nonlinear in B. Fluid turbulence is essentially more complex in this regard. To proceed in the simplest possible way, we suppose that τ_c is so small that not only (6.7) holds, but also $\tau_c |A_0| \ll 1$ (*i.e.* the effect of differential rotation over τ_c seconds can be ignored), so that $\exp(\pm s A_0) \simeq I$. Inserting this, and the explicit forms of A_0 and A_1 in (6.10), we get:

$$\partial_t B_0 = \nabla \times \left\{ u_0 \times B_0 - \eta \nabla \times B_0 + \int_0^\infty ds \ <u_1^t \times \nabla \times u_1^{t-s}> \times B_0 \right\} \tag{6.11}$$

Here, we have used that $\eta\nabla^2 = -\eta\nabla \times \nabla\times$, since $\nabla \cdot \boldsymbol{B}_0 = 0$. According to (6.4), $\nabla \times \int_0^\infty ds \cdots$ in (6.11) equals $< A_1\boldsymbol{B}_1 > = \nabla\times < \boldsymbol{u}_1 \times \boldsymbol{B}_1 >$, so that $\int_0^\infty ds \cdots$ in (6.11) equals $<\boldsymbol{u}_1 \times \boldsymbol{B}_1>$, apart from an irrelevant gradient term. Thus we have succeeded in expressing $<\boldsymbol{u}_1 \times \boldsymbol{B}_1>$ from Equation (4.4) in terms of $\boldsymbol{B}_0$ and the statistical properties of the turbulence $\boldsymbol{u}_1$:

$$
\begin{aligned}
<\boldsymbol{u}_1 \times \boldsymbol{B}_1> &= \int_0^\infty ds\, <\boldsymbol{u}_1^t \times \nabla \times \boldsymbol{u}_1^{t-s}> \times \boldsymbol{B}_0 \\
&= \int_0^\infty ds\, <\boldsymbol{u}_1^t \times \left\{ (\boldsymbol{B}_0\cdot\nabla)\boldsymbol{u}_1^{t-s} - \boldsymbol{B}_0(\nabla\cdot\boldsymbol{u}_1^{t-s}) - (\boldsymbol{u}_1^{t-s}\cdot\nabla)\boldsymbol{B}_0 \right\}>
\end{aligned}
\tag{6.12}
$$

We see that (6.12) will yield terms $\propto \boldsymbol{B}_0$ and $\propto \nabla_i B_{0,j}$, just as required for (4.4). We now work out the i-th component. Bypassing technicalities, the result is

$$
<\boldsymbol{u}_1 \times \boldsymbol{B}_1>_i = \alpha_{ik} B_{0,k} + \beta_{ik\ell}\nabla_k B_{0,\ell}
\tag{6.13}
$$

with

$$
\alpha_{ik} = \epsilon_{ip\ell}\int_0^\infty ds\, <u_p^t(\nabla_k u_\ell^{t-s})> - \epsilon_{ipk}\int_0^\infty ds\, <u_p^t(\nabla_\ell u_\ell^{t-s})>
\tag{6.14}
$$

$$
\beta_{ik\ell} = -\epsilon_{ip\ell}\int_0^\infty ds\, <u_p^t u_k^{t-s}>
\tag{6.15}
$$

To make the notation not too crowded, the index 1 on $\boldsymbol{u}_1$ is momentarily suppressed. At this point it is customary to suppose that the turbulence is isotropic. The tensors are of the first, second and third rank and, consequently, they must have the following form:

$$
\left.
\begin{aligned}
<u_p^t(\nabla_k u_\ell^{t-s})> &= a\epsilon_{pk\ell} &\rightarrow\quad <u_p^t(\nabla_k u_\ell^{t-s})> &= \tfrac{1}{6}<\boldsymbol{u}^t\cdot(\nabla\times\boldsymbol{u}^{t-s})>\epsilon_{pk\ell} \\
<u_p^t u_k^{t-s}> &= b\delta_{pk} &\rightarrow\quad <u_p^t u_k^{t-s}> &= \tfrac{1}{3}<\boldsymbol{u}^t\cdot\boldsymbol{u}^{t-s}>\delta_{pk} \\
<u_p^t(\nabla_\ell u_\ell^{t-s})> &= 0 & \text{(first rank tensor)}&
\end{aligned}
\right\}
\tag{6.16}
$$

To find the unknown constants a and b we have multiplied with $\epsilon_{pk\ell}$ and δ_{pk}, respectively, and contracted. Expressions (6.14) and (6.15) now become very simple; restoring the index 1 on $\boldsymbol{u}_1$:

$$
\alpha_{ik} = \alpha\delta_{ik} ; \qquad \alpha = -\tfrac{1}{3}\int_0^\infty <\boldsymbol{u}_1^t\cdot(\nabla\times\boldsymbol{u}_1^{t-s})> ds \simeq -\tfrac{1}{3}<\boldsymbol{u}_1\cdot\nabla\times\boldsymbol{u}_1>\tau_c
\tag{6.17}
$$

$$
\beta_{ik\ell} = -\beta\epsilon_{ik\ell} ; \qquad \beta = \tfrac{1}{3}\int_0^\infty <\boldsymbol{u}_1^t\cdot\boldsymbol{u}_1^{t-s}> ds \simeq \tfrac{1}{3}<u_1^2>\tau_c
\tag{6.18}
$$

This completes the proof of Equations (4.4) and (4.8) in their simplest form. Their validity requires a sufficiently short correlation time τ_c and isotropic turbulence. The remainder of this section deals with special topics, such as anisotropic turbulence (section 6.2), and the question of what happens if $\tau_c|A_0|$ is not small (section 6.3). Finally τ_c may be so large that FOSA breaks down (section 6.4).

6.2. ANISOTROPIC TURBULENCE

The actual practice in numerical modeling is to take position dependent scalars for α and β – in other words one assumes isotropic but *inhomogeneous* turbulence. This leads sometimes to contradictions. For example, let us compute $\nabla\beta$ assuming isotropy, that is $\beta \propto <u_k^t u_k^{t-s}>$, ignoring for the moment the integral in (6.18):

$$\nabla_i \beta \propto <u_k^t(\nabla_i u_k^{t-s})> + <u_k^{t-s}(\nabla_i u_k^t)> = 0 \tag{6.19}$$

Both correlation functions are first rank tensors and therefore zero due to isotropy. It follows that a scalar β (isotropy) must be position independent (homogeneity). Isotropy implies homogeneity. A position dependent β must also be a tensor, and this holds for α as well. The situation is clarified by writing

$$\beta_{ik\ell} = -\beta(\boldsymbol{r})\epsilon_{ik\ell} + \delta\beta_{ik\ell}(\boldsymbol{r}) \tag{6.20}$$

$\beta(\boldsymbol{r})$ is the locally isotropic part of $\beta_{ik\ell}$; $\delta\beta_{ik\ell}$ is everywhere small and therefore ignored. On the one hand, $\beta(\boldsymbol{r})$ is measured by $\frac{1}{3}\int_0^\infty ds <u_k^t u_k^{t-s}>$, but, on the other hand, $\boldsymbol{u}_1^t$ is only approximately isotropic, so that (6.19) is nonzero. Of course, we must require that $|\nabla\beta| \ll \beta/\lambda_c$; typically $\nabla\beta$ will be of the order β/R where R is the size of the dynamo (e.g. its radius).

In general the turbulence will be anisotropic and then α_{ik} and $\beta_{ik\ell}$ will have many nonzero components. This is a weak point of the theory, since we have usually no idea of their numerical value. The dynamo equation remains the same, except that $\alpha\boldsymbol{B}_0 - \beta\nabla \times \boldsymbol{B}_0$ in (4.8) is replaced by the right hand side of (6.13), i.e. by $\boldsymbol{\alpha}\cdot\boldsymbol{B}_0 + \boldsymbol{\beta} : (\nabla\boldsymbol{B}_0)$, in tensor form. This generalisation to anisotropic turbulence has been studied by many authors (e.g. Rädler 1983a,b; 1990, with many more references). Here, I restrict myself to showing that the antisymmetric part α_{ik}^A of α_{ik} is equivalent to an additional apparent mean flow. Since α_{ik}^A is antisymmetric, linear algebra tells us that $\alpha_{ik}^A B_{0,k} = (\boldsymbol{v} \times \boldsymbol{B}_0)_i$ with $\boldsymbol{v}$ given by $v_j = \epsilon_{ijk}\alpha_{ik}/2$. Therefore, the effect of α_{ik}^A is that in the dynamo equation, $\boldsymbol{u}_0$ is replaced by $\boldsymbol{u}_0 + \boldsymbol{v}$. To evaluate $\boldsymbol{v}$, we insert (6.14), dropping its second term for simplicity (incompressibility). With the help of $u_k^t\nabla_k* = \nabla_k(u_k^t*)$ we get

$$v_j = \frac{1}{2}\epsilon_{ijk}\epsilon_{ip\ell}\int_0^\infty ds <u_p^t(\nabla_k u_\ell^{t-s})> = -\frac{1}{2}\nabla_k \int_0^\infty ds <u_k^t u_j^{t-s}> \tag{6.21}$$

Relation (6.21) becomes very simple for isotropic turbulence:

$$\boldsymbol{v} = -\frac{1}{2}\nabla \frac{1}{3}\int_0^\infty ds <\boldsymbol{u}_1^t \cdot \boldsymbol{u}_1^{t-s}> = -\frac{1}{2}\nabla\beta \tag{6.22}$$

This expression shows that the mean field field is transported away from regions of intense turbulence, where β is large.

6.3. TWO-SCALE APPROXIMATION

The results obtained so far are based on (6.6) (FOSA), and on the assumption that τ_c is so small that $\tau_c|A_0| \ll 1$. If the latter is not true, then the effect of the operators $\exp(\pm sA_0)$ in (6.10) may become quite noticeable. An exact evaluation for arbitrary differential rotation is possible if $\eta = 0$ (Hoyng 1985). The result is that Equations (4.4) and (4.8) still hold, but that α and β become tensors even if the turbulence is isotropic. This is not unexpected since the differential rotation imposes a preferred direction. We shall now show that η seems to be sufficiently small for the dynamos of the Sun and the Earth, so that it is reasonable to take $\eta = 0$. The proof is simplest if there is no differential rotation. Equation (6.9) then looks as follows:

$$<A_1 B_1> = \nabla \times \int_0^\infty ds <u_1^t \times e^{s\eta\nabla^2}\nabla \times u_1^{t-s}> \times B_0^{t-s} \qquad (6.23)$$

The standard approach in the literature when $\eta \neq 0$ is to assume that $B_0^{t-s} \simeq B_0^t$ in (6.23). This is the *two-scale hypothesis*: B_0 is assumed to change on timescales much longer than τ_c, and over length scales much longer than λ_c.[†] Moffatt (1978, sections 7.8 and 7.9) and Krause and Rädler (1980, Chapter 7) show in detail how relation (6.23) can be elaborated further. Actually, they apply a Fourier transformation directly to (6.5), but that is a technical detail. Since the result of the calculation is well known, I shall resort to a short-cut which is highly nonrigorous but illustrates the essentials rather well.

Substitute $B_0^{t-s} \simeq B_0^t = B_0$ in (6.23) (two-scale hypothesis), and $\exp(s\eta\nabla^2) \approx \exp(-s\eta/\lambda_c^2)$ (think of a Fourier transform: $\nabla \to ik \approx i/\lambda_c$, since ∇ operates on u_1, which has a characteristic length scale λ_c). Assume next that $<u_1^t \times \nabla \times u_1^{t-s}> \simeq \exp(-s/\tau_c)\cdot <u_1 \times \nabla \times u_1>$, as is also implicitly done in (6.17). Noting that $<A_1 B_1> = \nabla \times <u_1 \times B_1>$, we deduce from (6.23) that

$$<u_1 \times B_1> \approx <u_1 \times \nabla \times u_1> \times B_0 \int_0^\infty \exp\{-s(\eta/\lambda_c^2 + 1/\tau_c)\}\, ds \qquad (6.24)$$

The integral is trivial and we may then proceed just as from (6.12) onwards. The end result is that τ_c is replaced by τ, as follows:

$$\left.\begin{array}{l} \alpha \approx -\tfrac{1}{3}<u_1 \cdot \nabla \times u_1> \tau \\[2mm] \beta \approx \tfrac{1}{3}<u_1^2> \tau \end{array}\right\} \qquad \frac{1}{\tau} = \frac{1}{\tau_c} + \frac{1}{\tau_d} \; ; \quad \tau_d = \frac{\lambda_c^2}{\eta} \qquad (6.25)$$

τ_d is the diffusion time for a typical eddy. Resistive effects reduce α and β since $\tau < \tau_c$. This is intuitively clear: the turbulence has less grip on B because the field lines are slipping. The exact result (Moffatt 1978, relations (7.78) and (7.98)) features a weighted integral over the so-called helicity and the energy spectrum of the turbulence instead of (6.25). The two limiting cases $\tau_c \gg \tau_d$ and $\tau_c \ll \tau_d$ are covered by Moffatt's (1978) expressions (7.81) and (7.85), respectively. The reader may verify that $\tau_d/\tau_c \gtrsim u_1 \lambda_c/\eta \equiv R_m$ (since $\tau_c \lesssim \lambda_c/u_1$); R_m is the magnetic Reynolds number of an eddy, which is known to be much larger than unity for the Sun as well as for the Earth (Moffatt 1978, section 4.5). We conclude that for these dynamos $\tau \simeq \tau_c$, and that the assumption $\eta = 0$ appears to be reasonable.[†] Differential rotation, finally, makes little difference. Equation (6.23) then contains the operator $\exp\{s(\eta\nabla^2 + \nabla \times u_0\times)\}$ instead of $\exp(s\eta\nabla^2)$. The typical magnitude of the two terms is $\tau_c\eta/\lambda_c^2 = 1/R_m \ll 1$ and $\tau_c\Delta\Omega_r$, respectively (see Table 1). It appears therefore that for the Sun and the Earth, the resistivity term in $\exp(\pm sA_0)$ is negligible, regardless the magnitude of the differential rotation term.

[†] There is little else one can do. For example, we may no longer put $B_0^{t-s} \simeq \exp(-sA_0)B_0^t = \exp(-s\eta\nabla^2)B_0$, as in (6.10). The reason is that the operator $\exp(-s\eta\nabla^2)$ is unbounded, due to the fact that the equation $\partial_t b = -\eta\nabla^2 b$ is ill-posed for advancing in time (the 'diffusion coefficient' $-\eta$ is negative). The situation is clearly unsatisfactory, since there may be a cascade down to spatial scales much smaller than λ_c. This point is connected to the so-called 'fast dynamo problem', see Childress (1992).

Moffatt's (1978, section 7.7) conjecture that *some* ohmic dissipation is a prerequisite for a nonzero α has been refuted by Drummond and Horgan (1986) for the case that $\tau_c \lesssim \lambda_c/u_1$, but is correct when $\tau_c \gg \lambda_c/u_1$. This limit, incidentally, is referred to as the *frozen turbulence* limit. It follows that systematic, non-stochastic flows (for which $\tau_c = \infty$) cannot produce a nonzero α unless $\eta \neq 0$, but random flows with $\eta = 0$ can (see also Kraichnan 1976).

6.4. LONG CORRELATION TIME

Based on laboratory experience and observations of the solar granulation we anticipate that $u_1\tau_c \approx \lambda_c$, so that *FOSA* may not be valid in the solar convection zone and in the Earth's core. The question of what happens if the correlation time is long is a major unsolved problem in dynamo theory. Van Kampen (1974a,b) has shown that when no terms in Equation (6.5) are ignored, the operator $\{\cdots\}$ in (6.10) becomes an infinite series involving so-called *ordered cumulants*. The n-th term contains n-tuple velocity correlation functions; its order of magnitude is $|A_1|^n \tau_c^{n-1} \approx (u_1\tau_c/\lambda_c)^n/\tau_c$ and it contains ∇^n as highest spatial derivative operating on B_0. If *FOSA* holds, only the first two terms remain, and (6.10) is recovered. Knobloch (1977, 1978) and Nicklaus and Stix (1988) have analysed the series for the case that $A_0 = 0$ (*i.e.* an α^2-dynamo) and homogeneous isotropic turbulence. Including terms up to $n = 4$ the dynamo equation takes the form:

$$\partial_t B_0 = \{(\alpha_2 + \alpha_3 + \alpha_4)\nabla \times +(\beta_2 + \beta_3 + \beta_4)\nabla^2 + \gamma_4\nabla^2\nabla \times +\mu_4\nabla^4\}B_0 \qquad (6.26)$$

α_2 and β_2 equal α and β from (6.17) and (6.18); α_3 and β_3 vanish because they contain only triple correlations $<u^{t_1}u^{t_2}u^{t_3}>$ which are all zero. For Gaussian turbulence u_1, Nicklaus and Stix (1988) argue that γ_4 and μ_4 vanish too. This result is important because it shows that in this (special) case the dynamo equation retains its mathematical structure, with *renormalised* coefficients α and β. Perhaps it is possible to generalise this property to arbitrary n or to other types of statistics. For $u_1\tau_c/\lambda_c \approx 1$, Nicklaus and Stix find that the corrections α_4 and β_4 are so large that they may reverse the sign of α_2 and/or β_2. This implies that *all* terms must be summed to determine the effective α and β, but that is unlikely to lead anywhere as the cumulant expansion may be divergent (Van Kampen 1976, § 6). The cumulant series approach could therefore be useful if $u_1\tau_c/\lambda_c$ is not too close to unity.

There are three techniques which are not based on a series expansion in $u_1\tau_c/\lambda_c$ and which can be used for *arbitrary* τ_c, *i.e.* even for frozen turbulence. Kraichnan (1976) and Drummond and Horgan (1986) have numerically evaluated the Lagrangian expressions for the effective α and β, for a velocity field consisting of a superposition of random waves. Drummond and Horgan (1986) found that the effective β is positive for all finite values of τ_c. In my view, this puts an end to speculations about the possibility and consequences of a negative β (Kraichnan 1976; Knobloch 1978; Parker 1979, section 18.4). For α the situation is less clear. Nonlinear convection simulations have shown that it is already very difficult to determine the *sign* of α (section 7). Secondly, Zel'dovich *et al.* (1984) and Dittrich *et al.* (1984) have analytically evaluated the behaviour of B in randomly renovating flows.

Finally there is the method of marginal averages (Van Kampen 1976, section 23) which to my knowledge has not yet been applied to the dynamo problem.

7. Open problems and recent developments

So far I have sketched a fairly rosy picture of mean field theory. I have, as Cowling (1981) puts it in his review, presented the subject very much from the standpoint of a believer. The emphasis has been on the physics, rather than on a discussion of the great variety of mean field dynamo models that exist for the Earth, the Sun, stars, the galaxy, accretion discs, *etc.* Instead, I have tried to convey an understanding of just one typical solar mean field model. Right from the early days of mean field theory, however, there have been a number of problems. Instead of disappearing in the course of time, they increased in number and severity, and as a result the theory finds itself in a kind of midlife crisis. This calls for a careful analysis of the patient's problems. I shall now review some of the major issues, from a somewhat broader perspective.

Two problems are inherent to all mean field dynamos, in any geometry, regardless of whether or not nonlinear effects are taken into account. These are the structure of the magnetic field in the dynamo, and the fact that the turbulence is likely to have long correlation time. This last problem has just been treated in section 6.4. The mathematical form of the dynamo equation is unknown if $u_1 \tau_c / \lambda_c \approx 1$, and this renders all predictions based on Equation (4.8) precarious. The problem of the magnetic field structure is that according to (4.1) the field in the dynamo is equal to $\boldsymbol{B}_0 + \boldsymbol{B}_1$, and that $\boldsymbol{B} \simeq \boldsymbol{B}_0$ if $\boldsymbol{B}_1$ is small, but this is not the case. An estimate from Equation (6.5) gives

$$B_1 \sim R_m^{1/2} B_0 \gg B_0 \qquad (7.1)$$

(R_m = magnetic Reynolds number of an eddy), see *e.g.* Krause and Rädler (1980, section 3.11) and Cowling (1981). This estimate does not allow for dynamical effects (influence of $\boldsymbol{B}_1$ on $\boldsymbol{u}_1$) and this will reduce B_1 considerably, but it is nevertheless likely that $B_1 > B_0$. The implication is that the field is chaotic and that there exists no ordered large-scale field as suggested in Figure 10. It is sometimes argued (*e.g.* Cowling 1981) that $B_1 > B_0$ invalidates the mean field approach, but the previous section shows that this is incorrect. It does follow, however, that mean field theory cannot explain why emerging active regions have a well-defined East-West orientation as observed.

7.1. NONLINEAR MEAN FIELD DYNAMOS

An obvious defect of Equation (4.8) is that it cannot predict the magnitude of $\boldsymbol{B}_0$ since it is linear. The reason is that the backreaction of $\boldsymbol{B}$ on $\boldsymbol{u}$ has not been accounted for; $\boldsymbol{u}_0$ and α and β (*i.e.* the statistical properties of $\boldsymbol{u}_1$) are taken to be independent of $\boldsymbol{B}_0$. Three types of nonlinearity have been considered:

(1). An extra term of the type B_0/τ is included on the right hand side of (4.8), to account for the loss of magnetic flux through magnetic buoyancy. The timescale τ is taken to be a decreasing function of B_0. This is of course phemomenological and, moreover, perusing the derivation of (4.8) in section 6, we see that there is no room for extra terms. What happens is that β changes, since buoyancy alters the radial transport properties of the convection. There is an extra large r-component in $\boldsymbol{u}_1$ which gives rise to extra components in the tensor $\beta_{ik\ell}$ in (6.15):

$$\beta_{ir\ell} = -\epsilon_{ir\ell}\int_0^\infty ds \; <u_{1,r}^t u_{1,r}^{t-s}> \;, \tag{7.2}$$

which depends on B_0. It follows that i, ℓ is either equal to θ, φ or to φ, θ. It is not difficult to see that this gives rise to enhanced radial diffusion ($\partial^2/\partial r^2$ – terms) of the θ and φ components of B_0 in (6.11). This idea has never been worked out; of course, if one is willing to replace $\partial/\partial r$ by $1/L$, one gets a term of the type B_0/τ.

(2). The coefficient α is taken to be a decreasing function of B_0, on the ground that a strong field reduces in particular the the the helicity of $\boldsymbol{u}_1$ (the correlation between $\boldsymbol{u}_1$ and $\nabla \times \boldsymbol{u}_1$). Expressions for $\alpha(B_0)$ have been derived (see Krause and Rädler 1980, Chapter 10). An indication that quenching of the α-effect indeed takes place is that linear mean field models require an α of the order of 10 cm s^{-1} if they are to have a period of 22 year, while mixing length estimates produce a figure which is typically a factor 100 larger, $\alpha \approx 10^3$ cm s^{-1}.

(3). The magnetic field changes (flattens) the profile of the differential rotation.

Many models have been constructed based on these concepts, and they show a bewildering variety of phenomena ranging from nonlinear periodic solutions to quasi-periodic, multi-periodic and chaotic behaviour. References can be found in Hoyng (1990) and Brandenburg and Tuominen (1991). Another line of attack has been to investigate the chaotic properties of dynamos. The underlying idea is as follows. By separating the r co-ordinate, the dynamo equation and the momentum equation (3.5) can be reduced to an infinite set of coupled *ordinary* differential equations for the expansion coefficients, which depend only on time. The dynamo describes a path in this infinite dimensional phase space, and the hope is that the orbit evolves in practice in a subspace of low dimension. It would then be possible to describe a nonlinear dynamo with a few coupled ordinary differential equations.

Ruzmaikin (1981) has shown that an $\alpha\Omega$-dynamo with α-effect quenching can be modeled by the Lorenz equations, which are known to possess chaotic solutions (Schuster 1988). The orbit in (3D) phase space is quasi-periodic, but lingers occasionally for a long time near the origin. This led to the suggestion that Grand Minima such as the Maunder Minimum might correspond to a strange attractor. A more elaborate model with up to seven variables has been formulated by Weiss *et al.* (1984). These studies can only give some indication of the behaviour of real dynamos because the spatial structure has been severely truncated. A major problem is that it has not been possible to demonstrate that the properties of the solution remain more or less unaffected as more variables are taken into account. In my view, there seems at present to be little reason to believe that the solar dynamo possesses an attractor of low dimension, neither from a theoretical nor from an observational point of view.

On the whole, the nonlinear theory of the Earth's dynamo appears to be in a more advanced state than its solar counterpart. A major flaw is that the nonlinearities are introduced phenomenologically, since dynamically consistent expressions for their dependence on B_0 are usually not available, and they may even not exist. Consider for example the effect of B on the mean flow u_0. This is given by the mean Lorentz force $<f>$:

$$<f_i> = \frac{1}{4\pi} <(\nabla \times B) \times B> = \frac{1}{8\pi} \nabla_j (2T_{ij} - T_{kk}\delta_{ij}) \qquad (7.3)$$

Hence, $<f>$ depends on a higher average $T_{ij} = <B_i B_j>$. One would hope that $<f> = (\nabla \times <B>) \times <B>/4\pi$, but that requires $B_1 = B - B_0 = B - <B>$ to be small, which is not the case. In other words, one cannot neglect the influence of the fluctuating field B_1. From the mathematical point of view, there is a *closure* problem, as $<B_i B_j>$ depends in turn on a higher average.

7.2. BOUNDARY LAYER DYNAMOS

Buoyancy is a problem for dynamos operating in the entire convection zone as in Figure 10, because it expels the field from the convection zone on a time scale of about one year (Parker 1975), and this impedes effective dynamo action. Moreover, for an equatorial migration of the activity belts, $\alpha \partial \Omega_r / \partial r$ should be negative in the Northern hemishpere. Since α is believed to be positive in the Northern convection zone, $\partial \Omega_r / \partial r$ should be negative, but helioseismology tells us that $\partial \Omega_r / \partial r \gtrsim 0$. Due to these and other problems, the solar dynamo is now believed to be located in the overshoot layer at the base of the convection zone (Spiegel and Weiss 1980; Galloway and Weiss 1981; Schüssler 1984; for a recent review see Zwaan 1992). The advantages are:

(1). The thermal stratification is subadiabatic there, which renders it much easier to suppress the buoyancy force for fields of the order of 10^4 G, the local equipartition value. An equatorward meridional flow may do so, too (Van Ballegooijen 1982, 1983; Pidatella and Stix 1986; Van Ballegooijen and Choudhuri 1988).

(2). Penetrative convection maintains some turbulence in the overshoot layer so that a turbulent dynamo may operate there. Moreover, α is likely to be *negative* in the Northern hemisphere. The preferred sense of rotation for sinking gas *in* the convection zone is anticlockwise in the Northern hemisphere, see Figure 9, but near the bottom the streamlines are forced to diverge rather than converge, whence the Coriolis force induces a clockwise rotation. This topological argument is supported by convection simulations of Glatzmaier (1985), Brandenburg et al. (1990) and Jennings (1991), who observe a sign reversal of the mean helicity along the radial direction.[†] The idea is then that equatorial migration is ensured due to a strong differential rotation $\partial \Omega_r / \partial r > 0$ in combination with $\alpha < 0$ (Northern hemisphere).

(3). Magnetic flux is expelled from the convection zone upward by buoyancy, but also *downward*. In the lower part of the convection zone the 'turbulent diamagnetic effect'

---

† However, Brandenburg et al. (1990) appear to predict a different overall sign for α than the other two authors.

(6.22) transports mean magnetic fields downward, since β is zero at the bottom, increases to a maximum somewhere in the convection zone and decreases again to zero at the surface. Parker's 'thermal shadows' may do likewise (Parker 1987b). Thus, magnetic flux accumulates in or near the overshoot layer.

Models of mean field boundary layer dynamos have been published by Schmitt and Schüssler (1989) and by Belvedere *et al.* (1990, 1991); see also Brandenburg and Tuominen (1991). The solar luminosity variation which is apparently in phase with the 11 year sunspot cycle (Hudson 1988) may be connected to a boundary layer dynamo. Suppose that the integrated luminosity deficit over one half-cycle of 5.5 year is used for a build-up of magnetic energy in a volume V in the overshoot layer (see Spiegel and Weiss (1980) for a more refined argument):

$$\int_{\text{half period}} \delta L(t)\, dt \;=\; \frac{B^2}{8\pi} V \tag{7.4}$$

Taking $\delta L(t)$ to be sinusoidal with an amplitude of $5 \times 10^{-4} L_\odot$ (Hudson 1988), we find that (7.4) equals 2×10^{38} erg. This corresponds to a magnetic field of 2×10^4 G in two equatorial belts of $10°$ latitudinal extent (roughly the width of a wing of the butterfly diagram), assuming that the magnetic field layer is 2×10^4 km thick.

The hope is that the prevailing strong differential rotation in combination with a relatively mild turbulence creates a strong East-West oriented toroidal field in the overshoot layer, of the order of a few times 10^4 G. How this is actually accomplished, and whether a mean field $\alpha\Omega$-dynamo can do it, are essentially open questions. Pending their solution, several authors have analysed under what conditions such toroidal fields may become unstable and rise to the surface to form active regions (Moreno-Insertis 1986; Parker 1987c; Chou and Fisher 1989; Choudhuri 1989; Choudhuri and D'Silva 1990). There remains, finally, the problem that the phase relations of the observed poloidal and toroidal field on the Sun indicate that α should be *positive* in the Northern hemisphere (Stix 1976b). However, it is not clear if this really applies to the overshoot layer, since the intervening convection zone may alter the phases. Indeed, Spruit *et al.* (1987), Durney (1988), and Van Geffen (1992) have suggested that there may be two dynamos operating in the Sun, a diffuse dynamo in the convection zone without a well-defined large-scale field, and a boundary layer dynamo responsible for the cycle.

7.3. EXTERNALLY FORCED DYNAMOS

Boundary conditions play an important role in shaping the solution of Equation (4.8). Normal practice is to assume that B_0 merges continuously into a potential field at the top of the convection zone, and that $B_0 = 0$ at the bottom of the convection zone (see Moffatt 1978; Krause and Rädler 1980). A nonzero boundary condition might be provided by a hypothetical relic field in the radiative interior of the Sun. Such a field will be smeared out by the turbulent convection, but there remains a visible effect: a zero frequency signal is effectively superposed on the cycle, and alternate 11 year sunspot cycles will have different amplitudes. These ideas have been pursued by Levy and Boyer (1982) and Boyer and Levy (1984). Sonett (1983) has derived in this way an upper limit for a fossil dipole field (the longest surviving component of a relic field) of the order of 1 G at the base of the convection zone. A periodic boundary condition might come about if the solar interior were to exert a torsional oscillation, with the magnetic field acting as a spring. Such ideas have been advanced by Piddington (1971, 1976) and Layzer *et al.* (1979), although a convincing model has never been worked out (*cf.* Cowling 1981). Apart from that, it is conceivable that a dynamo in the convection zone could be locked in to a coherent magnetic oscillation, if one were to exist in the radiative interior of the Sun. The phase stability of the cycle could then be very high, since it would be determined by an internal clock (which in this scenario would have a period of 22 years). For more details see Dicke 1978, 1988; Gough 1988, and section 2.3.

7.4. INTERNAL OR RANDOM FORCING

It is instructive to compare the dynamo equation, which describes turbulent transport of a divergence-free vector field, with turbulent transport of a scalar f. The starting point is the continuity equation $\partial_t f = -\nabla \cdot (\boldsymbol{u} f)$. The equation for $f_0 = <f>$ can be obtained by noting that Equation (6.10) is a general result. Any equation of the type $\partial_t f = \{A_0 + A_1(t)\} f$, as for example (6.1), leads to Equation (6.10) for the average, with $<\boldsymbol{B}>$ replaced by $<f>$. In the present case, $A_0 = -\nabla \cdot (\boldsymbol{u}_0 \cdots)$ and $A_1 = -\nabla \cdot (\boldsymbol{u}_1 \cdots)$. Take for simplicity $\tau_c |A_0| \ll 1$ so that $\exp(\pm s A_0) \to I$, and incompressible isotropic turbulence. The result is the ordinary diffusion equation:

$$\frac{\partial f_0}{\partial t} = \left\{ -\nabla \cdot (\boldsymbol{u}_0 f_0) + \int_0^\infty ds \, \nabla_k <u_k^t \nabla_p u_p^{t-s}> f_0 \right\}$$

$$= -\nabla \cdot (\boldsymbol{u}_0 f_0) + \nabla \cdot \beta \nabla f_0 \qquad (7.5)$$

The interpretation of (7.5) is straightforward: a patch of dye in the solar convection zone is advected and deformed by the shear flow $\boldsymbol{u}_0$, while turbulent diffusion spreads the patch over the entire convection zone on a turbulent diffusion timescale ($R_\odot^2 / \beta \sim 22$ years). Thereafter, f_0 becomes constant. The difference between Equations (4.8) and (7.5) is the absence of an α-term in the latter. The physical reason is that a vector (an advected arrow) senses the correlated lifting and twisting process depicted in Figure 9, but a scalar, which has no intrinsic direction, does not. Turbulent transport of a scalar leads to a featureless spreading of f_0 towards eventual homogeneity. There is a continuous loss of memory. Against this background, it seems unphysical that *the same process* for a vector results in a well-organised periodic wave pattern, which continues indefinitely without any degradation, as it does in Figure 10. The fact that we have neglected nonlinear effects cannot be the real answer: the 'mean field' of a collection of needles advected by the flow $\boldsymbol{u}$ is also described by the dynamo equation, but there is no Lorentz force.

This objection has led to a renewed interest in the statistical interpretation of mean field theory, which I shall briefly review in this section. The origin of the dilemma is that we have ignored the fluctuations in the turbulent convection. The way in which the effect of fluctuations shows up in the theory depends on what average is used. Things are conceptually easiest when $<\cdot>$ is interpreted as a longitudinal average. In that case

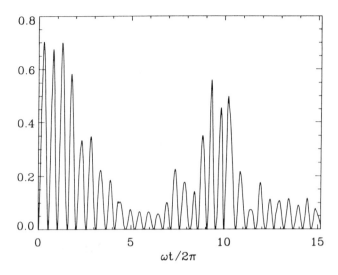

FIGURE 16. Irregular cycles of a simple $\alpha\Omega$-dynamo in the overshoot layer, with random forcing due to helicity fluctuations (Hoyng 1992). Vertical axis: the square of the amplitude of the toroidal field (arbitrary units); horizontal axis: time in units of the mean period.

α, β and u_0 are also longitudinal averages; for example, we have from (6.18) that $\beta = (\tau_c/3) \int u_1^2 \, d\varphi/2\pi$. These averages exhibit random fluctuations, because there is only a finite number of convection cells along a circle. In particular the fluctuations in α (*i.e.* helicity fluctuations) can be large. In recognition of this fact, Parker (1969) and Levy (1972) have studied the effect of a sudden jump in α and shown that it may induce a reversal of the geomagnetic field. Hoyng and Van Geffen (1992) have made a theoretical study of the effect of random fluctuations in a simple α^2-dynamo with $\alpha = \alpha_0 + \delta\alpha(t)$. It turns out that the dynamo operates slightly subcritically. Moreover, in addition to the fundamental mode, all higher eigenmodes are weakly stochastically excited. Choudhuri (1992) and Hoyng (1992) have studied the effect of helicity fluctuations in a simple $\alpha\Omega$-dynamo operating in the overshoot layer, see Figure 16. The effect of fluctuations in nonlinear dynamos has been investigated by Crossley *et al.* (1986) and by Meinel and Brandenburg (1990).

On the other hand, if $< \cdot >$ is interpreted as an ensemble average, then the effect of fluctuations is concealed in the fact that the 'energy tensor' $<\boldsymbol{B}\boldsymbol{B}>$ diverges to infinity, for marginal stable solutions of the dynamo equation (4.8). It follows that the trace of this tensor, $<B^2>$, becomes infinite, and therefore also the magnetic field $\boldsymbol{B}$ in the dynamo. The only remedy is to adapt the constants in the equation for $<\boldsymbol{B}\boldsymbol{B}>$ so that $<\boldsymbol{B}\boldsymbol{B}>$ remains finite – but then the dynamo equation (4.8) possesses only subcritical solutions. This means that $<\boldsymbol{B}>$ goes to zero exponentially in time, and this is in turn interpreted as a due to a loss of phase coherence between ensemble members, whose fields are all finite (Hoyng 1987b, 1988). The effect can be demonstrated by considering a simple α^2-dynamo, with constant α and β, and zero mean flow $\boldsymbol{u}_0$.[†] Assuming $\eta = 0$, zero resistivity, (4.8) becomes:

$$\frac{\partial \boldsymbol{B}_0}{\partial t} = \alpha \nabla \times \boldsymbol{B}_0 + \beta \nabla^2 \boldsymbol{B}_0 \qquad (7.6)$$

This equation describes the evolution of the mean field of a sphere of radius R, with homogeneous, isotropic helical turbulence. It is the only case for which the dynamo equation can be solved analytically in spherical geometry (see Krause and Rädler, 1980, Chapter 14). The mean energy density $\epsilon = <B^2>/8\pi$ of the magnetic field of this dynamo obeys a closed equation:

$$\frac{\partial \epsilon}{\partial t} = 2\gamma \epsilon + \beta \nabla^2 \epsilon \qquad (7.7)$$

with

$$\gamma \simeq \tfrac{1}{3} <|\nabla \times \boldsymbol{u}_1|> \tau_c \qquad (7.8)$$

A new dynamo coefficient γ appears, proportional to the mean *vorticity* of the turbulence. The α-effect plays no role in the energy balance. Now, if $\partial \epsilon/\partial t = 0$, then we find from (7.7) that $\gamma \approx \beta/R^2$. The dynamo number of this α^2-dynamo is given by $D_\alpha = \alpha R/\beta$ (*i.e.* the magnitude of $\alpha \nabla \times \boldsymbol{B}_0$ in (7.6) divided by that of $\beta \nabla^2 \boldsymbol{B}_0$ – see below (5.19)). Schwarz's inequality implies that $\alpha^2 \leq \beta \gamma$, *cf.* (4.5) and (7.8). It follows that $D_\alpha^2 = (\alpha R/\beta)^2 \leq \gamma R^2/\beta \approx 1$, and this implies that (7.6) has only subcritical solutions (Hoyng 1987b). These ideas have been applied by Van Geffen (1992) to a dynamo in the solar convection zone. One of the results is that the convection zone can only support a diffuse, aperiodic dynamo with a weak mean field $B_0 \lesssim 10^2$ G – otherwise there would be too much coronal heating.

7.5. THE PHYSICS OF TURBULENT DIFFUSION

The reader may have felt uneasy about the explanation of turbulent diffusion given in section 4. The two flux tubes of opposite polarity, indicated by an } in Figure 11 at (g), annihilate each other, and have disappeared in (h). This is allright for $\boldsymbol{B}_0$, but what happens to $\boldsymbol{B}$? The field in the two approaching tubes gets mixed. The (oppositely oriented) preferred direction for $\boldsymbol{B}$ which existed in each, is destroyed so that $\boldsymbol{B}_0 = 0$. But $\boldsymbol{B}$ is still there. This point is closely related to the previous section 7.4, and to the concept of *fast dynamo action* (Zel'dovich et al. 1990, Chapter 9; Childress 1992). To clarify what happens, we focus on the α^2-dynamo described by Equations (7.6) and (7.7), assuming that $<\cdot>$ is the ensemble average.

The resistivity η is zero, and therefore $\boldsymbol{B}$ cannot dissipate. Yet, the term $\beta \nabla^2 \boldsymbol{B}_0$ is equivalent to a very strong resistive dissipation. The explanation is that a clear distinction should be made between dissipation of $\boldsymbol{B}$ in any single dynamo of the ensemble (which does not happen since $\eta = 0$), and the decrease (or: 'diffusion') of the *probability that a certain $\boldsymbol{B}$ occurs in the whole ensemble* (which does take place). This underlines that only the mean field $\boldsymbol{B}_0$ (or rather: the probability distribution for $\boldsymbol{B}$, represented by the ensemble)

[†] Note that α and β are now defined as ensemble averages, and therefore they do not depend on time.

experiences turbulent diffusion, not $\boldsymbol{B}$ itself. That $\boldsymbol{B}$ is not dissipated, can also be seen by integrating (7.7) over the volume V of the dynamo. In this way, we obtain an equation for the mean energy balance:

$$\left\{\frac{\partial}{\partial t} - 2\gamma\right\} \int_V \epsilon \, d^3r = \beta \oint_{\partial V} (\nabla \epsilon) \cdot d\boldsymbol{\sigma} \qquad (7.9)$$

Since a volume term proportional to β is absent, it follows that turbulent diffusion does not create nor destroy mean energy (in spite of the fact that it causes the mean field to dissipate). It merely transports mean energy from one place to the other, keeping the total energy constant. According to (7.9), mean energy is created by random field line stretching, embodied in the term proportional to γ, and it is transported to the boundary ∂V (the local flux density is $-\beta \nabla \epsilon$), where it escapes into space – the escape rate being governed by the boundary condition on ∂V. In reality, η will be finite, and local dissipation will take place and provide an alternative sink of energy, in addition to escape at the boundary. Van Geffen (1992) has argued that such resistive effects are small in the case of the solar dynamo. For the dynamo of the Earth, on the other hand, resistive effects are expected to be the dominant sink of magnetic energy (even though they have little influence on the evolution of $\boldsymbol{B}_0$). Parker (1979, section 17.5) and other authors have argued that turbulent diffusion ultimately relies on resistive effects, which dissipate the small-scale fields. The present discussion shows that this is not the essential point: turbulent diffusion is caused by phase mixing between ensemble members (or: by dissipation of *probability*). Although a finite resistivity has some influence, *cf.* (6.25), it is not an essential ingredient for turbulent diffusion to work.

Acknowledgements: I am grateful to Drs. Joan Schmelz and John Brown for organising a very stimulating Summer School in Crieff. I also offer sincere thanks to Drs. Kanaris Tsinganos, John Brown and Jos van Geffen for their constructive criticism, which has been of considerable help to improve the presentation of this Chapter.

8. References

Alfvén, H.: 1950, *Tellus 2*, 74.

Backus, G.E.: 1958, *Ann. Phys. 4*, 372.

Barnes, J.A., Sargent, H.H. and Tryon, P.V.: 1980, in *The Ancient Sun*, eds. R.O. Pepin, J.A. Eddy and R.B. Merrill, Pergamon Press (New York), p. 159.

Belvedere, G., Pidatella, R.M. and Proctor, M.R.E.: 1990, *Geophys. Astrophys. Fluid Dynamics 51*, 263.

Belvedere, G., Proctor, M.R.E. and Lanzafame, G.: 1991, *Nature 350*, 481.

Boyer, D.W. and Levy, E.H.: 1984, *Astrophys. J. 277*, 848.

Braginskii, S.I.: 1965a, *Sov. Phys. JETP 20*, 726.

Braginskii, S.I.: 1965b, *Sov. Phys. JETP 20*, 1462.

Brandenburg, A. and Tuominen, I.: 1991, in *The Sun and Cool Stars: activity, magnetism, dynamos*, eds. I. Tuominen, D. Moss and G. Rüdiger, Springer-Verlag (Berlin), p. 223.

Brandenburg, A., Nordlund, Å., Pulkkinen, P., Stein, R.F. and Tuominen, I.: 1990, *Astron. Astrophys. 232*, 277.

Bullard, E.C. and Gellman, H.: 1954, *Phil. Trans. Roy. Soc. A247*, 213.

Bullard, E.C.: 1955, *Proc. Cambr. Phil. Soc. 51*, 744.

Childress, S.: 1992, in *Proceedings of the Workshop on Topological Fluid Dynamics*, Santa Barbara, eds. H. K. Moffatt and M. Tabor, to appear.

Chou, D.-Y. and Fisher, G.H.: 1989, *Astrophys. J. 341*, 533.

Choudhuri, A.R.: 1989, *Solar Phys. 123*, 217.

Choudhuri, A.R.: 1992, *Astron. Astrophys. 253*, 157.

Choudhuri, A.R. and D'Silva, S.: 1990, *Astron. Astrophys. 239*, 326.

Cowling, T.G.: 1934, *M.N.R.A.S. 94*, 39.

Cowling, T.G.: 1981, *Ann. Rev. Astron. Astrophys. 19*, 115.

Crossley, D., Jensen, O. and Jacobs, J.: 1986, *Phys. Earth Planet. Inter. 42*, 143.

Deinzer, W. and Stix, M.: 1971, *Astron. Astrophys. 12*, 111.

Dicke, R.H.: 1978, *Nature 276*, 676.

Dicke, R.H.: 1988, *Solar Phys. 115*, 171.

Dittrich, P., Molchanov, S.A., Sokoloff, D.D. and Ruzmaikin, A.A.: 1984, *Astron. Nachr. 305*, 119.

Drummond, I.T. and Horgan, R.R.: 1986, *J. Fluid Mech. 163*, 425.

Durney, B.R.: 1981, *Astrophys. J. 244*, 678.

Durney, B.R.: 1988, *Astron. Astrophys. 191*, 374.

Galloway, D.J.: 1986, *Adv. Space Res. 6, No. 8*, 19.

Galloway, D.J. and Weiss, N.O.: 1981, *Astrophys. J. 243*, 945.

Glatzmaier, G.A.: 1985, *Astrophys. J. 291*, 300.

Gough, D.O.: 1988, in *Solar-Terrestrial Relationships and the Earth Environment in the Last Millennia*, ed. G. Castagnoli-Cini, Soc. Italiana di Fisica (Bologna), p. 90.

Herzenberg, A.: 1958 *Phil. Trans. Roy. Soc. A250*, 543.

Hoyng, P.: 1985, *J. Fluid Mech. 151*, 295.

Hoyng, P.: 1987a, *Astron. Astrophys. 171*, 348.

Hoyng, P.: 1987b, *Astron. Astrophys. 171*, 357.

Hoyng, P.: 1988, *Astrophys. J. 332*, 857.

Hoyng, P.: 1990, in *Solar Photosphere: Structure, Convection, and Magnetic fields*, ed. J.O. Stenflo, Kluwer Academic Publishers (Dordrecht), p. 359.

Hoyng, P.: 1992, *Astron. Astrophys.*, in preparation.

Hoyng, P. and Van Geffen, J.H.G.M.: 1992, *Geophys. Astrophys. Fluid Dynamics*, to be submitted.

Hudson, H.S.: 1988, *Ann. Rev. Astron. Astrophys. 26*, 473.

Jennings, R.L.: 1991, in *The Sun and Cool Stars: activity, magnetism, dynamos*, eds. I. Tuominen, D. Moss and G. Rüdiger, Springer-Verlag (Berlin), p. 62.

Kiepenheuer, K.O.: 1953, in *The Sun*, ed. G.P. Kuiper, The University of Chicago Press (Chicago), p. 322.

Kiepenheuer, K.: 1959, *The Sun*, The University of Michigan Press (Ann Arbor).

Knobloch, E.: 1977, *J. Fluid Mech. 83*, 129.

Knobloch, E.: 1978, *Astrophys. J. 225*, 1050.

Kraichnan, R.H.: 1976, *J. Fluid Mech.* **77**, 753.

Krause, F.: 1967, Habilitationsschrift, Univ. Jena (translated in Roberts and Stix 1971).

Krause, F.: 1991, in *The Sun and Cool Stars: activity, magnetism, dynamos*, eds. I. Tuominen, D. Moss and G. Rüdiger, Springer-Verlag (Berlin), p. 3.

Krause, F. and Rädler, K.-H.: 1980, *Mean-Field Magnetohydrodynamics and Dynamo Theory*, Pergamon Press (London).

Larmor, J.: 1919, *Rep. Brit. Assoc. Adv. Sci. 1919*, 159.

Layzer, D., Rosner, R. and Doyle, H.T.: 1979, *Astrophys. J. 229*, 1126.

Levy, E.H.: 1972, *Astrophys. J. 171*, 635.

Levy, E.H. and Boyer, D.: 1982, *Astrophys. J. 254*, L19.

Libbrecht, K.G.: 1988, in *Seismology of the Sun & Sun-like Stars*, ed. E.J. Rolfe, ESA SP-286, p. 131.

Lortz, D.: 1972, *Z. Naturforsch. 27a*, 1350.

Meinel, R. and Brandenburg, A.: 1990, *Astron. Astrophys. 238*, 369.

Moffatt, H.K.: 1978, *Magnetic Field Generation in Electrically Conducting Fluids*, Cambridge U.P. (Cambridge).

Moreno-Insertis, F.: 1986, *Astron. Astrophys. 166*, 291.

Nicklaus, B. and Stix, M.: 1988, *Geophys. Astrophys. Fluid Dynamics 43*, 149.

Parker, E.N.: 1955, *Astrophys. J. 122*, 293.

Parker, E.N.: 1969, *Astrophys. J. 158*, 815.

Parker, E.N.: 1975, *Astrophys. J. 198*, 205.

Parker, E.N.: 1979, *Cosmical Magnetic Fields*, Clarendon Press (Oxford).

Parker, E.N.: 1987a, *Solar Phys. 110*, 11.

Parker, E.N.: 1987b, *Astrophys. J. 321*, 984.

Parker, E.N.: 1987c, *Astrophys. J. 321*, 1009.

Pidatella, R.M. and Stix, M.: 1986, *Astron. Astrophys. 157*, 338.

Piddington, J.H.: 1971, *Proc. Astron. Soc. Australia 2*, 7.

Piddington, J.H.: 1976, in *Basic Mechanisms of Solar Activity*, eds. V. Bumba and J. Kleczek, Reidel (Dordrecht), p. 389.

Ponomarenko, Yu.B.: 1973, *Prikl. Mech. Techn. Fiz. 6*, 47.

Rädler, K.-H.: 1983a, in *Stellar and Planetary Magnetism*, ed. A.M. Soward, Gordon and Breach (New York), p. 17.

Rädler, K.-H.: 1983b, in *Stellar and Planetary Magnetism*, ed. A.M. Soward, Gordon and Breach (New York), p. 37.

Rädler, K.-H.: 1990, in *Inside the Sun*, eds. G. Berthomieu and M. Cribier, Kluwer Academic Publishers (Dordrecht), p. 385.

Rikitake, T.: 1958, *Proc. Camb. Phil. Soc. 54*, 89.

Roberts, P.H. and Stix, M.: 1971, *NCAR-TN/IA-60*, (NCAR, Boulder, Colorado).

Ruzmaikin, A.A.: 1981, *Comm. Astrophys. 9*, 85.

Ruzmaikin, A.A., Sokoloff, D.D. and Shukurov, A.M.: 1988, *Nature 336*, 341.

Schmitt, D. and Schüssler, M.: 1989, *Astron. Astrophys. 223*, 343.

Schüssler, M.: 1984, in *The Hydromagnetics of the Sun*, ESA SP-220 (Noordwijk), p. 67.

Schuster, H.G.: 1988, *Deterministic Chaos*, VHC Verlagsgesellschaft (Weinheim).

Skaley, D. and Stix, M.: 1991, *Astron. Astrophys. 241*, 227.

Sonett, C.P.: 1983, *Nature 306*, 670.

Spiegel, E.A. and Weiss, N.O.: 1980, *Nature 287*, 616.

Spruit, H.C., Title, A.M. and Van Ballegooijen, A.A.: 1987, *Solar Phys. 110*, 115.

Spruit, H.C., Nordlund, Å., and Title, A.M.: 1990, *Ann. Rev. Astron. Astrophys. 28*, 263.

Stenflo, J.O. and Vogel, M.: 1986, *Nature 319*, 285.

Stenflo, J.O.: 1991, in *The Sun and Cool Stars: activity, magnetism dynamos*, eds. I. Tuominen, D. Moss and G. Rüdiger, Springer-Verlag (Berlin), p. 193.

Stepinski, T.F. and Levy, E.H.: 1990, *Astrophys. J. 362*, 318.

Stix, M.: 1976a, in *Basic Mechanisms of Solar Activity*, eds. V. Bumba and J. Kleczek, Reidel (Dordrecht), p. 367.

Stix, M.: 1976b, *Astron. Astrophys. 47*, 243.

Stix, M.: 1981, *Solar Phys. 74*, 79.

Stix, M.: 1983, *Astron. Astrophys. 118*, 363.

Van Ballegooijen, A.A.: 1982, *Astron. Astrophys. 113*, 99.

Van Ballegooijen, A.A.: 1983, *Astron. Astrophys. 118*, 275.

Van Ballegooijen, A.A. and Choudhuri, A.R.: 1988, *Astrophys. J. 333*, 965.

Van Geffen, J.H.G.M.: 1992, Thesis, Univ. Utrecht.

Van Kampen, N.G.: 1974a, *Physica 74*, 215.

Van Kampen, N.G.: 1974b, *Physica 74*, 239.

Van Kampen, N.G.: 1976, *Phys. Reports 24C*, 171.

Wang, Y.-M., Nash, A.G. and Sheeley, N.R.: 1989, *Science 245*, 712.

Weiss, N.O., Cattaneo, F. and Jones, C.A.: 1984, *Geophys. Astrophys. Fluid Dynamics 30*, 305.

Yoshimura, H.: 1975, *Astrophys. J. 201*, 740.

Zel'dovich, Ya.B., Ruzmaikin, A.A., Molchanov, S.A. and Sokoloff, D.D.: 1984, *J. Fluid Mech. 144*, 1.

Zel'dovich, Ya.B., Ruzmaikin, A.A. and Sokoloff, D.D.: 1990, *The Almighty Chance*, World Scientific (Singapore).

Zwaan, C.: 1981, in *The Sun as a Star*, ed. S. Jordan, NASA SP-450, p. 163.

Zwaan, C.: 1987, *Ann. Rev. Astron. Astrophys. 25*, 83.

Zwaan, C.: 1992, in *Sunspots: Theory and Observations*, eds. J.H. Thomas and N.O. Weiss, Kluwer Academic Publishers (Dordrecht), to appear.

6 THE MHD DESCRIPTION OF COSMIC PLASMAS

K. TSINGANOS
*Department of Physics, University of Crete and
Research Center of Crete
GR-71409, Heraklion, Crete
Greece*

ABSTRACT. The main goal of this Chapter is to provide a physical understanding of ideal Magnetohydrodynamics. *First,* its governing equations are derived from the full set of Maxwell's equations coupled with a kinetic model of the fully ionized gas, described by Boltzmann's equation for each of the two species of the plasma, electrons and ions. *Second,* we discuss the main assumptions and the region of validity of ideal MHD, in particular in the context of astrophysical and space plasmas. And *third,* some basic concepts in MHD which are useful in understanding the gross physical properties of solar hydromagnetic structures and magnetic activity are briefly presented, such as, magnetic field stresses – pressure and tension – the freezing condition of the magnetic fieldlines with the fluid, magnetic buoyancy, the virial theorem, and the expansive property of the magnetic field.

1. Introductory Concepts

The dynamics of plasmas is an extremely rich and relatively complicated subject, owing to the complexity of motions which can be performed by large numbers of charged particles under their mutual electromagnetic interactions. Discussion of these motions in terms of particle orbit theory proves useful in certain situations; however, it is adequate only when the particle density is low enough for their interaction to be ignored. At the other end of the spectrum is the magnetohydrodynamic (MHD) approach which we discuss in this Chapter and whose value lies in the fact that it is the best and most practical starting point for an understanding of the macroscopic equilibrium and stability physical properties of astrophysical as well as laboratory plasmas. The MHD model essentially describes how inertial, magnetic, gravitational, and pressure gradient forces interact within a fluid when some approximations in this interaction and the transport coefficients of the fluid are valid. The main goals of this Chapter are then three-fold. *First,* to derive the set of the MHD equations starting from a kinetic model described by the Boltzmann-Maxwell system of equations. *Second,* to discuss the various assumptions required for the MHD model, providing also a physical understanding of the applicability of the theory. And *third,* to outline some of the basic physical concepts in the MHD framework which play a key role in understanding the gross properties of magnetized fluids. Towards those goals, and in order to fully appreciate the physical content of MHD, we shall derive step by step the governing equations, starting from basic principles and keeping the discussion at a level as simple as possible. More detailed discussions can be found in the literature (see for example, Braginski 1965, Boyd and Sanderson 1969, Krall and Trivelpiece 1973).

J. T. Schmelz and J. C. Brown (ed.), The Sun, A Laboratory for Astrophysics, 139–154.
© 1992 *Kluwer Academic Publishers. Printed in the Netherlands.*

1.1. DISTRIBUTION FUNCTION

Assume that the plasma is fully ionized and consists of two species: electrons and ions. Denote the velocity distribution function for each species by $f_a(\vec{r}, \vec{v}, t)$ and their mass and electric charge by m_a and q_a, respectively, with $a = e, i$. The velocity distribution function, which in general depends on the spatial coordinates, the velocity and time, gives the number of particles species a in unit volume $d\vec{r}d\vec{v}$ of phase space. Then, the number of particles species a in unit volume $d\vec{r}$ is

$$n_a(\vec{r}, t) = \int f_a(\vec{r}, \vec{v}, t)d\vec{v}. \tag{1}$$

The mean velocity of the particles species a is then

$$\vec{V_a} = <\vec{v}> = \frac{1}{n_a} \int \vec{v} f_a(\vec{r}, \vec{v}, t)d\vec{v}, \tag{2}$$

while the instantaneous velocity $(\vec{v})$ is the sum of the mean velocity $(\vec{V_a})$ and the random velocity $(\vec{w})$, $\vec{v} = \vec{V_a} + \vec{w}$. Of course, since $\vec{w}$ represents the random motion of the particles, $<\vec{w}> = 0$.

As an example, consider the *isotropic* Maxwell velocity distribution for a system in thermodynamic equilibrium,

$$f_M = n \left(\frac{m}{2\pi k_B T}\right)^{3/2} e^{-mv^2/2k_B T}, \quad F(v)dv = dv \int_\Omega f_M(\vec{r}, \vec{v})d\Omega, \tag{3}$$

$$4\pi v^2 dv f_M = F(v)dv = 4\pi n \left(\frac{m}{2\pi k_B T}\right)^{3/2} v^2 e^{-mv^2/2k_B T} dv. \tag{4}$$

Then, for such a Maxwellian distribution, the mean velocities are,

$$<v_x> = 0, \quad <v_x^2> = \frac{k_B T}{m}, \quad <v> = \frac{8}{\pi} \frac{k_B T}{m}, \quad <v^2> = 3\frac{k_B T}{m}. \tag{5}$$

1.2. COLLISIONAL KINETIC EQUATION

Since the distribution function $f_a = f_a(x, y, z, v_x, v_y, v_z, t)$ is a function of seven variables, we may write for its total differential,

$$\frac{df_a}{dt} = \frac{\partial f_a}{\partial t} + \frac{\partial f_a}{\partial x}\frac{dx}{dt} + \frac{\partial f_a}{\partial y}\frac{dy}{dt} + \frac{\partial f_a}{\partial z}\frac{dz}{dt} + \frac{\partial f_a}{\partial v_x}\frac{dv_x}{dt} + \frac{\partial f_a}{\partial v_y}\frac{dv_y}{dt} + \frac{\partial f_a}{\partial z}\frac{dv_z}{dt}. \tag{6}$$

To derive an expression for df_a/dt note that it consists of two terms, one due to collisions and the other to free streaming: $df_a/dt = (\partial f_a/dt)_{coll} + (\partial f_a/dt)_{stream}$. But, if one moves with the streaming velocity in phase space, then the density f_a is constant, *i.e.*, $(\partial f_a/dt)_{stream} = 0$. Hence,

$$\left(\frac{\partial f_a}{\partial t}\right)_{coll} = \frac{\partial f_a}{\partial t} + (\vec{v} \cdot \vec{\nabla})f_a + \left(\frac{\vec{F_a}}{m_a} \cdot \vec{\nabla}_{\vec{v}}\right) f_a, \tag{7}$$

where $\vec{F}_a$ is the total force on the particle species a. This is the known *collisional kinetic equation*. If there are no collisions such that $(\partial f_a/\partial t)_{coll} = 0$ or, if the plasma is hot enough such that collisions can be neglected, Equation (7) reduces to

$$\frac{\partial f_a}{\partial t} + \left(\vec{v} \cdot \vec{\nabla}\right) f_a + \left(\frac{\vec{F}_a}{m_a}\right) \cdot \vec{\nabla}_{\vec{v}} f_a = 0 , \qquad (8)$$

the known *collisionless kinetic equation, or Vlasov's Equation.*

1.3. MAXWELL–BOLTZMANN EQUATIONS

In order to solve the collisional kinetic equation, we need to write an explicit expression for $(\partial f_a/\partial t)_{coll}$ in terms of f_a and the other variables on the right side of Equation (7). In this case, where we have assumed that the plasma is fully ionized and consists of two species, electrons and ions, we have two kinds of forces acting on the charged particles:

(i) long-range electromagnetic Lorentz forces, $\vec{F}_a = q_a[\vec{E} + (\vec{v}/c) \times \vec{B}]$ with $\vec{E}$ and $\vec{B}$ the electromagnetic fields created by the averaged electric charge and electric current density distributions in the plasma, and

(ii) short-range elastic Coulomb collisions between i–e, i–i and e–e.

If the collision term $C \equiv (\partial f_a/\partial t)_{coll}$ in the collisional kinetic equation (7) is given on the basis of the two-body scattering theory (Boyd and Sanderson 1969), then

$$(\partial f_a/\partial t)_{coll} = \int [f_a(1')f_a(2') - f_a(1)f_a(2)] \mid \vec{v}_1 - \vec{v}_2 \mid \sigma(\mid \vec{v}_1 - \vec{v}_2 \mid, \theta) d\vec{v}_2 d\Omega , \qquad (9)$$

and we obtain the *Boltzmann kinetic equation*. On the other hand, if there are many particles in the Debye sphere and, therefore, the collisions are not short range and binary, *i.e.*, if multiple Coulomb collisions are taken into account, we obtain the *Fokker–Planck equation*, a much more complicated equation.

The *self-consistent fields* $\vec{E}, \vec{B}$ depend, of course, on f_a and their evaluation requires a self–consistent solution of the kinetic equation for each species of particles together with Maxwell's equations,

$$\vec{\nabla} \times \vec{E} = -\frac{1}{c}\frac{\partial \vec{B}}{\partial t} , \quad \vec{\nabla} \times \vec{B} = \frac{4\pi}{c}\vec{J} + \frac{1}{c}\frac{\partial \vec{E}}{\partial t} , \qquad (10)$$

$$\vec{\nabla} \cdot \vec{E} = 4\pi\delta , \quad \vec{\nabla} \cdot \vec{B} = 0 , \qquad (11)$$

$$\delta(\vec{r}, t) = \sum_a q_a \int f_a(\vec{r}, \vec{v}, t)d\vec{v} , \quad \vec{J}(r, t) = \sum_a q_a \int f_a(\vec{r}, \vec{v}, t)\vec{v}d\vec{v}. \qquad (12)$$

The full set of the Maxwell–Boltzmann equations gives a complete description of the behavior of the plasma, combining the microscopic with the macroscopic behavior of the interacting particles. The charged particles generate electromagnetic fields through their charges and currents. In order to evaluate these fields, it would be necessary to know the position and velocity of every particle at all times. The motions of the charges themselves must be followed in the fields they generate as well as to those externally applied. It is evident, however, that this program is impossible to carry out due to the great number of particles involved and the enormous wealth of information it contains; the complexity of

Equations (7-12) makes impossible – even numerically – to solve them for any non–trivial geometry. This has required the development of relatively simpler models and MHD is one of them. In other words, instead of specifying the state of the plasma in terms of the state of each of its particles, a more macroscopic description can be followed wherein the emphasis is on its fluid nature. Let us define then the following useful macroscopic quantities,

$$\overleftrightarrow{\Pi}_a = n_a m_a < \vec{w}\vec{w} > \; , \quad P_a = \frac{1}{3}n_a m_a < w^2 > \; , \quad T_a \equiv P_a/n_a \, , \tag{13}$$

which denote the total pressure tensor, scalar pressure and temperature, respectively; the heat flux due to random motion,

$$\vec{h}_a \equiv \frac{1}{2}m_a n_a < w^2 \vec{w} > \; ; \tag{14}$$

the rate of mean momentum transfer between e–i due to the friction of their collisions,

$$\vec{K}_a \equiv \int m_a \vec{w} \left(\frac{\partial f_a}{\partial t} \right)_{coll} d\vec{w} \, ; \tag{15}$$

and the heating rate due to collisions between e–i,

$$Q_a \equiv \int \frac{1}{2}m_a w^2 \left(\frac{\partial f_a}{\partial t} \right)_{coll} d\vec{w} \, . \tag{16}$$

1.4. MOMENTS OF THE BOLTZMANN EQUATION

In order to proceed to the derivation of the MHD equations, we shall need to take the appropriate moments of the Boltzmann equation. The moment equations are obtained from the collisional kinetic equations (7) and (9) through multiplying them by various appropriate functions of the velocity, $\mu_a(\vec{v})$, and then integrating in velocity space. It is interesting that the particular choices $\mu_a(\vec{v}) = m_a$, $\mu_a(\vec{v}) = m_a\vec{v}$, and $\mu_a(\vec{v}) = m_a v^2/2$, correspond to mass, momentum, and energy conservation, respectively. In any case, the procedure is straightforward although tedious for higher orders. It effectively replaces an equation for the distribution function $f_a(\vec{r}, \vec{v}, t)$, by equations for quantities that are functions of $\vec{r}$ and t only.

First, integrate Equation (7) in $\vec{v}$-space to obtain the 0th–order moment equations,

$$\int \frac{\partial f_a}{\partial t} d\vec{v} + \int (\vec{v} \cdot \vec{\nabla}) f_a d\vec{v} + \frac{q_a}{m_a} \int \left(\vec{E} + \frac{\vec{v}}{c} \times \vec{B} \right) \cdot \vec{\nabla}_{\vec{v}} f_a d\vec{v} = \int \left(\frac{\partial f_a}{\partial t} \right)_{coll} d\vec{v} \, , \tag{17}$$

$$\Longrightarrow \quad \frac{\partial n_a}{\partial t} + \vec{\nabla} \cdot n_a \vec{V}_a = 0 \, . \tag{18}$$

The above equations are also known as the continuity equations since they express conservation of the number of particles of each species.

Second, multiply Equation (7) by $(m_a \vec{v})$, and then integrate in $\vec{v}$–space to obtain the 1st–order moment equations,

$$m_a \int \frac{\partial f_a}{\partial t} \vec{v} d\vec{v} + m_a \int (\vec{v} \cdot \vec{\nabla}) f_a \vec{v} d\vec{v} + q_a \int \left(\vec{E} + \frac{\vec{v}}{c} \times \vec{B} \right) \cdot (\vec{\nabla}_{\vec{v}} f_a) \vec{v} d\vec{v}$$

$$= \int \left(\frac{\partial f_a}{\partial t}\right)_{coll} m_a \vec{v} d\vec{v} = \vec{K}_a , \tag{19}$$

$$\Longrightarrow \quad m_a \frac{\partial}{\partial t}(n_a \vec{V}_a) + \vec{\nabla} \cdot m_a n_a < \vec{v} \cdot \vec{v} > - q_a n_a \left(\vec{E} + \frac{\vec{V}_a}{c} \times \vec{B}\right) = \vec{K}_a , \tag{20}$$

with $\vec{K}_a$ denoting the rate of the mean momentum transfer between electron–ion collisions as defined before, Equation (15).

Finally, multiply Equation (7) by $(m_a v^2/2)$, and integrate in $\vec{v}$–space to obtain the 2nd–order moment equations,

$$\frac{m_a}{2} \int \frac{\partial f_a}{\partial t} v^2 d\vec{v} + \frac{m_a}{2} \int (\vec{v} \cdot \vec{\nabla}) f_a v^2 d\vec{v} + \frac{q_a}{2} \int \left(\vec{E} + \frac{\vec{v}}{c} \times \vec{B}\right) \cdot (\vec{\nabla}_{\vec{v}} f_a) v^2 d\vec{v}$$

$$= \frac{1}{2} \int m_a v^2 \left(\frac{\partial f_a}{\partial t}\right)_{coll} d\vec{v} = Q_a \tag{21}$$

$$\Longrightarrow \quad \frac{\partial}{\partial t}\left(\frac{1}{2} m_a n_a < v^2 >\right) + \vec{\nabla} \cdot \left(\frac{1}{2} m_a n_a < v^2 \vec{v} >\right) - q_a n_a \vec{V}_a \cdot \vec{E} = Q_a . \tag{22}$$

The right hand sides of the above Equations (17-22) represent the rates of change *due to collisions* of the number, momentum, and energy densities, within the unit volume element $d\vec{r}$. If the protons within $d\vec{r}$ are gaining momentum (energy) by collisions at a rate $\vec{K}_i(Q_i)$, the electrons must be losing momentum (energy) by collisions at a rate $\vec{K}_e = -\vec{K}_i$ ($Q_e = -Q_i$),

$$\vec{K}_i = \int m_i \vec{v} \left(\frac{\partial f_i}{\partial t}\right)_{coll} d\vec{v} = -\int m_e \vec{v} \left(\frac{\partial f_e}{\partial t}\right)_{coll} d\vec{v} = -\vec{K}_e . \tag{23}$$

Altogether then, the first three moment equations together with Maxwell's equations are

$$\frac{dn_a}{dt} + n_a \vec{\nabla} \cdot \vec{V}_a = 0 , \tag{24}$$

$$n_a m_a \frac{du_a}{dt} = q_a n_a [\vec{E} + \frac{\vec{V}_a}{c} \times \vec{B}] - \vec{\nabla} \cdot \overleftrightarrow{\Pi}_a + \vec{K}_a , \tag{25}$$

$$\frac{3n_a k_B}{2} \frac{dT_a}{dt} + \overleftrightarrow{\Pi}_a : \vec{\nabla} \vec{V}_a + \vec{\nabla} \cdot \vec{h}_a = Q_a , \tag{26}$$

$$\vec{\nabla} \times \vec{E} = -\frac{1}{c} \frac{\partial \vec{B}}{\partial t} , \quad \vec{\nabla} \times \vec{B} = \frac{4\pi}{c}(q_i n_i \vec{V}_i + q_e n_e \vec{V}_e) + \frac{1}{c} \frac{\partial \vec{E}}{\partial t} , \tag{27}$$

$$\vec{\nabla} \cdot \vec{E} = 4\pi(q_i n_i + q_e n_e) , \quad \vec{\nabla} \cdot \vec{B} = 0 . \tag{28}$$

Note that the comoving derivative $d \cdot /dt = \partial \cdot /\partial t + (\vec{V} \cdot \vec{\nabla})\cdot$ in the above equations represents the temporal changes within an element moving with velocity $\vec{V}$. It is important to also stress that Equations (24-28) obtained so far are exact and no approximation has been made, except in the Boltzmann collisional integral (9).

2. The General Form of Ohm's Law and the Electrical Conductivity

It is instructive to consider for a moment the general form of Ohm's law in magnetized plasmas. To that purpose, multiply the ion momentum equation

$$\frac{\partial}{\partial t}(n_i m_i \vec{V}_i) + \vec{\nabla} \cdot m_i n_i (\vec{V}_i \vec{V}_i) + \vec{\nabla} \cdot P_i - q_i n_i \left[\vec{E} + \frac{\vec{V}_i}{c} \times \vec{B} \right] = \vec{K}_i, \qquad (29)$$

by q_i/m_i, the electron momentum equation

$$\frac{\partial}{\partial t}(n_e m_e \vec{V}_e) + \vec{\nabla} \cdot m_e n_e (\vec{V}_e \vec{V}_e) + \vec{\nabla} \cdot P_e - q_e n_e \left[\vec{E} + \frac{\vec{V}_e}{c} \times \vec{B} \right] = -\vec{K}_i, \qquad (30)$$

by q_e/m_e and then add. In terms of the electric current $\vec{J} = n_i q_i \vec{V}_i + n_e q_e \vec{V}_e$, we obtain

$$\frac{\partial \vec{J}}{\partial t} + \vec{\nabla} \cdot (n_i q_i \vec{V}_i \vec{V}_i + n_e q_e \vec{V}_e \vec{V}_e) + \vec{\nabla} \cdot \left(\frac{q_i}{m_i} P_i + \frac{q_e}{m_e} P_e \right)$$

$$-e^2 \vec{E} \left(\frac{n_i}{m_i} + \frac{n_e}{m_e} \right) - \frac{e^2}{c} \left[\left(\frac{n_i}{m_i} \vec{V}_i + \frac{n_e}{m_e} \vec{V}_e \right) \times \vec{B} \right] = e\vec{K}_i \left(\frac{1}{m_i} + \frac{1}{m_e} \right) \qquad (31)$$

Assuming that $m_i \gg m_e$, $P_i \simeq P_e = P/2$, $n_i = n_e = n/2$ and $V_i \simeq V_e = \vec{V}$ we get

$$\frac{\partial \vec{J}}{\partial t} - \frac{e}{2m_e} \vec{\nabla} P - \frac{\rho e^2}{m_e m_i} \left(\vec{E} + \frac{\vec{V}}{c} \times \vec{B} \right) + \frac{e}{m_e c} (\vec{J} \times \vec{B}) = \frac{e}{m_e} \vec{K}_i. \qquad (32)$$

Let $\vec{K}_i \propto (\vec{V}_e - \vec{V}_i) = -m_i \vec{J}/\rho e$. Choosing the constant of proportionality such that

$$\vec{K}_i = -\frac{\rho e}{\sigma m_i} \vec{J}, \qquad (33)$$

the above equation becomes,

$$\frac{m_e m_i}{\rho e^2} \frac{\partial \vec{J}}{\partial t} = -\frac{m_i}{e \rho c} (\vec{J} \times \vec{B}) + \frac{m_i}{2 \rho e} \vec{\nabla} P + \vec{E} + \frac{\vec{V} \times \vec{B}}{c} - \frac{\vec{J}}{\sigma}. \qquad (34)$$

2.1. ORDER OF MAGNITUDE OF VARIOUS TERMS IN OHM'S LAW

The order of magnitude of each term in the previous equation (34) is respectively,

$$\left(\frac{\omega}{\omega_p} \right)^2 \left(\frac{c}{V} \right)^2 : \left(\frac{\omega}{\omega_p} \right) \left(\frac{\Omega_{ce}}{\omega_p} \right) \left(\frac{c}{V} \right)^2 : \left(\frac{\omega}{\Omega_{ci}} \right) \left(\frac{c_s}{V} \right)^2$$

$$: 1 : 1 : \left(\frac{\omega}{\omega_p} \right) \left(\frac{\nu_{coll}}{\omega_p} \right) \left(\frac{c}{V} \right)^2 \qquad (35)$$

where ω_p is the electron plasma frequency, Ω_{ci}, Ω_{ce}, the ion and electron gyrofrequencies, c_s the ordinary sound speed in the gas, and, ν_{coll}, the collision frequency between elastic Coulomb collisions of electrons and ions. In the following, we examine the conditions under which some terms in this equation can be neglected such that (34) obtains a simpler form.
• The term $\partial \vec{J}/\partial t$ may be neglected if $(\omega/\omega_p)^2 \ll (V/c)^2$, or, if $L_D \ll L$

- The Hall term $\vec{J} \times \vec{B}$ may be neglected if $\omega \Omega_{ce}/\omega_p^2 \ll (V/c)^2$.
- The pressure gradient term may be neglected if $\omega/\Omega_{ci} \ll (V/c_s)^2$.
  If the previous inequalities are valid we are left with the last three terms,

$$\vec{J} = \sigma \left(\vec{E} + \frac{\vec{V}}{c} \times \vec{B} \right), \qquad (36)$$

- Finally, the condition that $\sigma \to \infty$ implies that $\omega \nu_{coll}/\omega_p^2 \ll (V/c)^2$.

2.2. SPITZER ELECTRICAL CONDUCTIVITY

In the following, we present a simple argument for an order of magnitude calculation of the electrical conductivity in the plasma. If τ_{coll} is the typical time between successive collisions, the electrical conductivity is given in terms of τ_{coll},

$$J_x = nq\dot{x} = nq\ddot{x}\tau_{coll} = nq\frac{qE}{m_e}\tau_{coll} = \sigma E, \qquad (37)$$

$$\implies \qquad \sigma = \frac{ne^2}{m_e}\tau_{coll}. \qquad (38)$$

On the other hand, τ_{coll} can by calculated from

$$\lambda_{coll} n \sigma_{cross} = 1, \qquad (39)$$

where λ_{coll} is the mean free path between successive collisions and σ_{cross} is the Coulomb scattering cross-section. Then

$$\tau_{coll} = \lambda_{coll}/V = 1/(nV\sigma_{cross}). \qquad (40)$$

Next, σ_{cross} can be calculated from the physics and geometry of a typical elastic Coulomb scattering which, in the Coulomb electric field of the target, $F = e^2/r_o^2$, has an impact parameter r_o, lasts a time $\tau_o = r_o/V$, and changes the momentum of the particle by

$$\Delta p = m_e V = F\tau_o = \frac{e^2}{r_o V}. \qquad (41)$$

It follows that

$$\tau_o = \frac{e^2}{m_e V^2}, \qquad (42)$$

and the scattering cross-section is

$$\sigma_{cross} = \pi r_o^2 = \frac{\pi e^4}{m_e^2 V^4}. \qquad (43)$$

Then the collision time is

$$\tau_{coll} = \frac{1}{nV}\frac{m_e^2 V^4}{\pi e^4} = \frac{m_e^2 V^3}{n\pi e^4}, \qquad (44)$$

and the electrical conductivity

$$\sigma = \frac{ne^2}{m_e}\tau_{coll} = \frac{ne^2}{m_e}\frac{m_e^2 V^3}{n\pi e^4} = \frac{m_e V^3}{\pi e^2}. \qquad (45)$$

With an electron speed of the order of $V \sim \sqrt{k_B T / m_e}$,

$$\sigma \simeq \frac{m}{\pi e^2} \left(\frac{k_B T}{m_e} \right)^{3/2} = \frac{m_e}{\pi e^2} \left(\frac{k_B}{m_e} \right)^{3/2} T^{3/2}, \qquad (46)$$

$$\implies \quad \sigma \simeq 10^7 T^{3/2}. \qquad (47)$$

At a typical astrophysical high temperature setting where T is of the order of $T \sim 10^6$ K (as is the case in the solar corona, for example), we obtain $\sigma \simeq 10^{16}$ sec^{-1}. This is an extremely high value of the electrical conductivity comparable to that of copper.

3. The Various MHD Approximations

Equations (24-28) are not very useful because there is a closure problem involving moments of higher order: there is always one more variable than there are equations. In the following, we shall discuss the MHD model which represents an approximate solution to this problem. The approximations involved in MHD are that *first*, all the high frequency and short wavelength information is eliminated. This is easily satisfied when considering the macroscopic behavior of most cosmic plasmas. And *second*, to assume the plasma to be either collision dominated or, in a strong magnetic field. In this last case the gyroradius of the ionized particles is small enough. Then, we may introduce single-fluid variables and approximate the higher moments from standard transport theory.

3.1. NEGLECT OF DISPLACEMENT CURRENT APPROXIMATION

Consider first the fact that most astrophysical flows are extremely nonrelativistic, *i.e.*, $V/c \ll 1$ where V is a characteristic bulk flow speed and c is the speed of light. Let also $\omega = 1/\tau = V/L$ a characteristic frequency with L a characteristic length of the system. We shall also use the fact that in most astrophysical plasmas, the temperature is such that the electrical conductivity is extremely high, $\sigma \to \infty$, as we discussed in the previous section. Thus, in this case of high electrical conductivity, (in order of magnitude) $E \sim (V/c)B$ and therefore,

$$\frac{\partial E}{\partial t} \simeq \left(\frac{1}{\tau} \right) \left(\frac{VB}{c} \right) \simeq \frac{V}{c} \left(\frac{VB}{L} \right) = \left(\frac{V}{c} \right)^2 \left(\frac{cB}{L} \right). \qquad (48)$$

Then, from Ampere's law, the conduction current J_{cond} is of order cB/L, while the displacement current J_{displ} is

$$J_{displ} \simeq \left(\frac{V}{c} \right)^2 \left(\frac{cB}{L} \right) = \left(\frac{V}{c} \right)^2 J_{cond} \ll J_{cond}, \qquad (49)$$

$$c(\vec{\nabla} \times \vec{B}) = 4\pi \vec{J}_{cond} + \frac{\partial \vec{E}}{\partial t} \implies \frac{\partial \vec{E}}{\partial t} \ll c(\vec{\nabla} \times \vec{B}), \qquad (50)$$

i.e., the displacement current is negligible if $V \ll c$ and $\sigma \to \infty$. Note that the condition $V \ll c$ is equivalent to the condition $\omega L \ll c$, or, $\omega \ll c/L$. In other words, the displacement current can be neglected in Ampere's law only for low frequency variations. It is important to remember that due to this assumption MHD is not used to describe high frequency plasma phenomena.

3.2. Charge-Neutral Approximation

Consider next Gauss law,

$$\vec{\nabla} \cdot \vec{E} = 4\pi\delta = 4\pi e(n_i - n_e),$$ (51)

with the electric field $\vec{E}$ always in a fixed frame of reference. The polarization current, J_δ, is also negligible because it is again of order $(V/c)^2 J_{cond}$,

$$J_\delta = \delta V \simeq \frac{E}{L} V \simeq \left(\frac{V}{c}B\right)\frac{V}{L} = \frac{V^2}{c^2}\left(\frac{cB}{L}\right) \qquad \Longrightarrow \qquad J_\delta \ll J_{cond}.$$ (52)

It follows that, to a first approximation, $n_i \simeq n_e = n$. This is the so-called *charge–neutral approximation* where the plasma is approximately neutral and deviations from neutrality occur only within the Debye sphere. The Debye length L_D determines the size of the region where deviations from neutrality become important while plasma electrical oscillations are characterized by the plasma frequency ω_p,

$$L_D = \left(\frac{k_B T}{8\pi n e^2}\right)^{1/2}, \quad \omega_p = \left(\frac{4\pi n e^2}{m_e}\right)^{1/2} \quad \text{for electrons}.$$ (53)

Therefore, in systems where $L \gg L_D$ and $\omega \ll \omega_p$ the charge–neutral approximation and the neglect of the displacement current in Maxwell's equations are valid. To get a quantitative feeling of these regimes, note that the Debye length L_D and the plasma frequency ω_p depend only on the number density n (in particles per cm^3) and the temperature T (in Kelvin),

$$L_D \simeq 4.9 \left(\frac{T}{n}\right)^{1/2} = 0.49 \text{ cm} , \quad \omega_p \simeq 2\pi 10^4 (n)^{1/2} = 2\pi 10^8 \text{ Hz},$$ (54)

for $T \sim 10^6$ K, $n \sim 10^8$ cm^{-3}.

Another approximation is to neglect the electron inertia, $m_e/m_i \ll 1$. This means that we consider time–scales long compared to those of the electron plasma frequency ω_p and the electron cyclotron frequency $\omega_{ce} = eB/m_e$. Similarly, the length scales must be long compared to the Debye scale L_D and the electron cyclotron radius $r_{ce} = V_e/\omega_{ce}$.

3.3. The Fluid Approximation

Assume that both electrons and ions are collision-dominated. Then, a given particle remains reasonably close to its neighbouring particles during time scales of interest. In this case, we may divide the plasma into small, identifiable fluid elements. To be more quantitative, define a characteristic time scale $\tau \sim L/V_{Ti}$ where V_{Ti} is the thermal speed of the ions, $V_{Ti} \sim \sqrt{k_B T_i/m_i}$ such that we may approximate $\omega \sim \partial/\partial t \simeq V_{Ti}/L$ and $\omega/k \sim V \simeq V_{Ti}$, with k the wavenumber.

In MHD time scales of interest, each species has sufficiently many collisions to make the distribution function nearly Maxwellian. The ions collide predominantly with ions, in a time scale τ_{ii}, while the electrons collide either with electrons in a time scale τ_{ee}, or with ions in a time scale τ_{ei}. Assume that electrons and ions have a common mean free path, $\lambda_{coll} = V_{Ti}\tau_{ii} = V_{Te}\tau_{ee}$ and that $T_e \simeq T_i$. Then it follows that $\tau_{ee} \sim (m_e/m_i)^{1/2}\tau_{ii}$ and the condition that ions are collision dominated is

$$\omega\tau_{ii} \simeq \frac{V_{Ti}\tau_{ii}}{L} \ll 1 \, , \tag{55}$$

while electrons are collision-dominated for,

$$\omega\tau_{ee} \sim \frac{V_{Te}\tau_{ee}}{L} \sim \sqrt{\frac{m_e}{m_i}}\frac{V_{Ti}\tau_{ii}}{L} \ll 1 \, . \tag{56}$$

Also, in order that the two species are collision–dominated, we must have $L \gg \lambda_{coll}$, a condition which is satisfied, however, if Equation (55) holds. Altogether, the following condition needs to be satisfied in order to have a collision–dominated plasma of ions and electrons,

$$\frac{V_{Ti}\tau_{ii}}{L} \sim \frac{V_{Te}\tau_{ee}}{L} \ll 1 \, . \tag{57}$$

If the above condition is satisfied, the viscosity term appearing at the right hand side of the momentum equation for ions and electrons is negligible. Therefore, if the plasma is collision dominated the momentum equation reduces to that of ideal MHD. However, note that in order to write the ideal MHD version of Ohm's law $\vec{J}/\sigma \simeq \vec{E} + \vec{V}/c \times \vec{B}$ and $\vec{J}/\sigma \ll \vec{V}/c \times \vec{B}$, many collisions are required, but not so many that the plasma is dominated by resistive diffusion.

Altogether, the main conditions required for the fluid–description are,

$$L \gg \lambda_{coll} \gg r_{ce} \gg L_D \, , \tag{58}$$

i.e., that $L \gg \lambda_{coll}$, but λ_{coll} must be large enough so most of the particles remain within the fluid element.

In *hydrodynamics*, the above conditions are assured by collisions; the mean free path λ_{coll} for a collision-dominated fluid is $\lambda_{coll} \ll L$.

In *hydromagnetics*, with elastic Coulomb collisions (see Equation 43)

$$\lambda_{coll} = V\tau_{coll} = \frac{1}{n\sigma_{cross}} = \frac{m^2 V^4}{\pi e^4 n} \simeq 3 \times 10^{-12}\frac{V^4}{n \ln \Lambda} \, . \tag{59}$$

Then, for $V \simeq 20$ km sec^{-1}, $n \simeq 10^9$ cm^{-3}, and $\ln \Lambda \simeq 20$, $\lambda_{coll} \simeq 30$ m, i.e., a very small value for astrophysical scales. Thus, in astrophysical systems like hydromagnetic structures in the inner solar corona with typical length scales of the order $L \sim 10^4$ km or more, the plasma is collision–dominated and MHD gives a very good description of the macroscopic physical properties of those systems. On the other hand, systems like the outer solar corona and the interplanetary solar wind, where the density drops several orders of magnitudes, may be regarded as collisionless plasmas because $L \sim \lambda_{coll}$. However, in strong magnetic fields, the cyclotron radius of the protons can be very small. In those cases, the magnetic field assumes the role previously played by collisions and binds particles tightly to the lines of force so that, in the motion of the fluid element, transverse movements of the particles are constrained within this element. This is the so–called guiding center approximation. The plasma is 2/3 of the way towards behaving like a fluid in this case.

In summary, although the behavior of real plasmas is extremely complicated, some of their basic properties may be captured by relatively simple macroscopic equations. These equations describe adequately those slower plasma effects which occur over long enough times and have larger spatial scales such that collisions and gyrations may establish sufficient coherence in the plasma to enable it to "act" as a single fluid.

3.4. MHD EQUATIONS AT THE SINGLE-FLUID APPROXIMATION

When the previous assumptions are valid, we may start defining single-fluid variables, such as the number and mass densities n, ρ,

$$n_i \simeq n_e \equiv n \,, \quad \rho = m_i n_e + m_e n_e = n(m_i + m_e) \simeq nm_i \,; \tag{60}$$

the bulk flow speed $\vec{V}$, which (since the momentum of the fluid is carried mostly by ions) is,

$$\vec{V} = \vec{V_i} \,; \tag{61}$$

the current density $\vec{J}$, which is proportional to the velocity difference between ions and electrons,

$$\vec{J} = en(V_i - V_e) = en(V - V_e) \,; \tag{62}$$

and, finally, the total scalar pressure P and temperature T,

$$P = P_e + P_i \,, \quad T = T_e + T_i \,. \tag{63}$$

The above equations relate the single fluid variables $\rho, \vec{V}, \vec{J}, P$ and T to the two-fluid variables $n, \vec{V_e}, \vec{V_i}, P_e, P_i, T_e$ and T_i.

Multiplying the ion mass conservation Equation (24) by m_i we obtain the MHD continuity equation

$$\frac{\partial \rho}{\partial t} + \vec{\nabla} \cdot \rho \vec{V} = 0 \,, \tag{64}$$

while adding the momentum equation for electrons and ions, Equations (25), we get

$$\rho \frac{\partial \vec{V}}{\partial t} + (\vec{V} \cdot \nabla)\vec{V} + \vec{\nabla}P - \frac{\vec{J} \times \vec{B}}{c} = - \vec{\nabla} \cdot [\overset{\leftrightarrow}{\Pi}_i{}^{visc} - \overset{\leftrightarrow}{\Pi}_e{}^{visc}] \,. \tag{65}$$

The lack of spherical symmetry of the pressure tensor $\overset{\leftrightarrow}{\Pi}_a$ is related to the frequency of collisions, or, the viscosity of the fluid. Therefore, the pressure tensor $\overset{\leftrightarrow}{\Pi}_a$ is usually broken into two parts, one traceless $\overset{\leftrightarrow}{\Pi}_a{}^{visc}$, the other diagonal $P_a \overset{\leftrightarrow}{I}$, i.e., $\overset{\leftrightarrow}{\Pi}_a = \overset{\leftrightarrow}{\Pi}_a{}^{visc} + P_a \overset{\leftrightarrow}{I}$. Then, if particle collisions are sufficiently frequent, i.e., if condition (57) is satisfied,

$$\frac{\overset{\leftrightarrow}{\nabla} \cdot \overset{\leftrightarrow}{\Pi}{}^{visc}}{\vec{\nabla}P} \sim \frac{V_{Ti}\tau_{ii}}{L} \ll 1 \,, \tag{66}$$

the viscosity term appearing at the right hand side of Equation (65) is negligible and the momentum equation reduces to that of ideal MHD. In this frequently used approximation the pressure term reduces to the gradient of a scalar (that is, $\vec{\nabla} \cdot \overleftrightarrow{\Pi}_a \to \vec{\nabla} P_a$).

Finally, for completeness, we should include the energy Equations (26) for electrons and ions which can be written,

$$\frac{d}{dt}\left(\frac{P_i}{\rho^\gamma}\right) = \frac{2}{3\rho^\gamma}(Q_i - \vec{\nabla} \cdot \vec{h}_i - \overleftrightarrow{\Pi}_i^{visc} : \vec{V}),\tag{67}$$

$$\frac{d}{dt}\left(\frac{P_e}{\rho^\gamma}\right) = \frac{2}{3\rho^\gamma}\left\{Q_e - \vec{\nabla} \cdot \vec{h}_e - \overleftrightarrow{\Pi}_e^{visc} : \left(\vec{V} - \frac{\vec{J}}{en}\right) + \frac{1}{en}\vec{J} \cdot \vec{\nabla}\left[\frac{P_e}{\rho^\gamma}\right]\right\}\tag{68}$$

In the single-fluid description, the two right hand sides of the energy equations (67-68) are neglected and we have $P \propto \rho^\gamma$. However, we may relax the constraints required to obtain this polytropic equation of state, if conserve energy by simply writing the first law of thermodynamics. We shall be using this approach in discussing MHD solutions appropriate for astrophysical outflows in Chapter 16.

4. The Frozen-in Approximation

Taking the curl of Equation (55) which gives the electric current density in the plasma and substituting the electric field from the induction Equation (10), we obtain

$$\frac{\partial \vec{B}}{\partial t} = \vec{\nabla} \times (\vec{V} \times \vec{B}) + \frac{1}{4\pi\sigma}\nabla^2 \vec{B}.\tag{69}$$

This induction equation describes how the magnetic field at a fixed point in space changes in time within the magnetized fluid of spatially uniform electrical conductivity σ, for a given velocity field $\vec{V}$ (Chapter 5).

The terms on the right hand side of (69) are of the order

$$|(\vec{\nabla} \times \vec{V}) \times \vec{B}| \sim \frac{VB}{L},\tag{70}$$

and

$$|\frac{1}{4\pi\sigma}\nabla^2 \vec{B}| \sim \frac{B}{4\pi\sigma L^2}.\tag{71}$$

It follows that the first term in (69) dominates if

$$\frac{B}{4\pi\sigma L^2} \ll \frac{VB}{L},\tag{72}$$

or,

$$R_m \equiv 4\pi\sigma V L \gg 1,\tag{73}$$

where R_m is the magnetic Reynolds number. Thus, for large scales L, large values of the electrical conductivity σ, and/or large velocities V, the diffusion of the magnetic field lines relative to the plasma can be neglected. In most astrophysical circumstances, the magnetic Reynolds number is very large. In the solar wind ($V \sim 400$ km sec^{-1}, $\sigma \simeq 10^{14}$ sec^{-1}), $R_m = 6\times 10^{14}$ across a distance of one Astronomical Unit. In the gaseous disk of the galaxy ($L \simeq$

100 pc, $V \simeq 10$ kms^1, $\sigma \simeq 10^{11}$ sec^{-1}), $R_m \simeq 10^{17}$. Even in the low electrical conductivity of the solar photosphere ($\sigma \simeq 10^{11}$ sec^{-1}) and for the slow convective velocities of $V \simeq 1$ kms^{-1} of granules ($L \simeq 1000$ km), $R_m \simeq 10^4$. In such cases (69) simplifies to

$$\frac{\partial \vec{B}}{\partial t} = \vec{\nabla} \times (\vec{V} \times \vec{B}). \tag{74}$$

This last equation has a simple physical interpretation which may be seen as follows. Consider an open surface S in the volume of the fluid bounded by a material contour C consisting of "marked" particles. At time t, the flux through S is

$$\Phi(t) = \int_S \vec{B}(\vec{r}, t) d\vec{S}. \tag{75}$$

Does the magnetic flux through this material contour C change due to the plasma motion? After some time δt, the original material contour C has been deformed to the new shape C' and has been transported to a new position. The change in flux during this motion from the shape C to the new shape C' in the time interval δt is

$$\delta \Phi = \int_S \frac{\partial \vec{B}}{\partial t} \cdot d\vec{S} \delta t + \int_C [(\vec{V} \delta t) \times d\vec{\ell}] \cdot \vec{B} = \delta \Phi_1 + \delta \Phi_2, \tag{76}$$

where $d\vec{\ell}$ is the infinitesimal length along C, $\delta \Phi_1$ represents the change of flux through S due to the temporal change of $\vec{B}$ only, while $\delta \Phi_2$ represents the change due to the escaping flux through the "walls" of the surface which the contour of the particles sweeps in moving from the initial configuration at C to the final at C'. Using Stokes' theorem, $\delta \Phi_2$ can be written

$$\delta \Phi_2 = \int_C (\vec{B} \times \vec{V}) \cdot d\vec{\ell} \delta t = -\int_S \vec{\nabla} \times (\vec{V} \times \vec{B}) \cdot d\vec{S} \delta t. \tag{77}$$

Adding $\delta \Phi_1$ and $\delta \Phi_2$ we obtain

$$\frac{\delta \Phi}{\delta t} = \int_S \left(\frac{\partial \vec{B}}{\partial t} - \vec{\nabla} \times (\vec{V} \times \vec{B}) \right) \cdot d\vec{S} = 0, \tag{78}$$

in view of Equation (74). It follows that the magnetic flux through a closed material contour moving with the fluid remains constant, in the limit of infinitely large electrical conductivity.

5. Magnetic Pressure, Tension and Buoyancy

The familiar electromagnetic stress tensor of an electromagnetic field

$$M_{ij} = -\delta_{ij} \left(\frac{E^2 + B^2}{8\pi} \right) + \left(\frac{E_i E_j + B_i B_j}{4\pi} \right), \tag{79}$$

reduces in the MHD limit to the Maxwell stress tensor

$$M_{ij} = -\delta_{ij} \frac{B^2}{8\pi} + \frac{B_i B_j}{4\pi}. \tag{80}$$

The magnetic forces, then, in a magnetofluid, $\partial M_{ij}/\partial x_j$, are the isotropic pressure, $B^2/8\pi$, in all three directions and a tension, $B^2/4\pi$, along the lines of force. The following simple argument helps us to understand the physical origin of this magnetic pressure. Consider a uniform magnetic field $\vec{B} = B_o \hat{z}$ in the half–space $y > 0$ with $\vec{B} = 0$ in $y < 0$ and $\vec{B} = (B_o/2)\hat{z}$, for continuity, in $y = 0$. From Ampere's law (10) (neglecting the displacement current), the associated electric current density has the form of a current sheet along $\hat{x}$ with the current δI_x per unit of length δz,

$$\frac{\delta I_x(y = 0)}{\delta z} = \frac{cB_o}{4\pi} . \tag{81}$$

¿From Laplace's law, this electric current $\delta I_x(y = 0)\hat{x}$ together with the magnetic field $\vec{B}(y = 0) = (B_o/2)\hat{z}$ exert a force $\delta\vec{F} = \delta F_y\hat{y}$ on the wall $y = 0$,

$$\delta F_y = -\frac{\delta x \delta I_x(y = 0)B_z(y = 0)}{c} , \tag{82}$$

while the magnetic force per unit area $\delta x \delta z$ on the surface $y = 0$ is

$$P_m = \frac{\delta F_y}{\delta x \delta z} = -\frac{B_o}{4\pi}\frac{B_o}{2} = -\frac{B_o^2}{8\pi} . \tag{83}$$

In other words, the magnetic field exerts a pressure on the surface $y = 0$ directed towards the half–space where $\vec{B} = 0$. Similar arguments can show that a magnetic flux tube behaves like a rubber band under tension. Equilibrium exists only when it is possible to balance the tension and pressure against each other. This magnetic stress M_{ij} is the cause of most of the activity observed in astrophysics because the magnetic field exerts strong forces on the fluid.

A direct consequence of the pressure of the magnetic field immersed in a gravitational field of an object is its *magnetic buoyancy*. Consider then a long magnetic flux tube with horizontal field lines which is immersed in a vertical gravitational field, (Parker 1979). Assuming that the field is in thermal equilibrium with the ambient gas (at a uniform temperature T_o), the condition of pressure equilibrium across the walls of the flux tube,

$$\vec{\nabla} P_g + \vec{\nabla} P_m = 0 , \qquad P_g = \frac{2k_B}{m_p}\rho T_o , \tag{84}$$

gives

$$\frac{2k_B}{m_p}T_o[\rho_{out} - \rho_{in}] = \frac{B^2}{8\pi} > 0 . \tag{85}$$

The inevitable result is that $\rho_{in} < \rho_{out}$ and the flux tube is buoyant. This basic property is responsible for the buoyant escape of the magnetic field from the solar/stellar interior where it is generated by the dynamo action (Chapter 5).

There is a point worth to mention quickly here on the basic assumption implicit in deriving Eq. (85). Evidently, Eq. (84) from which (85) follows, holds only when there is no tension. In other words, we have assumed that the magnetic field lines are stretched tight. For example, in a vacuum magnetic field (e.g. a dipole) where $P_g = 0$ we never have $\vec{\nabla} P_m = 0$, unless $\vec{B} = 0$ (Brown 1992).

6. The Virial Theorem and the Expansive Property of the Magnetic Field

To demonstrate the expansive property of the magnetic fields, consider a magnetized fluid with density ρ and pressure P subject to the magnetic stresses M_{ij} and a general external force $\mathcal{F}_i$ such that the momentum equation is

$$\rho \frac{\partial V_i}{\partial t} + \rho V_j \frac{\partial V_i}{\partial x_j} = -\frac{\partial P}{\partial x_i} + \frac{\partial M_{ij}}{\partial x_j} + \mathcal{F}_i \qquad (86)$$

and the continuity equation is

$$V_i \frac{\partial \rho}{\partial t} + V_i \frac{\partial}{\partial x_j} \rho V_j = 0 . \qquad (87)$$

Multiply the ith-component of Equation (86) by x_k

$$x_k \times \left| \frac{\partial}{\partial t} \rho V_i + \frac{\partial}{\partial x_j} \rho V_i V_j = -\frac{\partial P}{\partial x_i} + \frac{\partial M_{ij}}{\partial x_j} + \mathcal{F}_i , \right. \qquad (88)$$

and add to the kth-component of Equation (86) multiplied by x_i,

$$x_i \times \left| \frac{\partial}{\partial t} \rho V_k + \frac{\partial}{\partial x_j} \rho V_k V_j = -\frac{\partial P}{\partial x_k} + \frac{\partial M_{kj}}{\partial x_j} + \mathcal{F}_k . \right. \qquad (89)$$

The result is

$$\frac{\partial}{\partial t} \int_V \rho(x_i V_k - x_k V_i) d\vec{r} + \int_V x_k \left[\frac{\partial}{\partial x_j} \rho V_i V_j + x_i \frac{\partial}{\partial x_j} \rho V_k V_j \right] d\vec{r} =$$

$$\int_V \left[x_k \frac{\partial P}{\partial x_i} + x_i \frac{\partial P}{\partial x_k} \right] d\vec{r} + \int_V \left[x_k \frac{\partial M_{kj}}{\partial x_j} + x_i \frac{\partial M_{ij}}{\partial x_j} \right] d\vec{r} + \int_V [x_k \mathcal{F}_k + x_i \mathcal{F}_i] d\vec{r} . \qquad (90)$$

Define the inertia and kinetic tensors

$$I_{ij} \equiv \int_V d\vec{r} \rho x_i x_j , \quad T_{ij} \equiv \int_V d\vec{r} \frac{\rho V_i V_j}{2} . \qquad (91)$$

Then,

$$\frac{dI_{ij}}{dt} = \int_V d\vec{r} \rho(x_k V_i + x_i V_k) - \int_V d\vec{r} \frac{\partial}{\partial x_j} x_i x_k \rho V_j . \qquad (92)$$

Assuming that there is no fluid motion across the surface S of the volume V enclosing the fluid, apply Gauss theorem and obtain,

$$\frac{dI_{ik}}{dt} = \int_V d\vec{r} \rho(x_k V_i + x_i V_k) \implies \frac{d}{dt} \int_V d\vec{r} \rho(x_i V_k + x_k V_i) = \frac{d^2 I_{ik}}{dt^2} , \qquad (93)$$

$$\int_V d\vec{r} \left[x_k \frac{\partial}{\partial x_k} \rho V_i V_j + x_i \frac{\partial}{\partial x_j} \rho V_i V_k \right] = -4T_{ik} , \qquad (94)$$

$$\frac{d^2 I_{ik}}{dt^2} - 4T_{ik} = \int_V d\vec{r} [2\delta_{ik} P - 2M_{ik} + x_k \mathcal{F}_i + x_i \mathcal{F}_k]$$

$$+ \int_S dS_j [-P(\delta_{ij} x_k + \delta_{kj} x_i) + x_k M_{ij} + x_i M_{kj}] . \qquad (95)$$

This is the classical tensor virial equation (Parker 1979). Assuming an isolated system for which all motions and fields vanish on the enclosing surface S, set $i = k$ and summing over i,

$$\frac{1}{2}\frac{d^2I}{dt^2} = 2T + 3\int_V d\vec{r}P + \int_V d\vec{r}\frac{B^2}{8\pi} + \int_V d\vec{r}x_i\mathcal{F}_i\,, \tag{96}$$

where $I = I_i i$, $T = T_{ii}$, $M_{ii} = -\dfrac{B^2}{8\pi}$. The volume integral of the fluid pressure is positive and, for an ideal gas, equals twice its total thermal kinetic energy T_t,

$$3\int d\vec{r}P = \int_V d\vec{r}\rho < w^2 >= 2T_t\,, \tag{97}$$

while the volume integral of $B^2/4\pi$ is the total magnetic energy of the system M,

$$\int_V d\vec{r}\frac{B^2}{8\pi} = M\,. \tag{98}$$

We have then

$$\frac{1}{2}\frac{d^2I}{dt^2} = 2(T + T_t) + M + \int_V d\vec{r}x_i\mathcal{F}_i\,. \tag{99}$$

It is evident that since $T, T_t, M > 0$, all these terms contribute to expansion, $d^2I/dt^2 > 0$. External forces are needed in order that $d^2I/dt^2 = 0$. The magnetic field cannot contain its own stresses and is *expansive*.

In summary, the framework of the MHD theory is clear cut to provide us with confidence to work out answers to the several questions that remain unanswered, until eventually we obtain a satisfactory picture. In the meantime however, it is a fruitful area of academic research.

Acknowledgements. I am greatly indebted to the referee of this Chapter, Dr. Peter Hoyng, and the editors of the Volume, Prof. John Brown and Dr. Joan Schmelz for reading the manuscript and making several suggestions that led to a gradual improvement of the presentation. The research reported in this Chapter was partially supported by NATO grant 870221.

7. References

Boyd, T.J.M., and Sanderson, J.J. 1969, *Plasma Dynamics*, Thomas Nelson LTD, Melbourne.

Braginski, S.I. 1965, *Rev. of Plasma Phys.*, 1, 605.

Brown, J.C. 1992, (private communication).

Freiberg, J.P. 1982, *Rev. Mod. Phys.*, 65, No 3, 801.

Parker, E.N. 1979, *Cosmical Magnetic Fields*, Cambridge University Press, Oxford.

Krall, N.A., and Trivelpiece, A.W. 1973, *Principles of Plasma Physics*, McGraw–Hill Inc, New York.

7 SYMMETRIC AND NONSYMMETRIC MHD EQUILIBRIA

K. TSINGANOS
*Department of Physics, University of Crete and
Research Center of Crete
GR-71409, Heraklion, Crete
Greece*

ABSTRACT. The basic equations of Magnetohydrodynamics appropriate for an inviscid fluid of high electrical conductivity are considered for the case that the physical quantities are independent of one Cartesian coordinate, the angle ϕ in cylindrical coordinates or, in general for the case that there is one ignorable coordinate in a general curvilinear coordinate system. Several integrals are established for the case of steady (time-independent) equilibrium and the set of nine coupled differential equations is reduced to a pair of nonlinear equations for the magnetic flux density and the plasma density. This result allows a unified and systematic approach to the solution of problems involving steady hydromagnetic fields with a topological invariance in various curvilinear coordinates. Examples of the use of this method to construct such symmetric MHD equilibria are given in Chapter 16. Next, the question of the existence of nonsymmetric MHD equilibria is investigated. The Taylor-Proudman and Parker theorems, in hydrostatics and magnetostatics, respectively, are discussed, together with their equivalent in MHD. We explore the formal analogy between the field lines of symmetric equilibria and the surfaces of section of integrable Hamiltonian systems, concluding that spatially symmetric ideal MHD equilibria are topologically unstable to finite amplitude perturbations that do not share their symmetry properties. The resulting ergodicity of perturbed symmetric equilibria may have important consequences in the transport properties of magnetically dominated plasmas. It may be conjectured that this topological nonequilibrium is the basis for the continuous activity of the variable magnetic and velocity fields in astrophysics.

1. MHD Equations

As discussed in the preceding Chapter 6, under the conditions that the Debye length L_D, the cyclotron radius r_c, and the mean free path λ_{coll} are small compared to the characteristic scale L of a system, $L \gg \lambda_{coll} \gg r_c \gg L_D$, the plasma may be treated as a fluid. In many circumstances, the bulk flow speed V is non-relativistic, $V/c \ll 1$, the electrical conductivity of the gas extremely high, and the viscosity is low. Then, the governing equations of motion are the familiar equations of ideal magnetohydrodynamics (MHD) that describe the dynamical interaction of inviscid, compressible fluids of low electrical resistivity with magnetic and gravity fields,

$$\frac{\partial \rho}{\partial t} + \vec{\nabla} \cdot (\rho \vec{V}) = 0 \,, \tag{1a}$$

$$\vec{\nabla} \cdot \vec{B} = 0 \,, \tag{1b}$$

$$\frac{\partial \vec{B}}{\partial t} + \vec{\nabla} \times (\vec{V} \times \vec{B}) = 0 \,, \tag{1c}$$

J. T. Schmelz and J. C. Brown (ed.), The Sun, A Laboratory for Astrophysics, 155–172.

$$\rho \frac{\partial \vec{V}}{\partial t} + \rho(\vec{V} \cdot \vec{\nabla})\vec{V} = -\vec{\nabla}P + \frac{(\vec{\nabla} \times \vec{B}) \times \vec{B}}{4\pi} - \rho\vec{\nabla}\mathcal{V}, \tag{1d}$$

$$T = \frac{m_p}{2k_B} \frac{P}{\rho}, \tag{1e}$$

where $\vec{B}$, $\vec{V}$, $-\vec{\nabla}\mathcal{V}$ denote the magnetic, velocity and external gravity fields, respectively. The above system of equations may be closed by adding a detailed energy balance equation and Poisson's law for gravity. In most applications, however, the gravitational potential is given *a priori* (*i.e.*, self-gravitation of the plasma is neglected), a polytropic equation of state is assumed, and attention is focused on the restricted kinematic problem posed by the set of the nonlinear and coupled nine differential equations (1) for $\vec{B}$, $\vec{V}$, P, ρ, T.

The properties and general behavior of the solutions of the set (1) are not known due to its mathematical complexity. Thus, the time-dependent problem posed by equations (1) is formidable and no progress has been made so far towards its solution. On the other hand, the time-independent problem posed by equations (1) with $\partial \cdot /\partial t = 0$, is also equally difficult to solve, as it can be witnessed by the fact that no exact solution of (1) is available, with the exception of the trivial equipartition solution $\vec{V} = \vec{B}/(4\pi\rho)^{1/2}$, $\rho = \rho_o$ and solutions for symmetric configurations. In the following we shall discuss a method to generate families of symmetric solutions of the time-independent equations (1) and also investigate the question of the existence of nonsymmetric solutions of (1). Once these analytic equilibrium solutions are available, the unexplored so far question of their stability can be investigated, since the solutions are in closed forms.

2. Symmetric MHD Equilibria, $\partial \cdot /\partial t = 0$, $\partial/\partial x_3 = 0$

Equilibrium states that satisfy equations (1) with $\partial \cdot /\partial t = 0$ may be used to study the macroscopic behavior of laboratory fusion plasmas (Freidberg 1982), or astrophysical plasmas such as solar coronal loops and prominences, stellar winds, jets, etc. In this section, we shall develop a systematic method for constructing symmetric MHD equilibria. However, before proceeding with this construction, let us simply note that the requirement of equilibrium alone excludes space filling ergodic fieldlines; this property may formally be seen from the integral of the time-independent equation (1c), *i.e.*,

$$\vec{V} \times \vec{B} = \vec{\nabla}\Phi, \tag{2}$$

where Φ is the induction potential. An arbitrary hydromagnetic system in equilibrium should satisfy equation (2). However, this equation alone requires that magnetic lines and streamlines cannot be ergodic in 3-D space, since they are constrained to lie on 2-D surfaces. This follows immediately from Equation (2),

$$\vec{V} \cdot \vec{\nabla}\Phi = \vec{B} \cdot \vec{\nabla}\Phi = 0, \tag{3}$$

i.e., magnetic fieldlines, together with streamlines are constrained to lie on the surfaces Φ = constant. In the next section, we shall give the exact scalar differential equation that will define these surfaces in case the hydromagnetic system has a symmetry expressible through the existence of an ignorable coordinate.

2.1. THE THREE CLASSES OF SYMMETRIC MHD EQUILIBRIA

As is well known, to every symmetry of a physical system there corresponds a conservation law (Noether's theorem). It is therefore natural to start our study of hydromagnetics by seeking to isolate the conserved quantities of an MHD equilibrium system possessing some symmetry. On the other hand, it is also well known that there are three types of macroscopically symmetric physical systems in Euclidean space, namely those which are translationally symmetric, those which are rotationally symmetric and finally systems which are helically symmetric (a combination of translational and rotational symmetry).

Formally, a symmetry in a generalized nonorthogonal curvilinear coordinate system (x_1, x_2, x_3) can be revealed by a coordinate independence of the corresponding metric coefficients g_{ij} $(i, j = 1, 2, 3)$ (Edenstrasser 1980a,b),

$$\frac{\partial g_{ij}}{\partial x_3} = 0. \tag{4}$$

Then, in Euclidean space the corresponding unit vector $\vec{e}_3$ is a Killing vector and may be written in the form

$$\vec{e}_3 = \vec{a} + \vec{b} \times \vec{r}. \tag{5}$$

We have then three, and only three possibilities:

$\vec{a} = 0,$ corresponding to rotational symmetry,
$\vec{b} = 0,$ corresponding to translational symmetry and,
$\vec{a} \neq 0, \vec{b} \neq 0,$ corresponding to helical symmetry.

Thus, translational symmetry may be revealed by an independence of the physical quantities on the coordinate z, in the Cartesian coordinates $(x_1 = x, x_2 = y, x_3 = z)$ with $(h_1 = h_2 = h_3 = 1)$, or, cylindrical coordinates $(x_1 = R, x_2 = \phi, x_3 = z)$ with $(h_1 = 1, h_2 = R, h_3 = 1)$, etc. Rotational symmetry on the other hand may be revealed by an independence of the physical quantities on the azimuthal angle ϕ in cylindrical coordinates $(x_1 = z, x_2 = R, x_3 = \phi)$ with $(h_1 = 1, h_2 = 1, h_3 = R)$, or, spherical coordinates $(x_1 = r, x_2 = \theta, x_3 = \phi)$ with $(h_1 = 1, h_2 = r, h_3 = r\sin\theta)$, or, toroidal coordinates $(x_1 = u, x_2 = v, x_3 = \phi)$ with $(h_1 = h_2 = a/\sqrt{\cosh v - \cos u}, h_3 = a\sinh v/\sqrt{\cosh v - \cos u})$, or, oblate spheroidal coordinates $(x_1 = \xi, x_2 = \eta, x_3 = \phi)$ with $(h_1 = h_2 = a\sqrt{\cosh^2 \xi + \sin^2 \eta}, h_3 = a\cosh\xi\cos\eta)$, etc. Apparently the particular configuration of the system under study will indicate and the choice of the appropriate system of coordinates. For example, the toroidal geometries of tokamak devices indicate the use of toroidal coordinates (u, v, ϕ), while wind-type outflows, the use of spherical coordinates (r, θ, ϕ).

2.2. INTEGRALS OF SYMMETRIC MHD EQUILIBRIA

Consider next the integrals of the steady hydromagnetic equations (1) in such an orthogonal curvilinear coordinate system (x_1, x_2, x_3) with line elements $h_i(x_1, x_2)$ $i = 1, 2, 3$. Note first that an integration of the partial-derivatives differential equations (1) leads to arbitrary functions of the two relevant coordinates (x_1, x_2). Thus, we may integrate the mass and magnetic flux conservation laws (1a)-(1b) and define the magnetic and velocity fields on the plane $x_3 = $ constant through the magnetic potential $A(x_1, x_2)$ and the stream function $\Psi(x_1, x_2)$, respectively. Then, we may integrate the induction equation (1c) and the x_3-component of the momentum balance equation (1d), and introduce two more arbitrary functions, the induction potential function $\Phi(x_1, x_2)$, Eq. (2), and the total angular momentum function $L(x_1, x_2)$, respectively. The most interesting result however is that the corresponding surfaces, i.e., magnetic surfaces $A(x_1, x_2) = \text{const.}$, stream surfaces $\Psi(x_1, x_2) = \text{const.}$, equipotential surfaces $\Phi(x_1, x_2) = \text{const.}$, and total angular momentum surfaces $L(x_1, x_2) = \text{const.}$, all coincide, i.e.,

$$\Psi(x_1, x_2) = \Psi(A), \quad \Phi(x_1, x_2) = \Phi(A), \quad L(x_1, x_2) = L(A), \tag{6}$$

where the functions $\Psi(A)$, $\Phi(A)$, and $L(A)$ are arbitrary apart from some relationships among them that arise at the fixed points of the system of the original differential equations (1). For example, at the fixed point where the fields on the plane $x_3 = \text{constant}$ are related through the equipartition of energy relationship $\Psi_A^2 = 4\pi\rho$ (the familiar poloidal Alfvénic critical point), we require that

$$L = h_3^2 \Phi_A \equiv h_3^2 \Omega, \tag{7}$$

where the subscript A indicates derivative with respect to A and we have introduced the function $\Omega(A) \equiv \Phi_A(A)$ which has the dimensions of angular velocity.

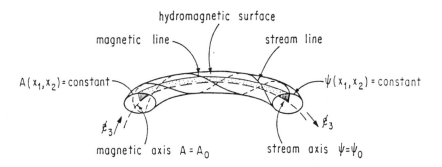

Figure 1. Schematic drawing of a toroidal hydromagnetic surface. The fields are independent of the coordinate x_3.

Using the above results we then obtain the following general expressions for the two fields, $\vec{B}$ and $\vec{V}$, passing over some algebraic calculations, (Tsinganos 1982a),

$$\vec{B}(x_1, x_2) = \vec{\nabla} A(x_1, x_2) \times \vec{\nabla} x_3 + \vec{\nabla} x_3 \frac{h_3^2 \Omega \Psi_A - L\Psi_A}{1 - \Psi_A^2/4\pi\rho}, \tag{8}$$

$$\vec{V}(x_1, x_2) = \frac{\Psi_A}{4\pi\rho} \vec{\nabla} A(x_1, x_2) \times \vec{\nabla} x_3 + \vec{\nabla} x_3 \frac{h_3^2 \Omega - L\Psi_A^2/4\pi\rho}{1 - \Psi_A^2/4\pi\rho}, \tag{9}$$

where $h_3 \vec{\nabla} x_3 = \vec{e}_3$ is the unit-vector corresponding to the ignorable coordinate x_3.

For each set of the three integral–functions $\Psi(A)$, $\Omega(A)$ and $L(A)$, the fields $\vec{B}$ and $\vec{V}$ are given in terms of the magnetic flux function $A(x_1, x_2)$ and the density $\rho(x_1, x_2)$. On the other hand, these two functions are related through

(i) the assumed equation for energy transport, or, equation of state and

(ii) the remaining two components of the momentum balance equation (1d).

In the following we discuss separately those two issues.

2.3. ENERGETICS AND EQUATIONS OF STATE IN SYMMETRIC MHD EQUILIBRIA

There are several ways in which we may close (in the usual formal sense) the system of equations (1). For each such way, there exists one more integral which is related to the energetics of the hydromagnetic system (Tsinganos 1981, 1982a). In the following we briefly outline the various equations of state, that introduce a corresponding fourth integral of equations (1).

(1) An incompressible atmosphere everywhere $\rho = $ constant, or along each streamline only, $\rho = \rho(A)$. The component of Equation (1d) along the streamlines of constant A can be integrated into the generalized Bernoulli relation,

$$P + \frac{1}{2}\rho|\vec{V}|^2 + \rho\mathcal{V} - \frac{\rho\Omega}{\Psi_A}h_3 B_3 = f(A),\qquad (10)$$

i.e., that the total energy density remains constant along a streamline. The invariance of f along each streamline is simply a consequence of the time-independence of the equations.

(2) A polytropic atmosphere, $P = P(\rho)$. Taking the plasma to be a polytrope of index, say, $\gamma \leq 5/3$, we may require that entropy is conserved,

$$(\vec{V} \cdot \vec{\nabla})(\frac{P}{\rho^\gamma}) = 0,\qquad (11)$$

which implies $P/\rho^\gamma = S(A)$, where the entropy S can be chosen to be an arbitrary function which is constant along each streamline. In that case, force balance along individual streamlines gives the following fourth integral of equations (1),

$$\frac{\gamma}{\gamma - 1}\frac{P}{\rho} + \frac{1}{2}|\vec{V}|^2 + \mathcal{V} - \frac{\Omega}{\Psi_A}h_3 B_3 = F(A),\qquad (12)$$

i.e., the sum of the thermal, kinetic, potential, and Poynting energy flux densities per unit of mass flux density is constant along each stream line, $F = F(A)$. This is Bernoulli's law for MHD.

(3) An atmosphere where the temperature $T = (m_p/2k_B)(P/\rho)$, is prescribed in consistency to a detailed energy–balance equation. In this case, momentum balance along streamlines results in a generalization of the barometric law for the drop of pressure with height in a gravitationally bound atmosphere, (Low 1975, Tsinganos 1982a),

$$P(A, Z) = P_o(A)\exp\int \frac{m_p dZ}{k_B T(A, Z)},\quad \text{with} \quad Z = \mathcal{V} + \frac{1}{2}|\vec{V}|^2 - \frac{\Omega}{\Psi_A}h_3 B_3.\qquad (13)$$

(4) An atmosphere where the density and pressure have the same radial profile along all streamlines A = constant (see Chapter 16, equations (2-3)).

2.4. EQUATION FOR MOMENTUM BALANCE ACROSS STREAMLINES

Momentum balance across the streamlines A = const. gives a second order, quasilinear differential equation of mixed elliptic/hyperbolic type (Tsinganos 1982a),

$$\left[A, \frac{\Delta A - \Psi_A \Delta \Psi}{4\pi \rho}\right] = \left[h_3^2 V_3^2, \frac{1}{2h_3^2}\right] - \left[h_3^2 B_3^2, \frac{1}{8\pi \rho h_3^2}\right] + \left[\frac{1}{\rho}, P\right], \tag{14}$$

where

$$\Delta A = -\vec{\nabla} x_3 \cdot \vec{\nabla} \times \vec{B}_3 = \frac{1}{h_1 h_2 h_3}\left[\frac{\partial}{\partial x_1}\left(\frac{h_2}{h_1 h_3}\frac{\partial A}{\partial x_1}\right) + \frac{\partial}{\partial x_2}\left(\frac{h_1}{h_2 h_3}\frac{\partial A}{\partial x_2}\right)\right], \tag{15}$$

and

$$\Delta \Psi = -\vec{\nabla} x_3 \cdot \vec{\nabla} \times \vec{V} = \frac{1}{h_1 h_2 h_3}\left[\frac{\partial}{\partial x_1}\left(\frac{h_2}{h_1 h_3}\frac{1}{4\pi \rho}\frac{\partial \Psi}{\partial x_1}\right) + \frac{\partial}{\partial x_2}\left(\frac{h_1}{h_2 h_3}\frac{1}{4\pi \rho}\frac{\partial \Psi}{\partial x_2}\right)\right], \tag{16}$$

are the electric current and fluid vorticity, respectively, in the direction of the unit vector $\vec{e}_3$ of the ignorable coordinate x_3. Also, the notation of Poisson bracket's has been used,

$$[f, g] = \frac{\partial f}{\partial x_1} \cdot \frac{\partial g}{\partial x_2} - \frac{\partial f}{\partial x_2} \cdot \frac{\partial g}{\partial x_1}. \tag{17}$$

Equation (14) is in a compact form and in order to make more transparent its properties, we need to write it explicitly for each of the previous cases of various equations of state that relate P and ρ. For example, in Cartesian coordinates and for a fluid of constant density $\rho = \rho_o$, the right hand side of (14) is zero and therefore,

$$\Delta A - \Psi_A \Delta \Psi = 4\pi \rho G(A), \tag{18}$$

with $G(A)$ an arbitrary function of A, a combination of $\Psi(A), \Omega(A)$ and $L(A)$ (Tsinganos 1982a). A solution of (18) gives $A(x, y)$ for given $\Psi(A), \Omega(A), L(A)$ and $f(A)$ while the pressure $P(x, y)$ is determined from (10) for a given $f(A)$.

On the other hand, when $P = P(\rho)$, only the last term in (14) is zero in non Cartesian (curvilinear) coordinates. Then, in terms of the three other integrals $\Psi(A), \Omega(A)$ and $L(A)$, as well as of the corresponding fourth integral (12) for the polytropic equation of state, (14) reduces to,

$$\Delta A - \Psi_A \Delta \Psi + \frac{1}{2(1 - \Psi_A^2/4\pi \rho)}\left[\frac{1}{h_3^2}\frac{d(L^2 \Psi_A^2)}{dA} + 4\pi \rho h_3^2 \frac{d\Omega^2}{dA} - 8\pi \rho \frac{d(L\Omega)}{dA}\right] +$$
$$\frac{1}{2(1 - \Psi_A^2/4\pi \rho)^2}\left[\frac{L^2 \Psi_A^2}{h_3^2} + 4\pi \rho h_3^2 \Omega^2 - 8\pi \rho L\Omega\right]\frac{1}{4\pi \rho}\frac{d\Psi_A^2}{dA} + 4\pi \rho \frac{dF}{dA} = 0. \tag{19}$$

Equation (19) should be integrated together with Equation (12), in order to give $A(x_1, x_2)$ and $\rho(x_1, x_2)$ for specified functions $\Psi(A), \Omega(A), L(A)$, and $F(A)$.

Finally, Equation (13) should be integrated together with

$$\Delta A - \Psi_A \Delta \Psi + \frac{1}{2(1 - \Psi_A^2/4\pi\rho)}\left[\frac{1}{h_3^2}\frac{d(L^2\Psi_A^2)}{dA} + 4\pi\rho h_3^2\frac{d\Omega^2}{dA} - 8\pi\rho\frac{d(L\Omega)}{dA}\right] +$$

$$\frac{1/4\pi\rho}{2(1 - \Psi_A^2/4\pi\rho)^2}\left[\frac{L^2\Psi_A^2}{h_3^2} + 4\pi\rho h_3^2\Omega^2 - 8\pi\rho L\Omega\right]\frac{d\Psi_A^2}{dA} + 4\pi\frac{\partial P(A,Z)}{\partial A} = 0 \qquad (20)$$

in order to give $A(x_1, x_2)$ and $\rho(x_1, x_2)$ for specified $\Psi(A)$, $\Omega(A)$, $L(A)$, and $P_o(A)$.

From Equations (19-20) it is also evident that they have singularities at the surfaces where the flow speed is equal to the Alfvén speed on the plane (x_1, x_2). This property separates the domain of interest in various subdomains where these differential equations can be of elliptic or hyperbolic type (Tsinganos and Trussoni 1991).

2.5. Streamline Coordinates for Equilibria Close to Magnetic Axis

The electric current and vorticity terms, ΔA and $\Psi_A \Delta \Psi$ that appear in Equations (18-20) above, involve second order derivatives in the two non ignorable coordinates x_1 and x_2. In this section we shall briefly outline a method for writing these differential expressions as first order derivatives. This is possible for a restricted class of MHD equilibria, namely those where the flux surfaces $A = $ constant are nested around some magnetic axis such that they are a monoparametric family of curves (Bacciotti and Chiuderi 1992).

Without loss of generality, consider a translationally symmetric system in the orthogonal coordinates x, y, z, with z being ignorable. Assume that the magnetic flux surfaces on the plane xy are nested around the axis $y = 0$. Defining the two new variables,

$$X = g(x) \qquad \text{and} \qquad A = \frac{y}{g(x)}, \qquad (21)$$

we want to determine $g(x)$ such that for any $A = A_o$, we obtain the shape of this flux surface on the plane $x - y$, $y = A_o g(x)$.

In switching from the old variables x, y to the new ones X and A we shall use the fact that,

$$\frac{\partial \cdot}{\partial x} = \frac{dX}{dx}\frac{\partial \cdot}{\partial X} - \frac{A}{X}\frac{dX}{dx}\frac{\partial \cdot}{\partial A} \qquad \text{and} \qquad \frac{\partial \cdot}{\partial y} = \frac{1}{X}\frac{\partial \cdot}{\partial A}. \qquad (22)$$

Therefore, the fields on the plane $x - y$ can be written in terms of the new coordinates (A, X),

$$B_x = \frac{F_o(A)}{X} \qquad \text{and} \qquad B_y = \frac{AF_o(A)}{X}\frac{dX}{dx}, \qquad (23)$$

and

$$4\pi\rho V_x = \frac{\Psi_A(A)}{X} \qquad \text{and} \qquad 4\pi\rho V_y = \frac{A\Psi_A(A)}{X}\frac{dX}{dx}, \qquad (24)$$

while the expressions for B_z and V_z are given in terms of the integrals $\Psi(A)$, $\Omega(A)$ and $L(A)$, as in Equations (8-9) with $x_3 = z$. For simplicity in the following we assume that $F_o(A) = 1$ and $\rho = \rho_o$, i.e., that the fluid is incompressible. The same method can be carried out if $F_o = F_o(A) \neq$ constant and for an incompressible fluid as well.

Substituting expressions (22) in the expression for ΔA, we obtain:

$$\Delta A = -\frac{A}{2X}\frac{dT}{dX} + \frac{2AT}{X^2} \quad \text{with} \quad T(x) = \left(\frac{dX}{dx}\right)^2. \tag{25}$$

Similarly for $\Delta\Psi$,

$$\Delta\Psi = -\frac{\Psi_A}{4\pi\rho}\frac{A}{2X}\frac{dT}{dX} + \frac{T}{X^2}\left[\frac{2A\Psi_A}{4\pi\rho} + A^2\frac{d}{dA}\frac{\Psi_A}{4\pi\rho}\right] + \frac{1}{X^2}\frac{d}{dA}\frac{\Psi_A}{4\pi\rho}. \tag{26}$$

Thus, Equation (18) becomes a first order differential equation for $T(x)$,

$$AX\frac{dT}{dX} - T\left[4A - \frac{A^2}{1-\Psi_A^2/4\pi\rho}\frac{d}{dA}\frac{\Psi_A^2}{4\pi\rho}\right] + \frac{1}{1-\Psi_A^2/4\pi\rho}\left[\frac{d}{dA}\frac{\Psi_A^2}{4\pi\rho} + 8\pi\rho GX^2\right] = 0, \tag{27}$$

and its solutions can be studied for the various functional forms of Ψ_A and $G(A)$. The same method can be applied to helically symmetric equilibria (Tsinganos 1982b).

3. Some Exact Solutions

In the following, we illustrate the use of the formulation of section 2 for deriving systematically classes of exact solutions of the steady MHD equations. In particular, we shall briefly present two families of solutions. In the first, the plasma has uniform density and gravity is neglected while, in the second, a nonuniform plasma density is in equilibrium in the Newtonian gravitational field of a central object. Both classes correspond to the following choice of the free functions : $\Psi_A = \text{const.}$ and $L, \Omega = 0$, corresponding to field–aligned flows, $(\rho\vec{V} = \lambda\vec{B})$.

(I) $\rho = \text{constant}$, $\mathcal{V} = 0$. In the three geometries of Cartesian, spherical, and cylindrical coordinates and corresponding physical situations, Equation (18) can be solved exactly for A to give,

   (i) Convective Rolls, in Cartesian coordinates, with $f = f_o + aA^2$ in Equation (10) (Tsinganos 1981),
   $$A(x,y) = A_o \sin k_x x \sin k_y y. \tag{28}$$

   (ii) An MHD vortex, in spherical coordinates, with $f = f_o + aA$, in Equation (10) (Tsinganos 1981),
   $$A(r,\theta) = (C_1 - C_2 r^2)r^2 \sin^2\theta. \tag{29}$$

In particular, the constant C_1 may be chosen such that, on the surface of the sphere $r = r_o$, the fields have only a tangential component,

$$B_r^{(\text{in})} = 2r_o^2 C_2(1 - r^2/r_o^2)\cos\theta, \quad B_\theta^{(\text{in})} = 2r_o^2 C_2(1 - 2r^2/r_o^2)\sin\theta. \tag{30}$$

Furthermore, by choosing $C_2 = -3B_o/4r_o^2$, we may match this interior solution to an asymptotically uniform field in the outside of $r = r_o$, (see Figure 2),

$$B_r^{(\text{out})} = 2B_o(1 - r_o^3/r^3)\cos\theta, \quad B_\theta^{(\text{out})} = -B_o(1 + r_o^3/2r^3)\sin\theta. \tag{31}$$

(iii) Toroidal Plasma Equilibrium, in cylindrical coordinates, r,θ,z with $f = f_o + aA^2$, wherein Equation (18) yields for A,

$$A = A_o \sin kz F_o(\eta, r^2),\tag{32}$$

$$B_r = -\frac{B_o r_o}{r}\cos kz F_o(\eta, r^2)\quad B_z = 2B_o \sin kz F_o\prime(\eta, r^2),\tag{33}$$

where F_o is the Coulomb wave function (Abramowitz and Stegun, 1970 p. 538) and η, k constants. The magnetic surfaces on the meridional plane are shown in Figure 3 to have two magnetic axes and an internal separatrix, i.e., the doublet configuration (Maschke 1973, Tsinganos 1981).

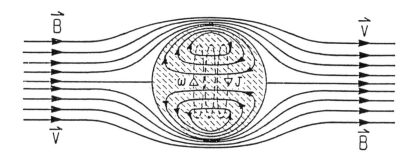

Figure 2. The MHD analog of Hill's spherical vortex. Outside the sphere $\vec{J}$ and $\vec{\omega}$ are zero.

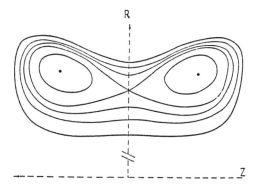

Figure 3. Toroidal plasma equilibrium in a tokamak of rectangular cross–section.

(II) Astrophysical outflows with $\rho \neq$ const., $\mathcal{V} = -GM/r$. Consider the following solutions pertinent to problems of astrophysical mass outflows,

   i) Stellar wind-type flows. With the ansatz,

$$\rho = \rho(r), \quad A = F(r)\sin^2\theta, \quad P = P_o(r) + P_1(r)\sin^2\theta, \tag{34}$$

   we may prescribe the radial function $F(r)$ in a way that results in a partially open magnetosphere (Low and Tsinganos 1986, Tsinganos and Low 1989), Figure (4), and solve Equation (1d) for the remaining radial functions $\rho(r)$, $P_o(r)$, and $P_1(r)$.

   ii) Jet-Type outflows. If, instead of assuming a spherical symmetric density, we set

$$\rho(r,\theta) = \rho(r)[1 + \delta\sin^2\theta], \tag{35}$$

   then we obtain collimated jet-type outflows of the form

$$V_r = V_r(r)\frac{\cos\theta}{(1 + \delta\sin^2\theta)^{1/2}}, \tag{36}$$

   with a high order Alfvénic singularity and an ordinary X-type singularity, (Tsinganos and Trussoni 1991 and Chapter 16).

   iii) Flows in Prominences. If we set $\mathcal{V} = gz$, a uniform gravitational field, and Ψ_A, L, Ω, and F_A constants, we obtain a solar prominence-like hydromagnetic structure with flows along the magnetic field and Alfvénic/sonic critical points (Tsinganos and Surlantzis 1992)

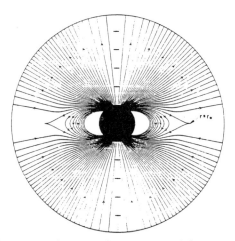

Figure 4. A partially open, axisymmetric, non-potential magnetosphere. The static field region around the equator has a tip at r_* where the separatrix fieldline has a cusp. The field is radial beyond r_*.

4. The Taylor-Proudman and Parker Theorems

Presently, two theorems exist on the necessary conditions for the steady, $\partial_t = 0$, equilibrium of hydromagnetic fields. The Taylor-Proudman theorem (Proudman 1916, Taylor 1922) states that any slow steady internal circulation $\vec{V}$ in a rotating fluid is invariant along the direction of the large scale angular velocity $\vec{\Omega}$. We may, without loss of generality, orient our coordinate system so that the z–axis lies along $\vec{\Omega}$, whereupon the necessary condition for equilibrium becomes

$$\frac{\partial \delta \vec{V}(x,y,z)}{\partial z} = 0. \tag{37}$$

The fluid motions are columnar in form, $\vec{V} = \vec{V}(x,y)$, a result which has been elegantly demonstrated experimentally by Taylor (1922).

On the other hand, for an equilibrium magnetic filed $\vec{B}_o(x,y)$ in a static ideal magnetofluid it has been shown (Parker 1972, 1979) that

$$\frac{\partial \delta \vec{B}(x,y,z)}{\partial z} = 0. \tag{38}$$

is the necessary condition for the equilibrium of a perturbing magnetic field which is analytic in the neighborhood of the equilibrium magnetic field $\vec{B}_o(x,y)$ that satisfies the magnetostatic equilibrium equations, and has a large-scale direction along $\hat{z}$. The conclusion is, therefore, that the topology of the magnetic lines of force must be invariant along the general direction of the field if there is to exist a static, well behaved, equilibrium state. In topologies lacking this invariance, the nonequilibrium takes the form of the dynamical neutral point dissipation and reconnection, destroying the nonivariant parts (Parker 1979).

Having well in hand the general formulation of the hydromagnetic equilibria, we are then led to ask whether there exist additional solutions of the steady MHD equations which are not included in our mathematical net of section 3. Clearly, if such solutions exist, they are lacking an invariance. To explore this possibility, we begin by looking for solutions in the neighborhood of some invariant states like those with an ignorable coordinate. Expanding an arbitrary hydromagnetic perturbation $\delta \vec{V}(x_1, x_2, x_3)$, $\delta \vec{B}(x_1, x_2, x_3)$ in ascending powers of some small parameter about an invariant equilibrium hydromagnetic field $\vec{V}_o(x_1, x_2)$, $\vec{B}_o(x_1, x_2)$, we find that the necessary condition for the perturbed field to be in equilibrium is (Tsinganos 1982c)

$$\frac{\partial \delta \vec{V}}{\partial x_3} = \frac{\partial \delta \vec{B}}{\partial x_3} = 0. \tag{39}$$

Thus, in the general case where there is no special direction within a hydromagnetic field other than the $x_3 = z$ direction (for example, the net magnetic mass fluxes are nonzero only in this z-direction), we are led to the conclusion that z-invariance of the pertubations is a necessary condition for equilibrium.

A physical way to demonstrate the nonequilibrium of nonsymmetric field topologies and show the consequences of the departure from equilibrium is the following: Consider two neighbouring Bernoulli surfaces $\Pi = P + (1/2)\rho V^2$ and $\Pi + \delta\Pi$, separated by an infinitesimal distance h, in an ideal magnetofluid (Figure 5).

The fields are assumed to have some large-scale direction and, without loss of generality, we may assume that they are parallel everywhere

$$\vec{V} = \lambda \frac{\vec{B}}{(4\pi\rho)^{1/2}},\tag{40}$$

where λ is constant along each flux tube. The constant magnetic flux δf_m crossing the infinitesimal area S formed by δh and the distance δw between two neighboring field lines on each Bernoulli surface is $\delta f_m = B\delta h\delta w$. On the other hand, denoting by $J_\perp$ the component of the current density perpendicular to $\vec{B}$, we obtain from the force balance equation,

$$\frac{\delta\Pi}{\delta h} = \frac{1 - \lambda^2}{4\pi} J_\perp B,\tag{41}$$

which, combined with $\delta f_m = B\delta h\delta w$, yields

$$J_\perp = \frac{\delta\Pi}{\delta f_m} \frac{4\pi}{1 - \lambda^2} \delta w.\tag{42}$$

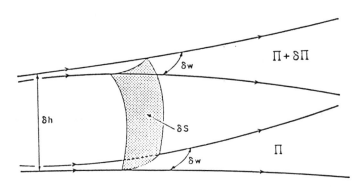

Figure 5. A sketch of two lines of force separated by a distance δw on each of two Bernoulli surfaces Π and $\Pi + \delta\Pi$ a distance δh apart.

Since $\delta\Pi$, δf_m, and λ are constants along the flux tube, $J_\perp$ is proportional to the separation δw of the two neighboring field lines as we move along the tube. In a field pattern which is not invariant along the mean direction of the field, δw changes along the fields. In most cases, somewhere in the field, δw increases without bound with the inevitable result that $J_\perp$ implies an unbounded increase in field gradients. Then, no matter how large the electrical conductivity is, dissipation and line cutting proceed (Parker 1979).

5. Hamiltonian Dynamics and the KAM Theorem

In this section, we briefly review the background of the Kolmogorov–Arnold–Moser (KAM) theorem relevant to the problem of MHD equilibria (Tsinganos *et al.* 1984). Suppose we are given an integrable Hamiltonian H_o, with H_o having no explicit time dependence. Hamilton-Jacobi theory then allows us to find, using a generating function S, cyclic coordinates whose conjugates are constants of the motion; we choose these to be the action-angle variables $(\vec{J}, \vec{\theta})$ such that $H_o = H_o(\vec{J})$ alone. If we consider a nonintegrable perturbation $\epsilon H_1(\vec{J}, \vec{\theta})$ with $\epsilon \ll 1$ such that

$$H(\vec{J}, \vec{\theta}) = H_o(\vec{J}) + \epsilon H_1(\vec{J}, \vec{\theta}), \qquad (43)$$

we may then ask whether there always exist (possibly new) action-angle variables $(\vec{J}', \vec{\theta}')$ such that $H(\vec{J}, \vec{\theta}) = H'(\vec{J}')$, *i.e.*, such that H' is a function of the action variable alone. The existence of these (possibly new) action-angle variables would guarantee that the perturbed Hamiltonian is still integrable (or, translating into the MHD domain, that the corresponding perturbed magnetic state still satisfies the MHD equations). Now, if such new action-angle variables exist, then there must exist a generating function $S(\vec{J}', \vec{\theta})$ with the property that $\vec{J} = \vec{\nabla}_{\vec{\theta}} S$; S must then satisfy the relation $H[\vec{\nabla}_{\vec{\theta}} S(\vec{J}', \vec{\theta}), \vec{\theta}] = H'(\vec{J}')$. One way to construct this generating function is to expand S in a perturbation series, say in powers of some parameter ϵ,

$$S = \vec{J}' \cdot \vec{\theta} + \epsilon S_1(\vec{J}', \vec{\theta}) \dots, \qquad (44)$$

when we obtain

$$H_o(\vec{J}' + \epsilon \vec{\nabla}_{\vec{\theta}} S_1 + \dots) + \epsilon H_1(\vec{J}' + \dots, \vec{\theta}) = H'(\vec{J}'). \qquad (45)$$

The lowest-order nontrivial result is obtained by equating coefficients of the terms linear in ϵ; thus, retaining terms to order ϵ, we have

$$H_o(\vec{J}') + \epsilon[\vec{\nabla}_{\vec{J}'} H_o(\vec{J}') \vec{\nabla}_{\vec{\theta}} S_1 + H_1(\vec{J}', \vec{\theta})] = H'(\vec{J}'). \qquad (46)$$

Now, since H_o is cyclic in $\vec{\theta}$, it follows that

$$\vec{\nabla}_{\vec{J}'} H_o(\vec{J}') = \vec{\omega}_o(\vec{J}') = [\vec{\omega}_{01}(\vec{J}'), \vec{\omega}_{02}(\vec{J}'), \dots], \qquad (47)$$

is the frequency vector of the unperturbed motions; furthermore,

$$H_1(\vec{J}', \theta) = \sum_{\vec{m} \neq 0} H_{1\vec{m}}(\vec{J}') e^{i\vec{m} \cdot \vec{\theta}}, \qquad (48)$$

because H_1 is a function of p and q, and these are periodic in $\vec{\theta}$; therefore S_1 is periodic in $\vec{\theta}$, with an arbitrary constant term we set to zero. Thus

$$S_1(\vec{J}', \vec{\theta}) = \sum_{\vec{m} \neq 0} S_{1\vec{m}}(\vec{J}') e^{i\vec{m} \cdot \vec{\theta}}. \qquad (49)$$

Equating Fourier coefficients, we obtain

$$\vec{m} \neq 0: \ S_{1\vec{m}}(\vec{J}') = +\frac{i H_{i\vec{m}}(\vec{J}')}{\vec{m} \cdot \vec{\omega}_o(\vec{J}')} + \dots$$

$$\vec{m} = 0: \quad H'(\vec{J}') = H_o(\vec{J}') + H_{10}(\vec{J}') + \ldots. \tag{50}$$

The generating function S is thus given by

$$S(\vec{J}', \vec{\theta}) = \vec{\theta}\vec{J}' + \epsilon \sum_{\vec{m} \neq 0} \frac{H_{1\vec{m}}(\vec{J}')}{m \cdot \vec{\omega}_o(\vec{J}')} e^{i\vec{m} \cdot \vec{\theta}} + \ldots. \tag{51}$$

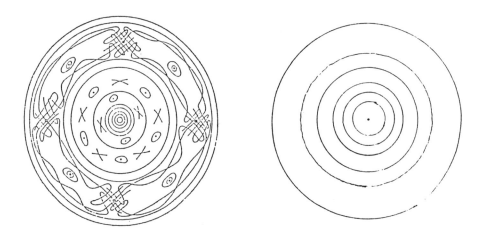

Figure 6. When perturbing an integrable Hamiltonian system, (a), every rational surface is destroyed; there remain only an even number of periodic orbits. For sufficiently small perturbations, most of the irrational surfaces still exist and retain their identity as invariant tori. Between these invariant tori are regions of irregular behavior of the orbits, (b).

Cursory inspection of Equation (51) immediately shows a great difficulty: if there exist motions such that the frequences $\vec{\omega}_o$ are commensurable, then there exist terms in the sum with $\vec{m} \cdot \vec{\omega}_o = 0$, so that the perturbation series diverges. In fact, even if the frequencies $\vec{\omega}_o$ are incommensurate, one can find values for $\vec{m}$ such that $\vec{m} \cdot \vec{\omega}_o$ is arbitrarily small. Hence,

Equation (51) can diverge even in the incommensurate case. This is the well-known problem of "small divisors" in classical mechanics (Arnold and Avez 1968); note, in particular, that the evident singular behavior does not reflect the behavior of the dynamical system under study but, rather, a failure of the method of solution (*i.e.*, the method of solving for the required generating function). The above discussion, which simply summarizes a classical result of dynamical theory, thus shows explicitly that a straighforward perturbation expansion of the corresponding MHD problem cannot answer the crucial question regarding the "structural" or topological stability of hydromagnetic steady MHD solutions.

The way out of this dilemma rests on the KAM theorem. The essence of the method underlying the KAM theorem is to replace the series expansion of Equation (51) by an iteration scheme (analogous to Newton's method) to calculate the form of the perturbed tori (Arnold and Avez 1968). That is, rather than expanding about the unperturbed Hamiltonian (as in Equations (50) and (51)) to calculate directly the generating function $S(\vec{J'}, \vec{\theta})$ of the perturbed tori, one instead calculates a sequence of generating functions by iteration. This procedure accelerates convergence for $\epsilon \ll 1$ for almost all unperturbed tori; for present purposes, the crucial point is that it can be shown that for a set of unperturbed tori of nonvanishing measure, convergence is not obtained. The key point here is that, unlike the case discussed above in connection with the straightforward perturbation leading to Equation (51), the lack of convergence for these exceptional tori is not a defect of the approximation scheme but, is instead, intrinsic to these particular sets of unperturbed tori. Thus, one has shown that these particular integral surfaces are destroyed by the symmetry-breaking perturbation (Figure 6). This is the basic result required in the following section.

6. 2-D MHD Equilibria and Hamiltonian Systems

Another approach to the problem of the existence of nonsymmetric MHD equilibria is based on an exact analogy between the flux surfaces of symmetric MHD equilibria and the surfaces of section of Hamiltonian systems. As discussed in sections 3 and 4, symmetric MHD equilibria are characterized by their flux surfaces $A(x_1, x_2) = $ const. which are 2-D surfaces in 3-D configuration space and on which the induction potential Φ, the stream function Ψ, the total angular momentum L, and the total energy flux density per unit of mass flux density F are constants. The topology of these flux surfaces is that of nested tori, if the field $B(x_1, x_2)$ is nonsingular and nonvanishing in some volume of space.

Consider for example a 1-D simple harmonic oscillator with a Hamiltonian

$$H^{(1)} = \frac{\alpha}{2}(p_1^2 + q_1^2), \qquad (52)$$

α being a constant. By making the correspondence

$$q_1 \to x \qquad p_1 \to y \qquad t \to z/B_z, \qquad (53)$$

and

$$H^{(1)} \to A = \frac{\alpha}{2}(x^2 + y^2), \qquad (54)$$

we obtain the following one to one correspondence between the simple harmonic oscillator and the magnetic field $B = [B_x(x,y), B_y(x,y), B_z]$ around an O-type magnetic neutral point,

$$\frac{dq_1}{dt} = \frac{\partial H^{(1)}}{\partial p_1} : \frac{dx}{dz} = \frac{B_x}{B_z}, \quad B_x = \frac{\partial A}{\partial y}, \tag{55}$$

$$\frac{dp_1}{dt} = -\frac{\partial H^{(1)}}{\partial q_1} : \frac{dy}{dz} = \frac{B_y}{B_z}, \quad B_y = -\frac{\partial A}{\partial x}. \tag{56}$$

The phase portrait of the simple harmonic oscillator in the phase space (q_1, p_1) is identical to the projection onto the xy–plane of the field lines of a twisted magnetic flux tube extending in the z-direction. Similarly, any symmetric MHD equilibrium with flux surfaces $A(x_1, x_2) =$const. may correspond to a 1-D Hamiltonian system with $H^{(1)}(q_1, p_1) = A(x_1, x_2)$.

On the other hand, it is well known that Hamilton's variational principle for a 1-D conservative dynamical system,

$$\delta \int_{t_1}^{t_2} L^{(1)}(q_1, p_1) dt = \delta \int_{t_1}^{t_2} [P_1 \frac{dq_1}{dt} - H^{(1)}(p_1, q_1)] dt = 0, \tag{57}$$

can be written in the following form by introducing some parameter (for example, the arc length measured along the phase space trajectory),

$$\delta \int_{\theta_1}^{\theta_2} [p_1 \frac{dp_1}{d\theta} - H^{(1)} \frac{dt}{d\theta}] d\theta = \delta \int_{\theta_1}^{\theta_2} [P_1 \frac{dq_1}{d\theta} - P_2 \frac{dq_2}{d\theta}] d\theta = \delta \int_{\theta_1}^{\theta_2} L^{(2)} d\theta = 0, \tag{58}$$

where we have set $q_2 = t$, $p_2 = -H^{(1)}$. But Equation (58) is simply Hamilton's principle of least action for a 2-D dynamical system with a Hamiltonian

$$H^{(2)}(q_1, q_2, p_1, p_2) = \frac{dq_2}{d\theta} [H^{(1)}(q_1, p_1) + p_2], \tag{59}$$

Thus, $H^{(2)}$ is by construction time-independent and has one ignorable coordinate, q_2. Hence, there are two integrals of motion and the dimensionality of the integral surfaces reduces from 4 to 2.

Altogether then, we have a one to one correspondence between the flux surfaces $A(x_1, x_2)$ = const. of a symmetric MHD system with one ignorable coordinate x_3 in equilibrium, and the 2-D invariant surfaces of a Hamiltonian system with 2 degrees of freedom (Tsinganos et al. 1984). It is straightforward then to apply all results of the KAM-theory to such MHD systems, as we shall do in the following section.

7. From MHD Equilibrium to Chaos

Consider now our original question: can MHD equilibria having some symmetry (expressible in the form of an ignorable coordinate) be subjected to perturbations which lead to states lacking any symmetry, but still satisfying the equilibrium equations?

To answer this question, we consider the equivalent question in the dynamical domain; that is, we consider a nonintegrable perturbation $\delta H^{(2)}$ of the equilibrium system defined by an integrable Hamiltonian $H_o^{(2)}$ corresponding (as discussed in section 6) to the given MHD equilibrium, with $\delta H^{(2)} \ll H_o^{(2)}$. The perturbed Hamiltonian then has the form

$$H^{(2)} = H_o^{(2)} + \delta H^{(2)}. \tag{60}$$

In the unperturbed system, the nested set of integral surfaces are 2–D level surfaces; in this nested topology, one can choose some parameter say, $a = A(x_1, x_2)$ labeling distinct surfaces. Now, as discussed in section 5, the KAM theorem shows that a finite measure of these integral surfaces of $H_o^{(2)}$ is destroyed by the application of the nonintegral perturbation; that is, for a range of a values in an interval of nonvanishing measure, "most" orbits associated with former integral surfaces labeled by this range of a values no longer remain on a 2–D surface. What does this imply about the corresponding MHD case? Translating from integral surfaces to magnetic flux surfaces, we recall that Ψ, Φ, L, and F are flux surface parameters which are like the flux function A (corresponding to a above), constant on flux surfaces, and that a nonintegral perturbation corresponds to a symmetry-breaking field/plasma perturbation (*viz.* one that violates the constraint $\partial_3 B = 0$). Such perturbations thus result in a nonvanishing measure of surfaces of constant Φ, Ψ, L, and F to be disrupted. Hence, symmetry-breaking perturbations of MHD equilibria necessarily result in the violation of the steady MHD equations of motion. This result corresponds precisely to the conclusions of section 4 with the restriction to infinitesimal perturbations removed, and hence the result is now generalized to arbitrary symmetry-breaking perturbations (Tsinganos *et al.* 1984).

Altogether, then, well behaved MHD equilibria seem to be available only to a limited number of systems for the following reasons. *First*, they have to satisfy the *necessary* conditions of coordinate invariance. *Second*, they need to satisfy the *sufficient* conditions of equilibrium, *i.e.*, all components of all forces must balance. And *thi rd*, they need to be mechanically stable. The class of systems satisfying all these requirements is expected to be rather small. These stringent requirements might account for the observed universal activity of the hydromagnetic fields (Parker 1979, Vaiana *et al.* 1981).

Acknowledgements. I am greatly indebted to Drs. Peter Hoyng, John Brown and Joan Schmelz for a critical reading of the manuscript and numerous suggestions that led to a gradual improvement of the presentation. The research reported in this Chapter was supported in part by NATO grant 870221.

8. References

Arnold, V.I., and Avez, A. 1968, in *Ergodic Problems in Classical Mechanics*, (New York Benjamin).

Abramowitz, M. and Stegun, I, 1965, *Handbook of Mathematical Functions*, (New York: Dover).

Bacciotti, F., and Chiuderi, C.: 1992, *Phys. Fluids B*, **4(1)**, 35.

Edenstrasser, J.W. 1980a, *J. Plasma Phys.*, **24(2)**, 299.

Edenstrasser, J.W. 1980b, *J. Plasma Phys.*, **24(3)**, 515.

Freidberg, P. 1982, *Rev. of Mod. Phys.*, **54 (3)**, 801.

Low, B.C. 1975, *Ap. J.*, **19**, 251.

Low, B.C., and Tsinganos, K. 1986, *Ap. J.*, **302** 163.

Maschke, E.K. 1973, *Plasma Phys.*, **1**, 535.

Parker, E.N. 1972, *Ap. J.*, **23** 746.

Parker, E.N. 1979, *Cosmical Magnetic Fields*, (Oxford : Clarendon Press).

Proudman, J. 1916, *Proc. Roy. Soc. London*, **A 92**, 408.

Taylor, G.I. 1922, *Proc. Roy. Soc. London*, **A 100**, 114.

Tsinganos, K. 1981, *Ap. J.*, **245**, 764.

Tsinganos, K. 1982a, *Ap. J.*, **252**, 775.

Tsinganos, K. 1982b, *Ap. J.*, **259**, 820.

Tsinganos, K. 1982c, *Ap. J.*, **259**, 832.

Tsinganos, K., Distler, J., and Rosner, R. 1984, *Ap. J.*, **278**, 409.

Tsinganos, K., and Low, B.C. 1989, *Ap. J.*, **342**, 1028.

Tsinganos, K., and Trussoni, E. 1991, *Astr. Ap.*, **231**, 270.

Tsinganos, K., and Surlantzis, G. 1992, *Astr. Ap.*, in press.

8 THE PHOTOSPHERE

H. ZIRIN
BBSO, Caltech
Pasadena, CA 91125
U.S.A.

ABSTRACT. The photosphere is the surface of the Sun that we see, and the direct source of its energy. It is dominated by granulation, supergranulation, and magnetic fields. We discuss the significance of limb darkening, the opacity, and model structure. The magnetic fields are dominated by the network, but the weaker fields are also of great interest. We discuss the Fraunhofer spectrum, and what it tells us about the surface, and finally, the newly-discovered emission lines at 12μ.

1. Nature of the Photospheric Surface

The photosphere is the surface of the Sun that we see in visible light, the layer from which most of the radiation reaches us. At that point, the opacity suddenly becomes negligible and all photons can now escape into space. As a result, there is a steep temperature decrease as the energy is lost. This continues for about 500 km until the unknown heating mechanism takes over in regions of stronger field.

The contrast between the solar and terrestrial atmospheres is interesting. The density at the photospheric level is 10,000 times less than that at the earth's surface. Why, then, is the solar atmosphere opaque? There are two reasons: While the scale height in the earth's atmosphere is only a few kilometers, that at the photosphere is about 150 kilometers, or 100 times greater. Second, the molecules in the earth's atmosphere are generally transparent in the visible. At the relatively high temperature of the photosphere, there is a considerable concentration of the negative hydrogen ion H^-, which absorbs radiation voraciously through most of the spectrum and produces the apparently opaque surface.

Because the photosphere is in hydrostatic equilibrium, the change in pressure from one level to the next is sufficient to sustain the weight of the overlying layers. In that case, the increase in pressure dP between point r and $r + dr$ is simply the weight of the added material. That weight is the gravitational attraction $g(r) = GM(r)/r^2$ of the mass of the star interior to r (the overlying material exerts no gravitational force on the material inside) on a cm³ of density ρ :

$$\frac{dP(r)}{dr} = -\frac{GM(r)}{r^2}\rho.$$ (1)

Substituting $P_g = nkT$, we have

J. T. Schmelz and J. C. Brown (ed.), The Sun, A Laboratory for Astrophysics, 175–190.

$$\frac{dP}{P} = -\frac{\mu m_H g(r)}{kT} dr = -\frac{dr}{H}, \tag{2}$$

where μ, the mean molecular weight, is about 0.6 for fully ionized material, but is about one in the photosphere where only the metals are ionized.

$$H = kT/\mu m_H g \tag{3}$$

is the scale height. At the surface, at 6000 K, H = 182 km. For short ranges where T and g change slowly, we can integrate, finding

$$P = P_0 \exp -(r - r_0)/H. \tag{4}$$

The scale height is the measure of any barometric atmosphere. An identical expression applies for density and particle number. Of course, the scale height from point to point is a function of temperature and local gravity, so we cannot simply integrate Equation (2) over large ranges. However, a piecewise application of (4) is satisfactory. An important characteristic of this solution is that the mass above any point is less than that in the next lower scale height, simply because each layer carries the overlying weight. So the dominant factor in the photosphere is the dramatic fall-off in pressure and density with a scale height of approximately 180 km.

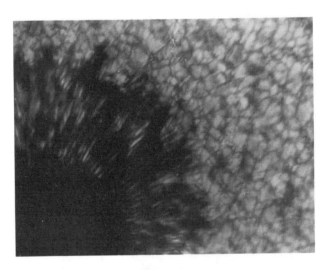

Figure 1. The granulation near a small spot (BBSO).

From the H^- absorption coefficient, we can determine the density of $\tau = 1$ in the photosphere, which, at 5000Å, is the official photosphere. This is easy to get if we use the table in section 42 of Allen (1981). The integrated absorption coefficient for H^- is

$$\tau = \int_{\infty}^{h_0} aN_H N_e kT dh, \tag{5}$$

where N_e, the electron density, is estimated as $10^{-3} N_H$ (we assume that most metals and a little hydrogen are ionized). Substituting a density from (3), we get approximately

$$\tau = 8.3 \times 10^{-16} aN^2 H/2. \tag{6}$$

With absorption $a = 3 \times 10^{-26}$ cm^{4}dyn^{-1} from Allen (1981), we find

$$\tau = 1.13 \times 10^{-34} N_H^2, \tag{7}$$

so the density at $\tau = 1$ is about 10^{17} atoms cm^{-3}. This is a generally useful procedure in atmospheres in hydrostatic equilibrium where the density falls with a scale height and everything happens in a single scale height.

Although there is a big convective zone below it and we see effects of the overshoot, the photosphere is in radiative equilibrium. This was concluded by Karl Schwarzschild when he first applied the Eddington approximation to the limb darkening. The Eddington approximation assumes a fixed ratio of three between the mean intensity J and the radiative pressure K:

$$J = \frac{1}{2} \int_{-1}^{1} I(\mu) d\mu = 3K = \frac{3}{2} \int_{-1}^{1} I(\mu)\mu^2 d\mu. \tag{8}$$

From the relation between flux F and K, we obtain (Mihalas 1978)

$$J(\tau) = \frac{3}{4} F \left(\tau + \frac{2}{3} \right),$$
$$B(\tau) = \frac{\sigma T^4}{\pi} = J(\tau) = \frac{3}{4} \frac{\sigma T_{eff}^4}{\pi} \left(\tau + \frac{2}{3} \right), \tag{9}$$

where B is the Planck function corresponding to the temperature at optical depth τ and is assumed to equal the source function S. If this is now inserted for the intensity I_λ, the result for the limb darkening is

$$\frac{I(0,\mu)}{I(0,1)} = \frac{3}{5} \left(\mu + \frac{2}{3} \right). \tag{10}$$

This gives a rather good approximation for the limb darkening, as we see in the table below (Zirin 1987). From this fit, Karl Schwarzschild first recognized that the photosphere was in radiative equilibrium. We can see that the Eddington approximation is really quite good up to $\mu = 0.2$ (μ is the cosine of the angle θ from Sun center). It is really very close to the optical limb. The deviations above this point are partly due to the roughness of the surface and partly due to the beginnings of the temperature reversal.

178

Table 1. The Eddington approximation for limb darkening.

μ	I/I_0(Eddington)	I/I_0(observed, 5000 Å)
1.0	1.00	1.00
0.8	0.88	0.88
0.6	0.76	0.74
0.5	0.70	0.68
0.4	0.64	0.64
0.3	0.58	0.52
0.2	0.52	0.42
0.1	0.46	0.32
0.05	0.43	0.20
0.02	0.41	0.14

Another way of understanding the dependence of temperature on position is the Eddington-Barbier relations (Barbier 1943):

$$I_\lambda(\tau = 0, \mu) \approx S_\lambda(\mu) \tag{11}$$

$$F_\lambda(0) \approx S_\lambda(2/3) \tag{12}$$

The first equation tells us that the intensity at the top of the atmosphere $(\tau = 0)$ as a function of the projection μ from disk center equals the source function at an optical depth equal to μ. The source function S is the ratio of emissivity to absorptivity at each point, generally equal to the Planck function. Thus, at Sun center, we see the source function at optical depth $\mu = 1$ and, at the edge of the Sun, we see the source function at optical depth $\mu = 0$. The limb darkening tells us that the temperature is decreasing upwards. The second Eddington-Barbier relation tells us that the emergent flux F at $\tau = 0$ is the source function at $\tau = 2/3$.

The variation of temperature in the photosphere can be derived from the limb darkening. When we look at the center of the Sun, we look much deeper than when we look at the limb. This variation is somewhat different at different wavelengths, depending on the variation of opacity. For example, the limb darkening is much stronger in the blue where a change in temperature produces a much greater change in the brightness. It is also different in the red where the Planck function effect is less. When we look in lines, we look higher in the atmosphere and see different gradients; for example, in the Hα line, we look above the temperature minimum, where the transparency is high and the temperature gradient is low; hence, there is very little limb darkening. A classic approach to the limb darkening problem was given by Pierce and Waddell (1961) who fitted the limb darkening and by comparing limb darkening at a range of wavelengths, were able to recover the wavelength dependence of the H$^-$ atom.

A much more extensive approach was made in the model of Vernazza, Avrett, and Loeser (1976), commonly known as VAL. This model was computed with detailed attention to deviations from LTE in the ions which affect the UV, and gives a good fit to both limb darkening and the variation of brightness in the continuum. Note that the height scale is measured from $\tau_\kappa = 1$ at 5000 Å, where the temperature is 6423 K and the density, $1.2 \times 10^{17} \text{cm}^{-3}$, is about 10^{-4} of the atmospheric density at the Earth's surface. The weakness of models like VAL is that they use observational data to extend the model from the low photosphere, where we probably understand what is happening (LTE, hydrostatic equilibrium), to the upper photosphere, where we do not understand what causes the upward increase in temperature and where the atmosphere is quite inhomogeneous. So, while these photospheric models are pretty good around $\tau = 1$, they would produce radio brightness far greater than observed (Gary *et al.* 1990; Zirin *et al.* 1990) and, hence, bear little resemblance to reality at greater heights where the model is governed by uncertain brightness measurements in the UV and IR or the extreme limb, as well as by unknown physics. Unfortunately, it is not necessarily true that the extensive calculations behind VAL make it right.

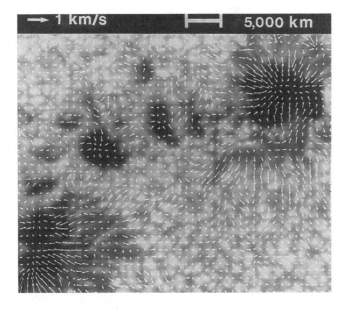

Figure 2. Motion in granulation and spots tracked by cross-correlation. Both small pores and large spots show strong inflow in the inner penumbra, but around the penumbra the Evershed effect produces outflow (Wang and Zirin 1991).

When we move out of the traditional visible range, the continuum changes and our data are somewhat inadequate to understand the physics. The H^- bound-free opacity falls with increasing wavelength but the free-free increases, so a minimum occurs near 1.7μ and we see perhaps one scale height deeper. Further into the IR and radio, the opacity rises with λ^2 and we see higher and higher. The brightness temperature first decreases to around 4000

K, then rises to about 8000 K in the chromosphere. There is no evidence for the fanciful high temperatures assigned by the VAL and other models which use theory, rather than observations, for this region. While UV lines from ions formed at higher temperature are observed, their intensity is far too weak (Zirin and Dietz 1962) to come from more than a tiny fraction of the region. The continuum from this region can be seen at 1600Å and around 200μ in the hidden IR. Although these regions are difficult for absolute measurements, the limited data argues strongly that the temperature is 4000-4500 K, rising to 6000 K at 4000 km and 8000 K at the top of the chromosphere. Eclipse measurements, further, show substantial density up to 4000 km; the limb at 2.6 mm is at least 5000 km above the photosphere. Imaging observations in the radio (Gary *et al.* 1990) and ultraviolet (Foing and Bonnet 1984) show a cool background with scattered high–temperature regions where the magnetic field is strong. While the temperature does rise above the temperature minimum, the important increase is in the magnetically enhanced areas. By contrast, Foukal *et al.* (1990) have reported that, at the opacity minimum near 1.5μ where we see one scale height deeper, the areas of stronger field are darker than the other regions, giving a backward extrapolation of the heating process.

The reason for the heating is unknown. The sound–wave heating model, pursued by theorists for years after the evidence of magnetic–related heating appeared, has finally been abandoned, and theorists are searching for ways in which the field might produce the heating.

2. The Fraunhofer Lines

Fraunhofer's discovery of the line spectrum was followed by Kirchhoff's (1859) more sophisticated laws governing the production of the spectrum:

1. The ratio of emissivity to absorptivity is independent of the composition of the material and depends only on the temperature and wavelength.
2. An opaque body radiates a continuous spectrum.
3. A transparent gas radiates an emission spectrum that is distinct for each chemical element.
4. An opaque body surrounded by a gas of low emissivity shows a continuous spectrum crossed by absorption lines at the same wavelength.
5. If the gas has a high emissivity, the continuous spectrum will be crossed by bright lines.

It is easy but important to understand the basis for Kirchhoff's laws. If we place two surfaces close together, they must each emit and absorb in the same ratio, or else one will rapidly get cooler than the other, and we will have built a magic refrigerator. This holds for any material, so the emissivity and absorptivity must depend only on temperature, so long as the surfaces are opaque. Because an opaque body is not transparent at any wavelength, it emits a continuous spectrum characteristic of that temperature.

A transparent gas, on the other hand, will be opaque only at the frequencies corresponding to the energy difference between the atomic or molecular energy levels, and emit only in those frequencies, giving an emission line spectrum. At each wavelength, we see the emissivity corresponding to the excitation temperature at the observed height. In the continuum we look deeper; in the line, higher. At each wavelength, we will see the local source function S_λ for that wavelength and the absorption line maps out the height variation of S_λ. If S_λ decreases outward, an absorption line results. Since S_λ cannot exceed the Planck function, we always have the maximum temperature at that point. With the line opacity, we can determine the exact height; in any event, the gradient of $dS_\lambda/d\tau_\lambda$ is determined.

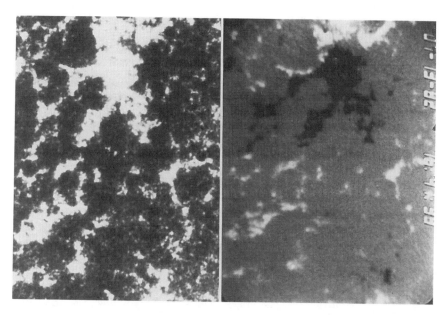

Figure 3. Comparison of the UV continuum at 1600Å (Foing and Bonnet 1984) with a videomagnetogram (BBSO) at right. There is almost perfect agreement between enhanced magnetic field and heating of this layer in the upper photosphere. The network is very prominent.

We can learn quite a bit about the photosphere by simply examining the general properties of the Fraunhofer lines which are common at 6000 K, where the thermal energy of each particle is about 0.5 volts and neutral and singly ionized atoms are abundant. But the dominant lines are from the 0.1% of the plasma made up of heavy elements; of those, the most abundant (C, N, O, Ne) are not observed, He is absent, and the H lines are weak. This is because these elements have excitation potentials above 10 volts and their strongest lines are in the UV. The strongest lines in the visible spectrum are the H and K (Fraunhofer's letters) lines of ionized calcium. Ca is easily ionized and mostly in the singly ionized state, and these lines are absorbed by ions in the ground, or lowest state. The sodium D lines are quite a bit weaker because, although they arise from the ground state, most of the Na is ionized and does not absorb. Along with the neutral Ca line $\lambda 4226$, these are the only resonance lines in the visible. In the UV, of course, we see the resonance lines of the ions formed in the hot elements of the upper photosphere and chromosphere. In both visible and UV, it does not take much material to make a line; the UV lines come from a tiny fraction of the surface, and the visible, from less abundant elements. The weaker Fraunhofer lines show a narrow Doppler core corresponding to the temperature with some broadening due to the mass motions. The stronger lines are optically deep through most of the profile, which is an inverted delta shape and maps the temperature distribution. In the center of these lines, there are narrow, very dark cores produced by pure scattering near the temperature minimum. As a result of these effects, the equivalent width of the lines is not directly proportional to the abundance, growing linearly at first, then as the square root. By plotting the "curve of growth" (Zirin 1988) of all the spectrum lines of an element produced by these effects, one can obtain a fit depending on abundance only and, from this, much of the basic determination of photospheric abundances has been carried out. This set of abundances occurs with great regularity throughout the universe, being seen in such diverse objects as quasars, meteorites, and new stars. However, measurements of solar energetic particles and the UV lines show that the abundance of elements of high first ionization potential such as C, N, O, Ne, Ar, is greatly reduced in the atmosphere above.

3. Granulation and Supergranulation

The granulation is the most striking characteristic of the photosphere, a "rice-grain" like structure occurring *everywhere* on the surface. The granules have a scale of about 1000 km and are separated by narrow dark lanes about 300 km across. They are thought to be the surface extrapolation of the upper convective zone, hot, bright columns rising to the surface, cooler material falling at the edges. That the bright granules truly rise can be confirmed by taking filtergrams in the red and blue wings of a weak or moderate absorption line. On the red side of the line, the image is brightened by upflow, the absorption line shifting to the blue, so the granular contrast is enhanced. On the blue side, the effect is the opposite: the darkening produced by the line shift cancels the granular brightness and produces low contrast. For this reason, videomagnetograms are made in the blue wing of photospheric lines.

The lifetime of the granulation is a bit confusing. Values like 5 or 10 minutes are reported, but these normally are autocorrelation times, the time for a granule to change substantially, usually decay. If we measure the entire lifetime from gradual formation of this upward flowing current to its dissolution, we get about 10 to 22 min (Zirin and Wang 1989), the bigger granules living longer. While most granules fade away, about 25% "explode." The granule breaks up and the smaller pieces flow out radially at about 1 km sec^{-1}. If we picture the visible granulation as the intersection of the convective column with the photosphere, this process tells us that the end of these granules is not a breakup by random motions, but development of overpressure and explosion that disrupts the structure. Further, all of the work on granule lifetimes has been devoted to the brightness field, but what we really want is the lifetime of the upflow, *i.e.* the Doppler lifetime. This is possible with modern techniques. Granules generally display only a small and random motion, except in the vicinity of sunspots. The Spacelab II team (Title *et al.* 1990) found a steady outflow in the moat surrounding symmetric spots where the Evershed flow is strong. But Zirin and Wang (1989) found an inflow before the appearance of small pores, continuing during their lifetime. One can explain such behavior with the speculation that the field becomes strong enough to reduce convection and cool the area, which reduces the lateral pressure, and the granules flow in until the field is strong enough to balance them. Again, since the granule must be a column, we do not know the three-dimensional picture. Do the granules lean in like the heads of kelp, or does the whole column translate? The physical basis of convection and granulation is thoroughly reviewed by Spruit *et al.* (1990), along with an extensive description of the Lockheed observational data.

The uniformity of the granulation is truly striking across the Sun. For field strengths below 1000 Gauss or so, the granular kinetic energy exceeds the magnetic pressure. But, for stronger fields, the granules are distorted, providing interesting evidence on the interplay between fields and convection. Two cases are known. The first is a reduction in size in the neighborhood of active regions found by Tarbell *et al.* (1988). This may be the incipient reduction of convection by the fields around the spot. Second is the elongation of granules in the center of emerging flux regions; these are the result of the emerging strong transverse fields. There is some evidence that the photospheric magnetic fields are concentrated in the intergranular lanes.

The wavelength dependence of granular contrast reflects the combination of Planck function and height effects. Granular contrast is lower in the red because brightness varies linearly with temperature in the Rayleigh–Jeans region while, on the blue side, it varies exponentially. Further in the blue, we find the granulation contrast little changed down to about 4100 Å, where a smaller–scale structure appears. This structure is also seen in the G band, a cluster of lines at 4305 Å; therefore, it must lie higher than the photosphere and may be identical to what we see in some absorption lines such as Na D. It is sinusoidal in form, with as much dark area as bright, smaller than the granulation, and without polygonal structures. It probably is the sum of the contributions of the many lines in the spectrum in the blue, since it appears weak in spectroheliograms made in continuum windows. This structure shows strong five-minute oscillation, and may be considered the upward extension of the granular structure. Velocity movies show a uniform cellular structure.

The granulation is convection with a velocity of ≈ 2 km sec^{-1}, a scale of 700 km, a lifetime about 18 min, and a temperature fluctuation greater than 100 K. The three-dimensional structure has been successfully modeled, most recently by Stein and Nordlund (1989). Spruit *et al.* propose that the exploding granules occur when the convection is inadequate to transport energy in the center of the granule, and note that this often occurs in three–dimensional simulations. That the granules are indeed rising bright elements is confirmed by filtergrams in the wings of photospheric lines.

The next scale size up is the so-called mesogranulation, a scale corresponding roughly to the exploding granules. Whether or not this exists is hard to say. Because there are many exploding granules, it probably is real and may be considered a larger–scale result of the granulation.

By contrast, the effects of the supergranulation are huge, ordering the quiet Sun magnetic fields as well as the structure of the chromosphere. This steady cellular flow, discovered by Leighton, Noyes, and Simon (1962), covers the entire Sun. On a Dopplergram of the whole Sun, the surface is covered by velocity cells of about 30,000 km across. This was a subtle effect when first discovered, but now is a gross feature that must be removed from Dopplergrams. In each cell, there is an outward flow of about 0.4 km sec^{-1}. No vertical flow has been detected. Magnetic fields are swept to this boundary, forming a network of fields on this scale. The supergranular cells last for a day or two as Doppler cells, but the fields concentrated on their edges live much longer. The heating of upper photosphere takes place on the edges of the cells.

4. Magnetic Fields

The photospheric magnetic fields occur in four patterns:

1. *Unipolar magnetic regions*, made up of an "enhanced network" where the magnetic network is relatively strong, complete, and of one polarity. In centerline Hα and CaK, the network is bright and, in the wing of Hα, a dark network of spicules is prominent. These are the accumulated remnants of active regions.
2. *Quiet magnetic regions.* The rest of the Sun, except for horizontal field regions, is covered by a pattern of weaker fields we shall call the quiet network. The magnetic elements are mixed in polarity and the network is incomplete. Sometimes (*e.g.* at the poles) these regions are unipolar.
3. *Horizontal field regions*, which separate unipolar regions of opposite polarity with filaments or horizontal fibrils and are devoid of network structure. The field lines may run parallel to or across the boundary.
4. *Ephemeral active regions*, small dipoles with a lifetime of about one day and no sunspots which erupt everywhere on the Sun. They are bright in chromospheric lines and X-rays, and have field strength similar to the network.

While the unipolar regions are the result of the breakup of active regions, the rest are present on the Sun at all times, regardless of the cycle. The first two consist of two parts:

1. The network edge elements have stronger fields, are longer lived and slower moving, and are the source of spicules and the hot component of the chromosphere. In the unipolar regions, they are all of the same sign; in the quiet regions, they are mostly the same sign and, in some regions, they are of mixed polarity. Their lifetime is unknown because they change shape before they merge or disappear, but they probably live for years.
2 The intranetwork fields (Zirin 1986) are little elements of mixed weak field which scurry hither and yon inside the network cells with velocity about 1 km sec^{-1}, much faster than the network. They are near the limit of resolution, but appear to be monopolar. Naturally, nothing is monopolar; this means each element is connected to an unknown sibling, which is reasonable in these mixed fields.

The unipolar fields, whose significance was first pointed out by Bumba and Howard (1965), clearly result from the decay of the active regions. But the quiet Sun network, weak as it is, has to be replenished locally. We have never really seen the elements die out, but assume they cannot last forever. The party line is that the little intranetwork field elements are swept to the boundaries by the supergranular flow and reform the network elements. Since these elements have random polarity, many must be swept before an element 10 or 100 times larger is formed.

Figure 4. A videomagnetogram of the quiet Sun, in a region dominated by positive (white) network polarity. The contours enhanced the dynamic range of the 8–bit display by wrapping the higher bits. Several bipolar areas occur here. The weak fields inside the network cells are mixed polarity (BBSO).

While the network elements appear to have a range of field strengths, there is a popular view that they represent fields nearly as strong as in sunspots, but are so small that they only fill a tiny fraction of the field of the magnetograph pixel and, thus, register as weak fields. The rationale is that field strengths measured in lines of different g-factor are not measured in the ratio expected. This is explained (Stenflo 1973) by saturation of the magnetograph in the strong fields. Since the magnetograph measures the average splitting in an area, these strong fields average with the weak or zero fields to give the low values measured. This cannot be confirmed directly since the Zeeman splitting is too small; for the widely–used 5250 Å line ($g = 3$), the splitting in a 10–Gauss field is only 3.2×10^{-4} Å. Only in sunspots could the Zeeman splitting be measured directly.

Because of hydrostatic equilibrium, the density at any height in the photosphere must be the same everywhere. In the horizontal direction, therefore, the magnetic pressure and gas pressure inside the spot must balance the gas pressure in the photosphere outside:

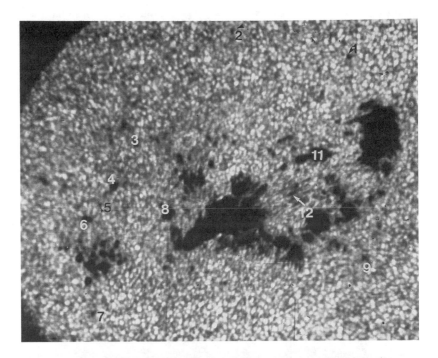

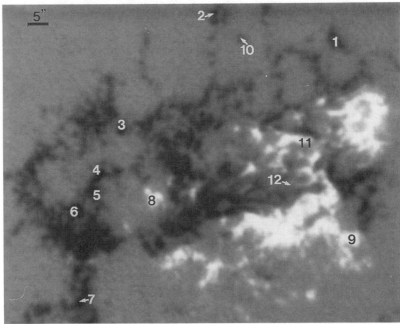

Figure 5. (Top) An emerging flux region surrounded by micropores (1, 2, 3, 7, 9). These can be matched against the magnetogram (Bottom). At 12 we see transient "spots" produced as the flux loops transit the surface (Zirin and Wang 1992).

$$nkT_{sp} + \frac{B^2}{8\pi} = nkT_{ph}. \tag{13}$$

The extra magnetic pressure inside can only be balanced if $T_{sp} < T_{ph}$ and, when the temperature is lower, we should see a sunspot. The tiny 1000–Gauss field regions would only be stable if they, too, were dark. However, if the 1000–Gauss field elements produce a magnetograph signal of 10 Gauss, they must only fill 0.01 of the resolution element and the quiet Sun spots will be invisible, which in fact they were. The published data apply only to fields stronger than the network, mostly enhanced network fields; the magnetographs used could not detect the intranetwork fields or even the quiet network. Furthermore, there is no guarantee that the line strengths should depend on the g-factor alone, since there are many subtleties in the measurements of magnetic fields.

Observatories around the world have been measuring the fields in sunspots for many years by the Zeeman splitting and commonly list spot fields of 500–600 Gauss for the smaller spots. If these many measurements are correct, then weaker fields clearly exist. It is said that the measured splittings in these small spots are too low because of scattered light, but splitting can scarcely be affected by scattered light. These results are still obtained at Mt. Wilson, Rome, and other places where spot fields are routinely measured.

The invisible spots have now been detected by Wang and Zirin (1991). It is found that almost all plages in active regions have small spots at their core. More important, the strongest elements of unipolar regions far from active regions have tiny micropores, about 300 km across. These are about 1/10 the area of the associated magnetic elements, with field strength about 300 Gauss and brightness about 80% of the photosphere at 5000Å. There may, of course, be yet smaller spots corresponding to the weaker elements of flux, but it should be remembered that the intranetwork fields are hundreds of times weaker than the fields marked by the "invisible spots" found by Wang and Zirin. Thus any micropores associated with the intranetwork field would have to be 30 km across, less than the mean free path of the radiation and, therefore, unstable because they could be easily heated from the side.

The structure of the photospheric fields is quite remarkable. We have carried out long integrations, and made even more sensitive measurements by superposing a number of magnetograms. We have been unable to find any general diffuse fields. So there is a stabilizing force, partly based on the cooling in micropores, that produces discrete elements, even if they are not so strong as expected.

Another question of interest is the tilt of the photospheric field lines. We have seen that, in emerging flux loops, this can be horizontal. When the initial fragmentation of the surface magnetic fields was recognized (Zirin 1966, p. 252), it was thought that the fields went straight up into the corona, or arched over to connect to the opposite pole. Because spicules showed a bush-like behavior, the concept of a canopy arching out from the field element grew. This concept may be studied quantitatively with modern magnetographs. If the field lines emerged isotropically into a hemisphere, there would be no loss of sensitivity as we go from center to limb. A completely vertical canopy would show a steady fall–off with $\cos\theta$. On the other hand, a conical bush, with lines no further than 45° from the radial, would show no drop–off until 45° from disk center, when some field lines have a line-of-sight component away from us. Recent measurements by Murray (1991) and Varsik and Zirin (unpublished) show that a cone with $\pm 60°$ best approximates the data. On high-resolution magnetograms near the limb, many clumps of field show a limbward companion of opposite sign due to the field lines pointing the other way.

5. The 12-micron Lines

A new set of solar emission lines in the 12-μ range was discovered by Murcray *et al.* (1981) and identified by Chang and Noyes (1983). The lines are due to high ($n = 7$ to $n = 6$) transitions in neutral Mg and Al and formed in LTE. Because the free-free absorption increases as λ^3, we see fairly high in the photosphere at this wavelength. The lines show absorption wings as the temperature continues to decrease upward; as we pass to the center of the line we see above the temperature minimum and there is an emission peak. Since the Zeeman splitting is proportional to the square of the wavelength, it is 600 times greater for these lines than for those in the visible. Thus, fields as weak as 200 Gauss may be measured directly, with filling factors irrelevant. Brault and Noyes (1983) found fields of 1600 Gauss in penumbrae and 200–500 Gauss in plages, but detected no splitting at all in the quiet Sun. Zirin and Popp (1989) obtained (Figure 6) the same result. The evidence supports strong fields (but only 500 Gauss) in plages, but quite weak fields elsewhere. It has been suggested that these relatively low values are the result of fields diverging with height, but plages near the limb do not give lower values, so it is probably not a height effect. Evidently, the plage fields are not sufficient to produce darkening, although pores seem to make an easy transition to plage and *vice versa*.

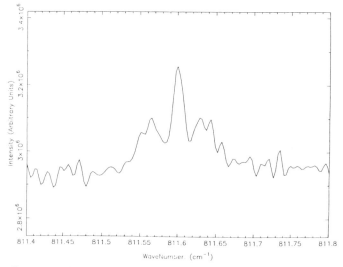

Figure 6. Zeeman splitting in a plage; the field is about 500G. The central spike is largely due to the quiet Sun (Zirin and Popp 1989).

The explanation of the reversal in the 12-μ line has recently been advanced by Chang *et al.* (1991) and by Carlsson *et al.* (1991). The highest bound levels of ions have b, the deviation from LTE, equal to one. By definition, $b = 1$ in the continuum, and there are so many collisions that equilibrium is established for these high levels. For lower levels, b decreases as electrons drain to the ground state faster than collisions from the continuum can replenish them. If we consider the Planck function (Zirin 1987, Equation 4.25 *et seq.*) with the b-values included:

$$B_\nu d\nu = \frac{2h\nu^3}{c^2} \frac{1}{(b_m/b_n)e^{h\nu/kT} - 1} d\nu \tag{14}$$

where n is the lower and m, the upper state. If $b_n > b_m$ and $kT \gg h\nu$, there can be significant increase in B_ν, producing the many emission lines from different elements. These lines show significant, in fact enormous, limb brightening, which the authors explain by the increasing departures from LTE with height. Deming (private communication) found, at eclipse, that the lines indeed originate in the upper photosphere below the temperature minimum, with a small contribution from higher up. The explanation of the limb brightening of the 12–μ lines is most important. The presence of reversal in these lines all over the Sun suggested the presence of a universal temperature increase, while the UV and radio data require the reversal to be limited to the magnetic regions. We now know that the latter is correct. Chang *et al.* also explain the complex structure of the infrared lines as manifestations of the quadratic Stark effect, which makes possible new density diagnostics of this important region. The Sun is truly an astrophysical laboratory and the "classical" photosphere seems to be much less classical than originally thought.

References

Allen, C.W. 1981. *Astrophysical Quantities, 3d ed.* London: Athlone.

Barbier, D. 1943. *Ann.d'Astrophys.* **6**, 113.

Brault, J.W. and Noyes, R. 1983. *Ap. J.* **269**, L61.

Bumba, V. and Howard, R.F. 1965. *Ap. J.* **141**, 1502.

Carlsson, M., Rutten, R.J., and Schchukina, N.G. 1991, to appear in Astronomy and Astrophysics.

Chang, E. S. and Noyes, R. W. 1983. *Ap. J.* **27**, L11.

Chang, E. S. *et al.* .1991. *Ap. J.*, 379.
   L79

Foing, B. and Bonnet, R. M. 1984. *Ap. J.* **279**, 848.

Foukal, P. *et al.* 1990. *Ap. J.* **353**, 712.

Gary, D., Zirin, H., and Wang, H. 1990. *Ap. J.* **355**, 321.

Kirchhoff, G. 1859, *Sitzungsber. Akad. Wiss. Berlin.* p. 783.

Leighton, R. B., Noyes, R. W., and Simon, G. W. 1962. *Ap. J.* **13**, 471.

Mihalas, D. 1978. *Stellar Atmospheres.* San Francisco: W. H. Freeman.

Murray, N. W. 1991. submitted to *Ap. J.*

Murcray, F. J. *et al.* 1981. *Ap. J.* **247**, L97.

Pierce, A. K. and Waddell, J. 1961. *Mem. Roy. Astro. Soc.* **68**, 89.

Spruit, H. C., Nordlund, Å, and Title, A. M. 1990. *Ann. Rev. Astr.* **28**, 263.

Stein, R. F. and Nordlund, Å 1989. *Ap. J.* **342,**, L95.

Stenflo, J. O. 1973. *Solar Phys.* **32**, 41.

Tarbell, T., Ferguson, S., Frank, Z., Title, A. and Topka, K.: 1988, *Tenth Sacramento Peak Summer Workshop*, Sunspot, NM .

Title, A. M., Shine, R. A., Tarbell, T. D., Topka, K. P. and Scharmer, G. B. 1990, *IAU Symp. 138*, p. 49.

Vernazza, J., Avrett, E. H. and Loeser, R. 1976. *Ap. J. Suppl.* **30**, 1.

Zirin, H. 1986. *Aust. J. Phys.* **38**, 961.

Zirin, H. 1966.

   , *The Solar Atmosphere* New York:Blaisdell

Zirin, H. 1988, *Astrophysics of the Sun* Cambridge: Cambridge Univ. Press, p.179.

Zirin, H. and Dietz, R. D. 1963. *Ap. J.* **138**, 664.

Zirin, H. and Popp, B. 1989. *Ap. J.* **340**, 571.

Zirin, H. and Wang, H. 1989. *Solar Phys.* **119**, 245.

Zirin, H. and Wang, H. 1990. *Solar Phys.* **125**, 45.

Zirin, H., Hurford, G. J. and Baumert, B. M. 1990. *Ap. J.* **370**, 779.

Zirin, H. and Wang, H. 1992, to appear in *Ap. J. Letters*.

9 SMALL-SCALE PHOTOSPHERIC MAGNETIC FIELDS

M. SCHÜSSLER
Kiepenheuer-Institut für Sonnenphysik
Schöneckstr. 6
D-7800 Freiburg
Germany

ABSTRACT. Most of the magnetic flux which is observed in the solar photosphere has a filamentary nature: structures with large magnetic flux density are embedded in a much less magnetized gas. A hierarchy of magnetic flux concentrations has been detected which ranges from large sunspots down to the smallest detectable features, the *magnetic elements,* which are the main topic of this lecture. We start with a description of observational methods and results concerning non-spot magnetic fields with particular emphasis on Stokes profile analysis and the line ratio method as well as on recent results from helioseismology. It follows a discussion of the theoretical approaches which have been used to describe small-scale solar magnetic fields, *i.e.* magnetoconvection, thin flux tubes, and numerical model calculations and simulations. In the remainder of the lecture, we go through the whole life cycle of magnetic elements and discuss observational and theoretical results concerning the various phases: formation by flux expulsion and thermal effects, magnetic and thermal structure in quasi-stationary equilibrium, dynamical processes and, eventually, instabilities, destruction, and decay. Observational and theoretical results are often presented in parallel reflecting fruitful interaction in this particular field of solar research.

1. Introduction

The discovery of the Zeeman effect in sunspots by G. E. Hale (1908) demonstrated that the Sun is a magnetic star. It took nearly 40 years until a magnetic field was detected on another star, the Ap star 78 Vir (Babcock 1947). In the intervening 40 years, it has been realized that magnetic fields are of key importance for basic astrophysical processes ranging from star formation to the acceleration of jets from active galactic cores. However, the Sun and the heliosphere still represent the "Rosetta Stone" for magnetic fields in astrophysics: only in the case of the Sun can we study the fundamental processes and interactions on their natural spatial and temporal scales which are observationally inaccessible for any other astronomical object.

Solar magnetic fields are organized in a hierarchy of structures which extends from large active regions with a size of 10^5 km down to the smallest magnetic flux concentrations with a diameter of the order of 10^2 km. In the course of the last 20 years, it became more and more clear that the bulk of the photospheric magnetic flux (outside sunspots, inside and outside active regions) is concentrated into small-scale structures with remarkably similar magnetic and thermal properties such that the term *magnetic elements* seems appropriate.

The concept of a unique magnetic structure as basis for non-spot solar magnetic fields stimulated theoretical work with the effect that research developed rapidly in successful

J. T. Schmelz and J. C. Brown (ed.), The Sun, A Laboratory for Astrophysics, 191–220.

interaction between observation and theory. In consequence, observational and theoretical aspects are often presented in parallel in this Chapter which is intended to give a short glimpse of our (or, at least, of the author's) present state of understanding of small-scale magnetic fields in the solar photosphere. The majority of topics is only mentioned cursorily, indicating relevant references in the literature; a few points are treated in somewhat more detail.

We shall concentrate on the physics of individual magnetic elements and, therefore, exclude topics like intranetwork fields, ephemeral active regions, and elementary bipoles whose structure and dynamics have been reviewed comprehensively by Martin (1990). By concentrating on the photospheric region, we omit a number of important topics, most notably the dynamics of filamentary magnetic field structures in the deep convection zone (with considerable implications for the dynamo problem; *cf.* Schüssler, 1984a, 1987a) and the effects of magnetic elements on the structure of the chromosphere and the question of atmospheric heating (see Ulmschneider *et al.* 1991, for an up-to-date overview).

We begin this Chapter by giving a short overview of the observational methods and some basic results in section 2. Stokes analysis and the line ratio method which are central for observational studies of magnetic elements are discussed in somewhat more detail. section 3 presents the theoretical approaches, *i.e.* studies of magnetoconvection, the magnetic flux tube concept, and numerical model calculations and simulations. The following sections discuss specific topics in relation to magnetic elements, *i.e.* formation of concentrated magnetic filaments (section 4), their magnetic and thermal structure (section 5), dynamics (flows, oscillations, and waves) in section 6, and the destruction of magnetic elements (section 7).

General information about many of these and related topics can be obtained in the textbooks by Parker (1979), Priest (1982), and Stix (1989), while more detailed discussions are given in the review papers by Spruit *et al.* (1991) and Solanki (1992) as well as in a number of conference proceedings (Schmidt 1985, Deinzer *et al.* 1986, Schröter *et al.* 1987, Stenflo 1990).

2. Observations

Since it is impossible to give a full account of the existing observational results concerning small-scale magnetic fields, we shall concentrate here on briefly summarizing the most important results and methods. More information can be found in the reviews by Solanki (1987abc, 1990, 1992) and Stenflo (1989).

2.1 HISTORICAL REMARKS

The first detection of distinct magnetic structures in the solar photosphere outside sunspots was reported by Hale (1922ab). He called them "invisible sunspots" since no trace of these magnetic features could be found in the visible light. With the development of the magnetograph which utilizes the circular polarization of the Zeeman components of spectral lines (Babcock and Babcock 1952, Kiepenheuer 1953), the measurement of magnetic fluxes in active regions and in the network outlined by the supergranular velocity pattern became feasible. A diffuse, quasi-homogeneous nature of these non-spot magnetic fields was assumed until the work of Sheeley (1966, 1967) indicated that the actual field strengths are larger than the average field strengths derived from low spatial resolution measurements under the tacit assumption of spatially resolved fields.

Beckers and Schröter (1968) were the first to detect small-scale features (1000–1500 km size) with strong magnetic fields (in excess of 1000 Gauss) in active regions; they called these structures (which, unlike pores and sunspots, are inconspicuous in the continuum light) *magnetic knots* and gave arguments that they comprise a large part of the magnetic flux of active regions. Comparing the magnetic response of different spectral lines, Howard and Stenflo (1972) showed that more than 90% of the magnetic flux outside sunspots is in the form of strong (kilogauss) fields. Since the average field strengths are much smaller, this indicated a filamentary nature of the magnetic fields in the solar photosphere: Small magnetic structures with large field strength are surrounded by much less magnetized plasma. Stenflo (1973, see also Frazier and Stenflo 1978) introduced a line-ratio technique to determine the intrinsic field strengths of unresolved magnetic filaments and found values in the range 1–2 kGauss. These results were confirmed later by a number of other authors (*e.g.* Harvey and Hall 1975, Tarbell and Title 1977, Wiehr 1978).

The last decade has seen a remarkable progress in our understanding of small-scale magnetic fields in the solar photosphere in spite of their spatially unresolved nature. This was made possible by the ingenious use of indirect spectroscopic methods and the development of sophisticated instruments, most notably the Kitt Peak Fourier Transform Spectrograph/Polarimeter (FTS) (Brault, 1978). The FTS is able to provide Stokes profiles (I, V, and Q) for a large number of spectral lines simultaneously. By comparison with synthetic Stokes profiles together with inversion techniques using theoretical models of magnetic structures, many properties of the filamentary magnetic fields could be determined (see reviews by Solanki 1987abc, 1990, 1992; Stenflo 1989, Spruit *et al.* 1991). The most important result is that, apart from a slight dependence on the area filling factor, almost all filaments seem to share basically the same properties (magnetic field strength, temperature, diameter; *cf.* Zayer *et al.* 1989, 1990). This led to the concept of a unique small-scale magnetic structure, the *magnetic element*, which constitutes the basic building block for magnetic fields in active regions and in the "quiet" Sun network.

Complementary to the FTS data which have low spatio-temporal resolution (but high spectral resolution and a large wavelength coverage) are two-dimensional, high-resolution magnetograms (*e.g.* Ramsey *et al.* 1977, Wang *et al.* 1985, Topka *et al.* 1986, Title *et al.* 1987, 1990). These data have clearly demonstrated the relation between magnetic, flow and intensity structures and the interaction of magnetic fields with convective and oscillatory velocities in the solar photosphere.

2.2 MAGNETOGRAMS, STOKES ANALYSIS, AND LINE RATIO METHOD

Magnetic fields in the solar atmosphere can be detected and measured by analyzing the polarization properties of the observed light since the Zeeman effect leads to a linear and/or circular polarization in the wings of spectral lines. It is customary to describe the polarization by means of the *Stokes parameters* I, V, Q, and U (*e.g.* Shurcliff 1962). Stokes I, the intensity of the unpolarized component, does not discriminate between magnetic and non-magnetic regions within the spatial resolution element of the observational device. The small-scale structures are spatially unresolved in spectra, thus measurements of spectral line profiles in Stokes I inevitably represent a mixture between profiles originating from magnetic and non-magnetic parts. Since the area fraction of magnetic elements (the "filling factor") generally is unknown, the interpretation of Stokes I profiles is difficult and often ambiguous.

On the other hand, Stokes V (the difference between left- and right-hand circularly polarized light) and Stokes U, Q (linearly polarized components) originate solely in the magnetic regions and, therefore, carry exclusive information about the magnetized parts of the atmosphere. Most studies have concentrated on Stokes V which is comparatively easy to measure; it depends mainly on the line-of-sight component of the magnetic field, *i.e.* its vertical component for observations near disc center. The classical magnetograph effectively is a detector for Stokes V in the wings of magnetically sensitive spectral lines. It must be kept in mind that, by definition, any polarimetric device measures polarization and *not* magnetic fields. To derive fields strengths and other magnetic information from polarimetric data is a non-trivial task and the results often depend on assumptions and models, especially if only one spectral line is used. We shall see below that, in the case of spatially unresolved fields, *i.e.* if the resolution element of the observational device contains a mixture between magnetic and non-magnetic parts, magnetograph observations in one spectral line effectively measure the magnetic flux within the resolution element (or the spatially *averaged* field strength) and not the *actual* value in the magnetic regions.

If more than one spectral line is used, properties of magnetic structures can be determined in remarkable detail, even if they are spatially unresolved. The so-called *line ratio method* (Stenflo 1973) is an example; it allows us to determine the actual magnetic field strength in spatially unresolved magnetic features. In what follows, we describe the method in its basic form.

For simplicity, we consider a spectral line which forms a normal Zeeman triplet. The magnetic field, B, is assumed constant and directed along the line of sight. The resulting Stokes V profile as a function of wavelength, λ, is given by

$$V(\lambda) = I_L - I_R = \frac{1}{2}\left[\Phi_m(\lambda + \Delta\lambda_B) - \Phi_m(\lambda - \Delta\lambda_B)\right] \tag{1}$$

where I_L, I_R are the intensities of left- and right-hand circularly polarized light, respectively, and $\Phi_m(\lambda)$ is the line intensity profile which would originate for $B = 0$ in otherwise the same atmosphere as in the magnetic structure. $\Delta\lambda_B = 4.67 \cdot 10^{-5} g \lambda^2 B$ [cgs] is the Zeeman splitting of the line which depends on its Landé factor, g.

Assume now a *two-component model*, *i.e.* the atmosphere in the resolution element is divided into a magnetic and a non-magnetic part with continuum intensities I_m and I_0, respectively, and with area fractions f and $1 - f$. The normalized signal S measured by a polarimeter is then given by the ratio of the spatial averages (denoted by angular brackets) of Stokes V and continuum intensity:

$$S(\lambda) = \frac{\langle V(\lambda)\rangle}{\langle I\rangle} = \frac{f \cdot \frac{1}{2}\left[\Phi_m(\lambda + \Delta\lambda_B) - \Phi_m(\lambda - \Delta\lambda_B)\right]}{f I_m + (1 - f) I_0}. \tag{2}$$

If the magnetic field is intrinsically weak in the sense that the Zeeman splitting is much smaller than the Doppler width of the line ($\Delta\lambda_B \ll \Delta\lambda_D$), the difference in the numerator can be approximated by the first term of a Taylor expansion. Additionally, for moderate spatial resolution and outside sunspots or pores one usually has $f \ll 1$. Hence, Equation (2) may be written approximately as

$$S(\lambda) \simeq f \cdot \frac{I_m}{I_0} \cdot \frac{1}{2}\left[\varphi_m(\lambda + \Delta\lambda_B) - \varphi_m(\lambda - \Delta\lambda_B)\right] \simeq f \cdot \frac{I_m}{I_0} \cdot \Delta\lambda_B \frac{d\varphi_m}{d\lambda}, \qquad (3)$$

where we have used the normalized line profiles $\varphi_m = \Phi_m/I_m$ (magnetic part) and $\varphi_0 = \Phi_0/I_0$ (non-magnetic part). We see from Equation (3) that, in this case, only the average field strength, $\langle B \rangle = f \cdot B$, multiplied by the ratio of the continuum intensities can be determined. The calibration of conventional magnetographs rests on the assumptions that $I_m = I_0$ and $\varphi_m = \varphi_0$; both are not fulfilled in the case of magnetic elements. Consequently, large errors in the determination of average fields and magnetic fluxes are possible (cf. Grossmann-Doerth et al. 1987), a problem which is of particular importance for the measurement of magnetic fields on active stars (e.g. Saar 1990).

The supposition of an intrinsically weak, spatially resolved field was implicitly made in all cases for which field strengths derived from magnetographic data were identified with the actual field strength. The line-ratio method, however, allows us to test whether this assumption is really valid. Take two spectral lines which have a different Zeeman-sensitivity (different Landé factors), but are otherwise "identical," i.e. are from the same chemical element, ionization level, and multiplet, and have similar excitation energies. A now classical example is the pair of iron lines Fe I 5247Å/5250Å with $g_1 = 2$, $g_2 = 3$, which were first used by Stenflo (1973). We take the ratio of their polarization signals, carry the Taylor expansion up to the third derivative and thereby obtain

$$\frac{S_1}{S_2} = \left(\frac{g_1}{g_2}\right)\left(\frac{\lambda_1^2}{\lambda_2^2}\right)\left(\frac{\varphi'_{m1} + \frac{1}{6}g_1^2\,\lambda_1^4\,B^2\,\varphi'''_{m1}}{\varphi'_{m2} + \frac{1}{6}g_2^2\,\lambda_2^4\,B^2\,\varphi'''_{m2}}\right) \qquad (4)$$

where the primes indicate derivatives with respect to λ. If the fields are intrinsically weak in the sense defined above, the third derivative terms are negligible and, since the intensity profiles $(\varphi_{m1}, \varphi_{m2})$ are the same, the ratio of polarization signals becomes equal to the ratio of the Landé factors (for $\lambda_1 \simeq \lambda_2$). If the measured ratio deviates from this, the assumption of intrinsically weak fields is invalid and, furthermore, Equation (4) now permits us to determine the *true* field strength. Observations of photospheric magnetic fields using this method have shown that more than 90% of the magnetic flux which permeates the solar atmosphere outside sunspots is in the form of intrinsically strong fields, i.e. the range of 1–2 kGauss. Moreover, the ratio of polarization signals is practically always the same, nearly independent of the polarization amplitude (filling factor) and the nature of the target (network, active regions, "quiet" Sun) (e.g. Frazier and Stenflo 1978, Stenflo and Harvey 1985). This means that the intrinsic field strength is always the same which suggests the idea of a unique magnetic structure, the *magnetic element*.

Note that, even if the true field strength is known, a determination of the filling factor, f, is not trivial since, in Equation (3), it is multiplied by the unknown ratio of continuum intensities. Assuming that this ratio is about unity, typical values of f for low spatial resolution observations (a few arcsec) range from about 1% in the network to between 10% and 20% in active region plages.

After the advent of polarized FTS spectra (Stenflo et al. 1984) with their enormous simultaneous wavelength coverage and spectral resolution (together with high signal-to-noise ratio), the line ratio method led the way for the development of a number of diagnostic techniques on the basis of Stokes profiles (mainly V, but also Q; for an overview, see Solanki 1992). Recently, the infrared spectral range found considerable interest (e.g. Stenflo et *al.* 1987b, Livingston 1991). Since the Zeeman sensitivity of spectral lines increases with wavelength, infrared lines open the possibility to detect intrinsically weak fields and to determine field strength distributions (Zayer *et al.* 1989; Rüedi 1991).

2.3 OBSERVATIONS WITHOUT POLARIZATION ANALYSIS

Observations in Stokes I, especially in white light or broad spectral bands, have the advantage that the number of available photons is large. This means that the image degradation due to seeing effects in the Earth's atmosphere can be minimized by using small exposure times. On the other hand, if the magnetic structures are unresolved, the Stokes I signal inevitably contains a mixture of information from magnetic and non-magnetic regions which makes quantitative analysis a very difficult task. Consequently, this kind of observation is best suited for morphological studies with high spatio-temporal resolution.

The first brightness structures found to be related to non-spot magnetic fields were white light faculae associated with active regions and enhanced network which are visible near the solar limb (*e.g.* Muller 1975, 1977; Hirayama 1978). At a spatial resolution of a few arc seconds, magnetic structures are also well correlated with core emission of the H and K lines of ionized Calcium (Skumanich *et al.* 1975) and with many lines in the UV.

Observations near the center of the solar disc which achieved very high spatial resolution revealed the existence of small-scale features (size less than 200 km, the resolution limit of existing solar telescopes) in the intergranular lanes, the downflow regions of the dominant convective flow pattern in the solar photosphere. These features appear bright in white light (*e.g.* Mehltretter 1974, Koutchmy 1978, Muller 1983, von der Lühe 1987, Auffret and Muller 1991), in spectral line cores (Sheeley 1967), and in line wings (Dunn and Zirker 1973). Some evidence exists that these bright features are related to concentrated magnetic field structures, *e.g.* a linear relation between the number of bright points and the magnetic flux measured in the same area (Mehltretter 1974), the cospatiality with chromospheric magnetic field indicators (Kitai and Muller 1984), and a close spatial association of bright features and polarization signal in high-resolution filtergrams/magnetograms (Title *et al.* 1990). However, the spatial resolution of presently feasible polarization measurements is insufficient to examine the magnetic nature of these structures directly. It is supposed that, due to geometric foreshortening effects and overlapping along the line of sight, the bright features observed near disc center are also responsible for the facular regions.

High-resolution spectra of Stokes I in facular regions have been obtained by Koutchmy and Stellmacher (1978) and by Stellmacher and Wiehr (1979) who used a two-component model with a bright magnetic element embedded in cool intergranular material to fit their data. Since Stokes I does not discriminate between magnetic and non-magnetic regions, many free parameters (filling factor, temperature structure of both components, magnetic field as function of height, velocities) have to be specified in such a model. If a polarization analysis is made (*e.g.* Koutchmy and Stellmacher 1987), this arbitrariness can be reduced considerably.

2.4 GLOBAL OSCILLATIONS

The global p-mode oscillations of the Sun become a more and more important diagnostic tool for the study of the solar interior (see Chapter 3) some profit from the wealth of these observational and theoretical studies presented in recent years. Magnetic elements and p-mode oscillations interact: On the one hand, observations of Stokes V-profiles have demonstrated that the magnetic structures participate in the global oscillations (Giovanelli *et al.* 1978, Wiehr 1985, Fleck 1991); on the other hand, the oscillation amplitudes are significantly reduced in regions with large filling factor of magnetic elements like active region plages (Tarbell *et al.* 1988). Similar to the case of sunspots which absorb oscillation energy by a yet unknown process (Braun *et al.* 1987), the reason for the amplitude reduction in plages is presently not well understood although proposals have been made (*e.g.* Bogdan and Knölker 1989). Once this problem is clarified, promising diagnostic opportunities will be opened.

Such a possibility has already materialized in the case of the dependence of the p-modes on the phase in the solar cycle and, more specifically, of the observed linear correlation between global oscillation frequencies and the total amount of magnetic flux in the photosphere (Woodard *et al.* 1991; see also section 6 of Chapter 3). Two models have been put forward to explain the influence of magnetic fields on the oscillation frequencies. On the one hand, Campbell and Roberts (1989, see also Evans and Roberts 1990, 1991) proposed a change in the atmospheric boundary condition caused by a variation of the chromospheric magnetic field. Following the results of Giovanelli and Jones (1982), they assumed a quasi-homogeneous "canopy" of mainly horizontal field in the upper photosphere or low chromosphere. This canopy modifies the boundary condition for the oscillations such that a frequency shift results which is proportional to the *square* of the strength of the canopy field while its sign depends on the assumed magnetic field gradient in the canopy. This result is intuitively plausible since the Lorentz force is quadratic in the magnetic field strength; it is not in agreement, however, with the frequency changes observed by Woodard *et al.* (1991).

In a recent paper, Evans and Roberts (1992) argue that the canopy effect nevertheless may lead to a linear relation between frequency shift and photospheric magnetic flux. Provided that the photospheric field is organized in concentrated, vertical flux tubes which expand like monopoles above a certain height, *i.e.* the canopy base, it can be shown that the mean square horizontal field at the base of the canopy (and, therefore, the frequency shift) is indeed proportional to the mean vertical photospheric field. It remains to be seen whether this property (and the required height dependence of the horizontal field) is in accordance with more realistic models of the chromospheric magnetic field (*e.g.* Solanki and Steiner 1990).

The other line of approach has been pursued by Bogdan and Zweibel (1987, see also Zweibel and Däppen 1989) and followed recently by Goldreich *et al.* (1991) who investigated the effect of a magnetic field organized in flux tubes ("fibrils") in the upper part of the convection zone. The influence on the oscillation frequencies results from an effectively enhanced rigidity or stiffness of the magnetized gas. Since the flux tubes do not fill the entire volume, the change $\Delta\omega$ of the oscillation frequency of a p-mode is proportional to the square of their (intrinsic) field strength, B, multiplied by the volume fraction filled by flux tubes, *i.e.* the filling factor f :

$$\Delta\omega \;\propto\; f \cdot B^2 \;\simeq\; \langle|B|\rangle\, B \,. \tag{5}$$

where $\langle|B|\rangle = f \cdot |B|$ is the (unsigned) average field strength which is proportional to the (unsigned) magnetic flux. If the variation of total flux is due to a variation in the number of flux tubes while their intrinsic field strength is unchanged, we see from Equation (5) that the frequency change is proportional to the average field or magnetic flux. This is in agreement with the observations of Woodard *et al.* (1991). *Quantitative* analysis of the

frequency shift in terms of flux tube structure, distribution, *etc.* is much more difficult and
depends critically on the choice of the averaging procedure (Bogdan and Cattaneo 1989).

We see that in both kinds of models the observed magnetic field dependence of the global
oscillation frequencies requires an intermittent (flux tube) structure of the magnetic field
in the photosphere. The oscillation results, therefore, support the conclusions drawn from
spectroscopic observations and, furthermore, may indicate that the filamentary structure
of solar photospheric magnetic fields extends into the upper parts of the convection zone.

2.5 SUMMARY OF OBSERVED PROPERTIES

The most important observed properties are briefly summarized in the following list. Some
of these points will be discussed in more detail within the appropriate sections below.
Small-scale magnetic flux concentrations in the solar photosphere

- comprise more than 90% of the magnetic flux outside sunspots,
- are located in the downflow regions of convective patterns,
- exhibit roundish ("tube") and elongated ("slab") geometry,
- show height-dependent field strengths between 1000 Gauss at $\tau_m = 0.01$ (continuum optical depth *within* the magnetic structure) and 2000 Gauss at $\tau_m = 1.$,
- are bright in continuum and in spectral lines and, therefore, hotter than the mean photosphere at the same *optical* depth,
- exhibit no significant stationary flows ("downdrafts") in their interior but show indications for strong non-stationary motions,
- are surrounded by rapid external downflows in their immediate environment,
- and, apart from slight differences between plage and network regions (different filling factor), their properties are remarkably uniform.

3. Theoretical Approaches

The observation of small-scale magnetic elements as the basic structure element of magnetic fields outside sunspots led to flourishing theoretical research. We can roughly divide
the concepts which have been used to describe the physics of magnetic elements and the
interaction with their environment into three basic approaches. These are described briefly
in the remainder of this section.

3.1 MAGNETOCONVECTION

The interaction between convection and magnetic fields is a topic which already found much
interest long before the existence of solar magnetic elements had been revealed and, in some
respects, it anticipated their detection. The two sides of this interaction are, on the one
hand, the advection and concentration of magnetic fields by convective flows (Parker 1963,
Weiss 1966) and, on the other hand, the modification of stability criteria and convective
patterns due to the action of the Lorentz force (*e.g.* Chandrasekhar 1961).

The focus of the studies performed along this line of research is not a detailed description
of magnetic structures in the solar atmosphere with all its complications but rather an isolation of the basic physical processes. Therefore, tractable problems are defined by assuming

idealized circumstances (*e.g.* closed box, constant diffusivities, no radiation, ...) and using convenient assumptions or approximations (incompressibility, Boussinesq approximation, two-dimensional systems). Reviews have been given by Proctor and Weiss (1982) and by Weiss (1991). A critical point of view has been put forward by Nordlund (1984a).

The *flux expulsion* mechanism found in the course of these studies (*e.g.* Weiss 1981ab, Hurlburt and Toomre 1988), a kind of "phase separation" between field-free, convecting plasma, and magnetic, almost stagnant regions, leads to concentrated magnetic structures in an electrically well-conducting plasma and is, therefore, of particular importance for the formation of solar magnetic elements.

3.2 THIN FLUX TUBES

A very useful approach, particularly for analytical studies, is the concept of a *magnetic flux tube*, *i.e.* a bundle of magnetic field lines (generally with cylindrical cross section), which is separated from its non-magnetic environment by a tangential discontinuity with a surface current. If the diameter of the flux tube is very small compared to all relevant spatial scales in the direction along its axis (scale height, radius of curvature, wavelength,...), the MHD equations can be cast in a very simple, quasi-one-dimensional form, the *thin flux tube approximation* (TFA) (*e.g.* Roberts and Webb 1978; Spruit 1981b; Ferriz-Mas and Schüssler 1989).

We describe the TFA here in its simplest form for an axisymmetric configuration with vertical axis. We use cylindrical polar coordinates (r, ϕ, z) and take $\partial/\partial\phi = 0$ for all quantities. If a physical quantity q (*e.g.* pressure, field and velocity components, ...) is regular at the axis $(r = 0)$, its dependence on the radial coordinate can be expanded in a Taylor series:

$$q(r, z, t) = q(0, z, t) + r\frac{\partial q}{\partial r}(0, z, t) + \frac{r^2}{2}\frac{\partial^2 q}{\partial r^2}(0, z, t) + \dots$$

$$\equiv q_0(z, t) + r q_1(z, t) + r^2 q_2(z, t) + \dots, \tag{6}$$

where t indicates the time-dependence. The abbreviations defined in the second line of Equation (6) simplify the writing of the formulae. It is a general property of the MHD equations under the given symmetry conditions (Ferriz-Mas and Schüssler 1989) that the expansion for scalar quantities (*e.g.* density ϱ, pressure p, temperature T, ...) and, for the z-components of vectors (vertical velocity v_z, vertical field component B_z), involves only *even* orders $(q_0, q_2, q_4, \dots)$ while, for the r- and ϕ-components of vectors $(v_r, v_\phi, B_r, B_\phi)$, we only have *odd* orders $(q_1, q_3, q_5, \dots)$. If we insert the expansions into the MHD equations and sort by expansion order, we obtain an infinite system of differential equations in the independent variables z and t.

Now take the radius, R, of the flux tube small enough compared to all scales of variation along the tube such that we may assume that all quantities vary weakly with r, except at the discontinuity at $r = R$. Under these circumstances, we may truncate the infinite system at low order such that a small number of differential equation remains. In the simplest case, the system is truncated at *zeroth* order; this is the TFA. We demonstrate the procedure with the simplest MHD equation, the solenoidality condition: $\nabla \cdot \mathbf{B} = 0$. In cylindrical coordinates, it reads:

$$\frac{\partial B_z}{\partial z} + \frac{1}{r}\frac{\partial}{\partial r}(rB_r) = 0. \tag{7}$$

Inserting the expansion given in Equation (6) leads to:

$$\frac{\partial}{\partial z}(B_{z0} + r^2 B_{z2} + r^4 B_{z4} + \ldots) + \frac{1}{r}\frac{\partial}{\partial r}(r^2 B_{r1} + r^4 B_{r3} + \ldots) = 0. \tag{8}$$

The zeroth order equation, therefore, is given by:

$$\frac{\partial B_{z0}}{\partial z} + 2B_{r1} = 0. \tag{9}$$

Note that zeroth order equations may well involve quantities of higher order like B_{r1} in Equation (9). This also shows that the TFA does *not* neglect the radial field component as is sometimes claimed. The same is true for the radial velocity component, v_r, as it becomes obvious from the full set of TFA equations which result from the procedure described above (primes denote $\partial/\partial z$, dots the time derivative, $\partial/\partial t$) :

$$\varrho_0 (\dot{v}_{z0} + v_{z0}v'_{z0}) + p'_0 + \varrho_0 g = 0 \tag{10}$$

$$\dot{\varrho}_0 + (\varrho_0 v_{z0})' + 2\varrho_0 v_{r1} = 0 \tag{11}$$

$$\dot{B}_{z0} + v_{z0}B'_{z0} + 2B_{z0}v_{r1} = 0 \tag{12}$$

$$\dot{p}_0 + v_{z0}p'_0 - \frac{\gamma p_0}{\varrho_0}(\dot{\varrho}_0 + v_{z0}\varrho'_0) = 0 \tag{13}$$

Together with Equation (9), we have 5 equations: the momentum equation (10) with gravitational acceleration g, the equation of continuity (11), the induction equation (12), and the energy equation (13, written here for adiabatic changes). The 6 unknown functions are: $\varrho_0, p_0, B_{z0}, B_{r1}, v_{z0}, v_{r1}$. We may combine Equations (11) and (12) to eliminate v_{r1}, *viz.*

$$\frac{\partial}{\partial t}\left(\frac{\varrho_0}{B_{z0}}\right) + \left(\frac{\varrho_0 v_{z0}}{B_{z0}}\right)' = 0, \tag{14}$$

but the system still is not closed. The missing equation is given by the condition of continuity of total pressure at $r = R$:

$$p(R, z, t) + \frac{B^2(R, z, t)}{8\pi} = p_e(R, z, t). \tag{15}$$

p_e is the (given) pressure in the external medium. To zeroth order, Equation (15) reads :

$$p_0 + \frac{B_{z0}^2}{8\pi} = p_e. \tag{16}$$

The step from Equation (15) to Equation (16) deserves a comment since it represents the transformation from a local condition (at $r = R$) to an equation which applies to the whole interior of the tube. This is justified if the time scale for the adjustment of pressure over the cross section of the tube (the crossing time τ_M of a magneto-acoustic wave) is small compared to the time scales of the processes we want to describe. Since $\tau_M \simeq 2R/v_M$, where v_M is the magneto-acoustic speed, it can be made arbitrarily small if the radius R of the tube is chosen small enough. For solar magnetic elements, we have $\tau_M \simeq 10$ sec.

Equations (9–13), together with Equation (16), form a closed system of equations, the zeroth order TFA. Apart from truncating at higher orders (*e.g.* Ferriz-Mas *et al.* 1989), the TFA equations may also be generalized to curved flux tubes (Spruit 1981c, see also Choudhuri 1990) and to include the effects of external flows (Parker 1975a), viscosity (Hasan and Schüssler 1985), and rotation (van Ballegooijen 1983, Ferriz-Mas and Schüssler 1992).

Let us discuss briefly the most elementary application of the zeroth order TFA equations, *i.e.* the static case in which the set of equations reduces to Equation (16), the pressure equilibrium condition, and

$$p_0' + \varrho_0 g = 0 , \tag{17}$$

which expresses hydrostatic equilibrium along the axis. Its solution is:

$$p_0(z) = p_0(0) \exp \left(- \int_0^z \frac{d\xi}{H_P(\xi)} \right) . \tag{18}$$

$H_P = \mathcal{R} T_0/\mu g$ is the scale height ($\mathcal{R}$: gas constant, μ: molecular weight). If the temperature inside the tube, $T_0(z)$, at each height is equal to the external temperature, $T_e(z)$, and if the external atmosphere is also in hydrostatic equilibrium, $p_e(z)$ has the same exponential factor as $p_0(z)$ in Equation (18). Therefore, we find for the plasma beta, *i.e.* the ratio between gas pressure and magnetic pressure, within the tube:

$$\beta \equiv \frac{8\pi p_0}{B_0^2} = \frac{p_0(z)}{p_e(z) - p_0(z)} = \frac{p_0(0)}{p_e(0) - p_0(0)} = \text{const.} \tag{19}$$

The plasma beta of a flux tube which is in temperature equilibrium with its environment is constant with height. From Equation (19), we find that, in this case:

$$B_0 \propto p_0^{1/2} \propto \exp \left(- \int_0^z \frac{d\xi}{2H_P(\xi)} \right) , \tag{20}$$

i.e. the magnetic field decreases with height and the corresponding scale height is twice the gas pressure scale height. Since magnetic flux conservation leads to $\pi B_0 R^2 = \Theta_m = \text{const.}$, we have

$$R \propto B_0^{-1/2} \propto \exp \left(\int_0^z \frac{d\xi}{4H_P(\xi)} \right) \tag{21}$$

which means that the scale height for the expansion of the flux tube is equal to four times the gas pressure scale height H_P. We may use the flux conservation condition again in order to determine the dependence of the expansion *rate* of a flux tube on its magnetic flux (or radius) for a fixed value of β:

$$\frac{dR}{dz} = \frac{d}{dz} \left(\frac{\Theta_m}{\pi B_0} \right)^{1/2} = \left(\frac{\Theta_m}{\pi} \right)^{1/2} \frac{d}{dz} \left(\frac{1}{B_0} \right)^{1/2} \propto \Theta_m^{1/2} \propto R . \tag{22}$$

We see that tubes with a large amount of magnetic flux (*e.g.* sunspots) expand faster with height than small tubes. This has important consequences for their stability (see section 7).

Among other applications, the TFA has been very successfully used for analytical and numerical studies of *convective collapse* (see section 4.2) and for the study of the properties of MHD wave modes in a flux tube geometry (see section 6). Moreover, thin flux tubes serve as a convenient and simple model for the raw interpretation of observations.

3.3 MODEL CALCULATIONS AND SIMULATIONS

A third theoretical approach is pursued with numerical MHD simulations attempting to represent or, at least, approximate solar conditions. The equations to be solved and the complications arising in this kind of approach have already been discussed in Chapter 4 (section 6).

Presently, calculations in two and in three spatial dimensions play a complementary role: While three-dimensional calculations (*e.g.* Nordlund 1986, Nordlund and Stein 1990) permit us to study the large-scale topology of magnetic fields and their interaction with granular motions, the spatial grid scale is limited by the immense requirements of computer time and storage. Consequently, the properties of individual, small-scale magnetic elements with thin boundary layers and the interaction with their environment cannot be studied. Two-dimensional calculations give this possibility since a much higher spatial resolution is possible, but they cannot simulate the interaction of magnetic elements with realistic, non-stationary granulation.

An adequate treatment of radiative energy transport is crucial for a sensible simulation of solar magnetic elements. Their thermodynamic structure is largely determined by a balance between cooling caused by radiative losses in the vertical direction and heating by horizontal radiative transport which is very efficient due to the small size of magnetic elements. Static models for magnetic elements with radiative transfer have been calculated by Steiner (1990, see also Steiner and Stenflo 1990) and by Pizzo *et al.* (1991) while time-dependent two-dimensional simulations have been presented by Grossmann-Doerth *et al.* (1989a), Shibata *et al.* (1990), and Knölker *et al.* (1991).

4. Formation of Concentrated Magnetic Structures

It is generally assumed that the magnetic flux which is observed in the solar photosphere originates in a dynamo process within (or immediately below) the convection zone from where it rises towards the surface due to the combined effects of buoyancy, instabilities, and convective flows (see Moreno-Insertis 1992). Simple estimates on the basis of conservation of mass and magnetic flux as well as numerical simulations of the eruption process suggest that the magnetic flux should arrive at the surface with a much lower field strength than the observed values of 1–2 kG (and even more in pores and sunspots). Two mechanisms have been studied in detail which are held responsible for the concentration of magnetic flux into structures of large field strength: *flux expulsion* and *convective collapse*. Although both processes are related, we discuss them in somewhat artificial separation in order to emphasize the basic effects.

4.1 FLUX EXPULSION

Parker (1963) showed first that, in an electrically well-conducting plasma with a stationary velocity field, a magnetic field is excluded from the regions of closed streamlines. We illustrate the process by a very simple example. Consider the following *kinematic* MHD problem: A two-dimensional velocity field

$$\mathbf{v}(x, z) = (v_x, v_y, v_z) = v_0 \left(-\sin\frac{\pi x}{L}, \ 0, \ \frac{\pi z}{L}\cos\frac{\pi x}{L} \right) \tag{23}$$

in Cartesian coordinates (x, y, z) is given in a box of dimension $L \times L$ in the xz-plane. The box is permeated by an initially homogeneous magnetic field $\mathbf{B} = (0, 0, B_0)$. The evolution of the field is determined by the MHD induction equation, *viz.*

$$\frac{\partial \mathbf{B}}{\partial t} = \nabla \times (\mathbf{v} \times \mathbf{B}) + \eta_m \nabla^2 \mathbf{B}, \tag{24}$$

where η_m is the (constant) magnetic diffusivity. A stationary state $(\partial/\partial t = 0)$ requires a balance between field advection, the first term on the right hand side of Equation (24), and diffusion, the second term. If the diffusivity is very small (the electrical conductivity very large) in the sense that the magnetic Reynolds number $R_m = v_0 L/\eta_m$ is large compared to unity, such a balance can only be achieved if the field is compressed ("expelled") in a narrow boundary layer near $x = 0$ such that the spatial gradient becomes sufficiently large. We may then apply standard boundary layer theory and write

$$v_x \simeq -\frac{v_0 \pi x}{L}; \quad B_z \simeq B(x) \tag{25}$$

near $x = 0$. The field $B(x)$ in the boundary layer is determined by the equation

$$-\frac{\pi v_0}{L}\frac{d}{dx}(xB) = \eta_m \frac{d^2 B}{dx^2} \tag{26}$$

which results in a Gaussian profile:

$$B(x) = B(0)\exp\left(-\frac{\pi R_m}{2L^2}x^2 \right). \tag{27}$$

Consequently, the extension of the boundary layer is inversely proportional to $R_m^{1/2}$ and, by flux conservation (which can be achieved by appropriate boundary conditions), we find:

$$B(0) = B_0(2R_m)^{1/2}. \tag{28}$$

We see that, for large values of R_m, high peak fields within the boundary layer are predicted, even more so in the cylindrical, axisymmetric case where, for geometrical reasons, we have $B(0) \propto R_m$. In reality, however, the back-reaction of the magnetic field on the velocity limits the field strength which can be reached in the course of this process: if the magnitude of the Lorentz force becomes of the order of the inertial force in the equation of motion (*cf.* Equation 23 of Chapter 4), *i.e.*

$$\frac{1}{4\pi}(\nabla \times \mathbf{B}) \times \mathbf{B} \simeq \varrho \mathbf{v} \cdot \nabla \mathbf{v}, \tag{29}$$

further amplification if the field ceases since the magnetic force suppresses the flow in the boundary layer. The order of magnitude of the corresponding field strength follows from Equation (29):

$$\frac{\mathbf{B}^2}{4\pi} \simeq \varrho \mathbf{v}^2 \tag{30}$$

which gives

$$B \simeq v(4\pi\varrho)^{1/2} \equiv B_{eq} . \tag{31}$$

Equation (30) expresses the equality of magnetic and kinetic energy density; the resulting field strength B_{eq} is, therefore, called the *equipartition* * field strength. A magnetic field amplified by flux expulsion cannot significantly exceed the equipartition field strength, unless additional effects come into play. One example is a plasma for which the kinematic viscosity is much larger than the magnetic diffusivity (Busse 1975). Such a situation, however, is not realized in a stellar convection zone or atmosphere. Other possibilities are thermal effects and superadiabaticity which are discussed in subsection 4.2 below.

Flux expulsion by convective flows has been investigated numerically in the kinematic as well as in the dynamic case (*e.g.* Weiss 1966, Galloway *et al.* 1978, Weiss, 1981ab). These and a number of other studies demonstrated that, in a convecting medium at high magnetic Reynolds number permeated by a magnetic field, the magnetic flux is concentrated into filaments between the convection cells. This effect is related to the phenomenon of "intermittency" for magnetic fields in turbulent flow (*e.g.* Kraichnan 1976, Orszag and Tang 1979, Meneguzzi *et al.* 1981). In a compressible stratified fluid, the magnetic flux concentrations formed by flux expulsion are found in the convective downflow regions (Nordlund 1983, 1986; Hurlburt *et al.* 1984; Hurlburt and Toomre 1988) in accordance with the observed patterns of magnetic fields in the solar photosphere: on the granular (Title *et al.* 1987) and on the supergranular scale (the "network") magnetic flux is predominantly located and concentrated in the downflow regions.

The simulations also show that the convective motions are excluded from strong flux concentrations and a kind of "phase separation" between field-free, convecting fluid and magnetic, almost stagnant regions evolves. This effect suppresses the convective heat exchange between the magnetic structure and its surroundings. Consequently, thermal interaction with the environment is reduced to radiative energy transport.

The well-known formal analogy between the MHD induction equation and the equation which determines the time evolution of *vorticity* supports the conjecture that the expulsion effect also operates for vorticity and leads to the formation of intense whirls or vortices. An example in cylindrical geometry has been given by Galloway (1978) while Schüssler (1984a) showed that a full analogy between *kinematic* expulsion of magnetic field and vorticity holds for two-dimensional flow in Cartesian geometry. Numerical simulations of turbulence (*e.g.* McWilliams 1984) and laboratory experiments in rotating, turbulent fluids (McEwan 1973, 1976; Hopfinger *et al.* 1982) have demonstrated vorticity expulsion for rotationally dominated flows. A similar effect occurs in the simulations of solar granulation carried out by Nordlund (1984b, 1985) who found that narrow granular downdrafts come into rapid rotation due to the concentration of angular momentum fluctuations (the "bathtub effect"). We expect, therefore, that in the surface layers of the Sun, magnetic field and vorticity are concentrated in the same locations such that magnetic flux concentrations become

---

* In the literature on stellar magnetic fields the term "equipartition field strength" is sometimes used for the field in a totally evacuated flux tube whose magnetic pressure balances the gas pressure in the atmosphere. This should not be confused with the conventional definition given in Equation (31).

surrounded by rapidly rotating, descending whirl flows. This has important implications for the stability properties of magnetic elements (see section 7 below).

In summary, we may say that the flux expulsion mechanism can qualitatively explain the observed patterns of magnetic fields in the solar photosphere and the concentration of fields in the convective downflow regions. The field strengths which can be achieved by this process alone, however, are too small: for the observable regions of the solar photosphere we have typical values of $v \simeq 2$ km sec^{-1} for the horizontal velocity and $\varrho \simeq 3 \cdot 10^{-7}$ g cm^{-3} which give an equipartition field of about 400 Gauss, much less than the observed fields between 1000 and 2000 Gauss. Consequently, an additional amplification mechanism is required,

4.2 THERMAL EFFECTS AND CONVECTIVE COLLAPSE

In the case of the solar (sub-)photosphere, magnetic fields can be concentrated beyond the equipartition limit of a few hundred Gauss by *thermal* effects. Since the horizontal flows of granular convection are responsible for both sweeping the magnetic field to the downflow regions and for carrying heat to those regions, the retardation of the flows by the growing magnetic field leads to a cooling of the magnetic region because the radiative losses can no longer be balanced by the throttled horizontal flow. This cooling effect causes an increase of the magnetic field since gas pressure in the magnetic region becomes smaller and it accelerates the downflow which gives rise to the *superadiabatic effect* (Parker 1978): An adiabatic downflow in a magnetic flux tube which is thermally isolated from its surroundings leads to a cooling of the interior with respect to the superadiabatically stratified surroundings and a partial evacuation of the the upper layers ensues. Pressure equilibrium with the surrounding gas is maintained by a contraction of the flux tube which increases the magnetic pressure. In this way, the magnetic field can be intensified locally to values which are only limited by the confining pressure of the external gas.

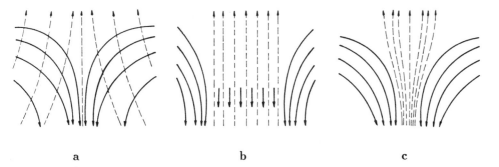

        a   b   c

Figure 1: Schematic sketch of the processes which lead to concentration of magnetic flux in the solar (sub)photosphere. **a:** The horizontal flows of granular convection sweep magnetic flux towards the downflow regions (full: stream lines, dashed: field lines). **b:** Magnetic forces suppress the convective motions when the magnetic energy density becomes comparable to the kinetic energy density. This throttles the energy supply into the magnetic region, the gas cools off due to radiative losses and a downflow is initiated. This triggers an instability caused by the superadiabatic stratification which strongly amplifies the flow. **c:** The downflow has evacuated the upper layers and the magnetic field has increased in order to maintain lateral force balance. A new equilibrium with a strong field has established itself.

It has been shown by a number of authors (Webb and Roberts 1978, Spruit and Zweibel 1979, Unno and Ando 1979) that the superadiabatic effect in the case of a flux tube which is initially in magnetostatic and temperature equilibrium with its environment drives a *convective instability* in the form of a monotonically increasing up- or downflow. Consequently, the initial downflow due to the radiative cooling will be enhanced by this effect leading to an even stronger amplification of the magnetic field, a process which is often referred to as *convective collapse*. Hasan (1983, 1984, 1985) and Venkatakrishnan (1983, 1985) have studied this process with aid of numerical simulations using the thin flux tube approximation. While they confirm the linear theory of convective instability of weak magnetic fields, the results for strong fields and for the nonlinear evolution of the instability are rather uncertain. They depend on the choice of boundary conditions as well as on assumptions concerning the initial state of the flux tube and its thermal interaction with the environment (for a more detailed discussion, see Nordlund 1984a and Schüssler 1990).

Presently, the final state of a magnetic element after convective collapse cannot be reliably predicted. It is even uncertain whether it is a static or an oscillatory state: although horizontal radiative transfer may, in principle, lead to overstable oscillations (Roberts 1976, Spruit 1979, Hasan, 1985, 1986; Venkatakrishnan 1985; Massaglia *et al.* 1989), a proper treatment of radiation far beyond the limits of the diffusion equation or "Newton's law of cooling" and an adequate level of spatial resolution in numerical simulations is necessary in order to decide whether overstable oscillations are excited in a realistic model of a magnetic element.

5. Magnetic and Thermal Structure

After the magnetic element has formed, it reaches a state which is stabilized against further convective collapse by the strong field and the (comparatively) low temperature of the gas in its interior. The upper limit of 250 m sec^{-1} for any systematic flow *within* the magnetic structures determined from the absolute shift of the Stokes V-profile zero crossings of spectral lines in spatially and temporarily unresolved FTS spectra (Stenflo and Harvey 1985, Solanki 1986; see section 6 below) indicates that magnetic elements reach a quasi-stationary state which does not deviate strongly from (magneto-)hydrostatic equilibrium. However, the existence of non-stationary velocities (oscillations, turbulence) is not excluded.

Most of the available observational information probably corresponds to this quasi-stationary phase of magnetic elements: a value of 90% for the fraction of magnetic flux in strong field form indicates that the time scales of the formation and destruction processes are much shorter than the duration of the stationary phase. The presently most sophisticated analysis of Stokes V data (which originate exclusively *within* the magnetic structures) has been given by Keller *et al.* (1990, see also Zayer *et al.* 1990) who carried out an inversion procedure on the basis of consistent two-dimensional magnetostatic equilibria for cylindrical flux tubes and synthetic Stokes profiles for many lines of sight. It is found that, at equal *geometrical* depth in the upper photosphere, the gas in the flux tube is 500-1000 K hotter than the non-magnetic atmosphere; the temperature difference decreases and reverses sign with increasing depth and, at the level of the formation of the continuum within the flux tube, the internal gas is about 2500 K cooler than the quiet atmosphere. On the *optical* depth scale for the continuum light, however, the flux tube atmosphere is everywhere hotter than the non-magnetic atmosphere: density and temperature deficit in the tube lead to small values of the continuum opacity (due to the H$^-$ ion) with the consequence that the whole optical depth scale is shifted downwards within the magnetic structure, similar to the well-known Wilson depression of sunspots.

The magnetic field strength as function of geometrical height as determined from Stokes inversion analysis turns out to be remarkably similar for a large range of filling factors (Zayer *et al.* 1990) while the temperature decreases somewhat for larger filling factor, *i.e.* magnetic elements are cooler in active region plages than in the quiet Sun network. It turns out that in the lower photosphere the height-dependence of the field is rather well described by the thin flux tube approximation discussed in section 3.2. As a function of continuum *optical* depth, τ_c, the field strengths range from typically about 1000 Gauss at $\tau_c = 0.01$, 1500 Gauss at $\tau_c = 0.1$, to about 2000 Gauss at $\tau_c = 1$.

As far as theoretical models are concerned, the objective is to describe the quasi-stationary phase of magnetic elements self-consistently including force balance and dynamics, energy transport by radiation and flows, and interaction with the environment. Some guidance as to the assumptions going into such models can be taken from simple considerations. For example, the strong buoyancy force of a magnetic element keeps it essentially vertical in the (sub-)photospheric layers (Schüssler 1986) such that the model of a vertical flux tube with a straight axis seems reasonable unless its diameter is much smaller than the scale height. Also, the assumption of an almost discontinuous transition (current sheet) between the flux concentration and its environment is supported by estimates of the widths of resistive and viscous boundary layers (Schüssler 1986).

The effects of finite resistivity and incomplete ionization are irrelevant: The cross-field flows due to finite resistivity are less than 10 m sec^{-1} for the photospheric regions while the drift velocities of neutral atoms are a few cm sec^{-1} at maximum (Hasan and Schüssler 1985). The small amount of mass influx into the tube caused by these effects is compensated by a moderate downflow with a velocity below 10 m sec^{-1}. This is very small compared to the sound speed such that the stratification practically stays hydrostatic.

Ultimately, magnetic element modeling calls for a comprehensive three-dimensional, time-dependent numerical simulation. However, this cannot be achieved with the computational facilities available at present or in the near future. For example, the simulations by Nordlund (1983, 1986) have a spatial resolution of about 250 km in the horizontal direction while a value of a few km is needed to resolve the boundary layer between a flux concentration and its surroundings which is crucial for a correct description of the energy balance. Consequently, the available three-dimensional simulations may describe the average motion of an *ensemble* of magnetic elements in a time-dependent granular velocity field but they give no information on the dynamics and the properties of individual flux concentrations. So it is necessary to consider models restricted to two spatial dimensions (axisymmetric flux tubes or infinite flux slabs) which allow a much better spatial resolution.

The small diameter of magnetic elements, their low internal opacity, and the suppression of convective energy transport by the strong magnetic field lead to a crucial role of radiative transfer for the energy balance of photospheric magnetic structures. The anisotropy of the radiation field near the inclined boundary layers between magnetic and non-magnetic gas and the necessity to describe both optically thick and optically thin regions demands a full treatment of the radiative transfer by integration of the transport equation. The present state of the art, therefore, is to avoid simplified procedures like the Eddington or the diffusion approximations but rather to integrate the full radiative transfer equation along many rays and angles, preferably in a non-grey model. Model calculations of this kind have been performed by Grossmann-Doerth *et al.* (1989a), Knölker *et al.* (1991) for time-dependent flux slab models with grey radiative transfer, by Pizzo *et al.* (1991) for static slabs with grey radiative transfer, and by Steiner (1990, see also Steiner and Stenflo 1990) for static cylindrical flux tubes with non-grey radiative transfer.

As an example of the results of such a model calculation, Figure 2 shows the stationary state reached in two-dimensional simulation of a magnetic slab carried out by Knölker *et al.* (1991). Since the convective energy transport is suppressed, the magnetic structure becomes cooler than its environment below $z = 0$ ($\tau_c = 1$ of the non-magnetic atmosphere) due to radiative losses at the surface. The temperature deficit (at the same geometrical height) reaches a maximum value of nearly 3000 K. On the other hand, the magnetic element is small and has a large transparency due to its low temperature and density. This sets up a *lateral* influx of radiation along the horizontal temperature gradient which heats the magnetic structure and, in fact, more than compensates for the absence of a significant convective heat flux: Energy is channeled into the magnetic elements which act accordingly as "heat leaks" in the uppermost layers of the convection zone and radiate more energy per unit area into space than the mean non-magnetic atmosphere (Zwaan 1967, Spruit 1976, 1977; Deinzer *et al.* 1984, Knölker *et al.* 1988, 1991). This inflow of heat balances the vertical radiative losses and limits the temperature deficit to a level which still leads to a temperature excess compared to the quiet Sun at equal *optical* depth for vertical incidence of the line of sight. These results are in qualitative agreement with semiempirical models determined from Stokes V-profiles (Solanki 1986, Keller *et al.* 1990) without the necessity to incorporate mechanical heating in the theoretical models.

The lateral energy influx into the small magnetic element is supplied by the surrounding atmosphere such that the spatially averaged energy flux is only slightly enhanced (a small part of the heat flux disturbance propagates into the thermally well-conducting convection zone which has a huge thermal inertia). This sets up a flow pattern outside the magnetic element which advects the heat which is radiated laterally into the magnetic structure. The thermally driven flow exhibits horizontal velocities of about 2 km sec^{-1} towards the magnetic element and a strong downdraft (up to 6 km sec^{-1}) immediately adjacent to the flux concentration. There are direct observational indications for both flow components (see section 6 below).

The lateral radiative heating and the shift of the opacity scale have the consequence that magnetic elements appear as bright structures in the continuum light. The model calculations predict values for the continuum intensity which are between 1.4 and 1.8 times larger than that of the quiet atmosphere; this is in accordance with the observations of bright structures in the intergranular spaces (*e.g.* Koutchmy 1978, Muller and Keil 1983, von der Lühe 1987). On the other hand, one might ask: Why are sunspots dark and emit much less energy than the quiet atmosphere ? The reason is their much larger size: Efficient radiative exchange is restricted to dimensions of the order of the photon mean free path which does not exceed about 100 km in the relevant regions. Consequently, a structure of that size can be heated efficiently by lateral influx of radiation while the much larger sunspots cannot benefit from this effect (except in a narrow boundary layer) and, therefore, cannot compensate for the reduction of convective energy flux brought about by their strong magnetic field.

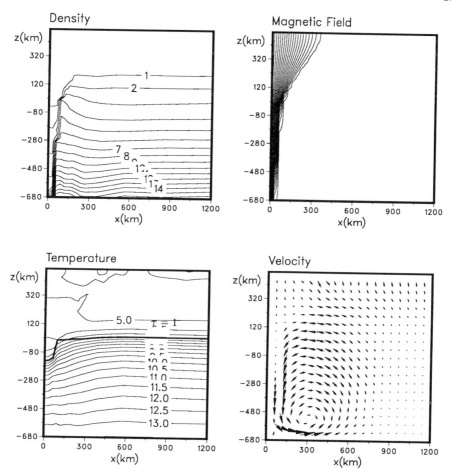

Figure 2: Result of a two-dimensional simulation of a flux slab in the solar (sub)photosphere. Only half of the symmetric structure is shown; the sheet extends infinitely in the direction perpendicular to the plane of the paper. The origin of the z-coordinate is the level of optical depth unity of the non-magnetic atmosphere (near to the visible solar "surface"). Densities are in units of 10^{-7}g·cm^{-3}, temperatures in 1000 K. While the magnetic structure is significantly cooler than the surroundings in the deep layers (due to suppression of convective transport), it is hotter in the upper parts (due to radiative illumination). The thick line in the isotherm plot (lower left frame) gives the location of optical depth unity (continuum, 5000 Å) for vertical incidence of the line of sight. The interior of the magnetic structure is almost static (except for an oscillation of the upper layers) and a convective cell circulates in the exterior which supplies the energy radiated into the magnetic element. The sizes of the flow velocity arrows scale with the modulus of the velocity; its maximum value is about 6 km sec^{-1}.

The model calculations which incorporate a full radiative transfer revealed another important effect, *i.e.* the heating of the upper layers within the flux concentration by *radiative illumination:* The material in the region above optical depth unity of a partially evacuated magnetic element is bathed in the radiation from the hot bottom (with a temperature of more than 7000 K at $\tau_c = 1$). It, therefore, reaches an equilibrium temperature which is larger than that of the gas at the same height in the quiet atmosphere which "sees" radiation from a layer of optical depth unity at a temperature of about 6400 K. As can be seen clearly in Figure 2, the result is the formation of a hot region with a temperature which is a few hundred degrees larger than that of the environment at equal *geometrical* depth (Grossmann-Doerth *et al.* 1989a, Kalkofen *et al.* 1989, Steiner 1990).

We conclude that the theoretical models are qualitatively in accordance with the empirical results concerning the temperature structure of magnetic elements as function of geometrical depth: a temperature depression in the deep layers is accompanied by a temperature enhancement in the middle and upper photosphere. Consequently, the models seem to have captured the basic physical processes which determine the energy balance.

The center-to-limb variation of the intensity contrast has been discussed in some detail by Schüssler (1987b; for a different point of view, see Schatten *et al.* 1986). Three effects contribute to the variation of the intensity contrast for finite inclination of the line of sight:

- The hot bottom of the magnetic element is obscured for rather small inclinations leading to disappearance of bright points (see Muller and Roudier 1984, Muller *et al.* 1989);
- the *bright wall* of the flux concentration becomes visible at large inclination angles and leads to a positive intensity contrast depending on the ratio between size and depth (Wilson depression) of the magnetic structure (Spruit 1976);
- the hot upper regions of the magnetic elements overlap near the limb and lead to a sharp contrast increase (Steiner and Stenflo 1990; see also Rogerson 1961).

These effects suffice to understand the various observational results if spatial resolution, selection effects, and the precise way of measurement are taken into account. High resolution observations near disc center are determined mainly by the hot bottom of the magnetic elements, while the selection of individual bright "facular granules" (*e.g.* Muller 1975) reveals the effect of the bright wall. The brightness evolution of individual faculae near the limb (Akimov *et al.* 1987), the observations of elevated faculae during eclipses (Akimov *et al.* 1982), and the measurements at the extreme limb (Chapman and Klabunde 1982, Lawrence and Chapman 1988) can be understood by the effect of overlapping hot regions.

6. Dynamics

Flows, oscillations, and waves are of particular interest for the study of magnetic elements. On the one hand, they give the possibility to compare with theoretical and simulation results and provide means to study the internal structure of flux concentrations while, on the other hand, it is suspected that magnetic elements act as guides for MHD waves carrying mechanical energy to heat the chromosphere and corona.

The major part of observational results on the dynamics of magnetic elements (see Solanki 1992 for a comprehensive overview) are based on analysis of Stokes V-profiles which can be used to determine the properties of spatially unresolved magnetic structures. The Stokes V-profiles of a spectral line in the presence of a magnetic field is shown schematically in Figure 3. Important parameters of observed Stokes V-profiles and the dynamical properties of magnetic elements which can be derived from them are:

- *zero-crossing wavelength* (λ_0): The wavelength shift between λ_0 and the rest wavelength of the line center can be used to measure systematic flows *within* magnetic elements. Solanki (1986) gives an upper limit of 250 m sec^{-1} for such flows. This rules out models of magnetic elements which rely on steady downflows and the Bernoulli effect.

- *wing width* $(\Delta\lambda_{b,r})$: Although strong systematic flows are not observed, the large width of the wings of Stokes V requires non-stationary ("turbulent") motions of a few km sec^{-1} (Solanki 1986, Keller *et al.* 1990). Since there is no strong center-to-limb variation of $\Delta\lambda$, both vertical and horizontal motions must be present (Pantellini *et al.* 1988).

- *asymmetry:* Observed Stokes V-profiles are asymmetric, *i.e.* for observations taken near the center of the solar disc, amplitude and area of the blue wing are generally larger than those of the red wing (Solanki and Stenflo, 1984). The *area* asymmetry can be used to study external downflows (see below) while the *amplitude* asymmetry indicates nonlinear oscillatory motions within the magnetic elements (Solanki 1989, Grossmann-Doerth *et al.* 1991).

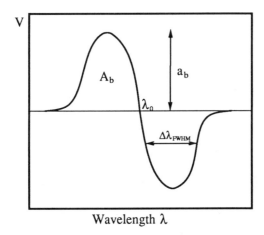

Figure 3: Schematic sketch of a Stokes V-profile. The quantities which contain information about dynamical properties of magnetic elements are: λ_0, the wavelength of the zero-crossing; $\Delta\lambda_{b,r}$: full width at half maximum of the wings; $a_{b,r}$: wing amplitudes, and $A_{b,r}$: wing areas – all for the blue (b) and red (r) wing, respectively.

The observed asymmetric and unshifted V-profiles presented a challenging riddle for a couple of years. While *amplitude* asymmetries arise easily (*e.g.* by superposition of symmetric, Doppler-shifted V-profiles for weak and strong fields), an *area* asymmetry, *i.e.* a net circular polarization of the whole profile, requires a combination of magnetic field and velocity gradients along the line of sight (Illing *et al.* 1975, Auer and Heasley 1978, Sanchez-Almeida *et al.* 1988). However, flows *within* the magnetic structure in a physically realistic configuration (*e.g.* magnetic fields decreasing with height) which reproduce the observed asymmetries lead to large shifts of the V-profile zero crossings which contradict the observations (Solanki and Pahlke, 1988).

Van Ballegooijen (1985) suggested that an area and amplitude asymmetry of the V-profile may also be caused by a downflow *outside* but in the immediate vicinity of a static magnetic flux concentration: Since the magnetic field flares out with height, lines of sight at the periphery traverse static magnetic (upper part of the atmosphere) and downflowing non-magnetic (lower part of the atmosphere) regions. It has been shown by Grossmann-Doerth *et al.* (1988, 1989b) that, quite generally, such a configuration leads to asymmetric V-profiles with *unshifted* zero crossings. The physical basis for the asymmetry is spectral line saturation: The magnetic field (in the upper part) and the downflow (in the lower part) shift the absorption coefficients for left- and right-hand circularly polarized light such that they are superposed in wavelength for one type of polarization and moved apart for the other. Line saturation has the consequence that the V-profile wing corresponding to the superposed absorption coefficients becomes weaker than the other.

Solanki (1989) was able to demonstrate that the observed V-profile area asymmetries of many spectral lines can be quantitatively reproduced by cool downflows adjacent to a magnetic structure which expands with height. Furthermore, since the downflows are fed by horizontal flows directed towards the flux concentration, this model, at the same time, accounts for the sign reversal of the asymmetry shown by observations near the solar limb (Stenflo *et al.* 1987a, Knölker *et al.* 1991, Bünte *et al.* 1991). Figure 4 illustrates the geometry of the magnetic element and the surrounding flow field which leads to the formation of asymmetric V-profiles.

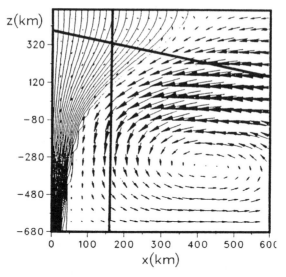

Figure 4: Velocity and magnetic field structure of a magnetic element model. Only half of the symmetric structure is displayed. Two representative lines of sight are indicated which traverse static, magnetic regions and non-magnetic moving gas leading to asymmetric Stokes V-profiles. The vertical line "sees" a flow away from the observer while the inclined ray cuts a flow towards the observer such that the resulting V-profile asymmetries have different signs (from Schüssler, 1990).

External flows near an expanding magnetic element lead to area and amplitude asymmetries of the same order of magnitude. Observed profiles, however, often show a much larger relative amplitude asymmetry than area asymmetry. It is suspected that oscillations in the interior of magnetic elements are responsible for this result (Solanki 1989). In fact, Grossmann-Doerth *et al.* (1991) were able to show that suitably chosen oscillatory internal flows together with external downflows can reproduce the observed amplitude *and* area asymmetries without exceeding the limit for the zero-crossing shift set by Solanki (1986).

The successful interpretation of asymmetric V-profiles in terms of external flows and internal oscillations supports the picture of magnetic elements which has emerged from analytical studies and numerical simulations. The external flow pattern which is deduced from the asymmetries is precisely the pattern predicted by the model calculations (*e.g.* Deinzer *et al.* 1984, Grossmann-Doerth *et al.* 1989a): the cool environment of the flux concentration supports a strong downdraft adjacent to the magnetic structure which is fed by a broad upflow and a horizontal flow which, at the same time, advects the energy radiated horizontally into the magnetic element. Internal oscillations have also been found (Knölker *et al.* 1991).

There is a large number of theoretical studies of oscillations and waves in magnetic flux tubes (*e.g.* Spruit 1981a, Herbold *et al.* 1985, Thomas 1985, Roberts 1986, 1991; Ryutova 1990). The flux tube geometry modifies the usual MHD *wave modes:* The slow mode develops into the longitudinal tube mode, while the linearly and circularly polarized Alfvén waves are transformed into the transversal and torsional tube mode, respectively. If higher orders in the slender flux tube approximation are included, a tube analogue of the fast mode also appears (Ferriz-Mas *et al.* 1989). Radiative energy exchange with the exterior medium can drive *overstable oscillations* (*e.g.* Hasan, 1986). The various wave modes of a flux tube may be excited in different ways:

→ the longitudinal mode by "buffeting" due to granules,
→ the transversal mode by "shaking" caused by horizontal flows,
→ the torsional mode by twisting due to whirl flows,
→ overstable oscillations by radiative heat exchange.

Oscillations and wave propagation in magnetic structures is a topic in its own right which probably has important implications for the structure of the solar chromosphere, both in "quiet" and in active regions. Since we have excluded the chromospheric layers from this lecture, we stop the discussion here and refer the interested reader to the reviews mentioned above and, in particular, to the proceedings edited by Ulmschneider *et al.* (1991).

7. Instability and Decay

The lifetime of individual magnetic elements is difficult to determine observationally. Muller (1983) found a mean lifetime of network bright points near disk center of about 20 minutes, but it is not clear whether this also represents the life span of the underlying magnetic structure. A crude estimate of a minimum lifetime of a magnetic element in strong-field form can be derived from the lower limit of 90% for the fraction of magnetic flux (excluding the "turbulent" flux) in strong-field form (Howard and Stenflo 1972) and the timescale of the convective collapse of 2 to 5 minutes (Hasan 1985, Nordlund 1986). Allowing for a quick destruction of the magnetic element by an instability (see below) within one minute, we find a minimum lifetime between 30 and 60 min for the quasi-equilibrium state.

Which processes can possibly destroy magnetic elements? In regions of mixed polarity, *re-connection* is important (Spruit *et al.* 1987). The result of the reconnection of two opposite polarity magnetic elements depends on the location of the reconnection point: If it is *below* the surface, a U-shaped loop forms which floats upwards due to magnetic buoyancy. It arrives there with a low field strength since the strong decrease of density with height and mass conservation leads to a significant expansion of the rising flux tube. Such a process is possibly a source of intrinsically weak magnetic field and might also be related to the "intranetwork fields" (Martin *et al.* 1985, Livi *et al.* 1985). If reconnection takes place *above* the surface, it forms a loop which can be drawn below the surface due to the action of magnetic tension forces if the footpoint separation is less than a few scale heights (Parker 1979, Chapter 8). Both possibilities lead to quite different observational signatures (see discussion in Spruit *et al.* 1991).

An individual flux concentration can be destroyed by dynamical processes, most efficiently by an instability. Besides the destabilizing influence of external flows related to the Kelvin-Helmholtz instability (*e.g.* Schüssler 1979, Tsinganos 1979, Bünte *et al.* 1992), the interchange or fluting instability is most important (Parker 1975b). While pores and sunspots can be stabilized by gravity (Meyer *et al.* 1977), small flux concentrations are stable with respect to fluting if they are surrounded by a strong whirl flow (Schüssler 1984b, Bünte *et al.* 1992). Such a structure is likely to form by advection of angular momentum towards the localized downflows by the familiar "bathtub effect." In fact, simulations of granular convection clearly show the formation of intense vortices (Nordlund 1985). Since the flux concentration cools its surroundings and thus enhances the converging downflows, magnetic structure and flow pattern can mutually stabilize each other: The strong thermal effects shape and stabilize the convective flow structure while this pattern, by means of advection of vorticity, stabilizes the magnetic element. The observed deformation of granules around bright points (Muller *et al.* 1989) and the prolonged lifetime of the granular pattern in plage regions (Title *et al.* 1987) support this conjecture.

If for some reason, *e.g.* because of a major reorganization of the pattern of convection, the supply of angular momentum becomes insufficient and the whirl decelerates, fluting instability sets in, and the flux concentration is disrupted typically within the Alfvén transit time of less than a minute (Schüssler 1986). The following evolution depends on the size of the fragments: If they are small enough (of the order of a few km), magnetic diffusion becomes relevant and the fragments tend to disperse into weak fields which may go through another flux expulsion/convective collapse cycle. Larger fragments may survive long enough to be reassembled by the granular flows and fuse into new flux concentrations without the necessity of another convective collapse.

In this way, a dynamical view of small-scale photospheric magnetic fields emerges. At any given time, most of the flux is in magnetic elements, but the individual elements sooner or later split into fragments. Small fragments diffuse and, together with rising U-loops, contribute to a weak-field component which partially becomes reconcentrated by flux expulsion and convective collapse. Larger fragments (with a diameter > 10 km, say) will rapidly heat up by radiation from the side leading to a decrease of the magnetic field strength. They are passive with respect to flows and may become severely distorted and inclined from the vertical direction before being assembled in intergranular downdrafts to form new magnetic elements. All processes of splitting, diffusion, expulsion, collapse and, accumulation of fragments operate in a timescale of minutes such that the majority of the flux at any given instant of time resides in the quasi-equilibrium strong-field form of magnetic elements.

8. Conclusion

The concept of small-scale fields consisting mainly of ensembles of similar structures (*magnetic elements*) which may be described by basic flux tube or flux slab models embedded in a non-magnetic environment has turned out to be remarkably successful. A consistent picture of magnetic elements begins to emerge based on analysis of spatially unresolved FTS data with their high spectral quality (in terms of resolution, noise, and wavelength range) and on comprehensive model calculations and analytical studies of idealized problems which have revealed the basic physical effects. Let us try to summarize this picture:

Expulsion of magnetic flux by strong horizontal granular flows leads to magnetic flux concentrations in the intergranular lanes. Radiative cooling and the large superadiabaticity of the uppermost layers of the convection zone cause a strong local intensification of the magnetic field by way of a partial evacuation (convective collapse). A quasi-equilibrium evolves which is characterized by the absence of systematic internal flows and the balance of the magnetic pressure by an internal gas pressure deficit. The magnetic flux density decreases with height in a way well described by the slender flux tube approximation.

This equilibrium is stabilized against further collapse by a temperature deficit of the layers below optical depth unity in the magnetic structure due to the suppression of convective energy transport. Heating by lateral influx of radiation, reduced density and the strong temperature dependence of the continuum opacity cause the magnetic element to be much hotter than the quiet atmosphere at equal *optical* depth and to reach a temperature above 7000 K at $\tau_c = 1$. Therefore, if observed with high spatial resolution near the center of the disc, the magnetic structure appears bright with a continuum intensity of about 1.5 times the value of the average Sun at 5000 Å. The "hot bottom and wall" of magnetic elements illuminate the upper layers of its atmosphere which becomes hotter than the environment even at equal geometrical depth.

The excess emission of magnetic elements is compensated nearly by an energy flux deficit in its environment such that only a small net excess flux is left. This cooling of the environment caused by lateral radiative energy flux into the magnetic element drives a thermal flow which supports, accelerates, and stabilizes the granular downflows next to the magnetic structure. The external flows are responsible for the asymmetry of the observed Stokes V-profiles. Conservation of angular momentum leads to rapid rotation of these downflows which stabilizes the magnetic element with respect to the interchange/fluting instability.

The quasi-equilibrium state seems to be well represented by the FTS spectra. However, spectroscopic observations of individual magnetic structures with large spatial resolution are highly desirable in order to have an independent check of the indirect methods which have been used to interpret the spatially unresolved FTS spectra. The challenging demand for high spatial resolution is compulsive for observational study of the formation and destruction processes, the dynamical interaction of the magnetic elements with convective flows, vortices, and p-mode oscillations, the excitation and propagation of oscillations and waves within magnetic structures, and the interaction with other magnetic elements.

Acknowledgements: Thanks are due to Oskar Steiner for useful comments on the manuscript. Parts of sections 7 and 8 have been taken from Schüssler (1990) with kind permission of Kluwer Academic Publishers.

216

References

Akimov, L.A., Belkina, I.L., Dyatel, N.P.: 1982, *Sov. Astron.* **26**, 334.
Akimov, L.A., Belkina, I.L., Dyatel, N.P., Marchenko, G.P.: 1987, *Sov. Astron.* **31**, 64.
Auer, L.H., Heasley, J.N.: 1978, *Astron. Astrophys.* **64**, 67.
Auffret, H., Muller, R.: 1991, *Astron. Astrophys.* **246**, 264.
Babcock, H.W.: 1947, *Astrophys. J.* **105**, 105.
Babcock, H.W., Babcock, H.D.: 1952, *Publ. Astron. Soc. Pacific* **64**, 282.
Beckers, J.M., Schröter, E.H.: 1968, *Solar Phys.* **4**, 142.
Bogdan, T.J., Cattaneo, F.: 1989, *Astrophys. J.* **342**, 545.
Bogdan, T.J., Knölker, M.: 1989, *Astrophys. J.* **339**, 579.
Bogdan, T.J., Zweibel, E.G.: 1987, *Astrophys. J.* **318**, 888.
Braun, D.C., Duvall, T.J., La Bonte, B.J.: 1987, *Astrophys. J.* **319**, L27.
Brault, J.W.: 1978, in G. Godoli, G. Noci, A. Righini (eds.): Proc. JOSO Workshop Future Solar Optical Observations - Needs and Constraints, *Osserv. Mem. Oss. Astrofis. Arcetri* No. 106, p.33.
Bünte, M., Steiner, O., Pizzo, V.J.: 1992, *Astron. Astrophys.*, in press.
Bünte, M., Steiner, O., Solanki, S. K.: 1991, in L. J. November (ed.): *Solar Polarimetry*, National Solar Observatory, Sunspot, p. 468.
Busse, F.H.: 1975, *J. Fluid Mech.* **71**, 193.
Campbell, W.R., Roberts, B.: 1989, *Astrophys. J.* **338**, 538.
Chandrasekhar, S.: 1961, *Hydrodynamic and Hydromagnetic Stability,* Clarendon, Oxford.
Chapman, G.A., Klabunde, D.P.: 1982, *Astrophys. J.* **261**, 387.
Choudhuri, A.R.: 1990, *Astron. Astrophys.* **239**, 335.
Deinzer, W., Hensler, G., Schüssler, M., Weisshaar, E.: 1984, *Astron. Astrophys.* **139**, 435.
Deinzer, W., Knölker, M., Voigt, H.H. (eds.): 1986, *Small Scale Magnetic Flux Concentrations in the Solar Photosphere,* Abh. der Akad. Wiss. Göttingen, Math.-Phys. Klasse, Folge 3, Nr. 38, Vandenhoeck und Ruprecht, Göttingen.
Dunn, R.B., Zirker, J.B.: 1973, *Solar Phys.* **33**, 281.
Evans, D.J., Roberts, B.R.: 1990, *Astrophys. J.* **356**, 704.
Evans, D.J., Roberts, B.R.: 1991, *Astrophys. J.* **371**, 387.
Evans, D.J., Roberts, B.R.: 1992, *Nature,* in press
Ferriz-Mas, A., Schüssler, M.: 1989, *Geophys. Astrophys. Fluid Dyn.* **48**, 217.
Ferriz-Mas, A., Schüssler, M.: 1992, *Geophys. Astrophys. Fluid Dyn.,* in press.
Ferriz-Mas, A., Schüssler, M. and Anton, V.: 1989, *Astron. Astrophys.* **210**, 425.
Fleck, B.: 1991, *Reviews in Modern Astronomy,* 4, 90 (Springer, Berlin).
Frazier, E.N., Stenflo, J.O.: 1978, *Astron. Astrophys.* **70**, 789.
Galloway, D.J.: 1978, in G. Belvedere, L. Paternó (eds.): *Workshop on Solar Rotation,* Catania, p. 352.
Galloway, D.J., Proctor, M.R.E., Weiss, N.O.: 1978, *J. Fluid Mech.* **87**, 243.
Giovanelli, R.G., Livingston, W.C., Harvey, J.H.: 1978, *Solar Phys.* **59**, 49.
Giovanelli, R.G., Jones, H. P.: 1982, *Solar Phys.* **79**, 267.
Goldreich, P., Murray, N., Willette, G., Kumar, P.: 1991, *Astrophys. J.* **370**, 752.
Grossmann-Doerth, U., Pahlke, K.-D., Schüssler, M.: 1987, *Astron. Astrophys.* **176**, 139.
Grossmann-Doerth, U., Schüssler, M., Solanki, S.K.: 1988, *Astron. Astrophys.* **206**, L37.

Grossmann-Doerth, U., Knölker, M., Schüssler, M., Weisshaar, E.: 1989a, in R.J. Rutten and G. Severino (eds.): *Solar and stellar granulation*, NATO ASI Series C Vol. 263, Kluwer, Dordrecht, p. 481.

Grossmann-Doerth, U., Schüssler, M., Solanki, S.K.: 1989b, *Astron. Astrophys.* **221**, 338.

Grossmann-Doerth, U., Schüssler, M., Solanki, S.K.: 1991, *Astron. Astrophys.* **249**, 239.

Hale, G.E.: 1908, *Astrophys. J.* **28**, 315.

Hale, G.E.: 1922a, *Proc. National Acad. Sci.* **8**, 168.

Hale, G.E.: 1922b, *Mon. Not. Roy. Astron. Soc.* **82**, 168.

Harvey, J.W., Hall, D.: 1975, *Bull. Amer. Astron. Soc.* **7**, 459.

Hasan, S.S.: 1983, in J.O. Stenflo (ed.): *Solar and Stellar Magnetic Fields: Origins and Coronal Effects*, IAU-Symp. No. 102, Reidel, Dordrecht, p. 73.

Hasan, S.S.: 1984, *Astrophys. J.* **285**, 851.

Hasan, S.S.: 1985, *Astron. Astrophys.* **143**, 39.

Hasan, S.S.: 1986, *Mon. Not. Roy. Astron. Soc.* **219**, 357.

Hasan, S.S., Schüssler, M.: 1985, *Astron. Astrophys.* **151**, 69.

Herbold, G., Ulmschneider, P., Spruit, H.C. and Rosner, R.: 1985, *Astron. Astrophys.* **145**, 157.

Hirayama, T.: 1978, *Publ. Astron. Soc. Japan,* **30**, 337.

Hopfinger, E.J., Browand, F.K., Gagne, Y.: 1982, *J. Fluid Mech.* **125**, 505.

Howard, R.W., Stenflo, J.O.: 1972, *Solar Phys.* **22**, 402.

Hurlburt, N.E., Toomre, J., Massaguer, J.M.: 1984, *Astrophys. J.* **282**, 557.

Hurlburt, N.E., Toomre, J.: 1988, *Astrophys. J.* **327**, 920.

Illing, R.M.E., Landman, D.A., Mickey, D.L.: 1975, *Astron. Astrophys.* **41**, 183.

Kalkofen, W., Bodo, G., Massaglia, S., Rossi, P.: 1989, in R.J. Rutten and G. Severino (eds.): *Solar and Stellar Granulation*, Kluwer, Dordrecht, p. 571.

Keller, C.U., Solanki, S.K., Steiner, O., Stenflo, J.O.: 1990, *Astron. Astrophys.* **233**, 583.

Kiepenheuer, K.O.: 1953, *Astrophys. J.* **117**, 447.

Kitai, R., Muller, R.: 1984, *Solar Phys.* **90**, 303.

Knölker, M., Schüssler, M., Weisshaar, E.: 1988, *Astron. Astrophys.* **194**, 257.

Knölker, M., Grossmann-Doerth, U., Schüssler, M., Weisshaar, E.: 1991, *Adv. Space Res.* **11**, 285.

Koutchmy, S.: 1978, *Astron. Astrophys.* **61**, 397.

Koutchmy, S., Stellmacher, G.: 1978, *Astron. Astrophys.* **67**, 93.

Koutchmy, S., Stellmacher, G.: 1987, in Schröter *et al.* (1987), p. 103.

Kraichnan, R.H.: 1976, *J. Fluid Mech.* **77**, 753.

Lawrence, J.K., Chapman, G.A.: 1988, *Astrophys. J.* **335**, 996.

Livi, S.H.B., Wang, J., Martin, S.F.: 1985, *Australian J. Phys.* **38**, 855.

Livingston, W.C.: 1991, in L. November (ed.): *Solar Polarimetry,* National Solar Observatory, Sunspot, p. 356.

Martin, S.F.: 1990, in Stenflo (1990), p. 129.

Martin, S.F., Livi, S.H.B., Wang, J.: 1985, *Australian J. Phys.* **38**, 929.

Massaglia, S., Bodo, G., Rossi, P.: 1989, *Astron. Astrophys.* **209**, 399.

McEwan, A.D.: 1973, *Geophys. Fluid Dyn.* **5**, 283.

McEwan, A.D.: 1976, *Nature* **260**, 126.

McWilliams, J.C.: 1984, *J. Fluid Mech.* **146**, 21.

Mehltretter, J.P.: 1974, *Solar Phys.* **38**, 43.

Meneguzzi, M., Frisch, U., Pouquet, A.: 1981, *Phys. Rev. Lett.* **47**, 1060.

Meyer, F., Schmidt, H.U., Weiss, N.O.: 1977, *Mon. Not. Roy. Astron. Soc.* **179**, 741.

Moreno-Insertis, F.: 1992, in J.H. Thomas and N.O. Weiss (eds): *Sunspots: Theory and Observations,* Nato ARW, Kluwer, Dordrecht, in press.

Muller, R.: 1975, *Solar Phys.* **45**, 105.

Muller, R.: 1977, *Solar Phys.* **52**, 249 .

Muller, R.: 1983, *Solar Phys.* **85**, 113.

Muller, R., Keil, S.L.: 1983, *Solar Phys.* **87**, 243.

Muller, R., Roudier, Th.: 1984, *Solar Phys.* **94**, 33.

Muller, R., Roudier, Th., Hulot, J.C.: 1989, *Solar Phys.* **119**, 229.

Nordlund, Å.: 1983, in J.O. Stenflo (ed.): *Solar and Stellar Magnetic Fields: Origins and Coronal Effects,* IAU-Symp. No. 102, Reidel, Dordrecht , p. 79.

Nordlund, Å.: 1984a, in T.D. Guyenne and J.J. Hunt (eds.): *The Hydromagnetics of the Sun,* ESA SP-220, p. 37.

Nordlund, Å.: 1984b, in S.L. Keil (ed.): *Small-scale Processes in Quiet Stellar Atmospheres,* Sacramento Peak Observatory, Sunspot, p. 174.

Nordlund, Å.: 1985, in Schmidt (1985), p. 1.

Nordlund, Å.: 1986, in Deinzer et al. (1986), p. 83.

Nordlund, Å, Stein, R.F.: 1990, in Stenflo (1990), p. 191.

Orszag, S., Tang, C.-H.: 1979, *J. Fluid Mech.* **90**, 129.

Pantellini, F.G.E., Solanki, S.K., Stenflo, J.O.: 1988, *Astron. Astrophys.* **189**, 263.

Parker, E.N.: 1963, *Astrophys. J.* **138**, 552.

Parker, E.N.: 1975a, *Astrophys. J.* **198**, 205.

Parker, E.N.: 1975b, *Solar Phys.* **40**, 291.

Parker, E.N.: 1978, *Astrophys. J.* **221**, 368.

Parker, E.N.: 1979, *Cosmical Magnetic Fields,* Clarendon, Oxford.

Pizzo, V.J., McGregor, K.B., Kunasz, P.B.: 1991, *Astrophys. J.,* in press.

Priest, E.R.: 1982, *Solar Magnetohydrodynamics,* Reidel, Dordrecht.

Proctor, M.R.E., Weiss, N.O.: 1982, *Rep. Progr. Phys.* **45**, 1317.

Ramsey, H.E, Schoolman, S.A., Title, A.M.: 1977, *Astrophys. J.* **215**, L41.

Roberts, B.: 1976, *Astrophys. J.* **204**, 268.

Roberts, B.: 1986, in Deinzer *et al.* (1986), p. 169.

Roberts, B.: 1991, in Ulmschneider *et al.* (1991), p. 494.

Roberts, B., Webb, A. R.: 1978, *Solar Phys.* **56**, 5.

Rogerson, J.B.: 1961, *Astrophys. J.* **134**, 331.

Rüedi, I.: 1991, *Measurement of solar magnetic fields with infrared lines,* Diplomarbeit, Institut für Astronomie, ETH Zürich.

Ryutova, M. P.: 1990, in Stenflo (1990), p. 229.

Saar, S.H.: 1990, in Stenflo (1990), p. 427.

Sanchez-Almeida, J., Collados, J., del Toro Iniesta, J.: 1988, *Astron. Astrophys.* **201**, L37.

Schatten, K.H., Mayr, H.G., Omidvar, K., Maier, E.: 1986, *Astrophys. J.* **311**, 460.

Schmidt, H.U. (ed.): 1985, *Theoretical Problems in High Resolution Solar Physics,* MPA 212, Max-Planck-Institut für Physik und Astrophysik, München.

Schröter, E.-H., Vázquez, M., Wyller, A.A. (eds.): 1987, *The Role of Fine-Scale Magnetic Fields on the Structure of the Solar Atmosphere,* Cambridge University Press.

Schüssler, M.: 1979, *Astron. Astrophys.* **71**, 79.

Schüssler, M.: 1984a, in T.D. Guyenne and J.J. Hunt (eds.): *The Hydromagnetics of the Sun,* ESA SP-220, p. 67.

Schüssler, M.: 1984b, *Astron. Astrophys.* **140**, 453.
Schüssler, M.: 1986, in Deinzer *et al.* (1986), p. 103.
Schüssler, M.: 1987a, in B.R. Durney and S. Sofia (eds.): *The Internal Solar Angular Velocity*, Reidel, Dordrecht, p. 303.
Schüssler, M.: 1987b, in E.-H. Schröter et al. (1987), p. 223.
Schüssler, M.: 1990, in Stenflo (1990), p. 161.
Sheeley, Jr., N.R.: 1966, *Astrophys. J.* **144**, 723.
Sheeley, Jr., N.R.: 1967, *Sol. Phys.* **1**, 171.
Shibata, K., Tajima, T., Steinolfson, R.S., Matsumoto, R.: 1990, *Astrophys. J.* **345**, 584.
Shurcliff, W.A.: 1962, *Polarized Light*, Harvard University Press, Cambrige (Ma.).
Skumanich, A., Smythe, C., Frazier, E.N.: 1975, *Astrophys. J.* **200**, 747.
Solanki, S.K.: 1986, *Astron. Astrophys.* **168**, 311.
Solanki, S.K.: 1987a, *The photospheric layers of solar magnetic fluxtubes*, Dissertation, Institut für Astronomie, ETH Zürich.
Solanki, S.K.: 1987b, in L. Hejna, M. Sobotka (eds.): *Proc. Tenth European Regional Astronomy Meeting of the IAU. Vol. 1: The Sun*, Publ. Astron. Inst. Czechosl. Acad. Sci., p. 95.
Solanki, S.K.: 1987c, in E.-H. Schröter *et al.* (1987), p. 67.
Solanki, S.K.: 1989, *Astron. Astrophys.* **224**, 225.
Solanki, S.K.: 1990, in Stenflo (1990), p. 103.
Solanki, S.K.: 1992, *Fundam. Cosm. Phys.*, in press.
Solanki, S.K., Pahlke, K.-D.: 1988, *Astron. Astrophys.* **201**, 143.
Solanki, S.K., Steiner, O.: 1990, *Astron. Astrophys.* **234**, 519.
Solanki, S.K., Stenflo, J.O.: 1984, *Astron. Astrophys.* **140**, 185.
Spruit, H.C.: 1976, *Solar Phys.* **50**, 269.
Spruit, H.C.: 1977, *Solar Phys.* **55**, 3.
Spruit, H.C.: 1979, *Solar Phys.* **61**, 363.
Spruit, H.C.: 1981a, in S. Jordan (ed.): *The Sun as a Star*, CNRS/NASA Monograph Series on Nonthermal Phenomena in Stellar Atmospheres, NASA SP-450, p. 385.
Spruit, H.C.: 1981b, *Astron. Astrophys.* **98**, 155.
Spruit, H.C.: 1981c, *Astron. Astrophys.* **102**, 129.
Spruit, H.C., Schüssler, M., Solanki, S.K.: 1991, in *The Solar Interior and Atmosphere*, W.C. Livingston and A.N. Cox (eds.), University of Arizona Press, in press.
Spruit, H.C., van Ballegooijen, A.A., Title, A.M.: 1987, *Solar Phys.* **110**, 115.
Spruit, H.C., Zweibel, E.G.: 1979, *Solar Phys.* **62**, 15.
Steiner, O.: 1990, *Model calculations of solar magnetic fluxtubes and radiative transfer*, Dissertation, Institut für Astronomie, ETH Zürich.
Steiner, O., Stenflo, J.O.: 1990, in Stenflo (1990), p. 181.
Stellmacher, G., Wiehr, E.: 1979, *Astron. Astrophys.* **75**, 263.
Stenflo, J.O.: 1973, *Solar Phys.* **32**, 41.
Stenflo, J.O.: 1989, *Astron. Astrophys. Rev.*, **1**, 3.
Stenflo, J.O. (ed.): 1990, *Solar Photosphere: Structure, Convection and Magnetic Fields*, IAU-Symp. No. 138, Kluwer, Dordrecht.
Stenflo, J.O., Harvey, J.W.: 1985, *Solar Phys.* **95**, 99.
Stenflo, J.O., Harvey, J.W., Brault, J.W., Solanki, S.K.: 1984, *Astron. Astrophys.* **131**, 333.
Stenflo, J.O., Solanki, S.K., Harvey, J.W.: 1987a, *Astron. Astrophys.*, **171**, 305.

Stenflo, J.O., Solanki, S.K., Harvey, J.W.: 1987b, *Astron. Astrophys.*, **173**, 167.

Stix, M.: 1989, *The Sun: An Introduction*, Springer, Berlin .

Tarbell, T.D., Peri, M., Frank, Z., Shine, R., Title, A.: 1988, in *Seismology of the Sun and Sun-Like Stars*, V. Domingo and E.J. Rolfe (eds.), ESA SP-286 (ESTEC Noordwijk, The Netherlands), p. 315.

Tarbell, T.D., Title, A.M.: 1977, *Solar Phys.* **52**, 13.

Thomas, J.H.: 1985, in Schmidt (1985), p. 126.

Title, A.M., Shine, R.A., Tarbell, T.D., Topka, K.P., Scharmer, G.B.: 1990, in Stenflo (1990), p. 49.

Title, A.M., Tarbell, T.D., Topka, K.P.: 1987, *Astrophys. J.* **317**, 892.

Topka, L.P., Tarbell, T.D., Title, A.M.: 1986, *Astrophys. J.* **306**,304.

Tsinganos, K.C.: 1980, *Astrophys. J.* **239**, 746.

Ulmschneider, P., Priest, E.R., Rosner, R. (eds.): 1991, *Mechanisms of Chromospheric and Coronal Heating*, Springer, Berlin.

Unno, W., Ando, H.: 1979, *Geophys. Astrophys. Fluid Dyn.* **12**, 107.

Van Ballegooijen, A.A.: 1983, *Astron. Astrophys.* **118**, 275.

Van Ballegooijen, A.A.: 1985, discussion remark in Schmidt (1985), p.177.

Venkatakrishnan, P.: 1983, *J. Astrophys. Astron.* **4**, 135.

Venkatakrishnan, P.: 1985, *J. Astrophys. Astron.* **6**, 21.

Venkatakrishnan, P.: 1986, *Nature* **322**, 156.

Von der Lühe, O.: 1987, in Schröter *et al.* (1987), p. 156.

Wang, J., Airin, H., Shi, Z.: 1985, *Solar Phys.* **98**, 241.

Webb, A.R., Roberts, B.: 1978, *Solar Phys.* **59**, 249.

Weiss, N.O.: 1966, *Proc. Roy. Soc. A*, **293**, 310.

Weiss, N.O.: 1981a, *J. Fluid Mech.* **108**, 247.

Weiss, N.O.: 1981b, *J. Fluid Mech.* **108**, 273.

Weiss, N.O.: 1991, *Geophys. Astrophys. Fluid Dyn.*, in press.

Wiehr, E.: 1978, *Astron. Astrophys.* **69**, 279.

Wiehr, E.: 1985, *Astron. Astrophys.* **149**, 217.

Woodard, M. F., Kuhn, J. R., Murray, N., Libbrecht, K. G.: 1991, *Astrophys. J.*, **373**, L81.

Zayer, I., Solanki, S.K., Stenflo, J.O.: 1989, *Astron. Astrophys.* **211**, 463.

Zayer, I., Solanki, S.K., Stenflo, J.O., Keller, C.U.: 1990, *Astron. Astrophys.* **239**, 356.

Zwaan, C.: 1967, *Solar Phys.* **1**, 478.

Zweibel, E.G., Däppen, W.: 1989, *Astrophys. J.* **343**, 991.

10 SUNSPOTS: A LABORATORY FOR SOLAR PHYSICS

J.B. GURMAN
Laboratory for Astronomy and Solar Physics
NASA Goddard Space Flight Center
Greenbelt, Maryland 20771
U.S.A.

ABSTRACT. After a brief discussion of sunspot morphology and the better-known aspects of the solar activity cycle, somewhat more detail is given on sunspot-to-photospheric continuum contrast, as a function of both wavelength and phase of the solar cycle. Umbral dots are a poorly understood phenomenon which may show the real limits on convection in the presence of a strong magnetic field. Simple, magnetostatic, and one-dimensional radiative transfer models of the observable layers of umbrae are introduced. Ultraviolet observations are used to determine whether the upper reaches of the umbral atmosphere are close to hydrostatic equilibrium, and models of umbral oscillations are discussed, as a paradigm of sunspots as "controls" for the less ordered parts of the solar atmosphere.

1. Introduction

Nearly four centuries after the first telescopic observations of the Sun, we understand neither how sunspots form nor why they persist for as long as they do. As Cowling (1946) pointed out, a convective instability should erase a uniform sunspot in a matter of hours, while, if convectively stable, the diffusion of its magnetic field should take hundreds of years, not several weeks, as observed. Repeated attempts to model the subsurface magnetic field of spots to obtain structures that could last for a few weeks have ended in frustration. The related problems of where, and in what form, the radiative energy missing from sunspots emerges have also intrigued theorists for some time. As Parker (1978) has said, "Sunspots remain an absurdity after so many years in that the structure remains unexplained by anything more than unproven conjectures." Only the heating of the corona and the acceleration of the solar wind rival the existence of sunspots as puzzles of long standing in solar physics.

For an observer, however, sunspots present a unique opportunity in the solar atmosphere. The atmosphere as a whole is characterized by disorder and multiple scales of features that often prevent us from distinguishing and understanding the various physical processes at work in those features. Most of the features in the Sun's atmosphere that catch our attention are transient, local increases in entropy over that of the quiet Sun: active regions, surges, eruptive filaments, coronal mass ejections, and flares. Sunspots, by contrast, while the most visible manifestation of the solar activity cycle, cause a *decrease* in the net solar irradiance and, for instance in the chromosphere, can represent an ordered structure found in no other sort of solar feature. Umbrae in particular are areas of lower entropy, and thus allow us to

221

J. T. Schmelz and J. C. Brown (ed.), The Sun, A Laboratory for Astrophysics, 221–243.

examine virtually all of the important processes at work in the solar atmosphere — without the plethora of surface features that complicate observation elsewhere. At a given

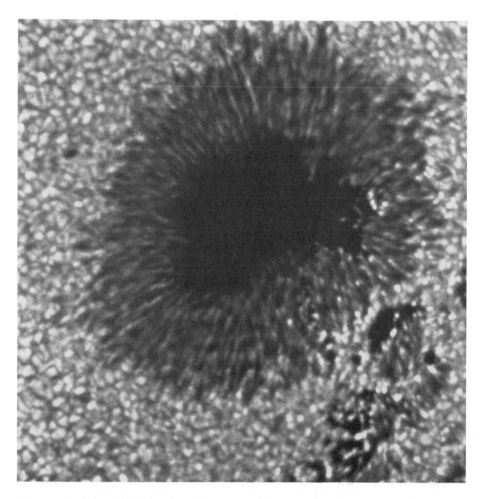

Figure 1. A broad-band, white-light image of a sunspot. The dark interior is known as the umbra (from the Latin for "shadow") and the mixed dark and light, filamentary, exterior is the penumbra. (Courtesy R. Shine, Lockheed Solar and Optical Physics Laboratory.)

lateral pressure, the umbra is cooler than the surrounding, quiet Sun; umbral dots last longer than photospheric granulation and appear not to explode; the strong, largely vertical magnetic field lasts from days to weeks in umbrae; the nonthermal broadening of spectral lines is less inside the umbra than outside, and so on. In many ways, the umbra represents a laboratory, or even a control, for what goes on in the more disordered parts of the solar atmosphere.

Sunspots also allow us to sample the quiet solar atmosphere and convection zone simply because they *are* different: sunspots float through the surrounding medium with magnetically aided buoyancy, they sometimes display large proper motions within a single spot group, they absorb global oscillations: they allow us, if only in principle as yet, the opportunity of examining how flux tubes — or rather agglomerations of them — respond to all the inputs of the solar plasma. What we learn about the behavior of sunspots is therefore of relevance to the entire solar atmosphere, threaded as it is with much thinner flux tubes.

2. Structure and Statistics

Chapter 22 addresses the formation of sunspots, so we will not deal with that issue in detail here. It is worth noting, however, that observers of many years' experience possess empirical knowledge much more extensive than our current theoretical understanding. An example is the observation that sunspots of leading and following polarity sometimes appear in the network (Born 1974) and separated by something like the size of a network cell — the supergranular scale. If the spot group turns out to be a particularly large, compound structure including several pairs of leading and following polarity spots, the longitudinal extent of such a structure can be ∼100 Mm (McIntosh and Wilson 1985). That turns out to be perhaps half the scale of the much sought-after, but as yet not conclusively identified "giant cells," which are associated with the depth of the second helium ionization. McIntosh and Wilson have even suggested that the curious arrangement of leader-polarity spots filling in an area much the same as that outlined by following polarity spots could be explained by a rather complex interaction of a variety of scales of convective cells' concentrating and dispersing flux tubes. The model itself has some substantial problems: Zirin (1991, private communication) has pointed out that the leading spots in the case discussed by McIntosh and Wilson emerged at a considerable distance from one another, and eventually coalesced into a single, large structure, but it shows schematically what essentially every sunspot model must: toroidal magnetic fields from below erupt into the convection zone and, being buoyant, rise to the surface and somehow coalesce to form sunspots.

2.1. Classification

Why sunspots form with the range of sizes they do, from pores a few arc seconds across to enormous complexes with convoluted polarity reversal ("neutral") lines, is still unclear. It is also unclear why most spots more than ∼10″ across are at least partially bounded with highly structured penumbra. Is there any reason to believe that morphological classification

224

will lead to a deeper understanding of the formation and evolution of sunspots? We should note that naturalists devised taxonomy long before genetics was understood. McIntosh's

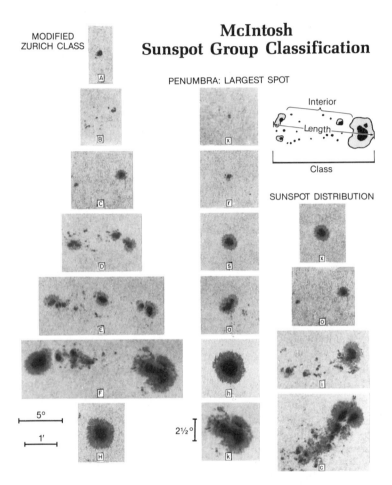

Figure 2. The modified Zürich sunspot classification scheme of McIntosh (1990). The penumbral and distribution descriptors were added to aid in flare prediction. (Courtesy of P. McIntosh, NOAA Space Environment Laboratory.)

(1990) extension of the Zürich sunspot classification scheme (Waldmeier, 1947; Kiepenheuer, 1953) is an attempt to define taxonomic classes of sunspot group size and complexity. The McIntosh scheme classifies spot groups according to three characteristics. A "modified Zürich class" describes whether the group is unipolar or bipolar, where in the group penumbra is present, and the heliographic extent of the group. The second descriptor indicates whether penumbra is present and, if so, its extent, while the third gives a measure of the density of spots between leader and follower. Although this scheme does not appear to tell us anything fundamental about sunspot formation or growth, it does turn out to be a useful predictor of large flares: unusually extended regions (modified Zürich classes E and F), large, asymmetric penumbrae (penumbra class k), and densely clustered spots (distribution class c) are much more likely to produce large X-ray flares than any other classes. As can be seen in Figure 2, this statistical result is simply a representation of the the fact that large flares are likely to occur in large active regions with highly complex magnetic fields. Still, as McIntosh (1990) points out, the predictive capability is so reliable that an artificial intelligence (AI)-based system for forecasting solar flares based on this classification scheme proved as good at prediction as the human experts at the U.S. Space Environment Services Center, and so was incorporated into their flare forecasting procedure. If flares are simply the most dramatic way for active region magnetic fields to relax, perhaps the McIntosh scheme could also predict quiescent region lifetimes.

2.2. THE SOLAR ACTIVITY CYCLE

The most familiar solar observational concept, even to the layperson, is the sunspot cycle.

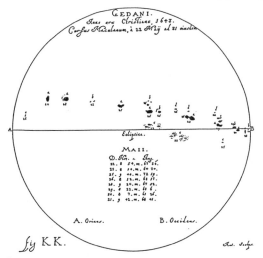

Figure 3. A time-lapse sunspot drawing by Hevelius, as reproduced in Eddy (1976).

Sunspot observations date back to Galileo and the earliest days of the telescope. Figure 3 shows a time-lapse drawing by Hevelius of the "course of the spots from the 22nd of May until the 31st thereof," 1643. The date is of interest because, within two years, the dynamo, or other mechanisms underlying the generation of sunspots, was suppressed for some 70 years.

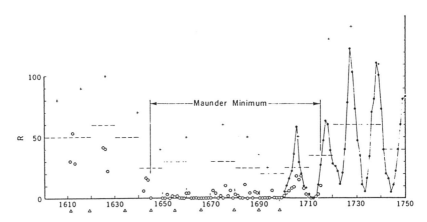

Figure 4. Sunspot number estimates, 1610 - 1750, including the Maunder minimum of 1645 - 1715 (Eddy 1976). The *open circles* are Eddy's best estimates of annual mean sunspot numbers R. (Reproduced with permission of *Science*.)

There were still occasional spots; indeed, a cycle may still have been going on, but the bottom dropped out of the overall sunspot number during this so-called Maunder minimum, first recognized in the 1890's by Spörer and Maunder (Eddy 1976). Records of eclipse observations from this period make no mention of coronal streamers, so the large-scale magnetic field may also have been absent. Finally, reports of aurorae were unusually sparse during the period. All in all, it seems that solar activity took a seven-decade vacation.

The sunspot "number," properly the Wolf or Zürich sunspot number R_Z, is a strange combination of raw sunspot numbers and the number of spot groups visible on a given day:

$$R_Z = k\,(10g + f),$$

where g is the number of spot groups visible on a given day, f the number of spots, and k a fudge factor to normalize results from different observers and observatories to Wolf's original figures. This is a peculiar sort of quantity, but when smoothed over 13 months, it correlates well with more obviously physical measures of solar activity, such as the 10.7 cm radio flux, which is produced by plage. The activity cycle, half of the cycle of general solar magnetic field reversals, has a mean of about 11.1 years, but with a variance of several months; several coniferous forests have been consumed for papers purporting to show longer term variations, of which the 80-year, or Gleissberg, cycle is probably the least unbelievable. It is important to note that we have only 13 cycles of really good sunspot number data (*i.e.*, since Wolf started observing regularly in 1848), so any attempt to infer longer-term periodicities, or to make purely statistical predictions of solar activity, must be depend on the statistics of rather small numbers.

SUNSPOT AREA COVERAGE

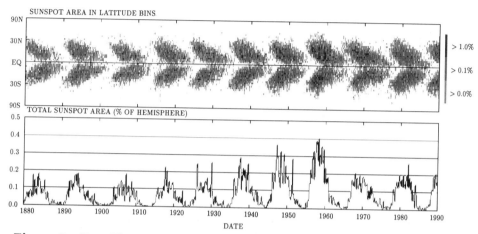

Figure 5. *Top:* The "butterfly" diagram showing the latitude of sunspot region appearance and *grey scale* the area of the spot regions, 1880-1990. *Bottom:* Fraction of the solar disk covered by sunspot regions, for the same period. (Courtesy D. Hathaway, NASA George C. Marshall Space Flight Center.)

The upper half of Figure 5 is the "butterfly diagram," so called because of the characteristic latitude migration of new spot groups toward the equator as the cycle progresses. Most sunspots, in fact, are confined to bands in each solar hemisphere between 28° and 7° from the equator. (There have also been many observational suggestions of "active" longitudes at which spot groups appear preferentially, but the evidence so far is marginal. Statistical evidence exists for preferred longitudes of large flare production, but this is simply because the majority of large flares in a given cycle tend to occur repeatedly in long-lived active regions.) In a given activity cycle, the polarity of leading spots in a given hemisphere (north or south) is opposite to that in the other hemisphere, and also opposite to that of the preceding or following cycle in the same hemisphere. This alternation of preceding and following polarity by hemisphere is known as the "Hale law."

2.3. HOW DARK ARE SUNSPOTS?

What are some typical parameters for individual spots? There is so much variation among spot groups that any mean must have a large variance assigned to it. It is, however, still worth examining a few of the parameters assembled by the late C.W. Allen (1973). The broadband continuum intensity of both umbra and penumbra increase with wavelength; the penumbra nearly disappears at wavelength around 4μm. Since the infrared continuum is formed below the visible surface, this behavior reinforces the view that the filamentary penumbral structures are geometrically higher than the umbra or quiet solar photosphere. At shorter wavelengths, umbral and penumbral continuum brightness fall off. Indeed, the shortest wavelength figures in Table 1 are quite uncertain, simply because of the difficulty of determining a continuum level in the presence of the large number of absorption lines below 400 nm. In the spectrum of the cool umbra, furthermore, there are many molecular absorption features not found in the photosphere. Maltby and his colleagues at Oslo have done some of the best work in this area (e.g. Ekman and Maltby 1974).

Table 1. Sunspot Contrast

Continuum Intensity Ratios
(after Allen 1973)

λ (nm)	300	400	500	600	800	1000	1500	2000	4000
$c_{u,ph}$	0.01	0.03	0.06	0.10	0.21	0.32	0.50	0.59	0.67
$c_{p,ph}$		0.68	0.72	0.76	0.81	0.86	0.89	0.91	0.94

$c_{u,ph}$: umbral to photospheric intensity ratio
$c_{p,ph}$: penumbral to photospheric ratio

They also discovered one of the most intriguing things about the solar cycle: in the near infrared, the umbra to photosphere contrast varies in the sawtooth fashion shown in Figure 6, in phase with the solar cycle (Albregtsen and Maltby 1978, Maltby et al. 1986). A comparison of $c_{u,ph}$ as a function of wavelength for three phases of the solar cycle is found in Table 2. Perhaps there is some property of the flux ropes that emerges at solar minimum that makes them less efficient in throttling convection than those that emerge at solar maximum, and the change is a drastic one when the new cycle spots start to appear. Skumanich (1991a) has suggested that this may be due not to variations in the spots, but to the continuum brightness of the Sun, which at least one group of observers believes may vary over the solar cycle, at least at some wavelengths.

Table 2. Solar Cycle Variation of Umbral Contrast

(after Maltby et al. (1986))

λ(nm)	579	669	876	12150	15400	16700	17300	20900	23500
$\phi = 0.1$	0.02	0.06	0.19	0.32	0.45	0.50	0.54	0.56	0.56
$\phi = 0.5$	0.06	0.09	0.21	0.34	0.49	0.54	0.57	0.58	0.58
$\phi = 0.9$	0.11	0.11	0.23	0.35	0.53	0.59	0.61	0.61	0.59

ϕ: fractional phase of the solar cycle

2.4. THE WILSON AND EVERSHED EFFECTS

The *Wilson effect* refers to the appearance of sunspots as depressions when close to the solar limb: the penumbra on the limbward side appears larger than that on the diskward side. There are some exceptions to this general rule, but the effect is easy to understand when one recalls that the primary source of continuous opacity at visible wavelengths is H^-; there is considerably more H^- in the photosphere than in the umbra, because the umbra is cooler and there are fewer free electrons about to combine with hydrogen atoms. The umbra is thus more transparent than the penumbra or photosphere. The Wilson depression describes how much deeper we can see into the umbra than in the photosphere (Bray and Loughhead 1964); it is usually taken to be 500 to 600 km (Giovanelli 1982).

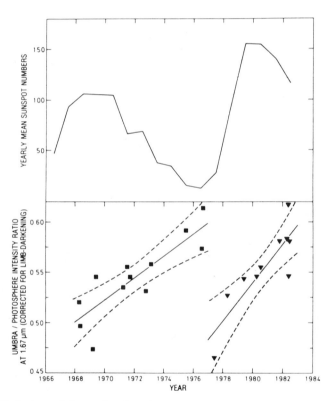

Figure 6. Variation of the ratio of umbral and photospheric continuum intensities at 1.67μm over more than a solar cycle (Maltby *et al.* 1986). (Reproduced courtesy of *The Astrophysical Journal*.)

When the continuum intensity contrast figures of Table 1 are integrated over wavelength, we find that something like a third of the photospheric flux is missing in sunspots. Where does it go? At one time, it was thought that spots were surrounded by bright "moats" where the excess radiative flux escaped, but moats really only surround large, unipolar spots and are visible primarily in chromospheric lines. Moats are in fact just facular material of the opposite polarity to the spot, where the field of a unipolar spot reenters the photosphere. Apparently, too, flux tubes regularly float outward from the penumbra into the moat, perhaps building it up by expelling opposite polarity flux tubes from the penumbra or as part of the general Evershed flow. This effect was first observed in 1909 by Evershed (at Kodaikanal) as a general outward flow in the penumbra in photospheric lines. Observations soon followed that showed the direction of the flow was reversed in the chromosphere.

Figure 7. Fe I 557.6 nm *(left)* and Hα *(right)* intensity *(top)* and line of sight velocity *(bottom)* images, 1990 June 21, obtained with the the Lockheed SOUP instrument at the Swedish Solar Observatory at La Palma. (Courtesy R. Shine, Lockheed Solar and Astrophysics Laboratory.)

Figure 7 demonstrates some of the properties of the Evershed effect. The bottom image in each case is a Dopplergram, obtained by differencing the signal in a narrow bandpass in the wings of an absorption line to obtain line-of-sight velocity information. Redshifts are displayed as bright areas, and blueshifts as dark. For a spot in the southern hemisphere, then, we expect that the photospheric, outward Evershed flow will be bright to the south and dark to the north, as it does here in the photospheric Fe I line at 557.6 nm. In the chromosphere, the flow is indeed in the opposite sense, as can be seen in the Hα image. Also of interest is the structure in Hα, where many of the fibrils seem to spiral into the penumbra.

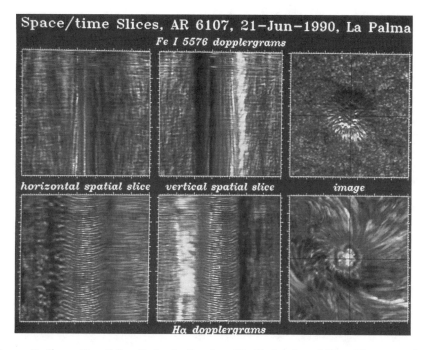

Figure 8. Space-time "slices" of velocity along North-South and East-West axes through the sunspot of Figure 7. (Courtesy R. Shine, Lockheed Solar and Astrophysics Laboratory.)

What is the structure of the penumbra? Broad-band, white-light images of sunspots (such as in Figure 1) show a penumbra consisting of long, filamentary structures. Some (*e.g.* Moore 1981) believe that the darker filaments actually overly the brighter ones, but it now appears from high-resolution observations (Title *et al.* 1992) that the magnetic field in the darker filaments is parallel to the surface, and that the outward Evershed flow is in fact confined to these horizontal structures. While the Evershed flow seems to indicate a general circulation, the impression from time-lapse movies of the Solar Optical Universal Polarimeter (SOUP) data taken at La Palma is that the penumbral bright features are all moving laterally, and probably represent areas of compression. Certainly there is plenty of activity in the penumbra. Despite the general inflow, travelling waves with periods on the order of 300 sec (Giovanelli 1973, Zirin and Stein 1972) pass *outward* through

232

the penumbra. Running penumbral waves usually cover only parts of the penumbra, but sometimes, most of the penumbra can be involved. Figure 8 shows the time history of two "slices," one vertical (*i.e.*, North-South) and one horizontal, through the spot of Figure 7. For the horizontal time slice, the penumbral waves appear as "puffs" every other period of an umbral oscillation, which shows up here as a herringbone pattern. (Umbral oscillations will be discussed below.).

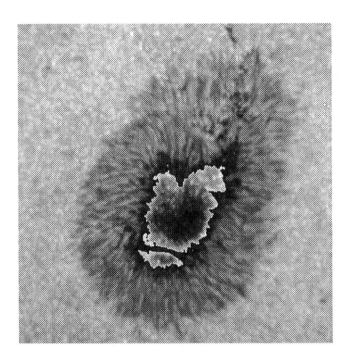

Figure 9. A broad-band, CCD image of a sunspot showing umbral dots in exquisite detail. This frame was chosen for good seeing. For a better reproduction, see Ewell (1992). (Courtesy of M.E. Ewell, Big Bear Solar Observatory, California Institute of Technology.)

And the umbra itself? Aside from *light bridges* of bright material in some sunspots, the umbra looks uniformly dark in conventional broad-band images, but there seemed to be some structure in the Dopplergrams of Figure 7, even in Fe I. Figure 9*(a)* shows an 8500 Å continuum image (Ewell 1992) of a a large spot with a light bridge; this is one of a set of CCD images chosen for particularly good seeing. Figure 9*(b)* shows the same data represented with an image transfer function of lower slope to bring out fainter features — a large number of small features known as umbral dots are now seen. In the darkest part of the umbra, no dots are visible, but perhaps a longer integration could have brought out dots there, as well. The dots appear to originate as bright grains in the penumbra and then move inward into the umbra at speeds of ~ 0.5 km sec^{-1}. Ewell estimates the most likely lifetime for the dots in this spot to be about 15 minutes, but some last over two hours. He finds rather more dots than previous authors, but that can probably be explained by the low noise level of the CCD, which makes it possible to distinguish dots only 2% brighter than the mean umbral intensity.

Several theories have been advanced to explain umbral dots; one popular one is that penetrative convection is able to push through the umbra. One problem, as Ewell (1992) points out, is that the dots would cool to umbral temperatures in a few seconds. Another possibility is oscillatory convection, which would have to be highly nonlinear for dots to last as long as they do but that in turn implies that there would be gross magnetic field changes in the dots on the time scale of the oscillation, and none is observed. In fact, high-resolution observations by Lites *et al.* (1991) at La Palma show no way to distinguish the dots from magnetic field data alone. They also found that upflows in the dots are considerably less than 1 km sec^{-1}, less than half what convective theories predicted. Better models are clearly necessary to explain the existence and lifetimes of umbral dots.

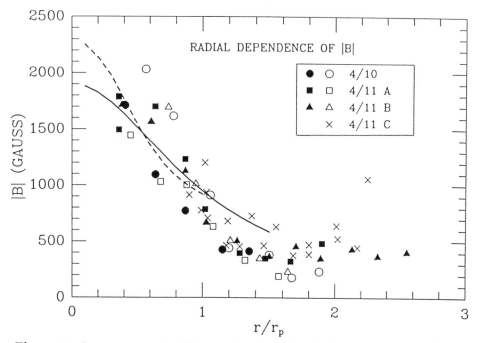

Figure 10. Sunspot magnetic field *strength* (not the longitudinal component of the field) as a function of radius in a roughly circular spot, as measured by the Zeeman splitting in the 12 μm line (Deming *et al.* 1988). The *solid* and *dashed* lines represent, repsectively, an earlier estimate based on $B_{\parallel}$ alone, and an analytical prediction. (Reproduced by courtesy of *The Astrophysical Journal*.)

Of course, the most striking thing about sunspots, after their appearance in the visible, is their intense magnetic field strength. Many authors have attempted to measure the radial variation of the magnetic field vector. Simply by using the word radial, we restrict ourselves to circular or nearly circular spots that are easy to cast in polar coordinates. Such spots are generally referred to as "theoreticians' spots" or "thesis spots." Stray light, instrumental polarization, poor spatial resolution, and poor detectors have all combined in the past to produce a bewildering variety of results for $\mathbf{B}(r, \phi)$. A popular empirical variation of $\mathbf{B}$ with radius is that of Beckers and Schröter (1969), in which the field strength at the outer penumbral boundary falls to only half the value at the center of the umbra. Several studies since that time, however, have shown that the field falls off faster; Figure 10 shows direct Zeeman splitting measurements of the 12μm Mg I line. Similar results have been obtained with the HAO Stokes I (Gurman and House 1981), Stokes II (Lites and Skumanich 1990), and the new Advanced Stokes Polarimeter (Lites *et al.* 1991). All agree that the field strength, essentially purely radial across the spot, drops to 500 or 600 Gauss at the outer edge of the penumbra.

3. Sunspot Models

Most sunspot models assume a geometry based on either a magnetic dipole embedded beneath the photosphere — which, of course, is equivalent to a current ring — or with a cylindrical current *sheet* of some vertical extent, which produces a more general solenoidal field on a sharp, outer boundary. The physics is deceptively simple: beyond cylindrical symmetry, one need only solve a boundary value problem (*i.e.*, at the radius of the cylindrical current sheet) involving the force balance equation,

$$\frac{1}{4\pi}(\nabla \times \mathbf{B}) \times \mathbf{B} = \nabla P + \rho \mathbf{g},$$

after making some assumptions about the behavior of the magnetic field $\mathbf{B}$. This magnetohydrostatic equation, however, is in general nonlinear, and most published solutions are for extremely restrictive, and not necessarily realistic, geometries. As Pizzo (1986) points out, solutions of the "thick" flux tube problem have also included "distributed-current" models that describe a continuous, radial variation of $\mathbf{B}$ and P across a spot; such models avoid the problem of discontinuities in $\mathbf{B}$ and P at the location of a current sheet. Pizzo instead shows numerical solutions for a more general class of thick-tube models in which the stratification of the gas is specified (from empirical models) as a function of radius and height. For reasonable values of the penumbral radius and Wilson depression, solutions are of the form seen in Figure 11. The $\tau = 1$ curve represents the visible surface; it is depressed over the umbra because of the Wilson effect — the opacity difference — described above.

Pizzo's model is particularly attractive because it is able to reproduce the $B(r, \phi)$ behavior shown by the 12μm and Stokes observations. (For an exhaustive review of magnetostatic sunspot models, see Skumanich (1991b)). How did Pizzo know what sort of stratification to put into his model? Just as in the the case of the quiet Sun, radiative transfer modellers have been able to produce non-LTE model atmospheres for sunspot umbrae that match the visible, infrared, and rather sparse ultraviolet line and continuum observations. Since such models are one-dimensional, they cannot hope to describe the complex structure of the penumbra. It does appear, however, that the plane-parallel assumption of the radiative transfer-based models is more appropriate for the umbra than anywhere else in the solar atmosphere, thanks to the strong magnetic field that spreads only slightly, all the way up to transition region heights.

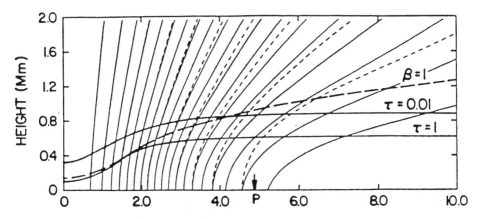

Figure 11. Magnetic field line solutions from the numerical magnetostatic treatment of Pizzo (1986). *Solid* lines: field lines without radiative constraint; *dashed* lines: field lines after requiring that no bright ring surround the sunspot. (Courtesy of A. Skumanich, High Altitude Observatory.)

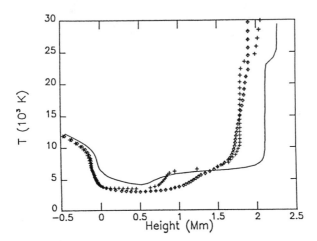

Figure 12. Temperature as a function of height for three sunspot models: the VAL3C quiet Sun model *(solid line)*, the Ondrejov umbral model *(diamonds)*, and the modified Sunspot sunspot model *(plusses)* (Gurman and Leibacher 1984).

The solid line in Figure 12 is the standard VAL3C quiet Sun model (Vernazza, Avrett, and Loeser 1981); the diamonds are the Ondrejov umbral model (Staude 1981), and the plusses are the model put together at a meeting in Sunspot, New Mexico a decade ago (Avrett 1981), with a modified transition region that does not appear on this plot (Gurman and Leibacher 1984). Below the height range shown here, all the models are assumed to merge with a standard convection zone model. If our ideas about how magnetic flux emergence varies over the solar cycle are correct, such an assumption is in error. But it is not clear by how much and, in any case, the lines and continua are by definition formed in the photosphere and above. Maltby *et al.* (1986) have since produced slightly better models, which take into account the solar-cycle variation of umbral contrast, but their gross features are similar to the other umbral models: the temperature minimum occurs at a lower temperature ($\sim 3300 - 3650$ K) than in the quiet Sun, and is more extended, while the chromosphere is shallower.

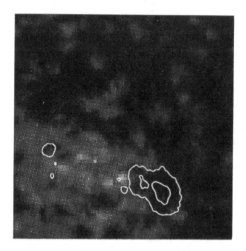

Figure 13. C IV 154.8 nm transition region image of AR 2502, from UVSP experiment 4393, 1980 June 15, 07:41 UT. Superposed are 309 nm umbral and penumbral contours from UVSP experiment 4398, 1980 15 June, 09:20 UT. Without the long-wavelength contours, sunspots are indistinguishable in transition region images.

4. The Upper Atmosphere of Sunspots

The topmost parts of the Sunspot sunspot model had to be modified because they predicted a denser, and therefore brighter, lower transition region ($T \sim 10^5$ K) than observed. The Ultraviolet Spectrometer and Polarimeter (UVSP; *cf.* Woodgate *et al.* 1980) on the *Solar Maximum Mission* (*SMM*) spacecraft produced a substantial number of spectroheliograms in transition region lines formed at $\leq 2.5 \times 10^5$ K. Figure 13 shows one such image, obtained in a 0.03 nm (second-order) bandpass in the 154.8 nm resonance line of C IV, which is formed at 10^5 K. Superposed on this transition region image are the umbral and penumbral contours from an image taken less than two hours later, with the same optics, but at a longer wavelength ($\sim$ 309 nm; first spectral order). The location of the spots relative to the plage and dark areas of the transition region is counterintuitive — the transition region over sunspots is nondescript. In a log-scaled image such as this, it is impossible to distinguish the sunspots without the longer-wavelength contours. In C IV, the spots are, on average, only a factor of three brighter than the mean quiet Sun, and much fainter than the plage.

The first-order detector allowed us to locate and restrict the raster pattern to sunspot umbrae. In one class of observations, Gurman and Athay (1983) examined line profiles taken within the umbra and time series of profiles in a single umbral pixel. They found that transition region line profiles are significantly narrower over umbrae than over the quiet Sun, as had already been noted in a single umbral spectrum obtained with the *Skylab* S082-B spectrograph by Cheng *et al.* (1976). A possible explanation for this difference lies in the ordered environment of the strong magnetic field in the umbra, as contrasted with the transition region in the quiet Sun, which is evidently made up of a multitude of small loops, possibly with random orientations. For a Gaussian profile, the line width is given by:

$$\Delta\lambda_D = \frac{\lambda}{c}\left(\frac{2kT_e}{m_i} + \xi_t^2\right)^{1/2}.$$

For C IV ($T_e = 10^5$ K) in the quiet Sun, $\langle\Delta\lambda_D\rangle \approx 150$ mÅ (Athay *et al.* 1982), so $\xi_{t,quiet} \approx$ 26.5 km sec^{-1}. Over sunspot umbrae, however, $\langle\Delta\lambda_D\rangle = 97$ mÅ, so $\xi_{t,umbra} = 14.6$ km sec^{-1}. If the $3'' \times 3''$ observing aperture averages over an ensemble of features in the quiet Sun that approximate a canonical ensemble with three degrees of freedom, but the strong, still nearly vertical magnetic field restricts nonthermal motion to one degree of freedom in the umbra, we would expect that

$$\xi_{t,umbra}^2 = \frac{1}{3}\xi_{t,quiet}^2,$$

or $\xi_{t,umbra} = 26.5$ km sec$^{-1}/\sqrt{3} = 15.2$ km sec^{-1}, in reasonably good agreement with the measured value (Gurman and Athay 1983).

How do we know that the umbral magnetic field is still strong at the height where C IV is formed? The field was measured with UVSP, the only instrument ever built, or currently planned, that could measure polarization in the ultraviolet. Zeeman broadening, which varies as λ^2, is obviously quiet small at this wavelength. Still, with sufficiently high signal-to-noise, the measurement can be made in features that do not vary significantly with time, as we surmise is the case with the umbral field. For an observation with $10''$ resolution over a spot and some adjacent plage, Henze *et al.* (1982) found that, over the umbra with a maximum photospheric field strength of about 2000 Gauss, the field was still over 1000 Gauss in the transition region. The most straightforward way of estimating the field gradient is simply to divide the difference between the photospheric and transition region fields by the height of the transition region above the photosphere predicted by a radiative transfer model. The field gradient, therefore, is about 0.4 Gauss km^{-1}. We can also use more sophisticated, if not more realistic, methods such as potential field extrapolation. The extrapolation methods all give a mean gradient of only 0.2 Gauss km^{-1}, or a transition region height of 4000 km (Hagyard *et al.* 1983).

This discrepancy should not be very troubling since limb observations of the chromosphere and transition region consistently show structure well above the hydrostatic scale height. But do we know if that applies to sunspot umbrae as well? One way to tell whether hydrostatic equilibrium is a valid assumption in the umbral atmosphere is to look for flows. The NRL HRTS instrument, which has been flown several times on sounding rockets as well as on the *Spacelab-2* Shuttle mission, images a 900″ long slit with an ultraviolet, stigmatic spectrograph. On the second HRTS rocket flight, the redshifts could be seen in all the transition region lines and in a number of chromospheric lines as well (Brekke *et al.* 1991). The downflow velocity was supersonic for most of the transition region lines and the fastest downflows, ∼200 km sec^{-1}, appeared to be independent of the temperature of maximum ionization concentration for the species producing the lines (Brekke *et al.* 1987). The same temperature independence also appears to characterize some slower, but still supersonic, components. Downflows were seen on the HRTS 1 flight, where the precise pointing was somewhat uncertain; on HRTS 2, over a strip of umbra and penumbra; and on *Spacelab-2*, over different areas of the penumbra and sometimes over the umbra as well, with variations from day to day. Kjeldseth-Moe *et al.* (1988) concluded that while the details of the flow pattern could change on time scales as short as five minutes, the transition region downflows were a "general phenomenon" over sunspots.

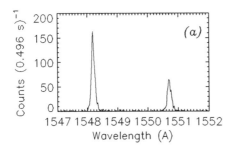 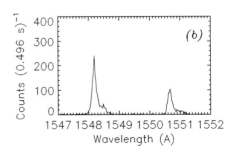

Figure 14. UVSP spectra of sunspot umbrae in *(a)* AR 4598, 1984 November 22 and *(b)* AR 4637, 1985 31 March. Both spectra were obtained with a 3″ × 3″ entrance aperture, 0.0030 nm exit slit, and 0.0025 nm sample size. Case *(b)* is one of the few obvious cases of supersonic umbral downflows obtained with UVSP; *(a)* is much more common.

The curious thing, however, is that strong downflows over sunspot *umbrae* were never seen with UVSP. Loops ending in the penumbra — the so-called superpenumbra — brighten and dim continually on time scales of two minutes or less. But when UVSP looked at 3″ × 3″ pixels corresponding to the darkest parts of umbrae in the visible, it never observed any redshifted C IV components stronger than those in Figure 14*(b)*; the profiles in Figure

14*(a)*are far more typical. UVSP observed several more spots than HRTS, but always
with restricted spatial coverage when there was good spectral bandpass, or limited velocity
bandpass when the images covered the entire spot region. Still, it was curious that HRTS
observed strong transition region downflows in three of three spots and UVSP saw none
in 20. It turns out that all three HRTS spots had light bridges, and there appears to be
a connection: the recent HRTS 7 flight observed two large spots *without* light bridges —
or conspicuous downflows (Cook 1991). This can be considered a cautionary tale about
how many sunspots makes a decent sample but, at the present writing, on the basis of
the UVSP data and the extended HRTS *Spacelab-2* observations (Maltby *et al.* 1990), it
appears that the dominant flow pattern over *umbrae* is one of blueshifts that are so weak
that hydrostatic equilibrium is a reasonable assumption. Over the penumbra, of course,
things are anything but static.

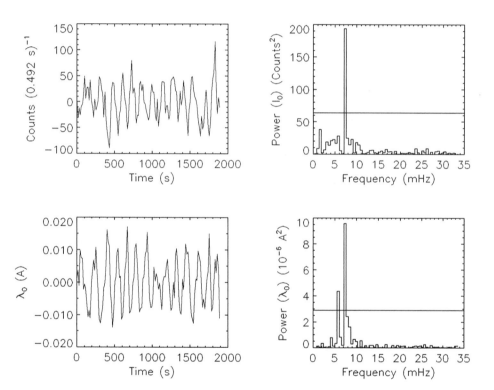

Figure 15. Time series of *(left)* central intensity *(top)* and line center position *(bottom)*,
and their power spectra *(right)*, for line profiles of the C IV resonance line at 1548.19 Å
the umbra of a large spot in AR 2522. The UVSP observations were obtained on 1980
June 26 with a 3″ × 3″ aperture, 30 mÅ exit slit bandpass, and a cadence of 14.88 s.

5. Umbral Oscillations

The assumption of hydrostatic equilibrium simplifies the modelling of the umbral oscillations which are visible in the time slices of Figure 8, particularly in Hα. The umbral oscillations are of higher frequency than the travelling penumbral waves, and typical periods fall in the 2 - 3 minute range. Maps of the spatial distribution of power within an umbra often show a good deal of fine structure, but the entire umbra appears to participate in the oscillations. They are visible at "photospheric" and chromospheric levels; in Ca II and Hα, the oscillations frequently steepen into umbral "flashes" that were first observed over 20 years ago (Beckers and Tallant 1969). In the upper chromosphere and transition region, the oscillations are still present, somewhat surprisingly as regular, nearly sinusoidal oscillations in intensity and Doppler shift in lines such as Mg II k (Gurman 1987), C IV, Si IV, and O IV (Gurman *et al.* 1982; see Figure 15). In these latter, optically thin transition region lines, Doppler shift and intensity oscillations are in phase, in the sense that maximum blueshift occurs at maximum brightness.

What, then, are the umbral oscillations? Since the waves are compressive and their periods are roughly the same in umbrae of all sizes and field strengths, it seems unlikely that they are fast-mode (Alfvén) waves. Instead, it seems likely that the mechanism must have something to do with the vertical stratification — the atmosphere. In the area where the plasma β, the ratio of magnetic pressure to gas pressure, becomes dominated by the magnetic field, one can approximate slow-mode MHD waves very well as pure acoustic waves, as long as the motion is along field lines (Roberts 1981). If the umbral field does not spread too rapidly, as indicated by both Pizzo's (1986) solutions and the measurements of transition region longitudinal magnetic field strength, we should be able to use this simplifying assumption, at least in the umbral chromosphere and transition region. But why is the transition region "tuned" to just one or two frequencies, and why in the range 5.5 to 8 mHz? After all, the umbral oscillations at photospheric heights have been interpreted as a trapped, subsurface, slow-mode MHD wave (Thomas 1985). In simultaneous observations of photospheric, chromospheric, and transition region oscillations in a sunspot umbra (Thomas *et al.* 1987), the transition regions always appear to have the highest quality Q. It appears that the umbral chromosphere acts as a resonant cavity for transmitting (magneto)acoustic waves (Žugžda, Locans, and Staude 1983, Gurman and Leibacher 1984), analogous to a Fabry-Pérot filter in classic wave optics: if an integral number of half wavelengths "fit in" to the cavity, constructive interference creates a resonance — a resonant transmission — at that wavelength. In a stratified atmosphere, the frequency or frequencies depend on the integrated sound travel time across the cavity where constructive interference takes place. In this case, the partially reflecting surfaces are provided (below) near the temperature minimum, where sound waves are turned when their frequency is below the local acoustic cutoff frequency, and (above) where they encounter a steep temperature gradient — and thus a sound speed gradient. Since (see Figure 12) umbral models have a somewhat shallower chromosphere than in the quiet Sun, we would expect a resonance with a shorter period than the 200 - 250 sec predicted for the quiet chromospheric cavity.

The acoustic resonant transmission model matches the observations well: the density perturbation one can derive straightforwardly from the optically thin, transition region line intensity oscillations and the velocity amplitude from the Doppler shift oscillations of the same lines coincide with what one would expect for a simple acoustic wave. The blueshift-intensity phase relation indicates that the waves are upwardly propagating, as one would expect for excitation from below. The phase relation in the optically thick, chromospheric lines is more difficult to interpret, but the frequency spectrum predicted for the modified Sunspot sunspot model agrees well with the observed power peaks. The remaining difficulty is to understand how wave energy and phase information are transmitted from the deep, slow-mode cavity through the $\beta \sim 1$ regime to the chromosphere.

6. The Sunspot Laboratory

The existence of a chromospheric resonant cavity in sunspot umbrae is a paradigm for how the ordered environment of the strong umbral magnetic field makes it possible to recognize physical processes with little of the noise that one encounters elsewhere in the solar atmosphere when one tries to isolate a single phenomenon from the welter of detail. It would be both facile and incorrect to say that the physics is easier in sunspots, but the entropy is lower than in the "quiet" Sun. Sunspots can help us understand the nature of the interaction of other processes with the ubiquitous magnetic flux tubes that populate the entire solar atmosphere. A good example is Braun, Duvall, and Labonte's (1987) discovery that the global, non-radial p-mode energy is absorbed in sunspots, particularly on spatial scales smaller than the umbral radius. The inversion of p-mode observations to obtain information about the convection zone will not be reliable until we have understood this process — which should shed light on the subsurface structure of sunspots as well. For all the mysteries still surrounding their birth, stability, and evolution, through their lower entropy, their stronger, more ordered magnetic fields, and their wealth of oscillatory phenomena, sunspots are able to give us unique insights into solar physics as a whole. Or, as the inscription on the sundial at York Minster puts it,

Lucem demonstrat umbra.

REFERENCES

Albregtsen, F. and Maltby, P. 1978, *Nature*, **274**, 41

Allen, C.A. 1973, *Astrophysical Quantities*, Athlone Press (University of London), pp. 181 - 186

Athay, R.G., Gurman, J.B., Henze, W., and Shine, R.A. 1983, *Ap. J.*, **265**, 519

Avrett, E.H. 1981, in L.E. Cram and J.H. Thomas (eds.), *The Physics of Sunspots*, Sunspot: Sacramento Peak Observatory, p. 235

Beckers, J.M. and Schröter, E.HM. 1969, *Solar Phys.*, **10**, 384

Beckers, J.M. and Tallant, P.E. 1969, *Solar Phys.*, **7**, 351

Born, R. 1974, *Solar Phys.*, **38**, 127

Braun, D.C., Duvall, T.L., and LaBonte, B.J. 1987, *Ap. J.*, **319**, L27

Bray, R.J. and Loughhead, R.E. 1964, *Sunspots*, Chapman and Hall (London), p. 99

Brekke, P., Kjeldseth-Moe, O., Bartoe, J.-D.F., and Brueckner, G.E. 1987, in *ESA SP-270: Proceedings of the 8th ESA Symposium on European Rocket and Balloon Programmes and Related Research*, Sunne, Sweden, p. 341

Brekke, P., Kjeldseth-Moe, O., Bartoe, J.-D.F., and Brueckner, G. 1991, *Ap. J. Suppl.*, **75**, 1337

Cheng, C.-C., Doschek, G., and Feldman, U. 1976, *Ap. J.*, **210**, 836

Cook, J.W. 1991, private communication

Cowling, T.G. 1946, *Mon. Not. Roy. Astron. Soc.*, **106**, 218

Deming, D.A., Boyle, R.J., Jennings, D.E., and Wiedemann, G. 1988, *Ap. J.*, **333**, 978

Eddy, J.A. 1976, *Science*, **192**, 1189

Ekman, G. and Maltby P. 1974, *Solar Phys.*, **26**, 83

Ewell, M.E. 1992, *Solar Phys.*, in press

Giovanelli, R.G. 1982, *Solar. Phys.*, **80**, 21

Gurman, J.B. 1991, in Byrne, B. and Mullan, D. (eds.), *Proceedings of the Armagh Observatory Colloquium on Surface Inhomogeneities in the Atmospheres of Late-Type Stars*, in press

Gurman, J.B. 1987, *Solar Phys.*, **108**, 61

Gurman, J.B. and Athay, R.G. 1983, *Ap. J.*, **273**, 374

Gurman, J.B. and House, L.L. 1981, *Solar Phys.*, **71**, 5

Gurman, J.B. and Leibacher, J.W. 1984, *Ap. J.*, **283**, 859

J.B. Gurman, Leibacher, J.W., Shine, R.A., Woodgate, B.E. and Henze, W. 1982, *Ap. J.*, **253**, 939

Hagyard, M.J. *et al.* 1983, *Solar. Phys.*, **84**, 13

Henze, W., Tandberg-Hanssen, E., Hagyard, M.J., Woodgate, B.E., Shine, R.A., Beckers, J.M., Bruner, M., Gurman, J.B., Hyder, C.L., and West, E.A. 1982, *Solar Phys.*, **81**, 231

Kiepenheuer, K.O. 1953, in G.P. Kuiper (ed.), *The Sun*, U. of Chicago Press, p. 322

Kjeldseth-Moe, O., Brynildsen, N. Brekke, P., Engvold, O., Maltby, P., Bartoe, J.D.-F., Brueckner, G.E., Cook, J.W., Dere, K.P., and Socker, D.G. 1988, *Ap. J.*, **334**, 1066

Lites, B.W. and Skumanich, A. 1990, *Ap. J.*, **348**, 747

Lites, B.W., Elmore, D., Murphy, G., Skumanich, A., Tomczyk, S., and Dunn, R.B. 1991, in L.J. November (ed.), *Solar Polarimetry*, Sunspot: National Solar Observatory, p. 3

Maltby, P., Avrett, E.H., Carlsson, M., Kjeldseth-Moe, O., Kurucz, R.L., and Loeser, R. 1986, *Ap. J.*, **306**, 284

Maltby, P., Brekke, P., Brynildsen, N., Kjeldseth-Moe, O., Bartoe, J.-D.F., and Brueckner, G.E. 1990, *Publ. Debrecen Obs.*, **7**, 244

McIntosh, P.S. 1990, *Solar Phys.*, **125**, 251

McIntosh, P.S. and Wilson, P.R. 1985, *Solar Phys.*, **97**, 59

Moore, R.L. 1981, *Ap. J.*, **249**, 390

Parker, E.N. 1978, in Eddy, J.A. (ed.), *The New Solar Physics*, Boulder: Westview Press, p. 1

Pizzo, V.J. 1986, *Ap. J.*, **302**, 785

Roberts, B. 1981, in L.E. Cram and J.H. Thomas (eds.), *The Physics of Sunspots*, Sunspot: Sacramento Peak Observatory, p. 369

Skumanich, A. 1991a, private communication

Skumanich, A. 1991b, in Byrne, B. and Mullan, D. (eds.), *Proceedings of the Armagh Observatory Colloquium on Surface Inhomogeneities in the Atmospheres of Late-Type Stars*, in press

St. John, C.E. 1913, *Ap. J.*, **37**, 322

Staude, J. 1981, *Astron. Astrophys.*, **100**, 284

Thomas, J.H.: 1985, *Australian J. Phys.*, **38**, 811

Thomas, J.H., Lites, B.W., Gurman, J.B., and Ladd, E.F. 1987, *Ap. J.*, **312**, 457

Title, A.M., Frank, Z.A., Shine, R.A., Tarbell, T.D., and Topka, K.P. 1992, submitted to *Ap. J.*

Vernazza, J.E., Avrett, E.H., and Loeser, R. 1981, *Ap. J. Suppl.*, **45**, 635

Waldmeier, M. 1947, *Publ. Zürich Obs.*, **9**, 1

Woodgate, B.E., Tandberg-Hanssen, E.A., Bruner, E.C., Beckers, J.M., Brandt, J.C., Henze, W., Hyder, C.L., Kalet, M.W., Kenny, P.J., Knox, E.D., Michalitsianos, A.G., Rehse, R., Shine, R.A., and Tinsley, H.D. 1980, *Solar Phys.*, **65**, 73

Zirin, H. and Stein, A. 1972, *Ap. J.*, **178**, L75

Žugžda, Y.D, Locans, V., and Staude, J.: 1983, *Solar Phys.*, **82**, 369

11   CHROMOSPHERIC STRUCTURE

J.B. GURMAN
Laboratory for Astronomy and Solar Physics
NASA Goddard Space Flight Center
Greenbelt, Maryland 20771
U.S.A.

ABSTRACT. A brief introduction is given to the variety of chromospheric features seen in Hα images. Despite their wealth of detail, images obtained in a single, narrow band in Hα also suffer from a surfeit of *missing* information. The most basic chromospheric structures are the supergranular network and spicules; these are visible in lines formed higher in the chromosphere, as well, which also show evidence for small-scale heating. The heating of even the lower chromosphere is still problematic, but may be explained by large-amplitude acoustic wave dissipation. A current challenge is understanding the apparent thermal bifurcation of the chromosphere into regimes of distinctly different scale heights. Finally, the possibility of chromospheric wave cavities is briefly discussed.

At the extreme solar limb the intensity of the continuous emission in the optical spectrum drops by a factor e over a region some 100 km in extent, thereby giving rise to the appearance of a sharp boundary. Immediately following this rapid drop, two marked changes occur in the character of the optical spectrum: (1) The spectrum changes from a strong emission continuum, broken by absorption lines, to a spectrum dominated by emission lines of neutral and singly ionized atoms superposed upon a weaker continuum. (2) For both lines and continuum, the emission scale height, which is the distance required for a decrease in intensity by a factor e, becomes much greater than the 100 km characterizing the continuum at the limb. These two characteristics of the emission persist for some 10,000 km, with only the strongest lines remaining visible throughout the entire region. This region, in which emission lines of neutral and singly ionized atoms are observed, is called the *chromosphere*. The first 5000 km, which presents a quasi-uniform appearance, is called the *lower chromosphere*. The upper 5000 km, where the emission is concentrated in small spike-like columns, is called the *upper chromosphere*. The columns are called *spicules*. The *classical chromospheric problem* is to find an explanation of the large scale heights of the chromospheric emission.

— R.N. Thomas (Thomas and Athay (1961))

1. Introduction

The "classical chromospheric problem" described by Thomas and Athay is still with us, a generation later. In the last chapter of the text, the authors admit that they have not solved the problem, even though they have laid the foundations of the elegant non-local-thermodynamic-equilibrium (non-LTE) line formation theory that would dramatically change the way we think about stellar chromospheres. It is still a problem, thirty years later. In some sense, there is not much to add for, while we have learned a great deal of detail in the last three decades, we still do not understand the chromosphere as a whole, perhaps because of the way it is observed.

J. T. Schmelz and J. C. Brown (ed.), The Sun, A Laboratory for Astrophysics, 245–259.

1.1. THE Hα ZOO

The traditional way of looking at the chromosphere is in the broad Balmer α line of hydrogen, or Hα. Generally, the images are obtained through a Lyot or other tunable filter with a bandpass of about a quarter of an Ångstrom, a line-center picture of the full disk at any time other than the very depth of solar minimum displays bright *plage* and dark sunspot umbrae in active regions, and long, dark *filaments*. When the filaments are visible over the limb, they appear in emission, although generally much weaker than the disk, and are then called *prominences*.

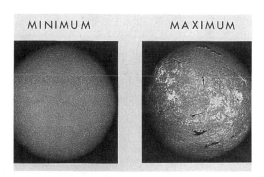

Figure 1. Full-disk Hα images obtained at solar maximum (left) and minimum (right). (Courtesy National Solar Observatory - Sacramento Peak)

Such features are representative of the variety of magnetic field-related phenomena that characterize solar activity. When the Sun is near the minimum of its activity cycle, there can be days without a trace of visible plage, and only the remnants of the so-called polar crown filaments are visible (*cf.* Figure 1). Nearer solar maximum, substantial areas adjacent to sunspots can go into emission in Hα during flares; often, this change from absorption to emission takes place along an entire filament as the chromosphere is compressed in a double-ribbon flare. Such events are discussed in detail in Chapters 22 and 23.

It is, however, worth noting that in the dark prominence material in the *spray* of Figure 2, obtained 0.5 Å out in the Hα line wing, we are observing line-of-sight velocities in excess of 40 km sec^{-1}. But the trick is understanding what such highly detailed images show.

In the quiet Sun, Hα is a very broad, pure absorption line, as one would expect from a photoionization dominated source function. The core is saturated and the line has damping wings several Å wide. If the entire line profile is observed, a substantial range of heights in the solar atmosphere can be explored.

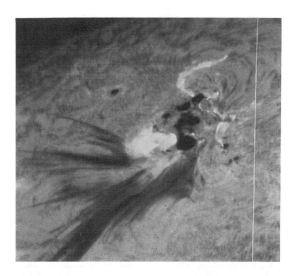

Figure 2. +0.5 Å Hα image of a flare spray. (Courtesy National Solar Observatory - Sacramento Peak)

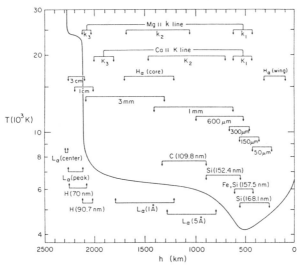

Figure 3. Temperature as a function of height in the photosphere and chromosphere of the standard reference atmosphere VAL3C (Vernazza, Avrett, and Loeser 1981). (Reprinted with permission of *The Astrophysical Journal*.)

In a standard model atmosphere such as that of Figure 3, the Hα wings are formed in the upper photosphere, below the temperature minimum, while the line core is formed in the lower chromosphere. An absorption feature, such as a dark spray or surge, that shows up in only one wing of the line profile, is not necessarily formed in the photosphere. The material is Doppler shifted — but there is no direct information on its transverse velocity. It is important to remember that an Hα filtergram is a convolution of optical depth and line-of-sight motions, especially when the filter is tuned 0.5 Å or more off line center.

Dopplergrams, velocity images formed by differencing simultaneous or nearly simultaneous images in each wing of a line, give much more than twice the information of a single line-wing image; time series of such images often allow us to differentiate between line shifts and opacity differences. In Dopplergram movies obtained with the Lockheed Solar Optical Universal Polarimeter (SOUP) instrument, for example, one can follow material travelling from one part of an active region to another as the absorption travels across the line profile. When these images are combined with photospheric magnetograms, the surging material in Hα can be seen to be travelling along closed magnetic features and their line-of-sight velocities are just components of a curved trajectory. If such an event is observed in just one wing of Hα, dark material mysteriously appears and disappears. Without the Doppler information from both line wings, it is impossible to obtain a realistic view of such dynamic events.

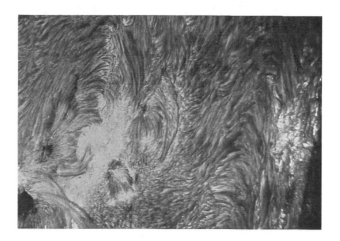

Figure 4. Hα line center image. Filamentary structures appear to map out magnetic field lines. (Courtesy National Solar Observatory - Sacramento Peak)

As a result, many ground-based observatories maintain patrols of active regions in which they obtain time series of images, cycling through the Hα line profile. It is wise to be skeptical when viewing Hα images at a single position in the line. While experienced observers may claim to know exactly what is happening in such images, there is simply not enough *usable* information provided. There is certainly information in such images, but there is also a tremendous amount of entropy $S = -k \ln I$, where I is a measure of information. Among the information missing from monochromatic Hα images: Is a dark feature cooler and higher than its surroundings, or is it just Doppler shifted? Is an asymmetric core profile excess emission shifted one way or absorption shifted the other? What is the total velocity vector of a feature seen in the line wing? Any monochromatic image of the Sun is full of entropy, but Hα is so rich in features that it has more entropy than lines that produce less spectacular images. Some observers (*e.g.* Zarro *et al.* 1988, Canfield *et al.* 1990) have obtained Hα *profiles* at every pixel in an image in an effort to understand the chromospheric dynamics of solar flares. Only by doing the hard work of line profile synthesis can we really claim to understand much of what is seen in Hα features, particularly transient features like flares.

Thus forewarned of the enormous amount of information *missing* from an Hα image, let us establish the classic picture of the Hα chromosphere. The line-center image of Figure 4 — formed in the low chromosphere — shows extended, often feathery filamentary structure in absorption near bright plage. The filamentary structures appear to map out mostly horizontal magnetic field lines — at some height — in the chromosphere. The plage regions are associated with concentrations of strong, usually more nearly vertical magnetic field. Filaments appear to be located along a neutral line of the magnetic field, however convoluted (Figure 5), and can also show evidence of horizontal velocity shear. Meanwhile, extending out from the sunspot penumbra, is a series of roughly radial, filamentary structures that are considered to form a "superpenumbra." (See Chapter 10 for a brief discussion of penumbral structure.)

Figure 5. Hα image showing a filament overlying the magnetic neutral line in an active region. (Courtesy National Solar Observatory - Sacramento Peak)

The primary structural constituents of the chromosphere, *spicules* and the *network*, can be seen in Figure 6. The large, cell-like structures of the supergranular network are more easily seen in spectroheliograms obtained in strong lines formed higher in the atmosphere than Hα (see below). Along the edges of the network cells appear the spicules: thin, elongated structures perhaps less than an arc second — 725 km — in width, and up to 10^4 km high. Many authors have compiled observations of the sizes, lifetimes, inclinations, and frequency of spicules. Vertical flows (*i.e.*, those aligned with the long axis of the spicule) generally appear to be on the order of 10 - 30 km sec^{-1}; most spicules are inclined from 10° to 30° from the vertical; and an estimated 10% of the solar surface is covered by spicules more than a few thousand km high (Zirin 1988). It should be noted that all these measurements are difficult to make, since the subarcsecond width of spicules makes them extremely difficult to resolve from the ground under any but the best seeing conditions.

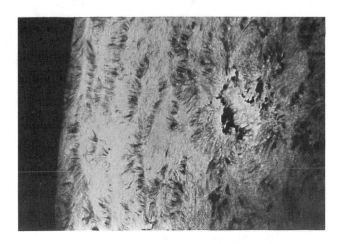

Figure 6. Hα image showing the "burning prairie" of spicules delineating supergranular network cells. (Courtesy National Solar Observatory - Sacramento Peak)

The distribution of inclinations may be a clue to the nature of several fine-scale Hα features. For *fibrils*, which are particularly thin, bent spicules at the edges of plage,the distribution is skewed toward the horizontal. Foukal (1990) described a grand unified view of spicules, fibrils, the bright mottles that make up plages, and the dark mottles that can be seen, at the right wavelength in the Hα wing, near the bases of spicules. The fibrils point toward areas of opposite polarity, and are presumably the footpoints of the large, active region loops seen in transition region and coronal lines. The spicules, on the other hand, are the lower parts of the weaker, network magnetic field that spreads out to fill the corona over the quiet Sun. As first observed in the EUV (Bohlin *et al.* 1975, Withbroe *et al.* 1976), "macrospicules," some tens of arc seconds in height, also apparently rise from the network and free fall back down, mostly near the poles. Since macrospicules appear to be related to microflares in emerging magnetic flux regions (Moore *et al.* 1977) their energy may be deposited in the corona; also, if there is a continuum of ever smaller microflares associated with spicules, they could actually heat the upper chromosphere and corona. Perhaps, too, as Mark Twain pointed out, science is wonderful precisely because such a wholesale return of speculation is obtained from such a minimal investment of fact. Well, let us speculate: the corona seems to be heated only in the closed loop structures in active regions — there is little white-light corona outside of the equatorial plane during solar minimum, and a nearly complete absence of X-ray and EUV loop structures. Let us speculate instead that the open, unipolar magnetic field structures into which spicules diverge are the source of the ubiquitous solar wind, by analogy to the high-speed streams originating in the open magnetic field structures of coronal holes. Observations with the *SOHO* spacecraft in the mid-1990s should prove or disprove this conjecture.

2. The Chromosphere in Ca II and Mg II

As can be seen in Figure 3, the central parts of the Ca II K and H resonance lines are formed everywhere from the temperature minimum, through the lower chromosphere, and on up to just below the beginning of the really steep temperature rise where the Mg II k and h resonance line cores are formed. The spectroscopic terminology for parts of the line profile almost approaches taxonomy: the K_1 features are the absorption minima just outside the emission core, the K_2 features are the emission peaks, and K_3 is the central reversal; V and R refer to violet and red. Conveniently, the K_1 minima are about 1 Å apart in plages, so Ca II filters are typically 1 Å (0.1 nm) wide to sample only the area above the temperature minimum.

A full-disk image obtained in the UV continuum at about 137.5 nm is shown in Figure 7 (center pane). The continuum there is formed at about the same height as the K_2 features, so it makes a good proxy. Cells of roughly 40″ to 60″ in diameter are seen over much of the surface (at least in larger reproductions of the image). This is the *supergranular network*, or network, as it usually called; its scale matches nicely with the depth of the first helium ionization in the convection zone, and in the photosphere, small magnetic elements are swept to the edges of supergranular cells at something like 0.5 km sec^{-1}. At high resolution, the structure resolves itself into small elements called *faculae*.

Figure 7. SELSIS Hα, *SMM* UVSP 137.5 nm, and National Solar Observatory photospheric longitudinal magnetograms for 1988 September 27. The diagonal structures in the northern acitivty belt represents the poleward diffusion of old plage.

The chromospheric network is particularly obvious in the active latitude belts and, in the northern hemisphere of Figure 7, there are eastward and poleward swept lanes of brighter network trailing behind the active regions. These consist of old plage that has been distorted by differential rotation and swept poleward by the large-scale flow. (See Wang and Sheeley (1991) for a discussion of the effects of both diffusion and meridional flows.) Darker lanes separate network of opposite polarity — formerly preceding and following polarity parts of the same active region (*cf.* the photospheric longitudinal magnetogram of Figure 7, right pane) — and between them are dark filament channels (as shown in the Hα image of Figure 7, left pane).

Since individual network cells appear to have coherence times of less than two days, it is this recirculation of old plage into the network that, in fact, replenishes network fields. As a result, the network is considerably harder to see at solar minimum than at solar maximum, in both Ca II spectroheliograms and photospheric magnetograms. Apparently, chromospheric emission lines — such as the K core — and their transition region counterparts are relatively weak at solar minimum because there is simply less magnetic field to dissipate energy in the network.

Figure 8. Spectroheliogram in the blueward emission peak of Ca II K (K_{2V}), obtained by Bruce Gillespie with the East Auxiliary of the McMath Solar Telescope. In this negative image, the K_{2V} bright points appear as small, dark dots. (Courtesy National Solar Observatory - Kitt Peak.)

It is possible to see spicules in the K line as well as in Hα. Figure 8 has been called "the best spectroheliogram ever taken." It was obtained in a very narrow bandpass about K_{2V}, the blueward emission peak of the K line, and, just in from the limb, shows lots of little black dots. This is a negative image, so the dots are really bright — they disappear near the extreme limb only because the wavelength of K_{2V} changes from center to limb. The curious thing is that these K_{2V} bright points, which have lifetimes on the order of a minute and recur 2 to 4 times at roughly three minutes intervals, occur in the *middle* of network cells, not in the network itself where there are strong magnetic fields and spicules. Rutten (1991) has explained the K_{2V} bright points as "piston kicks" that occur when the upward, compressive, shock-steepened phase of a 3-minute magnetoacoustic wave runs into the collapsing material from higher layers that had been driven upward in the previous cycle. The size of the bright points should then tell us something about the size of the driving piston, which must be some characteristic size for weak magnetic field elements in cell interiors. It is worth noting that this is going on in the quiet network cell interiors; on the cell boundaries, in the bright network, ultraviolet continuum observations with the HRTS and UVSP instruments have shown that 5 minute and longer period oscillations are the norm (Drake *et al.* 1989, Cook and Ewing 1991). Assuming that both 3- and 5-minute oscillations are compressive waves, is it possible that different features in the supergranular pattern have different scale heights that would lead to different characteristic sound travel times?

3. Thermal Bifurcation and the Modeller's Chromosphere

The quiet solar chromosphere has been the subject of several one-dimensional ("plane-parallel") models since the development of non-LTE radiative transfer to treat the synthesis of strong line profiles. It is essentially an iterative technique where the model atmosphere is adjusted to reproduce the line profiles observed at a variety of disk positions from center to limb. Enough good data are required, particularly in the infrared, microwave, and ultraviolet, to be able to model continuum source functions in much the same way. The core of H Lyα is formed in the plateau at ~ 2200 km (by convention measured from $h = 0$ at optical depth unity in the 500 nm continuum, $\tau_{500} = 1$). That plateau, at a temperature where about half the hydrogen is ionized, exists only to allow enough neutral hydrogen to produce the observed Lyα line strength. There have been arguments in the literature for several years over the reality of the plateau, but it is difficult to get around the requirement for neutral hydrogen in the upper chromosphere. Recently, however, Fontenla, Avrett, and Loeser (1990, 1991) have shown that ambipolar diffusion can continuously introduce enough neutral hydrogen at these heights that a plateau-less chromosphere can reproduce the observed Lyα line center intensities.

Another serious problem with one-dimensional, radiative transfer models is known as "thermal bifurcation." Ayres and collaborators have carried out a series of observations in the fundamental ($\Delta v = 1$) and first harmonic ($\Delta v = 2$) vibration-rotation bands of carbon monoxide at 4.8 μm in the infrared. They have found that the lines and nearby continuum are formed essentially in local thermodynamic equilibrium, but that the strong fundamental transitions show core brightness temperatures of only ≤ 3700 K. Ayres proposes that only $\sim 10\%$ of the Sun's surface is covered with areas where there is a chromospheric temperature rise and, whence, all the Ca II line core emission arises. The other $\sim 90\%$ has a much greater pressure scale height, with the temperature minimum occurring 400 - 500 km higher than in the chromosphere of the one-dimensional models.

To reproduce the observed CO band profiles, Ayres and Brault (1990) had to use a model atmosphere (labelled "0" in Figure 9) with a temperature minimum below 3000 K. The effect is to extend the CO band-forming region over what it would be in the standard model.

The idea of thermal bifurcation has provoked much discussion and further research. One of the most interesting pieces of work was done by Athay and Dere (1990), who used HRTS 1 data with $\sim 1''$ spatial resolution in a number of chromospheric lines of O I, C I, and Fe II. All of these optically thick lines are effectively thin in the chromosphere; that is, the upper levels of the transitions are populated primarily by collisions, and one can thus identify the range of temperature and density in which they are formed with some precision. Athay and Dere's analysis left both the size of the bright faculae and the fraction of the chromosphere covered as free parameters; their conclusion is that something like 90% of the Sun is covered by bright faculae which can only be regions of "normal" temperature minimum height and chromospheric temperature rise — exactly the opposite of the results from the CO work.

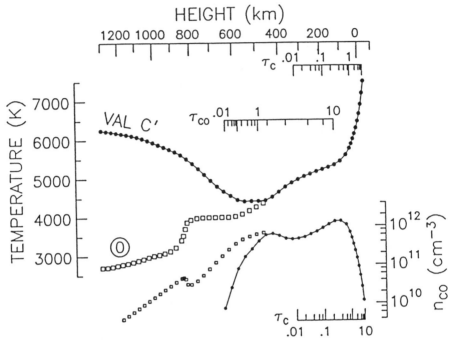

Figure 9. Thermal bifurcation: the quiet Sun model atmosphere of Maltby *et al.* (1986) (*filled circles*, model VAL C′) and the extended chromosphere model of Ayres and Brault (1990) (*open circles*, model 0). The deeper, cooler temperature minimum region of Model 0 is necessary to reproduce the observed CO first harmonic spectrum. (Reproduced courtesy of *The Astrophysical Journal*.)

Ayres has since attempted to explain the observations with so-called "cloud" models — either a cool, extended chromospheric cloud with small, bright, embedded elements, or a hot, standard chromosphere with embedded cool regions that can "thermally shadow" some of the hot regions at the limb; neither of these schemes can yet fit all the observations. With all we know about chromospheric structure, it is not enough to be able to understand these apparent contradictions.

4. Chromospheric Heating

The central mystery of the chromosphere, of course, is that it exists at all: if energy transport in the Sun's atmosphere were purely radiative, the temperature of the solar plasma would drop monotonically until it reached the temperature of the interplanetary medium, and solar physics would be considerably less interesting than it is. We also would know almost nothing of activity on other cool stars, since the best activity indicators are the strong chromospheric and transition region lines, and the presence or absence of a soft X-ray emitting corona.

Chromospheric heating is a difficult problem to address. It really breaks down into two questions: why is there a temperature reversal at all, and why is there a steep rise, to temperatures of tens of thousands of degrees, in the upper chromosphere? The plausible answers to the second half of the problem appear to be so different physically from those to the first, that they belong not here but in a discussion of coronal heating. Let us consider instead the simple, but as yet not convincingly answered question of just what heats the lower chromosphere.

In the early 1970s, the *OSO-8* spacecraft carried two high resolution, imaging spectrographs to determine whether acoustic waves could heat the outer solar atmosphere. With the benefit of hindsight, and the marvelous soft X-ray images of the corona obtained in the last two decades, it is now believed that the corona is heated primarily in active regions, where the magnetic field is strong, rather than ubiquitously, as one might expect with acoustic waves linked to the global, non-radial oscillations deeper in the solar atmosphere. In the lower chromospheric temperature plateau, however, the degree of ionization is quite low, so electromagnetic heating appears to be ruled out. The *OSO-8* instruments, therefore, were optimized to detect acoustic waves in both the chromosphere and transition region with periods shorter than the inverse of the acoustic cutoff frequency $\omega_{AC} = \gamma g/2s$ at the temperature minimum. In the linear theory at least, compressive waves with frequencies higher than $\omega_{AC}(T_{min})$ will propagate freely through the temperature minimum, while lower frequency acoustic waves will be turned. Quiet solar model atmospheres thus predicted that waves with periods shorter than about 200 sec could heat the chromosphere — but the power at such frequencies observed by *OSO-8* was far too low to heat the *corona* (Athay and White 1979, 1980). As a result, acoustic waves became unpopular as a candidate for heating any part of the outer solar atmosphere.

Recently, however, Anderson and Athay (1989, 1990) have constructed a model quiet solar atmosphere in a novel way. Instead of fitting a large number of ultraviolet, visible, and infrared lines and continua in the "semi-empirical" approach that Vernazza, Avrett, and Loeser (1981) used to produce standard models of the sort shown in Figure 3, Anderson and Athay started with a heat input that would reproduce that VAL3C model by including literally millions of lines and continua to see where the radiative losses occur. One of the fascinating things they learned was that the myriad of Fe II lines in the lower chromosphere are even more efficient radiative losses than the few strong lines — Ca II *H* and *K*, Mg II *k* and *h*, and the hydrogen lines. More importantly, they found that the *lower* chromosphere can be heated quite adequately with a flux of acoustic waves — presumably with high frequencies (which have not yet been observed) and an amplitude of about 5 km sec^{-1} — in a region where the sound speed is $\sim$ 7 km sec^{-1}. The ratio of those values is important because it clearly lies outside the domain of the theory of small, linear perturbations which is the basis for most of our modelling of acoustic waves in the Sun. Anderson and Athay quote values for the nonthermal broadening of lines formed in the lower chromosphere — usually called "microturbulence" — of 5.5 to 7 km sec^{-1}, but it is unclear whether such line broadening is caused by shocks.

In as steep a density gradient as that in the lower chromosphere (the density decreases by three orders of magnitude here over a few hundred km), compressive waves very rapidly steepen into shocks. As the density perturbation tries to remain the same, the pressure amplitude grows as the inverse of the density decrease and, as the work of Ulmschneider and his colleagues has shown, very efficient dissipation — heating — occurs. So while the interpretation of Anderson and Athay may be a bit oversimplified, it appears that the right flux of acoustic wave energy could heat the chromosphere below about 2000 km, or 7000 K. Of course, that is the one-dimensional, hydrostatic, radiative transfer model of the chromosphere — very different from the highly structured chromosphere we have seen here. It is difficult to determine how realistic any of our radiative transfer models may be, when the chromosphere is so highly structured, so extended, and so clearly not static.

5. Chromospheric Oscillations

The discussion in Chapter 3 of global oscillations concentrates on subphotospheric waves. As has been known for some time (Ando and Osaki 1977, Leibacher and Stein 1981), it is also possible for an acoustic wave cavity to exist in the chromosphere. The lower reflection is provided by the acoustic cutoff barrier and the upper internal reflection by the steep temperature gradient. There is also the possibility of a buoyancy wave cavity in the upper photosphere and low chromosphere, for frequencies below the Brunt-Väisälä frequency.

Let us first consider waves where pressure is the restoring force. The frequency of such waves is given by a relatively straightforward dispersion relation so, if there were a standing wave or mode, its period could be determined from the integrated sound travel time across the chromospheric cavity. What would such a mode look like on a $k - \omega$ diagram? The lower reflection is independent of scale size and the upper, total internal reflection, is nearly so because the temperature rise is so steep. So, rather than forming one of the parabolae seen on the familiar $k - \omega$ diagram, parabolae that reflect the essentially linear temperature gradient in the convection zone, we would expect a mode that would be nearly independent of frequency. Superposed on a set of deep, parabolic modes, the signature of chromospheric modes would thus be a series of "avoided crossings:" in the linear eigenvalue problem, no two solutions can have the same eigenvector k and ω: the linear solutions would avoid one another.

Since a spectral line formed everywhere in the photosphere and chromosphere has not been found, however, no observational $k - \omega$ diagrams look anything like this. In a part of the Ca II K line wings formed in the photosphere, Duvall et al. (1991) detected only the deep modes, even above the acoustic cutoff frequency at the temperature minimum. On the other hand, observations in a narrow band in the emission core of the K line (e.g. Kneer and von Uexküll 1983) show, in addition to a few of the lower-order parabolic modes, a very weak, horizontal feature in the $k - \omega$ diagram that is suggestive of the chromospheric mode at 4.1 mHz. Similar, horizontal features are visible in in the Hα $k - \omega$ diagram at a variety of frequencies. The 4.1 mHz feature shows up again, well below the 3σ level, in a CN bandhead $k - \omega$ diagram (Kneer, Newkirk, and von Uexküll 1982). There are also many observations obtained with limited spatial coverage and no spatial resolution that show $\sim$ 4 - 5 mHz oscillations in the K line core (Jensen and Orrall 1963), the 11 μm OH band (Deming et al. 1986), and the Mg II k line core (Gurman 1987).

One would like to speculate that all these weak results put together form almost convincing evidence for a chromospheric p-mode. One could even speculate that the multiple Hα frequencies were due to multiple vertical scales, each with a different sound travel time. The best observational $k - \omega$ diagrams obtained to date, however, show no statistically significant evidence for a chromospheric mode. One might expect that evidence to be strongest on the largest spatial scales — the smallest wavenumbers k, but that area of the $k - \omega$ diagram always seems to be noisy, while the evidence that does appear, always seems to be at larger wavenumbers — corresponding to small features. Perhaps, given the wealth of structure in the chromosphere, that is not too surprising: any 4 - 5 mHz oscillation in faculae, short spicules, or whatever is the expected response to forcing by 3.3 mHz oscillations tunneling up through the temperature minimum. When we instead observe a large area, all coherence is lost. If there is, in fact, a chromospheric p-mode, it is likely to be localized in individual structures, and possibly with different frequencies in different sorts of features. So why do we bother looking for a chromospheric p-mode? For exactly the reason that one *would* reflect the real height scales in the chromosphere. It might be the best probe of the real extent of the chromospheric material that produces the line profiles we observe.

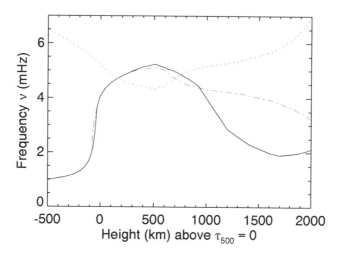

Figure 10. Comparison of the Brunt-Väisälä frequency *(solid line)* and the Lamb wave solution sk_x *(dashed line)* in the photosphere and low chromosphere (Gurman and Drake 1989). (The *dot-dashed line* indicates the misestimation of the Brunt–Väisälä frequency if the effects of ionization and microtubulence are neglected.)

There is, as already noted, another possible mode in the chromosphere, the buoyancy, or internal gravity mode. The solid line in Figure 10 is a fairly realistic run of Brunt-Väisälä frequency with height in part of the VAL3C model (Mihalas and Toomre 1981); the dashed line is just a scaled representation of the sound speed, to give an idea of the thermal structure: the peak in the Brunt-Väisälä frequency occurs at the temperature minimum. Deubner and Fleck (1989) observed 3-minute oscillations in chromospheric lines in network cell interior but ,in the network itself, the power was skewed toward lower frequencies — periods about 5 minutes. UV continuum intensity oscillations with 3 - 4 mHz frequencies have also been observed in the bright network at 1375 Å (Drake *et al.* 1989) and, more generally, at 1600 Å (Cook and Ewing 1991). It is difficult to see how such periods could correspond to a *p*-mode, unless the chromospheric height scale were greatly extended — which is certainly possible, but radiative transfer models of brighter features have smaller vertical scales than those for fainter features, thanks to the coronal "overpressure" necessary to produce the observed densities. The alternate possibility of a buoyancy wave is intriguing because such waves tend to "break," like waves on a beach, dissipatively — another way to heat the chromosphere.

6. Not Really a Conclusion

The reader may now feel, as does the author, that the the title of this Chapter is more than a bit redundant: it is very difficult, if not impossible, to talk about the solar chromosphere without getting into a discussion of individual structures, their lifetimes, their dynamics, their heating and cooling, their relationship with structures in the underlying photosphere and overlying transition region and corona, and so on. While there are still a number of contradictions in our physical description of the height scale of the chromosphere, exciting science is bound to come from trying to resolve such contradictions and understand as complex a phenomenon as the solar chromosphere. As Thomas and Athay (1961) said, in closing their book, thirty years ago,

> "although the corona manifests itself as a more sweeping phenomenon than the chromosphere, and the prominences as more spectacular phenomena than the spicules, we believe the key to the structure of the outer solar atmosphere, including the corona, lies in the combination chromosphere-spicules, even though we do not know which is the cause and which is the effect."

REFERENCES

Anderson, L.S. and Athay, R.G. 1989, *Ap. J.*, **336**, 1089

Anderson, L.S. and Athay, R.G. 1990, *Ap. J.*, **346**, 1010

Ando, H. and Osaki, Y. 1977, *Publ. Astron. Soc. Japan*, **29**, 221

Athay, R.G. and Dere, K.P. 1990, *Ap. J.*, **358**, 710

Athay, R.G. and White, O.R. 1979, *Ap. J. Suppl.*, **39**, 333

Athay, R.G. and White, O.R. 1980, *Ap. J.*, **240**, 306

Ayres, T.R. and Brault, J.W. 1990, *Ap. J.*, **363**, 705

Bohlin, J.D., Vogel, S.N., Purcell, J.D., Sheeley, N.R., Tousey, R., and VanHoosier, M.E. 1975, *Ap. J.*, **197**, L133

Canfield, R.C., Kiplinger, A.L., Penn, M.J., and Wülser, J.-P. 1990, *Ap. J.*, **363**, 318

Cook, J.W. and Ewing, J.A. 1991, *Ap. J.*, **371**, 804

Deming, D., Glenar, D.A., Käufl, H.U., Hill, A.A., and Espenak, F.: 1986, *Nature*, **322**, 232

Deubner, F.-L. and Fleck, B. 1989, *Astron. Astrophys.*, **228**, 506

Drake, S.A., Gurman, J.B., and Orwig, L.E. 1989, in Haisch, B.M. and Rodonò, M. (eds.), *IAU Colloquium No. 104: Solar and Stellar Flares (Poster Volume)*, Publ. Catania Astrophys. Obs., p. 235

Duval, T.L., Harvey, J.W., Jefferis, S.M., and Pomerantz, M.A. 1991, *Ap. J.*, **373**, 308

Fontenla, J.M., Avrett, E.H., and Loeser, R. 1990, *Ap. J.*, **355**, 700

Fontenla, J.M., Avrett, E.H., and Loeser, R. 1991, *Ap. J.*, in press

Foukal, P. 1990, *Solar Astrophysics*, John Wiley & Sons (New York), pp. 294 - 299

Gurman, J.B. 1987, *Solar Phys.*, **108**, 61

Gurman, J.B. and Drake, S.A. 1989, in Phillips, K.J.H. (ed.), *Proceedings of the Second Workshop on Thermal–Non-Thermal Interactions in Solar Flares*, Rutherford Appleton Laboratory, p. 2-4

Jensen, E. and Orrall, F.Q. 1963, *Ap. J.*, **138**, 252

Leibacher, J.W. and Stein, R.F. 1981, in S. Jordan (ed.), *The Sun as A Star*, NASA / CNRS, p. 273

Mihalas, B.W. and Toomre, J. 1981, *Ap. J.*, **249**, 349

Moore, R.L., Tang, F., Bohlin, J.D., and Golub, L. 1977, *Ap. J.*, **218**, 286

Kneer, F., Newkirk, G., von Uexküll, M. 1982, *Astron. Astrophys.*, **113**, 129

Kneer, F. and von Uexküll, M. 1983, *Astron. Astrophys.*, **119**, 124

Rutten, R.J. and Uitenbroek, H. 1991, in Ulmschneider, P., Priest, E.R., and Rosner, R. (eds.), *Mechanisms of Chromospheric and Coronal Heating*, (Berlin: Springer-Verlag), p. 48

Thomas, R.N. and Athay, R.G. 1961, *Physics of the Solar Chromosphere*, Interscience Publishers (New York), p. 1ff.

Vernazza, J.E., Avrett, E.H., and Loeser, R. 1981, *Ap. J. Suppl.*, **45**, 635

Wang, Y.-M. and Sheeley, N.R. 1991, *Ap. J.*, **375**, 761

Withbroe, G.L., Jaffe, D.T. Foukal, P.V., Huber, M.C.E., Noyes, R.W., Reeves, E.M., Schmahl, E.J., Timothy, J.G., and Vernazza, J.E. 1976, *Ap. J.*, **203**, 528

Zarro, D.M., Canfield, R.C., Strong, K.T., and Metcalf, T.R. 1988, *Ap. J.*, **324**, 582

Zirin, H. 1988, *Astrophysics of the Sun*, Cambridge University Press, p. 155ff.

12 SPECTROSCOPIC DIAGNOSTICS

A.H. GABRIEL
Institut d'Astrophysique Spatiale
C.N.R.S. - Université Paris XI
Bâtiment 121
Campus Universitaire
91405 Orsay Cede
France

ABSTRACT. A survey is presented of some of the available techniques for the interpretation of spectral intensities, in terms of the physical parameters of the emitting plasma. Following a discussion of the differences between LTE and "coronal" excitation conditions, the review is limited to "coronal" optically thin plasmas. The techniques include line ratio measurement of temperature and density, as well as the differential emission measure analysis, including continuum emission channels.

1. Introduction

This review presents a summary of some of the principal diagnostic methods based on the analysis of optically thin spectral line emission, as applied to the interpretation of solar plasma from the transition region, corona, active regions, and solar flares. It is not possible in a short article to include the derivation for all of the methods. For further detail, the reader is referred to Griem (1964), McWhirter (1965), Gabriel and Jordan (1972), and Gabriel and Mason (1982).

2. LTE versus Coronal Equilibrium

2.1. LOCAL THERMODYNAMIC EQUILIBRIUM

Local Thermodynamic Equilibrium (LTE) is a steady-state equilibrium condition in which upward and downward transitions are produced by electron impact. Although the plasma is optically thin so that the radiation escapes freely, this loss of energy is a small effect compared with the collisional transitions. In this condition, upward and downward transitions are exactly reverse processes, that is, they are equivalent in the case of time reversal. Figure 1 illustrates these processes for the case of a simplified two-level ion (states 1 and 2) and a simplified ionisation and recombination in which the ground-states only of adjacent ion stages q and $q + 1$ participate.

Ionisation occurs principally by electron impact collisions, while recombination is principally a three-body process. Thus we can write

$$N_q \, N_e \, S = N_{q+1} \, N_e^{\,2} \, R_3.$$ (1)

J. T. Schmelz and J. C. Brown (ed.), The Sun, A Laboratory for Astrophysics, 261–276.

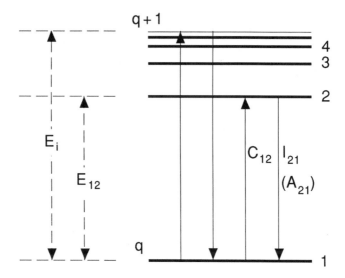

Figure 1. Energy level diagram showing, in simplified form, the transitions between the two lowest states 1 and 2 of the ion stage q, together with the ionisation and recombination processes involving the next higher ionisation stage $q + 1$.

From statistical mechanics, using the principle of detailed balance, we obtain from Equation (1) the familiar Saha equation

$$\frac{N_{q+1}N_e}{N_q} = \frac{2(2\pi mkT)^{3/2}}{h^3} \frac{U_{q+1}}{U_q} e^{-E_i/kT}. \tag{2}$$

For the case of transitions between two levels of the same ion, excitation and de-excitation are both two-body collision processes, so that

$$N_1 \, N_e \, C_{12} = N_2 \, N_e \, C_{21} \tag{3}$$

Using detailed balance to obtain the relation between C_{12} and C_{21} leads to the Boltzmann Distribution

$$\frac{N_2}{N_1} = \frac{g_2}{g_1} e^{-E_{12}/kT} \tag{4}$$

Note that, in the case of LTE, the ratio of population of adjacent ions depends on the electron density, whereas the ratio of level populations within one ion is independent of density.

2.2. CORONAL EQUILIBRIUM

This situation occurs in the limit of low density. In these conditions, radiative two-body recombination becomes more important than collisional three-body recombination. The equilibrium equation for ionisation balance then becomes

$$N_q \, N_e \, S = N_{q+1} \, N_e \, R_2, \tag{5}$$

so that

$$\frac{N_{q+1}}{N_q} = \frac{S}{R_2}. \tag{6}$$

Note that the ratio of adjacent ion stages is now independent of electron density and depends only on temperature through the quantities S and R. (In fact, higher order density dependent terms do occur but these are beyond the scope of this article.) Thus, definitive values of the ratios of ions as a function of temperature can be calculated and tabulated, as e.g. by Arnaud and Rothenflug (1985).

In the case of collisional excitation in the low density limit, this is balanced by the non-collisional process of spontaneous radiative decay, so that

$$N_1 \, N_e \, C_{12} = N_2 \, A_{21} = I_{21}, \tag{7}$$

where A_{21} is the radiative transition probability and I_{21} is the photon emission rate per unit volume. Effectively, the photon emission rate is simply equal to the collisional excitation rate in this simple two-level ion. From Equation (7), we see that the ratio of the excited state to the ground state population is

$$\frac{N_2}{N_1} = \frac{N_e \, C_{12}}{A_{21}}. \tag{8}$$

In the low density conditions of coronal equilibrium, this ratio is always very small compared with unity.

2.3. ELECTRON ENERGY DISTRIBUTION

Independent of the electron density, the transfer of energy and momentum to or from the free electrons is dominated by electron-electron collisions. This leads to an "LTE" relation for the electron kinetic energy, unless there are dominant outside non-thermal influences operating (such as a strong electric field). This thermal distribution of electron energies is the familiar Maxwellian distribution, and is a function only of the temperature.

3. Coronal Line Ratios for Electron Temperature Measurement

To observe the ratio of two resonance lines from the same ion, it is necessary to consider at least a three-level ion, as shown in Figure 2. Here the two excited states are each excited from the ground state (the only state in which there is an appreciable population), and each decay by radiation to the ground, emitting the two observed lines. Following Equation (7), we obtain

$$\frac{I_2}{I_1} = \frac{N_e\,N_g\,C_{g2}}{N_e\,N_g\,C_{g1}}$$
$$= \frac{C_{g2}}{C_{g1}}. \tag{9}$$

This ratio of atomic rate coefficients is simply a function of temperature. For reliable temperature measurement, it is necessary to use good atomic physics parameters. However, to examine the sensitivity, we can resort to simple approximations. Within such approximations, the form of a collision rate for an allowed transition is

$$C \approx \frac{Const.}{T^{1/2}} e^{-E/kT}, \tag{10}$$

so that

$$\frac{I_2}{I_1} \approx Const. \times e^{-(E_2 - E_1)/kT}. \tag{11}$$

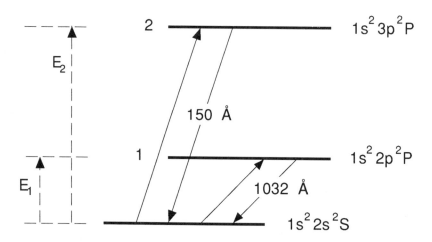

Figure 2. Temperature measurement by the ratio of lines from two excited allowed levels 1 and 2. The levels indicated are for the 3-electron lithium-like ion; in particular, the wavelengths indicated relate to the ion oxygen VI.

For the expression in Equation (11) to be most sensitive to temperature, it is necessary that $E_2 - E_1$ be large compared with kT. However, when this condition is satisfied, then the wavelengths of the two lines are widely separated, making the instrumental problem of relative intensity calibration very difficult. Furthermore, since the line ratio is very small, the weaker line can be too weak to measure. A further problem arises when the ratio becomes very sensitive, that small departures from the assumed isothermal nature of the source can lead to large errors in interpretation (see later).

3.1. DIELECTRONIC SATELLITES - A SPECIAL CASE

Figure 3 shows a simplified energy level system for a helium-like ion, together with the resonance levels of the type $1s2s2p$, which lie just below the first excited state $1s2p$. The resonant state can be excited by dielectronic capture C_s from the adjacent continuum. It can decay either by the reverse process of autoionisation A_a, or by radiative decay A_r to the ground state $1s^2 2s$ of the lithium like ion. Starting with the helium-like ion in its ground state $1s^2$, we consider the ratio of I_r, the intensity of the helium-like resonance line, to that of I_s, the satellite line resulting from decay of the 3-electron state. Excitation of the 3-electron state is equivalent to excitation of the helium resonance line, except that the impact electron has insufficient energy to escape and remains captured by the excited ion. Adapting Equation (7), we obtain for the satellite intensity

$$I_s = N_i \ N_e \ C_s \ \frac{A_r}{A_a + A_r}. \tag{12}$$

Here the final term represents the branching ratio, or the probability that the 3-electron state decays by radiation. In Equation (12), the coefficients C_s and A_a represent the forward and reverse identical processes. They are thus related by the detailed balance equation, in this case the Boltzmann-Saha equation

$$\frac{A_a}{C_s} = \frac{2(2\pi mkT)^{3/2}}{h^3} \ \frac{1}{g_s} \ e^{E_s/kT}. \tag{13}$$

Substituting in order to eliminate C_s, we get

$$I_s = N_i \ N_e \ \frac{h^3}{2(2\pi mkT)^{3/2}} \ g_s \ e^{-E_s/kT} \ \frac{A_a \ A_r}{A_a + A_r}. \tag{14}$$

Since for the helium-like resonance line (see Equation (10)), we have

$$I_r = N_i \ N_e \ C_r$$
$$\approx N_i \ N_e \ \frac{Const.}{T^{1/2}} e^{-E_r/kT} \tag{15}$$

Taking the ratio of Equations (14) and (15) then gives

$$\frac{I_s}{I_r} \approx \frac{1}{T} \ e^{(E_r - E_s)/kT} \ A_a \ \frac{A_r}{A_a + A_r} \tag{16}$$

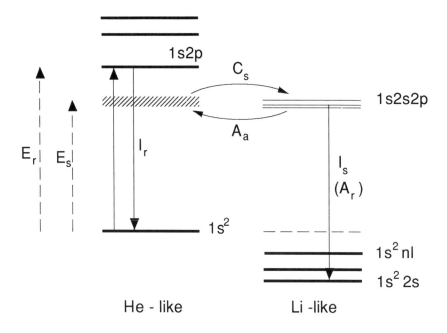

Figure 3. Energy level diagram, showing the processes determining the intensity of the lithium-like satellite lines adjacent to the helium-like resonance line.

Since the satellite and resonance line are very close in wavelength, there is no problem of intercalibration of the instrumentation between the two. Examination of Equation (16) shows how this ratio can be used to measure temperature. When A_a is small, the satellite is weak and the method does not work. Among the many lines in the satellite multiplets, it is necessary to choose one with an "allowed" autoionisation, that is, in the region of 10^{13} sec^{-1}. In this case, the ratio I_s/I_r will vary as A_r. Since A_r varies as Z^4, the satellite to resonance line ratio will increase rapidly with Z. Thus it is of the order 1% for oxygen, 10% for calcium and 50% for iron. Since the interval $E_r - E_s$ is small compared with kT, the main temperature dependence comes from the term T^{-1}. This useful sensitivity is not so large that it introduces serious problems in non-isothermal plasmas. Thus we have an excellent method for measuring temperatures, when they are high enough to produce helium-like calcium or higher, that is, in solar flares (Bely-Dubau *et al* 1982). Figure 4 shows such a spectrum of calcium obtained by the Bent Crystal Spectrometer on the *Solar Maximum Mission*.

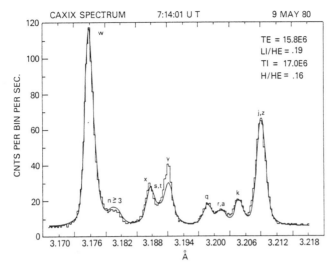

Figure 4. Soft X-ray spectra of calcium XIX and XVIII from a solar flare, showing the dielectronic satellite lines. Observations are represented by the stepped curve. The continuous curve is a theoretical spectrum.

4. Electron Density Measurement from Coronal Line Ratios

For the general case of the two-level ion shown in Figure 1, the excitation and de-excitation processes can be written as

$$N_1 \, N_e \, C_{12} = N_2 \, N_e \, C_{21} + N_2 \, A_{21} \qquad (17)$$

At high density, the first term on the right-hand-side dominates and we have the LTE relation of Equation (3). At low density, it is the second term which dominates and gives the coronal relation of Equation (7). However, even in the low density domain, normally characteristic of the coronal excitation conditions, the second term may not dominate totally if the value of the transition probability A_{21} is small, that is to say, for a metastable or forbidden transition. In such cases, for certain levels only, the two terms compete, so that $A_{21} \approx N_e C_{21}$. This introduces a different dependence on density for the level 2 than for other normal coronally excited levels.

To measure the electron density by this method, we choose a metastable level m such that $A_{21} \approx N_e C_{21}$ for the density region of interest. Ideally, the intensity of the line from m is compared with that from a normal coronally excited level 3 of about the same excitation energy. Such a scheme is shown in Figure 5. Thus for the allowed line, its intensity is given by

$$I_3 = N_1 \, N_e \, C_{13} \qquad (18)$$

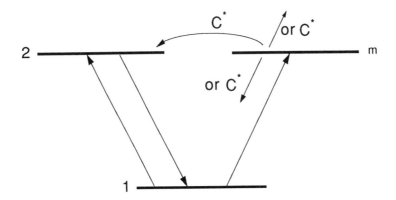

Figure 5. The processes leading to a density-sensitive line ratio.

However, for the metastable level, only that proportion of the excitation which decays by radiation contributes to the line intensity, giving

$$I_m = N_1 \; N_e \; C_{1m} \; \frac{N_m \; A_{m1}}{N_m \; A_{m1} + \Sigma \; N_m \; N_e \; C^*} \tag{19}$$

The ratio of the two observed intensities is then given by

$$\frac{I_m}{I_3} = \frac{C_{1m}}{C_{13}} \; \frac{A_{m1}}{A_{m1} + N_e \; \Sigma C^*} \tag{20}$$

The final term in Equation (20) introduces the density sensitivity into the observed intensity ratio.

The configuration shown in Figure 5 is only one of a number of alternatives possible. The collisional de-excitation of the level m can be towards the ground state, the level 3, or to unspecified other levels. The density-sensitive line can be a transition between two excited levels, not involving the ground. Furthermore, the metastable level can be either the upper or the lower level of the transition. Such variants modify considerably the equations, but we are always left with a merging of two density domains (and thus a useful density diagnostic) when $A^* \approx N_e C^*$. Here A^* is the decay rate of the metastable by radiation, and C^* is the rate coefficient for its collisional depopulation. Figure 6 shows a number of forms which this density-dependent line ratio can take, according to the precise energy-level structure applicable.

It is of interest to examine the variation along an iso- electronic sequence. The region of density-sensitivity occurs when $N_e \approx A^*/C^*$. C^* scales with the ion charge Z as Z^{-3}, while A^* for a forbidden transition scales as Z^6, Z^8, or even Z^{10}. Thus, for a given transition, the region of density sensitivity varies very steeply with the ion charge, *i.e.* between Z^9 and Z^{13}. This implies that there will be only one or two adjacent ion stages in a given iso-electronic structure that can be employed to measure the density of a given astrophysical medium.

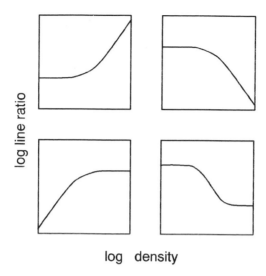

log line ratio

log density

Figure 6. Showing a number of different forms possible for a density-sensitive line ratio.

4.1. SPECIAL CASE OF THE HELIUM-LIKE ION

We refer again to the simplified model of Figure 5. If the two levels 3 and m have about the same energy, this tends to minimise the temperature sensitivity of the line ratio, whilst at the same time ensuring that the two lines fall in the same spectral region. A particular case arises for He-like ions, for which there are four $n = 2$ levels close together in energy, but having excitation energies which are a large proportion of the ionisation potential of the ion. Figure 7 shows a partial energy-level diagram, with the three spectral lines arising from three of these levels together with the two-photon continuum arising from the third. The three lines are designated using the notation w, y, and z defined by Gabriel (1972). The four levels, having very different values of transition probability, give rise to three different domains of density sensitivity (Gabriel and Jordan 1972).

The line w has a normal allowed transition probability. Then, in decreasing order, we have the intercombination line y, the two-photon emission, and the forbidden line z. At the lowest density, all four levels decay spontaneously to the ground. The first density sensitivity arises when collisional transfer from the $1s2s\ {}^3S$ level to the $1s2p\ {}^3P$ term competes with the forbidden line z. When this transfer is complete, the line z is no longer seen and the excitation of both z and y decay through the line y. At an intermediate density sensitive domain, the $1s2s\ {}^1S$ level transfers its population to the $1s2p\ {}^1P$ level. The effect here is to transfer the two-photon emission (difficult to observe) to augment the intensity of the line w. Then, at higher densities, we find the onset of collisional transfer from the y line

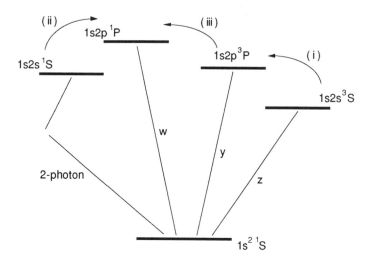

Figure 7. Simplified energy level diagram for the $n = 2$ states of the He-like ion.

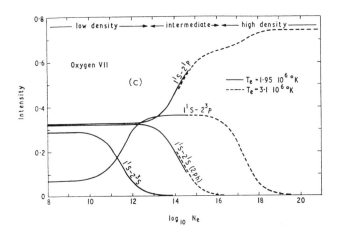

Figure 8. Showing the density sensitivity of the He-like oxygen line ratios, over a range of nearly ten orders of magnitude in density.

to the line w, so that finally all of the n=2 excitation decays through this line. The three domains are shown in terms of the relative line ratios for oxygen in Figure 8.

In reality, the equations are complicated by the fact that some of the $1s2p\ ^3P$ excitation decays by radiation to the $1s2s\ ^3S$ level. However, this does not affect the three domains, although it influences the magnitude of the line intensities.

In the Sun, it is possible to use the low-density region as a density diagnostic with He-like oxygen, neon, and magnesium, in active regions and flares. Observations for oxygen have been presented by McKenzie and Landecker (1982) and McKenzie (1987), and with a more recent theory by Gabriel *et al* (1991). The analysis of He-like neon spectra in flares has been reported by Wolfson *et al* (1983). It is possible to find other non-He-like ions to use in the transition region. However, no useful density-dependent line ratios have been found for temperatures exceeding 10^7 K.

5. Temperature from Ionisation Balance

In Section 2, it was shown that, in coronal equilibrium, the ratio of population of two adjacent ion stages N_q/N_{q+1} is a function only of temperature. Various authors have carried out computations of these ratios, and extensive tabulations exist in the literature (*e.g.* Arnaud and Rothenflug 1985). The functions are normally listed as the ratio of one ion stage to the total for the element: $N_q/\Sigma N_q$. An example of such a calculation, due to Jacobs *et al.* (1977) is shown in Figure 9.

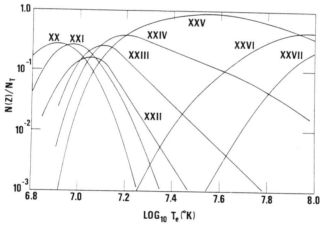

Figure 9. Ionisation balance curves for iron, according to Jacobs *et al.* (1977).

In reality, we measure not the abundance of an ion stage but the intensity of one of its resonance lines. The intensity of the line is given by

$$
\begin{aligned}
I_{21} &= N_q \, N_e \, C_{12} \\
&= \Sigma N_q \, N_e \, \frac{N_q}{\Sigma N_q} \, C_{12}\,, \\
&= \Sigma N_q \, N_e \, g(T)
\end{aligned}
\tag{21}
$$

where

$$g(T) = \frac{N_q}{\Sigma N_q} \, C_{12} \tag{22}$$

gathers together the two temperature dependent terms. The principal contribution to $g(T)$ comes from the ionisation balance, which has a very steep dependence on T in the wings of the function (see Figure 9). However, the exponential temperature factor in C_{12} enters also, and has the effect of displacing the function a little towards higher temperature, as compared to the ionisation balance.

Using this theory, the relative intensity of two lines, arising from adjacent ion stages, can be used to give a very sensitive determination of temperature. However, this analysis assumes a steady-state isothermal plasma, an assumption that will be examined in the next section.

6. Validity of the Assumptions

6.1. STEADY-STATE

Considering first the equilibrium between excited states, this is determined principally by the radiative decay times. Even for the metastable and forbidden decays, this time is normally less than 10^{-3} sec, which is much shorter than the time-scale for any disturbance in the solar atmosphere. We are therefore safe in assuming always steady-state excitation.

The time scale for ionisation equilibrium is somewhat slower and depends on the local electron density. It can be of the order 1 to 10 sec and, in these circumstances, we could envisage non steady-state effects. Attempts to exploit such effects for diagnostic purposes have concentrated on the most dynamic phenomena, *i.e.* solar flares. Only in exceptional cases have small departures from steady-state been detected. We therefore conclude that, in most other conditions, it is safe to assume steady-state ionisation.

6.2. THE ISOTHERMAL ASSUMPTION

This assumption is implicit in all of the methods so far discussed for measuring the temperature. Let us suppose that we have a modest distribution of temperatures around a mean value, that we are trying to measure. It is evident that, if a line ratio is sensitive to temperature, then relatively more of one line comes from the hotter region, and relatively more of the other from the cooler. This principle is equally valid whether one is considering the excitation of two lines from the same ion, or two lines from adjacent ions.

If the volume separation of emission of the two lines is not large, then a meaningful mean temperature can be deduced from the ratio. This condition can break down if either the spatial distribution of temperature is dramatic, or if the temperature-sensitivity of the line ratio is high. This is a strong argument for relying on excitation ratios in a single ion and, in particular, on the dielectronic satellite method (where applicable) with its modest T^{-1} sensitivity. With the ratio of adjacent ion stages, the situation is much more dangerous due to its much higher sensitivity. In an extreme case, the line ratio can say more about the ratio of two volumes containing hotter and cooler plasma, than be interpreted as a mean temperature.

The way around this problem is to use an analysis method such as the Differential Emission Measure, which does not assume isothermal conditions. It does, however, imply an assumption (normally valid) of steady state ionisation balance.

7. Analysis of Absolute Spectral Intensities

We consider in this section the use of the total intensity of a spectral line, rather than the application of line ratios. Nevertheless, we may choose to ignore some general factors common to all lines (such as the geometric collection efficiency of the spectrometer), and in that sense the analysis remains formally relative. We shall start with an isothermal plasma and then generalize the result to the multithermal situation.

In the coronal approximation, the photon intensity per unit volume for a line i is given by

$$I_i = \int N_q \, N_e \, C_i \, dV \tag{23}$$

Here C_i represents the effective collisional excitation rate coefficient for the line i, taking account of all processes in the ion. For a simple two-level ion, it is the excitation rate from the ground to the upper level, and this is often a sufficient approximation. For a more complex ion, it may be necessary to include both excitation and de- excitation by other routes. The integral is over the volume V which is observed. Normally this is limited in two dimensions by the resolution of the instrumentation. In the line-of-sight, for an optically thin plasma (an essential assumption), the totality of the plasma is involved.

Expanding Equation (23) gives

$$
\begin{aligned}
I_i &= \int \frac{N_H}{N_e} \frac{\Sigma N_q}{N_H} \frac{N_q}{\Sigma N_q} N_e{}^2 \, C_i \, dV \\
&= 0.8 \, Ab \, \frac{N_q}{\Sigma N_q} \, C_i \int N_e{}^2 \, dV \\
&= 0.8 \, Ab \, g_i(T) \, X,
\end{aligned}
\tag{24}
$$

where $X = \int N_e{}^2 \, dV$ is the isothermal emission measure and Ab is the abundance of the element relative to hydrogen. The factor 0.8 arises since, in a plasma of solar composition at coronal temperatures, there are more free electrons than protons, due to ionisation of the heavier elements present.

In order to generalize this expression for a non-isothermal plasma, it is necessary to define a Differential Emission Measure $Y(T)$ as

$$Y(T) = \frac{1}{\Delta T} \int_T^{T+\Delta T} N_e{}^2 \, dV \tag{25}$$

the spatial variation of N_e and T. It can be reduced to a simpler form

$$Y(T) = N_e{}^2 \frac{dV}{dT} \tag{26}$$

when there is a monotonic variation of N_e with T, as in some theoretical models of the solar atmosphere (see Chapter 13).

With this definition, we can now integrate the isothermal Equation (24) over the temperatures existing in the real plasma, to give

$$I_i = 0.8 \ Ab \int g_i(T) \ Y(T) \ dT \qquad (27)$$

In a practical situation, Equation (27) represents a matrix of equations, one for each spectral line i observed. The intensities I_i and the atomic functions $g_i(T)$ are known, and the problem rests in inverting the matrix to obtain the distribution $Y(T)$. It should be stressed that the function $Y(T)$ represents the totality of factors which determine the emitted intensities within the approximation assumed (coronal excitation, optically thin, steady-state plasma). If we wish to unfold Equation (25) in order to obtain N_e and T as a function of space, it is necessary to introduce further physical assumptions (*e.g.* hydrostatic equilibrium) regarding the source, as discussed in Chapter 13 for the solar corona.

It should be noted that a channel i, which gives rise to one of the Equations (27) of the matrix, represents in a completely general way one channel of the observing instrument. This can be a single spectral line, as assumed above, but it could also be a broad wavelength band containing continuum radiation, or even continuum overlaid with a number of spectral lines. In any case, it is necessary to understand the response of the channel sufficiently to calculate the value of $g_i(T)$, that is, the variation with temperature of the emission in the channel.

The inversion of the matrix of Equations (27) is "ill-conditioned" when the width of the $g(T)$ functions becomes comparable with either the structure in the function $Y(T)$, or the overall range of T observed. An ill-conditioned inversion implies that the uncertainty or "noise" level in the observed signals becomes much amplified in the inverted values of $Y(T)$. To avoid this ill-conditioned situation, it is preferable to choose $g(T)$ functions which are narrow, or a total range of temperatures which is very large. A large range of energies or temperatures implies that some of the higher channels contain principally continuum radiation. For such channels, the $g(T)$ functions can be readily defined, but are of a different shape from those for line spectra and in general much wider. Thus, the use of a larger temperature range will contribute to improving the conditioning of the inversion, but will be partially offset by the wider contribution functions of the continuum channels.

Many authors have discussed computational techniques for optimising the inversion. Basically, the technique is to iterate progressively a trial function $Y(T)$, until the best agreement is obtained between the calculated and observed set of values of the intensities I_i. The difference between the various methods rest in the choice of definition for the "best" fit. The various methods can also give a value for the variance, or the limits for the range of functions $Y(T)$ which are consistent, within the error bars of the observations and the atomic theory. It is also very instructive to perform the iteration by hand, using judgements based upon inspection of the fit, a technique which is by no means absurd for arriving at a sensible best fit. This is particularly useful in examining the range of different $Y(T)$ functions which are consistent with the observed data.

In general, the combination of inter-channel calibration and solar abundance errors imposes a limit of a factor 2 on the degree of fit expected. Other terms in the equation introduce smaller errors *e.g.* 50% for ionisation balance theory, 20% for excitation rates, and 10% for the dielectronic satellite theory. If we use existing computer-based iteration techniques, it is not possible to profit from these higher accuracies, all of the data being subjected to a common uncertainty of a factor 2. However, using a manual iteration technique, Gabriel *et al.* (1984) were able to fit wide energy range solar flare X-ray data, in a manner that enabled the inclusion of the 10% accuracy satellite line theory into an extended temperature differential emission measure analysis. This manual approach enable them to include also the spectroscopic effects of imbedded high-energy non-thermal electrons.

The method of differential emission measure is often presented as highly complex and obscure in interpretation. This misleading impression arises due to the complexity of the automatic iteration proceedures. The use of manual iteration demonstrates effectively the simplicity and directness of the technique. It is in effect no more than a quantitative application of the much used "ionisation temperature" method in which the existence of lines from various ions is used to deduce the presence of material at the temperature of maximum production of those ions.

The function $Y(T)$ has another particular advantage. It is possible to calculate this function for any theoretical model of an astrophysical plasma. It is possible, without introducing model assumptions to derive $Y(T)$ directly from observed spectra. This function then represents the ideal meeting point for comparing models with observations, a technique which is demonstrated in Chapter 13 for the solar corona.

8. Other Diagnostic Techniques

Due to the limitations of this short article, we have not dealt here with a number of other methods. For completeness, there follows a brief list.

8.1. NON-THERMAL EFFECTS

The existence within a plasma of streams of high energy electrons can give rise to some characteristic spectral line ratios (Gabriel *et al.* 1991) These can also be studied using a generalized version of the differential emission measure technique (Gabriel *et al.* 1984).

8.2. ABUNDANCE DETERMINATION

Evidence is accumulating that the relative abundance of the elements in solar plasma is not a constant, but varies with the type of feature and possibly with time. Spectroscopic studies and, in particular, the differential emission measure analysis allows us to study these effects.

8.3. WIDTH OF SPECTRAL LINES

Spectral lines in the chromosphere and transition region and, to a lesser extent, in the corona are found to have widths significantly larger than thermal doppler widths. This can be interpreted either as a local turbulence or due to the passage of an unresolved wave motion (Boland *et el* 1975, Saba and Strong 1991). (See also Chapter 13)

8.4. BULK PLASMA VELOCITIES

The observation from many sources of asymmetric or shifted spectral lines can give vital information on the dynamics of the plasma. Thus the blue-shifted resonance lines of He-like calcium and iron from solar flares have been interpreted as due to chromospheric evaporation; the rapid ejection of plasma when heated suddenly to coronal temperatures by the solar flare process (Antonucci *et al* 1982)

9. References

Antonucci, E., Gabriel, A.H., Acton, L.W., Culhane, J.L., Doyle, J.G., Leibacher, J.W., Machado, M.E., Orwig, L.E. and Rapley, C.G. (1982) *Solar Phys.*, **78**, 107.

Arnaud, M. and Rothenflug, R. (1985) *Astron. Astrophys. Suppl.*, **60**, 425.

Bely-Dubau, F., Dubau, J., Faucher, P., Gabriel, A. H., Loulergue, M., Steenman-Clark, L., Volonte, S., Antonucci, E. and Rapley, C. G. (1982) *Mon. Not. R. Astron. Soc.*, **201**, 1155.

Boland, B. C., Dyer, E. P., Firth, J. G., Gabriel, A. H., Jones, B. B., Jordan, C., McWhirter, R. W. P., Monk, P. and Turner, R. F., (1975) *Mon. Not. R. Astron. Soc.*, **171**, 697.

Gabriel, A. H. (1972) *Mon. Not. R. Astron. Soc.*, **160**, 99.

Gabriel, A. H., Bely-Dubau, F., Sherman, J. C., Orwig, L. E. and Schrijver, J. (1984) *Adv. Space Res.*, **4**, No. 7, 221.

Gabriel, A. H., Bely-Dubau, F., Faucher, P. and Acton, L. W., (1991) *Astrophys. J.*, **378**, 438.

Gabriel, A. H. and Jordan, C. (1972) 'Interpretation of Spectral Intensities from Laboratory and Astrophysical Plasmas' in E. W. McDaniel and M. R. C. McDowell (eds.), Case Studies in Atomic Collision Physics 2, North-Holland, Amsterdam, pp. 211-294.

Gabriel, A. H. and Mason, H. E. (1982) 'Solar Physics' in H. S. W. Massey and D. R. Bates (eds.), Applied Atomic Collision Physics, Vol 1, Academic Press, London, pp. 345-397.

Griem, Hans R. (1964) Plasma Spectroscopy, McGraw-Hill, New York.

Jacobs, V. L., Davis, J., Kepple, P. C. and Blaha, M., *Astrophys. J.*, **211**, 605.

McKenzie, D. L., (1987) *Astrophys. J.*, **322**, 512.

McKenzie, D. L. and Landecker, P. B., (1982) *Astrophys. J.*, **259**, 372.

McWhirter, R. W. P. (1965) 'Spectral Intensities' in R. H. Huddlestone and S. L. Leonard (eds.), Plasma Diagnostic Techniques, Academic Press, New York, pp. 201-264.

Saba, J. L. R. and Strong, K. T., (1991) *Astrophys. J.*, **375**, 789.

Wolfson, C. J., Doyle, J. G., Leibacher, J. W. and Phillips, K. J. H., (1983) *Astrophys. J.*, **269**, 319.

13   THE SOLAR CORONA

A. H. GABRIEL
Institut d'Astrophysique Spatiale,
C.N.R.S. - Université Paris XI,
Bâtiment 121,
Campus Universitaire,
91405 Orsay Cedex
France

ABSTRACT. The structure of the quiet solar corona is considered, excluding the effect of magnetic active regions. Starting with a simple one-dimensional geometry, models are derived both from *ab initio* theory and from analysis of spectral line intensities. To overcome the inconsistencies which arise, account is taken progressively of two-dimensional effects; the concentration of magnetic fields at the supergranule cell boundaries and the effect of coronal holes. The relationship is discussed between coronal holes, solar wind onset, the observation of fast streams, and the implication for other stars. The transition region is shown to be thicker than often assumed, thereby allowing the neglect of some of the more sophisticated non-Maxwellian effects which have been proposed in the past.

1. Introduction

Since the term "corona" is used in solar physics in a variety of different meanings, it is important to start with a definition. Here, the corona is that part of the solar atmosphere above the visible photospheric layer, which reaches temperatures of the order of half a million degrees or higher. This is in contrast to other definitions, which sometimes include cooler material (*e.g.* prominences) that occupy the space normally associated with the corona.

A basic question, which has eluded a detailed answer for half a century, concerns the mechanism for heating the solar corona. Since the corona is so much hotter than the underlying photosphere from which the energy is assumed to originate, it follows that the transfer mechanism cannot be thermal. Most theories involve some non-thermal mode of energy propagation, which transports some proportion of the available dynamic energy of photospheric motion to be dissipated high in the corona. This energy, once in thermal form, is dispersed in part by conduction back down to the chromosphere, in part in providing energy for the solar wind, and to a lesser extent by radiation in the form of optically thin spectral lines.

In the early days of coronal space observations in the 1960s the Sun was viewed as a star, *i.e.* without spatial resolution. With little other information, the data was interpreted in terms of spherically symmetric models. It was simple to separate the atmosphere into regions of chromosphere, transition region, and corona, with the supposition that some global physical principles dominated in each. As the techniques for spatial resolution of the

J. T. Schmelz and J. C. Brown (ed.), The Sun, A Laboratory for Astrophysics, 277–296.
© 1992 *Kluwer Academic Publishers. Printed in the Netherlands.*

corona have advanced, so has the realisation that there exist detailed structures having large local variations in intensity. We have now swung to the other extreme. It is fashionable to pretend that there exists no typical quiet corona, and that the only useful studies are detailed analyses of specific observable structures. I would claim that this approach has gone too far, and that we are in danger of "throwing the baby out with the bathwater,"to quote an old saying. By neglecting the possibility of drawing global conclusions regarding large long-lived regions, we risk ignoring a number of fairly obvious deductions concerning the energy and momentum balance of the solar atmosphere.

One of the most important departures from uniformity in the corona is due to the presence of the so-called active regions. Such regions, often but not always associated with sunspots, arise due to the emergence from the solar surface of intense magnetic field structures. Averaged over distances equivalent to the order of one arcminute, field strengths are between 100 and 3000 Gauss, as compared with the 1 to 10 Gauss typical of the remainder of the solar surface. Active regions are not always present on the visible half of the Sun, being especially rare at times of solar minimum. Nevertheless, the remaining hot quiet corona is always present, modulating only slightly in density and temperature between solar minimum and maximum. Thus we can conclude that the existence of a hot corona is not contingent on the presence of active regions, and that the physics of magnetic active regions can be considered as a separate question.

The next major departure from uniformity is due to the coronal holes. These are dark regions on the Sun, marked by well defined boundaries, that are assumed to be the regions having open field structures and the principal source of the solar wind. They exist in different forms at all phases of the solar cycle, and are therefore an essential feature of the global corona. It is clear however that, being much less intense, they contribute little to the global intensity of the Sun, seen as a star. Thus we may assume that the earlier derived spherical models of the Sun are representative of the closed field regions, rather than the open field corona.

Figure 1 shows a soft X-ray photograph of the solar corona. It is the finest and most recent in a long series of technological improvements for this type of imaging, and has been recorded by the Lockheed Soft X-ray Telescope instrument on the Japanese satellite *Yohkoh*, launched in August 1991. It shows in excellent resolution the wide range of intensities and structures visible at coronal temperatures. It presents vividly the complexity of the corona. However, one could also argue that it demonstrates clearly the categorization possible for coronal regions. We see the intense local active region emission. We see also the clearly defined dark coronal hole regions. What is left is a more or less uniform quiet corona, albeit crossed by structures indicating the form of the loop-shaped magnetic fields.

In this article we attempt to understand first the closed field and then the open field regions of the "quiet" solar corona, leaving aside totally the enhanced magnetic field active regions.

The approach taken is to start with the simple spherically symmetric model of the corona, and then to examine in turn the most important factors leading to a departure from this simplicity. The effect of superposing a fine structure on a global model can vary greatly. In some cases, we can expect the overall physics to behave in a manner which is simply an average of the detailed behaviour in the structures. In these cases, our averaged interpretation is *correct* in a limited or global sense. Other situations exist in which the systematic effects of fine structure lead to a reality which is significantly different from the average solution. It is most important to try to recognise such situations, and to apply the necessary corrections.

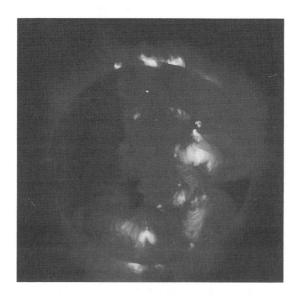

Figure 1. X-ray image of the solar corona, showing active regions, the quiet corona and coronal holes. This image was acquired by the Soft X-ray Telescope on the Yohkoh solar research spacecraft of the Japanese Institute of Space and Astronautical Science. The Soft X-ray Telescope experiment is a Japan/US collaboration involving the National Astronomical Observtory of Japan, the University of Tokyo, and the Lockheed Palo Alto Research Laboratory.

In general, there are two possible starting points for the analysis. We can use a theoretical modelling technique, *ab initio*, except for the choice of physically realistic boundary conditions, or we can construct models from the direct analysis of the observations. Both approaches are limited by the need to be sure that we are studying a single coherent region, and that the interpretation is not invalidated by fine structure effects. In reality, a more rapid advance in understanding can be achieved by exploiting a combination of the two proceedures.

2. Spherically Symmetric (or Plane-Parallel) Models

There is no difference in principle between these two descriptions. Both refer to one-dimensional models of the atmosphere, such as might be derived from observations of the Sun as a star, without spatial resolution. When the height range under consideration is small compared with the solar radius, the plane-parallel description may be a sufficient

280

approximation. However, since some terms vary as the square of the radius, it can be more precise to deal more correctly in spherical geometry. For our present purposes, the differences are not large, so that we will consider the plane-parallel case in order to simplify the presentation of the equations. It should be emphasised, however, that changing to the spherical case introduces no important changes in the physics, nor in the relative importance of the terms.

2.1. THEORETICAL MODELLING

The technique here is to establish steady-state equations for the conservation of energy and momentum flow, and then to solve them taking account of some assumed boundary conditions. The physics included is of course limited by the use of only those terms that are believed to be significant.

The divergence of the total energy flow must be equal to zero, if we include all significant transfer mechanisms. Thus

$$\nabla \vec{F} = 0. \tag{1}$$

$\vec{F}$ can be separated into its various components:

$$\nabla(\vec{F}_c + \vec{F}_r + \vec{F}_m + \vec{F}_k + \vec{F}_e + \vec{F}_g) = 0, \tag{2}$$

where, $\vec{F}_c$ is the conductive energy flux, $\vec{F}_r$ is the radiative energy flux, $\vec{F}_m$ is the mechanical (heating) energy flux, $\vec{F}_k$ is the kinetic energy flux, $\vec{F}_e$ is the enthalpy flux, and $\vec{F}_g$ is the gravitational energy flux.

The last three terms are zero, unless there is a flow of material, as for example due to the solar wind. In the plane-parallel approximation, for most of these terms the operator ∇ can be replaced by d/dh, and the flux itself by its vertical component. Specifically, we can write:
The conductive flux,

$$F_c = -a \cos^2\alpha \; \kappa T^{5/2} \frac{dT}{dh}, \tag{3}$$

where κ is the Spitzer conductivity 1.23×10^{-6} (in cgs units). The use of the Spitzer coefficient assumes collision dominated conduction. In spite of many contrary claims, we show later that this is indeed realistic. The area factor a and the inclination factor $\cos^2\alpha$ are introduced for ducted non-radial heat flow in this and the following equations. They will be referred to later, but for the present one-dimensional model, both are equal to unity.

The divergence of the radiative flux is equal to the energy radiated per unit volume :

$$\begin{aligned} \nabla F_r &= R \\ &= \Phi N_e^2. \end{aligned} \tag{4}$$

For optically thin radiation, Φ is a function of T only. It has been calculated and tabulated by a number of authors, e.g. Cox and Tucker (1969) and McWhirter et al. (1975).

The divergence of the mechanical energy flux is simply related to the energy deposition rate S per unit volume by

$$\nabla F_m = -S \tag{5}$$

The kinetic energy flux can be expressed as

$$aF_k = 1/2 \; aWv^2, \tag{6}$$

where aW is the mass flow rate per unit area (constant with height) and v is the local wind velocity. The enthalpy flux, F_e, is given by

$$F_e = \frac{5}{2} \; W \; \frac{kT}{AM_H}, \tag{7}$$

where A is the average atomic weight of the corona and M_H the mass of a hydrogen atom. The gravitational energy flux, F_g, is given by

$$\frac{d}{dh}F_g = Wg, \tag{8}$$

where g is the acceleration due to solar gravity.

The momentum equation in plane-parallel geometry can be written:

$$-\frac{dp}{dh} = \rho g + \frac{d}{dh}(\rho v^2) + \frac{d}{dh}p_m \tag{9}$$

where the right hand terms relate respectively to gravity, acceleration, and pressure due to the mechanical energy flux.

With the above set of equations, it is now possible to construct a theoretical model atmosphere, if we have access to certain boundary conditions. It is usual to adopt conditions at a certain height in the corona, a plasma which is well observed, and less complex than the base of the corona. A value of $\log T = 6.14$ and $\log p = 14.5$ (where $p = T \times N_e$) would be regarded as reasonably well established for a height in the corona of 30,000 km. However, it is also necessary to assume a value for the conductive flux at this height and values for the mechanical energy deposition S through the region under investigation. For these also, we must depend on deductions made from the observations.

2.2. MODELS BASED ON OBSERVATIONS

The use of satellite-borne ultraviolet spectrometers has given access to a wide range of spectral intensities emitted by highly-ionised atoms in the chromosphere and corona. The diagnostic techniques available using these lines include the direct measurement of electron temperatures and densities, as well as other more sophisticated studies. However, to unfold the temperature structure of the solar atmosphere, the most important methods involve the study of the distribution of ion stages formed at different temperatures. This technique, known as the differential emission measure analysis, has been developed by Pottasch (1963, 1964) and by Jordan and Wilson (1971). The idea is to use the absolute intensity of lines with well-understood excitation rates and temperatures of production in order to estimate the emission measure (or the volume integral of the density squared) as a function of temperature.

The intensity of a line i in a steady-state non-homogeneous plasma is given by

$$I_i = K_i A \int g_i(T) Y(T) d(\log T) \tag{10}$$

where $g_i(T)$ is the emission function for the line i, and the Differential Emission Measure (DEM) is defined as:

$$Y(T) = N_e^2 \frac{dh}{d(\log T)}. \tag{11}$$

With a wide range of emission line intensities available, it is possible to invert the set of Equations (10) in order to derive the value $Y(T)$ as a function of T. An example of such an analysis based upon the Sun-as-a-star data is shown in Figure 2. To turn this plot into a model of temperature and density versus height, it is necessary to make assumptions regarding the pressure, its law of variation with height, and a boundary value at some height. If we assume hydrostatic equilibrium *i.e.* the first term only of Equation (9), and a pressure in the corona at $logT = 6.2$ of $log(TN_e) = 14.5$, then a credible model of the atmosphere can be unfolded. This is shown as the curve A in Figure 3. The figure of 14.5 for the log pressure in the corona is based upon a number of independent measurements of coronal density made during eclipses. The model of Figure 3 is thus relatively free from arbitrary assumptions, other than that of the plane-parallel geometry.

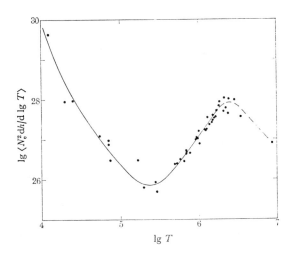

Figure 2. Differential emission measure for the quiet solar atmosphere, derived from observed spectral intensities.

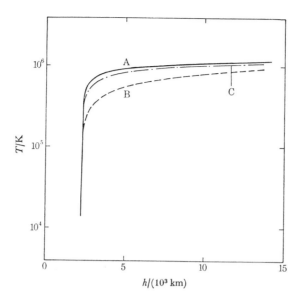

Figure 3. Temperature structure of the quiet solar atmosphere: A - plane-parallel model based on the intensities in Figure 1, B - an *ab initio* energy balance model, C - a hybrid network model.

If we return to the theoretical equations in section 2.1, it is possible to draw an interesting conclusion if we are prepared to make a number of approximations. These approximations can be shown to be valid if we limit the region under study to the temperature level of 3.0×10^5 K at the lower boundary and the height of 30,000 km for the upper boundary. We assume zero flow-rate (which is valid for the non-hole regions), constant pressure, and negligible radiated energy loss. Under these assumptions, manipulation of the equations predicts a straight line for the logarithmic plot of Figure 2, with a slope of +1.5, provided the conductive flux F_c is constant with height. Early workers, using the DEM technique noticed that the curve from the observations could be fitted by a straight line in the region $logT$ from 5.5 to 6.2, as can be seen from Figure 2. With the relatively large error bars due to uncertainties in the atomic physics and element abundances, they claimed that the slope measured was equal to 1.5. This they interpreted as showing that the conductive flux is indeed constant with height, and that the mechanical energy deposition in the region studied is consequently negligible. Although there are many factors which call into question the details of this interpretation, the basic idea of a constant conductive flux (apart from small variations due to the small radiated energy, the small departure from constant pressure, and the plane-parallel geometry) continues to hold favour today. The essential conclusion is that the mechanical energy flux, which must propagate upwards through this region, is not dissipated in the region considered up to 30,000 km, but at a greater and unknown height. Another important result to come out of this analysis is that the value of the conductive flux required by this model is around 3×10^6 erg cm^{-2} s^{-1}.

In fact, with more reliable data, the slope of the curve of Figure 2 can be shown to be somewhat steeper than 1.5. The explanation for this is given later in terms of failure of the one-dimensional model.

2.3. THE ROLE OF TURBULENCE

In our search for the non-thermal energy transport mechanism, there is one observable parameter which could be very important. This is the doppler width of the observed spectral lines. All of the conceivable modes of transport of energy in the solar atmosphere, be they coherent waves, shocks, or stochastic impulsive events, result in a movement of the ions, which should manifest itself by the doppler effect. Observations of lines formed in the transition region do indeed show an enhancement in the line widths over that expected from pure thermal broadening (Kohl, Parkinson, and Reeves 1973, Boland et al. 1973). In fact some of the lines are up to three times broader than expected, implying a kinetic energy component for these ions of ten times thermal.

To interpret these widths requires great care. The first step is to separate the non-thermal and thermal components. This involves a number of assumptions. The first is that the local electron and ion temperatures are equal, an assumption that is probably valid in the solar atmosphere, where gross departures from steady-state conditions are not observed. It is next assumed that the lines are formed in the electron temperature region determined by the maximum of their g(T) functions. With these two assumptions it is possible to predict the Gaussian profile of the true thermal doppler width of the line. We then assume that the kinetic energy due to the thermal and non-thermal components are additive, i.e. that the line-of-sight components of the velocity squared are combined by convolution of the profiles, in order to give the observed profiles. With these assumptions we can unfold the profile due to the non-thermal motions. For the non-thermal motions, it is assumed that all of the plasma in a local region moves coherently with the same velocity (unlike thermal motion in which the velocity varies inversely as the square root of the mass of the particle). This then enables a calculation of the local non-thermal energy density. The large widths are observed for heavy ions. Most of the solar plasma is made up of hydrogen and helium, for which the enhanced velocities are not large when compared with their own much higher thermal velocities. The increase in total local energy density compared with thermal is therefore much less than the factor of 10 indicated above, and varies from 50% low in the transition region to around 3% at coronal temperatures. Figure 4 shows a plot based on a collection of data from various sources, in which this ratio is presented in the form of the ratio of total pressure to thermal pressure.

The existence of an observed local velocity field might be associated with a propagating wave or disturbance (e.g. shock). It could also represent a standing wave structure or other non-propagating turbulence field. If the disturbance is propagating with a velocity c, then the mechanical energy flux will be given by

$$F_m = 2Ec \qquad (12)$$

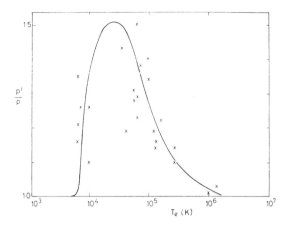

Figure 4. Ratio of total pressure to thermal pressure, based upon observed line profiles.

where E is the volume energy density derived as above. The value of c could correspond to the sound velocity, for acoustic waves, or to the Alfvén velocity for magnetic propagation of MHD waves. Choosing the larger of these (the Alfvén velocity for coronal conditions), we can safely state that Equation (12) represents the maximum value for the mechanical energy flux, consistent with the observed line profiles.

Boland *et al.* (1975) carried out this analysis, and derived for the plane-parallel atmosphere values of the mechanical energy flux of around 3×10^5 erg cm^{-2} s^{-1}.

2.4. CRITIQUE OF THE PLANE-PARALLEL MODEL

The most obvious difficulty arising from the above models lies in the value of the conductive flux required, which is an order of magnitude larger than the maximum permitted from the observed line profiles.

Another problem arises from the thickness of the transition region. This is often quoted in the literature as being extremely thin; of the order of 10 km. Such very steep temperature gradients would give rise to some strange physical effects. These arise when the mean free path for momentum transfer collisions between electrons (the collisions which establish thermal equilibrium) becomes comparable with the scale-length for temperature variation. Under these conditions a Maxwellian distribution is not maintained, the higher energy electrons moving freely across the region. The Spitzer thermal conductivity used in section 2.1 is then no longer valid, and large-scale electric fields are set up which oppose and limit the steepness of the gradient. In addition, various complex forms of ion diffusion occur (Delache 1967, Tworkowski 1976), which can have the effect of separating ions of different mass, and thus invalidating a basic assumption in all of our diagnostic analysis; that of constant chemical composition.

In reality, the model based upon DEM analysis gives thicknesses of 55 km for the range of $\log T$ between 4.5 and 5.5, or 29 km between 5.0 and 5.5. Models based on energy and momentum balance are somewhat steeper, becoming very steep in the region below 5.5, and it is these that are often used to justify the more extravagant claims. However these models are in error in neglecting local energy deposition in the transition region. We return later to this question of the thickness of the transition region.

3. Effect of the Super-Granular Network

At this point in the interpretation, it becomes necessary to ask the question, what is the most important effect which could lead to a breakdown in the above one-dimensional models? The clue to this came from the first high-resolution images of the transition region (Reeves, Vernazza, and Withbroe 1976). Figure 5 shows such images formed in a number of high chromosphere and transition region spectral lines.

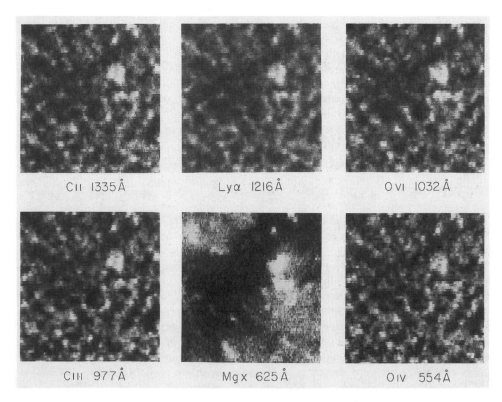

Figure 5. Images of the transition region network from Skylab, obtained in UV spectral lines covering a range of temperatures (Habbal and Withbroe 1981).

Clearly visible throughout the transition region is the network structure associated with the boundaries of the supergranulation cells. This structure, which was already well-known in the chromospheric ground-based Calcium K-line images, persists throughout the transition region, disappearing when we reach coronal temperatures around 1 million degrees. Moreover, the contrast seen in the transition region is high; the network boundaries being an order of magnitude brighter than the cell centres. It is clear that such a systematic ever-present pattern has serious consequences for the one-dimensional models. Not only do these fail to predict the pattern; they also falsely interpret the average intensities of the transition region lines.

3.1. SIMPLIFIED THEORETICAL MODEL

The supergranulation convection cells are of the order of 40 arcseconds or 30,000 km across and cover the entire surface of the Sun, being disrupted only by the very intense magnetic structure in active regions. The cell pattern evolves on a long timescale of the order 10 to 20 hours, and is fairly deep-lying below the photosphere; more so than the finer and shorter-lived granulation. Why then is it that the supergranulation rather than the granulation is manifested in the hotter overlying atmosphere? The answer lies in the magnetic field which transmits the influence to the higher layers. It has been long known from ground based magnetograms that the supergranulation flow has the effect of sweeping the quiet solar magnetic field into concentrations along the boundaries of the cells. It was clear that, higher in the corona, the plasma pressure would not be sufficient to maintain this concentration. This leaves the questions: how does the field concentration vary with height? and what is the effect of this field on the energetics of the atmosphere?

Even before the availability of the UV observations, Kopp and Kuperus (1968) tried to model these effects. There is an error in their analysis, which led to the prediction of an enhanced emission in the cell centre, the converse of what is observed. Later, Kopp (1972), drawing on some of the early observations, produced a model of the expansion of the field with height which gave some useful results. As he maintained the assumption of a quasi-vertical field throughout, he was unable to treat the cell centres. A model by Gabriel (1976a) offered a two-dimensional interpretation, which maintained energy and pressure balance throughout and included the magnetic field pressures. This model, shown in Figure 6, goes a long way to explaining many of the observed features.

It is assumed that both the upward mechanical energy flux and the downward conduction are constrained to flow along the inclined and tapering field lines. This modifies the equations of section 2.1, by including non-unity values for the quantities a and $\cos^2 \alpha$. The classical transition region, normal to the magnetic field, exists only in a strip 10 arcseconds wide overlying the cell boundaries, consistent with the enhanced intensities observed. The remaining cell centre areas have the field parallel to the transition region, thus reducing the thermal conductivity perpendicular to the region, which becomes much steeper, and has a consequently lower intensity. The model is obviously much simplified. It does not deal with the smaller scale dynamical structures, known to occur along the cell boundaries, nor with the effect of mixed fields of opposite polarity. It does however appear to give an explanation for some major defects of the one-dimensional models, and enables us to reconsider questions of total intensities and conductive fluxes. In approximate terms, since the boundaries occupy about one third of the surface, the local intensities are some three

288

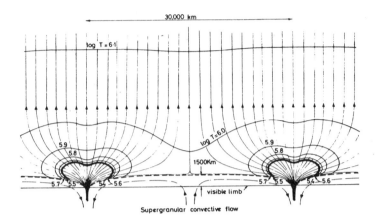

Figure 6. A model for the solar transition region, based upon the effect of the super-granular flow pattern, transmitted by the magnetic field to the overlaying atmospheric structure.

times higher than was assumed for the one-dimensional model, so that the transition region is some three times thicker and the local conductive flux three times smaller. Averaged over all of the surface, the mean conductive flux then decreases by three squared, *i.e.* by an order of magnitude to a figure of 3×10^5 erg cm^{-2} s^{-1}. This figure is now fully acceptable in terms of the non-thermal line broadening observed at the base of the corona. Being a two-dimensional model, it is not possible to produce a single curve for the variation of temperature with height. However, we can from this theoretical model compute the total (or average) value for the DEM as a function of temperature. This is shown superposed on the curve from the observations in Figure 7.

The agreement between the two curves shows that the theoretical network model is indeed consistent with the observations, for temperatures above $\log T$ of 5.3. We see also that the similar slopes for the two curves are significantly steeper than the figure of 1.5 derived for the one-dimensional model with constant conductive flux. This effect is due to the tapering of the field lines at lower heights. The divergence between the theory and observations below $\log T$ of 5.3 is interpreted as due to false assumptions made in the theoretical model, in particular that the mechanical energy flux deposition S is zero. Gabriel (1976a) showed that this difference is resolved if a mechanical energy of the order of 2×10^6 erg cm^{-2} s^{-1} is directly deposited in the region between $\log T$ of 4.3 and 5.3. This local effect at the network boundary is equivalent to a mean flux deposition of 6×10^5 erg cm$^{-2}s^{-1}$ over the solar surface.

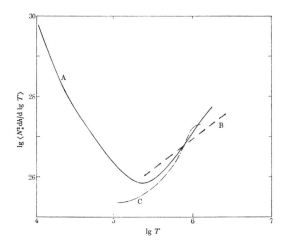

Figure 7. Differential emission measure, as a function of temperature: A - from observations, C - derived from the network model, B - a line of slope 1.5 for comparison (see text).

3.2. HYBRID NETWORK MODEL

To obtain a realistic atmosphere which extends below $\log T$ of 5.3, we could attempt to introduce this deposition into the energy equations. However this poses many problems of choice of functions. It is better to resort to observations and to integrate the observed DEM for the flux-tube geometry of Figure 6. For this we use an area factor a derived from the network model, and assume that all of the intensity observed comes from the network boundary. The resulting model is given in Table 1. It gives a thickness of 273 km for logT between 4.5 and 5.5, or 144 km for logT between 5.0 and 5.5. This model is also shown plotted in Figure 3. Both it and the observational plane-parallel model are capable of producing a temperature profile in the lower corona consistent with observed coronal temperatures. A typical *ab initio* energy-balance plane-parallel model (McWhirter *et al.* 1975) is however too cool in this region.

4. Coronal Holes

It has been known for some time that certain regions on the Sun, termed "M" regions (Bartels 1934), were mysteriously responsible for geomagnetic disturbances. These disturbances were found to occur at 27 day intervals, consistent with the apparent rotation rate of the Sun, although no obvious features on the solar surface were identified with this effect. With the SKYLAB mission, it was recognised that these features were what we now term Coronal Holes.

Table 1. Hybrid network model, based upon a theoretical network
and observed spectral intensities

$\log T$	$h(km)$	$\log N_e \ (cm^{-3})$
4.4	3384	10.45
4.6	3464	10.22
4.8	3518	10.01
5.0	3558	9.81
5.2	3595	9.60
5.4	3648	9.40
5.6	3807	9.20
5.8	4731	8.98
5.9	6403	8.86
6.0	10450	8.73
6.1	26360	8.51
6.15	37850	8.37

The introduction of coronal holes into our present discussion represents the second major divergence from spherical symmetry, which imposes a limitation on our simple one-dimensional models. This effect occurs higher in the corona. The transition region network is a uniform effect imposing an average influence on the solar physics, which has the effect of changing certain physical quantities by a factor of 10. However, the coronal holes impose a different kind of constraint, in effect dividing the solar atmosphere into two separate regions, hole and non-hole, each having a different balance between the physical processes.

4.1. OBSERVATIONS

From SKYLAB and subsequent observations, we can summarise a number of properties of holes. The holes are regions which are less intense by a substantial factor at temperatures above $\log T$ of the order 5.8. Since the network, which was discussed above, has contrast only below this temperature, we can display in an idealised manner on a single DEM plot (see Figure 8) the four different solar atmospheres that we must take into account; cell boundaries and cell centres, in hole and non-hole regions. For the present we will keep high in the corona, and consider only the properties of the holes as seen above one million degrees.

Holes cover between 20% and 30% of the Sun in the form of a small number (between 2 and 5) separate regions. Their most striking feature is one that is often neglected in their description : they appear to be relatively uniform from their centres up to a sharp well-defined boundary. At periods around solar minimum, there is a hole more or less symmetrically placed around each pole of the Sun. At periods near the maximum, the situation is more complex, with small holes appearing at any latitude. Correlation with solar magnetograms shows the holes to be uni-polar regions, and their complexity during the maximum forms part of the evolution of global magnetic field that results in the 11 year

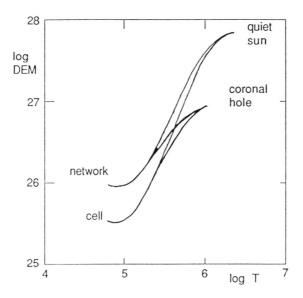

Figure 8. Idealised differential emission measure curves, showing the network contrast at lower temperatures and the coronal hole effects at higher temperatures.

field reversal. A combination of these various factors leads naturally to the conclusion that the holes are regions of the solar surface where the field lines connect to the interplanetary field, in contrast to non-hole regions where the field closes eventually onto the Sun. Since the question of closing or not is decided at some distance from the Sun, this provides an explanation for the sharp boundary and the resulting binary nature of the corona.

4.2. ORIGIN OF THE SOLAR WIND

It is logical to suppose that the loss of material in the form of solar wind is inhibited by closed field lines, and that the wind is thus generated solely from the hole regions. While this is now widely accepted, there exists also the possibility that high field active regions, whose complex field structures are predicted to have some open field lines, could also contribute to the wind. Leaving aside the question of active regions, it is interesting to examine whether the possibility for the wind to escape can explain all of the observed properties of holes, and to see what other predictions result from this hypothesis.

4.3. THE ATMOSPHERIC STRUCTURE IN HOLES

The approach used (Gabriel 1976b) is to solve with energy and momentum balance the full equations of section 2.1, in the network geometry of Figure 6, and with a realistic value for the mass flow rate W. An important problem is to establish appropriate boundary condition for the temperature and density in the corona at 30,000 km. SKYLAB data shows no lines hotter than 1 million degrees to be seen in the holes, leading to the conclusion that up to a height equivalent to one or two intensity scale heights we could take the temperature to have this value. For the density, Gabriel (1976b) adopted a value also deduced from SKYLAB data of a factor 3.6 lower than the normal quiet (closed field) corona. SKYLAB coronal hole observations at 1 million degrees are very near the limit of sensitivity, so that these assumptions could have large errors. Indeed we will return later to the question of whether the density in holes is in fact much lower.

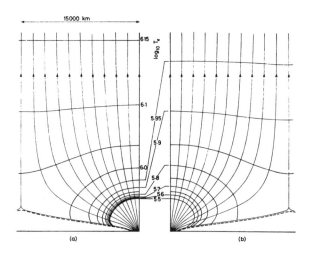

Figure 9. The transition region network model of Figure 6, recomputed for the wind flow in coronal hole regions.

Using these assumptions, a theoretical model was obtained which is shown in Figure 9, where it is compared with the non-hole model, derived in section 3.1. In the region shown by the isotherms plotted above $\log T$ of 5.5, the gradients of temperature are much shallower, by a factor 8. In the region below $\log T$ of 5.5, this factor increases to 12 (or the square of the density ratio of 3.6). Thus the intensity of the transition lines viewed normally to the solar surface remains the same in and out of hole regions, consistent with the observations. However the transition region is in reality dramatically changed. The density is lower by the factor 3.6 and the thickness greater by the square of 3.6. If this effect is hidden when viewing the disk normally, it should manifest clearly at the limb. Indeed, measurement of the absolute height of the Ne VII 465 Å line in a limb coronal hole from SKYLAB (Reeves, Vernazza, and Withbroe 1976) shows a height enhancement of 10 arcsec. This is comparable with the height expected from Figure 9 for a line formed between $\log T$ of 5.8 and 5.9. So that, in spite of the non-visibility of coronal holes on the disk in transition region lines, we have some firm observational evidence for the model proposed.

For the coronal hole model, we can examine the behaviour of the individual terms in equation (2), and their variation with height. Considering the three wind terms, the enthalpy is the most important in the region considered here. The downward conductive flux is rapidly dissipated in providing the enthalpy necessary to drive the wind. Thus, even in the assumption (as previously) of zero mechanical energy dissipation, the conductive flux is no longer constant with height. At 30,000 km the conductive flux has already dropped to 1.3×10^5 erg $cm^{-2}s^{-1}$, and continues to decrease rapidly with decreasing height. It is tempting to try to extrapolate the model to greater heights. This is not possible, because it is assumed that the real dissipation of mechanical energy occurs above the region in consideration, and we do not know how far above. What is clear is that the observations are consistent with a model in which the same mechanical heating flux of 3×10^5 erg $cm^{-2}s^{-1}$ is produced in hole and non-hole regions. In both regions this is converted initially into a downward conductive flux, starting at some unknown height greater than 30,000 km. However, in the hole regions, the majority of this flux is converted before it reaches the transition region into the enthalpy necessary to drive the solar wind. The lower flux remaining (less than 1×10^5 erg $cm^{-2}s^{-1}$ leads to the observed thickening of the transition region. For a more detailed consideration of these terms, the reader is referred to Gabriel (1976b).

4.4. TRANSITION REGION THICKNESS

The various models discussed above give thicknesses in the transition region which are significantly greater than often reported. These are summarised in Table 2. For the most representative models, the thickness between $\log T$ of 4.5 and 5.5 is 273 km and 3500 km respectively for the closed and open field regions. Under these circumstances, the more dramatic plasma effects due to very steep gradients are unlikely to occur. Furthermore, this helps to justify the use of Spitzer's classical thermal conductivity in the equations of section 2.1.

More recent studies of coronal hole densities from white light eclipse measurements (Koutchmy 1991, private communication) show that the relative densities hole/non-hole could be as little as 0.1. If this is confirmed, the model of Figure 9 should be re-evaluated, and the resulting transition region in holes will be found to increase even further in thickness.

TABLE 2. Thickness of the solar transition region between
$\log T$ of 4.5 and 5.5, according to the various models discussed.

MODEL	Thickness in km.
Theoretical 1-D model with no local dissipation	< 20
DEM analysis of observations, 1-D	55
DEM analysis in a network configuration (non-hole region)	273
DEM analysis in a network configuration (hole region)	3500

4.5. HIGH SPEED WIND STREAMS

The analysis of SKYLAB data has shown that the coronal holes are indeed the elusive
"M" regions responsible for the geocoronal disturbances. In particular, the coronal holes
which descend to lower latitudes have been correlated with the high-speed wind streams
observed in the vicinity of the earth. The problem is relatively straight-forward for such
near equatorial holes. It is less clear for the polar coronal holes, where it is necessary to
consider the curved magnetic paths in three dimensions between the Sun and the earth.
A discussion of such paths can be found in Gabriel (1988). Since the behaviour in the
out-of-ecliptic plane is not at all evident, there is uncertainty regarding the role of polar
coronal holes on the overall mass loss from the Sun.

4.6. THE IMPLICATION FOR STELLAR WINDS

Having advanced the understanding of the solar atmosphere several important steps beyond
the simple one-dimensional interpretation, it is instructive to reflect on the significance of
this work for stellar physics. For other stars there is little or no hope of resolving the surface
spatially, and only the integrated fluxes are available.

The first problem that will present itself for solar type stars is that of the coronal holes.
The comparatively low intensity of the hole regions implies that any model based upon
integrated intensities applies essentially to the non-hole regions, with no hope of deriving a
hole model from observations. However the flow rate of the wind depends entirely on the
physics of the hole regions. The problem faced is that the atmospheric structure deduced
from the integrated observations will have little or no relevance to the mass loss rate of the
star. This potentially serious difficulty has not yet been confronted correctly.

A second problem arises from the presence of active regions. Without spatial resolution, it is impossible to exclude active region spectra. Evidence exists from EXOSAT and other data that the Sun is atypical, and that other solar-type stars have a higher degree of activity. It is very questionable in these conditions whether the true "quiet corona" has ever been observed from other stars.

5. Forthcoming Space Observations

The many questions and paradoxes posed by the above considerations will be addressed by a number of space-borne instruments in course of preparation.

The Solar and Heliospheric Observatory (SOHO) is being prepared by ESA and NASA for launch in June 1995. SOHO will spend two years observing the Sun continuously, from its orbit at the $L1$ Legrangian point. It will carry a wide range of instruments for observing the solar atmosphere in a number of wavelength ranges, in addition to particle detectors to analyse the solar wind arriving at 1 AU. It will thus be able in a single mission to correlate the structures on the solar disk with the disturbances in interplanetary space. In particular, SOHO will carry the Coronal Diagnostic Spectrometer, a grazing incidence spectrometer capable of observing in the 150 to 500 Å range, in which the principal resonance lines of coronal temperature plasmas are to be found. In spite of the numerous solar satellites already launched, this important wavelength region has never been the study of an effective observatory-class satellite instrument. SOHO will offer for the first time ever the possibility of a systematic study of the physical conditions in the quiet solar corona.

Another ESA and NASA joint mission ULYSSES is a satellite carrying a range of particle and field instruments for *in situ* studies. This was launched in August 1990 in a trajectory that will make use of a Jupiter swing-by in order to pass out of the ecliptic plane, where it will overfly both poles of the Sun in 1994. By observing directly the solar wind at high ecliptic latitudes, it is hoped to have direct information on the wind that originates from the polar holes. It is at least conceivable that the polar wind is much stronger and faster than that which we observe in the ecliptic plane. The importance of such observations for the solar mass loss rate and stellar evolution will be obvious.

6. References

Bartels, J., (1934) *J. Geophys. Res.*, **39**, 201.

Boland, B.C., Engstrom, S.F.T., Jones, B.B. and Wilson, R., (1973) *Astr. Astrophys.*, **22**, 161.

Boland, B.C., Dyer, E.P., Firth, J.G., Gabriel, A.H., Jones, B.B., Jordan, C., McWhirter, R.W.P., Monk, P. and Turner, R.F., (1975) *Mon. Not. R. Astr. Soc.*, **171**, 697.

Cox, D.P. and Tucker, W.H. (1969) *Astrophys.J.*, **157**, 1157.

Delache, P., (1967) *Ann d'Astrophys.* **30**, 827.

Gabriel, A.H., (1976a) *Phil. Trans. R. Soc. Lond. A.*, **281**, 339.

Gabriel, A.H., (1976b) in 'The Energy Balance and Hydrodynamics of the Solar Chromosphere and Corona' I.A.U.Colloquium No 36 (eds Bonnet, R.M. and Delache, P.) Nice, p 375.

Gabriel, A.H., (1988) in 'Hot Thin Plasmas in Astrophysics' (ed.R. Pallavicini), Kluwer, p. 79.

Habbal, S. and Withbroe, G., (1981) *Solar Phys.*, **69**, 77.

Jordan, C. and Wilson, R.,(1971) in 'Physics of the Solar Corona' (ed. Macris), D.Reidel, Dordrecht, p. 219. .

Kohl, J.L., Parkinson, W.H. and Reeves, E.M., (1973) *Bull. Am. Astr. Soc.*, **5**, 274.

Kopp, R.A., (1972) *Solar Phys.*, **27**, 373.

Kopp, R.A. and Kuperus, M., (1968) *Solar Phys.*, **4**, 212.

McWhirter, R.W.P., Thonemann, P.C. and Wilson, R., (1975) *Astron. and Astrophys.*, **40**, 63.

Pottasch, S.R. (1963) *Astrophys.J.*, **137**, 945.

Pottasch, S.R. (1964) *Space Sci Rev*, **3**, 816.

Reeves, E.M., Vernazza, J.E. and Withbroe, G.L., (1976) *Phil. Trans. R. Soc. Lond. A.*, **281**, 319.

Tworkowski, A.S., (1976) *Astrophysical Lett.* **17**, 27

14   SOLAR RADIO OBSERVATIONS

G.J. HURFORD
Solar Astronomy
Caltech 264-33
Pasadena, CA 91125
U.S.A.

ABSTRACT. Solar radio observations of the quiet and active Sun are reviewed in the context of the properties of the emission processes. For plasma radiation, the emphasis is on interpretation of the frequency of the emission in terms of height. The roles of free-free bremsstrahlung in the quiet Sun, gyroresonance emission in active regions and thermal and nonthermal gyrosynchrotron emission in flares are discussed. Observational examples emphasize cases where the straightforward physics seems to work.

1. Introduction

As we are seeing, solar observations deal with a wide range of phenomena covering temperatures from a few thousand degrees in the photosphere to over 10^9 K in flares. Some Chapters are also suggesting the wide range of observing techniques, each with its perspective on a subset of these phenomena. In this spirit, solar radio observations have a slightly different role to play. Depending on wavelength and solar conditions, radio can provide some insights over this entire temperature range as well as provide access to the properties of nonthermal energetic electrons. The reason for this adaptability is that radio emission is generated by a number of mechanisms.

In this Chapter, our perspective will be to assume that we understand the physics of the radio emission processes. We will then investigate what radio observations can tell us about the Sun. (An alternative perspective would have been to use the solar radio emission to study the emission processes themselves.) Readers unfamiliar with the basic definitions as used in radio astronomy (flux density, brightness temperature, optical depth, etc.) or with the observing techniques used to acquire the observations may wish to review them in Chapter 19. In the observational examples in this Chapter, the emphasis will be on the simple cases, where the straightforward physics seems to work. The Sun is much more complex, however, and the literature is filled with puzzles that await explanation.

To keep from becoming lost in the diversity of radio phenomena, our most essential asset will be an awareness of what emission mechanism(s) are relevant in a given situation. The mechanisms that we will encounter in this Chapter include plasma radiation, free-free bremsstrahlung, gyroresonance, and thermal and nonthermal gyrosynchrotron. We will introduce these one at a time in the context of corresponding solar phenomena where they are often observed. The first mechanism to be considered, namely plasma radiation, is an example of "coherent emission" as opposed to incoherent emission where each electron acts

297

J. T. Schmelz and J. C. Brown (ed.), The Sun, A Laboratory for Astrophysics, 297–312.
© 1992 *Kluwer Academic Publishers. Printed in the Netherlands.*

independently. As we shall see, this is an important distinction which has profound effects on the character of the radiation and what it can tell us about the Sun.

2. Coherent Emission

The most important type of coherent emission for solar physics is plasma radiation, by which the energy of Langmuir waves is converted to electromagnetic radiation. (See Chapter 27). Observationally, the dominant characteristic of such radiation is that it is observed to be strongest at the plasma frequency (or at its second harmonic) at the site of emission. In Chapter 19, we recall that the plasma frequency depends only on the local electron density, so that the observed frequency of plasma radiation tells us the electron density at the site of emission. Brightness temperatures of plasma radiation can be very high ($> 10^{10}$ K) but are generally difficult to model on the basis of macroscopic solar parameters. Thus, the value of plasma radiation to the solar observer is threefold: first, its existence indicates the presence of some kind of a "disturbance" that initiated the plasma radiation; second, imaging can identify the line of sight to its location; and third, the frequency of the radiation indicates the local electron density at that site.

For a nominal model of electron density as a function of height in the solar atmosphere, Figure 1 shows how the plasma frequency varies with height. This relationship enables the observed frequency of plasma radiation sources to provide an indication of their height. Figure 2 shows observations of several types of radio bursts associated with plasma radiation. The observations are in the form of dynamic spectra in which an uncalibrated intensity is shown as a function of time and frequency. These were named in order of their historical classification as Type I, Type II, Type III, etc. Figure 3 shows a schematic view of the occurrence times and frequencies of such events in the context of a flare. Now let us briefly consider each one.

Type II events last for many minutes and are characterized by a slow drift downward in frequency. Type II emission frequently follows the impulsive phase of flares and often is seen simultaneously at two frequencies (or groups of frequencies) separated by about a factor of 2. Using Figure 1, the frequency drift suggests that the source is rising in the solar atmosphere. Converting the rate of change of frequency to a vertical velocity yields values typically in the range 400 - 2000 km sec^{-1}, velocities which exceed the Alfven velocities at the corresponding heights. Therefore, the interpretation is that the "disturbance" which initiates Type II emission is a flare-generated shock wave, propagating upward through the solar corona.

Type III events are often, but not always flare-associated, sometimes occur in groups of 10s to 100s, and are similar to Type IIs in that they show a downward drift in frequency with time. In this case, however, each Type III event lasts only for a few seconds, and the drift rate is much higher. Again interpreting the drift in frequency as a change of height, we can calculate source velocities of $\approx 10^8$ m sec^{-1} ($\approx c/3$). The interpretation in this case is that the burst is initiated by a beam of relativistic electrons. The coherent time structure of the burst over an order of magnitude in frequency (*viz* 2 orders of magnitude in coronal density or 4 coronal scale heights) implies a stability to such electron beams that poses a challenging problem to theoreticians.

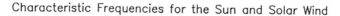

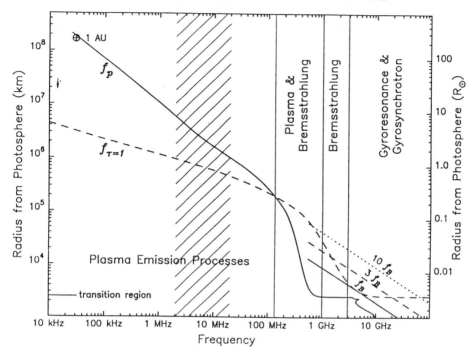

Figure 1. Plot relating the characteristic radio frequencies of the solar atmosphere. The solid curve shows the plasma frequency as a function of height for a nominal atmospheric model, The dashed curve shows the height at which the atmosphere reaches optical depth unity for overlying free-free bremsstrahlung. The curves labelled f_B correspond to the height of the indicated harmonic of the gyrofrequency above a sunspot. The hatched area denotes an observational gap. In all cases, the numerical values are model-dependent. From Gary and Hurford 1989.

Type I events (Figure 2a) can last for hours or days, are not necessarily flare-associated, and emit more uniformly over a wide range of frequencies, without the coherent time-drifts characteristic of Type II and Type III events. Type I bursts are shorter duration intensifications of the Type I events. These events are much less well understood, but probably are initiated by energetic electrons trapped for long periods of time in high coronal loops.

Type IV events are flare-associated, generally lasting a few minutes to a few hours after flares. Their frequency structure can range from fairly featureless to exhibiting a rich range of morphologies (Slottje 1981).

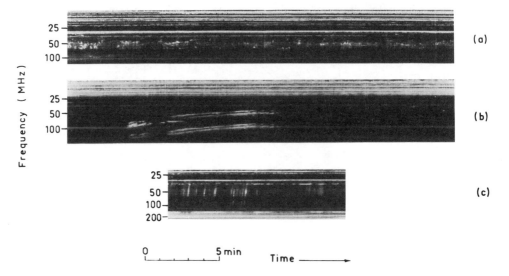

Figure 2. Examples of dynamic spectra for (a) Type I noise storm, (b) a Type II burst showing frequency splitting in addition to emission at the fundamental and first harmonic and (c) a group of Type III bursts. Adapted from McLean 1985.

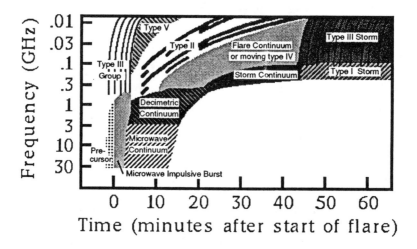

Figure 3. Schematic dynamic spectrum showing the time and frequency relationships among different types of radio emission during a large flare. Adapted from Dulk 1985.

The subject of coherent radio emission is a very rich one, in which our interpretation has been limited to the link between frequency and height of emission. Chapter 27 gives a introduction to the theoretical challenges such events can raise. For more details, including many examples of the rich range of frequency structure that is observed, the reader is referred to books such as those by Elgaroy 1977, Kruger 1979, and McLean and Labrum 1985.

3. Coherent *vs.* Incoherent Emission

We have seen how plasma radiation can be associated with some kind of a "disturbance," be it the presence of shock waves or energetic electrons. Let us now take the opposite circumstance and ask about solar radio emission for the idealized case of an "undisturbed," non-flaring Sun without even any magnetic fields. In such a case, the emission from the thermal plasma would be generated by free-free bremsstrahlung, namely collisions among the free electrons and ions. Such emission would depend on the number of such collisions and it can be shown that the resulting optical depth can be given by,

$$\tau = 0.2n_e^2 LT^{-1.5}\nu^{-2} \tag{1}$$

where the factor of 0.2 is an approximation for microwave frequencies at coronal temperatures and densities, n_e is the free electron density (cm^{-3}) (whose square is proportional to the collision frequency), L is the depth of the plasma (cm), T_e is the electron temperature (which determines the cross section) and ν (Hz) is the frequency.

Radio signals originating below a depth corresponding to $\tau = 1$ in the solar atmosphere will undergo significant absorption as they propagate outward. Thus, the $\tau = 1$ layer can be roughly interpreted as setting a lower limit to the height of the source of radio emission. Such a height is plotted in Figure 1 for a nominal solar atmosphere. It is particularly instructive to compare this height with that implied by the plasma frequency. The latter height can be interpreted as saying that radiation due to ANY mechanism originating below that height in unobservable because radiation at and below the plasma frequency does not propagate. Note that while the plasma frequency limit depends on the local value of electron density only, the free-free absorption limit depends on a combined *integral* of temperature and density.

Comparing the two curves in Figure 1, it is clear that at high frequencies, (> 1 GHz) emission at the plasma frequency would be absorbed by the overlying atmosphere and so could not be observed. Thus, the observation of plasma radiation is generally restricted to lower frequencies and, hence, necessarily associated with coronal phenomena. Thermal radiation can also be observed at lower frequencies, but its interpretation often cannot ignore the presence of layers where the plasma frequency is equal to the observing frequency.

4. The Quiet Sun

For the remainder of this Chapter, we will deal with the Sun at frequencies above 1 GHz where we can usually neglect the plasma frequency. Let us begin by considering the simplest possible situation of a quiet Sun with no flares or active regions to complicate our analysis. In such a circumstance, free-free emission from a thermal plasma is the operative mechanism. Since the optical depth is known as a function of the plasma parameters (Equation 1), the issue of how bright the quiet Sun is at radio frequencies should be a tractable problem.

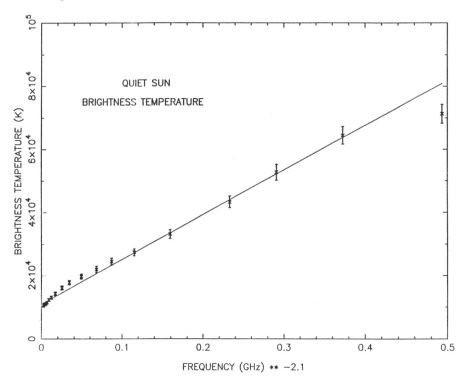

Figure 4. Observed brightness temperature of quiet Sun center, plotted as a function of $\nu^{-2.1}$ to illustrate the frequency dependence suggested by Equation 3. The index of -2.1 is used instead of -2 to compensate for a weak frequency dependence to the numerical coefficient in Equation 3. From Zirin *et al.* 1991.

One implication of the dependence of the optical depth on density and temperature in Equation 1 is that the nominal ×100 changes in temperature and density across the transition zone would result in a ×10^7 change in optical depth per unit distance for free-free bremsstrahlung. Putting numbers in, it is easy to show that the corona is optically thin and the chromosphere is optically thick at microwave frequencies. Viewing a chromosphere of temperature, T_{ch}, through an optically thin corona of temperature, T_{cor} and optical depth, $\tau_{cor} \ll 1$, the resulting brightness temperature would be

$$T_b = T_{ch} + T_{cor}\tau_{cor}. \tag{2}$$

Using Equation (1), this becomes

$$T_b = T_{ch} + 0.2n^2 L_{cor} T^{-0.5} \nu^{-2} \tag{3}$$

where L_{cor} is the coronal scale height.

Figure 4 shows some recent observations of how the surface brightness of the quiet Sun varies with frequency. The form of the plot confirms that the brightness temperature has the expected frequency dependence. The high frequency limit indicates a chromospheric temperature of about 10^4 K and the slope in Figure 4 implies a base coronal electron density of 3.2×10^8 cm^{-3} for a 10^6 K corona and scale height of 5×10^9 cm. Therefore, the observed brightness temperature of the quiet Sun is consistent with the broad run of temperatures and densities in the solar atmosphere.

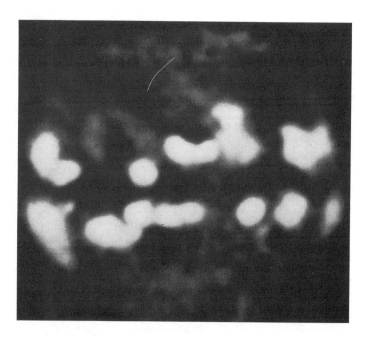

Figure 5. Full disk synthesis map of the Sun at 1.4 GHz. The bright features are associated with active regions. Courtesy of D. E. Gary.

The real Sun, of course, is not so simple. For example, a corollary of these arguments is that the Sun should be strongly limb brightened at microwave frequencies. This is because the path length in the corona and hence its contribution to T_b should be much larger toward the limb. Observations show that limb brightening of the expected magnitude is not observed. Figure 5 shows a full disk map at 1.4 GHz from the VLA showing a notable absence of limb brightening. Small scale quiet Sun structures at the few arcsecond level have also been observed and, in some cases, shown to be well-correlated optically.

5. Gyroresonance Opacity in Active Regions

Let us now introduce magnetic fields into the pristine quiet Sun that we have just been considering. How strong do such magnetic fields need to be before they become relevant? One measure of the importance of magnetic fields to radio emission from a thermal plasma is the ratio of the observing frequency to the gyrofrequency, the latter being the frequency with which an electron spirals in a magnetic field. The gyrofrequency depends only on the ambient magnetic field, independent of the electron velocity or plasma temperature and is given by:

$$\nu_g(GHz) = 2.8B(Kg). \tag{4}$$

If the observing frequency is large compared to the gyrofrequency (as is the case at microwave frequencies for fields of less than ≈ 100 Gauss), then the optical depth associated with free-free emission depends smoothly but weakly on magnetic field so as to increase or decrease the opacity for right- or left-circular polarized emission (depending on the direction of the field). This typically results in emission that is weakly polarized at the few percent level. Although the degree of polarization can be interpreted in terms of the field strength, to do so in practice requires a knowledge of gradients in temperature and density. (But see the interesting suggestion by Bogod and Gelfreikh 1980.)

The situation gets much more interesting for stronger fields, when the observing frequency is equal to a low harmonic of the gyrofrequency. At such frequencies, where $\nu = s\nu_g$, with s a low integer (usually 2, 3, or 4), gyroresonance provides a very strong source of opacity, so much so that under coronal conditions for low harmonics, τ can be $>>1$ within thin layers satisfying the resonance condition. (The dependence of τ on n_e and T_e is relatively weak.) However, each increment of s decreases τ by more than an order of magnitude so, under such circumstances, there is often a well-defined maximum value of s for which the corona is optically thick. Furthermore, τ_{rcp} and τ_{lcp} typically differ by an order of magnitude so there often exists a harmonic that is optically thick in one polarization yet optically thin in the other.

What are the implications of this interesting behavior for solar radio emission? First of all, to satisfy this resonance condition at microwave frequencies for low values of s, we require field strengths of several hundred Gauss in the upper chromosphere or corona. Let us consider the situation near an idealized sunspot (Figure 6). At a given frequency, the opacity of the million-degree corona is large on the isogauss surfaces satisfying the resonance condition. (Only the outer surface satisfying this condition matters, of course. Also, note that the chromosphere is optically thick due to free-free opacity.) Therefore, at a given frequency, the active region would appear as a million-degree brightness temperature source at the location of the strong coronal fields (with a diameter of the outermost optically thick shell at this frequency) surrounded by a 10^4 K chromosphere. Figure 7 shows an example of such intense cores of emission at 4.9 GHz. Other properties suggested by the cartoon are that the cores should have different diameters in right- or left-circular polarization, depending on the direction of the field. Furthermore, the diameters of the bright emission cores should systematically decrease toward higher frequencies, since stronger fields are required to satisfy the resonance conditions at higher frequencies and the corresponding

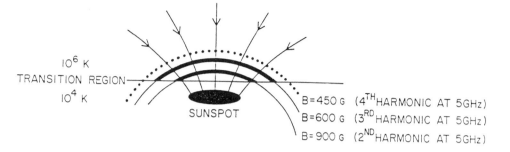

Figure 6. Schematic illustration of the magnetic field lines above an isolated sunspot, showing the isogauss contours at the indicated magnetic field values. In each case, 5 GHz corresponds to a low harmonic of the gyrofrequency so that gyroresonance may be effective. In such a case the heavily outlined surface would be optically thick and result in localized emission with coronal brightness temperatures.

diameter of the isogauss shells at the base of the corona is smaller. Figure 8 shows that this is, in fact, observed.

An alternative perspective on the effect of gyroresonance can be found by considering the brightness temperature spectrum as viewed along a single line of sight. Figure 6 suggests that we would see a million-degree temperature at low frequencies falling rapidly to 10^4 at a frequency corresponding to the isogauss shell which just passes through the transition zone. Furthermore, this should occur at a different frequency in the two polarizations. The left-hand panel in Figure 8 shows these features confirmed in a detailed model calculation while right-hand panel confirms this with corresponding observations. The sharp breaks in the spectrum occur at frequencies which can be directly related to the magnetic field strength at the base of the corona. This provides the basis for the hope that microwave observations with sufficient spatial and spectral resolution could function as a "coronal magnetograph," determining the magnetic field strength as a function of position at the base of the corona.

The goal of a coronal magnetograph has not yet been achieved. Partly because observations with the required combination of spatial and spectral resolution are not yet widely available and partly because the Sun is more complicated than these simple arguments suggest. In practice, temperature and density inhomogeneities complicate the picture as they affect both the gyroresonance emission and generate free-free emission on their own. In fact, as Figure 7 indicates, at frequencies below a few GHz, free-free emission continues to be important and can often dominate, so that the morphology of microwave maps can be dominated by the presence of loops. (In this case, there is hope that the spectra can disclose the temperature and density of these loops.)

There is an extensive literature of microwave observations of active regions and, in some of these cases, it would appear that one may need to invoke the presence of nonthermal particles or other phenomena such as "mode-coupling" to explain the observations. It remains to be seen whether more complete data will add to or resolve the complexity.

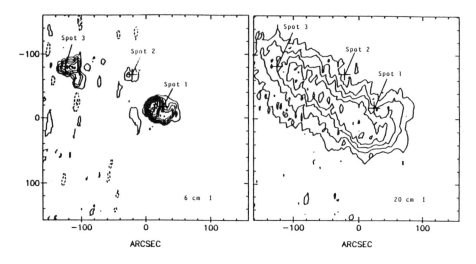

Figure 7. Active region maps made at the VLA at 4.9 GHz (left) and 1.4 GHz (right). The peak brightness temperatures are 1.1 and 1.3 $\times 10^6$ K respectively. At 4.9 GHz, the emission cores are located close to the sunspots. At 1.4 GHz, thermal bremsstrahlung is the dominant emission mechanism and more extended emission is associated with the overlying material. From Gary and Hurford, 1987.

6. Microwave Bursts and Gyrosynchrotron Emission

As discussed by Dulk (1985), one property of gyroresonance opacity is that as the temperature of the thermal plasma increases, the opacity falls off more slowly with harmonic number. As a consequence, at sufficiently high temperatures, the resonances associated with individual harmonics begin to blend so that the character of the frequency dependence of the opacity becomes less a resonance phenomena and more that of a continuum spectrum. In such cases, we label the emission as thermal gyrosynchrotron. Similar blending of the harmonic response occurs for nonthermal electron distributions as well, in which case the emission is labelled as nonthermal gyrosynchrotron.

The obvious context in which these high temperature thermal or nonthermal emissions can become relevant is in flare-associated microwave bursts. Figure 10 shows some representative microwave images, illustrating that, at high frequencies, microwave emission often appears as a compact source between H-alpha footpoints, presumably at the top of a loop. At lower frequencies, the microwave source is generally larger, perhaps either filling the loop or representing emission from a loop of larger size. Figure 10 also shows that microwave emission in the decay phase is also seen to have a different morphology than in the impulsive phase, presumably because it is by then dominated by thermal emission.

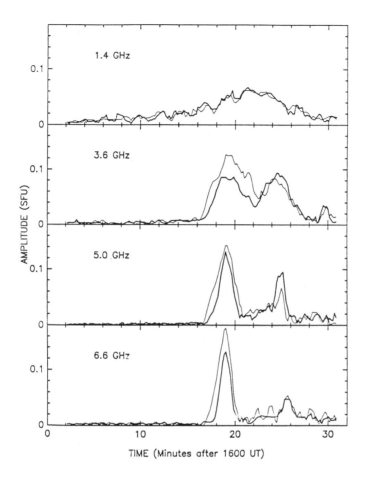

Figure 8. One dimensional profiles at several frequencies through the sunspots shown in Figure 7. The dark and light curves represent right- and left-hand circular polarization. The data were taken during an eclipse so that the time axis corresponds to spatial location. Note that the spot-associated source size decreases as frequency increases and that the two emission cores associated with opposite polarity sunspots have larger diameters in opposite senses of circular polarization. From Gary and Hurford, 1987.

For the impulsive phase, a conventional nonthermal picture is that accelerated electrons produce microwaves as they spiral in the ambient magnetic field while X-rays are produced when the electrons interact with the ambient plasma. There are significant differences, however, that should be kept in mind when interpreting and comparing microwaves and hard X-rays, even in circumstances where both are generated by nonthermal electrons. Hard X-rays are generally produced by electrons with energies greater than ≈ 30 keV; microwaves are associated with somewhat more energetic electrons, greater than ≈ 100 keV. The hard X-ray emission can be energetically very significant; microwaves are totally

308

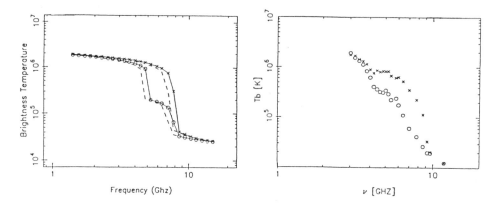

Figure 9. Representative brightness temperature spectra of microwave emission from a sunspot, from model calculations (left) and observations (right). The two symbols correspond to two senses of circular polarization. The steep falloff from coronal values occurs at harmonics of the gyrofrequency at the base of the corona. The dashed line in the model calculation shows the effect of reducing the field strength by 10 percent. From Hurford and Gary 1986.

insignificant. The locations at which hard X-ray emission occurs is weighted by the plasma density; for microwaves, magnetic field strength is critical at a given frequency. Hard X-rays are always optically thin; microwaves can be either optically thin or thick, a feature that makes their interpretation both more challenging and potentially more rewarding. Typical hard X-ray data consist of light curves and spatially-integrated spectra; typical microwave data consist of light curves and images. Hard X-ray observations are often statistically limited; microwaves are often limited primarily by systematic errors. It is clear that, even in situations where both microwaves and hard X-rays might be produced from the same population of nonthermal electrons, these data channels are complimentary and their comparison is often challenging.

The situation is complicated further by the fact that microwave emission can be thermally generated as well. In fact, it is quite possible for all four of the incoherent mechanisms we have discussed to dominate microwave emission at various locations and/or times during a flare. How does one make deal with such a potentially complicated situation? The key is to identify the dominant emission mechanism in each circumstance. One approach to this, widely used in the literature, is to examine all the possibilities, evaluate the relevant plasma/electron/field parameters that would be required in each case and hope that, for all but one mechanism, the parameters are unreasonable. The approach I personally favour, however, is based on exploiting the spectra.

Figure 11 shows the shape of the microwave spectra that would be expected for free-free bremsstrahlung and thermal and nonthermal gyrosynchrotron emission. Over a wide range of parameter space, the shape of the spectrum is independent of the values of the electron, field, and plasma parameters that govern the emission. In general, the effect of changing these parameters is to shift the peak frequency and/or the peak brightness temperature.

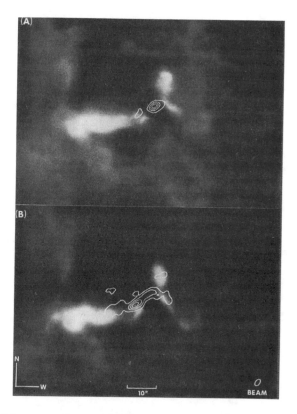

Figure 10. VLA maps of a small flare at 15 GHz, superimposed on Hα images. Between the impulsive and decay phases of the event shown in the top and bottom panels, the microwave emission has spanned the gap between the Hα footpoints. From Marsh and Hurford, 1980.

Thus, the shape of the brightness temperature spectra can be used to identify the emission mechanism and then the peak brightness temperature and frequency interpreted accordingly. Figure 12 shows a case where the spectrum was clearly thermal gyrosynchrotron.

In practice, the difficulty in using such an approach is that it must be done on a point-by-point basis, and so requires good maps, simultaneously acquired at many frequencies. The reason for this is that real microwave sources are not uniform and so simplifying assumptions based on single or dual frequency maps are not always valid. Clear evidence for this is shown in Figure 13 which shows the diameter of a burst source as a function of frequency. If the source parameters were spatially uniform, then the diameter of the source would be the same at all frequencies – independent of the emission mechanism! Because of the steepness of gyrosynchrotron spectra, as shown in Figure 13, even relatively modest changes in parameters can have significant effect on the brightness temperature when observed at a given frequency. Thus maps at a given frequency, in a sense, preselect a subset of parameter space at which the sources will be most readily seen. Such effects

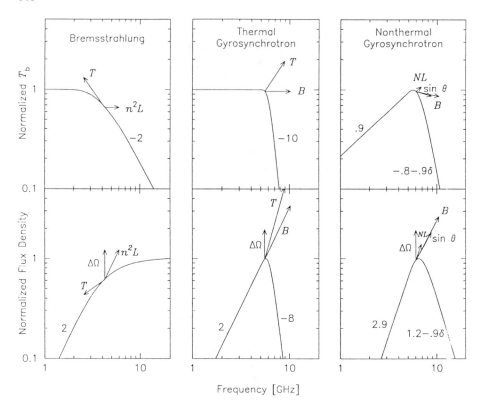

Figure 11. Universal spectra showing the form of microwave flux density and brightness temperature spectra for the a uniform source and the indicated emission mechanisms. These spectra retain their shape, but are shifted in frequency and intensity as the source parameters change. The effect of a factor of two increase in the parameters are shown for electron temperature T, ambient electron density n, path length L, magnetic field B, angle between the field and line of sight θ, electron power law index δ and source area $d\Omega$.

result in observations such as Dulk, Bastian, and Kane 1985 in which the sources appear at totally different locations when mapped at different frequencies.

Thus, we have this dichotomy whereby microwave observations at a fixed frequency or two can appear quite confusing, while in select circumstances the relatively straightforward physics seems to work fine. The resolution of this of course is that when observed with high spatial resolution, the Sun is quite nonuniform and microwave emission is sensitive to these nonuniformities. Perhaps, when observations can provide brightness temperature spectra at each location (to alleviate the nonuniform source problem), then the interpretation of

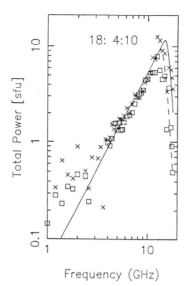

Figure 12. Flux density spectrum for a microwave burst in which the measured source size was independent of frequency. The two symbols correspond to the two senses of circular polarization. The curves are two-parameter fits for thermal gyrosynchrotron spectra of the form given in Figure 11. Adapted from Gary and Hurford 1989.

microwave data will become simple once again and we will have a powerful aid in interpreting the complexities of the real Sun.

7. References

Bogod, V. M. and Gelfreikh, G. B. 1980, *Solar Phys.*, **67**, 29.

Dulk, G. A. 1985, *Ann. Rev. Astron. Astrophys.*, **23**, 169.

Dulk, G. A., Bastian, T. S. and Kane, S. R. 1986, *Astrophys. J.*, **300**, 438.

Elgaroy, O. 1977, *Solar Noise Storms*, Pergamon Press.

Gary, D. E. and Hurford, G. J. 1987, *Astrophys. J.*, **317**, 522.

Gary, D. E. and Hurford, G. J. 1989, *Astrophys. J.*, **339**, 1115.

Gary, D. E. and Hurford, G. J. 1989, *Solar System Plasma Physics*, ed. Waite, J. H., Burch, J. L and Moore, R. L., A.G.U. Monograph 84, 237.

Hurford, G. J. and Gary, D. E. 1986, *NASA CP-2442*, 319.

Kruger, A. 1979, *Introduction to Solar Radio Astronomy and Radio Physics*, D. Reidel Publishing Co.

Marsh, K. A. and Hurford G. J. 1980, *Astrophys. J.*, **240**, L111.

McLean, D. J. 1985, *Solar Radiophysics*, McLean, D. J. and Labrum, N. R. ed, Cambridge University Press.

McLean, D. J. and Labrum, N. R. ed, 1985, *Solar Radiophysics*, Cambridge University Press, p37.

312

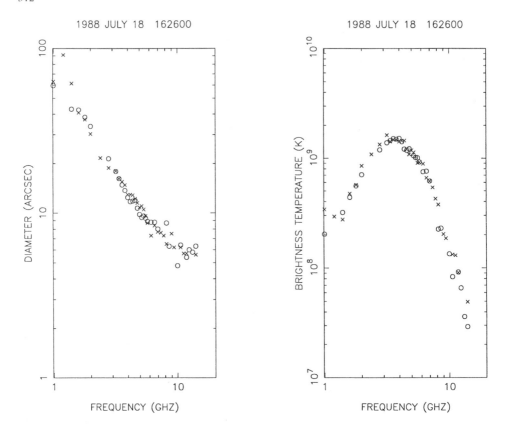

Figure 13. Observed diameter and central brightness temperature spectrum for a microwave burst source as observed with the Owens Valley frequency-agile interferometer. In this case the shape of the spectrum implies that the mechanism is nonthermal gyrosynchrotron.

Slottje, C. 1981, *Atlas of Fine Structures of Dynamic Spectra of Solar Type IV-dm and Some Type II Radio Bursts*, N.F.R.A. at Dwingloo.

Zirin, H., Baumert, B. M. and Hurford, G. J. 1991, *Astrophys. J.*, **370**, 779.

15 STELLAR CHROMOSPHERES, CORONAE, AND WINDS

R. PALLAVICINI
Osservatorio Astrofisico di Arcetri
Largo Enrico Fermi 5
50125 Florence
Italy

ABSTRACT. Chromospheres, coronae, and winds similar to those observed on the Sun have been detected in a variety of cool stars. In many cases, especially for main-sequence stars, there is good evidence that the processes leading to non-radiative plasma heating and wind acceleration are the same as for the Sun, although the radiative losses from stellar chromospheres and coronae are often some orders of magnitude larger. Rotation and, by inference, dynamo-generated magnetic fields are likely to be the key parameters controlling the level of non-radiative heating in outer stellar atmospheres. The heating mechanism, however, remains elusive and it is unclear how it is related to rotation, convection, magnetic fields, and age. Evolved stars of sufficiently late spectral types behave differently from the Sun. They apparently lack high-temperature coronal plasma and possess massive low-velocity winds that cannot be driven thermally. Although the disappearance of coronae and the onset of massive winds in these stars are not yet fully understood, there is some evidence that solar-type processes may be at work in this case, though under quite different physical conditions.

1. Introduction

The presence of chromospheres, coronae, and winds in late-type stars indicates the deposition of non-radiative energy and/or momentum in their outer atmospheres. In contrast to the solar case in which we observe a chromosphere, a corona, and a wind, not all stars later than F show all these phenomena. They show, however, at least one of these non-radiative processes. Thus, the fundamental questions we want to address are: Why significant differences exist in the outer atmospheres of stars of different spectral types and luminosity classes, and even among stars of the same spectral type? Could these differences be explained, at least partially, by observational selection effects? And, if not, what is the origin of non-radiative heating/momentum deposition for the various classes of stars? These questions will be the subject of this Chapter, in which I will present the observational evidence for stellar chromospheres, coronae, and winds, and I will discuss some of the theoretical ideas that have been put forward to explain the observations. It should be emphasized that our theoretical understanding of the mechanisms of chromospheric/coronal heating and wind acceleration in stars is still limited, and we are mainly at a stage where we are trying to make sense out of the large body of observational data obtained by space-borne and ground based facilities. Comparison with solar data is certainly useful, but we must also bear in mind that the properties of chromospheres, coronae, and winds in certain types of stars (*e.g.* in late-type giants) might be fundamentally different from those of the corresponding phenomena on the Sun.

J. T. Schmelz and J. C. Brown (ed.), The Sun, A Laboratory for Astrophysics, 313–348.
© 1992 *Kluwer Academic Publishers. Printed in the Netherlands.*

Nearly all we know about stellar coronae and a large part of what we know on stellar chromospheres and winds have been obtained over the past decade using space-borne instruments capable of observing the X-ray and ultraviolet radiation from stars. The *IUE*, *Einstein*, and *EXOSAT* satellites have been particularly instrumental in this respect, and new X-ray/UV data are now being obtained by *ROSAT* and the Hubble Space Telescope (*HST*). Chromospheres and, to a certain extent, winds can also be observed from the ground, and Ca II H and K observations of cool stars have a long tradition that extends back to the early sixties and even before. If some simplification can be accepted, our knowledge of stellar outer atmospheres can be summarized as follows. Chromospheres are present in cool stars of all spectral types and luminosity classes, though a substantial difference may exist between dwarfs (which show basically solar-type chromospheres of various activity levels) and very late giants and supergiants (which show evidence of geometrically extended chromospheres, comparable to and even larger than the stellar radius). X-ray coronae have been detected in all main-sequence stars of spectral type later than A, but there is no evidence of high temperature coronal plasma (at the present sensitivity levels) in red giants and supergiants. Winds from main-sequence stars have been observed only in the solar case, but their presence in all late-type dwarfs can be inferred from the decline of stellar rotation with age during main-sequence evolution. On the contrary, late-type giants and supergiants show evidence of "non-solar" winds and large mass losses.

These and other properties of the outer atmospheres of cool stars will be reviewed in some detail in the following sections. Since this volume is mainly intended for solar physicists, I will stress whenever possible the similarities and differences with respect to the solar case. Moreover, limits of space will not allow me to cover adequately all relevant problems. Thus, some important topics, such as radio emission from stellar coronae, detailed chromospheric modelling, radial pulsations in Mira-type giants, and many others, will not be discussed at all or will be mentioned only in a very cursory way.

2. Stellar Chromospheres

For the purpose of this Chapter, a stellar chromosphere is defined as that layer in the outer atmosphere of a cool star in which the temperature rises gradually to values of the order of $\approx 10^4$ K, before jumping abruptly (but not in all stars) to coronal values (*i.e.* to temperatures $> 10^6$ K). The properties of stellar chromospheres, as have been derived from optical and UV observations, have been reviewed, among others, by Ulmschneider (1979), Linsky (1980, 1985), Cassinelli and MacGregor (1986), Jordan (1986), Hammer (1987), Jordan and Linsky (1987), and Reimers (1989). Results from the extensive observations of chromospheric Ca II emission carried out at Mt. Wilson have been reviewed by Vaughan (1983), Baliunas and Vaughan (1985), Noyes (1986), and Hartmann and Noyes (1987).

2.1. SPECTROSCOPIC DIAGNOSTICS

There are many spectroscopic indicators of stellar chromospheres. These include the Ca II H and K lines at 3968 and 3934 Å, the Ca II infrared triplet at 8498, 8542, and 8662 Å, the Mg II h and k lines at 2803 and 2796 Å, the Hα line at 6563 Å, the 5876 and 10830 Å lines of He I, Lyα at 1216 Å, and other far-UV lines (such as lines of C I, C II, O I, Si II, and He

II observed with *IUE, cf.* section 3). Not all these lines are formed by the same process. A broad distinction can be made between lines that are collisionally dominated (such as the Ca II and Mg II lines and Lyα), which provide information on the physical conditions in the emitting region, and lines such as Hα that are photo-ionization dominated. The latter are not very sensitive to the physical conditions in the formation region, while being strongly dependent on the radiation field. A special case is that of the He lines. As shown by solar observations, these lines are sensitive to X-ray radiation from the upper coronal layers, and the He 10830 Å line is currently used to map the boundaries of solar coronal holes. Thus, although He lines are usually considered among chromospheric indicators, they are at least partially controlled by coronal radiation.

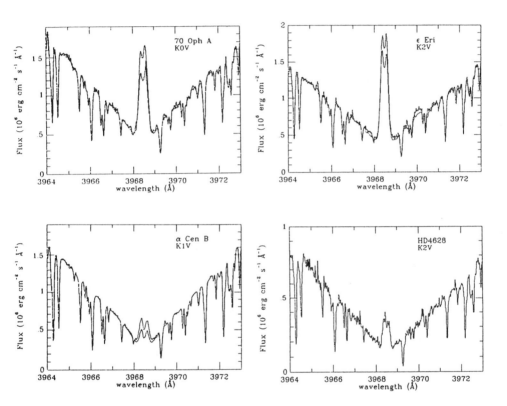

Figure 1. Spectra of cool stars in the Ca II H line (from Rebolo *et al.* 1989). For stars that show time variations in the Ca II H line, profiles are plotted for both the maximum and minimum emission levels.

The Ca II H and K lines are the most commonly used indicators of stellar chromospheres. They are easily observable from the ground and the blue sensitivity of early photographic plates explains why observers gave them an almost universal preference until recently. Figure 1 shows stellar spectra in the Ca II H and K lines. A central emission component is visible at the bottom of the deep Ca II absorption profile. While the latter is of photospheric origin, the emission component originates from chromospheric layers where the temperature inversion occurs. Ca II emission, therefore, is always seen against a photospheric background which is a strong function of the effective temperature of the star. As an example, we can consider the case of α Cen A (a G2 V star similar to the Sun) and α Cen B (which is much cooler, of spectral type K1 V). While the emission component is barely seen, even at high resolution, in α Cen A, it is easily observable in α Cen B. Yet the two stars have about the same radiative losses in the Ca II H and K lines ($\approx 4 \times 10^5$ erg cm^{-2} sec^{-1} at the star surface; see Pasquini, Pallavicini, and Pakull 1988).

High-resolution spectra of the Ca II lines calibrated in absolute flux units have been obtained for many late-type stars in recent years (Linsky *et al.* 1979b, Crivellari *et al.* 1987, Pasquini *et al.* 1988, 1989, Rebolo *et al.* 1989, Strassmeier *et al.* 1990). The conversion to absolute flux units is made by referring either to the pseudocontinuum at 3950 Å between the H and K lines (Catalano 1979, Pasquini *et al.* 1988), or to a 50 Å wide band encompassing the Ca II lines (Linsky *et al.* 1979b). In both cases, the reference point or band is calibrated in absolute units by means of narrow-band photometry. The chromospheric flux is measured by integrating the emission component above the zero level and between the K_1 (H_1) minima on both sides of the emission peak. The latter is referred to as the K_2 (H_2) component and its structure is typically doubled-peaked with a central self-reversal K_3 (H_3). An important correction that has to be made to the fluxes derived in this way is the subtraction of the photospheric contribution below the central emission peak. This is usually done by using grids of theoretically computed radiative equilibrium models (Kelch *et al.* 1988, 1989), though more empirical ways (*e.g.* extrapolation of the line wings to the line center, *e.g.* Blanco *et al.* 1974) have also been used.

It is instructive to compare the stellar data with solar Ca II K line data obtained by observing the Sun as a star (White and Livingston 1981, LaBonte 1986). Figure 2 shows Ca II solar data for the full-disk and for two spatially-resolved regions of different magnetic activity (a quiet Sun area and a plage region). As seen in the figure, the emission component is very small in the full-disk and quiet Sun spectra. Moreover, there are significant changes of the chromospheric component at different epochs during the solar cycle. This shows that systematic observations in the Ca II lines should allow the detection of stellar activity cycles, as stellar observations have indeed proven to be the case (Wilson 1978; see section 2.2 below). The other important point is that large differences exist between the quiet Sun spectrum and the spectrum of plages. Many solar-type stars that are younger and/or more rapidly rotating than the Sun have Ca II spectra that are very similar to those of solar plages. These stars are more active than the Sun, and the fraction of their surface covered by plage regions must be much larger.

The comparison between solar and stellar data shows other interesting features that are important diagnostics of the physical conditions in solar and stellar chromospheres. For instance, the separation of the K_1 minima increases, and the separation of the K_2 peaks decreases, as we move from quiet to plage areas (Shine and Linsky 1972). Similarly, in stars the separation of the K_1 minima increases, and the separation of the K_2 peaks decreases, as a function of the K_3 central intensity, *i.e.* for increasing stellar activity levels (Ayres 1981, Pasquini, and Pallavicini 1990). On the contrary, the FWHM of the central emission component in stars is independent of the activity level, while it is tightly correlated with the absolute visual magnitude of the star. This remarkable empirical correlation, whose theoretical interpretation is still a matter of controversy, is known as the *Wilson-Bappu effect* (Wilson and Bappu 1957) and is one of the most useful ways to derive absolute magnitudes (and, hence, distances) for late-type stars.

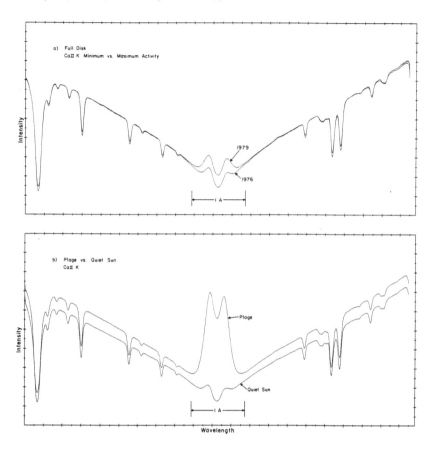

Figure 2. Spectra of the Sun in the Ca II K line. The upper panel shows the disk integrated Ca II K profile at maximum and minimum during the solar cycle. The lower panel shows a comparison of the Ca II K profiles in a quiet area and in a plage (from White and Livingston 1981).

The emission component at the center of the Ca II lines starts to appear around spectral type middle F. However, the fact that this component is seen against the photospheric background makes the detection of chromospheres in hotter stars increasingly more difficult. The Mg II *h* and *k* lines are more convenient in this respect, since the photospheric contribution, which decreases rapidly towards shorter wavelengths in cool stars, is usually negligible for these lines. This, and the larger abundance of Mg with respect to Ca, make the Mg II lines ideal diagnostics of stellar chromospheres. As a drawback, Mg II observations need to be carried out from space: many of these observations have been obtained over the past years with the *Copernicus* and *IUE* satellites (Basri and Linsky 1979, Stencel *et al.* 1980, Blanco *et al.* 1982, Crivellari *et al.* 1983, Doherty 1985, Vladilo *et al.* 1987).

With the Mg II data, we can study the onset of chromospheres among cool stars at spectral types earlier than those accessible to Ca II observations. The available data indicate that chromospheres first appear at late spectral type A. Altair (A7 IV-V) is often quoted as one of the earliest stars for which Mg II chromospheric emission (and also X-ray coronal emission) has been detected. The onset of chromospheres coincides, within the observational and theoretical uncertainties, with the appearance of significant subphotospheric convective zones in stars. This suggests a causal relationship between the deposition of non-radiative energy in outer stellar atmospheres and convective motions. The details of such causal relationship, however, are poorly known, though generation of acoustic or magnetoacoustic waves and/or of dynamo-generated magnetic fields are likely candidates (see discussion of heating mechanisms in section 5 below).

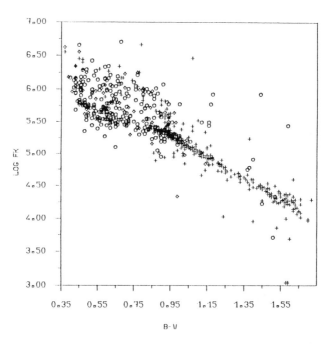

Figure 3. Ca II K line absolute surface fluxes as a function of colour index $B-V$. Fluxes are given in $erg\ cm^{-2}s^{-1}$. Different symbols refer to different luminosity classes (from Governini and Pallavicini 1986).

The Ca II and Mg II absolute fluxes obtained for stars of different spectral types and luminosity classes show a general decreasing trend towards lower effective temperatures (Linsky *et al.* 1979b, Basri and Linsky 1979, Governini and Pallavicini 1986, Schrijver 1987a and b, Pasquini, Pallavicini, and Pakull 1988; see Figure 3). A weak dependence on gravity may be present, but is not very prominent. Rather, stars of all luminosity classes show a large range of emission levels at each spectral type. It was suggested very early (Wilson 1963, Wilson and Skumanich 1964,and Kraft 1967) that different levels of chromospheric emission in dwarfs of the same effective temperature could be due to differences in rotation rate and/or age, and this has been confirmed by more recent studies (see section 2.2 and 5.2 below). The dependence on rotation, together with the clear association of chromospheric emission with convection, suggests that chromospheric heating may be due, at least partially, to the action of dynamo-generated magnetic fields. The strong similarity observed between Ca II and Mg II spectra of active stars and those of magnetic regions on the Sun (plages) is another argument in favor of this interpretation.

With the advent of solid state detectors that are more sensitive to the red part of the spectrum, the H I Hα line at 6563 Å has become an important diagnostic of stellar chromospheres (Zarro and Rodgers 1983, Herbig 1985, Bopp 1990, Pasquini and Pallavicini 1991). However, a quantitative estimate of chromospheric radiative losses in Hα is more difficult than for the Ca II and Mg II lines, at least for F to K stars. The reason is that chromospheric emission in Hα produces a much smaller contrast effect than in the Ca II and Mg II lines, and the chromospheric component appears as a (usually) subtle filling-in of the Hα profile, rather than as a distinct emission component (this also applies to the Ca II infrared triplet, *cf.* Linsky *et al.* 1979a). Although differential techniques (*i.e.* comparison of the observed Hα profiles with those of inactive stars of similar spectral type) can in principle allow a calibration of Hα data in terms of chromospheric radiative losses at the star surface (Herbig 1985, Pasquini and Pallavicini 1991), this is by no means a simple task and the errors are large. For most stars, Hα appears to be a qualitative rather than a quantitative indicator of chromospheres.

The situation is different for late K and M dwarfs, for T Tauri stars, and for at least part of the active binaries of the RS CVn type. In M dwarfs, Hα and, in general, all Balmer lines are often in emission (in the so-called dMe stars) and their relative contribution to the total chromospheric losses becomes increasingly more important with respect to that of other lines and continua. For instance, in the Sun and solar type stars, Ca II and Mg II lines contribute together some 40% of the total chromospheric radiative losses, while Hα and other Balmer lines contribute only a few percent (Linsky and Ayres 1978, Linsky 1991). In cooler stars, the contribution of the Balmer lines to the cooling of the chromosphere increases and they become eventually more important than the Ca II and Mg II lines (Linsky *et al.* 1982). Note that according to recent calculations Fe II lines are the most important cooling agents for the chromospheres of solar-type stars, contributing some 50% of the total radiative losses (*cf.* Linsky 1991). Moreover, the Ca II infrared triplet is typically more important that the Ca II H and K lines. Hα is in emission and very strong in T-Tauri stars (with equivalent widths that often may be as high as 50 Å); however, in these stars, Hα is not simply of chromospheric origin, but is largely produced in the dense circumstellar envelopes that surround T Tauri stars (see reviews by Kuhi and Cram 1989, Basri 1990, Hartmann 1990).

The behaviour of Hα emission in stars of similar spectral types but different levels of chromospheric activity is only partially understood. As for the Ca II and Mg II lines, Hα emission in F to K stars increases for increasing activity levels. However, the picture complicates substantially in M dwarfs as shown by theoretical models and observations (Stauffer and Hartmann 1986, Rodonò 1986, Mullan 1986, Robinson, Cram, and Giampapa

1990). M dwarfs with low activity levels have typically shallow Hα absorption profiles. As activity increases, the Hα absorption profile first becomes deeper and later, as activity increases further, the profiles gradually fill-in and eventually go into emission. In RS CVn binaries, Hα is usually filled-in but only the most active RS CVn binaries (like HR 1099) always have Hα in emission (Bopp 1990). Usually, the Hα profile in these active stars is highly variable with variations that are only partially related to the rotation of the star (Nations and Ramsey 1980, Collier, Cameron, and Robinson 1989). More often, the variations are chaotic and probably related to flare-like activity.

There are other spectroscopic diagnostics that are useful to derive information on the chromospheres of stars. Zirin (1976) and O'Brien and Lambert (1986) have obtained observations of the He I 10830 Å line, while the use of the He I 5876 Å line as a chromospheric diagnostic has been discussed by Danks and Lambert (1985) and Wolff, Boesgaard, and Simon (1986). Linsky et al. (1979a) and Foing et al. (1989) have discussed observations of the Ca II infrared triplet. Finally, far ultraviolet lines which originate in stellar chromospheres will be discussed in section 3 below.

2.2. THE MT. WILSON CA II H AND K DATA

Most of what we know at present about stellar chromospheres has not been derived from detailed high-resolution spectroscopic data (of the kind mentioned in the previous section), but more simply through spectrophotometric observations carried out at the Mt. Wilson Observatory over a period of nearly three decades (Wilson 1963, 1966, 1978, 1982, Vaughan and Preston 1980, Vaughan 1983, Baliunas and Vaughan 1985, Duncan et al. 1991). The Mt. Wilson photoelectric observations give an index S which is obtained by comparing the integrated flux in two 1 Å wide bands centered on the H and K lines of Ca II with the integrated flux in two 20 Å wide bands located on either sides of the H and K lines. As such, the S index is a contrast measurement that depends strongly on the effective temperature of the star. Although it can be converted to absolute flux units using appropriate calibrations (Noyes et al. 1984, Rutten 1984), it is not by itself a measure of chromospheric radiative losses in the Ca II lines. Therefore, it is not particularly suited to addressing questions such as the energy balance and heating processes in stellar chromospheres. Instead, it provides very precise relative measurements that are invaluable when comparing stars of the same effective temperature and luminosity class, or for investigating the time variability of chromospheric emission.

Mt. Wilson S indexes measured for main sequence stars in the solar neighbourhood are plotted as a function of the $B - V$ colour index in Figure 4, where the vertical bars indicate the range of values observed for the same stars at different times. The increasing trend towards cooler spectral types is an obvious consequence of the rapidly decreasing intensity of the 20 Å reference bands to which the Mt. Wilson H and K fluxes are referred (Ca II absolute surface fluxes actually decrease towards cooler stars, cf. Figure 3). It is clear that, at each spectral type, there is quite a large range of chromospheric emission levels. The Sun is close to the bottom of the observed distribution, but there are stars of the same spectral type as the Sun which have Ca II H and K absolute fluxes up to an order of magnitude larger. There is also some hint of the presence of a possible discontinuity in the observed S index distribution, with stars of either low or high chromospheric activity but no intermediate values (Vaughan-Preston gap). The reality of such a gap (which, if confirmed, would have important implications for our understanding of the stellar dynamo and its dependence on age) is, however, not convincingly proved (Hartmann *et al.* 1984). At any rate, both Figures 3 and 4 clearly demonstrate that chromospheric emission in stars does not depend only on the position of a star in the HR diagram, but also on other parameters such as rotation and age (Kraft 1967, Noyes *et al.* 1984; see discussion below and also section 4.2).

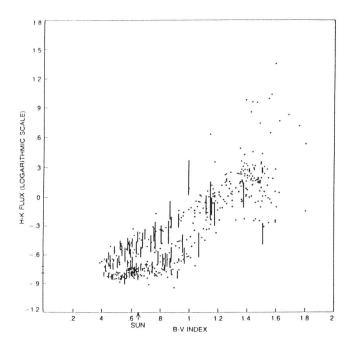

Figure 4. The Mt. Wilson Ca II S index as a function of the colour index $B - V$. Vertical bars indicate the range of values observed for some stars (from Vaughan 1983).

Probably the most significant result obtained by the use of the S index has been the discovery of solar-type cycles in stars other than the Sun (Wilson 1978). Figure 5 shows some of the observed cycles monitored over periods of nearly two decades (Baliunas and Vaughan 1985). While some of the observed cycles are similar to the solar cycle in both amplitude and duration, others show a quite different behaviour, with periods as short as a few years or apparently longer than the total monitored time. There is no obvious dependence of the amplitude and duration of the observed cycles on other parameters of the star. This is surprising since the characteristics of a stellar cycle would be expected to depend on rotation and on the properties of the convection zone. Unfortunately, the present

322

status of dynamo theory does not allow reliable predictions to be made about activity cycles in other stars (*cf.* Chapter 5 in this volume).

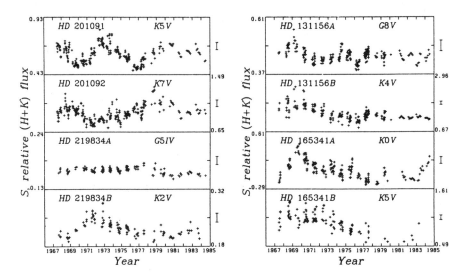

Figure 5. Stellar activity cycles as detected in the Ca II H&K lines over the period 1967 to 1985 (from Baliunas and Vaughan 1985).

The long-term systematic observations carried out at Mt. Wilson also show significant variations on time scales much shorter than those associated with stellar cycles. Some of these variations (over time scales of days to weeks) appear to be periodic and are best interpreted as due to the rotation of stars whose chromospheric emission is highly inhomogeneous (as might occur if bright plages are distributed irregularly over the stellar disk). This provides an excellent way to measure rotation rates for stars (such as the F and G) for which no optical continuum rotational modulation due to cool spots is typically observed. This method can measure rotational velocities that are lower than those obtainable at present with the best spectroscopic techniques and, in addition, the derived velocities are free of projection effects.

By using rotation rates derived from Mt. Wilson Ca II data, Noyes *et al.* (1984) have investigated the dependence of chromospheric Ca II emission upon rotation in dwarf stars, and Hartmann *et al.* (1984) have carried out a similar investigation for Mg II emission (see also Marilli and Catalano 1984). The results of Noyes *et al.* show that the Ca II absolute surface flux F'_{HK} (derived from the S index) is correlated with the star rotation period P_{rot}. There is some scatter, but no obvious dependence on spectral type. Ca II emission appears to decrease with increasing rotational periods, as could be expected from a dynamo mechanism which predicts – at least qualitatively – a greater efficiency of magnetic field generation in more rapidly rotating stars. Noyes *et al.* found, however, that the scatter in the correlation is somewhat reduced if a colour dependent Rossby number P_{rot}/τ_c, rather than simply P_{rot}, is used, where τ_c is the convective turnover time computed at the bottom of the convective zone (see Figure 6). This is more satisfactory from a theoretical point of view since one expects – again qualitatively – that the efficiency of the dynamo should depend on both rotation and convection, rather than simply on rotation.

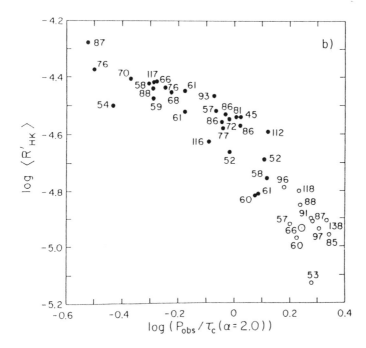

Figure 6. Normalized Ca II H&K line fluxes as a function of the Rossby number. Numbers indicate the colour index $B - V$ multiplied by 100. The position of the Sun is indicated (from Noyes *et al.* 1984).

While the above result strongly supports the interpretation of different chromospheric emission levels in stars as due to different degrees of magnetic activity, it remains unclear whether a formulation in terms of the Rossby number should be preferred to one in terms of the rotation period. In fact, the smaller scatter claimed by Noyes *et al.* is obtained by plotting the normalized flux $R'_{HK} = F'_{HK}/\sigma T^4_{eff}$ (a quantity which itself depends upon spectral type) rather than F'_{HK} (using the latter quantity the correlation becomes worse). Moreover, theoretical calculations of the convective turnover time are highly uncertain and depend critically on the assumed value of the ratio α of the mixing length to the pressure scale-height (the correlation improves only for a specific value of α). Finally, a detailed analysis carried out by Basri (1987) does not show any significant improvement by using Rossby numbers, rather than rotation periods. In conclusion, while there is little doubt that chromospheric emission depends at least partially on rotation (and, hence, presumably on dynamo-generated magnetic fields), it is unclear how the dynamo mechanism and chromospheric heating are mutually related. This is one of the major unsolved problems in stellar chromospheric research (see also section 5 below).

3. Stellar Transition Regions

As we know from the Sun, the temperature, after rising to values of the order of 2×10^4 K in the upper chromosphere, jumps abruptly to values of the order of $\approx 10^6$ K in the corona, through a narrow *transition region* whose width is much less that one pressure scale height. The properties of this region are best studied by using a number of optically thin far ultraviolet lines emitted by highly ionized atoms of C, O, N, Ne, Mg, Si, and S. The same occurs in main sequence stars of all spectral types later than A, in giants not cooler than early K, and in supergiants not cooler than middle G. The short-wavelength spectral range accessible to the *IUE* satellite (1100 - 1800 Å) has allowed the investigation of stellar transition regions in a variety of stars. The results have been reviewed, among others, by Ayres (1981), Linsky (1985), and Jordan and Linsky (1987). Here I will summarize some of the results for dwarf stars, while deferring the discussion of cool giants and supergiants (that show no evidence of material at $T \approx 10^5$ K or higher) to section 6.1.

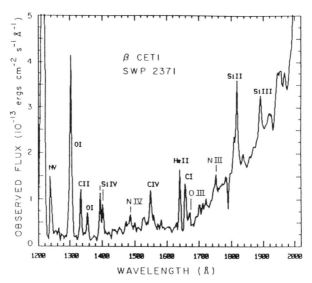

Figure 7. Short-wavelength, low-dispersion *IUE* spectrum of β Ceti. The most prominent emission lines are indicated (from Eriksson, Linsky and Simon 1983).

The short wavelength range of *IUE* contains, in addition to Lyα (which, however, is highly contaminated by interstellar absorption and geocoronal emission), lines of N V at 1241 Å, O I at 1304 Å, C II 1335 Å, Si IV at 1394 and 1403 Å, C IV at 1550 Å, He II at 1640 Å, CI at 1657 Å, and Si II at 1808 and 1817 Å. These lines are easily observable in low-resolution (≈ 6 Å) *IUE* spectra of sufficiently high signal-to-noise ratio (see Figure 7). Many other lines, including some which are important for density diagnostics, become visible in the high-resolution spectra that have been obtained for some of the brightest stars (see Jordan and Linsky 1987 for details). The observed lines form at different temperatures in the chromosphere and transition region up to temperatures of $\approx 2 \times 10^5$ K. For instance, the O I, C I, and Si II lines are produced deep in the chromosphere at temperatures of $\approx 7 \times 10^3$ K, C II is formed in the high chromosphere at $T \approx 2 \times 10^4$ K, Si IV forms in the transition region (at $\approx 7 \times 10^4$ K), and, finally, C IV and N V form at $T \approx 1 - 2 \times 10^5$ K. Fluxes in these lines allow the study of chromospheres and transition regions in various types of stars, and the construction of atmospheric models using techniques that have originally been developed for the Sun (differential emission measure analysis; *cf.* Jordan and Brown 1981, Jordan 1986, Jordan *et al.* 1987).

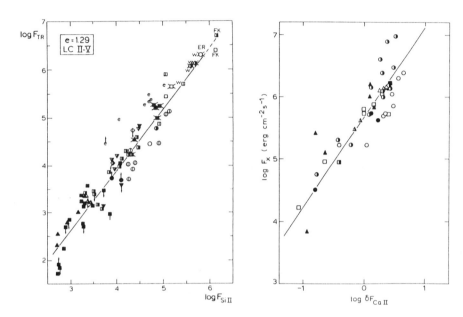

Figure 8. Flux-flux diagrams of different spectroscopic diagnostics of stellar chromospheres, transition regions and coronae. The left-hand panel (from Oranje 1986) shows transition region fluxes (in N V, C IV and Si IV) *vs.* chromospheric Si II fluxes. The right-hand panel (from Schrijver 1987) shows coronal X-ray fluxes *vs.* "excess" Ca II fluxes.

Not surprisingly, the *IUE* data show a large range of different emission levels even for stars of the same spectral type. The fluxes in different lines are well correlated with one another, as well as with Ca II and Mg II emission (Ayres, Marstad, and Linsky 1981, Oranje 1986, Cappelli *et al.* 1989; see Figure 8, left panel). There is very little dependence on spectral type and luminosity class in these flux-flux correlations, over nearly three orders of magnitude. Coronal X-ray fluxes are also correlated with transition region and chromospheric fluxes, with a slope that is somewhat higher than for low-temperature lines. All this indicates a very strong coupling between the different atmospheric layers, though it does not allow *per se* the identification of the relevant heating mechanisms. Schrijver (1987a,b) has argued that the flux-flux correlations improve significantly if a colour-dependent "basal" flux is subtracted from the Ca II and Mg II chromospheric fluxes, and the correlation is made in terms of an "excess" flux above this "basal" level (Figure 8, right panel). They have suggested that this "basal" flux represents that part of the chromosphere that is not heated by magnetic processes. I will come back to this point in section 5. At any rate, the transition region fluxes observed from stars of different emission levels appear to correlate well with the rotation rate (Simon and Fekel 1987), but there are indications of saturation effects at very high activity levels (Vilhu 1987).

Cappelli *et al.* (1989) have made a detailed comparison of *IUE* low-resolution spectra in stars of different activity levels with high-resolution spectra of spatially resolved solar regions obtained over the same wavelength range by the NRL Slit Spectrograph on board *Skylab*. They have shown that the solar data (which include spectra of quiet regions, active regions, and flares) are in good qualitative and quantitative agreement with the stellar data and obey the same flux-flux relationships with virtually the same slope and similar scatter. Cerruti-Sola, Cheng and Pallavicini (1991) have extended the comparison to Mg II data obtained by *IUE* and *Skylab*, again finding an excellent agreement. This is strong support of the notion that different levels of transition region and chromospheric emission in stars are due to different fractions of their surface being covered by magnetic regions. For very active stars (such as close binaries of the RS CVn type), the disk integrated transition region and chromospheric fluxes are comparable to those of the brightest solar plages and even of solar flares. Their surface must be entirely covered by magnetic regions as bright as the brightest solar plages or, alternatively, a smaller fraction of their surface must be covered by magnetic regions that are much brighter than those typically observed on the Sun.

4. Stellar Coronae

Until 1975, the Sun was the only star known to possess a high-temperature X-ray emitting corona. Then came the discovery of X-rays from the bright giant Capella (Catura, Acton, and Johnson 1975) and a number of detections followed in subsequent years, most of them being of active RS CVn binaries (Mewe 1979, Walter and Bowyer 1981). With the advent of more sensitive imaging telescopes on board the *Einstein*, *EXOSAT*, and *ROSAT* satellites, it is now well established that X-ray emission from high temperature low-density coronae is a general property of virtually all types of stars, the only exceptions being very late-type giants and supergiants and, perhaps, A-type dwarfs. The subject of X-ray emission from stars has been reviewed many times in recent years by, among others, Stern (1983), Rosner, Golub, and Vaiana (1985), Haisch (1986), Vaiana and Sciortino (1987), Schmitt (1988), Pallavicini (1988, 1989), Linsky (1990), Vaiana (1990), Rosner (1991). Here, I will present only the highlights of the results obtained so far, while referring to the above quoted papers for a more detailed discussion of the subject.

4.1. X-RAY EMISSION THROUGHOUT THE HR DIAGRAM

X-ray emission has been detected from both early and late-type stars. O-type stars of all luminosity classes are the brightest among stellar coronal sources, with X-ray luminosities ranging from $\approx 10^{31}$ to 10^{34} erg sec^{-1} (Chlebowski *et al.* 1989, Chlebowski 1989, and Sciortino *et al.* 1990). The X-ray luminosity of O and early B stars is well correlated with the bolometric luminosity ($L_x \sim L_{bol}^{-7}$) suggesting a possible connection with the massive radiatively driven winds of these stars. The presence of winds was revealed by UV observations with the *Copernicus* satellite and has been confirmed by *IUE* (see review by Cassinelli and MacGregor 1986). There is no correlation between the X-ray luminosities of early-type stars and their rotation rates (Pallavicini *et al.* 1981); indeed, no such correlation is expected since these stars have no subphotospheric convective zones and, hence, are unlikely to have dynamo generated magnetic fields.

The origin of X-ray emission in early-type stars is poorly understood. The favored mechanism (Lucy 1982) relies on the formation of density perturbations by radiative instabilities and subsequent shock heating throughout the wind. The presence of a thin corona at the base of the wind, confined by primordial magnetic fields (Cassinelli and Olson 1979, Cassinelli 1985), cannot be excluded, but is unlikely since the overlying dense wind should produce a noticeable absorption at low X-ray energies which is not observed. At any rate, the mechanisms of coronal formation and wind acceleration in early-type stars appear to be fundamentally different from those operating on the Sun and other late-type stars with convective envelopes. For this reason, I will not discuss early-type stars further.

The situation changes completely when we move along the main-sequence to stars that possess subphotospheric convective envelopes. All late-type dwarfs (from late A to middle M) have been detected as X-ray sources at luminosity levels ranging from $\approx 10^{26}$ to 10^{30} erg sec^{-1}. By comparison, the average X-ray luminosity of the Sun in $\approx 10^{27}$ erg sec^{-1}. The emission appears to be thermal with temperatures ranging from a few 10^{6} K to several 10^{7} K (Lemen *et al.* 1989, Schmitt *et al.* 1990). There is very little dependence of X-ray luminosity on effective temperature and bolometric luminosity, while there is a large range of emission levels (more than three orders of magnitude) at each spectral type (Vaiana *et al.* 1981, Maggio *et al.* 1987). This is similar to what is observed for chromospheric and transition region lines, but the range of emission levels at each spectral type is even larger. Note that the median X-ray luminosity is almost constant, to first approximation, from F to M stars and, hence, the X-ray surface flux increases towards cooler stars. This is the opposite of what is observed for chromospheric Ca II and Mg II emission and may have important implications for understanding the mechanisms of chromospheric and coronal heating (*cf.* section 5).

The parameter which has been shown to affect more significantly coronal emission in late-type dwarfs is, again, stellar rotation. An often quoted result relates the X-ray luminosity of middle F to middle M dwarfs to the square of the stellar rotation rate *Vsini* (Pallavicini *et al.* 1981). However, although a dependence on rotation is clearly present, the correlation is not as good as would be desirable (see Figure 9). There is a large scatter around the mean relationship (more than one order of magnitude in X-ray luminosity and at least a factor of two in rotation rate). The functional dependence of X-ray luminosity upon rotation is not well defined and many alternative L_X *vs.* V_{rot} laws have, in fact, been proposed (*e.g.* Walter 1982). Early F stars do not show any obvious dependence upon rotation and the correlation is not much better if a formulation in terms of the Rossby number (rather than simply rotation) is used (Schmitt *et al.* 1985). The active RS CVn binaries, as a class, are consistent with an increase of X-ray luminosity with rotation, but there is little or no dependence upon rotation within the class itself (Majer *et al.* 1986). Rapidly rotating K stars in the Pleiades are definitely less active in X-rays than predicted by the $L_x \sim (Vsini)^2$ law (Caillault and Helfand 1985). Clearly, rotation is important but probably is not the only relevant parameter.

X-ray emission is free of contrast effects such as those that plague the detectability of Ca II emission in hotter stars. Therefore, it is particularly suitable for investigating the onset of non-radiative heating among late-type stars. Analysis of *Einstein* data shows that X-ray emission, after dropping below detection threshold at spectral types from late B to middle A, rises again to values typical of late type stars (Schmitt *et al.* 1985). This is consistent with the onset of X-ray emission in coincidence with the appearance of appreciably deep subphotospheric convective zones. It is also consistent with the first appearance of Mg II emission at spectral types late A-early F. Unfortunately, this is the only clear indication of a possible relation between coronal X-ray emission and convective zone properties. Only at the other extreme of the main sequence, at spectral types later than M5, may we see another effect possibly related to convective zone properties. Bookbinder (1985) found a significant drop of X-ray detections in the *Einstein* sample for stars that are believed to be fully convective. If confirmed, this result would be a strong support of the dynamo theory and an indication that the dynamo operates very deeply in the convection zone, at the interface with the radiative core. This boundary layer is absent in fully convective stars.

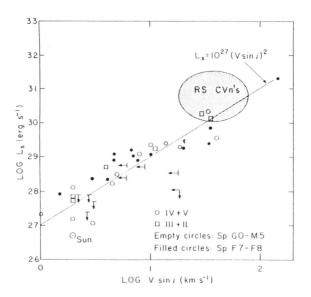

Figure 9. X-ray luminosity *vs.* rotation rates for stars of spectral type F7 to M5. The position of the Sun is indicated as well as the region occupied by RS CVn binaries (from Rosner, Golub and Vaiana 1985, based on results of Pallavicini *et al.* 1981).

The late-B and early-A dwarfs present an interesting problem. It is usually assumed, on the basis of *Einstein* results, that these stars do not possess coronae, consistent with current theoretical expectations for stars that have neither massive radiatively driven winds nor subphotospheric convective zones. Early reports of X-ray emission from A-type stars (including bright objects such as Vega or Sirius) were discarded either on the basis of possible detector contamination by UV radiation, or by the presence of unseen later-type companions (which would be the source of the observed X-ray radiation). However, there are a number of cases that cannot be easily discarded (Caillault and Zoonematkermani 1989, Pallavicini *et al.* 1990, Walter and Boyd 1991). Many A-type stars have recently been detected in the *ROSAT* All-Sky Survey (which is much less affected than the *Einstein* HRI by UV contamination). It will be interesting to see whether these detections can be explained by the presence of cooler companions or, rather, if new coronal mechanisms need to be considered for A-type stars.

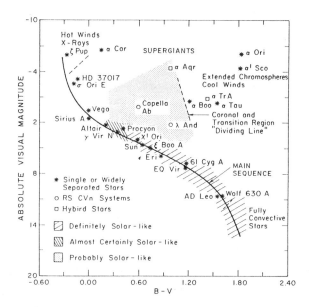

Figure 10. Summary of stellar activity throughout the HR diagram as derived from X-ray, UV, and optical data. The diagram shows schematically the position of the main-sequence as well as the regions of the HR diagram where massive winds (either hot or cool) occur. Some representative stars are indicated. Also shown is the position of the Coronal and Transition Region Dividing Line separating stars with hot coronae and solar-like winds (*to the left*) from stars with extended chromospheres and cool winds (*to the right*). Empty squares indicate the position of two "hybrid" stars which show evidence of both Transition Region material (at $T \approx 10^5$ K) and massive cool winds. Empty circles indicate RS CVn binaries. The regions of the HR diagram where solar-type phenomena are believed to occur are indicated (from Linsky 1985).

The dMe flare stars show other interesting peculiarities. Their X-ray luminosity appears to be largely independent of rotation, while it is well correlated with bolometric luminosity (Agrawal *et al.* 1986, Pallavicini, Tagliaferri and Stella 1990). This behaviour, which is at variance with that typically observed for late-type dwarfs, can be easily understood as a saturation effect, when the star becomes entirely covered by active regions. If this is the case, the dependence on bolometric luminosity would simply reflect a dependence on the size of the star. A similar dependence on radius has been found by Fleming, Gioia, and Maccacaro (1989) for M dwarfs detected serendipitously by *Einstein*. Since these stars represent the high luminosity tail of the M dwarf distribution, it is to be expected that their coronal emission occurs in a regime of saturation. Note that for the most active dMe stars, the ratio of X-ray to bolometric luminosity is on the order of $\approx 10^{-3}$. *i.e.* much larger than for both the Sun ($\approx 10^{-6}$) and for early-type stars ($\approx 10^{-7}$). The mechanism of coronal heating, whatever it is, must be extremely efficient in M dwarfs.

Among late-type evolved stars, the brightest ones are the RS CVn binaries, which have typical X-ray luminosities of $10^{30} - 10^{31}$ erg sec^{-1} (Majer *et al.* 1986, White *et al.* 1986, 1990). These close binaries have at least one component (typically of spectral type K0 IV) which is slightly evolved out of the main sequence. The high activity observed in these systems at X-ray, optical, and radio wavelengths is interpreted in terms of enhanced rotation and dynamo action (Linsky 1984). The enhanced rotation is a consequence of tidal interaction between the two components. Under these conditions, RS CVn binaries can provide important constraints on magnetic processes and dynamo action, but give little information on the behaviour of X-ray emission during post-main sequence evolution. For this, we have to turn to normal, single giants.

The observations with *Einstein* have shown that, among single evolved stars, the yellow giants are detected as X-ray sources, but the red giants (at spectral types later than $\approx$ K1 III) are not (Ayres *et al.* 1981, Maggio *et al.* 1990). The current upper limits are in the range $\approx 10^{28} - 10^{29}$ erg sec^{-1} but, in a few cases, long exposures have provided upper limits as low as $\approx 10^{27}$ erg sec^{-1}. Taking into account the much larger radius of giants with respect to the Sun, the derived upper limits imply surface fluxes that are equal to or lower than that of solar coronal holes (*cf.* Vaiana and Rosner 1978). Since late-type giants and supergiants also do not show evidence of material at $\approx 10^5$ K in *IUE* spectra but only at temperatures of $\approx 10^4$ K, it is likely that their coronae, if they exist, are extremely weak. A long exposure of the red giant α Boo with *ROSAT* has given an upper limit to its coronal emission as low as 3×10^{25} erg sec^{-1} (Ayres, Fleming, and Schmitt 1991). This corresponds to an X-ray surface flux $< 10^{-3}$ that of a solar coronal hole.

For the yellow giants that have been detected in X-rays, a wide range of X-ray emission levels is observed from less than $\approx 10^{27}$ erg sec^{-1} to almost 10^{31} erg sec^{-1}. This is even larger than the range of emission levels observed for G-type dwarfs. Some apparently single yellow giants have X-ray luminosities comparable to those of RS CVn binaries (Maggio *et al.* 1990). It is interesting to note that a similar behaviour is observed for Ca II H and K emission (Pasquini, Brocato, and Pallavicini 1990) and is most easily interpreted in terms of different masses and evolutionary histories. From a detailed comparison with evolutionary tracks, Pasquini *et al.* were able to demonstrate that giants with high chromospheric emission were rather massive stars that have not suffered magnetic braking while on the main-sequence. When they evolve out of the main-sequence and develop a subphotospheric convective zone, dynamo action starts to work thus causing the observed strong Ca II emission. On the contrary, less massive stars (with masses $\leq 2M_\odot$) have already suffered magnetic braking while on the main-sequence and significant dynamo action cannot operate anymore. The X-ray observations of yellow giants are apparently consistent with this picture, although more sensitive observations and a more accurate comparison with evolutionary tracks is required in order to confirm this scenario.

Figure 10 summarizes our present knowledge of non-radiative heating in outer stellar atmospheres as has been derived from X-ray, UV, and optical observations. Also indicated are the regions of the HR diagram where solar-type phenomena are likely to occur. The position of several stars mentioned above is also indicated.

4.2. CORONAL EMISSION AND AGE

In the course of this Chapter, I have mentioned several times a connection between chromospheric/coronal emission and age in late-type stars. The existence of such a connection was demonstrated more than twenty years ago, by comparing Ca II fluxes of solar-type stars in young open clusters (the Hyades and the Pleiades) with those of the Sun and field G stars (Kraft 1967). It was noticed that a similar decline with age also occurs for the average rotation rate of stars. Later, Skumanich (1972) formalized these findings by deriving a simple $t^{-1/2}$ decay law for both Ca II H&K emission and rotation rates. More recent studies along these lines include those of Duncan (1981) and Soderblom (1983).

The decline of rotation with age during main-sequence evolution is interpreted as a consequence of the magnetic braking that a star suffers under the action of outflowing magnetized stellar winds (Weber and Davis 1967). The magnetic fields imbedded in the wind cause corotation up to the Alfvén radius r_a and the loss of angular momentum is given by $\dot{J} \sim \Omega \, r_a^2 \, \dot{M}$, where Ω is the angular velocity of the star and $\dot{M}$ is the mass loss. Since $r_a \gg R_*$, the magnetic field acts like a lever arm to increase the angular momentum loss. On the other hand, the fact that Ca II emission also declines with age, with the the same law as rotation, suggests a physical connection between the two. The new X-ray observations made available from space allow further investigation of this dependence.

Several surveys of nearby open clusters have been carried out by *Einstein* and follow-up observations are now being obtained with *ROSAT* (Stern *et al.* 1981, Caillault and Helfand 1985, Micela *et al.* 1985, 1988, 1990, Schmitt *et al.* 1990b). The open clusters observed by *Einstein* include the Hyades, the Pleiades, and Ursa Major. Since the ages of these clusters range from $\approx 6 \times 10^7$ years for the Pleiades to $\approx 6 \times 10^8$ years for the Hyades, and the stars in each cluster represent a rather homogeneous sample, we have a powerful tool to investigate the evolution of coronal emission on the main sequence and to compare the properties of young stars with those of the Sun and other typical field stars (at ages of $\approx 5 \times 10^9$ years).

The integral luminosity functions derived for stars of different spectral types in each cluster show consistently that X-ray emission declines with age (Figure 11). However, the derived time dependence is not a simple $t^{-1/2}$ law like that derived from Ca II observations (see also Simon *et al.* 1985 for a similar discrepancy). An exponential decrease with a rapid fall off at ages later than that of the Hyades appears more consistent with the X-ray data. The decrease of X-ray activity with age occurs for all late type stars and there is some dependence on spectral type. This suggests that more than one parameter is involved in the process.

332

An important issue is whether the dependence on age is a primary effect or simply the result of the decrease of stellar rotation rate with advancing age. Stern *et al.* (1981) found that the higher X-ray emission of Hyades stars with respect to field stars is consistent with a dependence of X-ray luminosity upon $(V sini)^2$. The Pleiades data show a more complex pattern and a larger scatter. Rapidly rotating K stars in the Pleiades are less bright than predicted by the $L_X \sim (V sini)^2$ law, although the luminosity function of K stars in the Pleiades is shifted to higher luminosity values with respect to the Hyades and field stars (Caillault and Helfand 1985, Micela 1990). A dependence of X-ray luminosity upon rotation is also present within supposedly coeval stars in a cluster, but there are stars with similar rotation rates and widely different Ca II/X-ray emission levels (*e.g.* the late-type giants in the Hyades; Baliunas *et al.* 1983). Apparently, the relationship between coronal emission, rotation, and age is not a simple one and is likely to involve a complex interaction between convective motions, magnetic fields, mass loss through stellar winds, and the star angular momentum evolution.

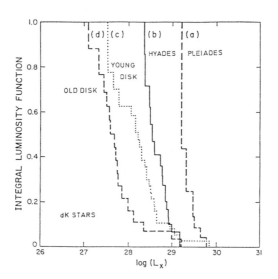

Figure 11. X-ray integral luminosity functions for K stars in open clusters of different ages (the Hyades and the Pleiades) and in field stars of both young and old disk populations (from Micela *et al.* 1990).

The dependence on age found for cluster stars extends to pre-main sequence objects (*i.e.* at ages $\leq 10^7$ years). *Einstein* made extensive observations of nearby star forming regions, including the Orion nebula, the Taurus-Auriga complex, the ρ Ophiuchi dark cloud, and the Chamaeleon dark cloud (Montermerle *et al.* 1983, Feigelson 1984, Walter *et al.* 1988, Feigelson and Kriss 1988). Even more extensive observations, covering the entire area of these regions, have been obtained recently during the *ROSAT* All-Sky Survey.

These observations have shown clearly that pre-main sequence stars are very strong coronal emitters, with typical luminosities in the range $\approx 10^{30}$ - 10^{31} erg sec^{-1}. These luminosities are larger than those typically observed in late-type field stars and are comparable to those seen in the very active RS CVn binaries.

A rather unexpected result was the finding that only a fraction ($\approx 1/3$ to $1/2$) of the X-ray detected stars in star forming regions are classical T-Tauri stars (Walter 1986, Walter et al. 1988). Most of the X-ray detected objects are stars that, in the optical, lack the extreme properties that characterize classical T-Tauri stars, i.e. strong Hα emission and infrared excess originating from dense circumstellar envelopes. These objects are indicated as Weak-lined T-Tauri stars. Some of them may be post-T Tauri stars, i.e. pre-main sequence objects that are older and more evolved than classical T-Tauri. However, many of them appear to be coeval with classical T-Tauri stars. In any case, they are apparently deprived of the circumstellar envelopes and dense winds of classical T-Tauri stars. Walter (1986) has named them Naked T-Tauri to emphasise this characteristic property. These young stars show enhanced chromospheric and coronal emission, are often variable in the optical and present a very strong Li I 6708 Å line indicative of young age (see reviews by Feigelson, Giampapa and Vrba 1991 and Montmerle et al. 1991).

The Sun must have been a Weak-lined T-Tauri star at same stage during its pre-main sequence evolution. By applying to the Sun what we have learned from X-ray and optical observations of Weak-lined T-Tauri stars, we can infer its properties at an age of $\approx 1 \times 10^6$ years (Feigelson, Giampapa, and Vrba 1991). At that age, the radius of the Sun was ≈ 2.5 times the present value and its optical luminosity ≈ 1.7 times higher. The effective temperature was lower by about 1500 K and the spectral type must have been around K5 IV. At that epoch, the Sun was rotating much more rapidly (at a rate of ≈ 25 km sec^{-1}) and the X-ray luminosity was three orders of magnitude higher. Presumably, also its flaring activity was much higher.

5. Chromospheric and Coronal Heating

The standard theory of chromospheric and coronal heating generally accepted up to the late seventies was based on the generation of acoustic waves by turbulent motions in sub-photospheric convective zones (Biermann 1946, 1948 and Schwarzschild 1948). The waves, propagating upwards in an atmosphere of rapidly decreasing density, steepen into shocks and dissipate, thus raising the temperature to values far in excess of the star's effective temperature. This theory has been challenged by both solar and stellar observations and it is now quite evident that a purely acoustic mechanism cannot be responsible for non-radiative heating in outer stellar atmospheres, although acoustic waves may still contribute to the heating of the chromosphere outside magnetic regions.

The early calculations of acoustic flux were based on the Lighthill (1952) and Proudman (1952) theory of sound wave generation and on the mixing-length theory of convection. There are many uncertainties on these calculations and the predicted fluxes must be considered with much caution. At any rate, the early calculations (see e.g. Renzini et al. 1977) gave a very steep dependence of the acoustic flux F_M on effective temperature ($\sim T_{eff}^{15}$) and a marked dependence on gravity as well ($\sim g^{-1}$). These predictions were not in agreement with the Mg II and X-ray data that became available by the end of the seventies (Basri

334

and Linsky 1979, Vaiana *et al.* 1981). The Mg II showed a much flatter dependence on effective temperature, a weak dependence on gravity, and, above all, a very large scatter. X-ray data showed even larger discrepancies, since the median X-ray flux increases, rather than decreasing, towards cooler spectral types.

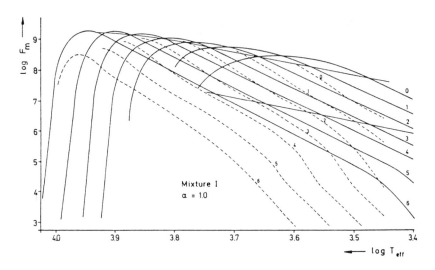

Figure 12. Theoretical calculations of acoustic energy fluxes for stars of different effective temperatures and gravities (from Bohn 1984). The full lines refer to the calculations of Bohn (1984), dashed lines are for the early calculations of Renzini *et al.* (1977).

New improved calculations of acoustic flux generation in late-type stars have been carried out by Bohn (1984; see also reviews by Ulmschneider 1991 and Musielak 1991). The new calculations, which include the effect of stratification in the subphotospheric convective zone, have substantially reduced the dependence on effective temperature, but the essential features remain unchanged (Figure 12). The acoustic flux is predicted to peak at spectral type F, to drop abruptly towards A-type stars (because convective zones become progressively thinner and eventually disappear in hotter stars), and to decrease for decreasing effective temperatures and increasing gravities at spectral type later than F (because the convective velocity decreases for cooler stars and larger gravities, and the acoustic flux depends on a very high power of the convective velocity). According to Bohn (1984), the dependence on effective temperature and gravity can be parameterized as:

$$F_M \sim g^{-0.5} \, T_{eff}^{9.75} \, \alpha^{2.8} \tag{1}$$

where α is the ratio of the mixing length to the pressure scale height. It is clear that a wave flux such as the one given above, which depends only on the position of a star in the HR diagram, cannot reproduce the observed large range of chromospheric and coronal fluxes observed for stars of the same effective temperature and/or gravity. Thus, the question is whether acoustic waves could possibly heat the non-magnetic part of stellar atmospheres or be relevant for those stars (such as old slowly rotating main-sequence stars or late-type giants) that cannot sustain a vigorous dynamo action.

Researchers at Utrecht (*e.g.* Schrijver 1987a,b, Rutten *et al.* 1991, Zwaan 1991), working with Mt. Wilson Ca II H&K and *IUE* data, have proposed that stellar chromospheric fluxes consist of two parts: a "basal" flux, which is independent of rotation and magnetic fields, and which could possibly be heated by purely acoustic waves, and a "magnetic" part which depends on rotation and age, and is heated by magnetic processes. Given the importance of this result for our understanding of chromospheric heating, it may be worth spending a few words about the definition of the "basal" flux, since its empirical determination is by no means straightforward.

The concept of a "basal flux" was introduced originally by noting that Ca II H&K fluxes (derived by converting Mt. Wilson Ca II S indexes to absolute units), when plotted as a function of colour index $B - V$, show a well defined lower boundary which decreases for decreasing T_{eff}. Above this lower boundary, there is a large range of Ca II fluxes at each spectral type, but no star apparently has a Ca II flux lower than a given value at each T_{eff}. In Ca II, most of this empirically determined lower boundary is photospheric light that falls in the pass-band of the Mt. Wilson detector, and which has not been subtracted from the derived Ca II absolute fluxes. The Utrecht researchers claim that, even when this correction is made (*e.g.* by using radiative equilibrium models), there is still an excess flux which must be considered as a "basal" chromosphere. A strong support of this interpretation comes from Mg II data which are much less affected by the photospheric background problem, and still show this lower boundary (Rutten *et al.* 1991).

The "basal" flux, therefore, is an intrinsically empirical quantity which, however, is not easily measured and must be extracted from the data by means of appropriate techniques. There are significant uncertainties in this process. The way adopted by the researchers at Utrecht is based on the use of flux-flux relationships of the type discussed in section 3 above. They assume that these flux-flux diagrams, when expressed in terms of "excess" fluxes above a given "basal" flux, are strictly power-laws with no dependence on colour. The "basal" fluxes are thus those which give the best power-law fits; they appear to be significant for Ca II and Mg II and even (but much less) for Si II. There is no evidence of a basal flux in X-rays.

The "basal" flux derived in this way for Ca II and Mg II decreases roughly as $\sim T^8$ (Zwaan 1991) and is weakly dependent on gravity: the "basal" flux for giants is slightly lower than for main-sequence stars. The dependence on effective temperature is not inconsistent with the theoretical acoustic flux given above (Equation 1), but the dependence on gravity is. However, the acoustic flux is computed at the top of the convective zone, while chromospheric Ca II and Mg II emissions refer to higher layers and propagation effects (*e.g.* radiation damping, Ulmschneider 1991) may be important. Moreover, the relative contribution of different cooling terms to the total chromospheric energy budget may change, in a still poorly known way, as a function of spectral type and luminosity class. Thus, while the concept of a "basal" chromosphere is an attractive and even plausible possibility, its existence, and interpretation in terms of acoustic heating, cannot be regarded as definitely proved.

The chromospheric flux above the "basal" flux, and all of the coronal X-ray flux, depend weakly on colour, while being a strong function of rotation (see sections 2.2 and 4.1 above). This is indirect evidence in favor of magnetic heating by dynamo-generated magnetic fields. However, which of the many proposed heating mechanisms (by either magnetohydrodynamic waves or electric currents) may be relevant for stellar chromospheres and coronae is not known. The microphysics of the heating process is poorly understood for the Sun, and it is unlikely that stellar observations may be of much help in this respect. Many of the proposed processes occur on spatial scales that are too small by orders of magnitude to be observed even in the solar case. Many plausible alternatives exist, but none of them can definitely be preferred. Comprehensive discussions on heating mechanisms can be found in Kuperus, Ionson, and Spicer (1983), Ulmschneider (1990), van Ballegoijen (1990), Narain and Ulmschneider (1990), Gomez (1990) and in the proceedings of the Heidelberg conference edited by Ulmschneider, Priest and Rosner (1991). Here I will briefly discuss only one of these mechanisms for the implications it has for stellar observations.

Parker (1983, 1988) has proposed that coronae may be heated by small-scale explosive events (micro-flares or nano-flares) that arise as a consequence of the continual shuffling and intermixing of the magnetic field lines by fluid motions. This concept has received considerable attention by solar and stellar physicists and claims have been made that stellar data (particularly those of M dwarf flare stars) provide observational evidence that X-ray coronae result from the cumulative effect of small short-lived microflares (Butler *et al.* 1986, Haisch 1986). There are three main arguments that have been presented in support of this claim: a) there is a good correlation (and possibly equality) between the X-ray luminosity of M dwarf flare stars and the time-integrated energy released in U-band stellar flares (Skumanich 1985, Butler *et al.* 1985, Whitehouse 1985); b) *EXOSAT* and simultaneous Hγ observations of some flare stars show a virtually continuous succession of short-lived impulsive events at both wavelengths (Butler *et al.* 1986); c) X-ray observations of stellar coronae show that rapid, low-level variability is ubiquitous among late-type stars (Ambruster *et al.* 1987).

While the three arguments above, taken together, support the idea of coronae powered by flare-like events, none of them by itself proves that this is actually the case. First, the existence of a correlation between X-ray luminosity and flare energy release rate may simply indicate a common driving agent (*i.e.* magnetic fields) rather than a causal relationship between the two. Second, arguments b) and c) are not really independent since the sample of *Einstein* data analyzed by Ambruster *et al.* is made almost exclusively by flare stars. Thus, the question becomes whether X-ray data really provide evidence for continuous short-lived microflaring activity. This question has been addressed by Pallavicini, Tagliaferri, and Stella (1990) using the entire *EXOSAT* sample of flare stars (more than 300 hours of data from 22 different sources) and applying several statistical techniques. The conclusion was that there is no evidence in the *EXOSAT* data for continuous low-level "microflaring" activity of the type reported by Butler *et al.* (1986). Variability is present, but occurs on longer time scales (> a few hundred seconds) and involves larger energies. The observed variability is more easily interpreted in terms of the occasional occurrence of "normal" flares and more gradual variations, rather than of continuous microflaring activity (see also similar conclusions based on smaller data samples by Collura, Pasquini, and Schmitt 1988). If stellar microflares exist, they are well below the present sensitivity levels.

6. Mass Losses and Winds in Cool Giants

As we proceed from the main sequence towards the giant branch, the properties of the outer atmospheres of cool stars change dramatically with respect to the solar case. High-temperature transition region and coronal plasmas apparently disappear and massive low-velocity winds take their place. The transition may not be as sharp as originally thought (since some intermediate cases do exist) but, nevertheless, the changes are sufficiently abrupt that we can talk of the existence of *dividing lines* in the cool half of the HR diagram. It is still a matter of controversy whether the disappearance of coronae and the appearance of massive winds across these dividing lines are causally related: if not, it a certainly a remarkable coincidence. Understanding the reasons of these atmospheric changes as we move from high to low gravity conditions can provide essential information on coronal heating and wind acceleration for all types of stars, including the Sun. Physical conditions and mass losses in cool giants have been discussed, among others, by MacGregor (1983), Brown (1984), Drake (1986), Dupree (1986), Cassinelli and MacGregor (1986), Dupree and Reimers (1987), Reimers (1989), and Simon and Drake (1989).

6.1. DIVIDING LINES IN THE HR DIAGRAM

The existence of dividing lines separating stars with and without high temperature coronae was first suggested on the basis of *IUE* data. Linsky and Haisch (1979) noticed that high-temperature lines of N V and C IV (which form at $\approx 10^5$ K) were apparently absent in the spectra of giants later than about early K and in supergiants later than about middle G. These stars show only lines formed at temperatures lower than a few times 10^4 K in their spectra (Simon, Linsky, and Stencel 1982; Haisch 1987, Haisch *et al.* 1990). Similarly, a coronal dividing line was found from *Einstein* observations separating stars with and without detectable X-ray emission (Ayres *et al.* 1981, Maggio *et al.* 1990; see also discussion in section 4.1 above). The transition region and coronal dividing lines coincide to within the observational uncertainties.

It was also realized quite early that the above dividing lines were in the general vicinity of other dividing lines separating stars with no detectable winds and stars which showed spectroscopic signatures that could be attributed to the presence of massive winds (Dupree 1986, Reimers 1989). There are several of these lines in the HR diagram as indicated schematically in Figure 13. They are not all coincident, since different sensitivity thresholds make the detections of winds more or less feasible using different spectroscopic diagnostics. Thus, we first have a Mg II dividing line separating stars with and without strong asymmetries in the Mg II line profiles, then a Ca II dividing line which refers to similar asymmetries in the Ca II lines, and finally a circumstellar dividing line separating stars with and without cool circumstellar absorption features. Progressively larger mass loss rates are required to detect winds by means of these various diagnostics.

The Sun has an extremely low mass loss rate ($\approx 10^{-14}$ M$_\odot$ year^{-1}) associated with its high velocity wind ($V_\infty \approx 400$-800 km sec^{-1}). This mass loss rate is too small to be detectable in other stars through spectroscopic signatures. Solar-type winds are probably present in all late-type dwarfs, but their existence can only be inferred indirectly from the progressive decrease of a star's rotation during main sequence evolution (*cf.* section 4.2 above). On

338

the contrary, the very detection of spectroscopic signatures of winds in late-type giants and supergiants means that these winds must be much stronger than the solar one. Mass loss rates as high as $\approx 10^{-6}$ $M_\odot$ year^{-1} have been derived for the very late M supergiants and values as high as 10^{-8} $M_\odot$ year^{-1} are common in all giants to the right of the dividing line. This is one million time larger than the mass loss rate observed from the Sun.

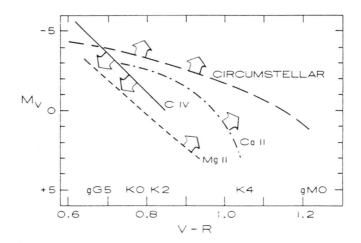

Figure 13. Dividing lines in the HR diagram between stars with and without massive winds. The various DL's have been determined from asymmetries in the Mg II and Ca II lines, and from blue-shifted absorption features indicative of circumstellar envelopes. The position of the Transition Region DL (as determined from IUE C IV spectra) is also indicated (from Dupree 1986).

In many giants, the presence of an outflowing massive wind can be inferred from the difference of the violet and red peaks of the Mg II and Ca II emission components. In main-sequence stars, the violet peak is comparable to or stronger than the red peak, while the ratio violet/red becomes less (and often much less) than 1 in giants and supergiants (Stencel and Mullan 1980). While violet/red ratios ≥ 1 are interpreted as due to downflows similar to those observed in the solar chromosphere (Linsky *et al.* 1979a), the strong depression of the violet peak in giants is attributed to the effect of an outflowing wind. The derived velocities are much less than the solar wind speed and of the order of a few tens km sec^{-1}. In other words, we have cool low-velocity winds as opposite to hot high-speed winds as in the Sun. The presence of these winds is also indicated by blue-shifted circumstellar absorption features that appear in the profiles of many optical and UV lines (Reimers 1977).

The winds from these giants cannot be thermally driven. In fact, for a Parker-type thermally driven wind (Parker 1963), the velocity at the base of the wind is given by

$$V_o \sim c_s \ r_s^2 \ exp(3/2 - 2r_s) \tag{2}$$

where $c_s = (\kappa T/\mu)^{1/2}$ is the sound speed and $r_s = GM_*/2c_s^2 R_*$ is the sonic point in units of the stellar radius R_*. By using solar parameters, $r_s = 4.7$, $V_o = 1.2$ km sec^{-1} and the mass loss rate $\dot{M} = 4\pi R_\odot^2 \mu N_o V_o \approx 1.2 \times 10^{-14}$ M$_\odot$ year^{-1} for base densities $N_o = 10^8$ cm^{-3} and mean mass per particle $\mu = 0.61 m_H$. By applying the same argument to a giant with $M = 16 M_\odot$, $R_* = 400 R_\odot$, and $T \approx 10^4$ K, one gets $V_o = 8 \times 10^{-13}$ km sec^{-1} and a vanishing small mass loss rate for all plausible base densities. As crude as this estimate may be, it is clear that the cool winds of late-type giants cannot be thermally driven. They also cannot be driven by radiation pressure (as the winds of early-type stars), since the lines and continua which are more relevant for absorption of photospheric radiation are in the visible and ultraviolet part of the spectrum, while the photospheric radiation of these cool stars peaks in the red and near infrared. Neither can they be accelerated efficiently by radiation pressure on grains, since this would occur at very large distances from the star (see discussion in MacGregor 1983 and Cassinelli and MacGregor 1986), while the asymmetries in the Mg II and Ca II lines indicate that the acceleration occurs in the chromosphere close to the star. Obviously, some other mechanism must be invoked to explain the cool winds of late-type giants.

Before discussing this problem, however, it is important to stress that the transition from solar-type conditions to giants with no coronae and massive low-velocity winds do not occur abruptly, since there are also intermediate situations. Hartmann, Dupree, and Raymond (1980, 1981) called attention to the existence of a class of stars which show C IV and N V transition region plasma, yet they have warm winds with mass loss rates of $\approx 10^{-8}$ M$_\odot$ year^{-1} and wind speed of the order of 100 km sec^{-1}. These have been called *hybrid stars*, of which two well-known examples are the G-type supergiants α Aqr and β Aqr. At least one hybrid star (α TrA, spectral type K2 IIb-IIIa) has been detected in X-rays by both *EXOSAT* (Brown et al. 1991) and *ROSAT* (Haisch et al. 1991), but others are apparently devoid of appreciable coronal emission. Clearly, hybrid stars are the best targets to investigate the disappearance of coronae across the dividing line.

A schematic summary of the present observational knowledge of the outer atmospheres and winds of stars in the cool half of the HR diagram is given in Figure 14.

6.2. ONSET OF WINDS AND DISAPPEARANCE OF CORONAE

The fundamental question we would like to answer is whether the onset of massive winds and the disappearance of coronae are causally related. At first sight, one could think of an analogy with solar coronal holes, where the energy that accelerates the wind is taken at the expense of the energy that heats the plasma in closed magnetic regions (Withbroe and Noyes 1977). However, this is a rather misleading analogy. The plasma in solar coronal holes is not much cooler than the coronal plasma in closed regions, while the amount of hot plasma per surface area in late-type giants may be up to several orders of magnitude less than in solar coronal holes (*cf.* section 4.1 above). Moreover, the wind speed in giants is much less than in the Sun, and the mass loss rates are many orders of magnitude larger. There is very little similarity between the conditions in solar coronal holes and those in cool giants and supergiants.

An attractive model to explain the nearly simultaneous disappearance of coronae and the onset of massive low-velocity winds was proposed by Hartmann and MacGregor (1980). They suggested that the winds could be accelerated by Alfvén waves, and explored the effects of changing gravity on the atmospheric structure. The key point is that the waves must deposit energy and momentum sufficiently close to the star to keep the terminal velocity low while, at the same time, lifting the gas out of the gravitational potential well. They found that mass loss rates of $\approx 5 \times 10^{-7}$ $M_\odot$ year^{-1} and terminal wind velocities of $V_\infty \approx 50$ km sec^{-1} could be obtained in low gravity stars provided the waves dissipate over a length scale comparable to the stellar radius. The plasma temperature remains low ($\approx 10^4$ K) up to distances of $\approx 2R_*$. When gravity increases, as in dwarf stars, the density at any given location also increases and the wave dissipation produces heating of the plasma, rather than wind acceleration. The problem is that the wave damping length must have a specific value ($\lambda_D \approx R_*$) and, more important, this condition needs to be verified remarkably precisely in order to have solutions with large mass loss rates and low terminal velocities (Holzer, Fla, and Leer 1983). It is unlikely that such stringent conditions can occur in reality.

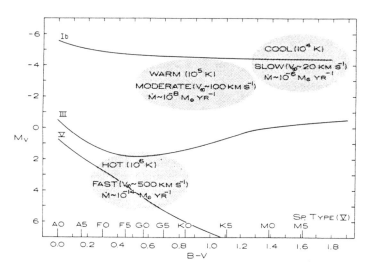

Figure 14. Mass losses and wind temperatures and velocities in different regions of the HR diagram (from Dupree 1986).

A completely different explanation was proposed by Antiochos, Haisch, and Stern (1986) in terms of magnetically confined loop structures. As shown by Antiochos and Noci (1986), a static loop model with a height H larger than the pressure scale-height H_g at the peak of the radiative loss function (*i.e.* at $T \approx 10^5$ K) allows only a hot ($T > 10^6$ K) solution. On the contrary, if $H < H_g$, two solutions exist, a hot one (with $T > 10^6$ K) and a cool one (with $T < 10^5$ K), but only the latter one is stable. In a star like the Sun, where magnetic loops can have a large range of heights, there will be a mixture of hot and cool loops (as observed). In a low gravity star, on the contrary, the pressure scale height H_g is much larger and only cool solutions may exist. In this picture, the disappearance of coronae and the onset of winds is regarded as a pure coincidence, and the wind acceleration remains to be explained.

Recently, Rosner et al. (1991) have made another attempt to explain in a unified way the change of atmospheric properties across the dividing line. They postulate that the dividing line marks a fundamental change in the topology of coronal magnetic fields, from prevalently closed structures (to the left of the dividing line) to prevalently open field lines (to the right of the dividing line). The reasons for this change in magnetic topology are not explained, but they could be caused by less efficient dynamo action and weaker magnetic fields among cool giants and supergiants. Rosner et al. argue that the open field topology is the cause of the absence of coronae among giants since any X-ray emitting coronal material needs to be magnetically confined in those stars in which the escape temperature is far less than typical coronal temperatures. In a G5 giant, the escape temperature is $\approx 3 \times 10^5$ K, while it is $\approx 10^7$ K for the Sun. Thus, while a several million degree corona not confined by magnetic fields could, in principle, exist for the Sun, this is not possible for giants. Any such material could expand freely at the sound speed. The other essential ingredient in the Rosner et al. model is the reflection of Alfvén waves that occurs when the wavelength of the waves becomes comparable to or larger than the local Alfvén speed scale height (defined as $h = V_A/(dV_A/dr)$ where $V_A = B/\sqrt{4\pi\rho}$ is the Alfvén speed). By example, they show that the reflection of Alfvén waves is much larger to the right of the dividing line than to the left (because of the lower temperature in the outer atmosphere of the former stars), thus leading to efficient wind acceleration in an open magnetic field topology. Although promising, the model needs to be worked out more quantitatively before one can claim that the long standing problem of coronae and winds in late-type giants has been solved.

7. Conclusion

In this paper I have summarized many new stellar data acquired from space over the past decade, together with some optical data obtained previously from the ground. From what I have presented, it emerges clearly that phenomena similar to the solar chromosphere, corona, and wind are characteristic of virtually all types of cool stars. The X-ray, UV, and optical emissions from these outer layers are often much larger in stars than in the Sun. Rotation appears to be a key parameter in determining the level of non-radiative heating in outer stellar atmospheres, thus suggesting that dynamo-generated magnetic fields may play a dominant role. However, the mechanisms of plasma heating and wind acceleration in different types of stars remain as elusive as for the Sun. It is still largely unclear how these mechanisms are related to convective zone properties, rotation, magnetic field, and age.

I have also shown that certain classes of late type stars, particularly the late-type giants and supergiants, present phenomena that, at first sight, appear completely different from those observed on the Sun. We do not yet understand the acceleration of massive winds and the absence of high temperature plasma in these stars. However, from the work that has been done so far, it appears that the physical processes responsible for these phenomena may not be too different from those operating on the Sun. Only the conditions (gravity and magnetic field strength or topology) under which the same processes occur may be different in the two cases. Understanding these processes in stars may lead to a better understanding of the physics of the Sun.

Acknowledgements. I thank Drs. J. Linsky and M. Machado for critical reading of the manuscript and useful comments.

References

Agrawal, P.C., Rao, A.R., and Sreekantan, B.V.: 1986, *Montly Not. Roy. Astron. Soc.* **219**, 225.

Ambruster, C.W., Sciortino, S., and Golub, L.: 1987, *Astrophys. J. Suppl.* **65**, 273.

Antiochos, S.K., Haisch, B.M., and Stern, R.A.: 1986, *Astrophys. J. Letters* **307**, L55.

Antiochos, S.K., and Noci, G.: 1986, *Astrophys. J.* **301**, 440.

Ayres, T.: 1981, in *The Universe at Ultraviolet Wavelengths: The First Two Years of IUE*, NASA CP-2171, p. 237.

Ayres, T.R., Fleming, T.A., and Schmitt, J.H.M.M.: 1991, *Astrophys. J. Letters* **376**, L45.

Ayres, T.R., Linsky, J.L., Vaiana, G.S., Golub, L., and Rosner, R.: 1981, *Astrophys. J.* **250**, 293.

Ayres, T.R., Marstad, N.C., and Linsky, J.L.: 1981, *Astrophys. J.* **247**, 545.

Baliunas, S.L., Hartmann, L., and Dupree, A.K.: 1983, *Astrophys. J.* **271**, 672.

Baliunas, S.L., and Vaughan, A.H.: 1985, *Ann. Rev. Astron. Astrophys.* **23**, 379.

Basri. G.S.: 1987, *Astrophys. J.* **316**, 377.

Basri, G.S.: 1990, in *High-resolution Spectroscopy in Astrophysics* (R. Pallavicini ed.), *Memorie Soc. Astron. It.* **61**, 707.

Basri, G.S., and Linsky, J.L.: 1979, *Astrophys. J.* **234**, 1023.

Biermann. L.: 1946, *Naturwiss.* **33**, 118.

Biermann, L.: 1948, *Z. Astrophys.* **25**, 161.

Blanco, C., Bruca, L., Catalano, S., and Marilli, E.: 1982, *Astron. Astrophys.* **115**, 280.

Blanco, C., Catalano, S., Marilli, E., and Rodonò, M.: 1974, *Astron. Astrophys.* **33**, 257.

Bohn, H.U.: 1984, *Astron. Astrophys.* **136**, 338.

Bookbinder, J.A.: 1985, PhD Thesis, Harvard University.

Bopp, B.W.: 1990, in *High-resolution Spectroscopy in Astrophysics* (R. Pallavicini ed.), *Memorie Soc. Astron. It.* **61**, 723.

Brown, A.: 1984, in *Cool Stars, Stellar Systems, and the Sun* (S.L. Baliunas and L. Hartmann eds.), *Lecture Notes in Phys.* **193**, 282.

Brown, A., Drake, S.A., van Steenberg, M.E., and Linsky, J.L.: 1991, *Astrophys. J.* **373**, 614.

Butler, C.J., Rodonò, M., Foing, B.H., and Haisch, B.M.: 1986, *Nature,* **321**, 679.

Caillault, J.-P., and Helfand, D.J.: 1985, *Astrophys. J.* **289**, 279.

Caillault, J.-P., and Zoonematkermani, S.: 1989, *Astrophys. J. Letters* **338**, L57.

Cappelli, A., Cerruti-Sola, M., Cheng, C.-C., and Pallavicini, R.: 1989, *Astron. Astrophys.* **213**, 226.

Cassinelli, J.P.: 1985, in *The Origin of Non-radiative Energy/Momentum in Hot Stars* (A.B. Underhill and A.G. Michalitsianos eds.), NASA CP-2358, p. 2.

Cassinelli, J.P., and MacGregor, K.B.: 1986, in *Physics of the Sun* (P.A. Sturrock ed.), Dordrecht: Reidel, p. 47.

343

Cassinelli, J.P., and Olson, G.L.: 1979, *Astrophys. J.* **229**, 304.

Catalano, S.: 1979, *Astron. Astrophys.* **80**, 317.

Catura, R.C., Acton, L.W., and Johnson, H.M.: 1975, *Astrophys. J. Letters* **196**, L47.

Cerruti-Sola, M., Cheng, C.-C., and Pallavicini, R.: 1991, *Astron. Astrophys..* in press.

Chlebowski, T.: 1989, *Astrophys. J.* **342**, 1091.

Chlebowski, T., Harnden, F.R.Jr., and Sciortino, S.: 1989, *Astrophys. J.* **341**, 427.

Collier Cameron, A., and Robinson, R.D.: 1989, *Mon. Not. Roy. Astron. Soc.* **236**, 57.

Collura, A., Pasquini, L., and Schmitt, J.H.M.M.: 1988, *Astron. Astrophys.*, **205**, 197.

Crivellari, L., Beckman, J.E., Foing, B.H., and Vladilo, G.: 1987, *Astron. Astrophys.* **174**, 127.

Crivellari, L., Franco, M.L., Molaro, P., Vladilo, G., and Beckman, J.E.: 1983, *Astron. Astrophys. Suppl.* **52**, 135.

Danks, A.C., and Lambert, D.L.: 1985, *Astron. Astrophys.* **148**, 293.

Doherty, L.R.: 1985, *Mon. Not. Roy. Astr. Soc.* **217**, 41.

Doyle, J.C., and Butler, C.J.: 1985, *Nature* **313**, 378.

Drake, S.A.: 1986, in *Cool Stars, Stellar Systems, and the Sun* (M. Zeilik and D.M. Gibson eds.), *Lecture Notes in Phys.* **254**, 369

Duncan, D.K.: 1981, *Astrophys. J.* **248**, 651.

Duncan, D.K., and other 17 authors: 1991, *Astrophys. J. Suppl.* **76**, 383.

Dupree, A.K.: 1986, *Ann. Rev. Astron. Astrophys.* **24**, 377.

Dupree, A.K., and Reimers, D.: 1987, in *Exploring the Universe with the IUE Satellite* (Y. Kondo ed.), Dordrecht: Reidel, p. 321.

Eriksson, K., Linsky, J.L., and Simon, T.: 1983, *Astrophys. J.* **272**, 665.

Feigelson, E.D.: 1984, in *Cool Stars, Stellar Systems, and the Sun* (S.L. Baliunas and L. Hartmann eds.), *Lecture Notes in Phys.* **193**, 27.

Feigelson, E. D., Giampapa, M.S., and Vrba, F.J.: 1991, in *The Sun in Time* (eds. C. P. Sonett and M. S. Giampapa), Tucson, AZ: University of Arizona Press, in press.

Feigelson, E.D., and Kriss, G.A.: 1988, *Astrophys. J.* **338**, 262.

Fleming. T.A., Gioia, I., and Maccacaro, T.: 1989, *Astrophys. J.* **340**, 1011.

Foing, B.H., Crivellari, L., Vladilo, G., Rebolo, R., and Beckman, J.E.: 1989, *Astron. Astrophys. Suppl.* **80**, 189.

Gomez, D.O.: 1990, *Fundamentals of Cosmic Phys.*, in press.

Governini, G., Pallavicini, R.: 1986, in *Cool Stars, Stellar Systems, and the Sun* (M. Zeilik and D.M. Gibson eds.), *Lecture Notes in Phys.* **254**, 67.

Haisch, B.M.: 1986, *Irish Astron. J.* **17**, 200.

Haisch, B.M.: 1987, in *Cool Stars, Stellar Systems, and the Sun* (J.L. Linsky and R.E. Stencel eds.), *Lecture Notes in Phys.* **291**, 269.

Haisch, B.M., Bookbinder, J.A., Maggio, A., Vaiana, G.S., and Bennett, J.O.: 1990, *Astrophys. J.* **361**, 570.

Haisch, B.M., Schmitt, J.H.M.M., and Rosso, C.: 1991, *Astrophys. J. Letters* , in press.

Hammer, R.: 1987, in *Solar and Stellar Physics* (E.-H. Schröter and M. Schüssler eds.), *Lecture Notes in Phys.* **292**, 77.

Hartmann, L.: 1990, in *Cool Stars, Stellar Systems, and the Sun* (G. Wallerstein ed.), San Francisco: Astron. Soc. Pacific, p. 289.

Hartmann, L., Baliunas, S.L., Duncan, D.K., and Noyes, R.W.: 1984, *Astrophys. J.* **279**, 778.

344

Hartmann, L., Dupree, A.K., and Raymond, J.C.: 1980, *Astrophys. J. Letters* **236**, L143.

Hartmann, L., Dupree, A.K., and Raymond, J.C.: 1981, *Astrophys. J.* **246**, 193.

Hartmann, L., and MacGregor, K.B.: 1980, *Astrophys. J.* **242**, 260.

Hartmann, L., and Noyes, R.W.: 1987, *Ann. Rev. Astron. Astrophys.* **25**, 271.

Hartmann, L., Soderblom, D.L., Noyes, R.W., and Burnham, J.N.: 1984, *Astrophys. J.* **276**, 254.

Herbig, G.H.: 1985, *Astrophys. J.* **289**, 269.

Holzer, T.E., Fla, T., and Leer, E.: 1983, *Astrophys. J.* **275**, 808.

Kelch, W.L., Linsky, J.L., Basri, G.S., Chin, H.Y., Chang, S.H., Maran, S.P., and Furenlind, I.: 1978, *Astrophys. J.* **220**, 962.

Kelch, R.P., Linsky, J.L., and Worden, S.P.: 1979, *Astrophys. J.* **229**, 700.

Kraft, R.P.: 1967, *Astrophys. J.* **150**, 551.

Kuhi, L.V., and Cram, L.E.: 1989, in *F, G, K Stars and T Tauri Stars* (L. E. Cram and L. V. Kuhi eds.), NASA SP-502, p. 99.

Kuperus, M., Ionson, J.A., and Spicer, D.S.: 1981, *Ann. Rev. Astron. Astrophys.* **19**, 7.

Jordan, C.: 1986, in *New Insights in Astrophysics - 8 years of UV Astronomy with IUE*, ESA SP-263, p. 17.

Jordan, C., Ayres, T.R., Brown, A., Linsky, J.L., and Simon, T.: 1987, *Mon. Not. Roy. Astron. Soc.* **225**, 903.

Jordan, C., and Brown, A.: 1981, in *Solar Phenomena in Stars and Stellar Systems* (R.M. Bonnet and A.K. Dupree eds.), Dordrecht: Reidel, p. 199.

Jordan, C., and Linsky, J.L.: 1987, in *Exploring the Universe with the IUE Satellite* (Y. Kondo ed.), Dordrecht: Reidel, p. 259.

LaBonte, B.J.: 1986a, *Astrophys. J. Suppl.* **62**, 229.

LaBonte, B.J.: 1986b, *Astrophys. J. Suppl.* **62**, 241

Lemen, J.R., Mewe, R., Schrijver, C.J., and Fludra, A.: 1989, *Astrophys. J.* **341**, 474.

Lighthill, M.J.: 1952, *Proc. Roy. Soc. London* **A211**, 564.

Linsky, J.L.: 1984, in *Cool Stars, Stellar Systems, and the Sun* (S.L. Baliunas and L. Hartmann eds.), *Lecture Notes in Phys.* **193**, 244.

Linsky, J.L.: 1980, *Ann. Rev. Astron. Astrophys.* **18**, 439.

Linsky, J.L.: 1985, *Solar Phys.* **100**, 333.

Linsky, J.L.: 1990, in *Imaging X-ray Astronomy: A Decade of Einstein Observatory Achievements* (M. Elvis ed.), Cambridge: Cambridge Univ. Press, p. 39.

Linsky, J.L., and Ayres, T.R.: 1978, *Astrophys. J.*, **220**, 619.

Linsky, J.L., Bornmann, P.L., Carpenter, K.G., Wing, R.F., Giampapa, M.S., Worden, S.P., and Hege, E.K.: 1982, *Astrophys. J.* **260**, 670.

Linsky, J.L., and Haisch, B.M.: 1979, *Astrophys. J. Letters* **229**, L27.

Linsky, J.L., Hunten, D.M., Sowell, R., Glackin, D.L., and Kelch, W.L.: 1979a, *Astrophys. J. Suppl.* **41**, 481.

Linsky, J.L., Worden, S.P., McClintock, W., and Robertson, R.M.: 1979b, *Astrophys. J. Suppl.* **47**, 47.

Lucy, L.B.: 1982, *Astrophys. J.* **255**, 286.

MacGregor, K.B.: 1983, in *Solar Wind Five* (M. Neugebauer ed.), NASA CP-2280, p. 241

Maggio, A., Sciortino, S., Vaiana, G.S., Majer, G.S., Bookbinder, J., Golub, L., Harnden, F.R.Jr., Rosner, R.: 1987, *Astrophys. J.* **315**, 687.

Maggio, A., Vaiana, G.S., Haisch, B.M., Stern, R.A., Bookbinder, J., Harnden, F.R.Jr., and Rosner, R.: 1990, *Astrophys. J.* **348**, 253.

Majer, P., Schmitt, J.H.M.M., Golub, L., Harnden, F.R.Jr., and Rosner, R.: 1986, *Astrophys. J.* **300**, 360.

Marilli, E., and Catalano, S.: 1984, *Astron. Astrophys.* **133**, 57.

Mewe, R.: 1979, *Space Science Rev.* **24**, 101.

Micela, G., Sciortino, S., Serio, S., Vaiana, G.S., Bookbinder, J., Golub, L., Harnden, F.R.Jr., and Rosner, R.: 1985, *Astrophys. J.* **292**, 172.

Micela, G.,Sciortino,S., Vaiana, G.S., Schmitt, J.H.M.M., Stern, R.A., Harnden, F.R. Jr., and Rosner, R.: 1988, *Astrophys. J.* **325**, 798.

Micela, G., Sciortino, S., Vaiana, G.S., Harnden, F.R. Jr., Rosner, R., and Schmitt, J.H. M.M.: 1990, *Astrophys. J.* **348**, 557.

Montmerle, T., Feigelson, E.D., Bouvier, J., and André, P.: 1991, in *Protostars and Planets III*, eds. E. H. Levy and J. I. Lunine, Tucson, AZ: University of Arizona Press, in press.

Montmerle, T., Koch-Miramond, L., Falgarone, E., and Grindlay, J.: 1983, *Astrophys. J.* **269**, 182.

Mullan, D.J.: 1986, in *The M-type Stars* (H.R. Johnson and F.R. Querci eds.), NASA SP-492, p. 455.

Musielak, Z.E.: 1991, in *Mechanisms of Chromospheric and Coronal Heating* (P. Ulmschneider, E.R. Priest, and R. Rosner eds.), Heidelberg: Springer-Verlag, p. 369.

Narain, U., and Ulmschneider, P.: 1990, *Space Science Rev.* **54**, 377.

Nations, H.L., and Ramsey, L.W.: 1980, *Astron. J.* **85**, 1086.

Noyes, R.W.: 1986, in *Physics of the Sun* (P.A. Sturrock ed.), Dordrecht: Reidel, p. 125.

Noyes, R.W., Hartmann, L.W., Baliunas, S.L., Duncan, D.K., Vaughan, A.H.: 1984, *Astrophys. J.*, **279**, 763.

O'Brien, G.T., and Lambert, D.L.: 1986, *Astrophys. J. Suppl.* **62**, 899.

Oranje, B.J.: 1986, *Astron. Astrophys.* **154**, 185.

Pallavicini, R.: 1988, in *Solar and Stellar Coronal Structure and Dynamics* (R.C. Altrock ed.), Sunspot, NM: Nat. Solar Obs., p. 19.

Pallavicini, R.: 1989, *Astron. Astrophys. Rev.* **1**, 177.

Pallavicini, R., Golub, L., Rosner, R., Vaiana, G.S., Ayres, T., and Linsky, J.L.: 1981, *Astrophys. J.* **248**, 279.

Pallavicini, R., Tagliaferri, G., Pollock, A.M.T., Schmitt, J.H.M.M., and Rosso, C.: 1990, *Astron. Astrophys.* **227**, 483.

Pallavicini, R., Tagliaferri, G., and Stella, L.: 1990, *Astron. Astrophys.* **228**, 403.

Parker, E.N.: 1963, *Interplanetary Dynamical Processes*, New-York: Interscience.

Parker, E.N.: 1983, *Astrophys. J.* **264**, 642.

Parker, E.N.: 1988, in *Solar and Stellar Coronal Structure and Dynamics* (R.C. Altrock ed.), Sunspot, NM: Nat. Solar Obs., p. 2.

Pasquini, L., Brocato, E., and Pallavicini, R.: 1990, *Astron. Astrophys.* **234**, 277.

Pasquini, L.. and Pallavicini, R.: 1990, in *High-resolution Spectroscopy in Astrophysics* (R. Pallavicini ed.), *Memorie Soc. Astron. It.* **61**, 737.

Pasquini, L., and Pallavicini, R.: 1991, *Astron. Astrophys.* **251**, 199.

Pasquini, L., Pallavicini, R., and Dravins, D.: 1989, *Astron. Astrophys.* **213**, 261.

Pasquini, L., Pallavicini, R., and Pakull, M.: 1988, *Astron. Astrophys.* **191**, 253.

Proudman, I.: 1952, *Proc. Roy. Soc. London*, **A214**, 119.

Rebolo, R.., Garcia Lopez, R.., Beckman, J.E., Vladilo, G., Foing. B., and Crivellari, L.: 1989, *Astron. Astrophys. Suppl.* **80**, 135.

Reimers, D.: 1977, *Astron. Astrophys.* **57**, 395.

Reimers, D.: 1989, in *F, G, K Stars and T Tauri Stars* (L.E. Cram and L.V. Kuhi eds.), NASA SP-502, p. 53.

Renzini, A., Cacciari, C., Ulmschneider, P., and Schmitz, F.: 1977, *Astron. Astrophys.* **61**, 39.

Robinson, R.D., Cram, L.E., and Giampapa, M.S.: 1990, *Astrophys. J. Suppl.* **74**, 891.

Rodonò, M. : 1986, in *The M-type Stars* (H.R. Johnson and F.R. Querci eds.), NASA SP-492, p. 409.

Rosner, R..: 1991, in *Mechanisms of Chromospheric and Coronal Heating* (P. Ulmschneider, E.R. Priest, and R. Rosner eds.), Heidelberg: Springer-Verlag, p. 287.

Rosner, R., An, C.-H., Musielak, Z.E., Moore, R.L., and Suess, S.T.: 1991, *Astrophys. J. Letters* **372**, L91.

Rosner, R.., Golub, L., and Vaiana, G.S.: 1985, *Ann. Rev. Astron. Astrophys.* **23**, 413.

Rutten, R.G.M.: 1984, *Astron. Astrophys.* **130**, 353.

Rutten, C.J., Schrijver, C.J., Lemmens, A.F.P., and Zwaan, C.: 1991, *Astron. Astrophys.* **252**, 203.

Schmitt, J.H.M.M.: 1988, in *Hot Thin Plasmas in Astrophysics* (R. Pallavicini ed.), Dordrecht: Kluwer, p. 109.

Schmitt, J.H.M.M., Collura, A., Sciortino, S., Vaiana, G.S., Harnden, F.R..Jr., and Rosner, R..: 1990a, *Astrophys. J.* **365**, 704.

Schmitt, J.H.M., Golub, L., Harnden, F.R..Jr., Maxson, C.W., Rosner, R., and Vaiana, G.S.: 1985, *Astrophys. J.* **290**, 307.

Schmitt, J.H.M.M., Micela, G., Sciortino, S., Vaiana, G.S., Harnden, F.R..Jr., and Rosner, R..: 1990b, *Astrophys. J.* **351**, 492.

Schrijver, C.J.: 1987a, *Astron. Astrophys.* **172**, 111.

Schrijver, C.J.: 1987b, in *Cool Stars, Stellar Systems, and the Sun* (J.L. Linsky and R.E. Stencel eds.), *Lecture Notes in Phys.* **291**, 135.

Schwarzschild, M.: 1948, *Astrophys. J.* **107**, 1.

Sciortino, S., Vaiana, G.S., Harnden, F.R..Jr., Ramella, M., Morossi, C., Rosner, R., and Schmitt, J.H.M.M.: 1990, *Astrophys. J.* **361**, 621.

Shine, R.A., and Linsky, J.L.: 1972, *Solar Phys.* **25**, 357.

Skumanich, A.: 1972, *Astrophys. J.* **171**, 565.

Skumanich, A.: 1985, *Austr. J. Phys.* **38**, 971.

Simon, T., and Drake, S.A.: 1989, *Astrophys. J.* **346**, 303.

Simon, T., and Fekel, F.C.Jr.: 1987, *Astrophys. J.* **316**, 434.

Simon, T., Linsky, J.L., and Stencel, R.: 1982, *Astrophys. J.* **257**, 225.

Soderblom. D.R.: 1983, *Astrophys. J. Suppl.* **53**, 1.

Stauffer, J.R., and Hartmann, L.W.: 1986, *Astrophys. J. Suppl.* **61**, 531.

Stencel, R.E., and Mullan, D.J.: 1980, *Astrophys. J.* **238**, 221.

Stencel, R.E., Mullan, D.J., Linsky, J.L., Basri, G.S., and Worden, S.P.: 1980, *Astrophys. J. Suppl.* **44**, 383.

Stern, R.A.: 1983, *Adv. Space Res.* **2**, No. 9. 39.

Stern, R.A., Zolcinski, M.-C., Antiochos, S.K., and Underwood, J.H.: 1981, *Astrophys. J.* **249**, 647.

Strassmeier, K.G., Fekel, F.C., Bopp, B.W., Dempsey, R.C., and Henry, G.W.: 1990, *Astrophys. J. Suppl.* **72**, 191.

Ulmschneider, P.: 1979, *Space Science Rev.* **24**, 71.

Ulmschneider, P.: 1990, in *Cool Stars, Stellar Systems, and the Sun* (G. Wallerstein ed.), San Francisco: Astron. Soc. Pacific, p. 3.

Ulmschneider, P.: 1991, in *Mechanisms of Chromospheric and Coronal Heating* (P. Ulmschneider, E.R. Priest, and R. Rosner eds.), Heidelberg: Springer-Verlag, p. 328.

Ulmschneider, P., Priest, E.R., and Rosner, R. (eds.): 1991, *Mechanisms of Chromospheric and Coronal Heating*, Heidelberg: Springer-Verlag.

Vaiana, G.S.: 1990, in *Imaging X-ray Astronomy: A Decade of Einstein Observatory Achievements* (M. Elvis ed.), Cambridge: Cambridge Univ. Press, p. 61.

Vaiana, G.S., and other 15 authors: 1981, *Astrophys. J.* **245**, 163.

Vaiana, G.S., and Rosner, R.: 1978, *Ann. Rev. Astron. Astrophys.* **16**, 393.

Vaiana, G.S., and Sciortino, S.: 1987, in *Circumstellar Matter* (I. Appenzeller and C. Jordan eds.), Dordrecht: Reidel, p. 333.

van Ballegooijen, A.A.: 1990, in *Cool Stars, Stellar Systems, and the Sun* (G. Wallerstein ed.), San Francisco: Astron. Soc. Pacific, p. 15.

Vladilo, G., Molaro, P., Crivellari, L., Foing, B.H., Beckman, J.E., and Genova, R.: 1987, *Astron. Astrophys.* **185**, 233.

Vaughan, A.H.: 1983, in *Solar and Stellar Magnetic Fields: Origins and Coronal Effects* (J.O. Stenflo ed.), Dordrecht: Reidel, p. 113.

Vaughan, A.H., and Preston, G.W. 1980: *Publ. Astron. Soc. Pacific* **92**, 385

Vilhu, O.: 1987, in *Cool Stars, Stellar Systems, and the Sun* (J.L. Linsky and R.E. Stencel eds.), *Lecture Notes in Phys.* **291**, 110.

Walter, F. M.: 1982, *Astrophys. J.* **253**, 745.

Walter, F. M.: 1986, *Astrophys. J.* **306**, 573.

Walter, F.M., and Bowyer,C.S.: 1981, *Astrophys. J.* **245**, 671.

Walter, F. M., and Boyd, W. T.: 1991, *Astrophys. J.* **370**, 318.

Walter, F. R., Brown, A., Mathieu, R. D., Myers, P. C., and Vrba, F.: 1988, *Astron. J.* **96**, 297.

Weber, E.J., and Davis, L.: 1967, *Astrophys. J.* **148**, 217.

White, N.E., Culhane, J.L., Parmar, A.N., Kellet, B.J., Kahn, S., van den Oord, G.H.J., and Kuijpers, J.: 1986, *Astrophys. J.* **301**, 262.

White, N.E., Shafer, R.A., Parmar, A.N., Horne, K., and Culhane, J.L.: 1990, *Astrophys. J.* **350**, 776.

White, O.R., and Livingston, W.C.: 1981, *Astrophys. J.* **249**, 798.

Whitehouse, D.R.: 1985, *Astron. Astrophys.* **145**, 449.

Wilson. O.C.: 1963, *Astrophys. J.* **138**, 832.

Wilson. O.C.: 1966, *Astrophys. J.* **144**, 695.

Wilson. O.C.: 1978, *Astrophys. J.* **226**, 379.

Wilson. O.C.: 1982, *Astrophys. J.* **257**, 179.

Wilson, O.C., and Bappu, V.: 1957, *Astrophys. J* **125**, 661.

Wilson, O.C., and Skumanich, A.: 1964, *Astrophys. J.* **140**, 1401.

Withbroe, G.L., and Noyes, R.V.: 1977, *Ann. Rev. Atron. Astrophys.* **15**, 363.

Wolff, S.C., Boesgaard, A.M., and Simon, T.: 1986, *Astrophys. J.* **310**, 360.

Zarro, D.M., and Rodgers, A.W.: 1983, *Astrophys. J. Suppl.* **53**, 815.

Zirin, H.: 1976, *Astrophys. J.* **208**, 414.

Zwaan, C.: 1991, in *Mechanisms of Chromospheric and Coronal Heating* (P. Ulmschneider, E.R. Priest, and R. Rosner eds.), Heidelberg: Springer-Verlag, p. 241.

16 EXACT 2-D MHD SOLUTIONS FOR ASTROPHYSICAL OUTFLOWS

K. TSINGANOS[1], E. TRUSSONI[2], C. SAUTY[3]

[1] Dep. of Physics, Un. of Crete, GR-71409, Heraklion, Crete, GREECE
[2] Osservatorio Astronomico di Torino, I-10025, Pino Torinese, ITALY
[3] Observatoire de Paris, DAEC, F-92195 Meudon Cedex, FRANCE

ABSTRACT. The systematic method outlined in Chapters 6 and 7 for obtaining and studying analytical solutions of the full MHD equations is illustrated here with a class of dynamical plasma equilibria, corresponding to the problem of steady outflows along the open magnetic fieldlines of nonpolytropic stellar atmospheres in the central gravitational field of stellar objects. Those solutions extend the familiar Parker model for a solar/stellar wind by including (i) a nonspherically symmetric magnetic field and outflow speed, (ii) nonspherically symmetric pressure and density distributions, (iii) a selfconsistent calculation of the heating required to drive the flow, and (iv) several novel hydromagnetic critical points in the flow which select a unique wind-type solution. In particular, the role of the poloidal and azimuthal magnetic fields and the Poynting energy flux to drive the outflow as in fast magnetic rotators is discussed. The rich topologies of these exact solutions present an interesting example of the mathematical complexity encountered in nonlinear physical systems. Such studies may be useful in efforts to understand the initial acceleration of axisymmetric hydromagnetic stellar winds and jets and should be taken into account in numerical modelling of those systems.

1. Introduction

Astrophysical winds have been studied to a considerable extent starting with the classical work of Parker (1958, 1963) which described a spherically symmetric outflow by using the basic equations of hydrodynamics. Since then, wind-type outflows have been investigated in connection with their central role in many astrophysical phenomena, such as fast solar wind streams and their association with coronal holes (Pneuman and Kopp 1971, Holzer 1977, Munro and Jackson 1977, Leer, Holzer and Flå 1982, Withbroe 1988, Wang and Sheeley 1990 and references therein), stellar winds and their association with stellar angular momentum loss (Temesváry 1952, Schatzman 1962, Weber and Davis 1967, Mestel 1968, Yeh 1976, Tassoul 1978, Belcher and MacGregor 1976, Sakurai 1985, 1987, 1990, Heyvaerts and Norman 1989), relativistic beams and their association with pulsar magnetospheres (Michel 1969, Goldreich and Julian 1970, Okamoto 1974, 1975, Phinney 1983, Camenzind 1989, and references therein), bipolar flows and their association with protostellar objects (Mouschovias 1977, 1981, Pudritz and Norman 1983, 1986, Mestel 1986, Mundt 1986, Shibata and Uchida 1986, 1987, Silvestro et al. 1987, Mundt et al. 1990 and references therein), cosmic jets and their association with compact galactic and extragalactic objects (Blandford and Rees, 1974, Ferrari et al. 1985, 1986, Lovelace et al. 1991, Owen et al.

J. T. Schmelz and J. C. Brown (ed.), The Sun, A Laboratory for Astrophysics, 349–376.
© 1992 Kluwer Academic Publishers. Printed in the Netherlands.

1989, Koupelis 1990, Hughes 1991 and references therein) and optically thick winds and their association to novae outbursts (Kato 1983, 1988, 1991, Orio *et al.* 1992, Cassatella and Viotti 1988 and references therein). Needless to say that the analogous accretion-type flows in various astrophysical objects have also been studied along the same lines.

It may be helpful to note in passing that the frequently quoted formulation of Weber and Davis (1967) that solved the one-dimensional problem on the equatorial plane, albeit a pioneering study for its time, nevertheless failed to take into account inertial and magnetic forces acting perpendicular to the equatorial plane and which are not unimportant, as will be made clear by this study. The studies of Sakurai (1985, 1990) proceeded a step further toward a more complete treatment by solving the two-dimensional problem in all space around the star. These solutions provided critical physical insight into the complexity of the physical problem, in particular the role played by critical points. However, they were obtained fully numerically, with the unavoidable result that a complete physical understanding of the global properties of the flow, as well as its parameter dependence, is not easily accessible. Sakurai's approach has been followed analytically by Heyvaerts and Norman (1989) who have shown that collimated outflows are the only physically acceptable solutions of the magnetohydrodynamic (MHD) equations that pass smoothly through the critical points. Finally, Lovelace *et al.* (1991) and Koupelis (1990), have been averaging across the direction of propagation of the outflow, thus allowing for a one-dimensional treatment valid close to the rotation axis.

All aforementioned studies were restricted to solutions corresponding to a polytropic equation of state with a *constant* index γ. The only reason for such an approach is the fact that the heating and cooling processes operating in the plasma under astrophysical conditions are usually poorly known. Therefore, the simplest way to close mathematically the system of hydromagnetic equations is by means of a polytropic relationship between the gas pressure and density. A consequence of this hypothesis is that in all previous studies the steady MHD solutions are always characterized by the presence of three critical points, namely the slow, fast and Alfvén critical points. In this way, some important insight into the solutions of the complete set of the hydromagnetic equations and their properties is gained, albeit by sacrificing equally interesting insight and information on the energetics of the outflows. Thus, it is well known that a polytropic equation of state with *any* constant γ cannot yield correct values of the physical parameters of the solar wind, both close to the Sun and at 1 A.U. (Parker 1963). For example, this inadequacy of the polytrope law as a means of describing the energy equation can be seen by the fact that when someone tries to reproduce the observed conditions at 1 A.U., he obtains unrealistically high velocities and unrealistically low densities near the solar surface (Weber and Davis 1967, Belcher and MacGregor 1976). Another related constraint in most of the previous studies is that the MHD equations are essentially solved along a streamline of the flow by using the generalized Bernoulli integral that exists under the polytropic assumption (Tsinganos 1982, Heyvaerts and Norman 1989). It would be interesting and complementary, then, to investigate analytically also other families of fully two-dimensional axisymmetric solutions. In such a way, by having the physical quantities of a global solution in closed forms, their behavior could be easily analysed.

In the approach that we review in the following, by means of an *a priori* physically sound choice of the angular dependence of the field and streamline pattern, the MHD equations are explicitly solved without imposing any polytropic relation. Then, through the energy equation, the form of the heating/cooling distribution along the flow is calculated *a posteriori* and *selfconsistently* with the assumed angular distribution of the field and flow structure. In other words, instead of replacing a solution of the energy equation by a single parameter, γ, we replace it by a function, $h(R, \theta)$ [*cf.* Equation (1d) below]. When compared with the approach of the polytropic equation of state, this approach has the advantage that by examining the energy function $h(R, \theta)$, we may decide if the assumed angular dependence of the physical quantities and the parameters used correspond to a physically acceptable solution (Tsinganos and Trussoni 1992). The analytical nature of the study will also enable us to have a detailed display of the various complicated topologies of the MHD solutions for magnetized winds, an experience useful in more sophisticated numerical studies. The interested reader may find a more detailed presentation in Low and Tsinganos (1986), Tsinganos and Low (1989), Hu and Low (1989), Tsinganos and Trussoni (1990, 1991) and Tsinganos and Sauty (1992a,b).

The structure of this Chapter is as follows. In the section 2, we outline the step by step construction of the solution by appropriately choosing the free integrals $\Psi(A)$, $\Omega(A)$, and $L(A)$ that were encountered in the formalism of Chapters 6 and 7. In section 3, the special case of a rotating radial flow embedded in a magnetic field with radial and azimuthal components is studied separately, with particular emphasis on two critical points, a star-type Alfvén critical point and an MHD X-type critical point, that are essential in choosing a unique wind-type solution. Then, in section 4, the nature of the acceleration in rotating magnetized outflows is analyzed with particular emphasis on the conversion of Poynting energy flux into kinetic energy of the flow. Finally, in section 5, the case of a nonrotating general poloidal flow with nonradial streamlines is examined with particular emphasis on the deduced shape of the streamlines and the associated multiple saddle and nodal critical points. The energetics of the flow and the parametric dependence of the topologies of the various solutions are briefly discussed also in these sections.

2. Specification of the MHD Integrals for Axisymmetric Flows

In the following, we write the basic MHD equations that govern axisymmetric outflows from the atmosphere of a central gravitating object and appropriately choose the free integrals $\Psi(A)$, $L(A)$, $\Omega(A)$ such that we obtain a physically interesting hydromagnetic field $(\vec{V}, \vec{B})$.

2.1. BASIC MHD EQUATIONS FOR AXISYMMETRIC FLOWS

As was explained in Chapters 6 and 7, our starting point is the following set of the MHD equations for the steady dynamical interaction of an inviscid but compressible fluid flow of high electrical conductivity with a magnetic field $\vec{B}$ in the spherically symmetric gravitational field surrounding a central gravitating object,

$$\vec{\nabla} \cdot \vec{B} = 0, \qquad \vec{\nabla} \cdot (\rho \vec{V}) = 0, \tag{1a}$$

$$\vec{\nabla} \times (\vec{V} \times \vec{B}) = 0, \tag{1b}$$

$$\rho(\vec{V} \cdot \vec{\nabla})\vec{V} = -\vec{\nabla}P + \frac{1}{4\pi}(\vec{\nabla} \times \vec{B}) \times \vec{B} - \frac{\rho G M (1 - \ell)}{r^2}\vec{e}_r, \tag{1c}$$

for the bulk flow speed $\vec{V}(r, \theta)$, the magnetic field $\vec{B}(r, \theta)$, the density $\rho(r, \theta)$, and the pressure $P(r, \theta)$ in spherical coordinates (r, θ, ϕ), with M the mass of the central body, G the gravitational constant and $\ell = L_r/L_{Edd}$, the constant ratio of the diffusive luminosity L_r and Eddington luminosity L_{Edd} (Ruggles and Bath 1979).

Equations (1) may be closed with a polytropic relationship between pressure and density, or with an energy equation. Since specification *a priori* of a constant polytropic index γ may not correspond to simultaneously correct conditions at the base of the outflow and at large distances from the central object (Weber and Davis 1967), as well as to artificial heating/cooling distributions along the flow, we shall follow the approach of allowing for a variable index γ and calculate the net heating rate required to support the deduced flow pattern. This heating rate can be deduced from energy conservation according to the first law of thermodynamics,

$$h = \frac{1}{\Gamma - 1}\left\{\vec{V} \cdot \vec{\nabla}P - \Gamma\frac{P}{\rho}\vec{V} \cdot \vec{\nabla}\rho\right\} \quad , \quad T = \frac{m_p}{2k_B}\frac{P}{\rho}, \tag{1d}$$

where m_p is the proton mass, k_B Boltzmann's constant, $\Gamma = 5/3$ the constant ratio of the specific heats for a proton gas and $h(r, \theta)$ the corresponding rate of total energy deposition in the unit volume of the fluid. We may for example, write for h

$$h = E_H - \rho^2\Lambda(T) + E_C, \tag{1e}$$

where E_H is the pure volumetric heating rate (dissipation of acoustic and MHD waves, viscous forces, etc), $\rho^2\Lambda(T)$ the radiation losses for an optically thin plasma with $\Lambda(T)$ a known function of T (Raymond and Smith 1977), and the last term includes the conductive energy losses. Note that in this approach of nonconstant γ, an effective variable γ can be calculated,

$$\gamma \equiv \frac{\partial \ln P}{\partial \ln \rho}\bigg|_{streamline} \neq const., \tag{1f}$$

and compared to some characteristic values of the constant γ approach, such as $\gamma = 1$ for an isothermal atmosphere, $\gamma = 3/2$ for the known heated Parker polytrope, or $\gamma = 5/3$ for an adiabatic gas.

2.2. ANGULAR DEPENDENCE OF DENSITY AND PRESSURE DISTRIBUTIONS

In the case of our Sun, it has long been established that the distribution of the plasma density in the solar atmosphere is clearly nonspherically symmetric and at least two-dimensional. For example, coronal holes from where the solar wind emanates are observed to be lower density regions than the surrounding streamers (Zirker 1977). In the case of other stars and in the presence of magnetic fields, it is also expected to have a meridionally anisotropic density distribution. For example, a latitudinal decrease of the density has also been inferred to exist in the chromospheres and cool envelopes of Be stars (Ringuelet and Iglesias 1991).

Theoretical arguments have long predicted that disks are a natural part of the stellar formation process and all current models of bipolar outflows require the existence of a pre-main sequence circumstellar disk (Königl 1982, Hartmann and MacGregor 1982). On the other hand, high angular resolution optical, infrared and radio images have accumulated observational evidence for the ubiquitous presence of axially symmetric circumstellar clouds around young stars (Torrelles *et al.* 1983, 1985, Harvey 1986, Cassen *et al.* 1986, Rydgren and Cohen 1986). With the above motivations in mind, it is inevitable to start our study by considering nonspherically symmetric density distributions. Thus, assume the following axisymmetric but meridionally anisotropic functional dependence of the density on R and θ

$$\rho(R,\theta) = \frac{\rho_o}{Y R^2}(1 + \delta f \sin^2\theta),$$ (2)

where ρ_o is the density at the polar base ($R = 1, \theta = 0$). The constant δ controls the meridional density asymmetry, such that for $\delta > 0$, the equatorial parts of the atmosphere are denser than the polar ones, at the same radial distance.

Expressions similar to Equation (2) have been used to model the density increase that we find as we move from the central axis of a polar coronal hole to the denser surrounding coronal streamer (Munro and Jackson 1977), the density distribution of the white light solar corona (Bagenal and Gibson 1991), and the density distribution of the low atmosphere of Be stars (Ringuelet and Inglesias 1991). Also, a similar angular variation of the density is thought to exist in accretion disks surrounding some astrophysical objects (Cassen *et al.* 1986).

In dealing with the r- and θ-components of the force balance equation (1b), we may separate the variables R and θ, if we assume that the pressure $P(R, \theta)$ has the following functional dependence on the two variables R and θ:

$$P(R,\theta) = P_o(R) + P_1(R)\sin^2\theta = \frac{\rho V_o^2}{2}\left[Q_o(R) + Q_1(R)\sin^2\theta\right].$$ (3)

Note that if $P_1(R) > 0$, the pressure has a polar deficit relative to its equatorial value at the same radial distance. This pressure distribution may be similar then to a *de Laval* nozzle type configuration where the pressure is lower along the polar direction of the steepest density gradient, than along the equatorial plane, as it has been assumed in models of bipolar outflows (Königl 1982).

2.3. Angular Dependence of Poloidal Magnetic Field

Magnetic fields around astrophysical objects are at *least* 2-dimensional, *i.e.*, axisymmetric, depending on the radial distance r and the colatitude θ. Then, such an axisymmetric but meridionally asymmetric magnetic field may also be constrained by requiring it to have the following three basic features: *first*, obviously to conserve magnetic flux, Equation (1a), *second*, to be radial along some magnetic axis ($\theta = 0$), and *third*, to have a radial field reversal at the magnetic equator ($\theta = \pi/2$), thus being associated with an electric current sheet there. The first requirement is identically satisfied by expressing the poloidal field

$\vec{B}_p(R,\theta)$, in the more general way, in terms of the scalar magnetic flux function $A(R,\theta)$ (Tsinganos 1982),

$$\vec{B}_p(R,\theta) = \vec{\nabla} \times \left[\frac{A(R,\theta)}{r \sin \theta} \vec{e}_\phi \right], \tag{4a}$$

such that $A(R,\theta) = $ constant defines a meridional magnetic field line. On the other hand, the simplest expression we could design for $A(R,\theta)$ which meets the second and third requirements is $A(r,\theta) = (B_o r_o^2/2) f(R) \sin^2\theta$. Note that the field is dipolar if $f(R) = 1/R$ (Low and Tsinganos 1986) while when $f(R) = R^2$ the field is uniform along the direction $\theta = 0$ (Uchida and Low 1981). The case $f(R) = 1$, that corresponds to conical fieldlines has also been studied in another connection (Tsinganos and Trussoni 1991). Thus, the poloidal magnetic field is,

$$B_r = \frac{1}{r^2 \sin \theta} \frac{\partial A}{\partial \theta} = \frac{B_o f(R)}{R^2} \cos\theta \;, \quad B_\theta(R,\theta) = -\frac{1}{r \sin \theta} \frac{\partial A}{\partial r} = -\frac{B_o}{2R} \frac{\mathrm{d}f}{\mathrm{d}R} \sin\theta. \tag{4b}$$

In this expression, $R = r/r_o$ is the dimensionless radial distance in terms of some base radius r_o corresponding to the stellar base and B_o is the magnetic field at this polar base ($R = 1$, $\theta = 0$).

Note that the general expression of the angular dependence of an arbitrary axially symmetric magnetic field [c.f. Equation (4a)], can be expanded on the orthogonal basis of the Legendre functions $P_l(\cos\theta)$, i.e, in an infinite series similar to the expansion in various multipoles (Jackson 1975). And, in Equation (4b) we have effectively taken the lowest order in such an expansion of the radial field $B_r(\cos\theta)$ on $\cos\theta$. Such an angular dependence of the magnetic field is not in disagreement with the solar poloidal magnetic field, for example, during periods of solar minimum as it is inferred from white light images of the K-corona (Bagenal and Gibson 1991), or, as inferred from interplanetary observations where to a first approximation is represented by a dipole component (Wang and Sheeley 1988).

On the other hand, we should keep in mind that this simple angular dependence of the magnetic field may not obviously be considered suitable for any general magnetic configuration. Magnetic patterns referred to specific applications may require the introduction of additional higher-order multipoles (Nash et al. 1988). For example, a $\cos^8\theta$ variation of the Sun's polar field during the period 1976-1977 was deduced empirically (Svalgaard et al. 1978). However, in such a case the solutions become necessarily more complicated. Therefore, in order to keep our study at an analytically manageable level, we shall confine ourselves to the 0th–order angular dependence of the magnetic field as given by Equation (4b).

2.4. ANGULAR DEPENDENCE OF POLOIDAL VELOCITY FIELD

Conservation of mass flux requires similarly from Equation (1a) that,

$$4\pi \rho \vec{V}_p = \frac{\Psi_A(A)}{r^2 \sin\theta} \frac{\partial A}{\partial \theta} \vec{e}_r - \frac{\Psi_A(A)}{r \sin\theta} \frac{\partial A}{\partial r} \vec{e}_\theta \tag{5}$$

where $\Psi_A(A) = \mathrm{d}\Psi(A)/\mathrm{d}A$ is an arbitrary function of A, as in Chapter 7.

We have suggested that a physically interesting choice of the arbitrary function $\Psi_A(A)$ is

$$\Psi_A(A) = \frac{4\pi\rho_o V_o}{B_o}\sqrt{1 + \delta\, f\sin^2\theta}\,, \tag{6}$$

which substituted in Equation (5), in combination with the chosen density functional dependence given in Equation (2), leads to the following poloidal velocity field,

$$V_r(R,\theta) = \frac{V_o Y f\cos\theta}{\sqrt{(1 + \delta\, f\sin^2\theta)}}\,, \quad V_\theta(R,\theta) = -V_o\frac{Y R}{2}\frac{df}{dR}\frac{\sin\theta}{\sqrt{(1 + \delta\, f\sin^2\theta)}}\,. \tag{7}$$

The constant δ controls the degree of velocity collimation around the axis of the flow ($\theta = 0$). Thus, $\delta = 0$ corresponds to a sinusoidally symmetric flow pattern, while $\delta \to \infty$ corresponds to a flow along the polar axis only, ($\theta = 0$). Take for example the highly–collimated bipolar flows and jets from young stars. Their majority have opening angles of about 3 to 10 deg (Mundt 1986), corresponding to large values of δ. Thus, for fixed R and $\delta \sim 100(1000)$, at $\theta = 10(3)$ deg the radial velocity in Equation (3) drops to less than one half its value at $\theta = 0$.

2.5. ANGULAR DEPENDENCE OF TOROIDAL MAGNETIC AND VELOCITY FIELDS

To express the toroidal components of $\vec{B}$ and $\vec{V}$, we also need to prescribe the remaining two free functions $\Omega(A)$ and $L(A)$. Thus, physically interesting expressions of B_ϕ and V_ϕ are obtained if we choose:

$$\Omega(A) = \frac{\lambda\beta^2 V_o f_\star}{r_o R_\star^2}\frac{1}{\sqrt{1 + \delta\, f\sin^2\theta}}\,, \quad L(A) = \frac{\lambda\beta^2 V_o r_o f\sin^2\theta}{\sqrt{1 + \delta\, f\sin^2\theta}}\,, \tag{8}$$

such that

$$B_\varphi(R,\theta) = -\lambda B_o\frac{f}{R}\sin\theta\left(\frac{1 - M_a^2 f_\star Y_\star/(fY)}{1 - M_a^2}\right)\,, \tag{9}$$

$$V_\varphi(R,\theta) = \lambda V_o\frac{R\sin\theta}{\sqrt{1 + \delta\, f\sin^2\theta}}\left(\frac{Y_\star f_\star - Y f}{1 - M_a^2}\right)\,. \tag{10}$$

In the above expressions, $f_\star = f(R_\star)$, $Y_\star = Y(R_\star)$ where $R_\star$ is the radial distance of the poloidal Alfvén transition of the flow, i.e., where $M_a = 1$, and the following definitions have been used,

$$V_o^a = \frac{B_o}{(4\pi\rho_o)^{1/2}}\,, \quad \beta = \frac{V_o^a}{V_o}\,, \quad M_a^2 = \left(\frac{V_r}{V_a}\right)^2 = \frac{4\pi\rho V_r^2}{B_r^2} = \frac{Y R^2}{\beta^2}\,, \tag{11}$$

Note that the mass efflux $\dot{m}$ (mass loss rate per unit solid angle $d\Sigma$) is

$$\rho V_r r^2 \equiv \frac{dM}{d\Sigma\, dt} = \dot{m}(R,\theta) = \rho_o V_o r_o^2 f\cos\theta\,(1 + \delta\, f\sin^2\theta)^{1/2}\,. \tag{12}$$

At each distance R, the mass efflux $\dot{m}(R,\theta)$ then has a peak between pole and equator at the angle $\theta_o = \cos^{-1}[1/\delta\, f(R)]/2$ and a strong equatorial concentration if δ is large ($\delta \gg 1$).

2.6. SET OF EQUATIONS FOR $Y(R)$, $f(R)$, $Q_o(R)$ AND $Q_1(R)$

Finally, in the remaining r- and θ-components of the force balance equation (1c), the variables r and θ can be separated with the previous angular dependences yielding the following three equations:

$$\frac{dQ_0}{dR} + \frac{2f}{R^2}\frac{d}{dR}(Yf) + \frac{\nu^2}{YR^4} = 0, \tag{13a}$$

$$\frac{dQ_1}{dR} - \frac{2f}{R^2}\frac{d}{dR}(Yf) + \frac{Yf}{R^2}\frac{df}{dR} - \frac{Y}{2R}\left[\frac{df}{dR}\right]^2 + \frac{\delta\nu^2 f}{YR^4} + \frac{\beta^2}{2R^2}\frac{df}{dR}\left[\frac{d^2f}{dR^2} - \frac{2f}{R^2}\right]$$
$$- \frac{2\lambda^2}{YR}\left[\frac{fY - f_\star Y_\star}{1 - M_a^2}\right]^2 + \frac{\beta^2\lambda^2}{R^2}\frac{d}{dR}\left[\frac{f - R^2 f_\star/R_\star^2}{1 - M_a^2}\right]^2 = 0, \tag{13b}$$

$$Q_1 - \frac{f}{2R^2}\frac{d}{dR}\left[YR^2\frac{df}{dR}\right] + \frac{Y}{4}\left[\frac{df}{dR}\right]^2 + \frac{\beta^2}{2R^2}f\left[\frac{d^2f}{dR^2} - \frac{2f}{R^2}\right]$$
$$- \frac{\lambda^2}{Y}\left[\frac{fY - f_\star Y_\star}{1 - M_a^2}\right]^2 + \frac{2\beta^2\lambda^2}{R^2}\left[\frac{f - R^2 f_\star/R_\star^2}{1 - M_a^2}\right]^2 = 0. \tag{13c}$$

where ν is the ratio of the escape speed and polar base speed, $\nu = (2GM(1 - \ell)/r_o V_o^2)^{1/2}$.

Equations (13) can be also derived by writing force balance along and across streamlines. For example, define the sum of the kinetic, gravitational, and Poynting energy flux densities per unit of mass flux density by E,

$$E \equiv \frac{1}{2}V_{total}^2 - \frac{GM}{r} - \frac{\Omega}{\Psi_A}r\sin\theta B_\varphi = \frac{V_o^2}{2}\frac{E_1(R) + E_2(R)A}{1 + \delta A}, \tag{14a}$$

where

$$E_1(R) = Y^2 f^2 - \frac{\nu^2}{R}, \tag{14b}$$

and

$$E_2(R) = Y^2 f\left[\frac{R^2}{4f^2}\left(\frac{df}{dR}\right)^2 - 1\right] + \lambda^2 Y^2 R^2 f\left[\frac{1 - Y_\star f_\star/Yf}{1 - M_a^2}\right]^2$$
$$- \frac{\delta\nu^2}{R} + \frac{2\lambda^2\beta^4 f_\star}{R_\star^2}\left[\frac{1 - M_a^2 f_\star Y_\star/(fY)}{1 - M_a^2}\right]. \tag{14c}$$

Then, force balance along an individual streamline $A(R,\theta) = $ const. gives,

$$\vec{B}_p \cdot [\vec{\nabla}E + (\vec{\nabla}P)/\rho] = 0, \tag{15a}$$

where $\vec{B}_p$ is the poloidal component of the magnetic field, or, equivalently,

$$\vec{B}_p \cdot \vec{\nabla}\left[\frac{E_1(R) + E_2(R)A}{1 + \delta A}\right] + \frac{YR^2}{\rho_o(1 + \delta A)}\vec{B}_p \cdot \vec{\nabla}\left[P_o(R) + \frac{P_1(R)}{f(R)}A\right] = 0, \tag{15b}$$

which gives Eqs (13a,b), while force balance across streamlines gives Equation (13c). Note that in this nonpolytropic approach, the enthalpy $\vec{\nabla}P/\rho$, cannot be written as the gradient of a scalar as is possible with the polytropic relation between P and ρ. Therefore, a generalized Bernoulli integral does not exist in this case. We want to note however that use of this generalized Bernoulli integral may be obscuring the true driving mechanism of thermal winds, which is the conversion of thermal energy added along the flow to kinetic energy. In the Bernoulli integral this added thermal energy does not appear explicitly as they appear the kinetic, gravitational and Poynting parts of the total energy. The existence of additional thermal heating is nevertheless implied by writing $P = K\rho^\gamma$ with $\gamma < 5/3$.

Equations (13) form a set of three nonlinear differential equations for the four unknowns $f(R)$, $Y(R)$, $Q_o(R)$, and $Q_1(R)$. We are then free to specify one of them. In the following we shall illustrate two possible sets of solutions: in the next section by specifying the streamfunction $f(R)$ while in section 5 the pressure-inhomogeneity function $Q_1(R)$ is related to $Q_o(R)$.

2.7. HEATING DISTRIBUTION, $h(R, \theta)$

Once Eqs. (13) that conserve momentum, magnetic flux and mass flux, as well as Faraday's law of induction have been solved, we may *a posteriori* determine the distribution of energy on the meridional plane required to maintain this flow pattern from the first law of thermodynamics, Equation (1d). Thus, if we write the heating distribution function $h(R, \theta)$ for a proton gas with the ratio of the specific heats $\Gamma = c_p/c_v = 5/3$, in the form

$$h(R, \theta) = [h_o(R) + h_1(R) \sin^2 \theta] \frac{\cos \theta}{(1 + \delta f \sin^2 \theta)^{1/2}}, \qquad (16a)$$

we obtain from Equation (1c) that h_o and h_1 need to have the following expressions:

$$h_o(R) = c_o \frac{f}{R^2} \left[Y R^2 \frac{dQ_o}{dR} + \frac{5}{3} Q_o R^2 \frac{dY}{dR} + \frac{10}{3} Q_o Y R \right], \qquad (16b)$$

$$h_1(R) = c_o \frac{f}{R^2} \left[Y R^2 \frac{dQ_1}{dR} + \frac{5}{3} Q_1 R^2 \frac{dY}{dR} + \frac{10}{3} Q_1 Y R \right], \quad \text{with} \quad c_o = \frac{3\rho V_o^3}{4r_o}. \qquad (16c)$$

The final criterion for accepting a given solution will be to verify that it corresponds to a physically acceptable form of the heating function $h(R, \theta)$. For example, a solution where the total heating rate required,

$$H_{tot} = \int_{Vol} h(r, \theta) dV, \qquad (16d)$$

is infinite, should be disregarded as unphysical (Tsinganos and Trussoni 1992).

Note that we may calculate a similar expression for the heating h for a proton gas where $c_p/c_v = 5/3$ and for the often used polytropic equation of state $P = K\rho^\gamma$, with $5/3 > \gamma =$ constant. Then, Equation (1d) gives the net heating rate per unit volume,

$$h = -\frac{5/3 - \gamma}{5/3 - 1} \frac{P}{\rho} \vec{V} \cdot \vec{\nabla}\rho, \qquad (16e)$$

where for $\gamma < 5/3$ and $\vec{\nabla}\rho < 0$ we have $h > 0$, *i.e.*, a net heating in the wind.

3. Helicoidal MHD Outflows

Setting $f(R) = 1$, we obtain the special case of a hydromagnetic field that is helicoidal without meridional components, $i.e.$, the configuration of a rotating radial outflow embedded in a magnetic field with radial and azimuthal components only.

3.1. DETERMINATION OF CRITICAL POINTS

Equations (13) simplify considerably when $f(R) \equiv 1$ yielding a single first order nonlinear equation for the Alfvén Mach number $M_a^2(R)$,

$$\frac{\mathrm{d}M_a^2}{\mathrm{d}R} = \frac{F(M_a^2, R; \delta \nu^2; \lambda; \beta; R_\star)}{G(M_a^2, R; \delta \nu^2; \lambda; \beta; R_\star)}, \tag{17a}$$

where

$$F = \frac{\delta \nu^2}{\beta^4 M_a^2} - \frac{4(1 - M_a^2)}{R^3} + \frac{4\lambda^2}{R(1 - M_A^2)}, \tag{17b}$$

$$G = \frac{2}{R^2} - \frac{\lambda^2}{(1 - M_A^2)^2} \left[\frac{(2M_A^2 - 1)}{M_A^4} \frac{R^4}{R_\star^4} - 1 \right]. \tag{17c}$$

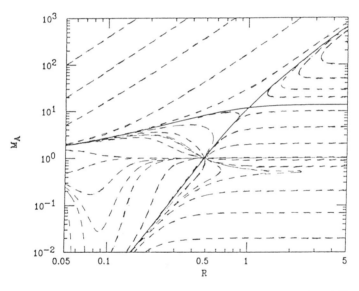

Figure 1. A typical solution topology of the Alfvén Mach number M_a^2 $vs.$ R for a strongly magnetized outflow, $\beta = 100$. Positions of Alfvénic, sonic and downwind X-type critical point are marked by star, square, and cross, respectively. The thick solid line corresponds to characteristic wind solution crossing all critical points and having asymptotically $M_a^\infty \propto R$. The other solid line corresponds to an $accretion$ $solution$ which, together with $breeze$-$solutions$ at lower part of graph, have $M_a^\infty \to const.$).

A topological analysis of Equation (17) shows that there are at most two critical points in the solutions, *i.e.*, there are two values of R ($R_\star$ and $R = R_x$) for which the numerator and denominator of Equation (17a) vanish. Those two critical points are introduced by the combined effects of rotation V_ϕ and the azimuthal magnetic field B_ϕ; for, if $\lambda = 0$ in (17) no such critical points exist. Also, note that of the three known polytropic critical points where the flow speed equals to one of the characteristic MHD modes for wave propagation in the plasma (slow, Alfvén, fast) (Heyvaerts and Norman 1989), the slow (sonic-like) critical point is not a critical point in this nonpolytropic approach. The situation is similar to radiation line driven outflows where the sonic point is not a critical point (MacGregor 1988).

3.2. STAR-TYPE ALFVÉNIC CRITICAL POINT

The first critical point is found at $R = R_\star$, where $M_a^2 = 1$ and $Y = Y_\star$: it is the familiar Alfvénic critical point, corresponding to the position where the poloidal flow speed reaches the poloidal Alfvén speed. This Alfvénic point corresponds mathematically to a third-order singularity, as compared to an ordinary first order X-type singularity. In fact, by Taylor expanding Equations (17) at $R \approx R_\star$, $Y \approx Y_\star$ up to the third order, it is found that all slopes are allowed at this Alfvénic point, contrary to the case of the X-type critical point wherein only two slopes are allowed. The Alfvén point marks the boundary between the inner magnetically–dominated region where angular momentum is predominantly transported by magnetic torques and the outer hydro–dominated region where the angular momentum is transported by the fluid. An easy demonstration of this fact can be given if we express the azimuthal components of the hydromagnetic field in terms of the radial flow speed at the Alfvén point, $V_R^\star$,

$$V_\phi(R,\theta) = \frac{\Omega r_o R \sin\theta}{V_R^\star} \frac{V_R^\star - V_R}{1 - M_A^2}, \tag{18a}$$

$$B_\phi(R,\theta) = -B_R \frac{\Omega r_o R \sin\theta}{V_R^\star R_\star^2} \frac{R_\star^2 - R^2}{1 - M_A^2} (1 + \delta \sin^2\theta)^{1/2}, \tag{18b}$$

where $\Omega(\theta)$ is approximately the angular velocity of the footpoints of the helical field lines at the base of the star if, as is usually the case, $\beta \gg 1$, $Y_\star \gg 1$

$$\Omega(\theta) = \frac{\lambda V_o Y_\star}{r_o} \frac{1}{(1 + \delta \sin^2\theta)^{1/2}}, \quad V_\phi(R = 1, \theta) = \frac{\lambda V_o \sin\theta}{(1 + \delta \sin^2\theta)^{1/2}} \frac{Y_\star - 1}{1 - 1/\beta^2}. \tag{19}$$

The asymptotic behavior of V_ϕ and B_ϕ can now be obtained. As can be seen in Figure 1, for $R \gg R_\star$, to a first approximation, $M_a \sim R$, while both V_ϕ and B_ϕ vary as $1/R$. On the other hand, for $R \ll R_\star$, where $V_R \ll V_R^\star$, they reduce to,

$$V_\phi(R,\theta) \approx \Omega(\theta) r_o R \sin\theta \times \left[1 - \frac{V_R^2}{V_R^{\star 2}} \left(1 - \frac{V_R^2}{V_R^{\star 2}}\right) + ...\right], \tag{20a}$$

$$B_\phi(R,\theta) \approx -B_R \frac{\Omega(\theta) r_o R \sin\theta}{V_R^\star} \times \left[1 - \frac{R^2}{R_\star^2} \left(1 - \frac{V_R^2}{V_R^{\star 2}}\right) + ...\right]. \tag{20b}$$

Near the base ($R = 1$), most of the angular momentum (L) is carried by the torque exerted by the magnetic field,

$$L = L_v^o + L_m^o \approx \Omega(\theta) r_o^2 \sin^2\theta \left(1 + R_\star^2\right). \tag{21a}$$

The relative contribution $R_\star^2$ of the magnetic term is dominant. On the other hand, further away from the stellar base, L_v and L_m add to

$$L = L_v^\infty + L_m^\infty \approx \Omega(\theta) r_\star^2 \sin^2\theta \left(1 + Y_\star / Y_\infty\right) \tag{21b}$$

where Y_∞ is the value of Y at some large distance R. Most of the angular momentum is carried now by the fluid since, in this case, the relative contribution $Y_\star / Y_\infty$ of the magnetic term is negligible.

3.3. X-type Magnetic Critical Point

The second critical point at $R = R_x = 1$ is present downstream of $R_\star$ and is a typical X-type point. The appearance of this second X-type critical point has to do with the mathematical nature of the Alfvén singularity that we referred above, namely that it is of the star-type, which means that there exist an infinite number of solutions that cross this point at $R_\star$. The boundary condition that the flow speed at infinity is finite for a wind solution – in order that the pressure and density there vanish – requires then the existence of this second X-type critical point at R_x, downstream of the Alfvénic critical point at $R_\star$, in order to filter a single solution. Thus, there is a unique wind-type solution that fulfils the requirement that $Y(R=1)=1$ and $M_a(R=1)=1/\beta$ with a continuous derivative at $R_\star$.

In order to understand the origin of the X-type critical point, it is instructive to consider for a moment the simpler case $\beta = 0$ and $\lambda \neq 0$, i.e., when there is only an azimuthal component of the magnetic field. In such a case, the conservation laws together with expression (7) for the radial velocity and (2) for the density, allow the following expressions for V_ϕ and B_ϕ,

$$V_\phi(R, \theta) = \frac{\lambda V_o}{R} \frac{\sin\theta}{(1 + \delta \sin^2\theta)^{1/2}}, \qquad B_\phi(R, \theta) = \frac{B_1}{YR}\sin\theta. \tag{22}$$

The r- and θ-components of the momentum balance equation, combined with the above expressions and with Equations (333) and (8), simplify considerably in this case and yield a single first order differential equation for the amplitude of the radial flow speed $Y(R)$:

$$\frac{dY}{dR} = \frac{Y}{R} \frac{\delta \nu^2 R - 6\lambda^2 + 4\Lambda^2 R^2 / Y}{2Y^2 R^2 + \lambda^2 - 2\Lambda^2 R^2 / Y}, \tag{23}$$

where $\Lambda = (B_1/\sqrt{4\pi\rho_o}V_o)$ is the ratio between the Alfvén speed associated with the azimuthal magnetic field B_1 and the radial flow speed at the polar base.

It is evident that, for some range of the parameters, the numerator and denominator of this differential equation vanish simultaneously. In other words, we have a critical point at $R = R_x$ and $Y = Y_x$. An analysis of Equation (23) around this critical point (see e.g. Holzer 1977), shows that the two slopes of $Y(R)$ at $R = R_x$ are one positive and one negative, resulting in the familiar X-type critical point geometry.

To further understand the nature of this X-type critical point, we may compare it to a similar situation that appears in the classical spherically symmetric hydrodynamic wind outflow (Parker, 1963). There, the attractive gravitational force and the expansive thermal pressure gradient force, when combined with the polytropic relationship between pressure and density, lead effectively to the formation of a gravitational/thermal-pressure nozzle at $R_P = [V_{esc}/2V_s]^2$ for an isothermal atmosphere with sound speed V_s. The boundary condition at the base, $V_R^{init} = V_o$, combined with the requirement that the pressure at infinity vanishes, constrains the free parameters of the problem such that the unique wind solution crosses the Parker X-type critical point at R_P with finite slope. Similarly in our case, the attractive component of the magnetic tension force and the expansive thermal pressure gradient and centrifugal forces, when combined with the law of angular momentum conservation, lead to the formation of a magnetic/(centrifugal-thermal pressure) nozzle at the critical point R_x. The boundary condition $Y(1) = 1$ and the requirement that the pressure at infinity vanishes or, equivalently, that the solution crosses the X-type critical point at R_x with a finite slope, lead to an eigenvalue-type condition among the parameters δ, λ, and Λ. It is interesting to note that for large λ, at the critical point R_x we have $V_r(R_x) \approx V_\phi(R_x)$, a relation that we shall encounter also in the most general case of fast magnetic rotators in the following section.

4. Nature of Acceleration in Rotating Magnetized Outflows

Rotating magnetized outflows are always hybrid winds, in which there is a "primary" wind mechanism that determines the mass loss in the absence of rotation and magnetic forces which can greatly modify the velocity structure and terminal speed (Cassinelli 1990). In the case discussed here, the "primary" mechanism is the thermal pressure gradients, while meridional magnetic forces and centrifugal forces together with magnetic rotator forces are modifying the acceleration of the wind. In the following, we show how part of the Poynting energy flux density per unit of mass flux density may be transfered to kinetic energy of the outflowing gas. Next, we apply the introduced concept of the Poynting magnetic rotator velocity to the model of previous section 3.

4.1. THE CHARACTERISTIC POYNTING MAGNETIC ROTATOR VELOCITY

The following argument illuminates the role of the magnetic rotator forces to drive the outflow and also clarifies the nature of the X-type critical point at $R = R_x$. In a rotating magnetized outflow, some energy deposition in the outflow may result from the transfer of energy from the rotating magnetic field. The flux of energy in the field depends on the Poynting vector $\vec{S}$,

$$\vec{S} = \frac{c}{4\pi}(\vec{E} \times \vec{B}) \qquad \text{for} \qquad \vec{E} = -\frac{\vec{V} \times \vec{B}}{c}, \qquad (24)$$

in the usual MHD approximation of large electrical conductivity $\sigma_e \to \infty$.

Since we are interested for the acceleration of magnetized collimated outflows in the radial direction, consider the form of the r–component of the Poynting vector $\vec{S}$ in the spherical coordinates (r, θ, ϕ),

$$4\pi S_r = (V_r B_\phi - V_\phi B_r)B_\phi + (V_r B_\theta - V_\theta B_r)B_\theta \,. \tag{25}$$

With (B_r, B_θ) and (V_r, V_θ) given by Equations (2a) and (3), respectively, the last term of the previous equation, $(V_r B_\theta - V_\theta B_r)$, is identically zero. On the other hand, with the general expressions of the azimuthal components B_ϕ and V_ϕ,

$$V_\phi = \frac{r\sin\theta\,\Omega - (L/r\sin\theta)\left(\Psi_A^2/4\pi\rho\right)}{1 - \Psi_A^2/4\pi\rho} \quad \text{and} \quad B_\phi = \frac{r\sin\theta\,\Omega\Psi_A - L\Psi_A/r\sin\theta}{1 - \Psi_A^2/4\pi\rho} \,. \tag{26}$$

for axisymmetric systems (Tsinganos 1982), the first term in Equation (25) divided by ρV_r gives the Poynting energy flux density per unit of mass flux density (energy per unit mass),

$$\frac{S_r}{\rho V_r} = -r\sin\theta\, B_\phi \frac{\Omega}{\Psi_A} = \frac{(r^2\sin^2\theta\,\Omega - L)\Omega}{M_a^2 - 1} \,. \tag{27}$$

Along a streamline ($A = $ constant), the term $r^2\sin^2\theta\,\Omega(A)$ is much larger than $L(A)$ for large R, while asymptotically $M_a \gg 1$. In order that the Poynting flux reduces with radial distance R, we should have M_a increasing faster than R. For example, in the present analytical solutions, equation (11), $M_a \sim RY^{1/2}$. Then, with $Y(R)$ increasing slowly with R, the Poynting flux term reduces as $1/Y(R)$ and part of this magnetic energy is transfered to the flow.

To see more quantitatively the conversion of Poynting energy flux to kinetic energy flux, write Equation (27) in terms of the magnetic flux $F_B(\theta) = B_r r^2$ and mass flux $F_M(\theta) = \rho V_r r^2$,

$$\left.\frac{S_r}{\rho V_r}\right|_\infty \simeq \frac{1}{V_r^\infty}\left[\frac{F_B^2}{4\pi F_M}\right]\Omega^2\sin^2\theta \,. \tag{28}$$

With $\Omega(A)$ constant along a streamline, we have that approximately,

$$\Omega(\theta) = \frac{V_\phi - M_a^2 B_\phi/\Psi_A}{r\sin\theta} \simeq \frac{V_\phi(r_o)}{r_o\sin\theta} = \Omega_o(\theta) \,, \tag{29}$$

where $\Omega_o(\theta)$ is the angular velocity of the footpoints of the streamlines at the stellar base r_o. Substituting the expression of $\Omega_o(\theta)$ from Equation (19) into Equation (29), we obtain for the Poynting energy flux density per unit of mass flux density at $R \to \infty$,

$$\left.\frac{S_r}{\rho V_r}\right|_\infty \simeq \frac{1}{V_r^\infty}\frac{F_B^2(\theta)\Omega_o^2(\theta)}{4\pi F_M(\theta)}\sin^2\theta \equiv \frac{V_M^3(\theta)}{V_r^\infty} \,. \tag{30}$$

In the last expression we have introduced the polar angle-depended Poynting velocity $V_M(\theta)$

$$V_M(\theta) = \left\{\frac{B_o^2 r_o^2 \Omega_o^2(\theta)}{4\pi\rho_o V_o}\cos\theta\sin^2\theta\sqrt{1 + \delta\sin^2\theta}\right\}^{1/3} \,. \tag{31}$$

A few comments on V_M are now in order. *First,* note that V_M is expressed exclusively in terms of quantities at the base of the wind: the angular velocity Ω_o, the magnetic field strength B_o, the density ρ_o and the flow speed V_o, all calculated at r_o. *Second,* V_M is proportional to $\sin\theta$, such that at the polar axis ($\theta = 0$) V_M is zero, a fact signifying the role of rotation on the magnetic rotator forces. *Third,* the amplitude of V_M is exactly equal to the so-called "Michel-velocity," which is introduced in one-dimensional magnetic winds (Michel 1969, Belcher and MacGregor 1976, Cassinelli 1990).

The velocity V_M plays a crucial role in rotating magnetic winds. If magnetic rotator forces are greater than thermal pressure forces, then there is enough energy in the Poynting flux such that a fraction of it is transfered to the flow. More quantitatively, if $V_M < V_P$, where V_P is the Parker velocity corresponding to an analogous thermally driven wind (see discussion below for the definition of V_P in the content of our model), magnetic rotator forces are insignificant and we have the so-called *slow-magnetic-rotator* (SMR). If, on the other hand, $V_M > V_P$, magnetic rotator forces are the dominant driving mechanism and we have the so-called *fast-magnetic-rotator* (FMR) (Belcher and MacGregor 1976, Cassinelli 1990).

4.2. Examples of Fast Magnetic Rotators

Let us apply the previous general considerations to our specific model for the angular dependence of the flow quantities. First note that the polar-angle dependence of the angular velocity of the footpoints of the streamlines, $\Omega_o(\theta)$, is

$$\Omega_o(\theta) = \frac{\lambda V_o \beta^2}{r_o R_\star^2} \frac{1}{(1+\delta\sin^2\theta)^{1/2}}, \tag{32}$$

Then in terms of our parameters, $V_M(\theta)$ is,

$$V_M(\theta) = V_o \left[\frac{\lambda^{2/3}\beta^2}{R_\star^{4/3}}\right] \left[\frac{\cos\theta\sin^2\theta}{(1+\delta\sin^2\theta)^{1/2}}\right]^{1/3}. \tag{33}$$

It turns out from Equation (17) that the amplitude Y_∞ of the asymptotic radial velocity V_r^∞ is

$$Y_\infty \approx \frac{V_M}{V_o}(6\ln R)^{1/3}, \tag{34}$$

i.e., equal to the velocity V_M, apart from the slowly increasing factor $(6\ln R)^{1/3}$. The corresponding expression of the Poynting energy flux is

$$\frac{S_r}{\rho V_r}\bigg|_\infty \simeq V_o^2 \frac{\lambda^2\beta^6}{R_\star^4}\frac{1}{(6\ln R)^{1/3}}\frac{\sin^2\theta}{1+\delta\sin^2\theta}. \tag{35}$$

It follows that the Poynting energy flux term decreases as $1/Y$ and therefore part of this energy is transferred to the flow.

In Figure 2, we plot the radial amplitudes $Y(R)$ of the radial flow speed $V_r(R)$ and of the azimuthal speed V_ϕ for various values of λ from 0 to 50 while keeping β fixed, $\beta = 40$. The effects of increasing the strength of the azimuthal components V_ϕ and B_ϕ by increasing λ while keeping the strength of the meridional magnetic field constant, are then three-fold. *First*, the higher is λ, the higher is the flow speed at $R \gg R_*$ and approximately $Y_\infty \approx (V_M/V_o)(6\ln R)^{1/3}$. On the other hand, at the X-type critical point the amplitude of the radial speed is $Y_x \approx V_M/V_o$. The amplitude of the thermal Parker speed Y_P in the absense of magnetic fields and rotation is (Tsinganos and Trussoni 1990),

$$Y(R) = \sqrt{1 + \delta \nu^2 \left(1 - \frac{1}{R}\right)}. \tag{36}$$

For the parameters used ($\delta = 4, \nu = 120$), we obtain $Y_\infty \equiv Y_P = 240$. Therefore, the amplitudes of the asymptotic flow speeds in Figure 2 are, for large λ, $Y_\infty \gg Y_P$. Then, we may say that for these large values of λ we are in the FMR regime. This is because the azimuthal component of $\vec{V}$ is proportional to λ and as we increase λ (for constant β) the Poynting flux available for the acceleration of the flow increases.

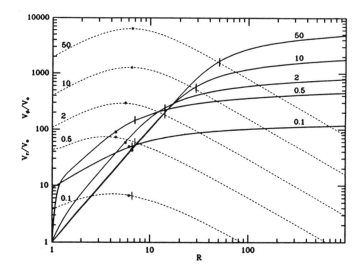

Figure 2. Radial dependence of the amplitude of radial and azimuthal flow speeds for $\delta = 4$, $\nu = 120$, $\beta = 40$ and $\lambda = 0.1, 0.5, 2, 10, 50$. The stars and vertical bars indicate the positions of Alfvén and X-type critical points, respectively.

Second, rotation itself has a decelerating effect on the flow close to the base. This is due to the fact that the centrifugal force is maximum at the equator and zero at the pole. Momentum balance then in the radial direction gives a decelerating net force. To see that consider force balance at the equator and at the pole at the same radial distance. At the equator, the pressure gradient together with the centrifugal force exactly balance the gravitational force. With this in mind, examine force balance at the polar axis and at the same radial distance. Now, there the force on the fluid is just the difference of the outwards pressure gradient and inwards gravity, since there is no centrifugal force at $\theta = 0$. But this net force is negative in view of the equatorial force balance conditions. Therefore, there is a decelerating force on the fluid, relative to the case with no rotation. This effect of pure rotation (*i.e.*, rotation which is not combined with B_ϕ to produce Poynting flux) is exactly the opposite to the accelerating effect of $\delta > 0$. This may be seen in Figure 2 where as λ increases, the strong initial acceleration due to the $\delta \nu^2$-effect is gradually diminishing. For strong enough λ however, some part of the Poynting flux is delivered close to the base and neutralizes the decelerating effect of pure rotation.

Third, in the same figure we have plotted the amplitude of the azimuthal speed, $V_\phi(R)$. Note that $V_\phi(R)$ has a maximum at $\sim R_\star$ and for large β we have $V_\phi(R_x) \approx V_r(R_x)$ at the X-type critical point.

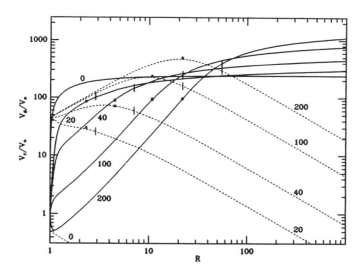

Figure 3. Radial dependence of the amplitude of radial and azimuthal flow speeds for $\delta = 4$, $\nu = 120$, $\lambda = 0.5$ and $\beta = 0, 20, 40, 100, 200$. The stars and vertical bars indicate the positions of Alfvén and Poynting critical points, respectively.

In Figure 3, we similarly plot the radial amplitude $V_r(R)$ of the radial flow speed for various strengths of the magnetic field, $\beta = 0 - 200$ and $\lambda = 0.5$. The effects of increasing the magnetic field may be seen to be similar to the previous effects of increasing rotation. Again, the higher is β, the higher is the flow speed at $R >> R_\star$ and approximately $Y_\infty \approx (V_M/V_o)(6 \ln R)^{1/3} >> Y_P$, *i.e.*, for large values of β we are in the FMR regime. This is because the azimuthal component of $\vec{B}$ is proportional to β and as we increase β (for constant λ) the Poynting flux available for the acceleration of the flow increases.

On the other hand, the meridional magnetic field has a decelerating effect on the flow close to the base. This is due to the $\cos\theta$ meridional dependence of B_r. Pressure equilibrium in the θ-direction requires a lower gas pressure at the polar regions because the magnetic pressure there is higher. Momentum balance then in the radial direction gives a decelerating force: since at the equator pressure gradient and gravity forces balance (neglecting centrifugal forces), as we move to the pole and at the same radial distance, the weight of the plasma is the same but the gas pressure is lower. The result is to have an inward force and deceleration. The effect of a pure meridional $\vec{B}$ (i.e., a meridional magnetic field which is not combined with rotation) is therefore exactly the opposite to the accelerating effect of $\delta > 0$ and the azimuthal magnetic field B_ϕ (Tsinganos and Trussoni 1992). This may be seen in Figure 3 where as β increases, the strong initial acceleration due to the $\delta \nu^2$-effect is gradually diminishing. For strong enough β, the velocity may even become zero.

Finally, in the same figure we have plotted the amplitude of the azimuthal speed, $V_\phi(R)$. Note that again $V_\phi(R)$ has a maximum at $\sim R_\star$ and for large β we have $V_\phi(R_x) \approx V_r(R_x)$ at the X-type critical point $R = R_x$.

4.3. Heating and Temperature Distribution

The selfconsistent energy distribution in the wind, Equation (16), implies a strong heating at the base for a weakly magnetized flow. In Figure 4, the heating distribution is shown for various values of the parameter β similar to those of Figure 3.

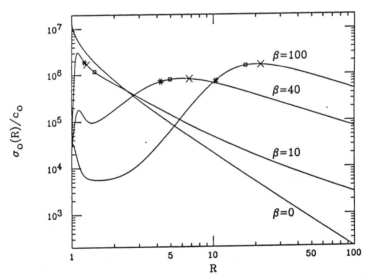

Figure 4. Radial dependence of the rate of heating per unit mass along the rotation axis expressed by the function $\sigma_o(R) \equiv h_o/\rho$, Equation (16b), in units of $c_o = 3V_o^3/4r_o$, for $\lambda = 0.5$ and various values of the magnetic parameter β representing the strength of the magnetic field.

In Figures 5a,b, the corresponding temperature distribution is shown for two extreme cases, $\beta = 0$ (a nonmagnetized) wind and $\beta = 100$, a highly magnetized outflow. For $\beta = 0$, the temperature reaches a maximum very close to the base and decreases quite strongly at larger distances, while the effective polytropic index reaches asymptotically the value $3/2$, the Parker polytrope. For strong magnetic fields however, the distribution of the heating and the temperature changes drastically. *First*, we have a region close to the base where the initial acceleration due to the effect of $\delta \nu^2$ corresponds to a heating and the temperature increases. *Second*, we have an intermediate region where the heating is drastically reduced and the temperature drops as in the case without azimuthal components. This is due to the decelerating effect of the poloidal magnetic field and in particular, its $\cos\theta$ dependence. The Poynting flux is still weak and does not transfer enough energy to the flow to accelerate it. *Third*, at $R \approx R_*$, the rotational velocity has increased enough so that part of the Poynting flux goes to acceleration, and the heating increases again together with the temperature. The temperature drops slowly further downstream and at large distances the gas can be considered isothermal. It is evident then that the heating distribution in a rotating outflow with a strong toroidal magnetic field is drastically different from that of a hydrodynamic outflow or, a magnetized meridional outflow.

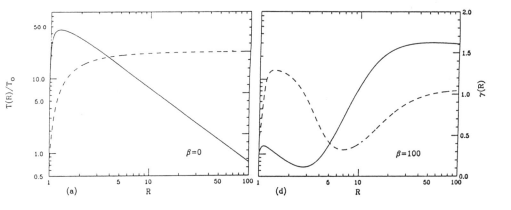

Figure 5. Temperature distribution $T(R)$ and effective polytropic index $\gamma(R)$ *vs.* R along polar axis, for $\lambda = 0.5$ and $\beta = 0, 100$ in (a) and (b), respectively. Note that for $\beta = 100$, asymptotically $\gamma \to 1$ and $T \to T_o = $ const. while, in the case of a pure hydrodynamic outflow, $\gamma \to 3/2$ and $T \to 0$.

We simply note in passing that a strong heating at the base of the wind can be related to high thermal conduction and dissipation of acoustic waves in the chromospheric zone. On the other hand, in a magnetized medium, Alfvén waves can heat the gas by means of dissipative processes that can be more effective along the plasma, far from the base. The results of this study seem to verify such an expected physical behavior, *i.e.*, hydrodynamic winds have the heating concentrated close to the base, while magnetized winds require a more extended heating.

5. Nonrotating Meridional Flows

A second class of solutions to Equations (13) is generated by setting $Q_1(R) = \kappa f(R)Q_o(R)$ and solving the system for $f(R)$, $Y(R)$, and $Q_o(R)$. This assumption of relating $Q_1(R)$ and $Q_o(R)$ essentially amounts to saying that the pressure, or the temperature, have the same basic radial profile along each streamline. We are dealing then with a class of solutions controlled by the five parameters κ, λ, β, $\delta\nu^2$, and $s = df(R = 1)/dR$. For given initial conditions, the set of Equations (13) can be solved for $Q_0(R)$, $Y(R)$, and $f(R)$ by standard numerical techniques. Note, however, that a nonlinear second order differential equation is obtained for $f(R)$, instead of the first order differential equation obtained in the previous case for M_a^2; and the solutions to this second order differential equation are controlled by several novel critical points that, for example, select again a unique wind-type solution, although the topology is entirely different in this case due to this higher order nature of the final equation.

An integral exists in this case as may be seen by treating the equations for force balance along each streamline. With $Q_1 = \kappa f(R)Q_o(R)$ in Equation (13c) we get

$$\vec{B}_p \cdot \vec{\nabla}E_1(R) + \frac{YR^2}{\rho_o}\vec{B}_p \cdot \vec{\nabla}Q_o(R) = -A\left\{\vec{B}_p \cdot \vec{\nabla}E_2(R) - \frac{\kappa YR^2}{\rho_o}\vec{B}_p \cdot \vec{\nabla}Q_o(R)\right\}. \quad (37)$$

This may be split into two equations, since the left hand side (LHS) depends only on R while the right hand side (RHS) is proportional to $A(R,\theta)$ times a function of R. Setting the LHS equal to zero we obtain Equation (13a) while setting the RHS equal to zero we obtain,

$$\vec{B}_p \cdot \vec{\nabla}[E_2(R) - \kappa E_1(R)] = 0. \quad (38)$$

Thus,

$$k \equiv \kappa E_1(R) - E_2(R) = Y^2 f\left[1 - \frac{F^2}{4} + \kappa f\right] + \frac{(\delta - \kappa)\nu^2}{R}$$
$$- \lambda^2\frac{R^2}{f}\left[\frac{fY - f_\star Y_\star}{1 - M_a^2}\right]^2 - 2\beta^4\lambda^2\frac{f_\star/R_\star^2}{f/R^2}\frac{f/R^2 - f_\star/R_\star^2}{1 - M_a^2}, \quad (39a)$$

has a constant value on all streamlines. To assign a physical meaning to k, note that in the special case $\kappa = \delta$, we have

$$k = [E(pole) - E(A)]\frac{1 + \delta A}{A}, \quad (39b)$$

i.e., the sign of k determines whether there is an excess of energy along the polar streamline as compared to the other streamlines (case $k > 0$) or a deficit of energy in the polar streamline as compared to the other nonpolar streamlines (case $k < 0$). Note that this integral will be used later in order to split the topological space into various domains according to the sign of this constant k.

5.1. THE SUBCLASS OF NONROTATING (MERIDIONAL) FLOWS

The simpler subcase $\lambda = \kappa = 0$ corresponding to nonrotating flows with a spherically symmetric pressure distribution, is interesting on its own and is studied in the following. The assumption $Q_1 = 0$ results in decoupling Equation (13a) from Equations (13b,c). Thus, we may first solve Equations (13b,c) and then integrate Equation (13a) for the pressure, with the constant of integration fixed such that the pressure be zero at infinity.

The formulation of the problem can be facilitated by the use of appropriate variables adapted to the particular situation. One such physical variable is M_a, which plays a significant role, even though there is no Alfvén critical point in the present nonrotating case. A second important physical variable is the radial variation of the geometrical shape of the streamlines, which we shall define for convenience as the logarithmic derivative of the streamlines function $f(R)$,

$$F \equiv \frac{R}{f}\frac{df}{dR} = \frac{d\ln(f)}{d\ln(R)}.$$

$\hfill (40)$

Besides its mathematical use for a rather simplified formulation, this quantity F represents physically the deviation of the streamlines from being radial. Thus, the three variables (R, F, M_a^2) should be the natural choice for a study of the solutions and especially their topology. By using these three variables, Equations (13b,c) can be written as a system of two equations for $F(R)$ and $M_a(R)$

$$R\frac{dF}{dR} = \left(1 - \frac{F^2}{4}\right)\frac{(F+1)(F-2) - FM_a^2 - \dfrac{\delta\nu^2 R^3}{2\beta^4 M_a^4 f}\dfrac{FM_a^2}{(1 - F^2/4)}}{M_a^2 - \left(1 - \dfrac{F^2}{4}\right)},$$

$\hfill (41\text{a})$

$$R\frac{dM_a^2}{dR} = \frac{M_a^2(F-2)}{8}\frac{F^2 + 4F + 8 - 2M_a^2(F+4) + \dfrac{4\delta\nu^2 R^3}{\beta^4 M_a^4 f}\dfrac{M_a^2 - 1}{F - 2}}{M_a^2 - \left(1 - \dfrac{F^2}{4}\right)},$$

$\hfill (41\text{b})$

while a third equation can be also obtained by simply dividing the two previous ones,

$$\frac{dM_a^2}{dF} = \frac{M_a^2(F-2)}{8(1 - F^2/4)}\frac{F^2 + 4F + 8 - 2M_a^2(F+4) + \dfrac{4\delta\nu^2 R^3}{\beta^4 M_a^4 f}\dfrac{M_a^2 - 1}{F - 2}}{(F+1)(F-2) - FM_a^2 - \dfrac{\delta\nu^2 R^3}{2\beta^4 M_a^4 f}\dfrac{FM_a^2}{(1 - F^2/4)}}.$$

$\hfill (41\text{c})$

Therefore, the above Equations (41) show that a three-dimensional representation of the solutions is possible by using (R, F, M_a^2) as the three independent coordinates.

5.2. CRITICAL POINTS AND TOPOLOGY IN THE CASE $\delta = 0$.

In order to get some insight, we shall consider first the case where the density is also spherically symmetric, $\delta = 0$. In this particular case Equation (41c) reduces to an ordinary first order equation for $M_a^2(F)$, such that the well known methods dealing with singular first order differential equations may be applied. Thus, the following critical points may be found on this (F, M_a^2)–plane:

(1) $(F_1, M_{a1}^2) = (-3 + \sqrt{5}, [3\sqrt{5} - 5]/2)$,   saddle point

(2) $(F_2, M_{a2}^2) = (-2, 1)$,   nodal point

(3) $(F_3, M_{a3}^2) = (-2, 0)$,   saddle point

(4) $(F_4, M_{a4}^2) = (-1, 0)$,   nodal point

(5) $(F_5, M_{a5}^2) = (+2, 0)$,   saddle point

(6) $(F_6, M_{a6}^2) = (+4, 5/2)$,   nodal point

In Figure 6, where the topology of the various solutions around the six physically interesting critical points on the plane (F, M_a^2) is plotted, critical solutions are indicated by solid lines and noncritical ones by dotted or dashed lines.

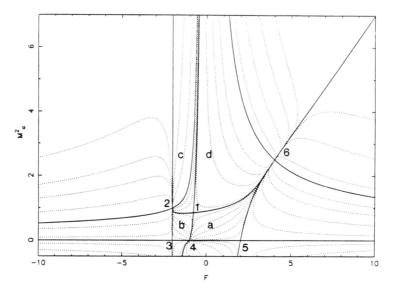

Figure 6. Topology of the solutions to equation (41c) for $\delta = 0$ on the plane–(F, M_a^2), containing three saddle points, (1,3,5) and three nodal points (2,4,6).

The next step is to draw the topology in the three-dimensional space which also implies finding the radial location of the critical points. First, note that only points (1,2,4,6) are true critical points while points (3,5) are projections of critical points appearing as individual critical points on the plane (F, M_a^2). Points (2,4) are located at the origin $R = 0$ while point (6) is ejected to $R = \infty$. The only critical point that remains at finite radius is point (1). Since its location depends on the normalization, we have chosen to fix the base at the location of this critical point (1) such that $R_1 = f(R_1) = Y(R_1) = 1$, $F_1 = s$, $M_{a1} = 1/\beta$. Note that this radial distance does not necessarily correspond to the stellar base. A full three-dimensional topology is presented at Figure 7. In this plot, critical solutions are indicated by solid lines and noncritical ones by dotted or dashed lines.

It is interesting to note that the three-dimensional space (R, F, M_a^2), Figure 7, is divided into three subdomains that do not communicate with each other, except at critical points. This can be understood by considering the generalized Bernoulli constant k, Equation (39), which is negative in $F < -2$, as well as in $F > 2$, while it is positive in the intermediate domain, $-2 < F < 2$. In each of these three subdomains, k has a constant sign, positive or negative. In the following, we shall consider separately these three domains which constitute the space of our solutions.

(i)   region $F < -2$.

All solutions in this domain terminate at finite R and, therefore, are not physical since all streamlines close at finite radii.

(ii)   region $-2 < F < 2$.

This is the more physically interesting region in the context of the present study because it contains the unique physical solution that may be related to astrophysical outflows. This is the critical solution that starts at the origin $R = 0$ at critical point (4), crosses the X-type critical point (1) and then ends asymptotically at $R \to \infty$ with $F \to 0$ (radial fieldlines, $M_a^2 \to \infty$). Moreover, it possesses the features of what we shall call the "critical coronal hole-type solution" in the more general case where $\delta \neq 0$.

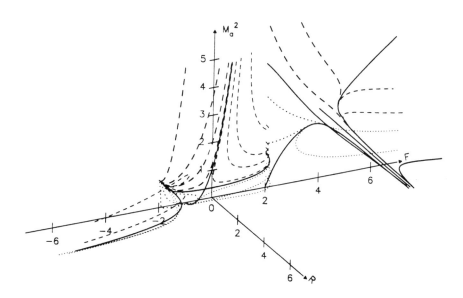

Figure 7. Topology of the solutions in the three-dimensional space (R, F, M_a^2) for $\delta = \kappa = 0$ and $\nu = 1$.

372

Of the other solutions in this subdomain, we note the dotted solutions both of which are trapped by the other critical branch that starts at $F = -2$ and ends at $F = 2$ after crossing critical point (1) (solid line). The other noncritical branches cannot reach low Alfvén Mach numbers and, therefore, may not be interesting since astrophysical flows usually start with a small M_a.

(iii) region $F > 2$.

At infinity, all solutions converge to the nodal point (6). The degenerate solution $F = 4$, $M_a^2 = 5/2$ is a straight line appearing as a single point, the critical point (6) on the plane (F, M_a^2). But the pressure drops fast , $P \sim const. - R^3$, and becomes negative at large radii. This property then excludes those solutions as unphysical.

5.3. CRITICAL POINTS AND TOPOLOGY IN THE CASE $\delta \neq 0$.

In the case $\delta > 0$, the situation is, in general, the same and it is interesting to note that the only physical domain again corresponds to a positive value of k, meaning that the energy supply is higher along the polar axis as compared to the other nonpolar streamlines.

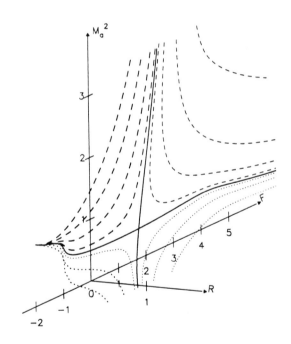

Figure 8. Topology of solutions in three-dimensional space of (R, F, M_a^2) for $\delta = 11.5$, $\nu = 1$.

In the following, we shall restrict our attention to this domain. The only critical point at finite radius is again similar to point (1) of the previous case $\delta = 0$, which we also choose as the base, $R_1 = 1$.

In Figure 8, the three-dimensional topology with $\delta = 11$ is shown and, evidently, it is analogous to the middle part of the topology in Figure 7. Thus, critical points (1) and (2) of the previous case $\delta = 0$ are the only ones that remain. Critical point (3) does not exist in this case of $\delta \neq 0$, while critical point (4) is ejected to the negative axis of M_a^2. Finally, critical points (5) and (6) do not exist in this case either because the topologies are constrained in the subspace where $k > 0$.

As in Figure 7, there are trapped solutions and noncritical solutions that do not reach low Alfvén Mach numbers. We also have the "critical coronal hole-type solution" where F is always slightly negative corresponding to faster than radial expansion of the streamlines and radial asymptotical behavior, as observed in flows of coronal holes. Note, however, that contrary to the case of a spherically symmetric density distribution ($\delta = 0$), the topology also exhibits some noncritical lines that reach zero Alfvén Mach number but with diverging F ($F = +\infty$). The corresponding solutions have fieldlines that start converging to the pole and form a nozzle before opening smoothly reaching a radial configuration. Those noncritical solutions may correspond to narrow opening angle jet-type outflows. Thus, in addition to the critical solution that has the characteristics of solar coronal hole-type outflows, there exist some noncritical solutions reminiscent of jet-type geometries.

6. Some Concluding Remarks from the Study of the Topologies.

In the previous analytical examples of magnetized outflows, it was found that, in all cases, there exists a unique and physically interesting solution corresponding to a characteristic astrophysical wind. The outflow in this case starts with subsonic and sub-Alfvénic speed and, in general, with a nonradial shape of the streamlines close to the star and ends (asymptotically) with super-Alfvénic/supersonic speeds, zero pressure, and, in the radial configuration, a situation reminiscent of solar/stellar winds. A *second* class of physically acceptable solutions was also found to exist in the case of meridional outflows with $f = f(R)$ and have a different streamline behavior. These solutions are again connected to the base with sub-Alfvénic/subsonic speeds and end up at infinity with super-Alfvénic/supersonic speed and zero pressure, but the streamlines initially converge toward the polar direction forming a nozzle, before extending conically to infinity, a situation reminiscent of the outflow configuration in astrophysical jets.

The nature and degree of acceleration of the outflow depends critically on:

• the density enhancement of the equatorial regions as compared to that of the polar regions, at each radial distance, δ. Note that this trend is in agreement with the observed density depletion inside the familiar solar polar coronal holes, as compared to the surrounding denser equatorial streamer, or, the denser accretion disk in the case of jet–type outflows from AGN and compact stellar objects.

• the magnitude of the poloidal magnetic field ($\beta \neq 0$) which decelerates the outflow, relative to an outflow without a meridional magnetic field, $\beta = 0$.

- the magnitude of azimuthal rotation ($\lambda \neq 0$) which also decelerates the polar outflow, relative to the case without rotation, $\lambda = 0$.
- the magnetic rotator forces (combined $\beta \neq 0$ and $\lambda \neq 0$) that accelerate the outflow transfering Poynting energy flux density to kinetic energy flux in the outflow.

In the case of rotating helicoidal outflows, the familiar Alfvén transition was shown to mark the boundary between an inner region where angular momentum is mainly carried by magnetic torques and an outer region, where angular momentum is carried by the fluid. The rotational flow speed attains its maximum value at the Alfvén point and decreases after to become equal to the increasing radial flow speed at the X-type critical point, for fast magnetic rotators. The heating and temperature distribution have been selfconsistently calculated and found to depend on the degree of magnetization, with the magnetized flows having a more extended heating and isothermal-like temperature distributions.

The main emphasis on this Chapter has been directed toward the problem of generation of exact solutions of the complete MHD equations pertinent to astrophysical outflows, a subsequent study of their topological properties and an analysis of the nature of the acceleration of rotating magnetized outflows. Such solutions have been shown to possess complicated topologies controlled by several critical points arising from the interplay of the magnetic, thermal pressure gradient, gravitational and inertial forces in the flow. The examples presented are indicative of the complexity encountered in the largely unexplored family of the two-dimensional solutions of the nonlinear MHD equations and should be taken into account in related numerical studies.

Acknowledgements. We are greatly indebted to Drs. Peter Hoyng, John Brown and Joan Schmelz for reading the manuscript and making numerous suggestions that led to a gradual improvement of the presentation. The research reported in this Chapter was supported in part by NATO grant 870221. This Chapter was written at the Osservatorio Astronomico di Torino and one of the authors (K.T.) acknowledges the kind hospitality of its director, Prof. A. Ferrari and the staff of the Osservatorio di Torino.

7. References

Bagenal, F., and Gibson, S. 1991, *J. Geophys. Res.*, (in press).

Belcher, J.W., and MacGregor, K.B. 1976, *Ap. J.*, **210**, 498.

Camenzind, M. 1989, in *Accretion Disks and Magnetic Fields in Astrophysics*, ed. G. Belvedere, Kluwer Academic, p. 129.

Cassatella, A., and Viotti, R.: 1988, *Physics of Classical Novae*, Springer Verlag L.N. in Physics 369.

Cassen, P., Shu, F., and Terebey, S. 1986, in *Protostars and Planets II*, eds. D. Black and M. Matthews, Univ. Arizona Press, Tucson, pp. 448-492.

Cassinelli, J.P. 1990, in *Angular Momentum and Mass Loss for Hot Stars*, eds. L. A. Wilson and R. Stalio, Kluwer Academic, pp. 135-144.

Ferrari, A., Trussoni, E., Rosner R., and Tsinganos, K. 1985, *Ap. J.*, **294**, 397.

Ferrari, A., Trussoni, E., Rosner R., and Tsinganos, K. 1986, *Ap. J.*, **300**, 577.

Harvey, P.M., 1986, in *Protostars and Planets II* (D. Black and M. Matthews eds.), Univ. Arizona Press, Tucson, p. 484-492.

Goldreich, P., Julian, W.H. 1970, *Ap. J.*, **160**, 971.

Hartmann, L., and MacGregor, K.B. 1982, *Ap. J.*, **259**, 180.

Heyvaerts, J., and Norman, C.A. 1989, *Ap. J.*, **347**, 1055.

Holzer, T.H. 1977, *J. Geophys. Res.*, **82**, 23.

Hu, Y.Q., and Low, B.C.: 1989, *Ap. J.*, **342**, 1049.

Hughes, P.A.: 1991, *Beams and Jets in Astrophysics*, Cambridge Astrophysics Series, Cambridge University Press, Oxford.

Jackson, J.D. 1975, *Classical Electrodynamics*, New York, Wiley, pp. 84–100.

Kato, M. 1983, *PASJ*, **35**, 33 and 507.

Kato, M. 1988, in *Physics of Classical Novae*, eds. A. Cassatella and R. Viotti, Springer Verlag L.N. in Physics 369, 236

Kato, M. 1991, *Ap. J.*, **369**, 471.

Hu, Y.Q., and Low, B.C. 1989, *Ap. J.*, **342**, 1049.

Königl, A. 1982, *Ap. J.*, **261**, 115.

Koupelis, T. 1990, *Ap. J.*, **363**, 79.

Leer, E., Holzer, T.E., and Flå, T. 1982, *Sp. Sci. Rev.*, **33**, 161.

Lovelace, R.V.E., Berk, H.L., and Contopoulos, J. 1991, *Ap. J.*, **379**, 696.

Low, B.C., and Tsinganos, K. 1986, *Ap. J.*, **302**, 163.

MacGregor, K.B. 1988, *Ap. J.*, **327**, 794.

Mestel, L. 1968, *M.N.R.A.S.*, **138**, 359.

Mestel, L.: 1986, in *Protostars and Planets II* (D. Black and M. Matthews eds.), Univ. Arizona Press, Tucson, pp. 320-339, Figs. 1,2.

Michel, F.C. 1969, *Ap. J.*, **158**, 727.

Mundt, R. 1986, in *Protostars and Planets II*, eds. D. Black and M. Matthews, Univ. Arizona Press, Tucson, pp. 414.

Mundt, R., Ray, T.P., Buehrke, T., and Raga, A.C. 1990, *Astr. Ap.*, **232**, 37.

Munro, R.H., and Jackson, B.V. 1977, *Ap. J.*, **213**, 874.

Mouschovias, T. Ch. 1977, *Ap. J.*, **211**, 147.

Mouschovias, T. Ch. 1981, in *Fundamental Problems in the Theory of Stellar Evolution.*, eds. D. Sugimoto, D.Q. Lamb and D.N. Schramm, Dordrecht, Reidel, pp. 27-42.

Nash, A.G., Sheeley, N.R., Jr. and Wang, Y.M. 1988, *Solar Phys.*, **117**, 359.

Okamoto, I. 1974, *M.N.R.A.S.*, **166**, 683.

Okamoto, I. 1975, *M.N.R.A.S.*, **173**, 357.

Orio, M., Trussoni, E., and Ögelman, H. 1992, *Astr. Ap.*, (in press).

Owen, F.N., Hardee, P.E., and Cornwell, T.J. 1989, *Ap. J.*, **340**, 698.

Parker, E. 1958, *Ap. J.*, **128**, 664.

Parker, E.N. 1963, *Interplanetary Dynamical Processes*, Interscience, New York.

Phinney, E.S. 1983, Ph.D. thesis, University of Cambridge.

Pneuman, G.W., and Kopp, R.A. 1971, *Solar Phys.*, **18**, 258.

Pudritz, R.E., and Norman, C.A. 1983, *Ap. J.*, **274**, 677.

Pudritz, R.E., and Norman, C.A. 1986, *Ap. J.*, **301**, 571.

Raymond, J.C., and Smith, B.R. 1977, *Ap. J. Suppl.*, **35**, 419.

Ringuelet, A.E., and Iglesias, M.E. 1991, *Ap. J.* **369**, 463.

Ruggles, C.L.N., and Bath, G.T. 1979, *Astr. Ap.*, **80**, 97.

Sakurai, T. 1985, *Astr. Ap.*, **152**, 121.

Sakurai, T. 1987, *Pub. Astr. Soc. Japan*, **39**, 821.

Sakurai, T. 1990, *Computer Physics Reports*, **12**, 247.

Schatzman, E. 1962, *Ann. Astrophys.*, **25**, 1.

Shibata, K., and Uchida, Y. 1986, *Pub. Astr. Soc. Japan*, **38**, 631.

Shibata, K., and Uchida, Y. 1987, *Pub. Astr. Soc. Japan*, **39**, 559.

Silvestro, G., Ferrari, A., Rosner, R., Trussoni E., and Tsinganos, K. 1987, *Nature*, **325**, 228.

Svalgaard, L., Duvall, T.L., Jr., and Scherrer, P.H. 1978, *Solar Phys.* **58**, 225.

Tassoul, J.-L. 1978, *Theory of Rotating Stars*, Princeton University Press.

Temesváry, S. 1952, *Zs. Naturforschung* **7a**, 103.

Torrelles, J.M., Rodriguez, L.F., Canto, J., and Carral, P. 1983, *Ap. J.* **274**, 214.

Torrelles, J.M., Canto, J., Rodriguez, L.F., Ho, P.T.P., and Moran, J. 1985, *Ap. J. (Letters)* **294**, L117-L120.

Tsinganos, K. 1982, *Ap. J.*, **252**, 775.

Tsinganos, K., and Low, B.C. 1989, *Ap. J.*, **342**, 1028.

Tsinganos, K., and Sauty, C.: 1992a, *Astr. Ap.*, (in press).

Tsinganos, K., and Sauty, C.: 1992b, *Astr. Ap.*, (in press).

Tsinganos, K., and Trussoni, E. 1990, *Astr. Ap.*, **231**, 270.

Tsinganos, K., and Trussoni, E. 1991, *Astr. Ap.*, **249**, 156.

Tsinganos, K., and Trussoni, E. 1992, *Mem. It. Astr. Soc.*, (in press).

Uchida, Y., and Low, B.C.: 1981, *J. Astr. Ap.* **2(4)**, 405.

Wang, Y.M., and Sheeley, N.R., Jr.: 1988, *J. Geophys. Res.* **93(A10)**, 11,227.

Wang, Y.M., and Sheeley, N.R., Jr. 1990, *Ap. J.*, **355**, 726.

Weber, E.J., and Davies, L. 1967, *Ap. J.*, **148**, 217.

Withbroe, G. 1988, *Ap. J.*, **325**, 442.

Yeh, T. 1976, *Ap. J.*, **206**, 768.

PART III
SOLAR INSTRUMENTATION

17   SOLAR OPTICAL INSTRUMENTATION

H. ZIRIN

BBSO, Caltech
Pasadena, CA 91125
U.S.A.

ABSTRACT. We discuss the various aspects of specialized instrumentation for solar observation. Telescopes are designed to minimize heating and turbulence. Because two–dimensional images are so important, two–dimensional monochromators are particularly important. Devices for magnetic field measurement and coronal observation are described, and the resolution of the various instruments is discussed.

1. Introduction

Observing the Sun is a tricky business, requiring common sense and realism. The most common danger is that most people do not distinguish sufficiently from night–time observations. If we look at a star, it does not disturb the local atmosphere; if we look at the Sun, we find ourselves in the midst of the winds and convective currents it excites. If we place a star on the telescope slit, we know that we see the integrated starlight; if we place the Sun on the slit, we do not know (without other data) which of the very different magnetic structures we are observing. For this reason, many observatories sit forlornly on top of sun–baked hills quite suitable for stellar work, but generating enough daytime turbulence to ruin solar observations.

The major differences between the requirements of solar and stellar observations are:

1. Site free of daytime convection
2. Telescope design
3. Focal plane instruments
4. Short, frequent exposures
5. Importance of two–dimensional data
6. High data rates

In the past, domes were often painted aluminum, and spectra were often published of unidentified regions of the Sun. Today almost all solar telescopes are evacuated to protect the optical path from solar heating. The domes are painted white, and slit–jaw cameras identify the regions under spectroscopic study.

379

J. T. Schmelz and J. C. Brown (ed.), The Sun, A Laboratory for Astrophysics, 379–393.
© 1992 *Kluwer Academic Publishers. Printed in the Netherlands.*

In my experience, the principal shortcoming in design of solar observing projects has been the lack of a clear plan and objectives (beyond the fact that we will look at the Sun). For example, one might prefer a site where occasional excellent frames might be obtained to one where continuous observations with good, but not superb, conditions could be carried out for long periods of time. For synoptic work, the highest resolution is not possible, so a site with the longest periods of clear skies would be preferred. The problem arises because the planners often have little actual experience observing.

2. Site Selection

High sky transparency and good seeing are both valuable but are not the same thing. Coronagraphs need high transparency and cannot function in a murky sky. But stable cirrus layers often provide good seeing. While mountain sites have the advantage of little overlying atmosphere, they often have the most turbulent air. All these factors must be considered. The problems of site surveys are addressed by Zirin and Mosher (1987). The principal lesson of that paper is that the physics of seeing is so poorly understood that one must rely on empirical, experimental results. This means comparative observations of the Sun through telescopes at different sites, preferably at the same height, on many days for many months.

Unfortunately there is little literature on the trade–offs of seeing for transparency, or short–term vs long–term seeing. Most site surveys have been too naive for such considerations; there has never, to my knowledge, been an objective test of whether one gains by going to a site of greater transparency but poorer seeing. All too often, a site is chosen for other considerations. There is nothing wrong with considerations of practicality of operation. If the weather conditions are severe or the travel distance great, it may indeed be worthwhile to compromise on a site of easier accessibility. One must always consider the total amount of work that will be accomplished with the funds and personnel at hand.

3. Basic Telescope Requirements

Solar telescopes must cover only $\pm 23.5°$ in declination and are limited by the poor daytime seeing, which makes it senseless to build systems greater than one meter diameter. Because the Sun is so bright, one can make very short exposures which may occasionally reach the resolution of the instrument. One may, therefore, use high–speed secondary guiders or simple computer re–registration to produce sequences that are close to the resolution limit. This has been done successfully for studies of granulation and other features requiring simple, direct images. More sophisticated measurements, however, are limited by the need for longer exposures with no image change. While the average image position can be maintained, wave front variation over the image plane varies the scale and position of individual elements. Therefore, spectroscopy, which requires longer exposures, is limited. For two–dimensional magnetography or Doppler maps, however, one requires a number of matching frames, and the wave–front variation over the image plane is so great that tedious de–stretching procedures are required. In this case, there is no guarantee that a second picture of the same quality can be obtained. For magnetograms of weak fields, a large number of frames are required, and this is probably not treatable by de–stretching. So, while strong–field magnetograms close to the resolution limit can be obtained, weak–field measurements are seeing–limited.

One solution is the use of active optics. In the application of Smithson *et al.* (1988), this consists of a 17–segment mirror which tilts the wave front from a segment of the pupil to achieve coherence at the center of the image. While the present device just barely works, one may assume that the continued progress of electronics will produce a more workable version. Recently, Dunn (unpublished) has fixed the Smithson device and obtained impressive improvement in mediocre seeing, with a field about one arc min. Whether further improvement occurs in good conditions is not yet known. A small spot must be in the field.

Observations with large apertures must average over a larger air column and may give a worse image. We find that telescopes 60 cm and larger will only achieve their limiting resolution a few days in the year. For this reason, continuous observation is needed to catch the best periods, which are hard to predict. However, for reasonable conditions at Big Bear Solar Observatory (BBSO), the 65 cm telescope always gives better images than the 25 cm. The maximum utilizable telescope size is not known, since present vacuum telescopes are not larger than 60–70 cm.

Obviously, the telescope should be carefully designed to have the best optics possible and avoid image distortion from internal heating.

Many other considerations occur in planning and carrying out solar observations. These are often ignored in the assumption that perfect conditions are possible. In fact, tradeoffs are always necessary. The dominant source of error is misunderstanding of the modulation transfer function (MTF).

Because funds, detectors, atmospheric quality, and other factors are always limited, the various elements of a system must be balanced to give the optimum resolution. Further, it is important to understand the actual resolution achieved. For example, it is common to display full–disk images claiming arc–second resolution. Anyone, even a theoretician, can place a ruler on the image, and, knowing that the Sun subtends 1900 arc sec, see if 1900 elements are resolved. Usually, they are not.

4. Modulation Transfer Function (MTF)

It is easy to think of mirrors as perfect, images as sharp points, and film as made up of very fine grains. But the image of a star is not a point but a spot, often blurred by atmosphere and poor optics, surrounded by a bright diffraction ring. The MTF is used to describe the performance of optical systems more quantitatively (a complete description may be found in Dainty and Shaw (1974)). The properties of the system can be described equally well by a point–spread or line–spread function, but the MTF is often more convenient. It is defined as the ratio of the output modulation of a sinusoidal form with spatial frequency ω to the input modulation at the same frequency. To measure it, a regular pattern (such as bars) of a certain spatial frequency is imaged through an optical system and a new distribution is produced, either on fine grain film or with a photoelectric scanner. The ratio of amplitudes is the MTF for that spatial frequency.

If the system is perfect, the MTF is one. The MTF for the overall system is the product of the MTFs of each component: objective, film, atmospheric seeing, and any intermediate optics. However, one must be careful where one element is designed to correct errors in another.

The MTF of an aberration–free circular dish is given by Zirin (1988) but a simple linear fit is adequate:

$$\text{MTF} \approx 1 - \frac{v}{v_0}. \tag{1}$$

where v is the spatial frequency in cycles mm^{-1}, and the limiting resolution v_0 corresponds to the Airy limit:

$$v_0 = \frac{\text{aperture}}{206265 \times \lambda} \quad \text{cycles arcsec}^{-1}. \tag{2}$$

The most important consequence of this form is that MTF = 0 if $v = v_0$ and there is no information at the limiting spatial frequency. The effect of aberrations is to reduce the MTF, especially for v/v_0, from 0.2 to 0. Central obscuration lowers the MTF around $v = v_0/2$, but actually increases it at v_0. If the aperture is covered except for two small holes at the edges, then we have a two–element interferometer, and the MTF is zero except for a spike around $v = v_0$. The MTF of film has a somewhat different form, fairly high for low spatial frequencies, then falling off steeply for higher frequencies near the grain size. But, just as v_0 corresponds to MTF zero, the film has little modulation near its resolution limit. The commonly used film Kodak 2415, popularly said to resolve 100 lines mm^{-1}, has an MTF of only 0.25 at that spatial frequency. So the image must be enlarged if arc–second resolution is desired. For a digital detector like a CCD, two pixels are necessary for each resolution element; four pixels are needed to distinguish two bars. Thus, if the scale is 1 arc sec per pixel, the limiting line separation is 2″, and the limiting spatial frequency is the inverse, 0.5 cycles arcsec^{-1}, at which the MTF is about 0.7. Thus, a full–disk picture with arc second resolution must have a detector with 3800 pixels across the diameter.

Since current CCDs are limited to 1024 pixels with slow readout or 720 with video, the real world requires a choice of field of view between low–resolution full–disk patrols or high–resolution observations with a limited field of view. The resolution, real or imagined, will never be as good as claimed or expected, because the optics and their alignment are never perfect, and the system MTF is the product of the MTFs of every component. For a 1024 × 1024 CCD full–disk telescope, we have 70% MTF at 3.7 arc sec ($v = .27$ cycles arcsec^{-1}); for a 15 cm aperture, the practical limit with a Lyot filter at 6563Å v_0 is 1.1 cycles arcsec^{-1}, so the overall MTF is 50% at 3.7 arc sec. That is the resolution of the best possible CCD systems today, no matter what is claimed.

5. Solar Telescopes

The primary aim of any solar telescope is to deliver a good image without distortion by the immense solar flux. Reflecting telescopes must be evacuated or possibly filled with helium. There is no need for this in refractors. Every non–optical surface exposed to the Sun must be painted titanium white to reduce heating. The vacuum window must be specially protected against heating from the edge. The mounting must be very stiff so that a servo guider can maintain accurate pointing to a high frequency. On older telescopes, a separate inside tower protected by a decoupled shell was used, but it has proved impossible to remove seismic coupling in such structures and the two separate structures are invariably weaker than a single massive one.

Because solar telescopes have long focal lengths and need only point at part of the sky, they are often fixed, light being reflected into them by movable flat mirrors. In the McMath telescope, a single–mirror heliostat reflects the beam to a large primary mirror at the bottom of the polar axis, and the reflected image is then brought to a vertical spectrograph. This simple arrangement produces a rotating image; in the coelostat, the face of the tracking mirror lies in and rotates about the polar axis, giving a non–rotating beam in a fixed direction. Since that beam is directed upwards, a second flat is needed to send it to the telescope. As the Sun moves across the sky, the second flat will shadow the coelostat, which must be moved to a different position. Since one wishes to avoid heat rising from the ground, the coelostat or heliostat is often placed atop a high tower, which also makes a long focal length and large image possible; however, the towers are usually small compared to the mountain and offer only partial relief. It is not easy to include a heliostat or coelostat in a vacuum system; some vacuum towers have been built with the mirrors outside but these do not appear to have been successful, perhaps because of mirror heating. In the Sacramento Peak tower, a computer–driven alt–azimuth system is used; this can be evacuated, but the image rotates and a rotating floor is required.

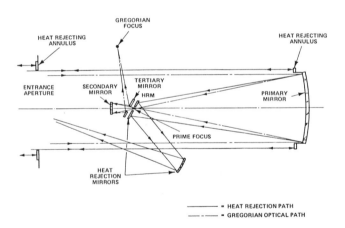

Figure 1. Scheme of the Gregorian solar telescope showing the beam and heat rejection paths. Only the primary sees full sunlight. The tertiary is thermally insulated from the heat rejection mirror.

Refractors in a closed tube usually do not need to be evacuated, probably because the light path is short and capped by the lens. Since the lens provides a window as well, it is easy to evacuate. For monochromatic observations, singlet lens systems may be ideal. Achromats, however, are needed if more than one wavelength, or a broad continuum, is to be observed. Yet achromats have a fatal flaw for solar work. Because achromatism is obtained by taking the difference in power between a strong positive and negative lens of different chromatism, a small thermal variation in the components makes a big difference in the focal length, and one must endlessly focus the lens, changing the image scale and interrupting the observations.

Lenses have other virtues. An error of $1/10$ wave in a mirror surface produces an error of $1/5$ wave in the resulting wave front because the beam traverses this distance twice; but a similar error in a refracting surface only produces an error of $1/30$ wave in the wave front, because it is the difference between glass and air paths $(n - 1)/n$ (where n is the index of refraction) that counts. Thus, for a given surface error, a lens is superior to a mirror and preferred for monochromatic work. Further, a lens reflects a few percent of the light incident on it, but absorbs almost none. Heating is thus minimized.

Equatorial systems have a non–rotating image and may be polarization–free. However, heavy secondary instruments are difficult to mount on the telescope and must be relegated to the coudé focus. For much solar observing, the Gregorian system (Figure 1) first introduced by ten Bruggencate (1950) at Locarno is particularly useful. Only the primary is exposed to a solar beam, forming an image on a cooled heat–rejection mirror. This only passes the field of view, the rest of the solar beam being reflected from the system. In a Cassegrain system, the whole beam falls on the secondary, with deleterious effects; in the Gregorian, all the optics beyond the heat rejection mirror get very little heating. The vertex of the Gregorian secondary must be carefully aligned to coincide with that of the primary; the field is limited and it is hard to change magnification, but the problem of solar heating of optics is solved. This scheme has been adopted for every proposed space solar telescope and is used in the BBSO reflector. Polarizing modulators must be located directly behind the heat stop.

If the telescope is equatorial, a dome is mandatory to reduce wind shake and heating; but turbulence around the dome slot degrades the image. It helps to use large fans to pull the air into the dome and exhaust it elsewhere; at the Pic du Midi, the front end of one telescope sticks out of the dome as in popular cartoons. From this point of view, the Sacramento Peak tower is ideal; the enclosed alt–azimuth mirror system at the top of the tower is relatively free of wind shake and the entire optical path is in vacuum. At the Huairou Observatory in China, the dome can be rolled away for domeless observing in calm weather.

While the main optics may be pointed at the Sun by a guider that tracks the four limbs, they are usually too massive to follow the rapidly dancing Sun. Therefore, secondary guiders have been introduced with light mirrors that can be rapidly pointed. Such guiders usually require a sunspot or some other distinguishing feature to lock onto. There has been some work with correlation trackers which may make it possible to guide on the granulation. The main telescope guider need not be complex; a good guider may be made with a short–focus (say 50 mm $f/10$) lens projecting an image on a silicon quadrant cell, with an occulting disk that covers much of the image. Even if the image is out of focus, any change in orientation will produce a detectable photon excess in the appropriate quadrants. The difference signal is then amplified and fed to the servos. Special precautions are required to compensate for passing clouds.

6. Spectral Resolution

Spectral resolution enables us to study the structure and velocities in the photosphere. Traditionally this was the province of the spectrograph but if two–dimensional observations are desired, the spectrograph functions must be replaced with narrow–band filters. A filter gives a two–dimensional image at a single wavelength, a spectrograph gives a one–dimensional spatial image at a range of wavelengths.

The spectrograph is the basic instrument to analyze light from the Sun. It consists of a slit to limit the entering light; a collimator, which makes the light diverging from the slit parallel; a grating, to disperse this light; and a camera to photograph the result. Various optical arrangements are used to achieve this result, depending on the goals of the particular spectrograph. All systems suffer from various off–axis aberrations, as the beams must go back and forth without hitting the other optics.

The simple Littrow spectrograph uses a single lens as collimator and camera. The slit is at the focus of the Littrow lens, which produces parallel light. This light is diffracted by the grating and refocussed by the lens on the film or plate. An image of the slit at each wavelength is produced. This system has the advantage of great simplicity, symmetry, and convenience in physical operation. Since the path length between the two ends of the grating is doubled, we get double retardation and higher dispersion. By combining BK7 and fluorite lenses, a nearly achromatic system with good UV transmission is obtained. Overlapping orders are either eliminated by filters, or separated and placed side by side by cross dispersers in the so–called echelle system. Good examples of the Littrow system are the thirteen–meter instrument at Sacramento Peak Observatory and the three–meter coudé at BBSO.

The critical part of the spectrograph is the plane grating, which diffracts light according to the formula

$$n\lambda = d(\sin\theta + \sin\phi), \tag{3}$$

where n is the order, d, the separation of the lines of the grating, λ, the wavelength, ϕ, the angle of incidence, and θ, the angle of diffraction. It is because n and λ occur together in the formula that the orders have to be separated; 6000 Å in the first order falls at the same point as 3000 Å in the second order. The inverse dispersion is the change in wavelength, $d\lambda$, per linear interval $fd\theta$ in the focal plane. Differentiating Equation (3) and using the Littrow condition $\theta \approx \phi$, and $\phi = $ constant gives

$$d\lambda = d\frac{\cos\theta}{n}d\theta = \frac{\lambda}{2}\cot\theta d\theta, \tag{4}$$

so the inverse dispersion (in Å/mm) is (for a Littrow system)

$$\frac{d\lambda}{fd\theta} = \frac{\lambda}{2f}\cot\theta. \tag{5}$$

The dispersion, expressed in mm/Å, depends only on the focal length of the spectrograph and the tangent of the angle of diffraction. Since the focal length is usually a fixed physical structure, the only way to get maximum dispersion (if desired) is to tilt the grating further and further over. Because the dispersion depends on the tangent of the angle of tilt, we get very large gains in dispersion at high angles. The factor two in Equation (5) is due to the autocollimation aspect of the Littrow system. As in other interference systems, the resolution depends on the retardation in wavelengths between light diffracted from the two ends of the system. A two–groove grating has the same resolution as a complete one; the function of the other grooves is to separate orders.

For a typical high–dispersion echelle–type grating, $\theta = 60°$, so the projected grating width is half the ruling width. To fit the spectrograph to a telescope of focal length F, one must match the slit width to the desired optical and spectroscopic resolution. We define RP(opt) $= 1/\Delta\Theta$ as the optical resolution, and RP(spec) $= \lambda/\Delta\lambda$ as the spectroscopic resolution. The smallest linear element observable is F/RP(opt), and this should match the slit size, which is $\Delta\lambda$ divided by the dispersion. Therefore,

$$\frac{F}{\text{RP(opt)}} = 2f\frac{\Delta\lambda}{\lambda}\tan\theta = \frac{4f}{\text{RP(spec)}}, \tag{6}$$

or,

$$\frac{F}{f} = \frac{4\text{RP(opt)}}{\text{RP(spec)}} \approx 4. \tag{7}$$

Since vertical angles at the slit must be equal, the telescope aperture in this case can be up to four times the projected grating width or twice the actual width. The factor four comes from the doubling of retardation in the Littrow mode and the $\tan\theta$ term. Because of the limitation of Equation (7), the speed of large telescopes for spectroscopy of point sources is limited by the size of the gratings. Wavelength calibration may be obtained by placing a sealed tube with a few iodine crystals ahead of the slit.

Unfortunately, the exposure time with such a system may be too long to obtain optimum RP(opt). Because of the low efficiency of gratings and the complex optics, spectroscopic exposures take seconds, while seeing degrades exposures longer than 1/30 second or so.

Other monochromators have been developed for more specialized applications: One important type is the Fourier Transform Spectrometer (FTS) which uses the reflection between two plates (a Michelson interferometer) to achieve high resolution, especially in the infrared. Unfortunately the FTS scans an even smaller area than the spectrograph, a single point, and the exposures can be many minutes.

Because of the limitation in examining a single point or line, two–dimensional monochromators were developed. The classic system is the spectroheliograph invented by George Ellery Hale in 1890 and used for his MIT B.A. thesis. The spectroheliograph makes an image in any line by scanning the slit across the Sun while the detector simultaneously moves across a second slit set to the desired wavelength. This gives a composite picture of the Sun in any chosen line. It is most flexible, but terribly slow, because many spectrograph images must be taken. But resolution is limited; changes in the sky transparency and guiding can produce a ragged image as the slit moves across the Sun. Leighton and co–workers pioneered the combination of images in different wavelengths to obtain Doppler, magnetic, and time–difference images. With modern electronic detectors, complex combinations of parts of the line profile are possible.

Because the spectroheliograph is bulky and slow, Lyot invented the birefringent filter, which permits instantaneous two–dimensional monochromatic images. If polarized light passes through a birefringent crystal of quartz or calcite with the axis of the crystal parallel to the face, the plane of polarization is rotated a different amount for each wavelength. A second polaroid at the back of the crystal passes only those bands of the right polarization. The resulting transmission is:

$$t_i = \cos^2 \pi n_i. \tag{8}$$

where n_i is the crystal retardation:

$$n = d(\epsilon - \omega)/\lambda. \tag{9}$$

ϵ and ω are the refractive indices for the two rays; for calcite, $\epsilon - \omega = -0.17$; for quartz, $\epsilon - \omega = 0.009$.

If we follow this sandwich with another half as thick, every other peak will be removed, and adding more, each again half as thick, the remaining peaks are removed so that a multilayer filter may be used to isolate the remaining peak.

Additional developments have made the Lyot filter the most powerful auxiliary for solar work. Lyot devised a wide–field version in which the element is split, the two halves rotated 90° and separated by a half–wave plate. This has the effect of making the optical axes symmetric and permits entrance of beams up to $f/15$ (4°wide). In addition, each element can be tuned in wavelength by placing a $\lambda/4$ plate before the second polaroid. Rotation of the polaroid shifts the wavelength between the normal peaks of the crystal sandwich. Evans (1949) showed that one could save a polaroid by placing one element, *sans* polaroid, between the split elements of a thinner one and crossing the end polaroids. The inner element is still wide field but the outer is not. However, if the outer is a thinner element, its field is wide anyway. It is possible to reduce the total polaroid number by four or five this way and nearly double the transmission. In this version the elements are no longer fully tunable.

Lyot filters have revolutionized the study of the Sun. The convenience of tuning, the wide field, and the clean pass–band have made it the monochromator of choice. However, optical calcite is rare and very expensive, so there have been a number of attempts to provide alternatives. So far, these have not really been successful.

One alternative has been the Fabry–Perot type filter: two coated plates with a thin separation, producing peaks separated by 3 or 4 Å. Rust et al. (1990) have used a single plate of lithium niobate, using an electric field to tune the plate. The problem is that the pass–band falls exponentially, while a Lyot filter element has real zeros in the transmission curve. Further, narrow–band multilayer filters are required to isolate the bands, and these are notoriously unstable and non–uniform.

The atomic resonance cell has been used by Cacciani and co–workers (Agnelli et al. 1975), employing the Macaluso–Corbino effect. Light passing through sodium vapor in a strong longitudinal magnetic field undergoes resonant scattering in the σ transitions and is circularly polarized. When the cell is placed between crossed polarizers, only the light absorbed and re–emitted in the σ transitions will have its plane of polarization rotated and pass through. Thus the filter isolates the wings of the Na D lines. A second cell placed in tandem is used as a selector cell to pick the blue or red wing.

Resonance cells may be made for the Na D lines or the potassium resonance line at 7699 Å. The passband is very narrow (about 0.03 Å) and free of sidebands. The cell is stable because the atoms always absorb their proper wavelength; on the other hand, that means it cannot be tuned easily to compensate for the Earth's diurnal and orbital motions or solar rotation. This is done by switching the selector cell or changing the magnetic field. For magnetic fields, a suitable modulator is placed in front of the cell which acts purely as a monochromator. For Doppler measurements, a modulator is placed in front of the second cell. There are excellent results for both longitudinal and transverse cells, and a single setup with two filter wheels can give longitudinal, transverse, and Doppler. The only drawback is that the system is somewhat long and the bandpass so narrow and pure that light levels are low. While the Cacciani cell has been used regularly both at Mt. Wilson and BBSO, it has otherwise been rejected by the solar physics community as "not invented here."

A new type of filter has been developed by Rakuljic (private communication), using reflection from a holographic pattern in $LiNbO_3$. This filter has low side bands and should be quite inexpensive; however, it is still in the development phase.

Finally, a warning: it is widely thought that extremely narrow filters are a necessity. While it is important to have a clean image free of variation and side bands, for many purposes an excessively narrow filter unduly extends exposures, complicates temperature stabilization, and gains little. For Hα patrols in the line center, 0.5Å is narrow enough. Videomagnetograms with 0.25Å are as good as, or better than, those with very narrow filters. And for active prominences, a few Å is adequate.

7. Magnetographs

Because of the dominant role of magnetic fields in solar activity, their measurement plays an extremely important role. George Ellery Hale was able to measure sunspot fields by the splitting of spectral lines, but this is only possible in sunspots. However, the direct measurement by splitting remains important as a calibration of magnetographs, since all of the photoelectric devices require independent calibration. Further, the strongest fields usually saturate magnetographs and can only be measured accurately by spectroscopic splitting, or some other device that separates the components in wavelength. To understand the magnetograph, we must understand the Zeeman effect, which is reviewed here.

In a magnetic field, the energy levels of an atom split according to the quantum number M_j (written as M for short), which is the projection of the total angular momentum J on the direction of the magnetic field. This is called the Zeeman effect (Figure 2). The line is split into a number of components, depending on the M, M' values of the two levels. There are three groups of components corresponding to what is called the "classical Zeeman triplet," for $J = 1 \rightarrow 0$. The unshifted line for $\Delta M = 0$ is called the π component, for "parallel"; it is seen when we look perpendicular to the magnetic field and is polarized *parallel* to the field. The two shifted components with $\Delta M = \pm 1$ are called σ_1 and σ_2 (for "*senkrecht*," or perpendicular), respectively; they are seen as left- and right-hand circularly polarized radiation along the direction of the magnetic field, and linearly polarized radiation *perpendicular* to the field vector when we look perpendicular to the field. The net polarization is zero in any direction; perpendicular to the field, the π component just equals the two σ components and, along the field, the two σ components cancel, giving no net polarization. The σ components are shifted in wavelength to the red and blue of the unperturbed line by

$$\Delta\lambda_H(\text{Å}) = 4.7 \times 10^{-13} g\lambda^2 H, \tag{10}$$

where H is the field strength in Gauss, g is the Landé g–factor averaged between the two states, and λ is the wavelength in Å. For a typical line, $g = 1$ and, at 6000 Å, the splitting at 3000 Gauss, typical for spots, is 0.05Å, barely detectable. Therefore, spectrograph measurements are made in the Fe 5250Å line, where $g = 3$, and the splitting is easily measured. For fields less than 1000 Gauss, the splitting is not easily measured and, therefore, photoelectric means are required.

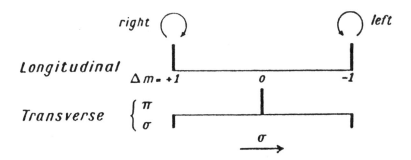

Figure 2. The Zeeman effect

The first magnetograph, developed by Harold Babcock in 1953, used an electro–optic KDP crystal ahead of the spectrograph. With the proper voltage, this crystal produces a retardation of $\pm\lambda/4$, which changes one or the other circularly polarized σ component into a plane component parallel to the axis of a polaroid placed behind it. A special slit separates the blue and red wings of the λ 5250 line; as the voltage is alternated, the absorption line shifts back and forth slightly in the spectrograph as one and then the other σ component comes through. Further refinements include radial velocity compensators and null–seeking systems. The Babcock magnetograph has operated at Mt. Wilson for many years.

The Babcock system measured the field at a single point. Further developments concentrated on measurements over an area. Leighton (1959) used simultaneous photographic spectroheliograms for two–dimensional magnetograms and Dopplergrams using the 60-ft tower at Mt. Wilson. For Dopplergrams, direct images in one wing are subtracted from the other; elements moving toward the observer are blue shifted and appear less intense in the blue; those moving away from us are brighter (darker on the negative). Superposition of the blue negative and red positive produces a uniform gray except where there is a Doppler shift. The superposition is necessary to eliminate the background brightness variation. In the case of a magnetic field, an analyzer is used to separate the Zeeman components and subtraction, as before, gives an image with density depending on the magnetic field. With these techniques, the five–minute oscillation and the structure of the chromospheric network were discovered.

With the coming of computers, the tedious matching of photographs has been replaced by electronic subtraction. Livingston *et al.* (1971) replaced the simple detector of the Babcock system with a double array of diodes along the slit of the Kitt Peak National Observatory spectrograph, so that the field at up to 512 points (now 2048) could be measured simultaneously. This powerful device has produced a matchless daily record of the solar magnetic field; it still produces the best full–disk magnetograms.

Leighton and Smithson (Smithson 1973) replaced the spectroheliograph with a Lyot filter and the diode array with a vidicon in the videomagnetograph of the BBSO. Two–dimensional magnetograms could then be obtained directly, without the slow process of adding up the linear results. A $1/4$ Å filter is used to limit the light to the wing or wings of a single line (usually 6439 Å or 6103 Å), with the initial polarizer removed (Ramsey 1971) to permit a double bandpass. Ahead of the filter, a potassium dihydrogen phosphate (KDP) crystal is modulated to give $\pm \lambda/4$. If it is positive, the left–circular Zeeman component σ_1 is linearly polarized in one plane and the σ_2 component is polarized orthogonally. When these two components enter the filter, they are passed at $\pm 1/2$ the passband (or ± 0.125 Å) with the result that a given magnetic polarity produces the same effect in either passband. The electro–optic crystals are difficult, but far superior to rotating wave plates, which usually shift the image, making subtraction impossible.

In the present system (Mosher 1976; Zirin 1986), the image is recorded by a CCD operating at video rates, digitized on–line, and accumulated in a memory that is set to half intensity at the beginning of the sequence. The KDP voltage is now reversed and a new exposure is obtained, with reversed signal. The process continues at the 30 frame sec^{-1} video rate; for active regions, good magnetic maps are produced in 32–128 frames, or 1–4 sec; for weak fields integrations of up to 4096 frames are used. At the same time, the brightness is accumulated in a separate register and divided into the signal. The addition of a circular polarizer (polaroid $+ \lambda/4$ plate) in the double pass–band mode alternates the wavelength between red and blue wings of the line, producing a Doppler analyzer. To obtain the transverse field, a circular analyzer is inserted which gives the linearly polarized σ component; this is measured at two positions, comprising the Stokes Q and U components. Finally, these are combined to give the transverse field direction and strength. Intensity is measured separately and the magnetogram is divided on-line by that signal.

The videomagnetograph is extremely sensitive because no scanning is involved and the images may be added up almost indefinitely. For active regions, 64 video frames are typically required, and, for the quiet Sun, as many as 4096. Because video rates are used, the cameras are fairly noisy, with a signal-to-noise ratio of about 50:1; for 4096 frames this gives 3200:1.

An alternative is to use a large–scale CCD. Since these have a large electron well, they can have a signal-to-noise ratio of 500:1 or more; thus, two frames are sufficient for an active region magnetogram. Unfortunately, the present chips can only be read out slowly, so the long interval between the subtracted images makes them much harder to match up. The two frames must be de–stretched, because image shifts vary across the field. The resulting magnetogram has higher resolution than the video type because two very good frames are exactly matched while, in a 32–frame videomagnetograph, the telescope wanders a bit. However, the two frames are inadequate for weak–field magnetograms and the many frames varied enough that destretching cannot be used. The many frames of the video system permit an even higher signal–to–noise ratio.

For deep magnetograms, the brightness variation of the granulation becomes a problem. It can be removed by using the filter in a single bandpass on the blue edge of the spectral line where the blue–shifted absorption due to the rising granule cancels its excess brightness and produces a more uniform brightness field.

This system is sensitive only to longitudinal magnetic fields but, if we insert a $\lambda/4$ plate the system, we alternately see the linearly polarized π and σ components emitted at right angles to the magnetic field. Modulation gives us the component amplitude of the transverse field in one direction. By rotating the $\lambda/4$ plate or modulating a KDP, the polarization component along another axis is obtained and, combining these two, gives the direction and magnitude of the transverse field. Brueckner (Hagyard 1982) developed a vector magnetograph at the Marshall Space Flight Center using such an arrangement. The sensitivity is not great and the reduction is tedious, but the data are much prized in flare studies. With modern image processors, these magnetographs should improve.

Because of the great interest in time variation and high spatial resolution in magnetic fields, almost all the activity in magnetography today is with two–dimensional systems. The point–by–point systems are best for full–disk observations where time resolution is not important. Recent years have seen migration to the near IR, where splitting is greater and seeing, less.

8. Coronagraphs

Astronomers, too impatient to wait for eclipses, long sought ways to observe the corona outside eclipse. George Ellery Hale climbed Mount Etna with a small telescope but volcanic dust and smoke stopped him. Since the corona is a million times fainter than the photosphere, a system free of scattered light is required. Lyot recognized that the limitations on observing close to the Sun were sky brightness and scattering from the objective. The first he eliminated by going to mountain altitudes. The second was eliminated by using a blemish–free singlet objective and an occulting disk or artificial moon. This does not eliminate diffraction from the circular aperture or double reflection at the center of the objective. So he placed a field lens behind the occulting disk which imaged the objective on a stop that removed the diffraction and internal reflection of the objective. Even such a system has scattered light of 10 millionths or more, while the K corona is 10 times fainter. So he used a spectrograph or Lyot filter to restrict observations to the coronal emission lines, which are 50 millionths of the disk intensity in their narrow 1Å ranges. This permitted observation of the coronal emission lines. Generals were persuaded that observations of the coronal lines would permit the prediction of solar activity, so coronagraphs built on this principle have been operated all over the world ever since. Although they were not much good at prediction, they did produce the first suggestion (Bell and Glazer 1954) that coronal holes were associated with magnetic activity.

The Lyot coronagraph is also unsurpassed for prominence cinematography and spectroscopy, since the prominences are often too faint for ordinary telescopes. The same is true for limb flares, where the weaker lines are most interesting. Modern enhancements using electronic cameras and narrow–band filters have permitted sky subtraction, and coronal lines may now be detected out to 0.5 $R_\odot$. Recently Koutchmy has reported progress with the reflecting coronagraph, which eliminates the severe chromatic aberration of the singlet lens in the Lyot system. One simply uses a superpolished off–axis primary mirror and a secondary with a hole in it; the disk goes through the hole while the corona goes back into the system.

The detection of the continuous coronal emission required development of the K–coronameter (Wlerick and Axtell 1957), which detects the electron corona by its high degree of polarization. Further development (Fisher *et al.* 1981) has made possible observation of the K corona to some distance from the Sun. The polarization is detected by an electro–optic modulator which, with a quarter–wave plate, gives alternately 0 or 1/2 wave retardation, which has the effect of choosing alternating orthogonal polarizations. These instruments must be exceptionally clean optically.

The scattering of sunlight in the objective limits further improvement on the ground, but the use of external occulting disks (Wagner *et al.* 1981) in space coronagraphs has made possible observations out to ten solar radii. Several disks must be used in tandem to block out the diffraction from the other disks.

9. *Gedankeninstruments*

Finally, there is the important class of *gedankeninstruments*, or virtual instruments. These are non–existent or improbable devices, normally proposed to government agencies or panels of theoreticians. These instruments exist only as dreams or promises; if ever completed, they normally have a duty cycle of 0%.

Gedankeninstruments are popular because they are easily funded (because they promise the world) and you do not have to observe with them. They are generally preferred to real instruments because people enjoy magic dreams, which need not be delivered because they never are placed in service. Real instruments, by contrast, always display some shortcomings and, therefore, appear inferior.

To plan a *gedankeninstrument*, you should first conceive a real need, then a non–viable solution, and arrange several workshops (preferably to audiences of theoreticians or NASA functionaries) on the device. Promise the world.

Examples are the *gedankenmagnetograph* (especially the *gedanken–vectormagnetograph* and its close relative the *gedankenStokespolarimeter*). In addition, the *gedankennarrow-bandfilter*, and of course the *gedankentelescope*. To avoid assassination, I will not mention specific examples.

10. References

Bell, B. and Glaser, H. 1957. *Smith. Contr. Astr.* **2**, 159.

Bruggencate, P. ten, Gollnow, H., and Jager, F. W. 1950. *Zs. f. Ap.* **27**, 223.

Cacciani, A., Varsik, J., and Zirin, H. 1990. *Solar Phys.* **125**, 173.

Dainty, J.C. and Shaw, R. 1974. *Image Science.* New York: Academic Pr.

Evans, J. W. 1949. *J. O. S. A.* **39**, 229.

Fisher, R. R. *et al.* 1981. *Appl. Opt.* **20**, 1094.

Hagyard, M. J. *et al.* 1982. *Solar Phys.* **80**, 33.

Leighton, R. B. 1959. *Ap. J.* **130**, 366.

Livingston, W. *et al.* 1971. *Publ. Roy. Obs. Edinburgh.* **8**, 52.

Lyot, B. 1930. *C. R. Acad. Sci. Paris.* **191**, 834.

Lyot, B. 1933. *C. R. Acad. Sci. Paris.* **197**, 1593.

Lyot, B. 1939. *M. N. R. A. S.* **99**, 586.

Mosher, J. M., 1976. *BBSO Preprint,* **No. 159** .

Ramsey, H. 1971. *Solar Phys.* **21**, 54.

Rust, D. M., Obyrne, J. W, and Sterner, R. E. 1990. *J-H APL-TEC* **11**, 77.

Smithson, R. C. 1973. *Solar Phys.* **29**, 365.

Smithson, R. C. Peri, M. L., and Benson, R. S.. *Applied Optics* **27**, 1615.

Wagner, W. *et al.* 1981. *Ap. J.* **244**, L123.

Wlerick, G. and Axtell, J. 1957. *Ap. J.* **126**, 253.

18 SOLAR ULTRAVIOLET INSTRUMENTATION

J.B. GURMAN
Laboratory for Astronomy and Solar Physics
NASA Goddard Space Flight Center
Greenbelt, Maryland 20771
U.S.A.

ABSTRACT. The ultraviolet is the only waveband in which the entire outer solar atmosphere, from chromosphere to corona, can be accessed. After examining the types of ultraviolet observations necessary to determine morphology and the state variables of the atmospheric plasma, I review very briefly some of the solar ultraviolet instrumentation of the last two decades. Among recent developments of note are changing detector technologies, multilayer coatings for high-resolution imaging at short wavelengths, and "solar-blind" optics.

1. Observables

With the exception of a few forbidden lines and the Thompson scattering K-corona in the visible, the upper solar atmosphere is observable only in the ultraviolet, soft X-rays, and at radio frequencies. Although competing radio emission mechanisms can often be discriminated (see Chapter 14), only the ultraviolet and soft X-rays offer unambiguous, optically-thin views of the upper solar atmosphere at all times. To test theories of the heating of the corona, understand the structure of the Sun's outer atmosphere, and discover where the solar wind is accelerated, therefor, we must observe at short wavelengths. To measure densities, temperatures, or magnetic fields in the transition region, we must observe the ultraviolet emission lines of the highly ionized species formed in the steep gradient between chromospheric and coronal temperatures.

What are the observables that give us information on the physical parameters we need to understand the great mass of detail in the solar atmosphere? Table 1 gives a simple-minded summary. The first "observable," morphology, is of particular interest for this part of the Sun's atmosphere, largely because the outer solar atmosphere *is* structure. Without some knowledge of the scales, lifetimes, and interactions of the structures that form the chromosphere, transition region, and corona, it is impossible to interpret reliably the measurements we may be able to make of more quantifiable, physical quantities. The early treatment of the transition region as a homogeneous, plane-parallel onion skin separating the chromosphere and corona (*e.g.* Pottasch 1964, Athay 1966) was not developed in ignorance of the highly structured nature of the chromosphere and, by inference at least, the transition region. Rather, these beautiful, plane-parallel models balancing downward conduction from the corona with radiative emission were able to explain the existing observations, because instruments of the day were unable to resolve any spatial or spectral features that would upset that simple picture. Thanks to observations of high spatial and spectral resolution,

J. T. Schmelz and J. C. Brown (ed.), The Sun, A Laboratory for Astrophysics, 395–410.

we now know that much of the transition region is concentrated in small, low-lying loops that are characterized by steady flows and sometimes violent jets, that active regions appear to vary on every spatial and temporal scale we have been able to measure, that transient flows characterize the transition region over sunspot penumbrae and regular oscillations characterize their umbrae, and so on. While we would like sub-arc second *spatial* resolution to determine the true sizes of the smallest features, very modest *spectral* resolution $\lambda/\Delta\lambda$ is required, particularly in the EUV, to isolate complexes of lines formed at roughly the same temperature.

Table 1. Solar Ultraviolet Observables

Information	Observable	$\lambda/\Delta\lambda$
Morphology	Imaging in lines and continua formed at various temperatures	≥ 10
n_e, T_e, emission measure	If optically thin, spectrally integrated line emission in various features	≥ 500
$v_\parallel$ **B**	Detailed line intensity profiles Detailed line intensity and polarization profiles	$\sim 3 \times 10^4$
$T(h)$, $p(h)$	If optically thick, detailed profiles of several lines (including strong lines such as Lyα, Mg II) and continua	$\sim 3 \times 10^4$

If all we want to know is the quasi-hydrostatic structure of a feature, density and temperature diagnostics, as well as differential emission measures, can be determined from integrated line profiles alone. The spectral resolution required for such observations is still rather low. If we want to be able to characterize subsonic velocities, however, we need to sample emission line profiles twice per Doppler width, or a resolution of $\sim 30,000$ for the C IV line at 155 nm. If we want to extract every bit of information possible out of line profiles, we can measure magnetic fields using the Zeeman effect. We then not only need to characterize the entire Stokes vector (linear and circular polarization as well as intensity), but we also require a very large number of photons to allow statistically reliable measurements of, say, the difference of left- and right-hand circular polarizations in the ultraviolet, since the λ^2-dependence of the Zeeman separation of the circularly polarized σ components is no great help at these wavelengths. Even the strongest magnetic fields in the transition region are difficult to measure without a light bucket of some size.

2. History

It is sometimes disconcerting to realize that solar UV instrumentation is less than 45 years old. As Richard Tousey of the U.S. Naval Research Laboratory, one of the pioneers of space solar astronomy, noted in his 1990 Hale Prize lecture, no one even knew if the Sun would have a measurable ultraviolet spectrum, or whether instead the Sun was a blackbody in the ultraviolet. That was a bit of an exaggeration, since it was already known that something was changing the ionization in the ionosphere on very rapid time scales in solar flares, but the beginning of our understanding of the outer solar atmosphere really dates from those first rocket flights. Most of the rocket flights in the first decade carried spectrographs but, by 1959, the NRL group was able to obtain Lyα images of the Sun, recorded on film, with about 1' resolution (Purcell *et al.* 1959). When integrated, photometric EUV spectra became available in the early 1960's, Pottasch (1964) and others used these observations to construct the first models of the transition region as a thin layer of nearly constant pressure. Only with the early *Orbiting Solar Observatory* spacecraft in the 1960's, however, was it possible to produce images of the transition region on a regular basis. The Harvard EUV instrument on *OSO-4* lasted only five weeks, but was able to produce 100 spectra in the range 30 - 140 nm, and some 4000 monochromatic, full-disk images at wavelengths in this range. With a spatial resolution ∼1', one could actually see individual active regions rotate across the disk in the activity belts.

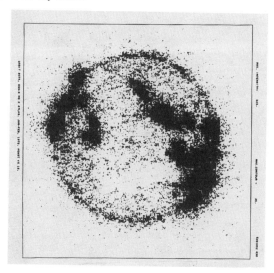

Figure 1. *OSO-6* Full-disk image in Mg X 62.5 nm. Note the coronal hole just south of the solar equator. From Gurman (1972).

The instrument that produced Figure 1 used normal incidence optics. Figure 2, in fact, shows the Harvard *OSO-6* instrument, but the only major changes were in the mirror figure — *OSO-6* employed an off-axis paraboloid, coated with iridium for EUV reflectance, while the earlier instrument had a spherical mirror — and the size of the entrance aperture, which gave approximately 35″ resolution. The detector was zero-dimensional: a photomultiplier with a bare tungsten photocathode — bare because coatings such as LiF, which transmits down to 105 nm, is opaque at shorter wavelengths. Tungsten was used because, due to its high work function, it is essentially "solar-blind;" that is, it is insensitive to photons much above 200 nm. (The zero-order detector, a photocell, was used as an optical reference to calibrate the cam motion that scanned the spectrograph grating.) The grating was spherical, ruled at 1800 lines mm^{-1}, and evaporation coated with gold. Spectral resolution was about 0.3 nm FWHM. The grating mount was driven by a stepper motor *via* an eccentric cam cut to provide linear step size in wavelength. The entire *OSO-6* instrument could be rastered to produce images. A 46′ × 46′ raster required about eight minutes; a 7′ × 7′ raster, some 30 sec. It was really with instruments such as these that it first became possible to measure densities and construct differential emission measure models that distinguished quiet and active regions (*e.g.* Dupree 1972, Withbroe and Gurman 1973). Indeed, it was with *OSO-4* observations that Munro and Withbroe (1972) were able to characterize the lower density and coronal temperature in coronal holes.

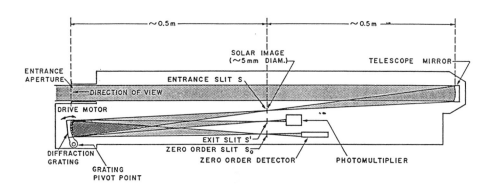

Figure 2. Optical layout of the Harvard EUV spectroheliometer on *OSO-6*.

The heritage of NRL spectrographs and spectroheliographs, and Harvard spectrometers, achieved its fullest flowering in the Apollo Telescope Mount (ATM) on board *Skylab*. In addition to a white-light coronagraph and a soft X-ray imager, the ATM carried four UV instruments. The S055 spectroheliometer was an upgrade of the earlier Harvard *OSO* designs, with a larger mirror, significantly better spatial resolution, and multiple photometric detectors to allow recording images in several lines simultaneously. The NRL S082-A instrument produced overlapping images of the solar disk with high spatial resolution and the S082-B, also from NRL, was capable of high spectral resolution over an extended wavelength range. Finally, the S020, infrequently mentioned, was originally designed to be hung out an airlock door but never did perform as designed — most probably due to contamination during EVA's. The environment of manned space vehicles is rather messy for instruments with thin filters that can get coated with hydrocarbons, or narrow slits that can be jammed with debris, and so on.

During the same period as *Skylab*, the *OSO-7* spacecraft was obtaining observations with a Wolter Type 2 telescope and spectrograph operating in the range 26 - 40 nm (Underwood and Neupert 1974). The paraboloidal-hyperboloidal mirror system was capable, in principle at least, of achieving 20″ spatial resolution, although the broad modulation transfer function, combined with the high contrast of solar features in some lines, made the actual resolution dependent on the spectral line observed. Spectral resolution of the 1152 line mm^{-1}, spherical grating spectrometer was ~500. The *OSO-7* payload, which also included soft X-ray filtergraphs and a hard X-ray polarimeter, was optimized for observing both flares and active regions, as well as coronal holes. In addition, strong lines such as Fe XV 28.4 nm were used to map the temperature distribution between 1.1 and 2 $R_\odot$ (Nakada *et al.* 1974).

Table 2. UV Instruments on *Skylab*

Instrument	Optics	Spatial Res. (″)	Spectral Res. (nm)	Spectral Range (nm)	Detector
S055	normal incidence	5	0.16 - 0.8	28 - 134	CEM's (7)
S082-A	normal incidence, Al filter	2	≥ 0.027	16 - 64	film
S082-B	normal incidence	2×60	0.006	117-195	film
S020	grazing incidence	none	0.004	1 - 20	film

While *Skylab* and *OSO-7* were flying, the Colorado and LPSP teams were building their spectrometers for *OSO-8*, and NRL was constructing its High Resolution Telescope and Spectrograph (HRTS). These were the first really high spectral resolution instruments which could begin to distinguish fine-scale features and describe their dynamics. The LPSP spectrometer obtained simultaneous, cospatial, high resolution profiles of several strong lines in any of a number of entrance slits arranged around a wheel in the spectrograph exit plane; the Colorado High Resolution Ultraviolet Spectrograph had a decker to adjust the entrance slit length and a wide range of accessible wavelengths. The *OSO-8* instruments are notable for two reasons. First, a large set of time series of chromospheric and transition region line profiles obtained with these instruments showed conclusively that acoustic waves provide insufficient energy to heat either the upper chromosphere or (by inference) the corona in the quiet Sun. Second, the rapid loss of sensitivity, particularly at shorter wavelengths, pointed out the dangers of exposure of magnesium fluoride-coated optics to ultraviolet emission! Both instruments lost a factor of 10^3 in sensitivity at Lyα during the first three months in orbit. (Figure 3). The throughput degradation of the Colorado instrument proceeded to longer wavelengths, dropping a factor of 100 at 155 nm in four months. In six months, the sensitivity at 280 nm decreased by over two orders of magnitude. In addition, the Colorado slit decker eventually malfunctioned and was left at a long setting.

What caused these drastic losses in throughput? Woodgate (1982) lists several reasons:

• Both instruments used Cassegrain telescopes that concentrated solar EUV on the secondary by a factor of 20 to 30.
• Neither instrument had a telescope door.

400

- An oil leak occurred during spacecraft vacuum testing.
- Oil-free vacuum systems were not used.
- Vinyl tubing was used for back-filling a vacuum chamber used for spacecraft testing.

Thus, the telescope design, the lack of an instrument door to allow outgassing before exposing the optics to the Sun, and a variety of sources of contamination with hydrocarbons combined to reduce instrument throughput drastically after only a few months. At the shortest wavelengths ($\lambda \leq 125$ nm), this loss occurred on a time scale of days to weeks. Exposure of MgF_2- or LiF-coated optics to solar EUV in the presence of free hydrocarbons appears to be a recipe for disaster.

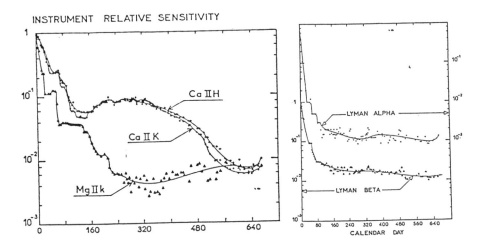

Figure 3. Sensitivity loss of the LPSP spectrometer on *OSO-8*, as a function of time, in days from launch. From Bonnet *et al.* (1978).

Meanwhile, HRTS was multiple times on sounding rockets — a recent flight was HRTS 7 — and for a week on the *Spacelab 2* mission of the Space Shuttle *Challenger* in 1985. The HRTS optical design (Figure 4) is quite elegant: it employs two concave Wadsworth gratings, not only to increase the total dispersion, but also to cancel out virtually all coma and astigmatism. The result is stigmatic spectra — spectra with spatial resolution of ∼1″ along the image of the 900″ long slit — obtained simultaneously over 70 nm of chromospheric, transition region, and even a few coronal lines (Bartoe and Brueckner 1975). The holographic gratings in the original, "symmetric tandem Wadsworth" design are of 2400 lines mm⁻¹. The only real drawbacks of the HRTS design are relatively minor. Despite what this diagram suggests, the HRTS focal plane is curved, requiring a film transport that can hold the film rigidly against a curved back. UV-sensitive film itself is a weakness, not for 5-minute rocket flights, but in the Shuttle payload bay environment. Thermal problems plagued HRTS on the *Spacelab 2* flight and the film, always sensitive to temperature, probably lost at least an order of magnitude in dynamic range due to in-flight baking. The length of the flight exacerbated the thermal problem since the Cassegrain telescope had never been exposed to solar irradiation for so long; the overheating of the entire package evidently led to some internal contamination. Once again, we must recognize that exposure of delicate instruments to solar UV, even for several days, is a stringent test of the robustness of their design and the completeness of contamination control during their construction. Nevertheless, the data returned by HRTS are nothing less than the best UV spectra of the Sun to date. Explosive events, sunspot flows, transition region downflows, and a variety of other small-scale phenomena have only been discovered because of the unique combination of spatial and wavelength coverage and resolution achieved by HRTS.

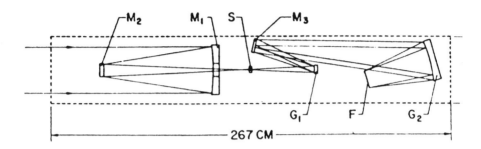

Figure 4. A schematic optical layout of a tandem Wadsworth spectrograph of the type employed in the NRL HRTS (Bartoe and Brueckner 1975). This design eliminates both coma and astigmatism.

The *Solar Maximum Mission* (*SMM*) was meant to carry both EUV (30 - 140 nm) and UV (115 - 360 nm) instruments but, due to financial constraints, only the longer-wavelength instrument actually flew. Although it was based closely on the Colorado *OSO-8* spectrograph, the Ultraviolet Spectrometer and Polarimeter (UVSP) on *SMM* was almost totally redesigned.

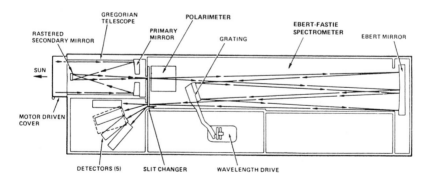

Figure 5. Schematic of the UVSP spectrometer on the *Solar Maximum Mission*. Novel features included the polarimeter, the wheel with both entrance and exit slits, and the door. From Woodgate *et al.* (1980).

Instead of a Cassegrain, the UVSP had a Gregorian telescope to reduce the solar flux incident on the secondary mirror. In addition, a field stop at the prime focus limited the beam to the $4' \times 4'$ field of view; and the stop was cooled by four copper legs connected to external heat pipes. The net effect of these two design changes was to reduce the flux on the secondary by a factor of 20 relative to the *OSO-8* design and, thus, reduce the chances of polymerizing hydrocarbons onto the secondary. Heaters were used to reduce condensation on the mirror surfaces, and the telescope door was opened only during spacecraft night for the first two weeks in orbit to allow outgassing. All these anticontamination measures worked quite well, and probably would have prevented all but the most minimal loss of short-wavelength sensitivity — if only the *OSO-8* spare wavelength drive unit were not used. Unbeknownst to the UVSP designers, when the instrument was kept warm enough to achieve best telescope and spectrograph focus, bleed valves in the wavelength drive gear train opened — and dispersed a hydrocarbon-based lubricant while solar UV was baking the mirrors and grating. As a result, UVSP throughput at Lyα dropped by a factor of 200 in the first three months in orbit. When the problem was understood, the instrument was operated at less than optimal focus and the reduced temperature apparently prevented further degradation. Fortunately, the sensitivity loss was largely confined to wavelengths below 130 nm. At C IV, throughput dropped by less than a factor of three in five years (for two and a half of these years, the instrument door was closed while waiting for the spacecraft attitude control system to be repaired).

The UVSP spectrograph was an Ebert-Fastie design, and the exquisite holographic grating scattered less than 10^{-4} of a line per Ångstrom into the surrounding spectrum. In this design, a spherical mirror acting as both collimator and camera removes all coma. The drawback is that the spectrograph is not stigmatic. (An excellent and entertaining review of the rediscovery of the principle of the Ebert spectrograph may be found in Fastie (1991).) This was not a difficulty at the time, however, because the only photometric, photoelectric detectors available in the mid-1970's were zero-dimensional: channel electron multipliers and photomultipliers. In order to sample more than one narrow bandpass at a time, UVSP had a slit wheel that paired entrance apertures ranging in size from $1'' \times 1''$ to $15'' \times 286''$ with exit slits that covered as little as 0.001 nm to as much as 0.33 nm. Many of the exit slits were bisected by beamsplitter prisms that allowed the short- and long-wavelength halves of transition region lines to be directed to different photomultipliers. When the secondary mirror was rastered, both intensity and line-of-sight velocity images ("Dopplergrams") could be constructed.

UVSP was also able to obtain high resolution spectra of superb quality: in first order, for example, the Mg II profiles had even better signal-to-noise than those obtained by the LPSP *OSO-8* spectrograph. Finally, UVSP was able to do something that no UV instrument before, since, or planned can: measure polarization. Measurements of several sunspots with UVSP have shown that the longitudinal component of the umbral magnetic field is still at least 1000 Gauss at the height of formation of C IV (Henze *et al.* 1982, Henze 1991). Force-free extrapolations from photospheric magnetograms imply that that height is some 4000 km above the visible surface, or about twice what hydrostatic model atmospheres would predict (Hagyard *et al.* 1983).

3. Where Are We Now?

While film is still used almost exclusively for sounding rocket flights, the increasing separation of manned space flight from scientific payloads means that film is unlikely ever to be used again as a detector in a long-duration mission. (In addition, Eastman Kodak, long the sole source of the UV-sensitive film used for solar space applications, has suspended production. Current stocks are not expected to last the decade.) In the last ten years, 1- and 2-dimensional photon-counting detectors have been introduced that are highly efficient in the ultraviolet; the most efficient are microchannel plates (MCPs) — arrays of curved channels that create a cascade of electrons for each photon that enters a given channel. Various anode structures, including the multianode microchannel array (MAMA) and wedge-and-strip, can be used to give precise positional information as well as time of arrival (which makes MCPs ideal for astrophysical applications). Bare MCPs, that is, without photocathodes, typically have quantum efficiencies of 10% from 10 to 100 nm, but are "solar-blind" at longer wavelengths. This is what makes MCPs so desirable as solar UV detectors (Rottmann 1990). If longer-wavelength response is desired, it is hoped that coating with an opaque photocathode material such as KBr can extend the MCP sensitivity. The *SOHO* Solar Ultraviolet Measurements of Emitted Radiation (SUMER) spectrograph will use MAMAs (see, *e.g.*, Timothy and Bybee 1986) with two different coatings, partly as a hedge against unexpected performance changes. MCPs have an important drawback: their dynamic range is limited to 10 - 100 counts sec^{-1} from an individual channel, since only so many electrons can cascade down a given channel. For solar applications, this means frequent readout to accommodate the broad dynamic range of transition region and coronal features.

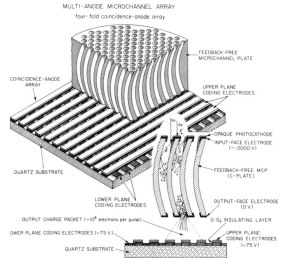

Figure 6. Schematic showing the amplification principal of the microchannel plate array, and the position encoding of the multiple anode multichannel array (MAMA).

Charge-coupled solid-state devices (CCDs) are perhaps the best known two dimensional detectors; because they are made of SiO_2 which absorbs strongly in the UV and EUV (Rottmann 1990), they are used infrequently in the ultraviolet. CCDs have excellent dynamic range — typically 10^5 — however, so they are a good match to the tremendous

Table 3. MAMAs and CCDs

Characteristic	MAMA	CCD
Intrinsic UV sensitivity	High	Low (can be UV flooded)
Solar blindness	Yes (bare, Cs_2Te, KBr photocathodes)	No
Dynamic range	≤ 100 count $chnl^{-1}$ sec^{-1}	10^5
Radiation damage	None	Well filling (raises background)
Operating temperature	No cooling required	-30 to -100 C
Hydrocarbon contamination	Less likely	Cold traps
Do 1024×1024 detectors exist yet?	?	Yes

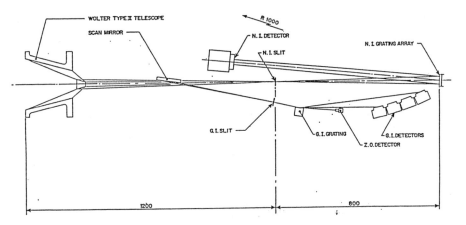

Figure 7. Schematic of the Coronal Diagnostic Spectrometer on *SOHO*. The grazing incidence detectors are one-dimensional microchannel plates, and the normal incidence detector system employs an EUV-activated phosphor and a visible-light-sensitive CCD. From Patchett *et al.* (1988).

dynamic range of solar UV emission. One approach to adapting CCD's to ultraviolet applications is to use the back side of a thinned CCD so that UV photons can reach the charge collection volume — the so-called "backside-thinned" CCD. Another approach is to design a UV detector that produces visible photons for a CCD back end. This is the scheme adopted by the Goddard Solar Physics group in building the normal-incidence detector for the *SOHO* Coronal Diagnostics Spectrometer (CDS). A UV-sensitive microchannel plate produces electrons that impinge on a phosphor that emits visible light, which is then imaged on the CCD by either a field lens or a fiber-optic bundle.

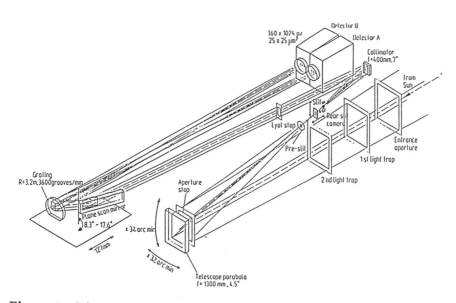

Figure 8. Schematic of the *SOHO* SUMER instrument (Wilhelm *et al.* 1988). The detectors are MAMAs with various photocathode treatments, including a bare section. The optics are fabricated of silicon carbide for visible light rejection.

The *SOHO* CDS and SUMER instruments are the latest culmination of the heritage of the various spectrographs already discussed here. The CDS has a grazing-incidence, toroidal, Wolter type 2 (coma-free) telescope. Part of the telescope beam is intercepted by a grazing-incidence scan mirror that directs the incident light to a Rowland-circle grating; the dispersed, astigmatic beam is detected by a series of four channel multiplier array (CMA) plate detectors with wedge-and-strip anode position sensing. The CMA is capable of processing on the order of 10^4 counts sec^{-1}. Since the grazing-incidence spectrometer is astigmatic — non-imaging — these detectors will in effect be one-dimensional in wavelength. Spatial coverage is achieved in one axis by scanning the grazing incidence mirror and, in the other, by stepping a 5″ × 5″ slit normal to the mirror rotation plane. Grazing incidence spectral coverage will run from 15 to 80 nm. The part of the beam not intercepted by the grazing incidence mirror is intercepted by a set of toroidal, and thus stigmatic, gratings, of

1800 to 3600 lines mm^{-1}. The two-dimensional detector thus records spatial information in one dimension and spectral in the other; a $2'' \times 2'$ entrance slit can be stepped to allow building up images with line profile information at each pixel. The spectral range here is 31.5 to 77 nm.

SUMER is a longer wavelength instrument, covering a range from 50 nm (overlapping CDS) at its short wavelength end to possible as high as 155 nm. Notice that this range includes the H Lyman series and continuum, as well as lines formed at the entire range of chromospheric, transition region, and quiet coronal temperatures. The SUMER detectors will be MAMAs; it is planned for them to bear different coatings to allow simultaneous access to the longer and shorter wavelength parts of this range. To keep even the coated MAMAs solar-blind, the SUMER mirrors have been fabricated of silicon carbide, which has a reflectance ranging from 0.2 at 60 nm to ≤ 0.4 at 100 nm. The three specular reflections (see Figure 8) alone yield over an order of magnitude attenuation at the longer wavelengths. The MAMA detectors will image space as one dimension and wavelength as the second; a scan mirror builds up the second spatial dimension.

Although *SOHO*'s objectives concern the quiet Sun — helioseismology and, in the case of CDS and SUMER, the acceleration of the solar wind — and thus do not usually require high temporal resolution, the raw data rate generated by such large-format, two-dimensional detector systems is substantial. The distance of the spacecraft, in a Lagrange L_1 orbit, however, restricts the total telemetry rate of the spacecraft, so much of the data will be digitally compressed. Some of the compression schemes are purported to be "lossless," while others, such as that currently planned by the SUMER team, simply omit downlinking featureless parts of the spectrum. The net accumulation rate on the ground of scientifically useful data will approach a gigabyte a week for SUMER alone. It is probably fair to say that no one yet knows how to deal with so much high spatial- and spectral-resolution data coming in every week for several years. Clearly, our observational tools have outstripped our ability to process data in a way that answers scientifically meaningful questions. One of the real challenges for solar physics in the next decade will be to provide the kind of analysis — not data reduction — capability that will allow us to exploit the richness of these large data sets.

The possibility for gargantuan data sets is heightened by another recent devlopment in XUV and soft X-ray optics. A grazing-incidence telescope, such as the *Skylab* S082-A, the *OSO*-7 EUV spectroheliometer, or the *YohKoh* Soft X-ray Telescope, is characterized by weak but extended sidelobes in its modulation transfer function. Any picture element is thus the convolution of a broad modulation transfer function and a number of features in the hottest parts of the Sun's atmosphere: no wonder that coronal structures in images produced with such instruments always looked hazy. This led to the belief that coronal features were quite diffuse. The modulation transfer functions of normal-incidence systems typically have much weaker wings, so that their resolution is correspondingly better, but normal incidence optics become very inefficient — reflectivities typically of only a few percent — below about 50 nm. In the last four to five years, however, an optical improvement has dramatically changed that picture. A conventional, normal incidence telescope coated with a partially reflecting layer of thickness some integer multiple of half wavelengths of an interesting bandpass acts as a Fabry-Pérot filter, or etalon, to improve the reflectance at that wavelength manyfold — typically to $\sim 40\%$. The result is a high resolution telescope with a bandpass $\Delta\lambda/\lambda$ of $\sim 10\%$ (Rottmann 1990). Groups at Lockheed, Stanford, and Harvard have flown rockets with multilayer coating, normal incidence telescopes in the XUV or soft X-ray range. In the superb 6.35 nm images obtained by Golub, one can see features down to a few arc seconds everywhere in the corona.

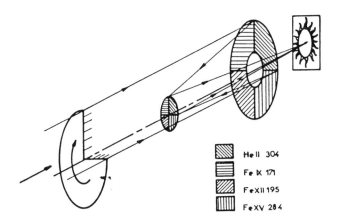

He II 304

Fe IX 171

F eXII 195

Fe XV 284

Figure 9. Schematic of the Extreme ultraviolet Imaging Telescope on *SOHO*. Both primary and secondary mirrors are coated in quadrants with multilayers tuned to one of the four wavelengths shown. From de la Boudiniere *et al.* (1988).

Although such telescopes are limited to a narrow bandpass, and are thus incompatible with broad-band spectroscopy, they are extremely valuable for what they can tell us about the morphology as a function of temperature. Indeed, information from several such bandpasses centered on lines formed at a variety of temperatures could prove a valuable diagnostic for coronal and transition region structures. One approach is 12-barrel multishooter for the U.S. Space Station, which since been "descoped" by, among other cost-cutting measures, jettisoning all scientific payloads. Another approach is that of the Extreme ultraviolet Imaging Telescope (EIT) on *SOHO*: rather than multiple telescopes, the four quadrants of the primary and secondary mirrors are coated with four different multilayers. A rotating, three-quadrant mask is used to prevent stray light from the other bandpasses entering the CCD camera. In this scheme, the multilayer coating and an Al prefilter provide solar-blindness for the detector.

If our aim is high-resolution spectroscopy, it is still possible to make use of multilayer coatings in grazing-incidence systems — by coating the grating rather than the telescope mirrors. This acts as a super blazing, giving enhanced throughput for the bandpass selected. The NASA Goddard Solar EUV Rocket Telescope (SERTS) has a long slit with two large ($5'$ × $8'$) lobes for imaging and a narrow center for spectroscopy. The performance of SERTS at the wavelengths around the chromospheric / transition region line He II 30.4 nm and the coronal line Fe XVI 33.5 nm (formed at $\sim 2 \times 10^6$ K) was optimized by applying a multilayer coating to the grating. This strategy improved the reflectivity of the grating by about a factor of 6 in the wavelength range of interest (Davila 1991).

4. Where Are We Headed?

A few months ago, solar ultraviolet observations appeared to be headed for a golden age; in addition to *SOHO*, NASA's Orbiting Solar Laboratory (OSL) was planned to include a "strap-on" version of HRTS, with several CCD detectors replacing film, as well as an Italian-U.S. XUV imager (XUVI) with two multlilayer-coated telescopes (each with four mirrors and corresponding filters). At this writing, it appears that the future of *OSL*, not for the first time, is in serious doubt. At the same time, however, there may be an effort to add an ultraviolet imaging component to the High Energy Solar Physics (HESP) mission, which was originally planned to include only a spinning spacecraft with γ-ray, hard X-ray, and soft X-ray detectors. The vast machine of study committees, instrument design teams, and space agency officials will no doubt turn out another mission concept; whether the concept will ever be placed in orbit is a matter of chance.

5. An Afterword: General Considerations for UV Instruments

Many of those who were students at the time of this Advanced Study Institute will be involved in the day-to-day operation of *SOHO* and perhaps the design of possible successor instruments. It thus may be worth enumerating a few points to consider in the design and operation of solar ultraviolet instruments.

Hydrocarbons that find a cold sink in unheated optical surfaces are baked onto flouride-coated surfaces when exposed to solar ultraviolet flux. Such contamination can have a devastating impact on instrument performance at wavelengths >110 nm, as can be seen from the *OSO-8* experience. Careful planning and execution of preventive measures throughout instrument construction, testing, launch, and operation are necessary to minimize the effects of such contamination.

The exterior of manned spacecraft can be an extremely harsh environment for instruments with fine slits, thin filters, and thermal control considerations. Despite the constant thirst for more and better data, we would do better to restrict our efforts to unmanned platforms, which can also be made far more stable.

To achieve "solar" (*i.e.*, visible) blindness, thin metal filters will pass short wavelengths (*e.g.*, $\lambda < 83.7$ nm for Al). Bare microchannel plates are solar-blind for wavelengths below ~ 100 nm. The red sensitivity of CCDs present a serious problem for their employment in longer-wavelength UV observations; currently, silicon carbide optics ("black mirrors") are the best solution.

I have not discussed the perennial problem of calibrating UV instruments, but it is clear that on-board calibration devices are optimal. An absolute wavelength comparison source that has been flown on a sounding rocket (Rottman *et al.* 1990) could be adapted to longer-lived spacecraft, but it might be prohibitively expensive to insure the long-term reliability of an on-board photometric calibration source. The traditional method of determining photometric response as a function of time is comparison with observations made by sounding rockets and calibrated before and after flight — usually with substantial extrapolation in the final phases of instrument lifetime. UVSP has been calibrated, at least partially, with observations of hot stars with well-determined ultraviolet fluxes. In any case, it is imperative to track accurately the *relative* photometric response, at all relevant wavelengths, of a solar UV instrument so that the relatively small number of comparisons with absolutely calibrated observations can be applied to the entire data set. While absolute calibration may not be necessary for line ratio diagnostics that yield density or temperature information, it is critical if one is interested in energetics or line profile synthesis. In any case, characterizing the variation of instrument throughput with time is crucial to instrument health, since it may provide early warning of thermal or contamination problems.

It would be facile to state that by the time instruments in the design phase actually fly, computer capabilities will have caught up with the expected flow of data. Serious thought needs to be applied to the problem of *how* we plan to analyze such large amounts of data.

Finally, a general consideration for all space-based instruments is to reduce the number and increase the reliability of mechanisms. UVSP had a door, a slit wheel, a polarimeter inserter and rotater, and a wavelength drive. Only one of these mechanisms had to fail to make our lives miserable.

There will always be additional considerations in the design of real instruments: how to balance science requirements with data rates, how to trade off capabilities *versus* mass and power consumption, and so on. All of the instruments described here — and many others — have been elegant compromises among such competing concerns. Their prodigious and scientifically revolutionary output has been all the more wondrous for being the product of such necessary compromise.

REFERENCES

Athay, R.G. 1966, *Ap. J.*, **145**, 784

Bartoe, J.-D.F. and Brueckner, G.E. 1975, *J. Opt. Soc. Am.*, **65**, 13

Bonnet, R.M., Lemaire, P., Vial, J.C., Artzner, G., Gouttebroze, P., Jouchoux, A., Leibacher, J.W., Skumanich, A., and Vidal-Madjar, A. 1978, *Ap. J.*, **221**, 1032

Davila, J.A. 1991, private communication

de la Boundiniere, J.P. *et al.* 1988, in *The SOHO Mission: Scientific and Technical Aspects of the Instruments*, ESA SP-1104, p. 43

Dupree, A.K. 1972, *Ap. J.*, **178**, 527

Fastie, W.G. 1991, *Physics Today*, 44, No. 1, 37

Gurman, J.B. 1972, "A Comparison of Photospheric Magnetograms and EUV Spectroheliograms," Undergraduate thesis, Harvard College, p. 69

Hagyard, M.J., Teuber, D., West, E.A., Tandberg-Hanssen, Henze, W., Beckers, J.M., Bruner, M., Hyder, C.L., and Woodgate, B.E. 1983, *Solar Phys.*, **84**, 13

Henze, W. *et al.* 1982, *Solar Phys.*, **81**, 231

Henze, W. 1991, in L. November (ed.), *Solar Polarimetry*, National Solar Observatory (Sunspot, New Mexico), p. 16

Munro, R.H. and Withbroe, G.L. 1972, *Ap. J.*, **176**, 511

Nakada, M.P., Chapman, R.D., Neupert, W.M., and Thomas, R.J. 1976, *Solar Phys.*, **47**, 611

Patchett, B.E. *et al.* 1988, in *The SOHO Mission: Scientific and Technical Aspects of the Instruments*, ESA SP-1104, p. 39

Pottasch, S.R. 1967, *Space Sci. Rev.*, **3**, 816

Purcell, J.D., Packer, D.M., and Tousey, R. 1959, *Nature*, **184**, 8

Rottman, G.J. 1990, *Physica Scripta*, **T31**, 199

Rottman, G.J., Hassler, D.M., Jone, M.D., and Orrall, F.Q. 1990, *Ap. J.*, **358**, 693

Timothy, J.G. and Bybee, R.L. 1986, *SPIE Ultraviolet Technology*, **687**, 109

Underwood, J.H. and Neupert, W.M. 1974, *Solar Phys.*, **35**, 241

Wilhelm, K. *et al.* 1988, in *The SOHO Mission: Scientific and Technical Aspects of the Instruments*, ESA SP-1104, p. 31

Withbroe, G.L. and Gurman, J.B. 1973, *Ap. J*, **183**, 279

Woodgate, B.E. *et al.* 1980, *Solar Phys.*, **65**, 73

Woodgate, B.E. 1982, presentation to Space Telescope Contamination Control Conference, Perkin-Elmer, Danbury, Connecticut

19   SOLAR RADIO INSTRUMENTATION

G.J. HURFORD
Solar Astronomy
Caltech 264-33
Pasadena, CA 91125
U.S.A.

ABSTRACT. The definitions of basic quantities such as flux density, brightness temperature and optical depth are reviewed in the context of solar radio astronomy. Current observational techniques and hardware are discussed in terms of the spatial and spectral resolution required for solar observations. The fundamentals of interferometry are outlined with consideration given to the limitations of interferometric imaging.

1. Introduction

Radio observations provide yet another perspective on a wide range of solar phenomena. As discussed in Chapter 14, the radio perspective is particularly useful for a number of purposes: for the study thermal and nonthermal processes occurring in the upper chromosphere and corona; for viewing nonthermal electrons and shock waves as they propagate through the corona; for providing unique observational data on magnetic fields in the lower corona; and for quantitative diagnostics of thermal and nonthermal flare emission.

In this Chapter, we review some basic quantities needed to understand radio data, and discuss current observational techniques and their limitations.

2. Basic Definitions

We first briefly review some of the basic definitions that will be indispensable in quantitatively describing solar radio phenomena. More complete developments can be found in texts such as Kraus (1986).

2.1. FREQUENCY, WAVELENGTH, AND POLARIZATION

Solar radio emission generally refers to the 7 decades of wavelength from about 1 mm to 10 km. Because of the closer link to the emission processes, it is often more useful to think in terms of the corresponding frequencies of 30 kHz to 300 GHz. In the Sun, radio emission is due to free electrons in a fully or partially ionized plasma. Neutral atoms can be neglected. In this respect, radio and H-alpha, for example, are viewing mutually exclusive components of the atmosphere.

J. T. Schmelz and J. C. Brown (ed.), The Sun, A Laboratory for Astrophysics, 411–422.

To orient ourselves within this seven decades of frequency, it is first useful to compare the observing frequency to the electron plasma frequency, whose value, given by ν [Hz] $= 9000 \, n^{-0.5}$ cm^{-3}, depends only on the ambient free electron density. Of major significance for solar radio emission is the fact that radio waves cannot propagate through a medium where the plasma frequency is greater than the radio frequency. For the solar atmosphere, the electron density and, hence, the plasma frequency generally decreases with increasing height. Thus, at any given observing frequency, there is a corresponding electron density and, hence, height from below which the radio waves cannot emerge. Thus, any detected radio emission must have originated above such a height which can be calculated with the use of an atmospheric model. (At 300 GHz, this corresponds to the middle chromosphere.) Higher frequency emission can (but need not) originate lower in the atmosphere than lower frequency emission.

It might be noted in passing that polarization in solar radio emission refers to circular rather than linear polarization. (This is not necessarily the case in nonsolar radio astronomy.) Linear polarization is absent in solar radio emission because it undergoes severe Faraday rotation as it propagates out of the solar atmosphere (Matzler 1973). Since observationally we must inevitably average over space, time, and frequency, the resulting differences in Faraday rotation suppress by many orders of magnitude any linear polarization originally present.

2.2. FLUX DENSITY

The strength of detected radio emission is described in terms of its flux density, which is the incident power per unit frequency passing through a unit area. MKS units of watts per square meter per Hz are inconveniently large for radio astronomy so that an important basic unit of 1 Jansky (Jy) $= 10^{-26}$ watts m^{-2} Hz^{-1} is widely used. However, solar radio astronomy, with its stronger signals, uses a unit of 1 solar flux unit (sfu) $= 10^4$ Jy $= 10^{-22}$ watts m^{-2} Hz^{-1}. Figure 1 shows the quiet Sun radio spectrum compared to some stronger non-solar sources; we see that, above $\approx$100 MHz, solar signals are much stronger than anything else in the sky, a fact which has implications, not all positive, for the observer.

A second implication of the solar spectrum in Figure 1 is readily apparent if one integrates under the curve. We see that solar radio emission is over 11 orders of magnitude below the solar constant of 1.3 kw m^{-2}. Energetically, solar radio emission is therefore totally insignificant.

2.3. BRIGHTNESS TEMPERATURE

In Chapter 14, we saw that the spectrum of surface brightness is a key diagnostic in microwaves. It is convenient to express the surface brightness at a given frequency in terms of a "brightness temperature," which is the temperature of opaque black body with the equivalent surface brightness.

We can relate the angular area, $d\Omega$ (sr) subtended by the source and the brightness temperature, T_b, of the source to the flux density, S, at frequency, ν:

$$S = 2kT_b d\Omega \nu^2 / c^2 \tag{1}$$

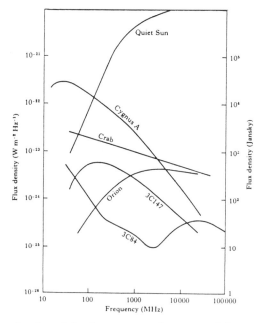

Figure 1. Flux density of the Sun compared to some of the stronger nonsolar sources.

where k and c are Boltzmann's constant and the speed of light respectively. In practical units, this is equivalent to

$$S(sfu) = 7.2 \times 10^{-11} T_b(K) d\Omega(arcsec^2) \nu^2(GHz) \qquad (2)$$

Comparing this with Figure 1, we see that, if the full Sun were a uniform source, solar brightness temperatures are 10^4 to 10^5 K, values which would not seem unreasonable for the upper chromosphere and above.

Part of the value of the concept of brightness temperature is that, for opaque thermal sources that are fully resolved, the brightness temperature is just equal to the electron temperature itself. In such cases, a measurement of brightness temperature provides a direct way to measure the plasma temperature.

Observationally, it is important to note that in order to measure T_b it is necessary not only to know the flux density, but also the angular size. Therefore it is necessary to have sufficient angular resolution to resolve the source. For transient sources, the emission must be resolved in time as well, or else the brightness temperature will be underestimated.

2.4. Optical Depth

If the source is not opaque, then the optical depth, τ, provides a convenient parameter to relate the brightness temperature, T_b, to the electron temperature, T_e, for thermal plasmas. Dulk (1985) provides a more complete development but, in this case for uniform thermal sources, we can write $T_b = T_e(1 - e^{-\tau})$. For optically thick sources, where $\tau \gg 1$, we have

$T_b = T_e$ as discussed above. For optically thin sources, where $\tau << 1$, then $T_b = T_e \tau$, and the optical depth can make the the observed surface brightness be significantly less than the electron temperature. The concept of optical depth is useful for many emission mechanisms and its value depends on plasma, field, and electron parameters. For nonthermal situations, T_e can be replaced by an effective temperature, T_f. Very important also is that the optical depth and, therefore, the brightness temperature (for non-opaque sources) depends on the observing frequency and polarization.

As discussed in Chapter 14, if we can observationally determine the brightness temperature as a function of frequency, we can extract these plasma, field, and electron parameters and have at our disposal a set of valuable diagnostic techniques.

3. Observational Techniques

Let us now consider the characteristics of hardware that is required to effectively observe solar radio radiation. These techniques employed are of course driven by the sensitivity, spectral, and spatial resolution that are needed to view the Sun through its solar radio emissions.

3.1. ANTENNAS AND RECEIVERS

The "optics" of a radio telescope often take the form of the familiar parabolic antenna. With typical diameters ranging from ≈ 1 to 30 meters, such antennas serve to focus the incoming radiation onto a "feed" which is located either a the primary focus or, for a Cassegrain configuration, at the secondary focus. Compared to solar optical telescopes, such antennas are very fast with f ratios (focal length to antenna diameter) typically of ≈ 0.4. Feeds are relatively small ($\approx$ a wavelength in size), passive structures whose geometry enables them to convert the focussed radio radiation to electrical signals in a cable or waveguide. Feeds can also serve to select which polarization(s) of the incoming radiation are observed.

A simplified block diagram of a generic "superhetrodyne" receiver is shown in Figure 2. The incident radio frequency (RF) energy is focussed by a parabolic antenna onto a feed. The output of the feed is amplified and fed into one input of a semiconductor "mixer." The other input of the mixer comes from an oscillator which is operating at the desired observing frequency. The mixer acts as a high speed switch, alternately opening and closing at the oscillator frequency. The output of the mixer is the sum and difference of its input frequencies. A low pass filter selects only the difference frequency which called the "intermediate frequency" or IF signal. Note that the strength (and phase) of this IF signal tracks that of the incoming RF signal at frequencies very near to the frequency of the oscillator. There is no response to RF signals at other frequencies. The mixer serves two functions. First, it limits the response of the receiver to the desired narrow range of frequencies. Second the mixer converts this signal to low frequencies where it is much easier to manipulate and where it can be carried by coaxial cables. In the simplest case, the IF signal would be carried into a convenient location where it would be rectified, detected by a diode whose output is proportional to the square of the voltage, and then averaged over the desired integration time. The resulting slowly varying voltage would be proportional to the square of the incident radio frequency electric field at the feed and, hence, to the incoming power at frequencies close to that of the oscillator.

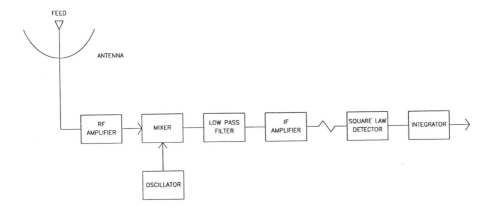

Figure 2. Schematic block diagram of a superhetrodyne receiver.

In terms of sensitivity, Figure 1 indicates that the signals are sufficiently strong that when the Sun fills the telescope beam, the receiver would see radiation levels corresponding to temperatures of 10^4 K and above. Typical receivers have equivalent noise temperatures of a few tens to a few hundred degrees K. Thus, the solar radio emission often dominates receiver noise. In fact, if you put an audio amplifier on the output of such a receiver, you can literally listen to solar flares. (Nothing dramatic, however, the solar "hiss" just gets louder.) Large signals are good, of course, but they do mean that care must be taken not to saturate amplifiers. Also, they can cause problems at facilities such as the Very Large Array (VLA) which were carefully designed for weak sources where the source signal is many orders of magnitude *below* the receiver noise (Bastian 1988).

Large signals also suggest the ability to achieve high signal-to-noise. For the frequently encountered situation where the solar signal dominates, the thermal signal to noise of an observation is given by $(Bt)^{1/2}$, where B is the bandwidth of the observation and t is the integration time. For typical values of $B = 50$ MHz and $t = 1$ sec, this suggests that s/n = 7000:1. What this illustrates is that, (unlike the typical situation in nonsolar radio astronomy) for solar radio astronomy thermal noise often is not a serious consideration - in practice the real s/n of solar radio observations may depend on systematic errors, some of which can be controlled, but many of which are difficult to quantify.

3.2. SPECTRAL RESOLUTION

Spectral observing requirements depend on the wavelength regime. Although there are no narrow spectral lines to consider, the spectral structure can be very rich in detail at low frequencies if observed with good time resolution. At microwave frequencies, with the possible exception of cyclotron lines, spectral resolution of $\approx 10\%$ is all that is required to characterize the continuum emission.

Spectral resolution can be achieved by a variety of techniques. At low frequencies, conventional dynamic spectra are obtained by sweeping the observing frequency continually, while arranging to display the intensity of the observed signal as a function of frequency and time. Conventional dynamic spectrographs, which used film as their recording media, provided good time and spatial resolution but only a very limited ability to measure the intensity of the recorded signals. Of course dynamic spectrographs with digital recording overcome this limitation. (*e.g.* Perrenoud 1982)

To get high spectral resolution over a limited frequency range, spectral analysis can be performed using a variety of hardware (such as filters, autocorrelators or acousto-optical spectrographs) on the IF signal. These can provide efficient approaches in terms of signal-to-noise but are more commonly used in non-solar work.

For solar work, the large signals make frequency-agile systems feasible (*e.g.* Perrenoud 1982, Hurford, Read, and Zirin 1984). With this technique, the oscillator frequency is rapidly changed so that different radio frequencies are sampled digitally in rapid succession. This has the advantage of flexibility in that trading frequency resolution, frequency range, and time resolution is determined by software, but is less efficient since each frequency is only observed for a small fraction of the time.

For low resolution spectroscopy, individual receivers (and antennas in some cases) can be used to independently determine the flux at several frequencies to provide an overview of the spectrum. This is the technique used by the RSTN network operated by the US Air Force. In addition to the limited number of spectral points, the disadvantage here is that each receiver/antenna has its own systematic errors which may degrade the quality of the resulting spectra.

3.3. REQUIREMENTS FOR SPATIAL RESOLUTION

In Chapter 14, we found that the size of burst sources showed a trend of increasing source size with decreasing frequency. Although this trend continues from microwave down to metric frequencies, at low frequencies, even intrinsically small sources would appear larger because of scattering by inhomogeneities in the solar corona. At high frequencies, the observed size (a few arcseconds at 10 GHz) is believed to be intrinsic to the source.

For solar radio astronomy, the motivation for achieving the spatial resolution suggested by these source sizes is greater than the natural desire to make nice images. As we have seen in the previous section, achieving sufficient resolution to determine the source size is an absolute requirement if we are to determine the brightness temperature of the source and, hence, to use radio data in a diagnostic fashion.

The challenge posed by such spatial resolution for radio observations becomes readily apparent when one recalls that the Rayleigh diffraction limit, θ, (radians) for a telescope of diameter D observing at wavelength λ, is given by $\theta = 1.22\lambda/D$. This suggests that the achievement of 10 arcseconds resolution at 5 GHz (6 cm) would require a telescope 1.5 km in diameter. Let us now turn to the issue of how to achieve such spatial resolution with more practical instrumentation.

3.4. REGIMES OF SPATIAL RESOLUTION

In general, there are four regimes of spatial resolution in solar radio astronomy. To achieve resolution of ≈ 10 degrees, one can use a feed or dipole antenna directly. Although the response of such a dipole extends over ≈ 100 degrees, mechanical scanning or spinning at least lets one determine the direction of the source centroid. Alternatively, a fixed feed viewing the sky can provide a very inexpensive way to observe time variable solar radio emission.

At higher frequencies, large parabolic antennas (*e.g.* Christiansen and Hogbom 1985) become increasingly more useful. In general, one does not have imaging detectors at the focus, so that imaging with a single antenna becomes a matter of mechanical scanning. This limits the time resolution of course but this is not serious for sources which are stable on timescales of an hour or so. The angular resolution, given by the diffraction limit of the antenna, is typically tens of arcseconds or a few arcminutes. While not sufficient to resolve the sources, it is adequate to identify larger qualitative features.

To achieve arcsecond-class resolution, it is necessary to use interferometry, and it to this topic that the next section will be devoted.

Before discussing interferometry in detail, however, we mention briefly the fourth regime of spatial resolution that is available to solar radio astronomy, namely that of Very Long Baseline Interferometry (VLBI) (*e.g.* Pearson and Readhead 1984). VLBI provides a proven technique for achieving milliarcsecond resolution imaging at radio wavelengths - that is resolution corresponding to ≈ 10 m on the Sun! Its successful use for solar work will probably be limited to nonthermal phenomena since very small sources would require very large brightness temperatures to be detectable. An additional requirement is that analyses be limited to very short integration times to avoid "smearing" the source. Finally, it should be noted that VLBI can only image sources. It cannot determine the location of small sources relative to the solar disk.

4. Interferometry

In this section, we review the principles of interferometric imaging. The emphasis will not be on mathematical rigour - there are several good sources in which the mathematics is carefully developed (*e.g.* Thompson and D'Addario 1982, Thompson, Moran and Swenson 1985) Rather, we will try to give a feel for how interferometry extracts its information about the source geometry with special emphasis on the limitations to interferometric imaging that are important to solar physics.

4.1. RESPONSE OF A TWO-ELEMENT INTERFEROMETER

The basic unit of an interferometer is a pair of antennas which simultaneously observe the same source at the same frequency (Figure 3). When the respective outputs from the two antennas are multiplied, signals that are common to the two antennas correlate to produce a nonzero averaged output. This is in contrast to signals that originate independently in each antenna whose correlation averages to zero. The correlated signal can be described by the amplitude of the correlation and its phase, with the latter, for example, being 0 or 180 degrees, depending on whether the peak of the wave which reaches one antenna was in or out of phase with the peak of the same wavefront that is seen by the other antenna. Let L be the antenna separation and θ be the angle between a vector joining the two antennas and a vector from an antenna to the source. Then the time delay for the detection of the wavefront at the two antennas is given by $L\cos\theta/c$ and a corresponding correlation phase (except for instrumental offsets) would be $L\cos\theta/\lambda \times 360$ degrees. For $\lambda = 6$ cm (5 GHz) and $L = 1.2$ km, a one arcsecond change in θ could change the observed correlation phase by up to 36 degrees, which is relatively easy to measure.

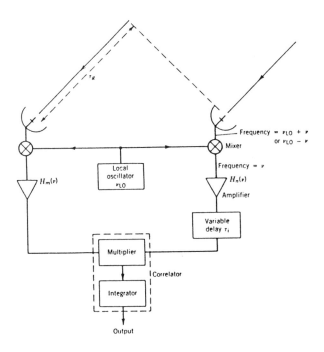

Figure 3. Schematic diagram of a 2-element interferometer. From Thompson, Moran and Swenson 1986.

In order to get a feeling for how the interferometer responds to different source geome-tries, let us consider a series of scenarios. Starting with a point source, for example, if we were to increase its strength, leaving its position in the sky unchanged, then the amplitude of the correlated signal would increase, but the phase of the correlation would be unaffected. Similarly, if the strength remained constant, but the source moved relative to the interfer-ometer, then the phase, but not the amplitude of the correlated signal could change. Note that for such a change to occur, the source displacement must affect the relative distance of the source to the two antennas so that the relative arrival time of the wavefront at the two antennas is changed. Furthermore, we note a positional ambiguity in that if the source were displaced so as to change the phase by a multiple of 360 degrees, the output of the interferometer would appear unchanged.

If we added a second source, then signals from each source would correlate independently, but the observed output would be the *vector* summation of the two correlated signals. The output amplitude could therefore increase, decrease or remain unchanged, depending on the relative location (viz. phase) of the two sources. A practical application of this is that if one is monitoring a single interferometer baseline, then the occurrence of a small flare might increase or decrease the observed amplitude, depending on the effective location of the flare relative to the preflare emission.

Finally, we consider the the effect of gradually increasing the diameter of the source, while leaving its total flux density unchanged. As the diameter is increased, contributions from various parts of the source have increasingly diverse phases. As a result, the correlated amplitude gradually decreases as the source size increases until, for sources large compared to the interferometer resolution, there is no correlated signal whatsoever. As Figure 4 suggests, this behavior provides the basis for an interferometer's ability to measure source size. Furthermore, the fact that an interferometer does not respond at all to extended sources, provides a dramatic practical advantage in suppressing signals from the extended background Sun.

These qualitative arguments are consistent with the rigorous result that the correlated signal from a two-antenna interferometer measures a single Fourier component of the source distribution.

Formally, it measures a complex visibility V with amplitude A and phase ϕ,

$$V(u,v) = A\exp^{i\phi} = \int (I(x,y)\exp^{-i(ux+vy)}\,dxdy \tag{3}$$

where the integral extends over the source distribution (as seen by the individual antennas) and u, v are the components of spatial frequency which depend on the baseline geometry, source location, and frequency.

In general, u and v vary as the hour angle evolves so that the resulting complex visibility $V(u,v)$ may also change. Thus, light curves of amplitude and phase may show variations even if the solar source itself is static.

420

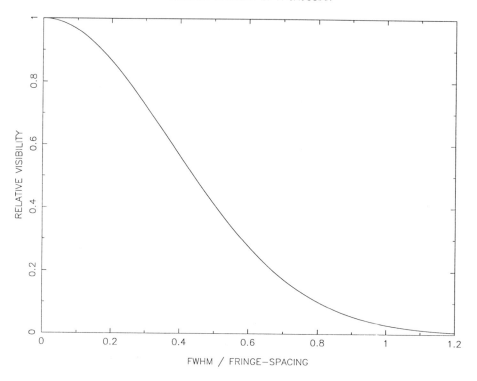

RELATIVE VISIBILITY OF A GAUSSIAN

Figure 4. Relative visibility amplitude of a gaussian source as a function of the ratio of the gaussian FWHM of the source to the interferometer 'fringe spacing' (angular source displacement corresponding to a 360 degree phase shift).

4.2. IMAGING WITH INTERFEROMETERS

Equation 3 forms the basis for interferometric imaging. It shows that the measured $V(u,v)$ is determined by the Fourier transform of the source distribution. If we make measurements of $V(u,v)$ at enough different values of u,v, we can perform the inverse Fourier transform,

$$I(x,y) = (2\pi)^{-1} \int V(u,v) \exp^{i(ux+vy)} du\,dv \qquad (4)$$

and recover the source image, $I(x,y)$.

To measure many values of $V(u,v)$, there are three things we can do. The first is to use many antenna pairs: N antennas can be grouped into $N(N-1)/2$ antenna pairs. For 27 antennas at the VLA, this corresponds to 351 u,v points. Figure 5 illustrates the u,v coverage and corresponding point response function or "synthesized beam" in this case. Second, we can repeat the measurements over the course of the day as a function of hour angle. This enables each antenna pair to make measurements at many positions in the u,v plane, not just at a single point. Although this is not useful for transient sources such as flares, it is quite effective for imaging slowly changing sources such as active regions. As a result, in such cases it is quite common to have tens of thousands of u, v points from which to reconstruct an image. Finally, if the interferometer is capable of observing at many wavelengths, then each antenna pair measures a different u, v point at each frequency. If we believe the source geometry does not change much with frequency (as with an optically thin source, for example) then this provides another way to measure additional Fourier components.

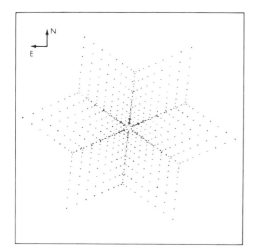
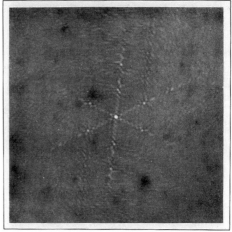

Figure 5. u, v coverage (left) and point response function (right) for a VLA snapshot map. Adapted from Ekers 1982.

Since one never observes a complete set of Fourier components, the production of maps often makes effective use of techniques such as CLEAN or maximum entropy algorithms.

4.3. LIMITATIONS OF INTERFEROMETER IMAGING

We close this Chapter with a discussion of the strengths and weaknesses of interferometer imaging. The purpose is to identify, in a generic sense, what can usually be trusted and what should be viewed more critically.

At microwave frequencies, measurements of the positions of solar sources using interferometric techniques are usually quite reliable. They are fundamentally measurements of time, in which the location of the source is determined to an arcsecond or two relative to a non-solar calibration source. At longer wavelengths, the positions are less well determined because of the effects of refraction in the ionosphere and in the solar corona.

Peak brightness temperatures are typically good to $\approx 10\%$ provided that the quality of the mapping is sufficiently high that good source sizes were measured. For transient sources, note that it is necessary that the source be resolved in time as well as in space.

422

The key limitation of interferometric mapping is the "dynamic range" of the map – namely, the surface brightness of the weakest believable source compared to that of the strongest. Thus, very weak sources should be viewed with the most caution.

Another important limitation in interferometric maps is that extended sources are just not shown. Their presence can be inferred only by comparing the integrated flux in the map and comparing it to a spatially integrated measurement. In this regard, it should be noted that most published maps are missing some of the flux. Whether it is a small or major fraction should strongly influence how the visible map is interpreted.

In general, any radio result that depends on a measurement at the $\approx 1\%$ level should be viewed with caution. For almost all radio telescopes, there are effects which cannot be compensated. For example, tracking jitter at the few arcsecond level or imperfect knowledge of telescope pointing can directly or indirectly affect results at that level.

The flux of very large events may not be as reliable as that from more modest events, primarily due to the potential for saturation effects.

4.4. SUMMARY

This brief look at solar radio instrumentation has suggested some of the strengths and weaknesses of solar radio observations. Broadly speaking, the solar radio observer often has the advantage of lots of signal, and fully adequate angular resolution. His problems lie more in the area of sometimes having to accept the limitations of hardware designed for the study of much weaker sources and the inherent limitations of interferometric imaging. While interferometry provides a very powerful glass through which to look at the Sun, its limitations of dynamic range and selectivity of spatial scales should be kept in mind in considering the observations outlined in Chapter 14.

5. References

Bastian, T. 1988 in *Synthesis Imaging in Radio Astronomy* Perley, R. A., Schwab, F. R. and Bridle, A. H. (eds.), Astron. Soc. of the Pacific.
Christiansen, W. N. and Hogbom, J. A. 1985, *Radio Telescopes*, Cambridge University Press.
Dulk, G. A. 1985, *Ann. Rev. Astron. Astrophys.*, **23**, 169.
Ekers, R. D. 1982 in *Synthesis Mapping*, Thompson, A. R. and D'Addario, L. R. 1982, ed, National Radio Astronomy Observatory, ch 12.
Hurford, G. J., Read, R. B. and Zirin, H. 1984, *Solar Phys.*, **94**, 413.
Kraus, J. D. 1986 *Radio Astronomy*, Cygnus-Quasar Books.
Matzler, C. 1973, *Solar Phys.*, **32**, 241.
Pearson, T. J. and Readhead, A. C. S, 1984 *Ann. Rev. Astron. Astrophys.*, **22**, 97.
Perrenoud, M. R. 1982, *Solar Phys.*, **81**, 197.
Thompson, A. R. and D'Addario, L. R. 1982, ed, *Synthesis Mapping*, National Radio Astronomy Observatory.
Thompson, A. R., Moran, J. M. and Swenson, G. W. 1986, *Interferometry and Synthesis in Radio Astronomy*, Krieger Publishing Company.

20   SOFT X-RAY INSTRUMENTATION

A. H. GABRIEL
Institut d'Astrophysique Spatiale
C.N.R.S. - Université Paris XI
Bâtiment 121
Campus Universitaire
91405 Orsay Cedex
France

ABSTRACT. The components of an X-ray spectroscopy configuration are identified by their function. Many physical components fulfil more than one of these functions. We review briefly the principles underlying a number of X-ray components, with emphasis on those of use in the soft X-ray region.

1. Introduction

The classical configuration for X-ray spectroscopy consists, as in the case of optical spectroscopy, of three components. The first is a collimator, or some component which serves to define the region in object space which is observed. The second is some form of dispersing element, while the third is the detector. However, in the case of X-rays, these three functions are not always accomplished by three separate components, it being often the case that one component serves two of the functions.

In Table 1, we summarise a number of components used in the X-ray region, and indicate their properties in respect of each of these three functions. Thus, it can be seen that a number of detectors are energy or wavelength sensitive, and serve also to select or measure the wavelength. Similarly, some dispersing elements serve also to collimate the radiation, whilst some forms of X-ray telescopes can select particular wavelengths.

In this short review, we shall consider the performance of some of these components, concentrating on those which are effective in the softer part of the X-ray spectrum.

2. Photographic Film

This excellent detector, in use since the early days of X-rays research, continues to have an important application, on account of its enormous data storage capacity, and its uses for imaging with high spatial resolution. However, the use of film for accurate intensity measurements can only be considered if elaborate calibration procedures are employed. The response is non-linear, and the film tends to vary in sensitivity from batch to batch and even within one batch.

J. T. Schmelz and J. C. Brown (ed.), The Sun, A Laboratory for Astrophysics, 423–434.
© 1992 *Kluwer Academic Publishers. Printed in the Netherlands.*

Table 1. The role of various components in an X-ray spectroscopy configuration

component	control of the geometry	wavelength selection	detection
photographic film	no	no	imaging
proportional counters	no	low resolution	non-imaging
position-sensitive propn. counters	no	low resolution	imaging or non-imaging
scintillators	no	medium resolution	non-imaging
solid state	no	high resolution	non-imaging
CCD	no	no	imaging
grating	no	high resolution (dispersion)	no
crystal	some (plus polarisation)	high resolution (selection)	no
grid collimator	selects point on object	no	no
grazing-incidence telescope	imaging	no	no
normal-incidence multi-layer	imaging	medium resolution	no

However, when compared to the visible region, the response of X-ray film is significantly more reliable. One photon absorbed is sufficient to sensitize one grain, on account of the higher photon energy, and thus effects such as inertia and reciprocity breakdown do not occur in the X-ray region. For very soft X-rays, the gelatin in the emulsion absorbs some of the photons, leading to a loss in sensitivity. Special low-gelatin emulsions are available for this region, but these are touch sensitive and often very variable in uniformity and sensitivity.

Evidently, emulsions are usable in space work only in those conditions where they can be physically recovered, *i.e.* in sounding rocket, balloon, or space shuttle borne instruments. The high quantum efficiency of film is not generally realizable in practice, due to the high detector background, which discourages grain counting techniques.

3. Proportional Counters

The classical proportional counter uses low pressure gas in a chamber of which the walls act as a cathode, with a thin central wire as the anode. A window in the side, allows the X-rays to enter and ionise the gas. The proportional counter acts in a region intermediate between the simple ionisation chamber with no gas gain and the Geiger counter with a catastrophic avalanche for each count. A controlled gas gain of between 10^3 and 10^5 is obtained close to the anode wire, where the principal acceleration process occurs.

The quantum efficiency of the counter can approach 100% when the gas is opaque and the window transparent. Since these properties are readily calculated or calibrated, the proportional counter can be used as an absolute detector.

The incident photon is absorbed in the gas and produces a fast photoelectron, which in turn creates a certain number of primary electron-ion pairs, depending on its energy. For argon, we get 37 pairs per kilovolt of the incident photon. The charge due to these pairs is then amplified by secondary ionisation, leading to a pulse size which is proportional to the original photon energy. However, the pulse width is limited by the Poisson statistics of the primary pair production. This leads to a resolution (FWHM) of

$$\frac{dE}{E} \approx \frac{0.14}{E^{0.5}}. \tag{1}$$

The time taken for the ions to drift to the cathode imposes a dead-time on the detector after each pulse, of the order 10 μ sec, which in turn limits the maximum useful counting rate of the detector.

The technology can be difficult for space applications, in particular concerning the sealing of the window. At low photon energies, thin windows must be used. These can allow the gradual leakage of the gas, which in some cases leads to the need for a continuous gas flow system.

4. Reflection of X-rays

Due to the generally increased absorption coefficient of most materials in the soft X-ray region, reflection coefficients are low. This leads to difficulties both for collimators and dispersing elements. Figure 1 shows how a single absorption feature (a) and a series of such features (b) affect the absorption and refractive index of a medium. The soft X-ray region, being beyond the major absorption features, normally results in a value of $\mu < 1$. In this situation, external reflection at the surface of the material is analogous to internal reflection in the case of visible light. A favourable reflection coefficient results when the grazing angle a is less than $\cos^{-1} \mu$, due to total "internal" reflection. In practice the cut-off is not sharp,

426

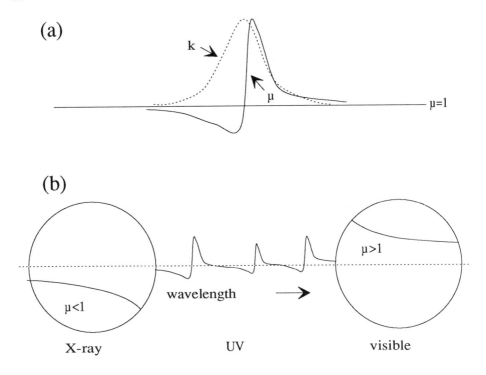

Figure 1. The effect of absorption features on the refractive index of a medium. (a) a single feature, and (b) a series of features.

and typically we find that $a = 4°$ is good down to 20 Å and $a = 2°$ down to 10 Å or lower, depending on the material.

5. The Diffraction Grating

The diffraction grating can be used for soft X-rays, but to ensure a good reflection coefficient, it is necessary to go to small grazing angles of incidence. However, the astigmatism, inherent in the Rowland Circle mounting, is a function of the angle of incidence. It can be negligible at normal incidence, but increases to a very large effect at grazing incidence. Figure 2 demonstrates that for small grazing angles there is little or no focussing effect in the plane orthogonal to the dispersion.

Very fine high-resolution spectra can be obtained with such a mounting. However, the angular dispersion of the grating is much smaller than that of a crystal. This then implies that, for the region in which the two techniques can be applicable, the amount of energy one can diffract without degrading the resolution more that a given figure is much larger for the crystal, which is therefore a more efficient spectrometer. At wavelengths longer

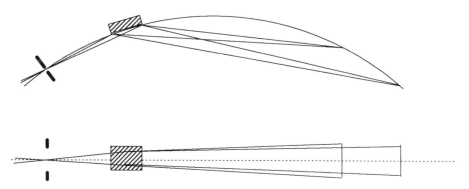

Figure 2. A Rowland Circle mounting in the grazing-incidence region, showing the effect of high astigmatism.

than say 50 Å, where the crystal suffers from absorption effects, the grating becomes more efficient. Grazing-incidence spectrometers have been flown effectively in space to study the spectrum of the solar corona (Breeveld *et al* 1988).

6. Crystal Diffraction

To understand the principle of the crystal, we consider first just the front plane of scattering centres, which act in the same way as a ruled diffraction grating. In the case of this element alone, the effect is shown in Figure 3(a). The complete dispersed spectrum is produced in the first and higher horizontal orders. In the zero horizontal order the spectrum is non-dispersed, all wavelengths being reflected together, with the angle of diffraction equal to the angle of incidence.

Figure 3(b) shows the effect of adding a number of similar layers of centres with a vertical displacement d. The dispersed first and higher horizontal orders now interfere destructively, and therefore disappear. The zero horizontal order also interferes destructively and vanishes, *except* when the Bragg condition

$$n \lambda = 2 d \sin \theta' \tag{2}$$

is satisfied. Thus we have a situation in which there is no dispersion, in the true sense of sending different wavelengths to different positions. However, the spectrum is indeed analysed by selectively reflecting only those wavelengths which satisfy the Bragg condition. With a plane parallel incident beam, the wavelength analysis is carried out by progressively rotating the crystal with respect to the beam. Since, at any one time, only one wavelength is reflected, the notion of efficiency is somewhat complex. For a single wavelength, the crystal is much more efficient than a grating, for the reasons indicated above. If the entire spectrum is required, this gain in efficiency is then lost, through the need to scan the spectrum sequentially.

428

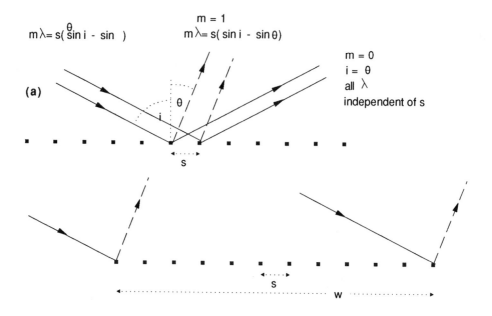

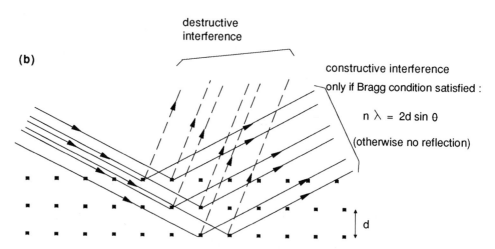

Figure 3. Diffraction from a crystal lattice. (a) a single layer, and (b) a three-dimensional lattice.

If it is required to record the entire spectrum simultaneously, it is necessary to resort to optical configurations in which the angle of incidence varies progressively across the face of the crystal. This can be achieved by having a divergent entrance beam, or with a parallel beam incident on a convex curved crystal. In this case, the overall efficiency is not increased, since any one wavelength is reflected from only a small fraction of the crystal surface. Both flat and curved crystals have been used in the Soft X-ray Polychromator (Acton *et al* 1980), a complex instrument for the study of solar flares and active regions from the Solar Maximum Mission satellite.

Consideration of Figure 3 can also be used to derive the resolving power of the crystal. For the elementary system shown in Figure 3(a), this is equal to the number of projected wavelengths in the total width of the crystal. In the case of the volume reflection, with zero horizontal order, the width of the crystal does not enter. Instead, the beams that interfere in the zero horizontal order are those from successive vertical layers. The resolving power is thus related numerically to the number of these layers which contribute. This is in turn related to the depth of penetration of the X-rays, *i.e.* to the degree of absorption. The softer the spectrum, the fewer the layers contributing, and thus the lower the resolution. It is this effect that limits the useful wavelengths for crystal diffraction to below 50 to 100 Å.

7. Grid Collimators

Mechanical collimators have been used since the earliest days in order to define the viewing direction and angle. At wavelengths shorter than around 1 to 5 Å, where reflection optics are no longer efficient, these are the only techniques feasible. Even at longer wavelengths, mechanical collimators can offer many advantages, depending on the situation. For astronomical observations, in which the source is effectively at infinity, the parallelism of the incident beam is not disturbed by such a mechanical collimator.

The earliest collimators were in the form of a number of closely spaced parallel foils, between which the X-rays passed on their way to the crystal. Such collimators were difficult to construct for high angular resolution and had a problem of parasitic grazing-incidence reflections from the surface of the foils. They have been largely superseded by grid collimators. The principle is shown in Figure 4. Two identical transmission grids are placed at a distance which determines, in conjunction with the grid period spacing, the resolution obtained. To block the second order transmission (*i.e.* radiation which passed one aperture in the first grid passing an adjacent aperture in the second), a third identical grid is placed at half the length of the collimator. To block fourth order transmission, a fourth grid is placed at the quarter position. Further grids are placed successively at half the remaining interval until the radiation allowed to pass at high order is sufficiently off axis to be beyond the field of view of the source. In the case of the Sun, this means beyond half an arcminute. Such collimators have been successfully flown on a number of missions, as for example on the Solar Maximum Mission (Acton *et al* 1980).

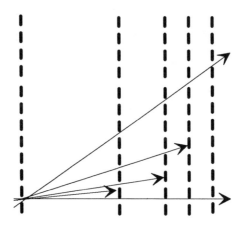

Figure 4. Showing the principle of the grid collimator

8. Grazing-Incidence Telescopes

As shown above, reflection optics can only be efficient in the X-ray region if they operate at grazing incidence. The first designs to accomplish this were developed many years ago by Wolter (1952) for use in X-ray microscopes. The basic concepts of Wolter's telescopes are used today for X-ray astronomy and have been flown many times in space. The excellent solar X-ray image shown in Figure 1 of Chapter 13 has been obtained using this technique.

The three principal designs of Wolter are shown in Figure 5. The first uses a single reflection from a paraboloid. It is exactly analogous to a normal-incidence paraboloid, except that the off-axis aberrations are exaggerated by the grazing-incidence. As a result, only the on-axis point gives a good image. Points off-axis are blurred by an amount equal or larger than the off-axis distance. It can be shown that, in a reflection optics system, off-axis aberrations can only be eliminated if we have an even number of reflections. In practice, this means two rather than one. The second Wolter design has two internal reflections; a paraboloid followed by a confocal hyperboloid. Such a system is capable of maintaining arcsecond resolution up to 5 arcminutes off-axis. The third system is an analogue of the second, but uses an external reflection for the hyperboloid. This Wolter type 2 telescope can be thought of as the X-ray analogue of a Ritchey-Chretien telescope. Like the Ritchey-Chretien it has a telephoto effect, with its focal length much longer than the focal distance.

The construction of such telescopes requires a very high precision and degree of polish, due to the very short wavelengths which must be reflected with low scatter. For non-solar astronomy, in which the fluxes can be very low, the effective aperture of a Wolter type 2 is often increased by nesting a number of confocal telescopes inside of each other.

Wolter type 0

Wolter type 1

Wolter type 2

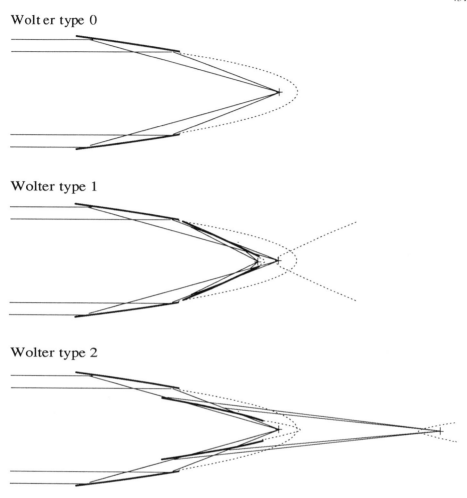

Figure 5. Grazing-incidence X-ray telescopes of the Wolter configurations.

9. Multilayer Reflection

A recent technique has been developed (Spiller 1972), which promises to bring to normal-incidence optics the high X-ray efficiency, hitherto only available at grazing-incidence. The potential advantages of such a development is enormous. Normal-incidence systems have much larger physical apertures for the same overall size, and provide more flexibility in design.

The principle of the technique is illustrated in Figure 6. A number of alternate layers of low- and high-Z material are deposited on a polished substrate. The reflections from each of the interfaces combine constructively to give an enhanced intensity in the reflected beam. The problem with this long-known principal is that high-Z materials are always good X-ray absorbers. The advance lies in designing the system so that the high-Z material is in very thin layers (a few angstroms), and is always located at the node of the standing wave pattern, where the energy density in the beam is locally reduced.

As always with a system based upon interferometer techniques, the resolution depends upon the number of reflecting layers employed. Figure 6 shows how a larger number of reflecting layers leads to a higher spectral resolution, or selectivity.

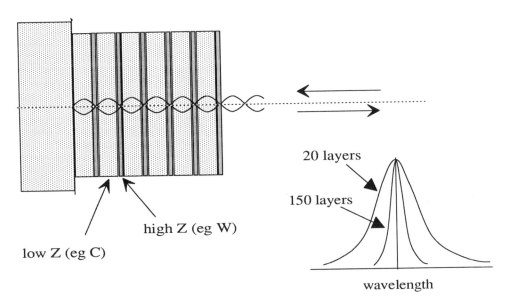

20 layers

150 layers

high Z (eg W)

low Z (eg C)

wavelength

Figure 6. Normal-incidence multilayer reflection for soft X-rays.

The principal difficulty lies not with the multilayer technique itself, but with the need to have an exceedingly high quality of super-polish in the substrate. In order to reduce scattering at each reflection, the local imperfections must be small compared with the X-ray wavelength. At grazing incidence, there is a large cosine factor which helps this requirement, but this is not available at normal incidence. So far, many good solar images have been obtained in the 150 to 300 Å region with multilayer telescopes, and some are now pushing down towards 50 Å.

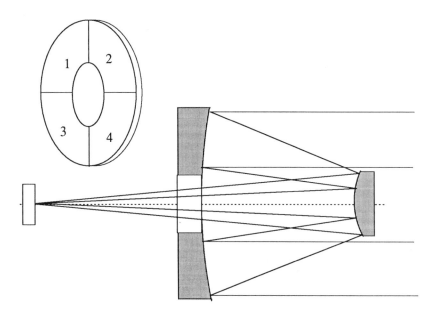

Figure 7. The telescope EIT for imaging the solar corona in specific XUV spectral lines.

An instrument, the EUV Imaging Telescope (EIT), is now under construction for flight on the future SOHO mission. EIT uses a Ritchey-Chretien telescope in which the two components are both multilayer reflectors (Delaboudiniere *et al* 1989). Figure 7 shows the construction of EIT. Each of the mirrors is coated with four different types of multilayer, sensitive at four different wavelengths, corresponding to different solar emission lines formed at different temperatures. A rotating mask is capable of uncovering each quadrant in turn, thereby recording at the same detector the four different images.

10. References

Acton, L. W., Culhane, J. L., Gabriel, A. H., Bentley, R. D., Bowles, J. A., Firth, J. G., Finch, M. L., Gilbreth, C. W., Guttridge, P., Hayes, R. W., Joki, E. G., Jones, B. B., Kent, B. J., Leibacher, J. W., Nobles, R. A., Patrick, T. J., Phillips, K. J. H., Rapley, C. G., Sheather, P. H., Sherman, J. C., Stark, J. P., Springer, L. A., Turner, R. F. and Wolfson, C. J., (1980) *Solar Phys.*, **65**, 53.
Breeveld, E. R., Culhane, J. L., Norman, K., Parkinson, J. H., Gabriel, A. H., Lang, J., Patchett, B. E. and Payne, J., (1988) *Astroph. Letts. and Comm.*, **27**, 155.

Delaboudiniere, J. P., Gabriel, A. H., Artzner, G. E., Millier, F., Michels, D. J., Dere, K. P., Howard, R. A., Kreplin, R. W., Catura, R. C., Stern, R. A., Lemen, J. R., Neupert, W. M., Gurman, J. B., Cugnon, P., Koeckelenbergh, A., Van Dessel, E. L., Jamar, C., Maucherat, A., Cauvineau, J. P. and Marioge, J. P., (1989), in "X-ray/EUV Optics for Astronomy and Microscopy," ed. Richard B. Hoover, *Proc. SPIE*, **1160**, 518.

Spiller, E., (1972) *Appl. Phys. Letts.*, **20**, 365.

Wolter, H., (1952) *Ann. der Phys.*, **10**, 94.

21   X-RAY INSTRUMENTATION

G.J. HURFORD
Solar Astronomy
Caltech 264-33
Pasadena, CA 91125
U.S.A.

ABSTRACT. The observational characteristics of solar hard X-ray and gamma-ray emission above 10 keV are reviewed to identify the sensitivity, spectral and spatial resolution desired for the corresponding X-ray instrumentation. The operating principles of non-imaging instrumentation are reviewed and their limitations discussed. Techniques for achieving high-sensitivity, high-spatial resolution hard X-ray images are outlined with emphasis on Fourier transform imaging.

1. Introduction

The purpose of this Chapter is to review the instrumentation and observing techniques associated with solar X-ray observations above energies of ≈ 10 keV. Such an energy cutoff is appropriate not only because because the observing techniques are distinctive, but because the observational requirements are driven by the fact that we are dealing almost exclusively with flare-associated solar phenomena. It also might be noted that above this energy there are nuclear, but no atomic lines of significance to solar physics. Our emphasis will not be to review existing observations, but rather to identify and discuss the observing techniques with the goal of conveying an appreciation of the the strengths and weaknesses of existing observations. Because of the likely significance of spatially-resolved observations in this energy range in the future, there will be some emphasis on imaging techniques for hard X-rays and gamma rays.

2. Characteristics of Solar X-ray and Gamma-Ray Emission

Flare-associated hard X-ray and gamma-ray emission has been the subject of many reviews, including those by Chupp (1984), Dennis (1985), and Emslie (Chapter 24). For our purposes here, we bypass the interpretation of these observations and ask what are the characteristics of the emission that we are setting out to detect. Figure 1 suggests that the energy range from 1 keV to 100 MeV encompasses several physically distinctive regimes. From ≈ 30 keV to ≈ 1 MeV, the hard X-ray continuum usually dominates. At low energies, the thermal bremsstrahlung continuum at ≈ 10 MK dominates and provides an observational link to spectroscopic techniques which are so useful at lower temperatures. At intermediate photon energies ($\approx 10\text{-}30$ keV), depending on the flare, either the thermal, non-thermal, or ≈ 30 MK superhot thermal emission can dominate.

435

J. T. Schmelz and J. C. Brown (ed.), The Sun, A Laboratory for Astrophysics, 435–445.

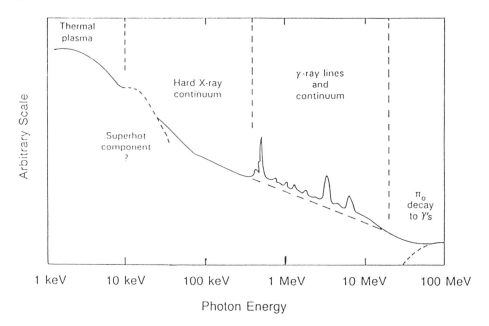

Figure 1. Schematic view of a flare X-ray and gamma ray spectrum, indicating the relevant spectral components. Courtesy of H. Hudson.

In terms of spectral lines, the 511 keV positron annihilation line can be a very useful diagnostic while at higher energies, high energy proton collisions with ambient nucleons produce a rich assortment of gamma ray lines which provide a direct observational window on the acceleration of nuclei in flares. The inverse reaction (high energy nuclei colliding with ambient hydrogen) makes a Doppler-broadened contribution to the gamma ray continuum. At the highest energies, pion decay to gamma rays can be relevant.

Despite the five decades of energy involved, high spectral resolution is desirable for two reasons. First, since thermal bremsstrahlung spectra can be as steep as E^{-9}, measurement of the spectral parameters can be severely distorted or indeed rendered unobservable if viewed with detectors with low spectral resolution. Furthermore, to fully exploit the information content of the gamma-ray lines, these lines must be resolved, not just detected, and this requires spectral resolution better than 1 percent.

Unlike the situation at radio wavelengths, where high signal-to-noise can often be taken for granted, for non-thermal X-rays and gamma rays, photon statistics is very often a limitation. As seen from earth, a typical once-a-day flare may provide only ≈ 4 photons cm^{-2} sec^{-1} above 30 keV. Therefore, an effective area $\approx 10^2$ cm^2 is required if we are to effectively observe hard X-rays. At high energies, sensitivity requirements become even more severe. On the other hand at low energies, the steep spectrum ensures there is often no shortage of photons. This also has implications for the dynamic range of the instrument, however, in that the hard X-rays and rare gamma rays often must be detected in the midst of a blizzard of low energy photons.

A time resolution ≈ 1 sec is a useful standard but there are occasional events that show significant structure down to $\approx 10^{-2}$ sec.

The need for arcsecond-class spatial resolution at lower thermal energies has been amply justified by *Skylab* (and the more recent *Yohkoh*) soft X-ray images. At hard X-ray energies, the HXIS experiment on *Solar Maximum Mission* saw structures both larger than and unresolved by its 8 arcsecond resolution. If we were to use the high frequency microwave data as a guide to the spatial distribution of the electrons, then arcsecond resolution would seem to be necessary.

3. Non-Imaging Observations

Having suggested the sensitivity, spectral and spatial resolution required for effective observation of solar hard X-rays, we now turn to the specifics of the instrumentation, beginning with observations with no spatial resolution.

Incident X-rays and gamma rays interact with matter through a variety of mechanisms. At low energies, where their energy is comparable to that of bound electrons, photo-ionization is important; at intermediate energies, where the electrons in the stopping material appear to be free, Compton collisions provide the primary interaction mechanism; at energies above ≈ 1 MeV, pair production becomes possible and begins to dominate. The stopping distances implied by Figure 2 show that at high energies, several cm of high-Z material are necessary to absorb the incident gamma rays.

Unlike the situation for solar optical or radio observations, the detection of hard X-rays and gamma rays is done one photon at a time. There are three types of detectors in common use: proportional counters, scintillation counters and germanium detectors. In each case, the incident X-ray photon interacts in the detector, depositing at least some (and hopefully all) of its energy in a form which is converted to an electrical pulse whose amplitude is proportional to the deposited energy. Such a pulse also gives an accurate indication of the arrival time of the photon.

At low energies, gas-filled proportional counters are commonly used. In such counters the incident X-ray ionizes a gas contained between high voltage electrodes. The electrons and ions drift toward the electrodes, accelerate as the field becomes stronger to further ionize the gas through collisions and so induce a detectable pulse. If necessary, the location of the X-ray interaction can be determined to an accuracy ≈ 1 mm.

At higher energies, where the stopping power of gas-filled proportional counters is no longer effective, scintillation detectors can be used. These consist of a scintillating material such as Cs-I in which the interacting X-ray produces a burst of optical photons which are then detected by a phototube. Figure 3 shows both a conceptual layout and practical implementation of a scintillation detector. Determining the location of the X-ray interaction is more difficult in this case. One approach uses multiple scintillators, each viewed by its own phototube or photodiode. Alternatively, a common scintillator can be viewed by several intercalibrated phototubes so that the ratio of their outputs implies a the locations of the initial interaction. Such techniques can achieve spatial resolution of ≈ 1 cm at best.

438

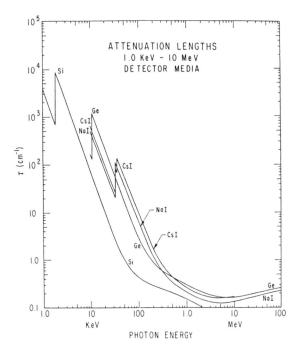

Figure 2. Attenuation lengths for common detector materials as a function of energy. The discontinuities are a result of photoionization effects.

Photons interacting in solid state germanium detectors create electron-hole pairs which create a current pulse due to a biassing electric field. Current germanium detectors are limited in size to $\approx$7 cm diameter, and must be cooled to liquid nitrogen temperatures, a significant practical complication in many cases. Techniques for achieving some spatial resolution are being developed for these detectors, but no single approach has been widely adopted as yet.

A key difference among these detector types is their respective capabilities for spectral resolution. One limitation to the spectral resolution of each detector type is the energy required to create each electron/ion pair, scintillation photon, or electron/hole pair. For example, a typical proportional counter requires $\approx$24 ev of energy loss for each primary electron-ion pair. Thus a 20 keV photon might create $\approx$ 800 electron-ion pairs. Although multiplication will subsequently take place, it is this number which fundamentally limits the spectral resolution due to Poisson statistics. (The statistical analysis needs to take into account that the electron-ion pairs are not created independently. This is described by a correcting "Fano factor" which is material dependent (*e.g.* Fraser 1989).)

For scintillators, the problem is even more serious for each detected photon represents $\approx$ 1 keV energy loss from the incident X-ray. The key advantage of germanium over scintillators is that each electon/hole pair in germanium requires only $\approx$ 3 ev so that well over an order of magnitude improvement in spectral resolution is achieved.

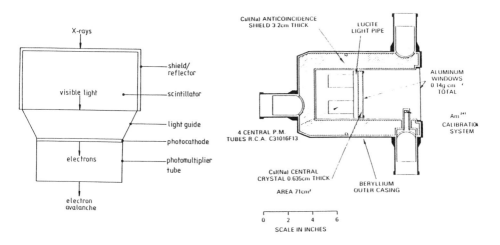

Figure 3. Schematic scintillation detector (left) and the HXRBS detector (right) on *SolarMaximumMission*. In the HXRBS detector, a separate CsI shield surrounds the central CsI detector and serves to minimize the effects of charged particles and photons that arrive from any direction except from the front. The central scintillator is viewed by 4 phototubes to enhance light collection efficiency. Adapted from Fraser 1989 and Orwig, Frost and Dennis 1980.

A typical instrument consists of one (or more) of these detectors along with electronics which measures the pulse height for each detected photon. The telemetry typically contains the number of counts per unit time interval detected in a number of pulse height intervals. For cases where telemetry is not a limitation, the arrival time and pulse height of each photon can be recorded on an individual basis.

Regardless of the type of detector system used, there are a number of potential limitations to which the interpreter of the data should be alert. First, poisson statistics of the number of detected photons often provides a fundamental constraint both for interpretation of the light curve and for the extraction of spectral parameters. Second, background is generated by variety of sources, including incident cosmic rays, residual radioactivity of the satellite, and diffuse X-rays of cosmic origin. The subtraction of background is usually straightforward, but this is not necessarily the case if the background is time varying as, for example, in a spacecraft in polar orbit. In such cases, the detection of weak, long-duration sources can be difficult. Third, the spectral response of some instruments can be complex because of limited spectral resolution and/or because the incident photon may not deposit all of its energy in the sensitive volume of the detector. Thus the task of converting a detected count rate spectrum to an incident photon spectrum can sometimes be quite challenging, and require a detailed understanding of the instrument response. Even when a corrected photon spectrum has been extracted, the conversion of such a spectrum to the electron spectrum at the Sun requires making assumptions such as thick or thin target models or for data with limited spectral resolution, making assumptions as to the thermal or nonthermal nature of the electron distribution.

Studies of X-ray emission under high rate conditions must be alert to the perils of dead-time and pulse-pileup effects. Dead-time is the interval (typically a few microseconds per photon) while the instrument is processing one photon during which it ignores any subsequent incident photons. Unless corrected, this tends to underestimate the intensity of large events. Pileup is a more insidious problem in which two or more low energy photons arrive sufficiently close together in time that they are indistinguishable from a single photon whose apparent energy is equal to their sum. Given the steep nature of the lower energy photon spectra, this can compromise both the apparent spectra and light curves at intermediate energies.

However, perhaps the most serious limitation affecting the interpretation of solar data from non-imaging hard X-ray and gamma ray instruments is the one that we have so long taken for granted, namely that any light curve or any spectrum that we so carefully extract represents a spatial average over the entire flare. In fact, almost everything we have learned about flares from hard X-rays and gamma rays has been based on data with no spatial resolution. While much has and will be learned from such data, this is a fundamental limitation indeed, given the strong evidence in all other spectral regimes that flares are a very spatially non-uniform phenomenon.

We now turn to the means by which this limitation may be overcome.

4. Hard X-ray Imaging

Imaging of hard X-rays is fundamentally constrained by the fact that there is no material or technique currently available for broadband reflection or refraction of light at wavelengths shorter than ≈ 2 Å (≈ 6 keV). Thus, imaging must use non-focussing optics, more specifically collimators, to selectively absorb photons, based on their angle of incidence.

Although there are many collimation techniques that might be considered, the list of options dwindles rather rapidly if, as implied in section 2, we want to image with both high spatial resolution (a few arcseconds) and high sensitivity (tens or hundreds of cm^2 effective area).

The simplest concept that might be employed is based on a medieval *camera obscura*, in which a single pinhole casts an inverted image of the bright object onto a screen (Figure 4). The size of the aperture determines the sensitivity and ratio of aperture diameter to screen/aperture separation limits the angular resolution. It is easily seen that to achieve both 10^2 cm^2 sensitivity and 10 arcsecond imaging with this basic concept would require a detector to be located ≈ 2 km from the screen.

A modern implementation of this idea uses a "pattern-redundant" array of thousands of small apertures which casts a distinctive multiple image of the source onto a detector plane. The use of multiple apertures retains high sensitivity while the use of small apertures addresses the resolution problem. While such approaches have enjoyed great success for non-solar observations with arcminute to degree class resolution, to achieve arcsecond-class resolution with a space-based instrument of practical length (≈ 10 m or less) would require both aperture size and detector resolution of $<<1$ mm. At the higher energies where

PINHOLE CAMERA

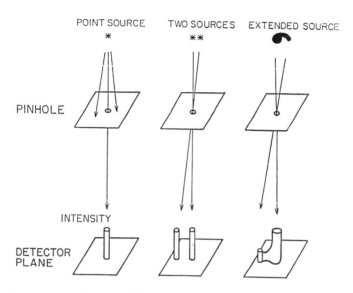

Figure 4. Schematic indication of the response of a simple pinhole camera to 1 or 2 point sources and to an extended source. Courtesy of T. A. Prince.

scintillators or germanium must be used, such spatial resolution cannot be achieved in current detectors.

To overcome the limitations imposed by low spatial resolution detectors, we must decouple the imaging and detecting functions. This can be accomplished by using multigrid collimators, a one-dimensional example of which is shown schematically in Figure 5. In this case, an X-ray opaque material of width, W, contains periodic slits and slats of width, a. Two such "grids" are present, separated by a distance, L with a detector located behind the rear grid. The transmission of such a grid pair as a function of incident angle is a sawtooth pattern whose FWHM angular resolution is $\theta_R = a/L$. Note that the angular resolution depends on the slit-slat width, and grid separation and is independent of the spatial resolution of the detector. The field of view, $\theta_F = W/L$, is determined by the grid size and separation. Figure 5 shows how inserting intermediate grids with the same pattern can isolate the response to one or more narrow maxima. In such a case, a detector without spatial resolution located behind the grids would respond only to X-rays coming from a narrow range of directions, θ_R. An array of such systems, each pointing and responding to a slightly different direction, can provide a basic imaging capability. This was basis of the HXIS imager (van Beek *et al.* 1980) which used 432 such subcollimators with apertures as small as 46 microns to achieve 8 arcseconds resolution.

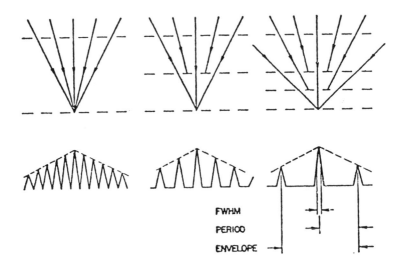

Figure 5. Schematic indication of the angular response of 2,3 and 4 grid collimators.

This approach has two serious problems, however. First, to get a multipixel image requires dividing the available instrument area into a very large number of subcollimators so that in an instrument of practical size, the width, W, available for each subcollimator must be quite small. HXIS, for example, had a high-resolution effective area of only about 0.07 cm^2 for X-rays coming from any particular direction. As a consequence its sensitivity was profoundly limited. Second, the practical task of setting up and maintaining mechanical alignment among multiple fine grids at the few-micron level becomes a very serious challenge. Therefore, we must look elsewhere for a technique to combine high sensitivity with even higher angular resolution (Prince *et al.* 1988).

A solution to the problem comes from a fresh look at the bigrid collimator shown in Figure 5. As Figure 6 suggests, its response as a function of source position varies periodically over the sky. Equivalently for a fixed source, if the collimator orientation is smoothly varied, then the count rate in the detector is modulated as a function of time. In fact, the amplitude and phase at the fundamental period of this modulation measures the amplitude and phase of a single Fourier component of the X-ray source. As we have seen in Chapter 19, this is exactly what is measured by a single baseline in a radio interferometer. Therefore, just as a multibaseline interferometer such as the VLA can reconstruct images by measuring many Fourier components, X-ray imaging can be accomplished by measuring many Fourier components. The mathematical equivalence is exact ((Makashima *et al.* 1977, Hurford and Hudson 1979).

Table 1 highlights the analogy between radio interferometry and Fourier Transform imaging. Note that the better image quality provided by earth-rotation synthesis in the radio regime can be easily obtained for X-rays by simply rotating the collimator. Therefore, a relatively small number of detectors can measure many Fourier components. This was the basis for the hard X-ray imager on *Hinotori* (Tsuneta 1984).

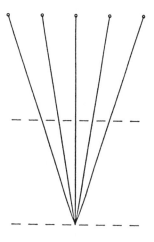
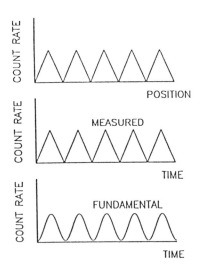

Figure 6. Schematic illustration of the response of a bigrid collimator. The left-hand panel shows the directions of maximum response of a bigrid collimator, indicating that such directions are periodically distributed. The count rate that would be detected as a function of source position is illustrated in the top-right graph. Equivalently, for a fixed source, if the orientation of the collimator were smoothly varied in time the count rate that would be measured is shown in the center-right graph. Given the periodic, measured count rate, if we ignore all but the fundamental period, the response would be reduced to that shown in the lower-right plot. It is the amplitude and phase of this curve that measures a specific visibility amplitude and phase of an X-ray source distribution.

TABLE 1
Analogy Between Radio Interferometry and Fourier Transform X-ray Imaging

	RADIO	X-RAY
Angular resolution:	wavelength/antenna separation	grid pitch/grid separation
	5cm/5km ==> 2 arcsec	50μm/5m ==> 2 arcsec
Field of view:	wavelength/antenna diameter	grid diameter/grid separation
	6cm/4m ==> 1deg	10cm/5m ==> 1deg
One Fourier component:	one antenna pair	grid pair + detector (one subcollimator)
Imaging requires:	many antenna pairs	many subcollimators
Rotational synthesis uses:	earth's rotation	rotating collimators

For X-rays, such Fourier transform imaging has a number of advantages compared to alternative schemes. High angular resolution can be achieved because the resolution is determined by the spacing of the apertures, not by the detector spatial resolution. Grids with 50 micron apertures separated by 5 m are practical and can achieve 2 arcsecond resolution. *uv* coverage is the same at all energies so that a common set of optics could image from 2 keV to 10 MeV. The energy range is limited at high energies by the need for thick grids to be opaque at gamma ray energies. At low energies the energy range can be limited by diffraction. High sensitivity can be achieved. For example, 10 subcollimators, each 10×10 cm in size would have a total effective area of 250 cm^2. By exploiting the mathematical analogy between radio and X-ray Fourier imaging, a great legacy of imaging tools and techniques developed for radio imaging is already available for application to X-ray imaging.

An important practical consideration is that even for an instrument designed to achieve X-ray images with arcsecond resolution, it is not necessary to align the "optics" to arcsecond tolerances. For example, moving grids parallel to their slits would have no effect while displacing grids perpendicular to their slits would merely *shift* the modulation pattern, not reduce it. In practice, it is necessary therefore to have knowledge, not control, of critical alignment parameters in such a telescope, a very important distinction.

Fourier transform imaging forms the basis of the hard X-ray imager on *Yohkoh* (Kosugi *et al.* 1991). This telescope does not rotate but, instead, measures the equivalent of 32 distinct points on the *uv* plane. It makes its measurements by a slightly different technique, equivalent to sampling the triangular modulation pattern at 2 fixed locations, and then determining the amplitude and phase by comparing these measurements with an independent measurement of spatially integrated flux.

A balloon-borne instrument called HEIDI (Crannell *et al.* 1991) will use grids with FWHM resolution as high as 11 arcseconds in a flight in the summer of 1992. The telescope, illustrated in Figure 7, modulates the X-ray flux by mechanically rotating two 5.2 m long collimators.

For the future, NASA is considering opportunities for a high resolution imager combining the high spectral resolution of germanium detectors with arcsecond class imaging. The straw-man instrument for such a mission would use 12 such detectors to make very high quality images.

Since Fourier-transform X-ray imagers share many of the advantages of radio interferometry, they also share their disadvantages as well. In particular, it will be very difficult to see low surface-brightness X-ray features in the presence of bright kernels of emission. Statistics will constrain the quality of X-ray images, particularly at the higher energies. Large diffuse X-ray sources will not be modulated and so will not be seen, just as interferometers do not respond to large diffuse radio sources. However in both cases, we can determine the amount of the "missing flux" by comparing the resulting images with the unmodulated signal levels.

Despite these limitations, the prospect of new instruments combining high sensitivity with high spatial and spectral resolution, in the next few years will certainly enrich our view of the high energy Sun.

HEIDI DUAL COUNTER–ROTATING RMCs

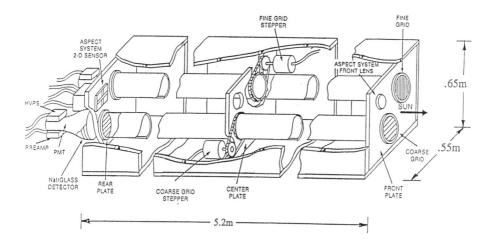

Figure 7. Schematic view of the HEIDI instrument. Fine and coarse grids in the front and rear plates are mounted at the ends of long tubes which are rotated by means of the motors in the center. The modulated X-ray flux is detected by a pair of scintillators mounted behind the grids at the left. Courtesy of C. J. Crannell.

5. References

Chupp, E. L. 1984 *Ann. Rev. Astron. Astrophys.*, **22**, 359.

Crannell, C. J. and 10 co-authors, 1991, in Proceedings of the AIAA International Balloon Technology Conference, AIAA-91-3653-CP.

Dennis, B. R. 1985, *Solar Phys.*, **100**, 465.

Fraser, G. W. 1989, *X-ray Detectors in Astronomy*, Cambridge University Press.

Hurford, G. J. and Hudson H. S. 1979, UCSD-SP-79-27 and BBSO preprint 0188.

Kosugi, T. and 11 coauthors, 1991, *Solar Phys.*, **136**, 17.

Makashima, K., Miyamoto, S., Murakami, T., Nishimura, J., Oda, M., Ogawara, Y. and Tawara, Y. 1977, in K. A. van der Hucht and G. Vaiana (eds.), *New Instrumentation for Space Astronomy*, Pergammon Press.

Orwig, L. E., Frost, K. J., and Dennis B. R., 1980 *Solar Phys.*, **65**, 25.

Prince, T. A., Hurford, G. J., Hudson, H. S. and Crannell, C. J. 1988 *Solar Phys.*, **118**, 269.

Tsuneta, S. 1984, *Ann Tokyo Astron. Obs.*, 2nd Series, **20**, 1.

van Beek, H. F., Hoyng,P., Lafleur, B., and Simnett, G. M. 1980, *Solar Phys.*, **65**, 39.

PART IV
SOLAR AND STELLAR ACTIVITY

H. ZIRIN
BBSO, Caltech
Pasadena, CA 91125
U.S.A.

ABSTRACT. The magnetic cycle lasts 22 years, reversing polarity in each 11–year half. Spots reach the surface in emerging flux regions (EFR) and grow in complexity. Ephemeral regions bring considerable flux to the surface, but play no great role. Diffusion models of the cycle are discussed, but the low observed diffusion constant makes them a poor fit to reality. We then turn to the properties of active regions, especially highly active regions producing many flares, and we discuss the properties of those flares.

1. The Cycle

The sunspot cycle is remarkable in its elegant complexity and regularity. The three major aspects of the cycle are the 11-year period of sunspot number, the Hale–Nicholson law of sunspot polarity, and the reversal of the general field. Superposed on this relentless regularity is a good deal of chaos.

As can be seen in Figure 5 of Chapter 10, each outbreak of spot activity begins steeply and declines slowly. The first spots of a cycle appear near 30° N and S latitudes, and the last ones near the equator. This variation is called Spörer's law, but its regularity is best demonstrated by Maunder's (1922) "butterfly diagram," Figure 1.

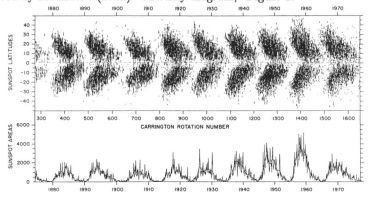

Figure 1. The butterfly diagram, with a single dot for each spot group in a number of cycles. The new–cycle spots form at high latitudes before the old–cycle spots disappear. The lower panel shows the sunspot number in each cycle.

449

J. T. Schmelz and J. C. Brown (ed.), The Sun, A Laboratory for Astrophysics, 449–463.

It is the *locus* of the successive sunspots that changes, while the individual spot groups typically last only one 27-day rotation or less and move very little relative to the surface. Note that the butterfly wings overlap, that is, old-cycle spots are still present at the equator when the new cycle breaks out, and each cycle outbreak is 12 or 13 years long. While the Zurich number is most frequently used to determine "sunspot maximum," the maximum *area*, which is more closely related to activity, lags behind the sunspot number maximum by about a year. The spot number in each hemisphere may be quite different, and one hemisphere may dominate for many years.

Hale and co-workers (1919) discovered that each successive outbreak of activity was marked by a distinct magnetic polarity. This is generally called the Hale-Nicholson (1938) law. The sunspot groups, which are all generally bipolar, appear strung out in the E-W rotation direction with the same magnetic polarity always leading in one hemisphere and the opposite in the other. The power of the Hale-Nicholson polarity rule is such that if a dipole comes up at right angles to the rotation direction, the spot of preceding (*p*) polarity immediately heads westward at 0.5-1 km sec^{-1}. If a big spot group has inverted polarity, it will be unusually active. The relation of dipole polarity to rotation reverses each 11-year semicycle, so the overall magnetic cycle is 22 years.

While some flux may be generated locally in random fashion, all stronger fields emerge through the sunspot process, break up, and diffuse across the surface to produce a bipolar global field. The polar field reverses sign after the peak of each semicycle, taking the sign of the following polarity in that hemisphere.

The current cycle has had major activity right through the number maximum. The huge March 1989 group appeared at 33°N, and the great June 1991 region at a similar latitude. Usually, the big spots are late in the cycle and near the equator.

The presence of a new cycle, even before spots erupt, has been suggested by Howard and LaBonte (1980) from torsional motions.

2. Flux Emergence

The emerging flux region is the characteristic form of sunspot emergence (Zirin 1972), clearly showing that magnetic flux tubes are emerging from below. In Hα, the EFR is marked by a bright patch (usually the brightest feature on the disk) crossed by dark arches, termed "arch filament systems" (AFS) by Bruzek (1967). In white light, the EFR appears as two clusters of small, growing pores about 20,000 km apart, too small to be a stable feature. More pores appear and gradually merge into spots with penumbra. Thin dark streaks appear between the spots, marking the flux arches penetrating the surface. The *p* spots run ahead, pulled by the Hale-Nicholson force, while the following (*f*) spots decay or remain fixed. In complex active regions, no AFS may be seen, and the emergence is marked by rapid spot emergence and separation. EFRs rarely grow to great size; great spots appear to form by merging of the results of several EFRs. Vrabec (1974) studied a growing active region with satellite *p* spots flowing into and merging with the leading *p* spot. He suggests a rising "tree trunk" model; the satellite spots represent the intersection with the surface of branches of a submerged tree which gradually merge as it rises. This process is more applicable to separate EFRs. In a single EFR, several pores come up and the space between them fills in. On the Sun, like fields often attract. The converse of this is that spots of opposite polarity never merge, but form a δ spot generally associated with high activity.

The arch filaments are not filaments at all, but connect opposite polarities and truly represent emerging flux loops. In the blue wing of Hα, only the top of the rising arch is seen; in the red wing, we see material draining downward. Thus, an Hα + 0.7Å (red) picture shows just the ends of the loops and Hα − 0.7Å (blue) shows the tops. As each arch passes through the surface, we briefly see dark fibrils in the photosphere. Thus each flux loop is discrete, with a discrete amount of material lifted, and that is why a single pore appears with each. The arches grow particularly dark when a spot is emerging and erupt when the spot appears. Chou and Zirin (1988) measured the radial velocities, finding the loop tops ascending with velocity 10–20 km sec^{-1}, while the feet showed material falling at $v = 20 - 40$ km sec^{-1}. Measurements of proper motions at the limb show the arches rising at about 15 km sec^{-1} for about 1000 sec, and erupting at heights above 15,000 km as the material drains out. The loss of material must produce an instability, leaving the loop buoyant and erupting upward. The downfall velocity exceeds the upward motion of the arch because of gravitational acceleration. There is a story that EFRs appear at nodes of the chromospheric network; unfortunately, the EFR is bright and can be misidentified in its early stages as a piece of network, so the evidence is unclear. The EFR is invariably bipolar.

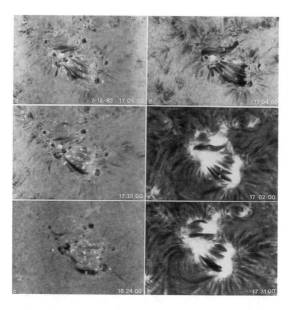

Figure 2. An emerging flux region, Hα blue wing left, with rising arches; red wing right top, with downflowing material; centerline lower right, arches and bright plage. The rapid growth here took only 80 min (Zirin 1987).

If the flux tube is to rise, the height gradient of the magnetic field must be steeper than the adiabatic. Because EFRs obey the Hale–Nicholson law, they must come from the global fields involved in the sunspot cycle; thus, the field may increase to great values far below the surface, at least down to $R = 0.8R_\odot$. Although the surface fields are in individual bundles, we may assume that their intensity increases at a rate greater than the density; this implies a field greater than 10,000 Gauss at $R = 0.8R_\odot$.

The EFR is a messenger from below the surface, but its story is far from clear. EFRs generally reach the surface in neat, organized bundles; the arch filament systems are generally parallel and apparently untwisted by subsurface convection. While the picture of a magnetic loop crossing the surface is supported by the measurements, there are problems. What happens to the loop after the spot group forms and dissipates? The answer may be that the loop sinks again. In some cases, we see this happen; the region shrinks and disappears. In others, most of the flux sinks and only a fraction remains at the surface. Why does all this happen at the surface? Possibly because the conductivity is at a minimum there and reconnection can occur.

Watching the spots grow is informative. When the first spots form penumbras, the arch filaments connected to them disappear but the p spot continues to move forward, leaving a limited arch filament system in the center so long as new flux continues to erupt. The spot begins to emit moving magnetic features as soon as the penumbra appears. The p spot moves forward $\left(1 \text{ km sec}^{-1}\right)$ in all cases; in most cases, the p spot grows and the f spot disappears. In a very few cases, the f spot dominates. Few single EFR dipoles ever exceed 50,000 km in length. Weart (1970) found that the majority of EFRs come up with the correct polarity; the small fraction with axes strongly tilted to the equator straighten out or die out early. Even if the axis is tilted, the p spot drifts straight west, often producing a hooked shape in the growing region.

A remarkable feature of EFRs is the frequent Ellerman bombs – bright points with very broad Hα wings (HW $\approx$ 5Å) that are low in the atmosphere so they are not visible in the Hα centerline. Severny (1957) called them *usi* or moustaches, because their spectra look like wide moustaches with a gap in the middle. It is hard to see how such broad wings are produced in such a tiny feature, and why they are tied to emerging flux. Shklovsky (1958) discussed the problems of confining such high velocities in tiny elements. They typically occur at the center of the EFR or on the edges of spots, places where the field is breaking the surface.

During the peak of the cycle, one or two EFRs per day emerge over the visible Sun outside active regions; but nearly as many emerge in the tiny fraction of the surface covered by active regions. Zirin and Liggett (1985) found a disproportionate probability that EFRs emerge near existing groups; we term these regions *centers of activity*.

Flux emergence plays an important role in many flares. The new flux must reach an accommodation with the old, forming a new magnetic structure. Since the lines of force are sticky and slow to change, the new flux pushes up or in until the field is sufficiently stressed to change to a more stable state. If this change, which usually involves reconnection, happens discontinuously, a flare results; if the adjustment is slow, we have simple heating. Without continuing flux emergence, the active region dies; the spot breaks up, the plages weaken and break up or submerge below the surface, and only enhanced network remains. The flux emergence also drives the spot motion so typical of the most active regions; as the spot moves, the field is sheared.

3. Ephemeral Regions

Ephemeral regions (ER) are small, spotless dipoles, which live for about a day. Their significance was first pointed out by Harvey and Martin (1973). The actual number of ER on the Sun is unknown because all the published counts lack calibration to a limiting magnitude. They are much less concentrated in latitude than the sunspots and their orientation is not really known. They show the typical separation rate of growing active regions – 1 km sec^{-1}. This pattern suggests that they are the result of local, rather than global, dynamo action. There may be some dependence of their number on the sunspot cycle – this might be due to the greater amount of flux spread around the Sun which can be intensified by local dynamos. Harvey (1985) concluded that the X-ray bright points are sites where ER are reconnecting with existing fields of opposite polarity. Gary *et al.* (1990) found no increased microwave emission from ephemeral regions compared to the network, but they may produce microflares and tiny radio bursts. The smallest observed are 9×10^{-4} sfu.

Figure 3. The reverse Evershed effect, in Hα blue; the limb is at left, and the material spiralling in is coming toward us (BBSO photo).

The ephemeral regions are readily identified as small bipolar structures in quiet Sun magnetograms, but the magnetograms vary in quality, strong active regions may mask them, and no way of setting a limiting magnitude has been found. Harvey *et al.* (1975) estimate 360 on the whole Sun in 1970, near sunspot maximum; and 178 in 1973, near minimum. The average flux is stated to be 10^{20} Maxwells (the average active region flux is 4×10^{22} Maxwells). The regions live for a day and typically give one tiny flare. Unlike growing active regions, they do not show arch filaments (although there may be a dark fibril or two) and they never develop pores.

Because the ER last such a short time, new ones have to erupt every day and, in fact, there are at least a thousand times more ER than active regions of all sizes. If these numbers are right, the total flux carried to the surface by ER exceeds that carried by sunspots by a factor of about five. So the contribution to the total magnetic flux is large although the net field is small.

Are the ephemeral regions just tiny active regions? We know that they do not obey the law of sunspot polarity or Spörer's law. Tang et al. (1984) studied 15 years of Mt. Wilson active region data and found an exponential dependence of the integral number of active regions down a size A of about three square degrees:

$$N(A) = 4788 \exp -(A/175).$$ (1)

This equation gives an asymptotic value of 4788 for the total number of active regions of all sizes in the 15-year period. By contrast, the number of ER during this period, using the Harvey et al. (1975) values, was 1000 times greater. Since Equation (1) gives an almost perfect fit to observations, a 1000-fold jump just below the Tang et al. threshold would be unreasonable. Combining this with the fact that they occur all over the Sun, the ephemeral regions may be a locally generated phenomenon.

4. Understanding the Cycle

Hale searched in vain for the general solar field for many years and thought he detected a 50 Gauss poloidal field. Not until the development of the magnetograph (Babcock and Babcock 1955) was it possible to confirm that there was a polar field ten times weaker than Hale found and, in 1959, H.D. Babcock detected its reversal in the new cycle. Babcock (1961) developed a regenerative model in which the fragments of f polarity migrated poleward to reverse the existing fields. Bumba and Howard (1965) showed how the spot fields broke up and spread into great unipolar regions which drifted toward the pole. The accumulation of f polarity reversed the polar field a few years after the onset of the new cycle.

Since all magnetic flux erupts on the Sun in dipoles, it is remarkable that the field can be organized into such large unipolar entities. Only the active regions are large enough to separate fields on such a scale. Leighton's (1964, 1969) kinematic model of the 22-year cycle was based on the random walk of fields diffusing from the spot groups. If individual elements of flux move a distance L before changing direction in a time τ, then their random walk will carry them a distance

$$r = L \left(\frac{t}{\tau}\right)^{1/2}$$ (2)

in a time t. The area filled by these elements will be

$$A \approx \pi L^2 t/\tau.$$ (3)

The spreading of the density n of magnetic elements is given by a diffusion equation

$$\frac{\partial n}{\partial t} = D \bigtriangledown^2 n, \tag{4}$$

where the diffusion constant D is given by:

$$D = \frac{1}{2}\frac{L^2}{\tau}. \tag{5}$$

Leighton chose size L and lifetime τ of a supergranule cell as the diffusion parameters. With L=15,000 km and $\tau = 7 \times 10^4$ sec (22 hours), he obtained a diffusion constant of 1600 km^2 sec^{-1}. He found the poleward drift rate of field could be matched by D = 1000 km^2 sec^{-1}. This behavior alone would not reproduce the polar field separation. Leighton had to introduce Joy's law (Zirin 1988) of spot tilt to explain the dominance of the f field; the slight poleward displacement of the f spots results in dominance of that polarity at the pole. Observations do not confirm Leighton's diffusion rate. Mosher (1977) and Zirin (1985) found $D = 300 \,\mathrm{km}^2$ sec^{-1}, one-third Leighton's value, from the actual motion of network elements. The enhanced network diffuses even more slowly.

Devore et al. (1985) and Devore and Sheeley (1987) produced excellent models of the evolution of magnetic regions by starting with an observed magnetogram and reproducing the field distribution observed a month later. They used the diffusion equation with $D = 300 \,\mathrm{km}^2$ sec^{-1}, Snodgrass's (1983) measured rotation, and a meridional flow rate varying significantly from case to case. Obviously, it helps to have many free parameters, but the diffusion model is close to predicting the evolution of fields, leaving only the flux emergence pattern, Joy's law, and the meridional flow to be explained.

Howard and LaBonte (1980), analyzing many years of velocity data from the Mt. Wilson magnetograph, found evidence for a torsional oscillation of the Sun in synchronism with the 11-year cycle. The amplitude is about 3 m sec^{-1} (0.167% of the rotational velocity) relative to the differentially rotating surface. Zones of faster rotation originate at the poles and drift to the equator in 22 years. The effect of these bands on the evolution of large–scale fields has not yet been explored.

The progress of the cycle is uneven: Even near maximum a completely quiet hemisphere can rotate into view. In the period 1955-70, there were far more spots in the northern hemisphere and, in the 1990 cycle, the southern hemisphere dominated. The two cycles peaking in 1946 and 1957 were the largest in history. Spots will often break out all over the Sun at once, supporting the idea that it is a global phenomenon. There also may be preferred meridians for their occurrence, but the evidence is weak.

While the sunspot cycle has been quite regular for some centuries, there is evidence for a period of low activity discovered by the English astronomer Maunder, who pointed out that very few spots were seen between 1650 and 1715. Although sunspots had been detected around 1600, there are few records of spot sightings in this period. Experienced observers reported the occurrence of a new spot group as a great event, mentioning that they had seen none for years. After 1715, the spots returned. The "Maunder minimum" was associated with a long cold spell in Europe, known as the Little Ice Age. Whether or not the Maunder minimum caused the Little Ice Age cannot be established, but there is little doubt that the sunspots dropped out during this period (although a weak 11-year period remained in the few spots present). There is evidence for other such periods at roughly 500–year intervals. When solar activity is high, the strong magnetic fields carried out by the solar wind block the high–energy galactic cosmic rays and less carbon 14 is produced. Measurements of the carbon 14 in dated tree rings thus confirm the low activity at this time. Still, the 11-year cycle was not detected until 1840, so observations were not that regular. For example, the solar corona was not mentioned until the latter part of the century, even though it surely must have been seen.

5. Highly Active Regions

Hale *et al.* (1919) introduced the Mt. Wilson magnetic classification, which remains the most significant classification of magnetic structure:

α: A single dominant spot, usually connected to plage of opposite magnetic polarity.
β: A pair of dominant spots of opposite polarity.
γ: Complex groups with an irregular distribution of polarities.
$\beta\gamma$:   Bipolar groups with no marked north-south inversion line.

In addition, the suffixes p and f are used when the preceding or following spot, respectively, is dominant. In the majority of groups, the p spot is dominant; Hale and Nicholson found 57% of the groups to be βp and αp, and only 13% with f dominant. Even if the polarity is inverted, the p spot is a big round one, moving westward.

Early attempts to predict flares by this classification were quite unsuccessful. Of course, it was recognized that sunspots were connected with flares (although the data were so poor that some papers can be found claiming flares are unrelated to sunspots!), but many big sunspots gave no flares. This is because the classification totally disregarded magnetic mixing. Künzel (1960) showed that none of these classes had an excess of flares and defined a new δ configuration that did:

δ: Umbrae of opposite polarity in a single penumbra.

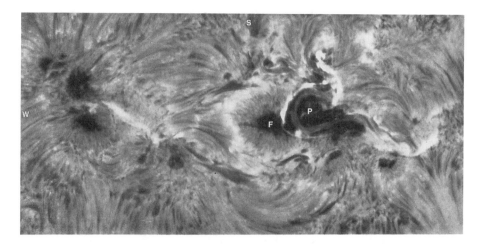

Figure 4. An energetic δ spot, July 1974. The complex spot at right, which emerged in this form, contains a large p spot surrounded by f spots (BBSO photo).

He showed that the δ groups produced far more flares than the other Mt. Wilson classes, a conclusion soon confirmed by others. The δ configuration can occur within any of the older classes; it simply requires two umbrae butted up against one another.

The high activity associated with δ spots should come as no surprise. Two poles with strong opposite vertical fields are pushed together; their field cannot be directly connected. Instead the connecting field lines run parallel to the boundary in a sheared fashion. This has been confirmed (Zirin, in preparation) with vector magnetograph observations and with spectrographic measurements of the Zeeman splitting. The field in the shear area is just as great as in the umbrae.

After Künzel's discovery, his ideas were confirmed by observations. Rust (1968) pointed out that δ configurations formed by satellite spots of opposite polarity near a large umbra were a common source of flares. Tanaka (1975, 1980) showed that 90% of δ groups with inverted polarity (relative to the Hale–Nicholson law) were associated with great activity. δ spots are almost always large; conversely, large spot groups are more likely than smaller ones to show the δ property.

Zirin and Liggett (1987) found that δ spots form in three ways:

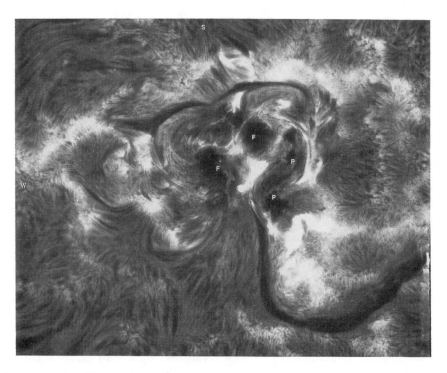

Figure 5. The great August 1972 region. While it rotated to the left, the f, or following, spots were ahead of the p spots; polarities are completely interchanged, and all the spots are tightly clustered together (BBSO photo).

1. A single complex emerges at once with dipoles intertwined and polarities reversed from the Hale–Nicholson rules (Figure 4). This gives the most active regions.
2. Large satellite dipoles emerge close to existing spots so that the expansion of the EFR pushes a p spot into an f spot or *vice versa*.
3. A growing bipolar spot group collides with another dipole so that opposite polarities are pushed together. Examples of these are given by Künzel *et al.* (1961) and Tang (1983).

The third type is the most frequent, occurring whenever an EFR comes up in the wrong (or right) place. It only forms from emerging umbrae, not plage. If the dipole only expands into plage, modest flares occur without δ spot formation. If it collides with an umbra of opposite polarity, the δ spot forms and bigger flares occur. If it collides with an umbra of the same polarity, the two do not merge, but coexist peacefully separated by a light bridge. The biggest flares, however, occur in the first class, particularly in what is called the "island delta," a large, round spot or spots surrounded by bright plage of opposite polarity.

Zirin and Liggett (1987) studied properties of highly active regions observed at Big Bear Solar Observatory (BBSO) and summarized them as follows (modified by more recent experience):

1. The most important activity occurs in "island δ" spots with high magnetic shear and rapid spot motion.
2. δ spots rarely last more than one rotation and are shorter-lived than other spots of the same size, but new δ spots may emerge in the same complex (April-May 1984, Mt. Wilson 24030, 24057).
3. The polarity is generally inverted as compared to the Hale-Nicholson law.
4. Most δ spots are locked in a deadly embrace from which they never escape; but a small fraction eject an umbra which produces many flares as it rapidly separates.
5. Components of δ spots are not connected by direct lines of force, but by sheared magnetic field lines running parallel to the inversion line.

The epitome of the island delta was the August 1972 group (Figure 5) which produced some of the greatest flares in history (Zirin and Tanaka 1973). It emerged with a large p spot behind a ragged f plage; this evolved on the far side of the Sun into a completely inverted δ configuration, with a large new p spot pushing through its center. There was no place for the p spot to go and great shear built up as it pushed forward, producing several great flares.

The sheared inversion line of the δ spot is the scene of most of the energy release, although smaller flares often occur in the surrounding tangle of spots. It is marked in Hα by elongated fibrils parallel to a neutral line, leading out to a filament. Often, the Hα emission covers the umbrae, so we know that the field turns 90° between the vertical umbral state and the horizontal fibrils. In the continuum, the penumbra has been sheared off by the δ stress, and even elongated umbrae (which never found a place in spot classifications but are a sure portent of flares) often appear.

The eruption of the island δ raises the question: Why did it not simplify below the surface? It is possible that all active regions are formed complex and most are simplified when they erupt at the surface. Or, conversely, they were formed simple and these regions got twisted up in the convective zone. The fact that they occur away from the number maximum and that they are typically quite large suggests they were produced in the complex form. No matter, the Hale–Nicholson force pulls the p spots forward.

Zirin and Marquette (1990) listed the following criteria for expecting high activity and issuing a BEARALERT, a warning from BBSO:

- Large "island δ" sunspots.
- Elongated umbrae.
- High shear in transverse field or steep gradients in longitudinal. Penumbral fibrils aligned with the inversion line.
- A δ configuration with bright Hα obscuring the umbrae.
- An emerging p spot that is moving into an f-polarity region, or an umbra expelled from a δ spot.
- A filament curled around a sunspot or overlying a plage.
- Rapid spot movement along the inversion line.
- Intense surging or flaring at the east limb, or continuous loop eruption on the disk
- Big spots have big flares.

All of these characteristics represent sheared fields, flux emergence, *etc.* The BEAR-ALERTS are considerably better than random, but still miss important activity. The principles are probably correct: We can usually find a reason for the flare after it occurs but the prediction is still difficult.

While activity can be predicted from observable spots, the eruption of great spot groups and even the eruption of new EFRs in old regions are still the secrets of the solar interior.

6. Observing Flares

Optical observations are, of course, the classical technique for observing flares. The high resolution, short exposures, and possibilities for magnetic and Doppler measurements make it the prime technique. Also, a major part of the energy comes out in optical frequencies. However, the high-energy phase, so important to the flare process, can only be detected by inference and the relation between "the flare" and the optical emission is not obvious. The classical technique for observing flares is with the Lyot filter in Hα or another line. This is certainly the most sensitive line to atmospheric heating by the energetic particles that result from the flare. Hα is particularly valuable because it shows the magnetic background of the flare, especially key aspects such as the neutral line filaments that play a key role in the energy release. Typically, the Hα emission goes up with the hard X-rays (and microwave) and decays with the soft. The initial rise represents emission from the elevated flare source itself as well as heating and excitation of the underlying chromosphere. The soft X-ray emission is usually in the form of the well–known two–ribbon flare, where conductive heating of the surface intersection of a cylindrical shell takes place. This heating is always in the shell in which loop prominences form; the rest of the coronal cloud appears to have no effect.

Kernels of the flare where the line profile is broadened by great heat or motion are observed in the wing of Hα. Because the line center is optically thick, the wing gives a new view, seeing down to the underlying sunspots which are often covered by emitting material. Emission at 5Å or more from line center is the exceptional case but most flares are only an angstrom or two wide.

Flares may be observed in other lines, each with something to offer. The helium lines, D3 and 10830, appear on a structureless background and are always optically thin. Thus, they present a true picture of the flare without background confusion. However, much of these observations can be duplicated by tuning off-band in Hα. Emission appears wherever the collision rate exceeds the radiative de–excitation rates, usually $N_e = 10^{12.7}$ cm^{-3}. Less dense regions where He is excited by thermal or radiative means appear in absorption; this is usually true of surges. Thus, we have a good density diagnostic. Many flares have been observed in D3; we have only recently begun 10830 monitoring. BBSO typically observes in 5 or 6 wavelengths: Hα ± 0.7Å, K-line or He, longitudinal and transverse magnetic field, and Doppler. All of these contribute to the analysis. Flares may be observed in other lines: Hβ, further out on the Planck curve, is more sensitive than Hα, but ejecta are less visible. The K line has the same properties. Lines like Na D and Mg b are less interesting, showing the upper photosphere itself.

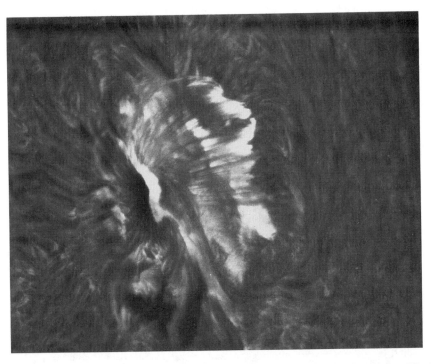

Figure 6. Post–flare loops following the great flare of 1974 September 10. The loop tops, seen at right projected against the chromosphere, are brighter than the loop legs, which are in absorption. The loop tops are denser and hotter (BBSO photo).

What do we see? Only a movie can show the various complex elements. Every flare is different. Exceptionally energetic flares show the famous flash phase, an explosion, doubling brightness or area every 10 sec or so. If they are on the outer edge of the big spot, an eruption and shock wave will appear, with very intense hard X-rays and low soft X-rays. These flares are underrated by the present soft X-ray (*X3, etc.*) classification. All flares have two ribbons to some extent, simply because all closed field lines intersect the surface in two points, and energetic particles and thermal energy flow easily along them. Sometimes, quite distant points brighten simultaneously; they are simply the terminus of field lines originating in the flare. If the flare is really brilliant, the surrounding chromosphere will be illuminated. Neither the peak Hα intensity nor the soft X-ray flux from a big flare is known since all the observations are saturated. We have measured 3× the continuum; at this wave length this means the minimum flare chromospheric temperature must be 18,000 K. Mary Rowe, a Caltech undergraduate, studied 122 flares to see if they were bright beforehand; 111 were. Thus the brightening occurs in existing poles, not in random places.

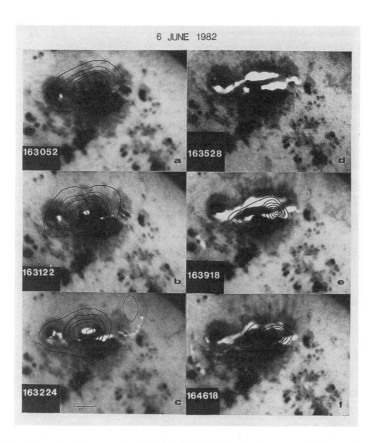

Figure 7. Flare development in He D3. This X13 flare began with a number of bright points along the neutral line of the δ spot, elongating into two bright ribbons which then separated. Contours mark the hard X-ray emission measured by Hinotori (Tanaka and Zirin 1985).

Many, maybe most, flares originate in a filament along the inversion line; the filament is usually activated before the flare. In the impulsive phase of a big flare, we see tiny bright points, usually pairs of opposite polarity, along the neutral lines. Presumably, these are distinct, quite flat flux loops which are individually activated. In a big flare, the growth continues, energy is released over a wider area, and a violent ejection of material may occur. As the peak is reached, loop prominences occur connecting the two ribbons, which then spread apart. The loops *always* connect the two ribbons, they *always* form further and further up, they *always* form at the loop top, and they *always* rain down. Because heating of the ribbons only occurs in the loop shell and X-ray images similarly show bright loops, we must assume that the rest of the volume is either cooler or far less dense. Probably runaway cooling in the loops produces compression and the peak density at the loop top.

The soft X-ray thermal cloud is commonly thought to be due to "chromospheric evaporation," material rising from the heated chromosphere. Such evaporation is not observed in Hα, but the material may be so highly ionized that it is invisible in Hα. The only material seen rising is the erupting filament or spray material, which usually pass through the loop area and move much further out. They may very well be a source of material, but the first flare loops occur at a low height.

Many different flare morphologies are seen; all produce high–energy particles in varying amounts. One common flare, usually small, is marked by the brightening of two footpoints, nothing in between. Presumably something has happened in the corona at the top of the connecting flux loop. Another common flare, sometimes much larger, has a filament twisting and rolling, followed or accompanied by footpoint brightenings at both ends. Eruption does not always disrupt the filament.

When the shear is high, one sees a channel in Hα along the inversion line, with alternating dark and bright fibrils. In white light, we see sheared penumbra, or even elongated sunspots, which have strong transverse fields (which can be measured with a spectrograph). For emitting material to cover the spot, the field must take a sharp turn from the normal vertical spot field. We find the most active spots show narrow alternating channels of opposite polarity, moving in opposite directions.

All these occurrences are part of the flare story. There is much we can learn by examining it.

7. References

Babcock, H. W. 1961. *Ap. J.* **133**, 572.
Babcock, H. D. and Babcock, H. W. 1955. *Ap. J.* **121**, 349.
Bruzek, A. 1967. *Solar Phys.* **2**, 451.
Bumba, V. and Howard, R. F. 1965. *Ap. J.* **141**, 1502.
Chou, D.-Y. and Zirin H. 1988. *Ap. J.* **333**, 420.
DeVore, C. R. *et al.* 1985. *Aust. J. Phys.* **38**, 999.
Devore, C. R. and Sheeley, N. R. 1987. *Solar Phys.* **108**, 47.
Gary, D., Zirin, H., and Wang, H. 1990. *Ap. J.* **355**, 321.
Hale, G. E. *et al.* 1919. *Ap. J.* **49**, 153.
Hale, G. E. and Nicholson, S. B. 1938. *Carnegie Inst. Wash. Publ.* **49**, 1.
Harvey, K. L. and Martin S. F. 1973. *Solar Phys.* **32**, 389.

Harvey, K. L. *et al.* 1975. *Solar Phys.* **40**, 87.

Harvey, K. L. 1985. *Austr. J. Phys.* **38**, 875.

Howard R. and LaBonte, B. J. 1980. *Ap. J.* **239**, L33.

Künzel, H. 1960. *Astron. Nachr.* **285**, 271.

Künzel, H., Mattig, W., and Schröter, E. H.: 1961. *Die Sterne* **9/10**, 198.

Leighton, R. B. 1964. *Ap. J.* **140**, 1547.

Leighton, R. B. 1969. *Ap. J.* **156**, 1.

Liggett, M. A. and Zirin, H. 1985. *Solar Phys.* **97**, 51.

Maunder, E. W. 1922. *M. N. R. A. S.* **82**, 534.

Mosher, J. M. 1977. Ph. D. thesis, California Institute of.

Severny, A. B. 1957. *A. J. U. S. S. R.* **34**, 328.

Shklovsky, J. S. 1958. *Soviet Astronomy* **2**, 786.

Snodgrass, H. B. 1983. *Ap. J.* **270**, 288.

Tanaka, K. and Zirin, H. 1985. *Ap. J.* **299**, 1036.

Tang, F. 1983. *Solar Phys.* **89**, 43.

Tang, F., Howard, R., and Adkins, J. M. 1984. *Solar Phys.* **9**, 175.

Vrabec, D. 1974. *IAU Symp.* **56**, 201.

Weart, S. R. 1970. *Ap. J.* **162**, 887.

Zirin, H. 1972. *Solar Phys.* **22**, 34.

Zirin, H. 1985. *Australian Journal of Physics* **38**, 961.

Zirin, H. 1987. *Solar Physics* 114, 239.

Zirin, H. and Liggett, M. A. 1987. *Solar Phys.* **113**, 267.

Zirin, H. and Tanaka, K. 1973. *Solar Phys.* **32**, 173.

Zirin, H. and Marquette, W. H. 1991. *Solar Phys.* **131**, 149.

23 OVERVIEW OF SOLAR FLARES

A.G. EMSLIE
Department of Physics
The University of Alabama in Huntsville
Huntsville, AL 35899
U.S.A.

ABSTRACT. We discuss the energetics of, and characteristic emissions associated with, solar flares, and various flare "classification schemes" that have been devised. It is argued that flares derive their energy from current-carrying, but force-free, magnetic fields; observations of the magnetic "shear" associated with these currents are presented. We discuss the basic principles of flare modeling, both "forward" (*i.e.*, starting with a prescribed initial configuration and solving the relevant equations) and "inverse" (*i.e.*, fitting observed emissions to a self-consistent model of energy release). We close with a brief discussion on the effects of flares and the importance of flare research.

1. Introduction

Solar flares are associated unquestionably with the rapid release of stored magnetic energy. A large flare can release up to 10^{32} ergs of energy in a few minutes, over an area of under 10^{18} cm^2 (roughly 10 arc-seconds square when viewed from the Earth). Although this is an impressive amount of energy, it corresponds to only one-fortieth of a second of the normal solar radiative output, so that, bolometrically, solar flares represent an insignificant (and, so far, unobservable) increase in the total luminosity of the Sun. They are, therefore, not nearly as energetic as some of their stellar counterparts (see Chapter 26), whose occurrence does show up in the light curves of the star.

This release of energy is sometimes associated with (note, not necessarily the *cause* of) the ejection of up to 10^{15} g of material at speeds in excess of 100 km sec^{-1}. The energy is stored over a period of hours to days in the stressing of magnetic field configurations in so-called *active regions*, although it is not yet determined whether this stressing occurs before or after the flux emerges from below the solar surface.

Thanks to Hal Zirin *et al.*, we know that flares are complicated three (or four!) dimensional animals, evolving in a chaotic and difficult-to-predict manner. Aside from morphological studies of individual events, however, flare modelers typically resort to "cartoons," whose basic building block is the concept of an isolated magnetic flux tube, or "loop." Flare modeling typically proceeds by one of two paths. In the "direct" approach, a pre-flare magnetic field/plasma configuration is adopted and the plasma physical equations are solved to predict the subsequent evolution of the configuration and its modes of energy release. By contrast, aficionados of "inverse" modeling attempt to infer the characteristics of the energy release through analysis of flare observations. We shall review both of these approaches.

J. T. Schmelz and J. C. Brown (ed.), The Sun, A Laboratory for Astrophysics, 465–474.
© 1992 *Kluwer Academic Publishers. Printed in the Netherlands.*

466

In addition, we will discuss attempts at arranging flares into observational "classification schemes."

2. Flare Emissions

Flares produce enhanced emission across most of the electromagnetic spectrum. A brief summary of the types of emission seen, and the exciting agent responsible for the emission, follows.

• γ-Rays

γ-ray emission in the quiet Sun is totally absent, so that any feature in γ-rays is an emission feature associated with the flare. Both lines and continuum are seen; the lines are due to both prompt nuclear processes (e.g., deexcitation of excited states of nuclei such as Carbon and Oxygen), and delayed features such as the 511 keV e+/e- annihilation line and the 2.223 MeV deuterium formation line, produced by absorption of slow neutrons by ambient protons. The continuum is due to bremsstrahlung of relativistic electrons, and is typically power-law in shape. Figure 1 (from Murphy et al. 1991) shows a typical solar flare γ-ray spectrum. Gamma-ray emission is discussed further in Chapter 24.

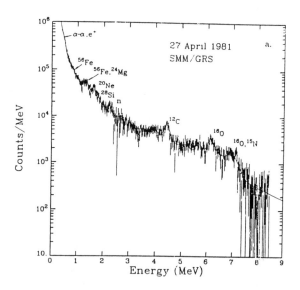

Figure 1. Gamma-ray spectrum from a flare on 27 April 1981 (after Murphy et al. 1991). Note the conspicuous line emissions superimposed on the broad (bremsstrahlung) continuum. The features at 4.5 and 6.1 MeV are due to prompt deexcitation of excited nuclei; by contrast, the feature at 2.2 MeV is due to deuterium formation and is delayed because of the time required to slow the accelerated neutrons to a velocity at which the cross-section for capture on ambient protons is significant.

● Hard X-rays

X-rays are "hard" in the energy range $10 \simeq 300$ keV. Below this energy they are "soft," and above it they are considered, by solar physicists at least, as γ-rays. There is little doubt that they are produced by bremsstrahlung of comparable energy electrons on ambient protons; they are the canonical signature of the impulsive phase of the flare and are discussed at length in Chapter 24.

● Soft X-rays

This wavelength range contains a substantial fraction of the radiated energy in flares (see, *e.g.*, Canfield *et al.* 1978). Largely in recognition of the transition to "softer" emission, these are characterized more by wavelength than by photon energy. Soft X-ray emission spans the range $1 \simeq 10$ Å, and consists of an admixture of bremsstrahlung (free-free) continuum, bound-free continuum, and spectral lines, typically due to highly ionized species of metals, *e.g.*, Ca^{18+} and Fe^{24+}. The electrons responsible for exciting both the lines and continuum are typically thermal electrons with a temperature around 10^7 K. Such temperatures are typical of flare coronas; hence, soft X-ray lines in particular provide a valuable diagnostic of conditions (temperature, density, velocity) in flaring atmospheres (see Chapter 25.

● EUV

Thermal emission from temperatures of order $10^4 - -10^5$ K, we again see a mixture of lines and continuum. Two major components are relevant: broadband (10 – 1000 Å) emission from the dense upper chromosphere (a valuable diagnostic of energy input to this region – see Chapter 25) and optically thin line emission from the thin chromosphere/corona transition region. This intensely studied region of the spectrum provides diagnostics on the rapidly changing temperature and density structure of flaring atmospheres.

● Optical

Here we find a plethora of lines and continuum, constituting a substantial fraction of the total energy released, and, for the first time, we encounter emission from lines (*e.g.*, $H\alpha$) normally seen in absorption on the quiet Sun. Imaging of flares at optical wavelengths is, of course, possible with large ground-based telescopes, and has resulted in a vast amount of morphological data on flare structure and evolution (see Chapter 22). Originating in optically thick regions of the atmosphere, the line shapes are complex functions of the global structure of the flare chromosphere; as a result, they offer valuable diagnostics of atmospheric conditions there. Historically, the intensity and/or area of the flare in $H\alpha$ was the only way to classify flare size; our understanding of "big" versus "small" flares has since evolved considerably, leading to other methods of flare classification (Section 3).

• Radio

Radio waves constitute a low energy radiation field that nevertheless provides us with a diagnostic of some of the highest energy phenomena in the flare. Totally insignificant energetically (although it is humbling to note that complete conversion of the rest mass of a human into energy might *just* power the radio flare), radio emission is nevertheless rich in detail and complexity, and it affords us diagnostics not only of energetic electrons (up to several hundred keV) but also of magnetic field strengths, densities, *etc.* The principal mechanisms of emission are gyrosynchrotron radiation, free-free emission (bremsstrahlung), and collective plasma processes and are elaborated in Chapter 14.

3. Flare Sizes and Classifications

As mentioned in the previous section, flares have been classified historically by their appearance in the Hα lines of hydrogen. A two-parameter classification scheme (see, *e.g.*, Svestka 1976) has been devised; the first parameter (1-4) refers to the area of the bright Hα area patch, the second to the intensity of the enhanced emission (faint, normal, or bright). Thus, for example, a small, very bright Hα region might be classified as a 1b flare.

With the advent of space instrumentation, it was soon discovered that this time-honored classification scheme was not always the best indicator of the quantity of energy release. On the grounds that a large fraction of the radiated energy in flares appears as soft X-rays, radiated by the hot coronal plasma produced by the flare process, a flare classification based on the energy flux in the $1-8$Å waveband, as synoptically observed by the GOES satellite series, was devised. This scheme uses a letter to denote the order-of-magnitude flux and a number to denote the multiple of the base amount. The broad classification is as follows:

$$B \qquad 10^{-4} \text{erg cm}^2 \text{sec}^{-1}$$

$$C \qquad 10^{-3} \text{erg cm}^2 \text{sec}^{-1}$$

$$M \qquad 10^{-2} \text{erg cm}^2 \text{sec}^{-1}$$

$$X \qquad 10^{-1} \text{erg cm}^2 \text{sec}^{-1}$$

Thus, for example, a C5 flare has a flux of 5×10^{-3}erg cm^{2}sec^{-1} in the 1-8Å channel.

Hard X-ray flares have been classified into three main classes (*e.g.*, Dennis 1985). Unlike the previous classification schemes, this classification is primarily concerned not with the geometrical extent or intensity of the event, but with the duration and shape of the light curve. Flares are categorized as follows:

A: "Thermal" flares, which are presumably compact, low altitude, events and which exhibit a smoothly varying flux *vs.* time profile, presumably corresponding to the energization of a single loop.

B: "Impulsive" flares, generally associated with larger loops, exhibiting a "spiky" flux-*vs.*-time profile.

<u>C</u>: "Gradual" flares, believed to occur in very long ($\geq 50,000$ km) loops. These presumably involve particles in a lower density coronal trap decaying slowly through Coulomb collisions.

The distribution of flare powers ranges over several orders of magnitude, with less intense events being more frequent than brighter ones. In fact, the distribution of flare frequency F (e.g., number of events in a given energy range per day) with size of event S (e.g., counts sec^{-1} in hard X-rays) follows a well-defined power-law form, with a spectral index equal to 1.8 (Figure 2). The turnover at low energies is undoubtedly a selection effect ("the world's smallest flares"); however, a cutoff at *high* energies is *required* by the fact that otherwise the total energy release rate $\int S\, F(S)\, dS$ would be infinite. It has been suggested that there is a natural explanation for this power-law behavior in terms of an analogy with avalanche sizes in growing sandpiles (Lu and Hamilton 1991).

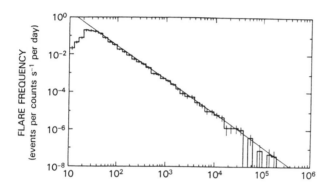

Figure 2. Distribution of flare frequency *vs.* size of event (after Dennis 1985).

4. Energy Buildup and Storage

There is little doubt that the source of flare energy lies in stressed magnetic fields, if for no other reason than the fact that other candidate energy sources fail by orders of magnitude to account for the quantity of energy release that is observed. For example, the thermal energy content in the corona above an active region is of order $E_{th} = (3nkT)V$, where n is the density, T the temperature, and V the volume. Setting $nV \approx 10^{38}$, $T = 10^6$ gives $E_{th} \approx 10^{28}$ ergs, some three to four orders of magnitude smaller than the energy released in a large flare (the situation is worse if one questions, with the Second Law of Thermodynamics in mind, how the conversion of this thermal energy leads to *hotter* temperatures!). The gravitational energy $E_{grav} = nVgh$ is also $\approx 10^{28}$ ergs; this is not a coincidence but rather a direct consequence of the virial theorem applied to an atmosphere in hydrostatic equilibrium. However, the magnitude energy stored is of order $(B^2/8\pi)V$, which, for $B \approx 10^3$ Gauss is $\approx 10^{32}$ ergs. Not only is this *sufficient*, but its closeness to the actual amount of liberated energy is surely too significant to be dismissed as coincidence.

Not all of the energy in the magnetic field can be released. The "ground state" of the field is the potential (current-free) state, and only the excess energy above this state is available for conversion into other forms. We conclude that flares must arise in regions of non-potential field, *i.e.*, whenever currents, or magnetic shears, are present.

Consider the momentum equation for a single fluid plasma, in the form

$$\rho\frac{d\mathbf{v}}{dt} = -\nabla p + \frac{1}{4\pi}(\nabla\times\mathbf{B})\times\mathbf{B}, \tag{1}$$

where ρ is the density, $\mathbf{v}$ the velocity, p the pressure, and $\mathbf{B}$ the magnetic field. The last term on the right hand side is the Lorentz force associated with charges moving perpendicular to field lines (the current density $\mathbf{J} \simeq \nabla\times\mathbf{B}$) and, for currents substantially non-parallel to the field lines, results in a force density of order $\nabla(B^2/4\pi)$. Since $B^2 >> 4\pi p$ for solar coronal conditions, it follows that (1) cannot hold unless the currents and fields are aligned, *i.e.*,

$$(\nabla\times\mathbf{B})\times\mathbf{B} = \mathbf{0}. \tag{2}$$

Fields that satisfy (2) are known as *force-free* and have been extensively studied because of their importance in defining the possible magnetic field configurations of solar active regions. In order of complexity, the solutions to (2) are $\mathbf{B} = \mathbf{0}$ (trivial), $\nabla\times\mathbf{B} = \mathbf{0}$ (the "ground-state" potential case with no energy available for release),

$$\nabla\times\mathbf{B} = \alpha\mathbf{B}, \tag{3}$$

the linear force-free case, and

$$\nabla\times\mathbf{B} = \alpha(\mathbf{r})\mathbf{B}, \tag{4}$$

the general case, where $\alpha(\mathbf{r})$ is now a function of position, but is nevertheless constant along a field line, as can be seen by taking the divergence of (4):

$$0 = \nabla\cdot(\nabla\times\mathbf{B}) = \alpha\nabla\cdot\mathbf{B} + (\mathbf{B}\cdot\nabla)\alpha \tag{5}$$

and noting that $\nabla\cdot\mathbf{B} = 0$ so that any gradient in α must be perpendicular to $\mathbf{B}$.

There is a wide diversity of both analytic and numerical solutions of (4), which will not be reviewed here. The essential component of such solutions, for the purposes of solar flare physics, is the value of α (the "shear" of each field line relative to the potential state). Direct measurement of $\nabla\times\mathbf{B}$ from observations of the transverse component of $\mathbf{B}$ is, however, subject to considerable uncertainty. Another measure of "shear" (that is less susceptible to uncertainties introduced by seeing, the solar atmospheric model used to relate the emergent polarization Stokes vector to the components of $\mathbf{B}$, *etc.*), is the angle between the observed transverse field and that deduced from a potential extrapolation of the observed line of sight field. (Since $\nabla\cdot\mathbf{B} = 0$ and $\mathbf{B} = \nabla\Omega$ for a potential field, determination of the potential field reduces to solving Laplace's equation $\nabla^2\Omega = 0$ in a region where the normal component $\nabla\Omega$ is known on a boundary, and is a straightforward problem.) Figure 3 (after Hagyard *et al.* 1984) shows the observed transverse field for solar active region number 2372 (April 1980) together with the corresponding potential transverse field, superimposed on a map of the line of sight neutral line ($\mathbf{B}_{\parallel} = 0$). In the potential field calculation, the transverse field crosses the neutral line at right angles, while the observed (non-potential) field in general does not. In some regions (*e.g.*, between points labelled 30 and 40 in Figure 3(a)), the transverse field is sheared by a full 90° and points *along* the neutral line, rather than across it. This highly sheared state is presumably conducive to flare activity and, indeed, a flare did occur in the very location under discussion.

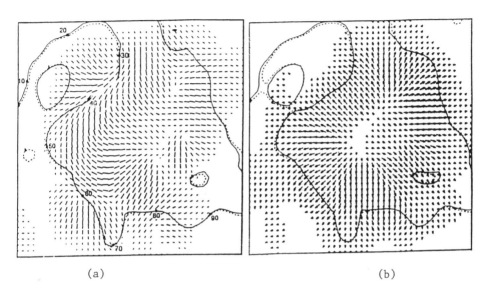

(a)   (b)

Figure 3. Observed (a) and equivalent potential (b) fields in an active region prior to flaring activity. The hash marks delineate the direction and strength of the transverse field; the continuous line delineates the line-of-sight neutral line. Note the strong shear (transverse field significantly non-perpendicular to the neutral line) in the top left corner of (a).

Theoretical calculations of the buildup of energy when footpoints of magnetic structures are sheared relative to each other have been carried out by a number of authors (*e.g.*, Wu *et al.* 1984). The results generally show that most of the mechanical energy of the shear is stored in magnetic energy, rather than kinetic, thermal, or gravitational. This is hardly surprising, since the gas pressure is negligible compared to the magnetic pressure in the first place. However, these calculations do show that the rates of magnetic energy buildup are sufficient to accumulate significant amounts of energy, verifying the viability of photospheric shearing motions as a mechanism for energy buildup. Alternative mechanisms involve the stressing of fields into nonpotential configurations before they emerge from below the photosphere. In principle, measurement of the time evolution of action region currents prior to flaring activity should be capable of distinguishing between shearing *before* and *after* flux emergence. However, there are considerable uncertainties in the transverse field data from vector magnetographs, especially considering that different instruments in different geographic locations are used at different times and the fact that the location of the active region on the solar disk (and, hence, the projection of a given field into line-of-sight and transverse components) varies with time. Hence, a reliable measurement of shearing (or lack thereof) prior to flaring has yet to be made.

5. Energy Release

As will be discussed in Chapter 25, there is considerable theoretical difficulty accounting for the rapid release of energy that takes place during a flare. The fundamental reason for this difficulty is the immense volume over which energy release occurs, which gives it both a high inductance L, and a low resistance R, both of which drive the (L/R) inductive energy release timescale to values of the order of years, rather than minutes or seconds. Modelers of the energy release process in flares have typically (and presumably out of frustration and desperation) ignored this rather glaring issue, and have instead invoked a variety of models which are scaled (through $e.g.$, a ten-order-of-magnitude increase in the resistivity) to fit observed numbers. There are basically two approaches to flare modeling:

- *The "direct" approach*

Here, an initial plasma/magnetic field configuration of some form is assumed, and the relevant equations of resistive magnetohydrodynamics are solved to yield the evolution of the configuration. Apart from the fact that it is a highly complicated problem (in general, a set of coupled, nonlinear, partial differential equations involving 16 equations in 16 unknowns – Chen 1974), its proper implementation requires knowledge of the preflare structure on size scales orders of magnitude lower than provided by observation. It does, however, remove the need for the "magic wand" approach, whereby the rapid heating of several billion tons of plasma to temperatures in excess of 10^8K, or the continuous acceleration of a significant percentage of the electrons to energies an order of magnitude above the thermal energy, are simply assumed.

- *The "inverse" approach*

Favored by many modelers, this approach starts with observations of flare manifestations over a variety of wavelengths, and attempts to infer the characteristics of the energy release ($e.g.$, heating $vs.$ acceleration, electron $vs.$ proton energization) through the interrelationships amongst these observations. The main difficulty inherent in this approach is that the physics of energy *release* is inextricably coupled to the physics of energy *transport* over the source of the observed emission. Further, the energy transport process of necessity increases the entropy above that corresponding to the instant after energy release; thus, information is irretrievably lost. Further discussion of "inverse" modeling can be found in Chapter 25.

6. Conclusions

Solar flares have been studied for well over a century. While our *observational* knowledge of flares has dramatically increased over the last few space-age decades, our theoretical understanding has manifestly failed to keep pace. This is not a slight against theoreticians (like myself!): the size-scales over which candidate physical processes occur are orders of magnitude smaller than the observers can ever hope to resolve and, as has been remarked on a number of occasions, the applicable physics is subtle and far from straightforward to apply. Nevertheless, we pursue studies of solar flares in the hope of understanding not only the role of similar processes in other domains (*e.g.*, flare stars, active galactic nuclei, *etc.*), but also the physical processes themselves. Solar flare physicists, therefore, have close ties not only to the astronomical community, but also to the communities of plasma physicists and atomic physicists, for which the Sun represents a wonderful extension of their laboratory into size scales that are clearly impossible to duplicate on the Earth.

Other reasons for studying flares are also the dramatic (and occasionally terrifying) effects that they can have on the Earth and its immediate environment. Consider that during a recent series of large flares, aurorae could be seen as far south as the tropics, radio noise storms paralyzed communication networks over continents, and that electrical transformers in the Canadian province of Quebec literally exploded as a result of the immense voltages ($\int \mathbf{E} \cdot \mathbf{d}\boldsymbol{\ell} = d\Phi/dt$) that were imposed upon them. Even domestic air travellers were not immune: the occupants of Concorde flying at 53,000 feet or so over the North Atlantic were exposed to a dosage rate equivalent to a medical chest X-ray machine until the pilots were able to reduce the altitude of the plane to a safer level. The fatality rate for space station occupants during such events has been estimated to be as high as 50%.

So much for the rationale for studying solar flares - onto the rationale for coming to an NATO Advanced Study Institute on the topic. Those U.S. residents (and visitors) who are avid followers of the David Letterman late-night television show will recognize the format of the following "top ten" list:

TOP TEN REASONS TO COME TO A NATO ADVANCED STUDY INSTITUTE

10. To gain 15 lbs.
9. To model your latest fireproof nightwear at 3 a.m.
8. To discover that those lawn bowl things aren't *supposed* to roll straight.
7. To try and take Steve Fullerton's last £10 in a sunset-sunrise poker game.
6. To provide a much-needed boost to the Scottish whisky industry.
5. To reduce the average age of the Crieff Hydro residents by half an order of magnitude.
4. To catch Hal Zirin in a rare duet performance with the 1980 Scottish Pub Entertainer of the Year.
3. To test out your gas mask on the smoking bus.
2. To prove yourself to be a fine ambassador for your country through the exercise of diplomatic relations of the Hydro golf course.
1. To fend off hordes of rampaging senior citizens trying to gatecrash the bagpipe concert.

References

Canfield, R.C. *et al.* 1978, in *Solar Flares - A Monograph from Skylab Solar Workshop II*, ed. P.A. Sturrock (Colorado: Colorado Associated Press).

Chen, F.F. 1974, *Principles of Plasma Physics* (New York: McGraw Hill).

Dennis, B.R. 1985, *Solar Phys.*, **100**, 465.

Hagyard, M.J., Smith, Jr., J.B., Teuber, D., and West, E.A. 1984, *Solar Phys.*, **91**, 115.

Lu, E., and Hamilton, R.J. 1991, *Ap.J.*, **380**, L89.

Murphy, R.J., Ramaty, R., Kozlovsky, B., and Reames, D.V. 1991, *Ap. J.*, **371**, 793.

Svestka, Z. 1976, *Solar Flares* (Dordrecht: Reidel).

Wu, S.T., Hu, Y.Q., Krall, K.R., Hagyard, M.J., and Smith, J.B., Jr. 1984, *Solar Phys.*, **90**, 117.

24 HIGH-ENERGY FLARE EMISSIONS

A.G. EMSLIE
Department of Physics
The University of Alabama in Huntsville
Huntsville, AL 35899
U.S.A.

ABSTRACT. For almost a century after their discovery in 1859, solar flares were viewed almost exclusively through their optical (*e.g.* Hα) signature. The advent of space-based observations, with their ability to observe across the entire electromagnetic spectrum (including the X-rays and γ-rays blocked by absorption in the Earth's atmosphere), led to a dramatic rethinking of the flare process. It was discovered that most flares have an impulsive signature in hard X-rays (and even γ-rays) coincident with the optical flash, and that this high energy manifestation has profound implications for the modes of energy release and transport in the flare. A considerable amount of effort, both observational and theoretical, has been expended on elucidating the nature of the process(es) responsible for high energy emissions in flares. In this Chapter, I will review the various components of high energy emission and the mechanisms responsible for them. A disproportionately large part of the discussion will, due to personal bias, be concerned with hard X-ray emission – its spectrum, polarization, spatial structure, and temporal characteristics. Both "nonthermal" and "thermal" models for hard X-ray bursts will be reviewed, with the main thrust of the argument aimed at laying this historically significant, but physically questionable, dichotomy to rest. The current status of our understanding, together with a prognosis for the future, will conclude the Chapter.

1. Observational Characteristics

A brief review of the various observed manifestations of a flare appears in the preceding Chapter. *High energy emissions* on this list include, of course, γ-rays and hard X-rays, but somewhat surprisingly, a mention of radio emission is also appropriate here inasmuch as the electrons responsible for the emission of (very low energy) radio waves are themselves mildly relativistic and, hence, constitute evidence for a high-energy physical process.

Figure 1 shows the time histories of several simple impulsive flares, showing the synchronism between the hard X-ray emission and the microwave (radio) emission. Clearly the process which accelerates the electrons responsible for the (bremsstrahlung) X-rays and the (synchrotron) microwave emission does so simultaneously over a wide energy range. A more dramatic example is shown in Figure 2, where, over a series of seven successive bursts from an active region (and quite possibly the same loop), the hard X-ray, microwave, and γ-ray light curves peak within seconds of each other. This shows that, not only are 40-500 keV electrons accelerated essentially simultaneously, but so also are the deka-MeV protons responsible for exciting the nuclear deexcitations in the 4-6 MeV range (see below).

J. T. Schmelz and J. C. Brown (ed.), The Sun, A Laboratory for Astrophysics, 475–488.

476

Direct evidence for acceleration of particles in flares comes from detection of electrons, protons, and neutrons in interplanetary space. The charged particles are observed to have broken power-law spectra, with electron spectral indices typically around unity for $E \leq 100$ keV, steepening to $\sim$ 3-4 above this energy. Proton spectra are similar, with the spectral break occurring around 1 MeV. The times of arrival of neutrons at the Earth imply very high (relativistic) energies, indicating that GeV protons are also accelerated rapidly in the flare process. *Indirect* evidence comes from many different high-energy radiation fields, the basics of which we now summarize.

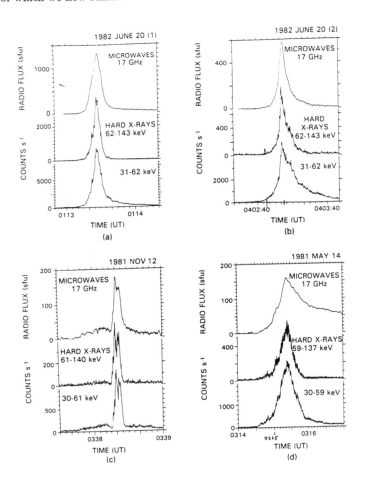

Figure 1. Light curves for low and high energy X-rays, and microwaves, for several simple impulsive events. Since the microwave emission is produced by mildly relativistic electrons, we see that energization of electrons over a wide range of energies occurs essentially simultaneously.

1.1. γ-RAYS

Continuum γ-rays are produced by bremsstrahlung of relativistic electrons on protons; the spectrum typically takes the form of (surprise, surprise!) a power-law (see Murphy *et al.* 1991). Superimposed on this broadband continuum are a large number of lines formed by deexcitation of nuclear species excited by proton and heavy ion bombardment. Since the deexcitation timescales are short, these lines are prompt and, hence, they provide a real-time diagnostic on the acceleration of protons above about 30 MeV, the excitation threshold. A list of lines and their calculated approximate relative intensity (based on photospheric element abundances) can be found in Ramaty (1986); two of the most prominent are (i) the line at 4.43 MeV, produced by decay after a combination of $^{12}C(p,p')^{12}C^{*4.43}$, $^{12}C(\alpha,\alpha')^{12}C^{*4.43}$, and $^{16}O(p,p\alpha)^{12}C^{*4.43}$ reactions, and (ii) the line at 6.14 MeV with the excited state formed through $^{16}O(p,p')^{16}O^{*6.14}$ and $^{16}O(\alpha,\alpha')^{16}O^{*6.14}$ reactions.

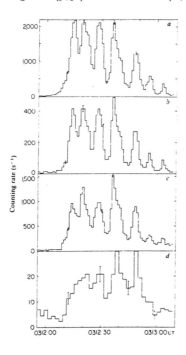

Figure 2. Time profiles of a multiply impulsive burst in the energy ranges (a) 40 - 80 keV, (b) 80 - 140 keV, (c) 330 - 380 keV, and (d) 4.1 - 6.4 MeV (after Forrest and Chupp 1983). The first three energy ranges correspond to electron bremsstrahlung, the last is principally due to nuclear line emission. Note the synchronism of the emissions over a wide range of energies.

Other important nuclear γ-ray lines are the e^+e^- annihilation line at 0.511 MeV, the positrons being produced in a variety of proton-impact reactions (see Ramaty 1986), and the deuterium formation [n(p,γ)D] line at 2.223 MeV. This latter feature is especially note-worthy in that capture of neutrons can only occur after they have slowed down considerably. The line is, therefore, not only delayed relative to other impulsive phase emissions, but is also formed at great depths in the atmosphere. This leads to strong limb-darkening of the feature, which can be used to place constraints on the neutron acceleration process (see Ramaty 1986).

1.2. MICROWAVE RADIATION

A more thorough review of this process can be found in Chapter 14. Electrons spiral around magnetic field lines at the *cyclotron frequency*

$$\nu(Hz) = \frac{eB}{2\pi mc} = \frac{2.8 \times 10^6 B(gauss)}{\gamma}, \tag{1}$$

where γ is the relativistic Lorentz factor. The radiation pattern from relativistic electrons is highly beamed in the direction of motion of the emitting electron (an aberrational ef-fect) and, hence, the intensity-time profile of the radiation in a given direction exhibits a "spiky" pattern. Fourier analysis of this profile reveals significant power at high harmonic frequencies; these high harmonics give rise to emission at very high frequencies and, with modest spectral resolution, blend into a "continuum" of microwave emission.

Unlike most other "high energy" emissions, microwave radiation can be optically thick at some frequencies (typically the first few harmonics - *e.g.* Chanmugam and Dulk 1981), with significant absorption and reemission due to gyroresonance and collective plasma effects (the gyro frequency Ω and plasma frequency ω_{pe} have similar magnitudes in flaring coronal plasmas). The intensity seen by an observer is also a sensitive function of the angular distribution of the emitting electrons and of the strength of the magnetic field. This has both a plus and a minus side. On the negative side, clean diagnostics of, for example, the number and spectrum of the accelerated electrons is difficult due to the other factors influencing the emergent radiation field. On the other hand, however, microwave emission offers us about the only opportunity to estimate directly the strengths of coronal magnetic fields, to be compared with, for example, force-free extrapolations of observed photospheric fields. Convenient analytical approximations for the opacity κ as function a of magnetic field strength, density, and pitch angle, for various ratios of (ω_{pe}/Ω) have been derived (Dulk and Marsh 1982).

The total power in the *emergent* microwave radiation is a relatively small part of the flare budget; it has been suggested, however, that the amount of power emitted and absorbed within the flaring volume can be considerably larger, amounting to a substantial fraction of the total energy released. Since photons, unlike charged particles, are not obliged to follow magnetic field lines, microwave emission and absorption has been suggested as an important mode of cross-field energy transport in flares (Melrose and Dulk 1982).

1.3. HARD X-RAYS

Discussion of this invaluable diagnostic radiation field forms the bulk of the ensuing discussion and, consequently, merits a separate section.

2. Hard X-Ray Emission

2.1. BREMSSTRAHLUNG EFFICIENCY

The bremsstrahlung ("braking streaming") radiation process is very straightforward; an electron undergoes a high-angle-deflection (low impact parameter) collision with an ambient proton, giving off a photon of energy comparable to the initial electron energy. The cross-section for this process is a function of the energies and outgoing directions of both the emitted photon and the scattered electron, and of the polarization of the emitted photon. It has been calculated in considerable detail for both nonrelativistic and relativistic regimes (Koch and Motz 1959). For studies of total yield in hard X-rays below a few hundred keV, it is sufficient to use the non-relativistic Bethe-Heitler cross-section, which is an average over the *directions* of the emitted photon and outgoing electron and of the polarization state of the photon, but remains differential in the energy ϵ of the emitted photon. The cross-section is

$$\sigma_B = \frac{K}{\epsilon E} \ln \left[\frac{1 + (1 - \epsilon/E)^{1/2}}{1 - (1 - \epsilon/E)^{1/2}} \right] \text{ cm}^2 \text{ erg}^{-1}, \tag{2}$$

where the logarithmic factor drives σ_B to zero at $\epsilon = E$, and $K = (8\alpha/3)r_0^2 m_e c^2 = 7.9 \times 10^{-25} \text{cm}^2$ keV (r_0 being the classical electron radius, α the fine structure constant, m_e the electron mass, and c the speed of light). The rate of bremsstrahlung production by electrons of energy E (erg sec^{-1}) in a target of density n is then given by

$$P_x = \int_{\epsilon=0}^{E} \sigma_B(\epsilon, E) \; nvE \; d\epsilon. \tag{3}$$

If the electrons interact with an atmosphere consisting of relatively cold target particles, then the rate of energy loss (ergs sec^{-1}) due to low-angle (high impact parameter) inelastic collisions with other electrons is

$$P_c = \sigma_E \; nvE, \tag{4}$$

where $\sigma_E = 2\pi e^4 \Lambda/E^2$ is the collision cross-section, e being the electronic charge and Λ a logarithmic factor which incorporates the fall-off of energy loss with impact parameter under the long range Coulomb force (see, *e.g.* Spitzer 1962, Emslie 1978).

The *efficiency* η of such a cold source of hard X-ray production is simply the ratio of the bremsstrahlung production rate (3) to the collisional energy loss rate (4), *i.e.*

$$\eta = \frac{\int_0^E \sigma_B \epsilon d\epsilon}{E\sigma_E}. \tag{5}$$

Neglecting logarithmic factors, we obtain

$$\eta \approx \frac{KE}{2\pi e^4 \Lambda} = \frac{4\alpha}{3\pi\Lambda} \left[\frac{E}{m_e c^2} \right]. \tag{6}$$

480

For $E \approx 30\text{keV}$ (typical of the energy of the bulk of the hard X-ray emission in flares), we obtain, with $\Lambda \approx 20$, $\eta \approx 10^{-5}$. This shows that bremsstrahlung in a cold target is a very inefficient process, requiring of order 100,000 times more energy input to target heating than to bremsstrahlung emission. To maximize the bremsstrahlung production in such a scenario, we make the target *thick*, *i.e.* we require that the electrons decay to the threshold energy $E = \epsilon$ in the source, rather than escaping with excess energy intact. In a *thin* target, the electron energy loss and, hence, the bremsstrahlung production rate is even lower, thereby requiring an even larger energy of injected electrons to produce an observed hard X-ray flux.

Of course, if the target is warm, then Equation (4) is replaced by

$$P_e = 0, \tag{7}$$

reflecting the fact that there is, an average, no secular charge in the energy of the bremsstrahlung-emitting electrons as a result of collisions with electrons of comparable energy. Such a *thermal* source is, in principle, 100% efficient at converting electron energy into photon energy; however, in practice, other energy losses, such as mass motions and thermal conduction, reduce this efficiency somewhat (*e.g.* Smith and Auer 1980).

2.2. BREMSSTRAHLUNG SPECTRA - THERMAL AND NONTHERMAL MODELS

There exists a simple relationship between the energy spectrum of the emitting electrons and the spectrum of the emitted hard X-rays. For details, we refer the reader to Brown (1971) and Tandberg-Hanssen and Emslie (1988); for now we present a simple argument that emphasizes the salient features. In a *thin* target, the emitted photon spectrum is a function of the instantaneous electron spectrum in the source. Using (2) and neglecting logarithmic factors, we find that the hard X-ray spectrum $I(\epsilon)$ (photons keV^{-1} sec^{-1}) is given by

$$I(\epsilon) \sim \int_\epsilon^\infty F(E) \frac{1}{\epsilon E} dE, \tag{8}$$

where $F(E)$ (electrons keV^{-1}sec^{-1}) is the instantaneous electron flux. If we characterize this flux by a spectral index $\delta = -\partial \ln F / \partial \ln E$, then

$$I(\epsilon) \sim \frac{1}{\epsilon} \int_\epsilon^\infty E^{-\delta} \frac{dE}{E} \sim \epsilon^{-(\delta+1)}, \tag{9}$$

so that the characteristic exponent $\gamma = -\partial \ln I / \partial \ln \epsilon$ of the hard X-ray spectrum is related to its electron counterpart by

$$\gamma = \delta + 1. \tag{10}$$

If the target is *thick*, then the instantaneous electron spectrum on the source is not the same as the (physically more interesting) *injected* spectrum. These are related by

$$F(E) \sim F(E_0) \, \tau \, (E_0), \tag{11}$$

where $\tau(E_0)$ is the collisional lifetime of an electron of energy E_0, and is the appropriate weighting factor to apply to the injected spectrum to obtain the instantaneous spectrum. From (4), we see that

$$\tau(E_0) \sim \sigma_E^{-1} \sim E_0^2, \tag{12}$$

so that

$$F(E) \sim E_0^{2-\delta}, \tag{13}$$

where δ now refers to the characteristic exponent of the injected flux. Substituting (13) in (8) gives

$$\gamma = \delta - 1 \tag{14}$$

two powers flatter (i.e. harder) than for the thin target case (10).

For a thermal distribution of electrons with a Maxwellian distribution characterized by a temperature T,

$$F(E) \sim \frac{E \, \exp(-E/kT)}{T^{3/2}}, \tag{15}$$

so that, by (8),

$$I(\epsilon) \sim \frac{e^{-\epsilon/kT}}{\epsilon T^{1/2}}. \tag{16}$$

While this appears to have a fundamentally different form than the power-laws discussed above, one must recall that these spectral shapes are basically arbitrary. For example, (16) has a local "spectral index"

$$\gamma = \frac{\epsilon}{kT} + 1. \tag{17}$$

Further, as first shown by Brown (1974) and later elaborated upon by Brown and Emslie (1988), a suitably inhomogeneous thermal source can result in a wide range of bremsstrahlung spectra. Defining the *differential emission measure* ("radiating power per unit temperature") function

$$\xi(T) = n^2 \frac{dV}{dT}, \tag{18}$$

the emission from a general source may be written

$$I \sim \int \frac{e^{-\epsilon/kT}}{\epsilon T^{1/2}} \, \xi(T) \, dT. \tag{19}$$

With $s = (kT)^{-1}$ and $\hat{\xi}(s) = k^{-1/2} s^{-3/2} \epsilon^{-1} \xi(1/ks)$, we find that

$$I(\epsilon) \sim \int e^{-s\epsilon} \hat{\xi}(s) \, ds = \mathcal{L}[\hat{\xi}(s), \epsilon], \tag{20}$$

the Laplace transform of the emission-measure-related function $\hat{\xi}(s)$. Since, for example, the inverse Laplace transform of a power-law $I(\epsilon)$ is another power-law $\hat{\xi}(s)$, it follows that spectra alone *cannot distinguish between hard X-ray emission models*. (One can, however test the viability of an assumed model through analysis of $I(\epsilon)$ and its various derivatives [Brown and Emslie 1988]: Emslie *et al.* (1989) have used this technique to show consistency of the high spectral resolution hard X-ray spectra of Lin *et al.* (1981) with a purely thermal source.) We must, therefore, seek simultaneous information on the spatial structure, temporal variability, anisotropy (directly), and polarization of the hard X-ray radiation in order to determine which model is applicable.

At this point, the reader may well ask why the determination of the appropriate model is so important. The answer lies in the energetic efficiency of the three scenarios with, in general

$$\eta(thin\ target) < \eta(thick\ target) << \eta(thermal). \tag{21}$$

As will be discussed in the next Chapter, the energetic requirements associated with a thick target interpretation are severe, requiring a large fraction of the flare energy to be released in the form of accelerated electrons, and probably requiring a highly filamented structure for the energy release region. The constraints imposed by the (lower efficiency) thin target model are even more stringent. A thermal model poses significantly less constraints on the energy release mechanism, with regard to both the quantity and quality (entropy) of the released energy. It is, therefore, vital to the overall problem of flare energetics to determine the extent to which each of the scenarios discussed above is relevant.

Of course, some hybrid of these models is not only possible, it is inevitable. The high energy electrons in an otherwise collisionally relaxed Maxwellian have very long mean free paths and, therefore, act like a nonthermal beam impinging on a cold target. Conversely, the collisional heating effected by the passage of a "nonthermal" electron beam produces hot plasma and corresponding thermal emission. Hence, I submit that the "thermal/nonthermal" controversy that has raged in the solar physics community for over two decades may be a question with comparatively little physical basis. What is really required is an accurate determination of the electron phase-space distribution function f($\mathbf{r}$,$\mathbf{v}$,t), since then the ratio of bremsstrahlung to collisional losses (both functionals of f($\mathbf{r}$,$\mathbf{v}$,t)) can be determined and the observed hard X-ray emission used to place reliable constraints on the primary energy release process.

Ideally, knowledge of f($\mathbf{r}$,$\mathbf{v}$,t) could come from hard X-ray observations with sufficient spatial, energy, and temporal resolution that f($\mathbf{r}$,$\mathbf{v}$,t) does not vary significantly over either the observed pixel size or the observational integration time. In such a case, a point-by-point, time-by-time measurement of the hard X-ray spectrum is possible. Such observations are indeed planned for the *High-Energy Solar Physics (HESP)* mission scheduled for the next solar maximum in the year 2000. In the meantime, reasonable constraints have been placed using the limited resolution observations available at present. These observations will now be discussed, albeit *via* somewhat of a return to the thermal/nonthermal dichotomy alluded to above.

2.3. SPATIAL STRUCTURE

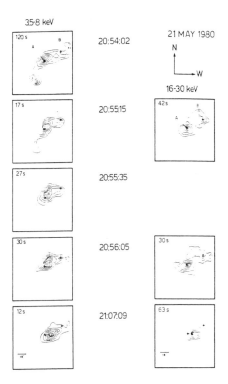

Figure 3. Images in soft (3.5 - 8 keV) and hard (16 - 30 keV) X-rays for an event on May 21, 1980 (after Hoyng *et al.* 1981). Note the relatively amorphous structure in soft X-rays throughout the event, and the "double footpoint" structure in hard X-rays early in the event.

Spatially resolved hard X-ray observations from flares have been obtained by two techniques – direct collimation, as used on the *Solar Maximum Mission (SMM)* Hard X-ray Imaging Spectrometer (HXIS) and *via* a rotating modulation collimator. In the latter design, used on the Japanese *Hinotori* satellite and part of the design for the proposed *HESP* mission, a pair of metal plates, each with a regular array of slits, is rotated so that the shadow of the source showing through the top grid, coupled with the slit pattern on the bottom grid, produces a fringe pattern that provides information on the spatial Fourier component oriented perpendicular to the slits. By rotating the whole assembly, a Fourier map of the source is obtained which can be inverted by standard techniques to obtain the source structure. The reader is referred to Campbell *et al.* (1990) for details. Other evidence on the spatial structure of hard X-ray emission has come from rare events in which two different spacecraft observed a flare, with one detector's view of the flare partially occulted by the solar limb (Kane *et al.* 1980). Such observations have shown that, at least in the events studied, only a small ($< 1\%$) fraction of the hard X-ray emission at energies from a few tens of keV upward originates from heights $> 25,000$ km in the corona.

Early results from HXIS indicated a "two-footpoint" structure early in the event, consistent with the impingement of an electron beam (or beams) on the thick chromospheric target. Later in the event (Figure 3), the emission became dominated by a more amorphous structure with a centroid in between the original footpoints, again consistent with a thick-target model in which heated chromospheric material is driven upward into the corona – see Chapter 25. However, not only was the time resolution of statistically significant HXIS data so poor that this supposed "evolution" could not be reliably tracked, McKinnon *et al.* (1985) have pointed out that the emission shown in Figure 3 may only represent a small fraction of the total emission, the remainder being spread over many weak pixels not transmitted by the *SMM* satellite because of telemetry limitations. *Hinotori* observations, which reflect *all* the photons arriving at the instrument, tended to favor a more extended coronal source, but were at a spatial resolution sufficiently poor that it is debatable whether a footpoint structure could have ever been seen. In short, a clear picture of the spatial structure of hard X-ray emission, and its implications for the emission model, must await observations with more resolution and sensitivity than available at present.

2.4. POLARIZATION

Here we find a remarkable example of a situation where the number of theoretical predictions vastly outnumber the number of reliable data points. Theoretical simulations of both thick-target and thermal hard X-ray polarization, using more complicated cross-sections, differential in the direction and polarization of the outgoing photon, have been carried out by a number of authors, dating back to Elwert and Haug (1970) and Brown (1972), and continuing through Leach and Petrosian (1983). Thick-target models typically exhibit a few tens of percent polarization (for a simple source geometry!), with the exact value dependent on the assumptions regarding the angular distribution of the injected electrons, the variation in the guiding magnetic field strength, and the role of photospheric backscattered radiation. Somewhat surprisingly, thermal models also exhibit some degree of polarization (Emslie and Brown 1980), due to temperature gradients in the source. Expected values run to only a few percent, however.

Observationally, one of my favorite graphs shows the variation of polarization with time, where "time" refers to the epoch at which the observations were made. Early observations claimed high polarization, consistent with the available thick target predictions. However, after much debate regarding detector cross-calibration, noise, and contamination of the detectors, claims in recent years have been more modest, culminating in the marginally significant (consistent with zero within $1 - \sigma$ error bars) polarization reported by Tramiel *et al.* (1984) from a Space Shuttle Experiment. As a result, a meaningful comparison of observation with theory is next to impossible at present.

The prognosis for such observations is not healthy. No plans currently exist for better hard X-ray polarization observations of flares. This may well be due to the disturbing possibility that any measurement will fail to yield definitive results because of any source inhomogeneity that may be present. Nevertheless, a definitive observation of a high degree of polarization would be a very significant observation because of its implications for the nature of the distribution function of the bremsstrahlung-emitting electrons.

2.5. DIRECTIVITY

A similar situation exists here. Theoretical predictions abound, usually made concurrently with predictions of the polarization. Observationally, one can either study the center-to-limb brightness distribution of flares and use a statistical argument to infer directivity, or use two spacecraft at different viewing angles to more directly ascertain the directivity in suitable events.

At hard X-ray energies, center-to-limb studies are inconclusive. Datlowe *et al.* (1977) found no evidence for limb brightening (or darkening) of events; however, they did find evidence that spectra of flares on the limb tended to be harder, showing an energy-dependent directivity effect. At γ-ray energies, the center-to-limb brightness distribution is remarkable: all events observed by *SMM* showing γ-ray emission above 10 MeV occurred within a few degrees of the limb (Figure 4). This result clearly indicates a directionality in the emission process: for downward-propagating electrons, the most favored direction for emission is downward (and hence never seen), followed by emission at right angles (seen only for limb events), and last, emission in the outward (*i.e.* backward) direction (as would be seen only for flares on the disk). Note also that more recent observations from the *Gamma-Ray Observatory* do show > 10 MeV flares on the disk; this may be due to nonvertical field lines in the active region, and can in principle be tested using vector magnetograph observations (see previous Chapter).

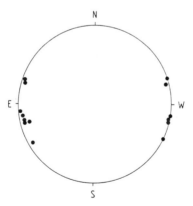

Figure 4. Locations of *all* events producing emission above 10 MeV, as observed by the *Solar Maximum Mission*. The striking clustering of sources at both limbs is indicative of a strong directionality of emission at these energies.

Direct stereoscopic observations of hard X-ray flares, using detectors on both the *ISEE-3* satellite at the inner Sun-Earth Lagrange point and the *Pioneer Versus Orbiter* have been reported by Kane *et al.* (1988). No measurable difference in the deka-keV flux observed in the detectors was evident, over viewing angle differences ranging from $-90°$ to $+90°$ (Figure 5).

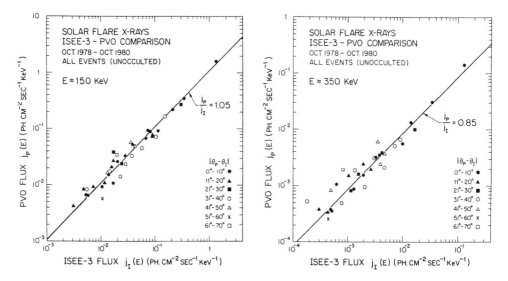

Figure 5. Hard X-ray flux at 150 and 350 keV observed by the Pioneer Venus Orbiter (PVO) and International Sun-Earth Explorer 3 (ISEE-3) satellites, for different heliocentric angular differences between the two detectors (after Kane *et al.* 1988). No systematic directionality in the emission is evident.

2.6. TEMPORAL VARIATIONS

Statistically significant fluctuations in hard X-ray bursts have been observed on timescales as short as a few tens of milliseconds (Kiplinger *et al.* 1983, Figure 6). Such rapid rise times point to a rapid acceleration and/or heating process, while the similarly rapid decay times place significant constraints on source lengths, densities, *etc.*, in order for the collisional decay of the electrons to occur sufficiently rapidly (see Kiplinger *et al.* 1983). More significantly, they indicate that all flares may be a superposition of such very short bursts, only resolvable when they do not significantly overlap. A similar idea, but invoking a fundamental timescale of 5-10 seconds, was proposed by de Jager and de Jonge (1978) — they termed the building blocks "elementary flare bursts." However, results from the Burst and Transient Source Experiment (BATSE) on the *Gamma-Ray Observatory*, with a detector about 100 times more sensitive than any other flown, tend to confirm the sub-second timescale as being the more fundamental. A model which naturally account for these timescales as the transverse magnetohydrodynamic expansion times of exploding coronal "filaments" has been proposed by Roumeliotis and Emslie (1991).

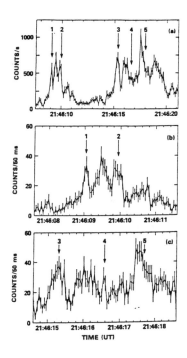

Figure 6. Rapid fluctuations in hard X-ray burst time profiles (after Kiplinger *et al.* 1983).

3. Summary and Conclusions

High energy emissions provide information on the energy released in a flare at a time when the emitting particles, not yet having had time to thermalize in the ambient plasma, most closely reflect conditions in the primary energy release process. We know that flares are indeed efficient particle accelerators, and may in fact release most of their energy in the form of electrons of energy ~ 10 keV and upward, which manifest themselves through hard X-ray bremsstrahlung. Protons of MeV and even GeV energies are accelerated essentially simultaneously with the electrons. However, the value of these radiations in determining the nature of the energy release/particle acceleration process is limited by the lack of spatial, temporal, and spectral resolution currently available. Definitive diagnostics require that one sample the spectrum produced by the emitting particles on size and timescales of the same order as those associated with changes in the particle energies. It is straightforward to estimate the size and timescales: for example, the mean free path and collision time for a 20 keV electron in a medium of density $n \approx 10^{11} \text{cm}^{-3}$ (typical of the flaring corona) are 400 km ($\sim 5''$ on the Sun) and 0.05 seconds, respectively. Observations of statistically significant spectra over these lengths and timescales are anticipated from the forthcoming *HESP* mission.

References

Brown, J.C. 1971, *Solar Phys.*, **18**, 489.

Brown, J.C. 1972, *Solar Phys.*, **26**, 441.

Brown, J.C. 1974, in IAU Symposium 57, *Coronal Disturbances* (ed. G.A. Newkirk, Jr.), p. 105.

Brown, J.C. and Emslie, A.G. 1988, *Ap.J.*, **351**, 554.

Campbell, J.W., Davis, J.M., and Emslie, A.G. 1990, in Proc. SPIE 34th Annual Symposium.

Chanmugam, G. and Dulk, G.A. 1981, *Ap.J.*, **244**, 569.

Datlowe, D.W., O'Dell, S.L., Peterson, L.E., and Elcan, M.J. 1977, *Ap.J.*, **212**, 561.

de Jager, C. and de Jonge, G. 1978, *Solar Phys.*, **58**, 127.

Dulk, G.A. and Marsh, K.A. 1982, *Ap.J.*, **259**, 350.

Elwert, G. and Haug, E. 1970, *Solar Phys.*, **15**, 234.

Emslie, A.G. 1978, *Ap.J.*, **224**, 241.

Emslie, A.G. and Brown, J.C. 1980, *Ap.J.*, **237**, 1015.

Emslie, A.G., Coffey, V.N., and Schwartz, R.A. 1989, *Solar Phys.*, **122**, 313.

Forrest, D.J. and Chupp, E. 1983, *Nature*, **305**, #5932, p. 291.

Hoyng, P. *et al.* 1981, *Ap.J.*, **246**, 1155.

Kane, S.R., Anderson, K.A., Evans, W.D., Klebesadel, R.W., and Laros, J.G. 1980, *Ap.J.*, **239**, L85.

Kane, S.R., Fenimore, E.E., Klebesadel, R.W., and Laros, J.G. 1988, *Ap.J.*, **326**, 1017.

Kiplinger, A.L., Dennis, B.R., Emslie, A.G., Frost, K.J., and Orwig, L.E. 1983, *Ap.J.*, **265**, L99.

Koch, H.W. and Motz, J.W. 1959, *Rev. Mod. Phys.*, **31**, 920.

Leach, J. and Petrosian, V. 1983, *Ap.J.*, **269**, 715.

Lin, R.P., Schwartz, R.A., Pelling, R.M., and Hurley, K.C. 1981, *Ap.J.*, **251**, L109.

McKinnon, A.L., Brown, J.C., and Hayward, J. 1985, *Solar Phys.*, **99**, 231.

Melrose, D.B. and Dulk, G.A. 1982, *Ap. J.*, **259**, 844.

Murphy, R.J., Ramaty, R., Kozlovsky, B., and Reames, D.V. 1991, *Ap.J.*, **371**, 793.

Ramaty, R. 1986, in *Physics of the Sun*, Vol II. (eds. P.A. Sturrock et al.), Chapter 14.

Roumeliotis, G. and Emslie, A.G. 1991, *Ap.J.*, **377**, 685.

Smith, D.F. and Auer, L.H. 1980, *Ap.J.*, **238**, 1126.

Spitzer, L.W., Jr. 1962, *Physics of Fully Ionized Gases* (New York: Interscience)

Tandberg-Hanssen, E., and Emslie, A.G. 1988, *The Physics of Solar Flares* (New York: Cambridge University Press).

Tramiel, L.J., Chanan, G.A., and Novick, R. 1984, *Ap.J.*, **280**, 440.

25 ENERGY RELEASE AND TRANSPORT IN FLARE PLASMAS

A.G. EMSLIE
Department of Physics
The University of Alabama in Huntsville
Huntsville, AL 35899
U.S.A.

ABSTRACT. We discuss the fundamental issues involved in the dissipation of stored magnetic energy during solar flares, and the various magnetic field topologies and geometries that have been invoked to enhance the energy release rate. Transport of the released energy, particularly by energetic electrons, is reviewed, noting especially the global electrodynamic consequences of such a scenario. Results from hydrodynamic simulations of electron-heated flare atmospheres are compared critically with recent observational data.

1. Overview

As discussed in Chapter 23, solar flares occur in regions of stressed magnetic field, *e.g.*, in the vicinity of sunspots. It was also shown there that there is sufficient energy $1/8\pi \int B^2 dV$ in the magnetic field to power the observed (radiative + kinetic) manifestations of the flare. However, a major unanswered question is just how this magnetic energy is dissipated quickly enough to account for observed flare timescales. The reason for this difficulty lies in the extremely large inductance and low resistance of the appropriate volumes of coronal plasma, and it forces upon us the idea that flare energy release occurs in extremely thin regions of field reversal, such as neutral sheets. Some basic topological and geometrical scenarios illustrating the possible configurations will be presented, and the fundamental equation describing the evolution of such configurations, the *magnetic diffusion equation*, will be derived and discussed.

Once the energy has been released, it is transported throughout the flare atmosphere to produce enhanced temperatures throughout the corona and chromosphere and, in turn, enhanced radiation signatures across the electromagnetic spectrum. Various mechanisms have been proposed to effect this transport of energy, most notable amongst them being (i) the propagation of accelerated suprathermal electrons and (ii) the (relatively slower) diffusion of heat from the region of primary energy release. As noted in the previous Chapter, there is considerable observational evidence that nonthermal electrons indeed play a significant role. However, there are significant unresolved theoretical issues raised by this; these issues will be presented and discussed.

J. T. Schmelz and J. C. Brown (ed.), The Sun, A Laboratory for Astrophysics, 489–508.

The last decade has seen a dramatic increase in the use of high-speed computers as a modeling tool. As a result, numerous investigations into hydrodynamic response of the solar atmosphere to various forms of energy deposition have been carried out. These numerical simulations yield the temperature, density, and velocity response of the atmosphere and, hence, can be used to predict a wide variety of observational signatures such as UV line intensities, soft X-ray fluxes and line profiles, and optical emissions such as Hα. Comparison of these predictions with observation can tell us much about the mode of energy deposition and, hence, about the mode of energy release and the environment in which it occurs.

We will carry out such a comparison for the particular case of heating by nonthermal electrons. Comparison of synthetic soft X-ray line profiles with observation leads us to the conclusion that a simple one-dimensional geometry (i.e., uniform guiding field strength), together with a preflare environment typical of active region conditions, does not yield satisfactory agreement between theory and observation. However, other observations, such as hard X-ray/EUV flux correlations, and the behavior of UV line intensities do add support to the basic electron-heated concept. We conclude that some form of coronal temperature and density enhancements prior to the impulsive phase, and/or the presence of significant inhomogeneities in the guiding field strength, may be necessary to account for all the impulsive phase manifestations studied.

2. The Magnetic Diffusion Equation

Recall a simple high-school laboratory experiment, in which a wire is moved between the poles of a horseshoe magnet, and generates a voltage ($\mathbf{E} = \mathbf{v} \times \mathbf{B}/c$) along the wire as a result. In the absence of a load (resistance) to support this voltage, it is clear that motion of the wire across the lines of force is not possible. Thus, in the limit of zero resistance, the wire and the field are "frozen" together; only when a resistance is present can slippage between the wire and the magnetic field occur, leading to ohmic dissipation of the currents produced.

Since flares derive their energy from dissipation of currents in stressed magnetic fields, it is clearly important for the resistance of the dissipating region to be large enough for the energy to dissipate on observable timescales. This can be done in one of two ways: by increasing the *resistivity* η (e.g., through some form of turbulence) or by decreasing the cross-sectional area through which the current flows. The formal equation describing the above physics is known as the *magnetic diffusion equation*, which we shall now derive:

We start with Ohm's Law in the steady-state form

$$\mathbf{E} + \frac{\mathbf{v} \times \mathbf{B}}{c} = \eta \mathbf{j} \tag{1}$$

where $\mathbf{E}$ and $\mathbf{B}$ are the electric and magnetic fields, $\mathbf{v}$ the velocity of the fluid relative to the observer, $\mathbf{j}$ is the current density, and η is the resistivity (assumed to be a scalar). Note that the total electric field on the left hand side is a superposition of any external electrostatic or inductive field present, plus the induced electric field due to motion of the fluid across the magnetic field lines.

We take the curl of (1), and use Faraday's law $\nabla \times \mathbf{E} = -(1/c)\partial \mathbf{B}/\partial t$ to eliminate $\mathbf{E}$. This results in

$$-\frac{1}{c}\frac{\partial \mathbf{B}}{\partial t} + \frac{1}{c}\nabla\times(\mathbf{v}\times\mathbf{B}) = \eta\nabla\times\mathbf{j}, \tag{2}$$

where we have assumed a uniform η. Next, we eliminate the current density via the steady-state Ampere's law: $\nabla\times\mathbf{B} = (4\pi/c)\mathbf{j}$. This gives us an equation involving only the vector fields $\mathbf{v}$ and $\mathbf{B}$:

$$\frac{\partial \mathbf{B}}{\partial t} = \nabla\times(\mathbf{v}\times\mathbf{B}) - \frac{\eta c^2}{4\pi}\nabla\times(\nabla\times\mathbf{B}). \tag{3}$$

Expanding the last term, and using the solenoidal condition $\nabla\cdot\mathbf{B} = 0$, we obtain

$$\frac{\partial \mathbf{B}}{\partial t} = \nabla\times(\mathbf{v}\times\mathbf{B}) + \frac{\eta c^2}{4\pi}\nabla^2\,\mathbf{B}. \tag{4}$$

This is the desired equation. It expresses the rate of change of magnetic field as a combination of (i) advection of field lines carried by the moving plasma and (ii) diffusion of the field through the plasma. If $\eta \to 0$, the latter contribution vanishes and we have the "frozen-in" condition discussed qualitatively above. We can formally prove, using (4), that the magnetic flux through a surface comoving with the fluid is a constant for $\eta = 0$ (see, e.g., Parker 1975 and Chapter 6).

Each term on the right side of (4) expresses a characteristic timescale associated with the change in magnetic field; the first term defines the *advection timescale*

$$\tau_a = \frac{L}{v}, \tag{5}$$

while the second defines the *diffusion timescale*

$$\tau_d = \frac{4\pi L^2}{\eta c^2} \tag{6}$$

Here L is the characteristic scale over which the magnetic field strength changes. The ratio of these timescales is called the *magnetic Reynolds number* (or *Lundquist number*):

$$S = \frac{\tau_d}{\tau_a} = \frac{4\pi\,vL}{\eta c^2}. \tag{7}$$

For classical resistivity, $\eta = 10^{-7}T^{-3/2}$ (*e.g.*, Spitzer 1962); thus

$$\tau_d \approx 10^{-13}T^{3/2}L^2; \quad S \approx 10^{-5}T^{3/2}L. \tag{8}$$

At $T \approx 3 \times 10^6$ K (typical of preflare coronal structures) this gives

$$\tau_d \approx 10^{-3}L^2; \quad S \approx 10^5\,L, \tag{9}$$

so that for $L \approx 10^{8-9}$ cm (a typical scale length for flaring regions) τ_d is of the order of millions of years, with $S \approx 10^{13}$. The plasma is thus frozen-in to the field to a very high degree, and the diffusion (or energy dissipation time) is many orders of magnitude larger than observed flare timescales. Note that this is *not* really a consequence of the conducting properties of coronal material (the resistivity of which is about the same as that of copper wire), but rather of the huge scale lengths involved. For a laboratory copper wire

492

$(\eta \approx 10^{-18}$ e.s.u.$)$ of characteristic thickness $L \approx 1$ mm, the corresponding values of τ_d and S are 10^{-5} sec and 10^4, respectively. The field thus diffuses (*i.e.*, the current disappears) rather quickly when the wire is removed from an externally imposed e.m.f. In a sense, we can view the e.m.f. as primary and the current and magnetic field as consequences of this applied voltage. By contrast, in astrophysical situations, the *magnetic field* is primary; currents and voltages are, respectively, consequences of sheared magnetic fields, and the presence of resistivity rather than the cause of the magnetic field itself.

3. Energy Dissipation in Flares

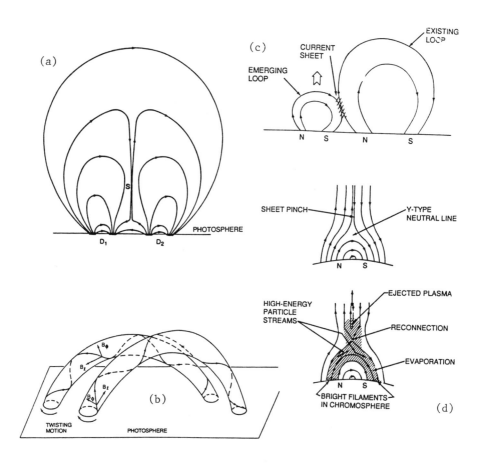

Figure 1. Schematic modes of magnetic energy dissipation in solar flares (for details see Tandberg-Hanssen and Emslie 1988). (a) Merging bipole model, due to Sweet; (b) Merging loop model of Gold and Hoyle; (c) Emerging flux model; (d) Sheet reconnection model due to Sturrock.

In order to dissipate flare magnetic energy on observable timescales, we must significantly reduce L, the scale length for variation of the magnetic field. This requires the proximity of two oppositely directed magnetic fields. Possible scenarios for accomplishing such a steep magnetic field gradient ("current sheet") are shown in Figure 1. For further discussion see, *e.g.*, Tandberg-Hanssen and Emslie (1988).

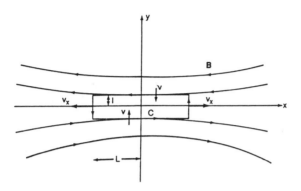

Figure 2. Neutral sheet model. The magnetic field in the x-direction reverses sign along the line $y = 0$. Material advects inward in the y-direction and outward along the x-axis.

Under the assumption that the magnetic field structure "spontaneously" releases energy (*i.e.*, is not driven by external agents), the rate of energy release in a current sheet is straightforward to calculate (Parker 1975). Using the geometry shown in Figure 2, we write down the equations of continuity, momentum, and energy, together with Ampere's law relating $\mathbf{j}$ and $\mathbf{B}$. This results in the system of equations

$$vL = v_z \ell \tag{10}$$

$$\frac{B^2}{8\pi} = p_i = p_0 = \frac{1}{2}\rho v_z^2 \tag{11}$$

$$(\eta j^2)(4L\ell) = \frac{B^2}{8\pi} v \, (4L) \tag{12}$$

$$B(4L) = \frac{4\pi}{c} j \, (4L\ell) \tag{13}$$

Equation (11) gives

$$v_z = \left[\frac{B^2}{4\pi\rho}\right]^{1/2} = V_a, \tag{14}$$

the Alfvén speed. Substituting in (10) gives

$$\frac{v}{\ell} = \frac{V_a}{L}. \tag{15}$$

494

Eliminating (j/B) between (12) and (13) yields

$$v\ell = \frac{\eta c^2}{2\pi}.$$ (16)

Finally, (15) and (16) give

$$v = \left(\frac{2}{S^{1/2}}\right) V_a$$ (17)

$$\ell = \left(\frac{2}{S^{1/2}}\right) L,$$ (18)

where $S = (8L\, V_a/\eta c^2)$ is the Lundquist number for the sheet (of length $2L$), see Equation (7). Equations (17) and (18) are the desired rate of field merging, and the width of the sheet, respectively. The rate of energy release per unit length of sheet (z-direction in Figure 2) is

$$P = \frac{B^2}{8\pi}4Lv = \frac{B^2\, V_a\, L}{\pi\, S^{1/2}}\text{erg cm}^{-1}\text{ sec}^{-1}$$ (19)

Inserting typical numerical values, $viz\ B \approx 10^3$ Gauss, $V_a \approx 10^8$ cm sec^{-1}, $L \sim 10^9$ cm, $S \sim 10^{14}$, we obtain $P = 3 \times 10^{15}$erg cm^{-1}sec^{-1}, so that even if we adopt a width of 10^9 cm for the sheet (*cf.* geometries of Figure 1), we can still only extract $\approx 10^{24}$ erg s^{-1} from the field, a value several orders of magnitude too low to account for flare energy release rates, which can exceed 10^{30} erg sec^{-1} in large flares (*e.g.*, Sturrock 1980).

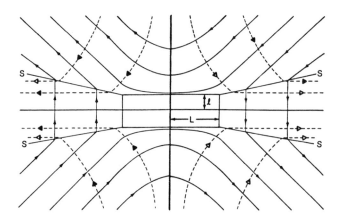

Figure 3. Petschek reconnection topology. As the material advects toward the x-axis, it undergoes a discontinuous change in velocity at the standing shocks S. Note that the connectivity of the field lines is dramatically different from Figure 2; this is true magnetic *reconnection*

In an attempt to enhance the rate or energy dissipation without the need to introduce an *ad hoc* order of magnitude enhancement in the resistivity (and thereby reduce both τ_d and S), various authors have taken a closer look at the process of fluid flow in the vicinity of a current sheet. Petschek (1964) pointed out that, near the origin in Figure 2, several interesting things happen. According to the simple picture presented above, the velocity changes abruptly from a gradual inflow at a speed $<< V_a$ to a rapid outflow at speeds $\approx V_a$. He also noted that, since the field and plasma are still frozen together to a high degree of approximation, the fluid inflow velocity is also $\approx v$. Consider then a parcel of fluid with initial horizontal extent Δx and vertical extent Δy. By continuity, for incompressible flow, $\Delta x \Delta y$ is conserved, so that $\Delta x \to \infty$ as $\Delta y \to 0$, *i.e.*, as the parcel advects toward the current sheet. Since the flux in the x-direction (across the ends of the parcel) is also conserved (by the frozen-in condition), $B_x \to \infty$ as $\Delta y \to 0$ also. These unphysical behaviors are both prevented by the formation of standing shocks (S in Figure 3), at which the fluid flow changes abruptly, without having to pass near the origin. The small region around the origin becomes a *stagnation point*, and the topology of the field lines changes dramatically as shown. .

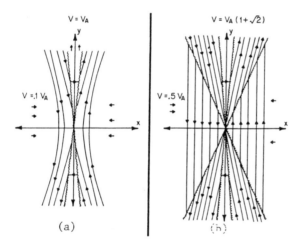

Figure 4. A comparison of the Petschek (a) and Sonnerup (b) modes of magnetic re-connection. The Sonnerup mode involves two standing shocks per quadrant; the Petschek mode only one. (After Forman *et al.* 1985.)

Note that in the Petschek scenario, the field lines truly *reconnect* to other, previously distinct, lines; this is indeed "magnetic reconnection," as opposed to simple *dissipation*. Due to the associated change in field topology, there is now a very strong magnetic tension force to the right and left (*i.e.*, along the x-axis), which propels the frozen-in fluid at Alfvénic velocities. Most of the energy is thus released through relaxation of this magnetic tension and appears as *kinetic* energy of the propelled ("slingshot") material.

Other more complicated topologies, such as that considered by Sonnerup (1970), which involves *two* sets of standing shocks, are also possible. Figure 4 compares and contrasts the Petschek and Sonnerup models. Priest and Forbes (1988) have reviewed the magnetic dissipation and reconnection processes in terms of a comprehensive framework, of which the Parker, Petschek, and Sonnerup models are special cases. More recently, Craig and McClymont (1991) have explored, both analytically and numerically, the physics of driven reconnection near an X-type neutral point. The numerical simulations of Biskamp (1986) suggest that it is difficult to sustain a Petschek topology and that it will in fact relax back to the neutral sheet topology discussed above. In summary, no clear picture as to the mode of dissipation of magnetic fields in flares exists; furthermore, the fundamental problem of the high magnetic Reynolds number, and its implications for the rate of energy dissipation possible, remains.

4. Energy Transport - The "Secondary" Flare

The magnetic energy $1/8\pi \int B^2 dV$ having been released, it is now transported through the remainder of the flaring region in order to create the multi-faceted signatures that characterize the flare. The relationship amongst these various signatures (*e.g.*, hard X-ray, soft X-ray, EUV) depend critically on the form that the released energy takes, for example bulk heating versus acceleration of particles. In the absence of a good understanding of magnetic dissipation/reconnection, two schematic models of the energy release process have been employed traditionally. In one, a "magic wand" is waved and the magnetic energy is converted into the appearance of a large mass of $\sim 10^8$ K material; on the other, we wave a somewhat different wand to create a stream of very energetic ($\gtrsim 10$ keV) particles, notably electrons. We know that some combination of these two processes must take place, because of the ubiquitous hard X-ray bursts ($\epsilon \gtrsim 10$ keV) associated with flares; however, as we shall see, the extent to which each represents the truth is still far from a resolved issue.

At this point, it is well worthwhile to make a few very useful (if somewhat unconventional) definitions concerning the modes of energy release and hard X-ray production in flares:

Thermal A source of hard X-rays is said to be *thermal* if the mean energy of the bremsstrahlung-producing electrons is comparable to the mean energy of the entire ensemble of electrons in the source. It follows that there is no secular loss of energy from the bremsstrahlung-producing electrons, and, in principle, that such a source could be close to 100% efficient at hard X-ray production. Note that the velocity distribution function need *not* be perfectly Maxwellian for a source to be considered "thermal."

Non-Thermal By contrast, an ensemble of hard X-ray producing electrons are said to be *nonthermal* if their average energy greatly exceeds that of the background electrons in the source. The bremsstrahlung-producing electrons now lose a considerable fraction of their energy in collisions with ambient, slower, electrons, and only a small fraction (typically 0.001%) of their energy is lost in small impact parameter, large scattering angle, bremsstrahlung-emitting collisions with ambient protons. The source efficiency is, therefore, typically very small, requiring many more ergs of electron energy than that manifested in the hard X-ray production itself. Note that, aside from the requirement that the distribution function be substantially enhanced over the Maxwellian high-energy tail, there is no requirement that the electron spectrum fit any particular form (*e.g.*, power-law, Bessel function, *etc.*).

5. The Electron-Heated Thick-Target Model - A Critical Overview

As discussed in the previous Chapter, observations of hard X-ray bursts provide considerable evidence pointing to nonthermal electrons as having a primary role in flare energy transport. Due to the relatively low efficiency of such nonthermal electrons at producing bremsstrahlung, this interpretation requires that a considerable fraction of the released energy be deposited in suprathermal electrons. These electrons propagate along the guiding field lines and are ultimately stopped in the flare loop. The loop, therefore, acts as a "thick target" (Brown 1971). Standard formulae exist to calculate the required electron spectrum for a given hard X-ray yield: the logarithmic slope (spectral index) δ of the electron spectrum is typically one larger than the slope γ of the corresponding photon spectrum ($\delta = \gamma + 1$), and the integrated rate of injection of electrons above, say, 20 keV can be $\gtrsim 10$ sec^{-1} in large events. (Note that the choice of 20 keV as a reference energy is basically arbitrary. Since we are ultimately interested in total energy requirements, however, the appropriate choice of low energy cutoff is in fact that below which a "nonthermal" description (see above) is no longer valid. This "transition energy" is, despite some heroic efforts, not well determined observationally at present.)

5.1. BEAM DYNAMICS

Apart from the theoretical problem of how to accelerate these electrons (with a total power requirement of order $10^{28} - 10^{29}$ erg sec^{-1} in large events), the above large number of electrons provides some other interesting insights. (i) The total number of electrons in a flare loop is of order 10^{37} (10^{10}cm$^{-3} \times 10^{27}$cm^3), so that all the electrons would be depleted in a few seconds, unless there is a real-time supply of fresh electrons to the acceleration region. Consequently, models which invoke the steady accumulation, during the energy buildup phase, of high-energy electrons in some sort of low-density trap, followed by their subsequent release during the impulsive phase of the flare, are not valid. (ii) Tagging each electron with its 4.8×10^{-10} units of charge results in a current of some 10^{27} e.s.u. Application of Ampere's circuital law at a point 10^9 cm from this axis of the loop then gives an azimuthal magnetic field of $B = 4I/c\ell \approx 10^8$ Gauss. Apart from the rather obvious fact that field strengths of this magnitude have never been observed on the Sun, the associated energy in this field $1/8\pi \int B^2 dV$ of order 10^{42} ergs, some ten orders of magnitude larger the flare process that allegedly accelerated the electrons themselves. (iii) The voltage drop created by displacing even one second's worth of electrons (some 10^{18} electrons cm^{-2}) a distance of $\ell = 10^9$ cm is $V = 4\pi\sigma \ell \approx 10^{19}$ statvolts (σ = charge density); the associated electrostatic energy is $1/8\pi \int E^2 dV = 10^{46}$ ergs, an even more embarrassing amount.

The issues in the preceding paragraph are all alleviated if we admit a *return current*, cospatial with the beam and consisting of ambient electrons dragged along by the electrostatic and inductive fields associated with the change separation and beam current respectively (see, *e.g.*, van den Oord 1990 and references therein). This return current replenishes the acceleration region, neutralizes the beam current, and cancels the change separation electric field. However, it clearly can exist only *outside* the acceleration region itself (since, otherwise, electrons would be moving opposite to the direction of the applied electric force); thus, replenishment of the acceleration region remains a rather interesting and unresolved (Holman 1985) problem.

An alternative to the return current involves current closure in a circuit (much like in the laboratory). This replenishes the acceleration region and solves the charge separation problem; however, the unneutralized beam current remains. As previously pointed out, all 10^{36} electrons sec^{-1} flowing in one direction creates an unacceptably large induced magnetic field; thus this method of current closure requires many oppositely directed beams (imagine a bundle of arrows, half pointing upward and half pointing downward) and, hence, a highly structured energy release mechanism. Holman (1985) has estimated the number of separate energy release regions as at least 10,000.

If, on the other hand, the energy release process involves non-directed (*e.g.*, stochastic) acceleration of particles, then there is no preferred direction for the beam current and, hence, no return current is required. However, the nature of stochastic processes indicates that a majority of the unreleased energy would go into high entropy energization (*i.e.*, heating), rather than the much lower entropy condition of the acceleration of a small percentage of the ambient electrons to many times the mean thermal energy.

In summary, the global electrodynamics of the large number of electrons required for hard X-ray burst production in a nonthermal model pose significant theoretical problems that have not yet been adequately resolved. Nevertheless, the abundant observational evidence pointing to such a large-scale acceleration process offers us a powerful diagnostic of the flare energy release process. This is because, as remarked above, nonthermal electrons are a low entropy (high information) product of the energy release, offering us the chance to probe its physics to a level of detail not possible with high entropy (low information) thermal emissions. We look forward to analysis of observations at high spatial and/or temporal resolution commensurate with the physical processes (*e.g.*, Coulomb collisions) affecting the electrons. High-time-resolution ($\sim 10^{-3}$ sec) observations are now available, with the successful launch of the *Gamma-Ray Observatory (GRO)*, while high-spatial resolution (arc-second) observations are part of the proposed *High Energy Solar Physics (HESP)* mission objectives.

5.2. ATMOSPHERIC RESPONSE

Notwithstanding the above remarks, thermal emissions from flares can still provide valuable information as to the mode of energy release. For example, in an electron-heated model, we can derive rather precise formulae for the heating rate as a function of position and time. Using this as a source term in the hydrodynamic (or magnetohydrodynamic) equations describing the response of a fluid to a deposition of energy within it leads to predictions as to the variations of density (zeroth moment of the distribution function), velocity (first moment), and temperature (second moment) with position and time. These hydrodynamic results can in turn be used to predict observable quantities such as line strengths and profiles, emission measures, *etc.* It turns out that these diagnostic predictions are sufficiently dependent on the assumed mode of energy release to permit discrimination amongst candidate models.

5.2.1. *Heating by Nonthermal Electrons.* The variation of energy E with depth of an electron injected into an ionized hydrogen target is given by (Emslie 1978)

$$E^2 = E_0^2 - 2KN \qquad (20)$$

where E_0 is the initial energy, $K = 2\pi e^4 \Lambda$ where e is the electronic charge, Λ is the Coulomb logarithm), and N is the column density traversed by the electron $= \int n(s)ds$. Here, s is the arc length traversed by the electron. Due to the changes in direction suffered by the electron in each collision, the distance travelled along the guiding magnetic field line is, on average, $(2/3)$ of the total path length (Emslie 1978); for simplicity, we shall ignore this factor here.

The heating rate $\left(\text{erg cm}^{-3} \text{ sec}^{-1}\right)$ effected by an injected flux $F_0(E_0)$ (electrons $\text{cm}^{-2}\text{sec}^{-1}$ per unit E_0) is

$$I_B = n \int F_0(E_0) \left| \frac{dE}{dN}(E_0, N) \right| dE_0. \qquad (21)$$

Assuming that the heating rate is dominated by Coulomb collisions in a uniform guiding field (i.e., neglecting collective plasma processes and field line convergence), we may substitute, from (20), $|dE/dN| = K/E$, and hence easily obtain the heating rate as a function of column depth N, given the initial spectrum $F_0(E_0)$. For a power-law with cutoff (total energy flux $\Im$ erg $\text{cm}^{-2}\text{sec}^{-1}$):

$$F_0(E_0) = \begin{cases} (\delta - 2)\Im/E_c^2 \left[E_0/E_c\right]^{-\delta} & ; E_0 \geq E_c, \\ 0 & ; E_0 < E_c, \end{cases} \qquad (22)$$

expression (21) reduces to

$$I_B = (\delta - 2)\Im E_c^{\delta - 2n} \int_{E^*}^{\infty} \frac{E_0^{-\delta} \, dE_0}{(E_0^2 - 2KN)^{1/2}}. \qquad (23)$$

The lower limit on the integral is $E^* = \max(E_c, (2KN)^{1/2})$, reflecting the fact that at large column depths the lower cutoff in the spectrum (22) is irrelevant, since electrons of energy less than E_c cannot penetrate to such depths anyway. Thus, for $N > E_c^2/2\,K$, we obtain the relatively simple formula

$$I_B = \frac{1}{2}Kn(\delta - 2)B\left[\frac{\delta}{2}, \frac{1}{2}\right] \frac{\Im}{E_c^2}\left[\frac{2KN}{E_c^2}\right]^{-\delta/2}, \qquad (24)$$

where $B(u, v) = \int_0^1 x^{u-1}(1 - x)^{v-1}dx$ is the beta function. (Inclusion of scattering terms changes the $(2KN)$ into $(3KN)$ and the second argument of the beta function to $(1/3)$ - Emslie 1978.)

Equation (24) shows that the heating per particle (I_B/n) falls off like a simple power-law, with index $(\delta/2)$. It is, therefore, a straightforward matter to calculate the total heat deposited in a given region of the atmosphere: one simply integrates Equation (24) over the appropriate range of column density. An application of this simple technique is presented in the next subsection.

500

5.2.2. *Enhancement of EUV Radiation.* EUV radiation in the range $10 - 1030$ Å is detected throughout the Sudden Frequency Deviations in radio signals bounced off the Earth's ionosphere. These frequency deviations are Doppler shifts induced by the motion of the ionosphere in response to the arrival of the EUV radiation (see, *e.g.*, Kane and Donnelly 1971). The radiation originates in the upper chromosphere $(T \leq 10^4$ K$)$ and is the mechanism by which the atmosphere loses the energy deposited by the flare energy transport process.

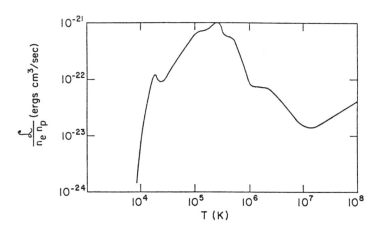

Figure 5. Radiative loss function for an optically thin ionized gas with solar abundances. The radiated power per unit volume is given by multiplying this function by the square of the local density.

The ability of an optically thin plasma to radiate energy is a complicated function of temperature, depending on the ionization/excitation balance of the various species present. For solar abundances, the radiative loss function $\varphi(T)$ has been calculated by several authors (*e.g.*, Cox and Tucker 1969, Raymond, Cox, and Smith 1975). These calculations (*e.g.* Figure 5) show that, in general, $\varphi(T)$ *decreases* with T in the interval 5×10^4 K $\leq T \leq 10^7$ K, due to the decrease in bound-free and bound-bound radiation. (Above $T \approx 10^7$ K, the radiative loss rate again increases due to the growing role of free-free (bremsstrahlung) emission.) Consider then, plasma in this interesting range, where $\partial\varphi/\partial T < 0$. If the plasma is heated somewhat, then the resulting reduction in radiative losses leads to an even greater imbalance between heating and cooling. The gas, therefore, becomes hotter, the radiative losses become even less, and so on. This leads to a runaway increase in temperature known as *radiative instability.* For practical purposes, it means that it suffices to heat the chromospheric plasma to $T = T_{crit} \approx 5 \times 10^4$ K; above this critical temperature radiative instability rapidly drives the gas to coronal temperatures.

In the case of electron heating, the energy deposition rate per particle is a decreasing function of column depth N (Equation (24)). Thus, for some N_{crit}, the gas at depths $N > N_{crit}$ is only mildly heated to temperatures $T < T_{crit}$. It thus remains relatively cool, and emits the EUV radiation mentioned above. However, for $N < N_{crit}$, the gas is heated to $T > T_{crit}$ and, aided by the radiative instability mechanism, is rapidly heated to coronal temperatures, at which time it principally emits soft X-rays. The brightness of the cool EUV source is thus equal to the electron power deposited in layers with $N > N_{crit}$ ($n > n_{crit}$), where

$$n_{crit}^2 \varphi(T_{crit}) = \frac{1}{2} K n_{crit} (\delta - 2) B \left[\frac{\delta}{2}, \frac{1}{2} \right] \frac{\Im}{E_c^2} \left[\frac{2K N_{crit}}{E_c^2} \right]^{-\delta/2} \tag{25}$$

For heating sufficiently rapid that no significant mass motions occur, the density n is given by the hydrostatic preflare relation, *viz*

$$N_{crit} m_H g = 2 n_{crit} k T_{crit}. \tag{26}$$

Here m_H is the proton mass, k Boltzmann's constant, and g the solar gravity. (Equation (26) is valid for ionized atmosphere of pure hydrogen; partial ionization and/or the introduction of other elements introduces only a multiplicative constant.) With this, (25) reduces to

$$N_{crit} = C \Im^{\frac{2}{B\delta + 2}}, \tag{27}$$

where $C = [K(\delta - 2) B [\delta/2, 1/2] \Im k T_{crit} / m_H g \, \varphi(T_{crit}) E_c^2]^{2/(\delta+2)}$ is a constant.
The electron power deposited below $N = N_{crit}$ is

$$I_{EUV} = A \int_{N_{crit}}^{\infty} I_B \left[\frac{dN}{n} \right] \mathrm{ergsec}^{-1}, \tag{28}$$

where A is the flare area (cm^2); using (24) this evaluates to

$$I_{EUV} = C' \, A \Im \, N_{crit}^{1-\delta/2}, \tag{29}$$

where $C' = K B [\delta/2, 1/2] E_c^{\delta-2} (2K)^{-\delta/2}$. Substituting from (27) into (29) gives, to within a constant of proportionality,

$$I_{EUV} \sim A \, \Im^{\frac{4}{2+\delta}}. \tag{30}$$

Now the injected beam flux $\Im$ is related to the hard X-ray bremsstrahlung yield I_x by $\Im A \sim I_x$; thus

$$\left[\frac{I_{EUV}}{I_x} \right] \sim \Im^{\frac{2-\delta}{2+\delta}} \sim \left[\frac{I_x}{A} \right]^{\frac{2-\delta}{2+\delta}}. \tag{31}$$

502

For a given flare area A, this result says, interestingly enough, that (I_{EUV}/I_x), the ratio of EUV field to hard X-ray intensity, *decreases* with the hard X-ray intensity of the flare. Physically, this corresponds to a higher beam flux, hence, a larger value of N_{crit} and a lower fraction of the energy deposited in the cool, radiatively stable, EUV-emitting plasma. Equivalently, for a given hard X-ray yield I_x, the EUV intensity increases with the flare area, since the electron power is now diluted, leading to most of the atmosphere being in a "simmering," EUV-emitting, situation, rather than in an explosive heating to soft X-ray emitting temperatures.

For fixed area A, Equation (31) provides a test of the electron-heated model, with its characteristic heating-versus-depth profile. A plot of observed (I_{EUV}/I_x) values vs. I_x does indeed yield a power-law relationship, with a best-fit power-law index equal to -0.4 (Figure 6, from McClymont and Canfield 1986). This is consistent with Equation (31), for $\delta = 14/3 \approx 5$. Not only is such a value of consistent with that inferred from hard X-ray spectra (the spectral index of which is $\gamma = \delta - 1$ - Brown 1971), but the excellent correlation of Figure 6 shows that the area A does not vary significantly from event to event. Carrying through all the numerical constants, McClymont and Canfield found that the area A corresponding to the best-fit line of Figure 6 is $\approx 2 \times 10^{16} \text{cm}^2$.

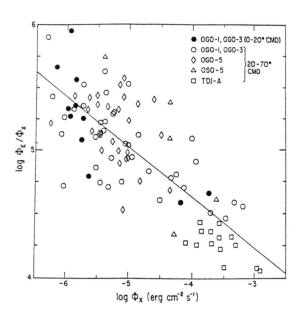

Figure 6. Ratio of EUV to hard X-ray flux *vs.* hard X-ray flux for a variety of events, observed with various satellites. Note that the EUV to hard X-ray flux ratio decreases with the size of the flare.

This conclusion (that all flares have a universal area) is quite remarkable - big flares are not so much "big" as "intense." Furthermore, the "universal area" above is extremely small - it corresponds to a square, two arc seconds on a side, much smaller than the area of, say, Hα kernels. Such an area appears totally incompatible with the requirements for *stellar* flares (see Chapter 26). Even for the solar case, it suggests that the filling factor (the fraction of the flare area over which the beam is injected) is quite small, and may even lend support to the "multiple spaghetti strand" model mentioned in section 5.1. LaRosa and Emslie (1988) have further pointed out that such low areas are completely untenable for beam injection areas in large events, inasmuch as they correspond to physically unacceptable beam fluxes and, hence, they suggest that flare loops must have a much lower cross-sectional area in the chromosphere than they do in the corona, thereby allowing the beam injection area to be much larger than the chromospheric precipitation area. We shall return to this concept of a tapered loop geometry in the next section.

5.2.3. Hydrodynamic Response of the Atmosphere. The electron power depicted in the atmosphere gives rise to a variety of physical effects. The initial effect is, of course, a heating of the ambient atmosphere. This sets up pressure gradients, leading in turn to hydrodynamic flows and, ultimately, to density variations.

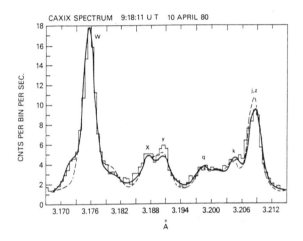

Figure 7. Ca XIX spectrum observed by Antonucci *et al.* (1980). Note the blue "shoulder" on the resonance ("w") line, indicating an upward motion of the emitting material; the principal component is, however, at the laboratory rest wavelength.

Despite the apparent simplicity of the above physics, calculation of the actual hydrodynamic response of the solar atmosphere to a prescribed form of energy input actually requires a solution of a rather formidable set of coupled, nonlinear, partial differential equations, the source terms of which are themselves nontrivial to compute. The complexity of this problem is underscored by the large range of answers to the same problem found by modelers using different hydrodynamic codes (the "benchmark" test of Kopp *et al.*, 1986). Nevertheless, the qualitative physical description of the atmospheric response is similar in the many to modeling attempts carried out to date (*e.g.*, Mariska *et al.*, 1989, and references therein): the strong chromospheric heating affected by the electron beam, aided by the radiative instability mechanism, gives rise to a rapid increase in temperature, a two-to-three order of magnitude increase in the gas pressure, and, consequently, a rapid upward motion of the heated material. This process is known universally as "chromospheric evaporation" (somewhat of a misnomer since no change of state occurs) and leads to a characteristic Doppler shift in the soft X-ray emission produced by the heated material.

These Doppler shifts should be detectable in the spectra of the highly ionized species appropriate to soft X-ray temperatures. A notable example is the Helium-like complex of lines around 3.2 Å due to CaXIX. This complex has been observed in a number of flares by the Bent Crystal Spectrometer (BCS) experiment on the *Solar Maximum Mission (SMM)* and by the *P78-1* satellite (before it became a target for an Air Force missile test!). Figure 7 shows a typical CaXIX spectrum; its most prominent feature is the resonance line (labelled "w" in the alphabetical ordering scheme of Gabriel 1972) at the short wavelength end of the complex. In flares, this line typically shows two components, one at the rest wavelength and one blueshifted by a few mÅ, corresponding to line-of-sight velocities of order a few hundred km sec^{-1} toward the observer.

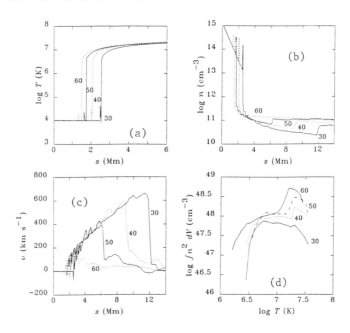

Figure 8. Hydrodynamic simulation results for an electron- heated solar atmosphere (for details, see Mariska *et al.* 1989). (a) temperature *vs.* height; (b) density *vs.* height; (c) velocity (positive = upward) *vs.* height; (d) emission measure *vs.* height. The curves are labelled with the time in seconds since the start of electron injection. Note the upward velocities and resulting coronal density enhancements throughout the event.

An obvious test of the electron-heated thick target model is its ability to account for observed features such as soft X-ray line profiles. Consequently, there has recently been a considerable amount of modeling of the hydrodynamic response of the atmosphere to a heat input of the form (24), and the resulting spectroscopic diagnostics. The basic model adopted in these studies is as described by Nagai and Emslie (1984) and Mariska *et al.* (1989). Electrons with a spectrum similar to that in (22) are injected at the apex of a semicircular magnetic loop; the electrons propagate along guiding field lines which, because of the high magnetic field strength appropriate, form rigid "conduits" for the electrons (the requirement for this assumption to be valid, namely that the gas pressures generated by the heating are everywhere less than the confining magnetic pressure $B^2/8\pi$ can be tested *a posteriori*). The temporal profile of the electron injection is that of a triangle, with a linear rise to a maximum flux around $t = 30$ sec, followed by a symmetrical fall to zero. The heating effected by the electrons is given by (21), and the atmospheric response is typically treated in a single fluid hydrodynamic approximation .

Temperature, velocity, and density profiles for a typical case ($F = 5 \times 10^{10}$erg cm^{-2}sec^{-1}, $E_c = 15$ keV, $\delta = 4$) are shown in Figures 8a through 8c. The initial atmosphere consists of a loop of length 10^9 cm in hydrostatic equilibrium with a coronal temperature of 2×10^6 K; the chromosphere is modelled as isothermal at $T = 10^4$ K. This is not, of course, a terribly accurate picture of the real solar chromosphere; however, due to the fact that rapid heating to temperature in excess of 10^7 K occurs in a few tens of seconds (Figure 8a), the details of the initial structure are largely irrelevant to the subsequent evolution.

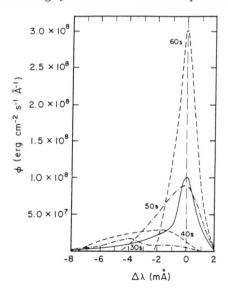

Figure 9. *Dashed lines:* synthesized Ca XIX "w" line profiles, using the hydrodynamic results of Figure 8, at various times throughout the event. *Solid line:* line profile averaged over the first 60 seconds. Compare with the observed profile of Figure 7. (After Li *et al.* 1989.)

A few tens of seconds into the heating, a large fraction of the preflare chromosphere has been heated to coronal temperatures (Figure 8a). As a result, the narrow chromosphere/corona transition region "moves downward" into a region of high density; one should stress, however, that this is not a *physical* downward movement but rather the downward motion of a constant temperature phase surface. The true physical velocity is *upward* (Figure 8b), driven by the large pressure gradient produced by the three order-of-magnitude temperature enhancement in the preflare chromospheric layers. Upward velocities saturate at around 500 km sec^{-1}, the sound speed for the 2×10^7 K material in question. As a result of this upward velocity an order-of-magnitude density enhancement moves up the loop toward the apex (Figure 8c); due to the assumed symmetry of the flare loop, these upward moving density enhancements exactly cancel each other's momentum at the loop apex, going rise to a hot, dense, stationary component there some 30 seconds into the simulation.

The hot CaXIX-emitting material, therefore, has two components, an upward-moving one half-way up the leg of the loop and a stationary one developing later near the loop apex. The spatially-integrated CaXIX line profile, therefore, evolves in a manner shown in Figure 9. A time average of the first 60 sec of the flare (solid line in Figure 9) is directly comparable to the observations of Figure 7, which are for a similar integration time.

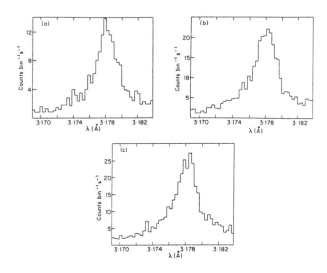

Figure 10. Ca XIX profiles observed (a) at the beginning of the impulsive phase; (b) near the peak of the impulsive phase; (c) in the decay phase (after McClements and Alexander 1990). Despite the superior time resolution (< 10 seconds) of this data, the observed profiles still have a principal component that is at the laboratory rest wavelength. This observation is in contradiction to the theoretical profiles of Li *et al.* for the electron-heated model (see Figure 9).

Due to the evolving nature of the profile, a more critical test of the model is the line profile during the first ten or twenty seconds, before the stationary component has had a chance to develop. The model quite definitively predicts that the line profile during this phase should be blueshifted by about 3 − 4 mÅ (see Li *et al.* 1989 for more details); however, observations at the required time resolution (Figure 10) again reveal a line profile with a principal component that is *stationary*, inconsistent with the model predictions. This inconsistency points to a serious deficiency in the thick-target model as presented to date. Various attempts at resolving the discrepancy are in progress: they include the use of a nonuniform loop cross-sectional area and/or the introduction of a "preheating" phase which enhances the coronal density through weak (and, hence, unobserable) evaporation. Both of these lead to a reduction in the electron power reaching the chromosphere and, hence, presumably in the amount of evaporation and blue-shifted line emission, produced. However, one must not be tempted to make premature judgements as to the viability and/or consistency of these approaches. The introduction of a tapered loop may indeed reduce, through the magnetic mirror force (*e.g.* Chen 1974), the precipitating electron flux; however, the upward-driven motions see a Lavall nozzle geometry and, hence, may be readily driven to supersonic flow. A dense, preheated corona will also lead to reduced electron precipitation and to a heavier "piston" suppressing upward evaporative flow; however, a high-pressure corona is only consistent with the EUV/hard X-ray flux relationship of Figure 6 if δ is unreasonably small (McClymont and Canfield 1986).

0.0.1. Summary. As a result of our brief exploration of the physics of electron-heated flare atmospheres, we may conclude that the model is not without its strengths, yet fails, on its present (undoubtedly oversimplistic) form to reproduce satisfactorily (or even qualitatively) certain crucial impulsive phase manifestations. Further, the model itself rests on some very shaky theoretical ground, in that no self-consistent explanation to date has been offered for either the acceleration or the global electrodynamics of the electron beam. Research into these latter areas (with due recognition for the available observational constraints) remains one of the most challenging aspects of solar physics today.

References

Antonucci, E. *et al.* 1980, *Solar Phys.*, **78**, 107.

Brown, J.C. 1971, *Solar Phys.*, **18**, 489.

Biskamp, D. 1986, *Phys. Fluids*, **29**, 1520.

Chen, F.F. 1974, *Principles of Plasma Physics* (New York: Plenum)

Cox, D.P. and Tucker, W.H. 1969, *Ap.J.*, **157**, 1157.

Craig, I.J.D. and McClymont, A.N. 1991, *Ap.J.*, **371**, L41.

Emslie, A.G. 1978, *Ap.J.*, **224**, 241.

Forman, M.A., Ramaty, R., and Zweibel, E. 1985, in *Physics of the Sun*, eds. P.A. Sturrock *et al.*, Chapter 13. (Dordrecht: Reidel).

Gabriel, A.H. 1972, *MNRAS*, **160**, 99.

Holman, G.D. 1985, *Ap.J.*, **293**, 584.

Kane, S.R. and Donnelly, R.F. 1971, *Ap.J.*, **164**, 151.

Kopp, R.A. *et al.*, in *Energetic Phenomena on the Sun*, NASA CP-2439, Chapter 7.

LaRosa, T.N. and Emslie, A.G. 1988, *Ap.J.*, **326**, 997.

Li, P., Emslie, A.G., and Mariska, J.T. 1989, *Ap.J.*, **341**, 1075.

Mariska, J.T., Emslie, A.G., and Li, P. 1989, *Ap.J.*, **341**, 1067.

McClements, K.G. and Alexander, D. 1990, *Solar Phys.*, **123**, 161.

McClymont, A.N. and Canfield, R.C. 1986, *Ap.J.*, **305**, 936.

Nagai, F. and Emslie, A.G. 1984, *Ap.J.*, **279**, 896.

Parker, E.N. 1975, *Cosmical Magnetic Fields* (Oxford).

Petschek, H.E. 1964, in *Physics of Solar Flares* (ed. W.N. Hess), AAS-NASA, p. 426.

Priest, E.R., and Forbes, T. 1986, *J. Geophys. Res.*, **91**, 5579.

Raymond, J.C., Cox, D.P., and Smith, B.W. 1976, *Ap.J.*, **204**, 290.

Sonnerup, B.U.O. 1970, *J. Plasma Phys.*, **4**, 161.

Sturrock, P.A. 1980 (ed.) *Solar Flares - A Monograph from Skylab Solar Workshop II* (Colorado: Colorado Associated Press).

Tandberg-Hanssen, E. and Emslie, A.G. 1988, *The Physics of Solar Flares*(New York: Cambridge).

van den Oord, G.H.J. 1990, *Astr. Ap.*, **234**, 496.

26 STELLAR FLARES: OBSERVATIONS AND MODELLING

R. PALLAVICINI
Osservatorio Astrofisico di Arcetri
Largo Enrico Fermi 5
50125 Firenze
Italy

ABSTRACT. Increasing observational evidence shows that stellar flares, especially those on M dwarf stars, have many similarities with solar flares. Their energies, however, can be many orders of magnitude larger. Observations of stellar flares at all wavelengths from X-rays to the radio are reviewed stressing the similarities and differences with respect to the solar case. Empirical models developed in the framework of the solar analogy are also discussed and it is shown that they provide only a very crude description of the observed phenomena. Major advances in the observations of stellar flares are required before reliable models can be developed and the fundamental question answered whether stellar flares are simply scaled-up versions of solar flares.

1. Introduction

Flare-like brightenings similar to those observed on the Sun have been detected from many different types of stars (Pettersen 1989). These include classical UV Ceti-type flare stars, RS CVn and Algol-type binaries, pre-main sequence stars (both classical T Tauri and weak-lined T Tauri stars) and a few other individual objects (*e.g.* an X-ray flare has been detected from the G0 V star π^1 UMa, Landini *et al.* 1986, and another from the A-type visual binary Castor A+B, Pallavicini *et al.* 1990a). Not all these flares share the same observational characteristics, and not all of them are as similar to solar flares as the events observed from UV Ceti-type stars. The latter are a subgroup of late K and M dwarfs in the solar neighbourhood with Balmer lines in emission (dKe-dMe stars). Many of them show periodic photometric variations in continuum optical light, that are attributed to the passage of dark spots across their surface (Bopp and Fekel 1977, Bopp and Espenak 1977).

Flares from UV Ceti-type stars have been studied for more than forty years, mostly in optical continuum radiation. The available data show that there is a strong analogy between flares on these stars and solar flares, but there are also some significant differences. First, and most important, the energy released may be up to three orders of magnitude larger than in the *largest* solar flares. The long duration events that are typically observed at X-ray and radio wavelengths from RS CVn and Algol-type binaries are even more energetic and may reach values as high as five orders of magnitude more than the largest solar flares. Clearly, if this energy derives from the same mechanisms thought to be responsible for solar flares (*i.e.* dissipation of magnetic energy in stressed non-potential configurations), these events must involve much larger volumes and stronger magnetic fields than solar flares.

J. T. Schmelz and J. C. Brown (ed.), The Sun, A Laboratory for Astrophysics, 509–533.

An as yet unsolved fundamental question is whether the larger energies and the other differences that unquestionably exist between solar and stellar flares are due to the different environments in which the same physical processes occur or, rather, if the environments are so different that even the basic flare mechanisms are not the same. This is not an easy question, and there are neither observational nor theoretical arguments at present that can convincingly discriminate between these two possibilities. Thus, while early studies of flare stars emphasized the differences rather than the similarities (*e.g.* Gurzadyan 1980), there is now a large consensus about the interpretation of stellar flares as analogues of solar flares (*e.g.* Haisch and Rodonò 1989).

It must be remembered that stellar properties change dramatically across the HR diagram and even stars that have the same effective temperature and gravity may differ, for instance, in chemical composition, rotation rate, magnetic fields and age. It is not expected, therefore, that flares on different types of stars, or on stars in different evolutionary phases, should all be the same, even if produced by similar mechanisms. An interesting case is that of the X-ray flare seen by *EXOSAT* on the G0 V single star π^1 UMa (Landini *et al.* 1986). This flare released in X-rays $\sim 10^{33}$ erg, *i.e.* a factor ten more than the *total* energy released by the largest solar flares. Yet, the internal structure and global properties of this star cannot differ significantly from those of the Sun, except for a younger age and more rapid rotation. Apparently, the latter two properties are sufficient to produce a more vigorous magnetic activity (through enhanced dynamo action) and larger flares. However, the possibility of other more exotic flare mechanisms cannot be excluded at least for certain types of stars (see Chapter 27).

Rather than trying to answer the above basic questions, in this Chapter I will present an overview of the stellar flare phenomenon, stressing the similarities and differences with respect to the solar case. I will present the observational data as well as various solar-type models that have been developed over the past few years in order to interpret the data. For reasons of space, only a summary of the main problems in stellar flare research will be given, while referring to a number of recent conference proceedings and review papers for a more extensive treatment of the subject and complete lists of references (see, *e.g.* Byrne and Rodonò 1983, Gondhalekar 1986, Havnes *et al.* 1988, Haisch and Rodonò 1989, Mirzoyan *et al.* 1990, Haisch *et al.* 1991, Pettersen 1991).

2. Stellar Flare Energetics and Diagnostics

2.1. ENERGETICS

We can make some simple considerations about stellar flare properties if we assume, as a working hypothesis, that the flare energy derives, as for solar flares, from the dissipation of stressed magnetic fields. As will be shown in the next section, the largest flares on UV Ceti type stars have energies of the order of a few times 10^{34} erg in the optical U or B band (Gershberg 1989, Shakhovskaya 1989). Energies of the same order are also derived in soft X-rays for the largest flares on dMe stars (Pallavicini *et al.* 1990b). Taking into account that the total bolometric radiative losses from a stellar flare are estimated to exceed those measured in the optical U band by a factor 20-30 (Pettersen 1988), the total energy released in the largest flares on dMe stars is a few times 10^{35} erg. Note that this is a lower limit since it does not include the unknown contribution from mass motions. Even so, it is three orders of magnitude larger than the total energy released in the largest solar flares.

If this energy has to come from dissipation of magnetic fields, the energy ΔE released by the flare must be related to the magnetic field strength B and its variation ΔB by

$$\Delta E = \frac{B(\Delta B)}{4\pi} V \tag{1}$$

where V is the flare volume and we have assumed that only a small fraction of the magnetic field B is dissipated (things would not be different, to within a factor of 2, in the unlikely circumstance that the entire field is annihilated).

For the Sun, $\Delta E \sim 10^{32}$ erg and the volume of a large 2-ribbon flare is $V \sim 10^{29}$ cm^3 (e.g. Svestka 1976). Hence $B(\Delta B) \sim 10^4$. If $B = 300$ Gauss (a reasonable value for solar flares), a variation $\Delta B \sim 30$ Gauss is sufficient to account for the flare energy release. This is not the case for stellar flares for which $\Delta E \geq 10^{35}$ erg. In fact, if the volume were the same as in the solar case, $B(\Delta B)$ would be $\sim 10^7$ and, in this case, one should dissipate entirely a magnetic field of several thousand Gauss to do the job. More likely, if we want to maintain the same ratio 1 to 10 that we have assumed above for the solar case, one has to postulate magnetic fields as high as $\sim 10^4$ Gauss in dMe stars and variations $\Delta B \sim 10^3$ Gauss. Although we cannot exclude the existence of magnetic fields of $\sim 10^4$ Gauss in localized regions at the surface of flare stars (cf. Mullan 1984), the available evidence suggests that average fields are, at most, a factor three higher than the solar ones (Saar 1990).

Alternatively, the flare volume could be much larger than in the solar case. For instance, if the volume were an order of magnitude larger, $B(\Delta B) \sim 10^6$ and one needs to dissipate ~ 300 Gauss in a field of ~ 3000 Gauss. Note that a volume of 10^{30} cm^3 or larger implies a linear size for the flare that is comparable to the stellar radius in dMe stars, since these stars have typical radii $R_* \approx 0.3 R_\odot$. In either case, it is clear that the *largest* flares on UV Ceti-type stars require substantially higher magnetic fields and/or larger volumes than solar flares. However, the range of observed flare energies overlaps between the solar and stellar case. The smallest flares detected on dMe stars are, in fact, comparable to, and even smaller than, the strongest solar flares.

The above considerations apply only to flares on dMe stars. The intense long-duration events that are typically observed from RS CVn and Algol-type binaries and from PMS stars may reach energies as high as $\sim 10^{37}$ *erg* in X-rays (cf. Linsky 1991) and, perhaps, one order of magnitude more over the entire electromagnetic spectrum. Application of the solar analogy to these cases requires even larger volumes, if the magnetic fields are not much higher. Volumes as large as $\sim 10^{32} - 10^{33}$ cm^3 are required for the largest flares. Again, this implies flare linear sizes comparable to the stellar radius since these stars have typical radii $R_* \sim 3 R_\odot$.

2.2. DIAGNOSTICS

Another important difference between the solar and stellar case is in the range of observational diagnostics available. As mentioned above, flares on dMe stars have been most commonly observed in optical continuum emission. This is in striking contrast with the solar case, for which "white-light" flares are a relatively rare phenomenon and most observations have been made in Hα. With the advent of more sensitive techniques in ground-based observations, and the opening of the space era to stellar observations, new diagnostics have become accessible for stellar flares. These include high-resolution spectroscopy of optical lines, the detection of ultraviolet line and continuum emission, the study of soft X-rays at energies $\leq 10 \; keV$, and radio continuum emission (including the possibility of obtaining radio dynamic spectra). Although this has considerably reduced the gap between solar and stellar observations, and has contributed enormously to our understanding of stellar flares, there are still many other diagnostics that remain completely inaccessible. For instance, we do not have any *direct* information on the spatial structure of stellar flares, and we have no measurements of hard X-rays, γ-rays or particle emission. Our knowledge of the dynamics of the stellar flare plasma is also quite limited, while lack of sufficient sensitivity and spectral resolution has prevented the application of plasma diagnostic techniques at X-ray wavelengths. These fundamental limitations need to be taken into account when comparing solar and stellar flares and when trying to determine how successful solar-type models are in explaining flares on stars.

3. Stellar Flare Observations

Stellar flares, like solar flares, are extremely complex phenomena that involve all atmospheric levels, from the photosphere to the corona, at various phases throughout the flare process. Coordinated multiwavelength observations are required if we want to grasp the essential physics involved. Unfortunately, these are not easy to arrange, and only a few coordinated campaigns have been successfully carried out up to now. Among the most successful are the observations of Kahler *et al.* (1982) of a flare on YZ CMi (25 Oct 1979), those of Rodonò *et al.* (1989) of a large flare on AD Leo (28 March 1984), and those of Hawley and Pettersen (1991) of an even larger flare on the same star (12 April 1985). In the following discussion, I will refer mostly to the results of these coordinated campaigns (see Figures. 1-3). Other multiwavelength campaigns were more limited in scope and usually involved simultaneous observations at only two different wavelengths (*e.g.* Haisch *et al.* 1983; Bromage *et al.* 1986; de Jager *et al.* 1986, 1989; Doyle *et al.* 1988a,b; Kundu *et al.* 1988). Several additional campaigns on both UV Ceti-type and RS CVn stars have been organized recently involving instruments such as *ROSAT*, *HST*, and a variety of ground based optical and radio facilities.

3.1. OPTICAL CONTINUUM EMISSION

Flares on UV Ceti-type stars have been most extensively studied in broad-band optical continuum emission (see reviews by Byrne 1983, Pettersen 1988, 1989; Gershberg 1989, Shakhovskaya 1989, Butler 1991). Since stellar flares are blue while dMe stars are intrinsically red, the flares are best seen in the U and B bands where the flare amplitude may reach values as high as five or six magnitudes. Only recently, observations have been carried out at longer wavelengths, such as the V, R and I bands (Doyle *et al.* 1989b, Hawley and Pettersen 1991). Figure 3a shows observations of a large flare on AD Leo obtained simultaneously by Hawley and Pettersen (1991) in the U, B, V and R bands. Lack of contrast with respect to the underlying photosphere explains why it is much more difficult to detected optical continuum flares in solar-type stars or in RS CVn binaries, since these stars are typically hotter and brighter than dMe stars. However, a 0.6 magnitude U-band flare has been detected recently from the RS CVn binary HR 1099 (Foing *et al.* 1991).

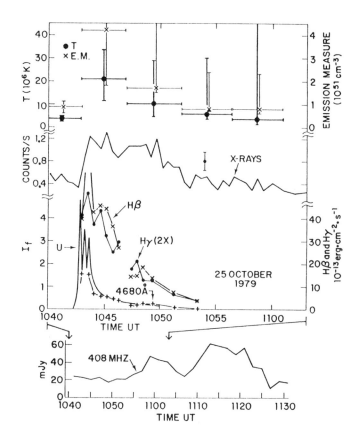

Figure 1. Multiwavelength observations of a flare on YZ CMi on 25 October 1979. From top to bottom: coronal temperature and emission measure, soft X-ray emission, Balmer line and continuum emission, radio emission at 408 MHz (from Kahler *et al.* 1982).

Statistical studies over large samples (Lacy *et al.* 1976, Gershberg and Shakovskaya 1983, Pettersen *et al.* 1984) indicate that the energy emitted in the U band is only slightly larger than in the B band ($E_U \sim 1.2\ E_B$), while the total energy released at optical wavelengths $E_{opt} \sim 3 - 5\ E_U$. For the flare on AD Leo studied by Hawley and Pettersen (1991), the total energy in the optical continuum was ~ 6 times that in U band, while the U and B bands contributed almost equally to the total energy budget. The total energy released throughout the electromagnetic spectrum is estimated to be $E_{bol} \sim 5 - 6\ E_{opt}$. Since flares on dMe stars are observed to involve energies from $\sim 10^{28}$ erg to $\sim 10^{34}$ erg in the U band, the total radiative energies may exceed these values by more than one order of magnitude.

The intrinsically brighter among UV Ceti-type stars flare more frequently than the fainter ones and produce more energetic flares. While the latter result may be due at least partially to a selection effect (since stellar flares are seen against the integrated emission from the star), the higher flare frequency of intrinsically brighter stars suggests a dependence on the stellar radius, as might occur if larger areas of the star are covered by active flaring regions. Although periodicities have occasionally been reported, there is no convincing evidence that stellar flares repeat regularly in time. Surprisingly enough, no clear periodicities related to the stellar rotation and/or to stellar cycles have yet been found. One possibility is that the large activity of these stars at all times, and the random occurrence of flares, reduce the amplitude of any such modulation.

The light curves of optical flares (see, *e.g.*, Figures 1-3) present a large variety of different shapes with typically impulsive, often structured, profiles. The observed rise times are in the range $\sim$ 1-100 sec and the observed decay times are on the order of $\sim$ 10-1000 sec. Large flares tend to have longer time scales than small flares, as also observed for solar 2-ribbon flares. Moreover, flares observed in the optical continuum are typically more impulsive and of shorter duration than flares observed in optical Balmer lines or in soft X-rays. This is similar to what is observed in solar flares, where white-light emission has a much shorter lifetime (and is localized in much smaller areas) than Hα emission. In solar flares, there is a very good temporal correlation between white-light flares and hard X-ray emission (Neidig 1989). If the same were true for stellar flares, we could use the optical continuum emission as a proxy for hard X-rays which are otherwise unobservable at the present sensitivity levels. Although there is no observational support for this claim, the solar analogy suggests that optical continuum emission could indeed be used to derive information on the primary energy release and particle acceleration in stellar flares. Unfortunately, the origin of optical continuum emission in stellar (and solar) flares is poorly understood, the most likely candidates being hydrogen free-bound emission and H^- radiation (Neidig 1989, Mauas *et al.* 1990).

3.2. OPTICAL LINE EMISSION

Optical line emission (in the hydrogen Balmer lines, in the H and K lines of Ca II and in He I and He II lines) is enhanced during stellar flares (see reviews by Giampapa 1983, Worden 1983, Foing 1989, Butler 1991). Typically, line emission peaks later than optical continuum emission, has a more gradual profile, and a much longer duration (Bopp and Moffett 1973, Moffett and Bopp 1976; see Figures 1-3). While an impulsive component is often present in the Balmer lines (in addition to the later gradual phase), the Ca II lines show a smoother behaviour and are further delayed with respect to Balmer and continuum emission. By contrast, He I and He II lines behave rather impulsively, similarly to the continuum flare. Although more prominent in the later flare phases, optical line emission is always a small fraction (of the order of 10%) of the optical continuum emission at all flare phases. For instance, in the large flare on AD Leo studied by Hawley and Pettersen (1991), the contribution of line emission to the total optical emission increased from 4% during the impulsive phase to 17% during the subsequent gradual phase. The electron densities estimated from the Balmer line decrement or from the highest order Balmer lines detectable are in the range 10^{13} to $10^{15} cm^{-3}$, somewhat larger than for Hα solar flares. This is consistent with the larger densities inferred for the chromospheres of quiescent dMe stars (*e.g.* Mullan 1977) although there are large uncertainties on the derived values given the difficulty of correctly identifying and measuring the highest members of the Balmer series.

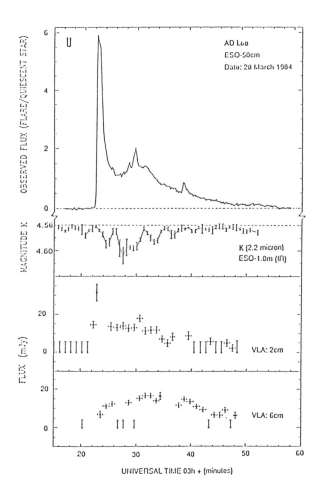

Figure 2. Multiwavelength observations of a flare on AD Leo on 28 March 1984. From top to bottom: optical continuum emission in the U band, infrared emission at 2.2μ, radio emission at 2 and 6 cm (from Rodonò *et al.* 1989).

Typically, the Balmer lines are strongly broadened during the early flare phases and the broadening decreases with time during flare evolution. The Ca II lines are not appreciably broadened and this suggests that the broadening may be largely due to the Stark effect to which the Ca II lines are much less sensitive (Byrne 1989, Robinson 1989). However, Doyle *et al.* (1988a) found it difficult to fit both the line core and the wings with the Stark effect for lines observed in a flare on YZ CMi. Red asymmetries of the Balmer lines are sometimes

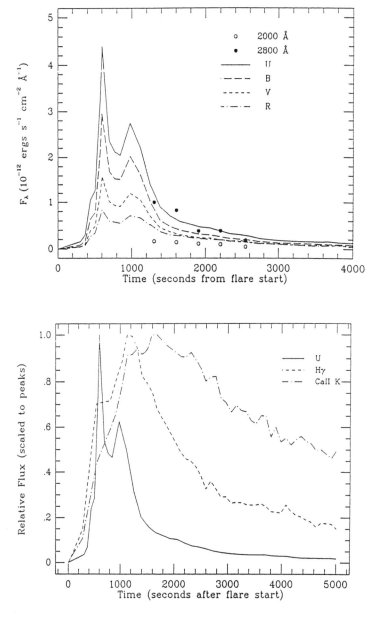

Figure 3. Light curves of a very large flare on AD Leo observed on 12 April 1985. (a) light curves in the optical U, B, V and R bands and at two UV wavelengths; (b) light curves in U-band continuum emission, Hγ and the Ca II K line. Fluxes in (b) have been normalized to their peak value (from Hawley and Pettersen 1991).

observed during the early flare phases (Byrne 1989, Foing 1989), as they are for the Sun (*e.g.* Zarro *et al.* 1988). The red-asymmetries and the broadening (if the latter is due to the Stark effect) both suggest that optical line emission might originate as a consequence of the formation of a compressed high-density region moving downward through the stellar atmosphere (see section 4.1. below).

3.3. INFRARED EMISSION

There are very few infrared observations of stellar flares. One of the best cases is shown in Figure 2 for the 28 March 1984 flare on AD Leo. In the infrared K-band at 2.2 μ, a *negative* flare was detected at the same time as the optical U-band flare. Negative flares have also been reported in the optical, especially in the red I-band (Bruevich *et al.* 1980, Grinin 1983) and are usually interpreted as due to an increase of H^- opacity caused by enhanced ionization of metals (Hénoux *et al.* 1990). However, the H^- opacity interpretation predicts very short-lived negative flares prior to the optical event: this is consistent with the short-lived pre-flare dips usually seen in the optical I-band, but is at variance with the 2.2μ negative flare seen on AD Leo since this lasted for almost 15 min. Other explanations, such as inverse Compton effect on infrared photons to produce the optical continuum flare (Gurzadyan 1988) also have problems since, in this case, one would expect the infrared flare to be the mirror image of the optical U-band flare. Unfortunately, there are virtually no observations of solar flares in the infrared to test whether similar phenomena occur on the Sun.

3.4. ULTRAVIOLET EMISSION

Ultraviolet lines formed in the chromosphere and in the transition region between the chromosphere and the corona are strongly enhanced during stellar flares (Bromage *et al.* 1986, Byrne 1989, Doyle *et al.* 1989a, Linsky *et al.* 1989, Neff 1991). They include Lyα, the Mg II h and k lines at 2800 Å, and UV lines of C II, C IV, N V, Si II, Si IV and He II. The UV observations of stellar flares obtained so far with *IUE* were severely affected by lack of adequate sensitivity for time-resolved observations. In most cases, especially for dMe stars, the exposure times used were comparable with the flare life-time, with the consequence that only a rough comparison could be made between flaring and quiescent conditions. This situation is expected to improve significantly with the new data that are becoming available from *HST*. In addition to UV lines, a strong continuum in the far ultraviolet is also observed during stellar flares (Bromage *et al.* 1986), similar to what is observed in solar flares, but several orders of magnitude stronger. Most likely, this continuum is due to recombination radiation of neutral Si excited by intense UV line emission (Phillips *et al.* 1991).

Despite the limited knowledge of UV stellar flares so far obtained, the available evidence suggests that the total energy radiated by a flare in transition region line and continuum emission is roughly comparable to the total energy emitted in the optical (Pettersen 1988, Byrne 1989). Both are also comparable to the energy emitted in soft X-rays. We may expect, however, that significant differences occur from flare to flare and even in the course of the same event.

3.5. X-RAY EMISSION

Flares from UV Ceti-type stars have been observed at soft ($\leq$ 10 keV) X-ray energies by both the *Einstein* and *EXOSAT* Observatories. By contrast, no hard X-ray observations of stellar flares have yet been possible, and there is very little hope that the situation may change in the near future. In solar flares the ratio of hard to soft X-rays is typically $\sim 10^{-5}$. If the same ratio applies to stellar flares, detector areas of $\sim$ 1 km^2 would be required to obtain observations of the same quality as those obtainable at present for solar flares. Even allowing for the possibility of larger hard X-ray fluxes and/or flatter spectra in the case of stellar flares, we need major technological advances before hard X-ray observations may become possible for stars.

Figure 1 shows an example of a soft X-ray flare observed from YZ CMi with the *Einstein* satellite. The X-ray emission was much more gradual and lasted longer than the optical continuum emission, as typically observed in solar flares. The flare coronal temperature reached a peak value of $\sim 2 \times 10^7$ K and the emission measure was a few times 10^{51} cm^{-3}. The early *Einstein* observations of stellar flares were reviewed by Haisch (1983). With the launch of *EXOSAT*, our knowledge of X-ray flares on all type of stars has increased enormously thanks to the long uninterrupted observations allowed by the highly eccentric orbit of the satellite (see review by Linsky 1991).

The *EXOSAT* observations of flares on UV Ceti-type stars have been discussed most extensively by Pallavicini *et al.* (1990b). There is a large variety of total energies and time scales for stellar X-ray flares. The time scales range from 10^2 to 10^4 sec and are similar to the typical time scales for solar X-ray flares. Also the coronal temperatures are similar to solar values and typically in the range $2 - 4 \times 10^7$ K. On the contrary, the emission measures and the total X-ray energies are much larger than in solar flares. The total radiative energies in X-rays range from 10^{30} to 10^{34} erg and the emission measures are in the range 10^{51} to 10^{53} cm^{-3}. There are no direct measurements of flare coronal densities (as could be derived from ratios of density sensitive lines). Indirect estimates, based on the radiative cooling time, suggest electron densities in the range 10^{11} to 10^{12} cm^{-3}.

The *EXOSAT* observations have suggested the existence of two different classes of stellar flares similar to solar *compact* and *2-ribbon* flares (Pallavicini *et al.* 1990b). Although this classification is based only on morphological evidence (*i.e.* the time scales and total energies are different in the two cases), it is tempting to speculate that these two classes may also correspond to different physical mechanisms, as believed to occur for solar flares (Pallavicini *et al.* 1977, Poletto *et al.* 1988, Poletto 1989). The observational evidence in favor of the existence of stellar analogues of solar 2-ribbon flares, and the modelling of such events in disrupted magnetic configurations that relax back to closed structures (Kopp and Pneuman 1976, Kopp and Poletto 1984), are discussed in Pallavicini (1991).

Some additional comments need to be made on X-ray flares from RS CVn and Algol-type binaries and from PMS stars. The flares seen from these stars in X-rays, UV, optical line emission, and radio wavelengths are typically much stronger and longer-lived than flares from UV Ceti-type stars; also the inferred coronal temperatures are often higher (White *et al.* 1986, Tagliaferri *et al.* 1988, Pallavicini and Tagliaferri 1989, Tsuru *et al.* 1989, Linsky 1991). This might suggest that flares on RS CVn binaries and pre-main sequence stars are fundamentally different from flares on dMe stars and on the Sun, as has been suggested on several grounds (*e.g.* Byrne 1989, Mullan 1989). This conclusion, however, may be affected at least partly by selection effects. The quiescent X-ray emission of RS CVn binaries and of pre-main sequence stars (typically of the order of $\sim 10^{30} - 10^{31}$ erg) is, in fact, much higher than the quiescent luminosity of most dMe stars. Thus, small short-lived X-ray events, if they exist, could easily go undetected against the much higher quiescent luminosity of the star. In spite of these possible selection effects, it is clear, however, that flares on RS CVn and Algol-type binaries and on pre-main sequence stars can reach intensities much higher than for UV Ceti-type stars. Total X-ray energies of $10^{35} - 10^{36}$ erg are commonly observed (Linsky 1991), and an X-ray energy of $> 10^{37}$ erg has been reported for a very long flare on the RS CVn binary UX Ari (Tsuru et al. 1989).

3.6. RADIO EMISSION

Radio flares from UV Ceti-type stars have been observed from centimetric to metric wavelengths (see reviews by Gibson 1983, Dulk 1985, Kuijpers 1989, Lang 1990, Bastian 1990, and Bookbinder 1991). Although energetically unimportant (the ratio of microwave to X-ray flux being typically on the order of $10^{-7} - 10^{-8}$), microwave observations of stellar flares possess great diagnostic capabilities since the emission originates (at least partly) in the region of primary energy release and particle acceleration. Unfortunately, interpretation of the radio data is usually not straightforward, since the emission mechanism is often not uniquely identified, the observed emission may be strongly modified by propagation effects, and the geometry and physical parameters of the flare region are largely unknown. Despite these difficulties, substantial progress has been made in our understanding of stellar radio emission, especially with the use of sensitive arrays such as the Very Large Array.

An example of simultaneous observations of dMe stellar flares at centimetric wavelengths and in optical continuum radiation is shown in Figure 2 for the 28 March 1984 event on AD Leo. At 2 cm, there is a good correspondence between the radio flare and the optical continuum flare, while this is much less evident at 6 cm. The good correspondence found at 2 cm is reminiscent of the strong correlation found for solar flares between microwave and hard X-ray bursts (if we take the stellar optical continuum emission as a proxy for hard X-rays, as explained above). However, it would be wrong to generalize this conclusion to all flares observed from UV Ceti-type stars. Most stellar flare observations have been obtained so far at 6 cm and 20 cm, rather than at 2 cm, and lack of correlation between radio flares and optical (and soft X-ray) flares appears the rule rather than the exception (Nelson et al. 1986, Kundu et al. 1988).

The large body of data collected at the Very Large Array (Gary et al. 1982, Lang and Willson 1986, White et al. 1986, Kundu et al. 1988, Jackson et al. 1989), as well as data from Arecibo (Lang et al. 1983, Lang and Willson 1988), have shown that most microwave flares from UV Ceti-type stars are narrow-band ($\Delta\nu/\nu \leq 0.1$), have very high-brightness temperatures ($T_b \sim 10^{12} - 10^{15}$ K) and a very high degree of circular polarization (approaching 100%). These characteristics can only be explained by a *coherent* emission mechanism that could be either electron cyclotron maser or some form of plasma radiation (Dulk 1985, Kuijpers 1989; see Chapters 14 and 27). This is a fundamental difference with respect to solar microwave flares: the latter in fact are most commonly produced by an incoherent radiation process that is believed to be non-thermal gyrosynchrotron emission. The poor correlation that is typically found between stellar microwave flares and optical/X-ray flares is consistent with the coherent nature of most stellar radio flares since optical and

X-ray emissions are certainly incoherent. On the other hand, the importance of coherent processes is expected to decrease at shorter wavelengths, and this may explain why a much better correlation is found at 2 cm rather than at 6 cm (*cf.* Figure 2).

In contrast to microwave bursts, observations of stellar flares at metric wavelengths have provided much less information, especially because the early observations were heavily affected by radio frequency interference. Interestingly enough, the metric flares were often delayed with respect to optical flares (Spangler and Moffett 1976). An example of this behaviour, as observed at 408 MHz from Arecibo, is shown in Figure 1: the decimetric burst appears delayed by 17 min with respect to the optical continuum flare. This suggests an analogy with Type II and Type IV solar bursts, which are due to disturbances moving outward through the corona. As for the Sun, dynamic radio spectra (obtained by observing radio bursts as a function of both frequency and time) could provide essential information for identifying the radio emission mechanism and for studying propagating disturbances. Such dynamic spectra are starting to become available in the stellar case at decimetric wavelengths (Bastian and Bookbinder 1987, Bastian *et al.* 1990).

Finally, a few words about radio flares from RS CVn binaries and PMS stars. Contrary to dMe stars, microwave emission from RS CVn and PMS stars does not typically require a coherent mechanism. The observed emission, which is unpolarized or only weakly polarized and has a far lower brightness temperature, is best interpreted as due to non-thermal gyrosynchrotron emission of high-energy electrons spiralling in magnetic fields of 10-100 Gauss (Feigelson and Montmerle 1985, Mutel *et al.* 1987, White *et al.* 1991). *VLBI* observations of some RS CVn binaries and PMS objects (Mutel *et al.* 1985, Lestrade 1988, Lestrade *et al.* 1988, Phillips *et al.* 1991) indicate the presence of very large radio emitting structures which extend to several stellar radii and may possibly include interconnecting loops in binary systems. However, it is unclear whether these large regions represent the flare source itself or, rather, are due to the injections of accelerated particles from the flare site into more stable magnetic structures.

4. Stellar Flare Models

Models developed so far for stellar flares are largely based on the solar analogy and on the application of solar-type concepts to the often different stellar conditions. On the Sun, most flares are believed to originate in the corona, at the top of one or more magnetically confined loop-like structures (see Chapter 25 in this volume). Energy may be released either in the form of non-thermal particles that are accelerated to high energies, or as direct Joule heating of the plasma. In both cases, energy is transferred to lower levels by either particle beams or heat conduction. The response of the atmosphere to the various forms of energy deposition and transfer is responsible for the variety of flare phenomena observed at different wavelengths.

In the popular *thick-target* model of solar flares (Brown 1971) the accelerated particles are mainly low-energy electrons with power-law spectra $F(E) \sim E^{-\delta}$ and a low-energy cut-off of $E_c \sim 20$ keV. The electrons, accelerated in the corona, are channelled by the magnetic field towards the dense loop footpoints and produce hard X-rays by non-thermal bremsstrahlung (while the tail of the distribution generates microwave emission by non-thermal gyro-synchrotron). Most of the electron energy, however, is deposited at the loop footpoints by collisional losses, thus producing the observed optical and UV line and continuum emission. Collisional heating at the footpoints, in turn, causes evaporation of high-temperature plasma which fills the structure and is ultimately responsible for the observed soft X-ray thermal emission (see Dennis and Schwartz 1989 and Chapter 25 in this volume for a more detailed discussion and complete lists of references).

In the *thermal* model, the impulsive flare energy release produces mainly heating in the coronal portion of the loop. Energy is transferred down by a conduction front and the lower atmospheric layers (from which optical and UV emission originate) are heated by thermal conduction rather than by non-thermal particles. In this model, hard X-rays are produced thermally in a very hot ($\sim 10^8$ K) region created by the initial energy release. In this case too, the large heat flux deposited at the loop footpoints causes heating and evaporation of chromospheric material and the filling of the loop with high-temperature soft X-ray emitting plasma.

The two models sketched above are by no means mutually exclusive and it is likely that both thermal and non-thermal processes occur in the same events. What is not known even in the solar case, however, is the relative importance of thermal *vs.* non-thermal mechanisms in the overall flare energy budget. In the stellar case, we may assume that the same basic physical phenomena are present (*i.e.* magnetic confinement, particle beams, thermal conduction, evaporation of chromospheric material, etc.), but their relevance to the flare are even less understood than in the solar case. Moreover, these physical processes operate under conditions that are usually quite different from the solar ones. Thus, a major challenge to the present stellar flare models is to understand how the different conditions prevailing in stars affect those processes that are believed to be relevant for solar flares.

A comprehensive stellar flare model that treats all atmospheric levels from the photosphere to the corona, including all relevant physical effects, does not yet exist. Models have been developed so far either for the optically thick lower atmospheric layers (with the purpose of modelling optical flares), or for the optically thin coronal part (with the purpose of modelling X-ray flares). In the rest of this section, I will discuss these two classes of models, and then briefly mention an attempt towards a more unified picture of the whole flare phenomenon.

4.1. THE LOW-TEMPERATURE FLARE

A first attempt to model optical stellar flares by means of the solar analogy was made by Katsova *et al.* (1981) and Livshits *et al.* (1981). They assumed that the optical flare (both continuum and lines) originates in the chromosphere (at levels where the density is of the order of $10^{14} - 10^{15}$ cm^{-3}) as a consequence of the hydrodynamic response of the atmosphere to energy deposition by non-thermal particles. They assumed that a large energy flux ($F_o = 10^{12}$ erg cm^{-2} s^{-1}) is deposited in the upper chromosphere by non-thermal electrons with a power-law energy distribution and a low-energy cut-off $E_c \sim 20$ keV. The assumed spectrum and cut-off energy are similar to those derived for solar flares from hard X-ray observations (Dennis and Schwartz 1989).

According to Katsova *et al.* (1981), the high pressure region produced by the collisional heating generates two perturbations, one upward (which drives chromospheric evaporation and leads to the formation of the soft X-ray emitting region; *cf.* next section), and one downward toward deeper chromospheric layers. Energy is transported downward by a conduction front and by a shock wave which precedes the thermal wave. As shown in Figure 4, a narrow ($\Delta L = 10$ km) high-density ($n \sim 10^{15}$ cm^{-3}) region at $T = 10^4$ K is predicted to form behind the shock front, and the emission from this compressed region could well be responsible for the optical flare. An interesting prediction of this model is that the Balmer lines should be Stark-broadened and should have red-asymmetries, in agreement with some stellar flare observations.

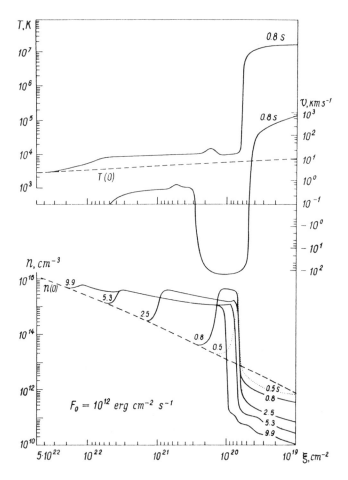

Figure 4. Gasdynamic simulation of a stellar flare heated by non-thermal electrons showing the development of a downward moving compressed region in the chromosphere (from Livshits *et al.* 1981).

In contrast to the above picture, Mullan (1989) argued that electrons with low energy cut-off $E_c \sim 20$ keV should be unable to penetrate as deep as the upper chromosphere because of the larger densities and longer loop lengths typically expected in dMe stars. The penetration depth of an electron with energy E_{keV} is given by:

$$\xi_p(E_{kev}) \sim 7 \times 10^{19}(E_{keV}/20)^2 \;\; cm^{-2} \tag{2}$$

while the expected column density of a flaring loop is $\xi_L = <n> L = 10^{21} - 10^{23} cm^{-2} \gg \xi_p(20)$. Hence, low energy electrons are likely to be stopped in the corona and energy will be transferred down by thermal conduction (Mullan 1976, 1977; Kodaira 1977). If this is the case, the ratio of X-ray to optical flare luminosity L_{opt}/L_x would be a measure of the relative importance of conduction vs. radiative losses since $L_{opt}/L_x \sim \tau_r/\tau_c$. Typically this ratio is ~ 1 in stellar flares (Byrne 1989) but large variations occur from flare to flare and even in the course of the same event (e.g. Kahler et al. 1982).

It seems unlikely that a purely conductive model can explain the *impulsive* emission of stellar flares. Temporal variations as short as 0.1 sec have been observed in optical continuum flares (Gershberg 1989), while the shortest time variations expected from a conductive model should be on the order of L/c_s, where L is the loop length and c_s is the sound speed. Since c_s is of the order of 500-1000 km s^{-1} in typical flare conditions, one expects variations on time scales longer than several tens to a few hundred seconds if the optical flare is heated by thermal conduction. Another way to discriminate between thermal conduction vs. non-thermal particles is to look at the profiles of Balmer lines. Heating by non-thermal electrons produces Stark-broadened Balmer lines with central reversals (Cram and Woods 1982, Canfield et al. 1984), while heating by conduction would produce narrow Balmer lines with weak or no central reversals. Solar observations (Canfield et al. 1984, Zarro et al. 1988) favour a non-thermal model during the impulsive phase of flares, while the observed broadening of Balmer lines during the initial phase of stellar flares (Robinson 1989, Doyle et al. 1989a) also argues against a purely conductive model. However, thermal conduction may be an important energy transfer mechanism during the later gradual phase of both solar and stellar flares.

The above considerations need to be modified if the optical continuum emission in stellar flares originates from much deeper layers than those assumed by Katsova et al. (1981) and Livshits et al. (1981) and/or if the non-thermal particles accelerated in stellar flares have much higher energies than those responsible for solar hard X-ray bursts. From an analysis of optical continuum flares Grinin and Sobolev (1977) concluded that stellar flares originate in layers where the density is $n = 10^{15} - 10^{17}$ cm^{-3}, much higher than assumed by Katsova and Livshifts. At such high densities the optical thickness in the continuum beyond the Balmer jump is $\gg 1$ and the observed emission has a quasi black-body spectrum (Mochnacki and Zirin 1980, Kahler et al. 1982, de Jager et al. 1986.). This is a significant difference with respect to solar flares for which the optical thickness is always less than unity. In these conditions, neither thermal conduction nor shock fronts nor low-energy particles can be very efficient in transferring energy to the lower layers. Instead, Grinin and Sobolev (1988, 1989; see also Grinin 1991) have proposed that optical continuum flares originate from direct heating of dense photospheric layers by non-thermal particles with energies much larger than those on the Sun.

Two possibilities have be considered: either electrons with energies ≥ 100 keV or protons with energies ≥ 5 MeV (protons have also been proposed on other grounds by van den Oord 1988 and Simnett 1989). The calculations of Grinin and Sobolev (1988, 1989) show that protons with a power-law energy spectrum $\sim E^{-3}$ at energies ≥ 5 MeV can reproduce the observed colour indices of optical flares provided the energy flux carried by the protons is $F_0 \sim 10^{12}$ erg cm^{-2} s^{-1}. Note that, in this model, the heating of deep atmospheric layers is produced by both protons and the optical and UV photons created as the beam penetrates the dense atmospheric layers. The heating due to photons becomes increasingly important in deeper layers. The optical flare could also be produced if a similar energy input is deposited by electrons with energy $E_e = E_p(m_e/m_p)^{1/2} \sim E_p/50$. Unfortunately, it is not possible at present to discriminate between heating by electrons or protons. This is difficult for solar flares (*e.g.* Brown *et al.* 1990, Simnett 1991) and is even more difficult for stellar flares for which no observations of hard X-rays and γ-rays are available.

4.2. THE CORONAL FLARE

With the advent of space observations that have allowed the study of the high temperature coronal part of stellar flares, great interest has arisen recently in the development of hydrodynamic models that treat the time dependent response of a coronal loop to various heating perturbations. A large number of such numerical codes have been developed for solar flares (*e.g.* Nagai 1980, Cheng *et al.* 1983, 1985, Pallavicini *et al.* 1983, Nagai and Emslie 1984, MacNeice *et al.* 1984, Fisher *et al.* 1985abc, Peres *et al.* 1987). It is natural, therefore, to extend similar calculations to the stellar case. In a sense, this approach is complementary to that of Katsova *et al.* (1981) and Livshits *et al.* (1981), which were concerned with the low-temperature optically-thick layers. Coronal models, instead, treat the chromosphere only as a boundary layer and a mass reservoir, while concentrating the attention on the phenomena occurring in the optically-thin part of the loop.

A first attempt to model a stellar X-ray flare by means of hydrocodes was made by Reale *et al.* (1988) using the Palermo-Harvard one-dimensional single-fluid code (*e.g.* Peres 1989). The flare chosen for modelling was a long-duration event observed on Prox Cen by the *Einstein* Observatory (Haisch *et al.* 1983). By assuming a circularly symmetric loop of length L, and a localized heating at the top of the loop, they tried to reproduce the observed light curve in the *Einstein* IPC band. The heating was assumed to be constant throughout the flare rise and to stop abruptly after 700 sec. The loop length was treated as a free parameter, searching for the value that gives the best fit to the flare light curve. Since the number of free parameters was limited by choosing a particular form of the heating function, the decay of the flare is determined almost uniquely by the loop length, longer loops producing more gradual decays. The best fit is obtained with a loop with $L = 1.4 \times 10^{10}$ cm. When compared to the Sun, this is a much longer loop than for confined solar flares. Since the radius of Prox Cen is only $0.14R_\odot$, the inferred length of the flaring loop is in fact comparable to the stellar radius, in contrast to typical solar compact flares which have loop lengths of $\approx 0.1R_\odot$ or less.

How sensitive are the conclusions of Reale *et al.* (1988) to the assumed heating function? In order to answer this question, Cheng and Pallavicini (1991) have carried out extensive modelling of flares on dMe stars using the NRL one-dimensional two-fluid code (Cheng *et al.* 1983). In all cases intense heating was assumed to occur close to the top of the loop. Rather than trying to fit a particularly well observed event, they built a grid of different models by changing loop length, preflare conditions, flare heating rate, and spatial and temporal dependence of the heating function. By exploring the parameter space, it is thus possible to get better insights into the flare process, and to test how sensitive the models are to the assumed input parameters.

The basic hydrodynamic results are similar to those obtained by applying the same codes to the solar case (Pallavicini *et al.* 1983; Cheng *et al.* 1983, 1985; Peres *et al.* 1987). The major difference is due to the higher gravity and smaller pressure scale height of dMe stars. If energy is deposited at the loop top, a high temperature region forms there. Energy is rapidly conducted down towards the loop footpoints and the transition region between the chromosphere and the corona steepens and moves downwards. If the energy conducted to the chromosphere is larger than can be radiated away, a high pressure region forms at the top of the chromosphere. The heated chromospheric material expands upwards at velocities of several hundred km s^{-1} filling the coronal portion of the loop ("chromospheric evaporation") while, at the same time, the high pressure region acts like a piston compressing the underlying chromospheric material. The compressed region that forms at the loop footpoints is similar to that found by Katsova *et al.* (1981) and Livshits *et al.* (1981) when heating the upper chromosphere by non-thermal electrons. However, because of the larger gravity and smaller pressure scale height of dMe stars, the compression region encounters a very steep density gradient that limits the amplitude and time duration of the compressed region. For plausible energy input rates (*i.e.* those which appear to be required to fit the X-ray observations), it seems unlikely that the compressed region that forms in a thermal flare can contribute significantly to the flare optical emission in dMe stars.

The results of Cheng and Pallavicini (1991), and their limitations, are best summarized in Figure 5 which shows a comparison of predicted and observed emission measures and total X-ray energies for flares observed by *EXOSAT* and for 4 of the 10 different models run by them. Models 3 and 9 are both for a small loop with $L = 2 \times 10^9$ cm, but the total energy released in model 9 is one order of magnitude larger than in model 3. Models 5 and 8, instead, are for a large loop (with $L = 8 \times 10^9$ cm) and, again, differ by one order of magnitude in energy input. The energy deposited in models 9 and 5 is about the same, but the two models differ in the loop length. As can be seen from the figure, small X-ray flares with energies of the order of $\sim 10^{31}$ erg and emission measures of the order of $\sim 10^{51}$ cm^{-3} (similar to the Prox Cen flare discussed above) can be reproduced by a small loop (model 3), without the need of the large loop comparable to the stellar radius required by Reale *et al.* (1988). Only very large flares (with X-ray energies of the order of $\sim 10^{34}$ erg and emission measure of $\sim 10^{53}$ cm^{-3}) require large loops (model 8), unless unrealistically high magnetic fields are invoked to provide for both energy release and confinement. Intermediate size flares (with energies of the order of $\sim 10^{32}$ erg and emission measures of $\sim 10^{52}$ cm^{-3} can be reproduced equally well with both a small loop (model 9) and a large loop (model 5). With a suitable choice of the time dependence of the heating function, it is always possible to fit the observed X-ray light curve but, without further constraints, we are unable to discriminate between alternative models.

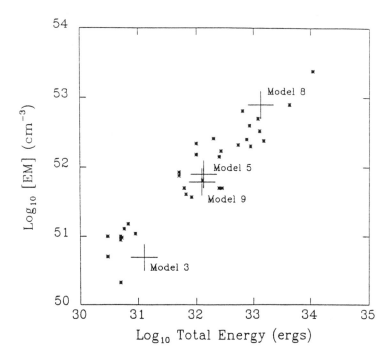

Figure 5. Results of numerical simulations of stellar flares and comparison with *EXOSAT* observations. Different models differ in the loop length and/or energy input (from Cheng and Pallavicini 1991).

4.3. TOWARDS A GLOBAL FLARE MODEL

So far we have treated the low temperature (optical) flare and the high temperature (X-ray) flare as separate entities. Recently, an ambitious attempt has been made to build a global model to account for the entire flare atmosphere in a self-consistent way (Hawley and Fisher 1991, Hawley 1991). Not surprisingly, this broader scope has been reached at the expense of drastic assumptions that were required to make the physical problem mathematically manageable. Thus, while the proposed model is certainly a quite promising approach, it is far from being a detailed numerical solution of the complete set of non-linear time-dependent hydrodynamic and radiative transfer equations. Such a complete model would be a formidable task to build and does not yet exist even for solar flares.

In the Hawley and Fisher approach, the time evolution of the coronal parameters is obtained by means of a simplified algorithm that reduces the solution of the time dependent non-linear partial differential equations of mass, energy, and momentum conservation (*i.e.* the equations that are solved numerically in the codes discussed above) to the solution of an ordinary differential equation (Fisher and Hawley 1990). This is possible if one considers time scales much longer than the loop sound transit time. Thus the model, which also neglects completely non-thermal particles, is suitable for describing the gradual phase of flares, but is inadequate for the impulsive one. Comparison with detailed hydrodynamic simulations (*e.g.* Pallavicini *et al.* 1983) shows, however, that the simplified model is sufficiently accurate for describing the coronal properties during the gradual phase.

If energy is deposited in the coronal part of a loop, heat conduction will make the transition region much steeper and will move it to lower levels. The detailed temperature structure in the chromosphere and temperature minimum region, however, does not depend so much on heat conduction (which is ineffective at such low temperatures), but is rather determined by irradiation by soft X-ray and UV photons from the upper part of the loop (for early suggestions of the possible relevance of X-ray irradiation for the temperature structure of the chromosphere see also Somov 1975, Machado *et al.* 1978 and Cram 1982). Under these assumptions, and with the further simplification of adopting an equivalent static loop to simulate the coronal flare, Hawley and Fisher (1991) were able to construct a series of chromospheric models that can be parameterized in terms of the coronal temperature at the loop apex. Higher coronal temperatures move the transition regions to higher mass column densities in the atmosphere while increasing the chromospheric temperature at any given column mass (see Figure 6a). Radiative transfer calculations allow the intensities of optical chromospheric lines to be calculated as a function of the coronal apex temperature. Thus, from the observed behaviour of optical lines as a function of time, one can infer the time dependence of the coronal apex temperature during the flare and compare it with the one predicted by the coronal model. For self-consistency, the two temperature profiles should be equal, and both should be consistent with soft X-ray observations of the flare.

Hawley and Fisher (1991) have applied their model to the large flare on AD Leo observed on 12 April 1985 (see section 3 above and Figure 3). For this particular event, no X-ray observations were available, so no direct comparison could be made with the predictions of the coronal model. For the latter, a heating function that follows the optical emission was assumed. A comparison between the coronal temperature predicted by the model and that inferred from the optical lines is shown in Figure 6b for Hγ and the Ca II K line. There is a quite good agreement between predictions and observations for Hγ and other Balmer lines during a large part of the flare evolution. However, in the late flare decay, the coronal model predicts a much steeper temperature decrease suggesting that additional heating should be provided during the decay of this long-duration event. More disturbing is the fact that the coronal temperature inferred from the Ca II K line is substantially different from that inferred from the Balmer lines, *i.e.* a given coronal model is unable to reproduce simultaneously all the optical lines observed. This, as well as the large discrepancy found between the predicted and observed blue continuum, suggests that additional effects, such as consideration of non-thermal particles, should be added to the model to make it more consistent with observations.

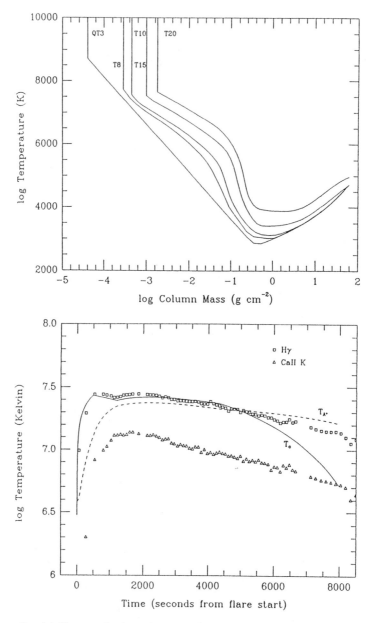

Figure 6. (a) Chromospheric models of stellar flares for different values of the coronal apex temperature. (b) Comparison of the time evolution of the coronal temperature inferred from Hγ and Ca II K line emission and that computed from a loop coronal model (full line). The dashed line represents the temperature of the equivalent static coronal loop used in computing chromospheric models (from Hawley and Fisher 1991).

5. Conclusions

In this Chapter, observational data on stellar flares have been summarized, and the far more limited models so far developed using the solar analogy have been discussed. From what has been presented, one could easily argue that what is really needed at this point is a major theoretical effort aiming at interpreting the data already obtained. One could even argue that the problem whether stellar flares are simply scaled-up versions of solar flares has already been solved. Neither of these two conclusions is probably right.

The empirical models discussed above, although crude and only partially successful in explaining the observational data, are nevertheless able to reproduce many of the observed characteristics of stellar flares. The problem, however, is that the observations themselves are insufficient to effectively constrain the models and to discriminate between alternative scenarios. Paradoxically, fitting a given model to stellar flares is, to a certain extent, simpler than for solar flares. Good examples are the hydrodynamic simulations of X-ray flares. They show quite a good agreement with observed light curves and global parameters such as average coronal temperatures and emission measures. However, without other observational constraints, such as line profiles and shifts or density diagnostics, it would be difficult to use the numerical simulations to obtain information on the mechanisms of energy release and the geometry of the flare.

The examples given in the previous sections also show that there are many uncertainties in the interpretation of the data. For instance, the origin of optical emission (both continuum and lines) is poorly understood, and the role played in these processes by non-thermal particles, heat conduction, gasdynamic effects and X-ray irradiation is still unclear. We would also like to understand why coherent processes apparently play a much larger role in stellar microwave emission than in solar flares. And, finally, we would like to know whether the larger energies involved in stellar flares are simply due to larger volumes and larger magnetic fields or, rather, are the consequence of processes that are fundamentally different from those operating on the Sun.

These, and many other questions, are likely to remain on the stage for many years to come. However, if any progress has to be made in this area, I believe that this can only be obtained through major advances on the observational side. Improved spectral and temporal resolution, and improved sensitivity, at optical, X-ray, UV, and radio wavelengths are clearly required. VLBI observations have the potential of providing information on the spatial structure of stellar flares, at least in those cases, such as RS CVn binaries, in which the source is large enough to be resolved. And, finally, observations of hard X-rays and γ-rays, although unlikely in the near future, would open a completely new perspective, thus allowing, for the first time, fundamental questions in stellar flare research to be addressed.

Acknowledgements. I thank Drs. G. Emslie and B. Pettersen for critical reading of the manuscript and useful comments.

References

Bastian, T.S.: 1990, *Solar Phys.* **130**, 265.

530

Bastian, T.S., and Bookbinder, J.A.: 1987, *Nature* **326**, 678.

Bastian, T.S., Dulk, G.A., Bookbinder, J., and Davis, M.: 1990, *Astrophys. J.* **353**, 265.

Bookbinder, J.A.: 1991, in *Stellar Flares* (B.R. Pettersen ed.), *Memorie Soc. Astron. Ital.* **62**, 321.

Bopp, B.W., and Espenak, F.: 1977, *Astron. J.* **82**, 916.

Bopp, B.W., and Fekel, F.Jr.: 1977, *Astron. J.* **82**, 490.

Bopp, B.W., and Moffett, T.J.: 1973, *Astrophys. J.* **185**, 239.

Bromage, G.E., Phillips, K.J.H., Dufton, P.L., and Kingston, A.E.: 1986, *Montly Not. Roy. Astron. Soc.* **220**, 1021.

Brown, J.C.: 1971, *Solar Phys.* **18**, 489.

Brown, J.C., Karlicky, M., MacKinnon, A.L., and van den Oord, G.H.J.: 1990, *Astrophys. J. Suppl.* **73**. 343.

Bruevich, V.V., Burnashev, V.I., Grinin, V.P., Kilyachkov, N.N., Kotyshevv, V.V., Shakovskaya, N.I., and Shavchenko, V.S.: 1980, *Izv. Krym. Astrofiz. Obs.* **61**, 90.

Butler, C.J.: 1991, in *Stellar Flares* (B.R. Pettersen ed.), *Memorie Soc. Astron. Ital.* **62**, 243.

Byrne, P.B.: 1983, in *Activity in Red-dwarf Stars* (P.B. Byrne and M. Rodonò eds.), Dordrecht: Reidel, p. 157.

Byrne, P.B.: 1989, *Solar Phys.* **121**, 61.

Byrne, P.B., and Rodonò, M. (eds.): 1983, *Activity in Red-dwarf Stars*, Dordrecht: Reidel.

Canfield, R.C., Gunkler, T.A., and Ricchiazzi, P.J.: 1984, *Astrophys. J.* **282**, 296.

Cheng, C.-C., Karpen, J.T., and Doschek, G.A.: 1985, *Astrophys. J.*, **286**, 787.

Cheng, C.-C., Oran, E.S., Doschek, G.A., Boris, J.P., and Mariska, J.T.: 1983, *Astrophys. J.* **265**, 1090.

Cheng, C.-C., and Pallavicini, R.: 1991, *Astrophys. J.* **381**, 234.

Cram, L.E.: 1982, *Astrophys. J.* **253**, 768.

Cram, L.E., and Woods, T.: 1982, *Astrophys. J.* **257**, 269.

Dennis, B.R., and Schwartz, R.A.: 1989, *Solar Phys.* **121**, 75.

de Jager, C., and 19 other authors: 1986, *Astron. Astrophys.* **156**, 95.

de Jager, C., and 19 other authors: 1989, *Astron. Astrophys.* **211**, 157.

Doyle, J.G., Butler, C.J., Byrne, P.B., and van den Oord, G.H.J.: 1988a, *Astrophys. J.* **193**, 229.

Doyle, J.G., Butler, C.J., Callanan, P.J., Tagliaferri, G., de la Reza, R., White, N.E., Torres, C.A., and Quast, G.: 1988b, *Astron. Astrophys.* **191**, 79.

Doyle, J.G., Byrne, P.B., and van den Oord, G.H.J.: 1989a, *Astron. Astrophys.* **224**, 153.

Doyle, J.G., van den Oord, G.H.J., and Butler, C.J.: 1989b, *Astron. Astrophys.* **208**, 208.

Dulk, G.A.: 1985, *Ann. Rev. Astron. Astrophys.* **23**, 169.

Feigelson, E.D., and Montmerle, T.: 1985, *Astrophys. J. Letters* **289**, L19.

Fisher, G.H., Canfield, R.C., and McClymont, A.N.: 1985a, *Astrophys. J.* **289**, 414.

Fisher, G.H., Canfield, R.C., and McClymont, A.N.: 1985b, *Astrophys. J.* **289**, 425.

Fisher, G.H., Canfield, R.C., and McClymont, A.N.: 1985c, *Astrophys. J.* **289**, 434.

Fisher, G.H., and Hawley, S.L.: 1990, *Astrophys. J.* **357**, 243.

Foing, B.H.: 1989, *Solar Phys.* **121**, 117.

Foing, B.M., and 15 other authors: 1991, in *Eruptive Solar Flares* (B.V. Jackson, M.E. Machado, and Z. Svestka eds.), Heidelberg: Springer-Verlag, in press.

Gary, D.E., Linsky, J.L., and Dulk, G.A.: 1982, *Astrophys. J. Letters* **263**, L79.

Gershberg, R.E.: 1989, *Memorie Soc. Astron. It.* **60**, 263.

Gershberg, R.E., and Shakhovskaya, N.I.: 1983, *Astrophys. Space Sci.* **95**, 235.

Giampapa, M.S.: 1983, in *Activity in Red-dwarf Stars* (P.B. Byrne and M. Rodonò eds.), Dordrecht: Reidel, p. 223.

Gibson, D.M.: 1983, in *Activity in Red-dwarf Stars* (P.B. Byrne and M. Rodonò eds.), Dordrecht: Reidel, p. 273.

Gondhalekar, P.M. (ed.): 1986, *Flares: Solar and Stellar*, Abingdon: Rutherford-Appleton Laboratories.

Grinin. V.P.: 1983, in *Activity in Red-dwarf Stars* (P.B. Byrne and M. Rodonò eds.), Dordrect: Reidel, p. 613.

Grinin, V.P.: 1991, in *Stellar Flares* (B.R. Pettersen ed.), *Memorie Soc. Astron. Ital.* **62**, 389.

Grinin, V.P., and Sobolev, V.V.: 1977, *Astrofizika* **13**, 587.

Grinin, V.P., and Sobolev, V.V.: 1988, *Astrofizika* **28**, 355.

Grinin, V.P., and Sobolev, V.V.: 1989, *Astrofizika* **31**, 527.

Gurzadyan, G.A.: 1980, *Flare Stars*, Oxford: Pergamon Press.

Gurzadyan, G.A.: 1988, *Astrophys. J.* **332**, 183.

Haisch, B.M.: 1983, in *Activity in Red-dwarf Stars* (P.B. Byrne and M. Rodonò eds.), Dordrecht: Reidel, p. 255.

Haisch, B.M., Linsky, J.L., Bornmann, P.L., Stencel, R.E., Antiochos, S.K., Golub, L., and Vaiana, G.S.: 1983, *Astrophys. J.* **267**, 280.

Haisch, B.M., and Rodonò, M. (eds.): 1989, *Solar and Stellar Flares*, Dordrecht: Kluwer.

Haisch, B.M., Strong, K.T., and Rodonò, M.: 1991, *Ann. Rev. Astron. Astrophys.* **29**, 275.

Havnes, O., Pettersen, B.R., Schmitt, J.H.M.M., and Solheim, J.E. (eds.): 1988, *Activity in Cool Star Envelopes*, Dordrect: Kluwer.

Hawley, S.L.: 1991, in *Stellar Flares* (B.R. Pettersen ed.), *Memorie Soc. Astron. Ital.* **62**, 271.

Hawley, S.L., and Fisher, G.H.: 1991, *Astrophys. J.*, in press.

Hawley, S.L., and Pettersen, B.R.: 1991, *Astrophys. J.* **378**, 725.

Hénoux, J.-C., Aboudarham, J., Brown, J.C., van den Oord, G.H.J., van Driel-Gesztelyi, L., and Gerlei, O.: 1990, *Astron. Astrophys.* **233**, 577.

Jackson, P.D., Kundu, M.R., and White, S.M.: 1989, *Astron. Astrophys.* **210**, 284.

Kahler, SD., and 30 other authors: 1982, *Astrophys. J.* **252**, 239.

Katsova, M.M., Kosovichev, A.G., and Livshits, M.A.; 1981, *Astrofisika* **17**, 285.

Kodaira, K.: 1977, *Astron. Astrophys.* **61**, 625.

Kopp, R.A., and Pneuman, G.W.: 1976, *Solar Phys.*, **50**, 85.

Kopp, R.A., and Poletto, G.: 1984, *Solar Phys.* **93**, 351.

Kuijpers, J: 1989, *Solar Phys.* **121**, 163.

Kundu, M.R., Pallavicini, R., White, S.M., and Jackson, P.D.: 1988, *Astron. Astrophys.* **195**, 159.

Lacy, C.H., Moffett, T.J., and Evans, D.S.: 1976, *Astrophys. J. Suppl.* **30**, 85.

Landini, M., Monsignori-fossi, B.C., Pallavicini, R., and Piro, L.: 1986, *Astron. Astrophys.* **157**, 217.

Lang, K.R.: 1990, in *Flares Stars in Star Clusters, Associations and Solar Vicinity* (L.V. Mirzoyan *et al.* eds.), Dordrecht: Kluwer, p. 125.

Lang, K.R., Bookbinder, J., Golub, L., and Davis, M.: 1983, *Astrophys. J. Letters* **272**, L15.

Lang, K.R., and Willson, R.F.: 1986, *Astrophys. J.* **305**, 363.

Lang, K.R., and Willson, R.F.: 1988, *Astrophys. J.* **326**, 300.

Lestrade, J.-F.: 1988, in *The Impact of VLBI on Astrophysics and Geophysics* (M.J.Reid and J.M. Moran eds.), Dordrecht: Kluwer, p. 265.

Lestrade, J.-F., Mutel, R.L., Preston, R.A., and Phillips, R.B.: 1988, *Astrophys. J.* **328**, 232.

Linsky, J.L.: 1991, in *Stellar Flares* (B.R. Pettersen ed.), *Memorie Soc. Astron. Ital.* **62**, 307.

Linsky, J.L., Neff, J.E., Brown, A., Gross, B.D., Simon, T., Andrews, A.D., Rodonò, M., and Feldman, P.A.: 1989, *Astron. Astrophys.* **211**, 173.

Livshits, M.A., Badalyan, O.G., Kosovichev, A.G., and Katsova, M.M.: 1981, *Solar Phys.* **73**, 269.

Machado, M.E., Emslie, A.G., and Brown, J.C.: 1978, *Solar Phys.* **58**, 363.

MacNeice, P., McWhirter, R.W.P., Spicer, D.S., and Burgess, A.: 1984, *Solar Phys.* **90**, 357.

Mauas, P.J.D., Machado, M.E., and Avrett, E.H.: 1990, *Astrophys. J.* **360**, 715.

Mirzoyan, L.V., Pettersen, B.R. and Tsvetkov, K.K. (eds.): 1990, *Flare Stars in Star Clusters, Associations and Solar Vicinity*, Dordrecht: Kluwer.

Mochnacki, S., and Zirin, H.: 1980, *Astrophys. J. Letters* **239**, L27.

Moffett, T.J., and Bopp, B.W.: 1976, *Astrophys. J. Suppl.* **31**, 61.

Mullan, D.J.: 1976, *Astrophys. J.* **207**, 289.

Mullan, D.J.: 1977, *Solar Phys.* **54**, 183.

Mullan, D.J.: 1984, *Astrophys. J.* **279**, 746.

Mullan. D.J.: 1989, *Solar Phys.* **121**, 239.

Mutel, R.L., Lestrade, J.F., Preston, R.A., and Phillips, R.B.: 1985, *Astrophys. J.* **254**, 641.

Mutel, R.L., Morris, D.H., Doiron, D.J., and Lestrade, J.-F.: 1987, *Astron. J.* **93**, 1220.

Nagai, F.: 1980, *Solar Phys.* **68**, 351.

Nagai, F., and Emslie, A.G.: 1984, *Astrophys. J.* **279**, 896.

Neff, J.E.: 1991, in *Stellar Flares* (B.R. Pettersen ed.), *Memorie Soc. Astron. Ital.* **62**, 291.

Neidig, D.F.: 1989, *Solar Phys.* **121**, 261.

Nelson, G.J., Robinson, R.D., Slee, O.B., Ashley, M.C.B., Hyland, A.R., Touhy, I.R., Nikoloff, I., and Vaughan, A.E.: 1986, *Montly Not. Roy. Astron. Soc.* **220**, 91.

Pallavicini, R.: 1991, in *Eruptive Solar Flares* (B.V. Jackson, Machado,M.E., and Z. Svestka eds.), Heidelberg: Springer-Verlag, in press.

Pallavicini, R., Peres, G., Serio, S., Vaiana, G.S., Acton, L., Leibacher, J., and Rosner, R.: 1983, *Astrophys. J.* **270**, 270.

Pallavicini, R., Serio, S., and Vaiana, G.S.: 1977, *Astrophys. J.* **216**, 108.

Pallavicini, R., and Tagliaferri, G.: 1989, in *Solar and Stellar Flares - Posters Papers* (B.M. Haisch and M. Rodonò eds.), Catania: Astrophysical Observatory, p. 17.

Pallavicini, R., Tagliaferri, G., Pollock, A.M.T., Schmitt, J.H.M.M., and Rosso, C.: 1990a, *Astron. Astrophys.* **227**, 483.

Pallavicini, R., Tagliaferri, G., and Stella, L.: 1990b, *Astron. Astrophys.* **228**, 403.

Peres, G.: 1989, *Solar Phys.* **121**, 289.

Peres, G., Reale, F., Serio, S., and Pallavicini, R.: 1987, *Astrophys. J.* **312**, 895.

Pettersen, B.R.: 1988, in *Activity in Cool Star Envelopes* (O. Havnes *et al.* eds.), Dordrecht: Kluwer, p. 49.

Pettersen. B.R.: 1989, *Solar Phys.* **121**, 299.

Pettersen, B.R. (ed.): 1991, *Stellar Flares*, Florence: Società Astronomica Italiana.

Pettersen, B.R., Colemann, L.A., and Evans, D.S.: 1984, *Astrophys. J. Suppl.* **54**, 375.

Poletto, G.: 1989, *Solar Phys.* **121**, 313.

Poletto, G., Pallavicini, R., and Kopp, R.A.: 1988, *Astron. Astrophys.* **201**, 93.

Phillips, K.J.H., Bromage, G.E., and Doyle, J.G.: 1991, *Astrophys. J.*, in press.

Phillips, R.B., Lonsdale, C.J., and Feigelson, E.D.: 1991, *Astrophys. J.*, in press.

Reale, F., Peres, G., Serio, S., Rosner, R., and Schmitt, J.H.M.M.: 1988, *Astrophys. J.* **328**, 256.

Robinson, R.D.: 1989, in *Solar and Stellar Flares - Poster Papers* (B.M. Haisch and M. Rodonò eds.), Catania: Astrophysical Observatory, p. 83.

Rodonò, M., and 9 other authors: 1989, in *Solar and Stellar Flares - Poster Papers* (B.M. Haisch and M. Rodonò eds.), Catania: Astrophysical Observatory, p. 53.

Saar, S.H.: 1990, in *High-resolution Spectroscopy in Astrophysics* (R. Pallavicini ed.), *Memorie Soc. Astron. Ital.* **61**, 559.

Shakhovskaya, N.I.: 1989, *Solar Phys.* **121**, 375.

Simnett, G.M.: 1991, in *Stellar Flares* (B.R. Pettersen ed.), *Memorie Soc. Astron. Ital.* **62**, 359.

Somov, B.V.: 1975, *Solar Phys.* **42**, 235.

Spangler, S., and Moffett, T.: 1976. *Astrophys. J.* **203**, 497.

Svestka, Z.: 1976, *Solar Flares*, Dordrecht: Reidel.

Tagliaferri, G., Giommi, P., Angelini. L., Osborne, J.P., and Pallavicini, R.: 1988, *Astrophys. J. Letters* **331**, L113.

Tsuru, T., and 12 other authors: 1989, *Publ. Astron. Soc. Japan* **41**, 679.

van den Oord, G.H.J.: 1988, *Astron. Astrophys.* **207**, 101.

White, N.E., Culhane, J.L., Parmar, A.N., Kellet, B.J., Kahn, S., van den Oord, G.H.J., and Kuijpers, J.: 1986, *Astrophys. J.* **301**, 262.

White, S.M., Kundu, M.R., and Jackson. P.D.: 1986, *Astrophys. J.* **311**, 814.

White, S.M., Pallavicini, R., and Kundu, M.R.: 1991, *Astrophys. J.*, in press.

Worden, S.P.: 1983, in *Activity in Red-dwarf Stars* (P.B. Byrne and M. Rodonò eds.), Dordrecht: Reidel, p. 207.

Zarro, D., Canfield, R.C., Strong. K.T., and Metcalf, T.R.: 1988, *Astrophys. J.* **324**, 582.

27. PHYSICS OF FLARES IN STARS AND ACCRETION DISKS

J. KUIJPERS[1]
Sterrekundig Instituut
Rijksuniversiteit Utrecht
P.O. Box 80 000
3508 TA Utrecht
The Netherlands

ABSTRACT. This is a review of the physics of magnetic flares in the Sun. The details of the energy release processes in a solar flare are, at present, not clear and do not form a subject of this Chapter. Rather, we point out fundamental aspects of the conversion of kinetic into magnetic energy. We show the physical conditions required for magnetic flares to appear. Particular attention is paid to force free equilibria, energy storage, the amount of liberated energy, resistivity, and magnetic helicity. These findings are then applied to other stars, in particular to accretion disks around magnetic neutron stars.

1. Introduction

Solar flares (Haisch and Rodonò 1989; Haisch, Strong, and Rodonò 1991) are eruptive phenomena in the solar atmosphere. They consist of an explosive release of stored magnetic energy in a magnetically dominated environment, the solar corona. The magnetic structure which holds the free energy is anchored in a dense streaming plasma below the photosphere. Typically, the energy release occurs on a timescale of a few times the Alfvén crossing time of the relaxing field structure. The released amount of energy can be comparable to that of the unperturbed magnetic structure although often its value is much less.

Eruptive phenomena are observed in other astrophysical objects as well and in many of them magnetic fields are at play. In particular, powerful bursts of X-rays and intense radio emission have been observed to occur in close binaries, with and without compact objects, in pre-main sequence stars, and in active galactic nuclei (Kuijpers 1989). The question then arises whether these explosions are governed by similar underlying physics as solar flares.

We shall first summarize our understanding of solar flares. Investigation of the physical building blocks allows us to extrapolate the magnetic flare phenomenon to other objects in

---

[1]Also at CHEAF (Centrum voor Hoge Energie Astrofysica), P.O. Box 41882, 1009 DB Amsterdam and at Discipline Natuurkunde, Katholieke Universiteit Nijmegen, Toernooiveld 1, 6525 ED Nijmegen.

J. T. Schmelz and J. C. Brown (ed.), The Sun, A Laboratory for Astrophysics, 535–597.

the universe. We shall see that, from a theoretical point of view, magnetic explosions are the natural course of events in a number of quite different objects which have the following elements in common:

1. The system is stratified: it consists of a *dense driver* and an adjacent *dilute magnetic corona*;

2. The following ordering of time scales exists: The time scale for structural changes at the boundary between driver and corona t_v is less than the *Ohmic decay time* of the system t_η and larger than the travel time of magnetic signals across the coronal structure t_A:

$$t_\eta \gg t_v \gg t_A. \tag{1}$$

Here the *flow crossing time* t_v is determined by differential motions at the boundary:

$$t_v \equiv \ell/\Delta v, \tag{2}$$

ℓ is the characteristic dimension of the structure at the boundary and Δv is the corresponding velocity difference. The Ohmic decay time t_η is large because of the large extent of the astrophysical object (and not because of the value of the collisional conductivity which is only that of copper at room temperature):

$$t_\eta \equiv \mu_0 \ell_B^2/\eta, \tag{3}$$

where η is the resistivity ($\eta = \sigma^{-1}$, σ is the conductivity) and ℓ_B is the characteristic dimension of the structure (in general $\ell_B > \ell$). The *Alfvén time* t_A is defined as

$$t_A \equiv \frac{\ell}{c_A}, \tag{4}$$

where c_A is the characteristic Alfvén speed;

3. The *driver is dynamically dominated by matter.* Because of the relatively large Ohmic decay time in comparison with the flow crossing time, kinetic energy is converted into magnetic energy. In other words, the driver can act as a dynamo;

4. The *corona is magnetically dominated:* Its gas pressure p_{gas} is much less than the magnetic pressure $p_{mag} = B^2(2\mu_0)^{-1}$. In terms of the plasma beta, where

$$\beta \equiv \frac{p_{gas}}{p_{mag}}, \tag{5}$$

this is expressed as

$$\beta \ll 1. \tag{6}$$

Alternatively, this condition can be rephrased as a highly supersonic Alfvén speed. In a static situation, pressure gradients across the coronal magnetic field are then balanced by a small deviation of the electric current direction from the magnetic field direction. To a first approximation, the corona is, therefore, *force free*

$$\vec{j} \times \vec{B} = 0;$$ (7)

5. There is a continuing power transfer from the driver to the corona in the form of a Poynting flux. This *"electrodynamic" coupling* is caused by the perturbing action of the driver on the coronal magnetic fields. It implies a large dynamic pressure in the driver in comparison with the magnetic pressure in the corona.

As we shall see, these conditions set up a *relaxation type behaviour:* The interplay between driver and corona causes a continuous energization of coronal magnetic structures while the magnetic properties of the corona ensure a periodic *catastrophic release of magnetic energy in the form of heating, particle acceleration, expulsion of plasmoids, and waves.* We shall first demonstrate the necessity of the above conditions in turn and then summarize our understanding of the flare phenomenon and the remaining problems.

2. Conversion of Kinetic into Magnetic Energy

If the *driver moves slowly* through an ambient (low-beta) plasma in comparison with the Ohmic time scale, the perturbed magnetic field has ample time to diffuse. The motion, therefore, gives rise to resistive heating and *not much energy is stored in ambient fields.*

If, on the other hand, *the driver moves fast enough through the corona,* initially resistive dissipation cannot keep pace with the power input from the driver (the first inequality in Equation (1)). Then, *kinetic energy is converted into electromagnetic energy.*

Let us derive an expression for the power input at the boundary in the ideal, non-resistive, case. The temporal change of the electromagnetic energy density

$$W_{EM} \equiv \frac{\epsilon_0 E^2}{2} + \frac{B^2}{2\mu_0}$$

in a non-moving volume V with bounding surface S is described by *Poynting's theorem*

$$\frac{dW_{EM}}{dt} = -\int_V \vec{E} \cdot \vec{j} \, dV - \oint_S (\vec{E} \times \vec{B}) \cdot \frac{d\vec{S}}{\mu_0},$$ (8)

where $\vec{E}$ is the electric field and $\vec{B}$ the magnetic field. Equation (8) follows directly from Maxwell's equations by taking the inner product of Ampère's law with $\vec{E}$, and subtracting from it the inner product of Faraday's law with $\vec{B}$. The first term on the right-hand side of Equation (8) gives the rate of change in electromagnetic energy by the work of the electric field on the electric currents inside the volume. The second term represents the flow of electromagnetic energy or Poynting flux,

$$\vec{F}_{EM} = \frac{\vec{E} \times \vec{B}}{\mu_0}$$ (9)

through the surface. In the ideal mhd approximation, valid for low-frequency motion, the electric energy can be neglected in comparison with the magnetic energy: $W_{EM} \approx W_M \equiv \int_V B^2(2\mu_0)^{-1}dV$ and we can put $\vec{E} = -\vec{v} \times \vec{B}$ so that

$$\frac{dW_M}{dt} = -\int_V (\vec{j} \times \vec{B}) \cdot \vec{v}dV + \oint_S \frac{\vec{B} \cdot \vec{v}\vec{B} \cdot d\vec{S}}{\mu_0} - \oint_S \frac{B^2\vec{v} \cdot d\vec{S}}{\mu_0}. \tag{10}$$

The first term on the right-hand side corresponds to the first term in Equation (8), and is the work done by the electric field on the current. In the ideal mhd case, it consists entirely of the reversible work done by the Lorentz force; in the resistive case, there is also a contribution from irreversible Ohmic dissipation. The second term shows how shear on the boundary causes transfer of energy provided the magnetic field extends through the boundary and provided the field has a component in the direction of the velocity component along the boundary. The final term shows how the emergence of new magnetic flux through the boundary changes the electromagnetic energy. At a corona-driver interface, characteristically all three terms are important, as both effects (shear and emergence) occur. The form of the electromagnetic energy now depends on the ratio of the Alfvén crossing time (Equation (4)) to the flow crossing time (Equation (2)).

If the *Alfvén crossing time is relatively large* the moving object excites

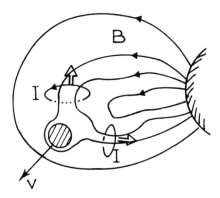

Figure 1: Radiation of mhd waves by a well-conducting body moving quickly $(t_A > t_v)$ through a magnetosphere.

magnetohydrodynamic disturbances in the surrounding plasma (Neubauer 1980; Barnett and Olbert 1986; Dobrowolny and Veltri 1986; Scheurwater and Kuijpers 1988) (see Figure 1).

If, on the other hand, the reverse is true and the second *inequality in Equation (1) is satisfied, the corona has ample time to adjust itself* to the new position of the field lines at the boundary as dictated by the driver. Then, instead of generating mhd waves or *Alternating Currents*, the motion sets up a *Direct Current pattern*. Since the coronal plasma is low-beta, it will be characterized by a slowly-changing sequence of states, each of which, in the limit of very long flow-crossing times (slow motion), approaches a *force free equilibrium* (Equation (7)). (Note that the static condition cannot be applied to arbitrarily large altitudes where the signal travel time becomes larger than the flow crossing time at the boundary. There, Lorentz forces must be balanced by inertia. The question then arises whether for a steady flow at the boundary the corona asymptotically approaches a steady state where the power input from the

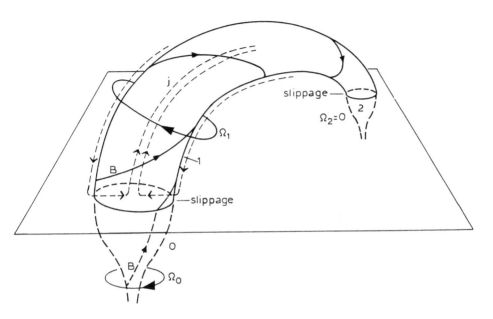

Figure 2: Hypothetical steady rotating loop, driven by rotation ($\Omega_2 > \Omega_1$) of the subphotospheric portion of the tube at one footpoint. For simplicity, the subphotospheric part at the other end is kept fixed ($\Omega_2 = 0$). Finite resistivity at the transition region allows the current density $\vec{j}$ (dashed lines) to close across the field (oblique drawn line) and causes slippage between parts 1 and 0 and between 1 and 2. The current is distributed throughout the tube volume.

driver is balanced by dissipation elsewhere in the electric circuit (see Figure 2). We shall show that such a steady state is unrealistic for a large variety of astrophysical objects provided they have a force free corona.

3. Force Free Equilibria, the Virial Theorem and Geometromagnetics

A static corona in force free equilibrium can only store a finite amount of magnetic energy which is dictated by the magnetic field distribution at its bounding surface (Aly 1985; Low 1986). This can be seen from application of the scalar magnetic virial theorem (see Chapter 6 for its derivation) to the force free part of the magnetic structure. Let us take two concentric spheres, one just above the photosphere in the corona (see Figure 3) and the other with radius going to infinity. We now integrate the virial theorem over the volume between the two spheres.

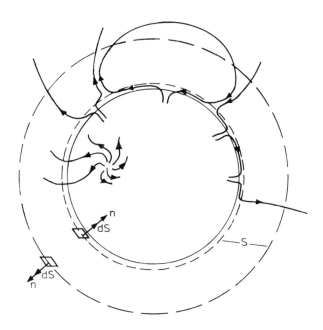

Figure 3: The figure shows the spheres of integration (dashed) used in Equation (11) in a force free corona surrounding a spherical driver, some field lines (drawn) rooted in the driver and extending into the corona, and surface elements $d\vec{S}$ with normal vector $\vec{n}$.

Note that we can neglect the non-magnetic terms: the time derivative becomes immaterial as we consider an equilibrium; the kinetic energy can be neglected as the equilibrium is static (Note that this is not true for a wind); the gravitational energy in the dilute corona is negligible in comparison with the magnetic energy; similarly, in a low-beta plasma, the internal energy of the gas does not contribute and, finally, the pressure term at the bounding surfaces can be neglected if we choose the spheres to lie inside the force free corona. Further, the magnetic contribution from the upper surface varies as $B^2(r)r^3$ and vanishes for a static equilibrium when r goes to infinity. The result then takes on the very simple form:

$$\int_V \left\{ \frac{B^2}{2\mu_0} \right\} dV =$$

$$= \oint_{S(R)} \left\{ \left(\frac{B^2}{2\mu_0} \right) \vec{r} \cdot d\vec{S} - \frac{(\vec{B} \cdot \vec{r})(\vec{B} \cdot d\vec{S})}{\mu_0} \right\}. \tag{11}$$

This tells us that the total magnetic energy inside the corona W_M can be expressed as the surface integral of a simple function of the magnetic field:

$$W_M = \frac{2\pi R^3}{\mu_0} (< B_n^2 > - < B_t^2 >), \tag{12}$$

where B_n is the field component normal to the bounding surface, B_t is the tangential field component, and the angular brackets denote averaging over the lower surface. (Note: In plane geometry, one finds, similar to Equation (12), the expression $W_M = \int(xB_x + yB_y)B_z \, dx \, dy/\mu_o$ where the integral extends over the infinite plane $z = 0$. One is not justified in limiting the integration to a finite observed area only (Hofmann 1990).)

Below we shall draw two important conclusions from Equation (12), one concerning the *inevitability of energy storage* in a force free corona, the other concerning the *maximum amount of energy which can be liberated* in a single concerted magnetic explosion. In passing, we note that work has also been done on non-force free static equilibria where pressure forces are important (Zwingmann 1987). Further, steady dynamic equilibria, where Lorentz forces are balanced by inertial forces, have not yet received much attention.

3.1. ENERGY IS STORED IN THE CORONA

Because magnetic energy is a positive quantity, *the tangential magnetic field component can never exceed the average normal field component* on the bounding surface. This is the result of the force free nature of the field: As one stresses the field at the surface, energy is transported into the corona. Now inside the corona the shape of the field lines is determined by the force free condition. This implies that the angle of the field at the surface, as given by the ratio B_t/B_n, and, therefore, the shear of the field at the surface, cannot be imposed arbitrarily but is determined self-consistently by the force free condition. As the energy content of the structure increases, the field lines expand (see Figure 4). In the

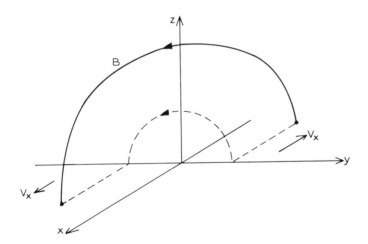

Figure 4: Energization and expansion of a coronal ($z \leq 0$) magnetic arcade by relatively slow subphotospheric velocity shear.

ideal case (in the absence of resistivity), the field lines open up asymptotically towards infinity as sketched in Figure 5 with current sheets (Aly 1991). As the surface fields open, the contribution from $\vec{B} \cdot \vec{v}$ vanishes along the boundary surface (see Equation (10))

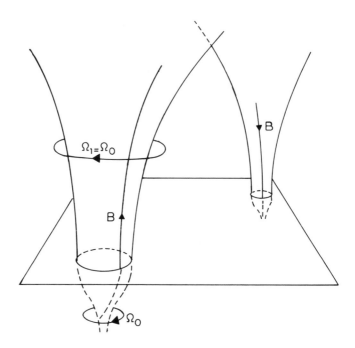

Figure 5: Realistic asymptotic state for continuing rotation of a coronal flux tube at one end. Instead of Figure 2 the fields open up to infinity. The front tube is rotating with the applied frequency Ω_0 and the tube at the back is standing still (assuming infinite inertia below that end). A current sheet occurs between both tubes in the high corona.

and no further energy is transmitted by the driver into the corona. (Note that the condition of Equation (1b) cannot be satisfied at arbitrarily large altitude when the structure becomes infinite. Therefore, some energy continues to be transmitted to the corona in the form of Alfvén waves and the quantity $\vec{B} \cdot \vec{v}$ does not vanish entirely on the boundary.)

In earlier calculations (Heyvaerts *et al.* 1982), a catastrophe appeared during a slow but large increase of the value B_t/B_n at the boundary. As we have seen, one does not have the freedom to impose large values for this quantity at the boundary of a force free atmosphere. Therefore, these initial results, however interesting, cannot be applied to solar flares.

The property of force free flux tubes to expand when energized at one end is also responsible for the *poor feedback of stresses* to the other end of a flux tube below the photosphere. The rate of angular momentum flow per unit area communicated by magnetic stresses in a magnetic flux tube rotated from below is proportional to the magnetic stress $B_z B_\phi$ (z is the vertical coordinate and ϕ the azimuthal angle). If the coronal field were not force free but could be constrained by sufficiently strong body forces, the coronal part of the flux tube would become strongly twisted so that the stresses at the other footpoint become equal (but opposite) to those applied, as is sketched in Figure 6a. Then the

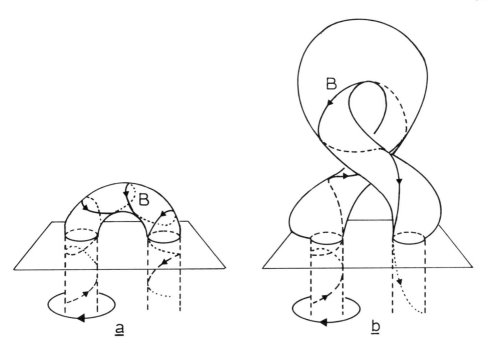

Figure 6: The tightly wound tube in (a) is not force free. In a force free corona, a flux tube expands upon being twisted at one end as in (b). As a result, the returned stress at the other footpoint remains less than the applied one.

angular momentum flux traveling up on one side of the tube would completely go down at the other end and be returned to the subphotospheric plasma. In the force free case, however, (Figure 6b) a flux tube becomes kink unstable if twisted too much. As a result, the *average stress remains limited* to a value $B_z B_\phi \leq B_z^2$. A similar result is valid for the momentum transport in a sheared arcade.

This force free limit on the twist has interesting consequences for the observations of twists in magnetic fields. The limit on the perturbed field component applies to any force free surface, in particular to the top of a flux tube where the fields are weakest. The torque free condition requires the angular momentum transport ($\propto B_\phi B_z r^3$) along the force free part of the tube (radius r) to remain constant. Conservation of flux ($\propto B_z r^2$) then implies that the twist as defined by $B_\phi (B_z r)^{-1}$ is constant along the loop. The shear at the base, therefore, has an upper limit

$$\left. \frac{B_\phi}{B_z} \right|_{pho} \lesssim \frac{r_{pho}}{r_{cor}}, \tag{13}$$

where the r's are radii of the tube in the photosphere and corona respectively (Figure 7). The right-hand side

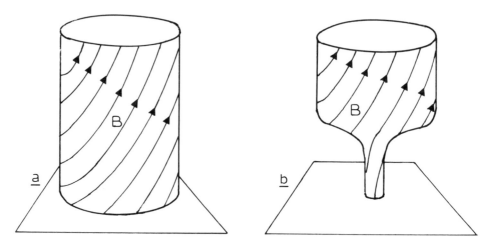

Figure 7: Because of the torque free condition in a force free flux tube, the value B_ϕ/B_z decreases upon constriction of the tube (b) and is directly related to the filling factor at the line-tying "surface." In (a) the filling factor is unity.

is the square root of the *filling factor* of the solar photosphere with magnetic fields (which is a few per cent). Therefore, an upper limt to the shear of rotating tubes at photospheric levels is of order 0.1 in the solar case. (For arcades, a similar reasoning shows that the shear can go up to unity.)

3.2. THE MAXIMUM LIBERATED ENERGY

It follows from Equation (12) that the *maximum amount of magnetic energy which can be stored* in a force free corona is of order

$$W_M \lesssim \frac{2\pi R^3}{\mu_0} < B_n^2 > . \tag{14}$$

The right-hand side is of the order of the energy in a potential field which surrounds a sphere and has a strength B_n on the surface. Physically, this means that the maximum amount of stored energy is entirely determined by the field values at the shearing surface and by the geometry. The best the shearing motions can do is to build up the field (to a value of order of the boundary value) in a sheath surrounding the "anchoring" surface with a scale height comparable to the linear dimension of the (finite) object. For larger altitudes, the geometry becomes spherical and the force free nature ensures that the fields expand radially so that the stored energy in these far-out regions stays negligible. This result suggests that the *energy which can be stored* above an isolated photospheric field area S of average field strength B at the photosphere, is

$$W_M = S H \frac{B^2}{2\mu_0}, \tag{15}$$

where the scale height H is of order $H \approx S^{1/2}$. If the magnetic area is not isolated but surrounded by regions of similar field strength, the expansion of the coronal fields above the specific area is hindered by the neighbouring structures. Then the scale height to be used in the estimate Equation (15) may go up. However, a firm upper limit is always given by the radius of the star or, more generally, the linear dimension of the driver by which the fields are disturbed.

Numerical calculations of two-dimensional magnetic arcades driven by shear at the footpoints show that an ideal instability can occur when the stored energy is a few times the potential energy (Biskamp and Welter 1989; Biskamp 1989; Mikić, Barnes, and Schnack 1988; Mikić, Schnack, and Van Hoven 1990; Schnack, Mikić, and Barnes 1990), quite consistent with the above estimates. However, the (ideal) instability only appears if the sheared structure is embraced sideways either by other dipolar structures (see Figure 8) or by an exactly periodic pattern. This is consistent with analytical considerations (Aly 1990) which show that an isolated arcade just opens up asymptotically but does not become ideally unstable (It may, however, become resistively unstable). Finally, as the energy release is driven by an ideal mhd instability, the *time scale is a number of Alfvén travel times* consistent with the observations of solar flares.

Van Tend and Kuperus (1978) have developed a simple flare model based on a *line current* suspended magnetically. In this limit, the system is again force free. As the current increases above a certain value, a *catastrophe* appears. This early model, which has been elaborated by Kaastra (1985), van den Oord (1988), Martens and Kuin (1989), and Forbes and Isenberg (1991), shows a similar behaviour to the numerical calculations of the spatially distributed force free systems: As the line current increases, the structure *expands*. Further, the characteristic free energy per unit length at the moment of instability is $H^2 B^2(H)/2\mu_0$ in agreement with Equation (15).

At present, it is not settled whether the energy release in solar flares is always caused by an ideal mhd instability or, for some class of flares, the trigger is a resistive instability (Sakurai 1989; Aly 1990) or if a loss of equilibrium is at play (Klimchuk and Sturrock 1989). We also mention that non-force free models for flares exist (Zwingmann 1987). We note, however, that, in view of the extreme topological complexity of solar magnetic fields, ideal instabilities can occur on a variety of scales and could produce various kinds of observed flares such as large (two-ribbon) flares, smaller (compact) flares, miniflares, microflares, and nanoflares (Schadee, de Jager, and Svestka 1983; Schadee 1986; Dere 1990; Vlahos 1990). The different geometries of the fields completely determine the flares and their energies (see Equation (15)). If this is true, the global evolution of flares can be described by simple *"geometromagnetics."*

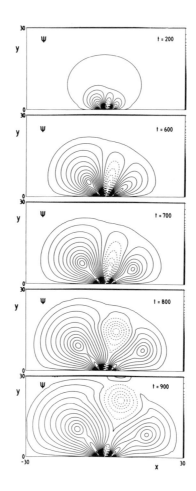

Figure 8: Evolution of a triple arcade configuration. The Ψ contours are the field line projections on a vertical plane (Biskamp and Welter 1989).

3.3. EMERGING FLUX

Most of the theoretical studies consider shearing or rotating flows and their effect on existing flux tubes. However, new emergence of magnetic flux is known to play an important role in the Sun. Upon emergence, flux tubes are seen to be swept up along the boundaries of supergranular flows. Often, flares are known to be associated with emergence of magnetic flux (Zirin 1983). It has, therefore, been suspected for a long time that such an emerging flux region causes the existing field structure to become ideal mhd unstable. While these effects must be included, they extend, but do not invalidate, the importance of the above considerations.

It is sometimes (Melrose 1990) claimed that storage of energy in the corona is not important for flares. In our opinion, a robust argument exists against such a claim. If storage were unimportant, the liberated energy should be delivered during the flare from subphotospheric layers. If the energy transport is along preexisting magnetic field channels, the energy propagation speed is limited to the Alfvén speed in a low-beta plasma. If, on the other hand, the energy is in rising flux tubes, the upward velocity is limited to a (hybrid) Alfvén speed. The reason for the latter limitation is that a rising flux tube below the photosphere experiences an aerodynamic drag in the surrounding gas of order ρv^2 per unit area perpendicular to the rising tube surface which counteracts the buoyancy lift $B^2(2\mu_0)^{-1}$. Therefore, subphotospherically stored magnetic energy can never appear in the corona at a rate substantially exceeding its characteristic magnetic energy density times the surface area times the minimum Alfvén speed along its path. For an individual rising flux tube of radius 500 km, field strength 0.1 T, and Alfvén speed of 50 km sec^{-1} in the photosphere, an upper limit to the energy flow by emergence into the corona is of order 10^{20} W. This power is smaller than the observed power from a large solar flare, 10^{22} W during 10^3 sec.

3.4. CONVERGING FLOWS

Possibly converging flows are more important for the storage of magnetic energy in the solar corona than the shearing or rotating flow patterns (Kuperus and van Tend 1981; Machado et al. 1988). In the case of prominences, accurate observations (Martin 1990) show that prominences are situated above regions of converging flow while shear flow, on the other hand, is not always observed. In fact, a very simple argument (Kuijpers 1990b) demonstrates that the *free energy in a flux tube continuously increases*

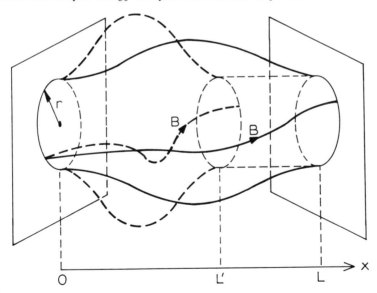

Figure 9: Shortening of a twisted flux tube with rigid boundary wall conditions. The tube is surrounded by untwisted field and the plasma is ideal.

if one tries to shorten the tube length. Figure 9 demonstrates that, for an ideal plasma, the free energy inside a line-tied current-carrying flux tube varies as

$$\frac{B_\phi^2}{2\mu_0}\pi r^2 L \propto B_z^2 \frac{r^4}{L} \propto \frac{1}{L} \tag{16}$$

because of flux freezing. The free energy, therefore, increases as the tube (radius r, length L) shortens (irrespective of its thickening). Figure 10 shows how upwelling flux

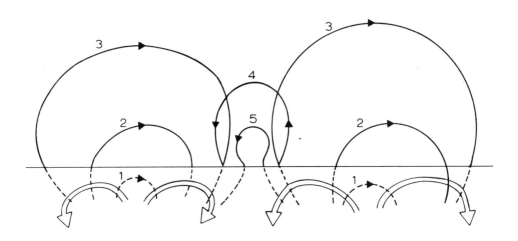

Figure 10: The thick arrows indicate convection below the surface. Magnetic fields rise to the surface $(1 \rightarrow 2)$, then converge and reconnect $(3 \rightarrow 4)$ and finally become more and more energized $(4 \rightarrow 5)$ provided the tube contains a net current at stage 4.

tubes become distorted, reconnect (resistivity needed !), and are progressively shortened as their footpoint motion is determined by subphotospheric flows. The free energy of the loop at stage 5 exceeds that at stage 4 (Equation 16) and continues to increase if no flaring occurs.

The above results are only valid in an *ideal or infinitely conducting non-viscous plasma.* We have entirely *neglected resistive effects* on the basis of the first inequality in Equation (1). Moreover, ideal instabilities can only lead to conversion of magnetic into kinetic energy but not to heating and particle acceleration. In the following sections, we shall consider resistive effects.

4. The Impossibility of Steady Slipping Circuits

We are now in a position to decide if the situation of Figure 2 with slippage is realistic in the corona. As a first correction to the global ideal evolution, we shall consider the effect of a *realistic quiescent resistivity.* The resistivity in the Sun is largest in a shallow surface layer near the temperature minimum. At this region, the ions become demagnetized (the ion-electron collision frequency becomes larger than the ion cyclotron frequency) so that the particles can follow potential drops across the magnetic field. At larger altitudes, charged particles are still glued to the field lines. Below this region, the conductivity increases because of the rising temperature. In the coolest parts of sunspots, the electric resistivity is estimated at 1 ohm m (Kovitya and Cram 1983).

We, therefore, introduce a thin resistive layer at the photosphere. In such a layer, the fields are no longer frozen into the plasma but can slip with respect to the gas. We want to investigate if such a resistive layer can prevent the expansion of the loop in Figures 5 and 6b. In particular, we want to *find out if a steady state exists with a moderate value of the electric current* (small expansion, small value of B_ϕ/B_z) as sketched in Figure 2.

In such a steady state, one submerged end of the tube is rotating at the prescribed angular frequency Ω_0. This is the end where the "external" torque is applied. Since there is a layer of finite resistivity between the submerged and the coronal portion of the tube, there is *slippage* between these parts at footpoint 0. The coronal part rotates at a speed $\Omega_1 < \Omega_0$. At the other footpoint, there is again slippage between the coronal and the submerged part. For simplicity, we take the second submerged part to be static $\Omega_2 = 0$, corresponding to an infinite submerged mass. We further make the simplifying assumption that the shape of the flux tube is not changed by the electric currents and that the tube is in "solid body" rotation around its axis. This, of course, amounts to relaxing the force free condition in the coronal part and to making an appeal to *unspecified body forces* to keep the shape fixed. The aim of our gedanken experiment is to find out how large the electric current and its associated magnetic field are if one insists on a stationary state. If the field created by the current is small compared to the potential field, a steady force free configuration exists resembling the non-force free tube with prescribed shape. *If, however, the magnetic field from the "steady state" current is much larger than the potential field, the resistivity is too small (for the given dimensions and velocities) to prevent the tube from substantial expansion.*

In a steady state, the electric current and the rotation frequency Ω_1 are completely determined by Faraday's equation and Ohm's law, given the external rotation frequency Ω_0, the geometry of the background field, and the resistivity. The coronal electric current system consists of a body current through the tube which crosses the magnetic field in the resistive layer at one footpoint and then returns as a surface current to the other footpoint where it closes across the field in the resistive layer. In a steady state, any coronal magnetic field line is an electric equipotential since the resistivity along the static coronal field can be neglected in comparison to the resistivity of the surface layer. Further, in a steady state, Faraday's law implies that the tangential electric field is continuous at footpoints. The potential difference between the axis and the surface of the cylinder is, therefore, the same, whether measured over the resistive layer at footpoint 0, in the corona, or over the resistive layer at footpoint 2:

$$V_0 = V_1 = V_2. \tag{17}$$

In Equation (17), one recognizes Kirchhoff's law for electric circuits. Using Ohm's law in an isotropic plasma (for simplicity) $\vec{E} = -\vec{v} \times \vec{B} + \vec{j}/\sigma$ one has

$$V_0 = \Phi\Omega_0(2\pi)^{-1} - IR_0, \tag{18}$$

$$\mathcal{V}_1 = \Phi\Omega_1(2\pi)^{-1},$$
$$\mathcal{V}_2 = I R_2,$$

where Φ is the magnetic flux contained in the flux tube, I is the total current crossing the flux tube in each resistive layer at its outer radius, a formal total resistance is defined by $R_i \equiv (4\pi\sigma_i h_i)^{-1}$ for $i = 0,2$ with h_i the height of the resistive layer i, and we have used solid body rotation for the tube around its axis. From Equations (17) and (19) follows the total current

$$I = \frac{\Omega_0\Phi}{2\pi(R_0 + R_2)} = \tag{19}$$

$$= \frac{(\Omega_0 - \Omega_1)\Phi}{2\pi R_0} = \tag{20}$$

$$= \frac{\Omega_1\Phi}{2\pi R_2}, \tag{21}$$

and the rotation speed in the corona

$$\frac{\Omega_1}{\Omega_0} = \frac{R_2}{R_0 + R_2}. \tag{22}$$

Finally, from Equation (19) and Ampère's law for a steady state, we find that the magnetic field created by the current at the surface of the tube is

$$\frac{\delta B}{B} = \frac{\mu_0\Omega_0 r}{4\pi(R_0 + R_2)}, \tag{23}$$

where r is the tube minor radius. Substituting $\Omega_0 r = 0.1$ km sec^{-1} and $R_0 + R_2 = 10^{-7}$ Ohm, one finds $\delta B/B \approx 10^2$, which is far too large for a force free structure (compare Figure 6). We, therefore, conclude that the introduction of a *realistic time-independent resistivity does not prevent the expansion of flux tubes* when they are energized by differential motions at the footpoints.

5. Resistivity and the Flare Products

In the previous section, we have considered the (small) effect of a realistic quiescent resistivity on the ideal mhd evolution. The real problem, however, is that during the energy release in a flare, the *effective resistivity quickly increases* far above the classical collisional value and is determined by plasma turbulence and by microinstabilities. This the heart of the difficulty behind the extremely small collisional resistivity in the corona. A measure of the collisional resistivity is the *Lundqvist number*,

$$S \equiv \frac{\ell_B c_A \mu_0}{\eta}. \tag{24}$$

A typical value is $S \approx 10^{13}$ in the corona whereas in present-day numerical calculations it is only of order 10^4. Because of the extreme smallness of the collisional resistivity, it is no longer sufficient to use an mhd description with a scalar resistivity, even if a computer could handle a realistic Lundqvist number. Rather, the smallness of the collisonal resistivity means that new, *collisionless or collective*, phenomena appear. The collisional dissipation time is extremely large in comparison with the Alfvén time on which global changes in the magnetic structure occur during the flare. As the magnetic field structure changes quickly in comparison with the slow storage phase, *electric fields are large during a flare*. Therefore, ample time exists for the development of *beams of accelerated particles and other non-Maxwellian particle distributions* under the influence of electric fields. Consequently, an mhd description can never handle the dissipation phase – the energy release – in the flare correctly. In particular, one would like to know what fraction of the released energy is converted into *bulk motion by Lorentz forces*, what fraction into *accelerated particles by electric fields* and what fraction into *heating by collisions* (see Chapters 14 and 24). At present, this problem of the *energy partitioning in a flare* is not solved and forms the most important gap in our understanding the flare phenomenon.

From a global "mhd" point of view, these collisonless effects reflect themselves in a quickly increasing effective resistivity. However, not only is the energy conversion highly *time-dependent*, it is also extremely *inhomogeneous*. The effective resistivity is only large in very small volumes, at first perhaps in a very thin layer around an X-type neutral point. For instance, in the calculations by (Biskamp and Welter 1989; Mikić, Schnack, and Van Hoven 1990) an ideal mhd instability drives "reconnection" in a thin layer, creating a topologically separated gas blob or plasmoid, which is ejected at the local Alfvén speed (Shibata and Uchida 1986). Following the trigger, the region of energy conversion may be spreading to a *multitude of small dissipation sites*, as is indicated by observations of radio emission in flares (Slottje 1978; Kuijpers, van der Post, and Slottje 1981; Tapping *et al.* 1983; Allaart *et al.* 1990; Kuijpers 1990a).

The term *magnetic reconnection* (Kadomtsev 1987) is used for the topological change of the magnetic field structure in a nearly ideal plasma (where the field lines are frozen in and can be followed in time by the motion of the plasma). Resistivity in a small region (see Figure 11) liberates part of the magnetic energy. This energy is converted into kinetic energy by Lorentz forces of the reconnected state and into particle acceleration by electric fields in the reconnecting layer. A first description of steady reconnection is given in (Sweet 1958; Parker 1957; Petschek 1964) and since then most research on reconnection has gone into finding stationary solutions (Priest 1982). However, reconnection in the laboratory and in the cosmos may have little to do with these idealized models. Observations indicate *reconnection to be very time-dependent* and to lead to turbulent motions and turbulent magnetic fields in the plasma (Dere *et al.* 1991; Dubois 1985; Biskamp 1985; Strauss 1988). Recently, three-dimensional resistive calculations of reconnecting structures have been performed for the earth magnetosphere/solar wind interface (Otto 1990; Lee 1990).

The crucial property of reconnection is that it can release part of the free energy of a magnetic structure on an Alfvén crossing time. To see this, consider two thin flux tubes as in Figure 11a. After reconnection, the

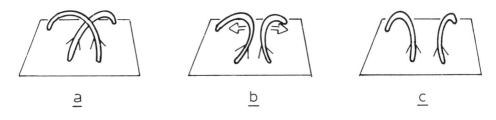

Figure 11: Reconnection of two thin flux tubes. Resistivity is needed only in a small region of contact between the tubes in (a). The relaxation in (b) is ideal and (c) shows the final state when the energy excess has been dissipated.

subsequent evolution is completely ideal (Figure 11b,c). The total relaxation time of the tube structure is, therefore, of the order

$$t_{rel} \approx \frac{\mu_0 d^2}{\eta} + \frac{\ell_B}{c_A},$$

and approaches the Alfvén crossing time if the individual tube diameter d is sufficiently small. In Figure 12a we have

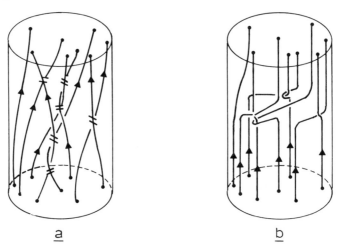

Figure 12: (a) shows a braided force free structure with the reconnection sites indicated by double slashes. (b) is topologically equivalent to (a), but now the knots (and the regions of reconnection) have been moved to a thin slice at the middle. Note that internal reconnection can remove only part of the linkage of the field lines. Some global twist inside the tube remains because the footpoints are frozen in and are in general not opposite to each other.

sketched a braided field structure inside a larger tube. Figure 12b demonstrates its topological equivalence to a (non force free) field where the "linkage" is largely confined to a thin disk. Since the disk volume can be made vanishingly small, the *entire structure of thin strands can relax on the Alfvén crossing time*. The same is, therefore, true in the real flux tube on the left-hand side provided resistivity operates efficiently and nearly simultaneously at the required locations (with vanishingly small total volume). Exactly how reconnection arises is still an unsolved problem (Sakai and de Jager 1991).

6. Energy Release and Magnetic Helicity

In section 3, we derived an upper limit on the energy which can be stored under force free conditions. This maximum energy can be compared with the energy in the potential field with the same distribution of the normal field component on the boundary surface. We have seen that the upper limit on the energy excess is of the order of the energy in this potential field. It is, however, not at all clear if relaxation to the potential field structure is possible on a short "flare" time scale. With the use of *Taylor's hypothesis* (Taylor 1974, 1986), based on earlier work by Woltjer (Woltjer 1958), one can calculate the energy which is liberated when a specific magnetic field structure relaxes by reconnection. *The hypothesis states that a magnetic field structure inside a magnetically closed volume V which relaxes under the only constraint that the total magnetic helicity is conserved, relaxes to a linear force free field inside V* (a field with $\nabla \times \vec{B} = \alpha\vec{B}$ where α is constant in space). The hypothesis permits calculation of the available free energy in an actual situation when the complete magnetic vector field at the surface is determined from the observations with sufficient accuracy. We shall first sketch the derivation of the hypothesis and the conditions for its validity.

A magnetically closed volume is a region of space enclosed by magnetic field lines (the field lines themselves are closed since $\nabla \cdot \vec{B} = 0$). The quantity $\vec{B} \cdot \vec{n}$ vanishes therefore on its boundary, with normal vector $\vec{n}$.

The magnetic helicity K inside V is defined by

$$K \equiv \int_V \vec{A} \cdot \vec{B} dV, \tag{25}$$

with $\vec{A}$ the vector potential corresponding to $\vec{B}$ ($\vec{B} = \nabla \times \vec{A}$).

The helicity inside a volume V is a measure of the amount of *linkage* of the magnetic field lines within that (magnetically closed) volume. To see this, consider two

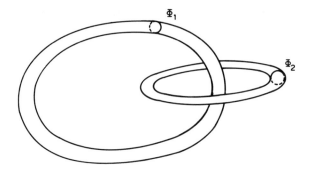

Figure 13: Two flux tubes of flux Φ_1 and Φ_2 have a magnetic helicity $K = 2\Phi_1\Phi_2$.

thin flux tubes which are linked as in Figure 13. The magnetic flux of tube i ($i = 1,2$) is Φ_i, its volume V_i, its cross-section $\Delta\sigma_i$, and A_i is an arbitrary surface spanning loop i. Then, the helicity of the entire structure (Equation (25)) can be written as

$$K_{12} = \sum_{i=1,2} B_i \Delta\sigma_i \oint_{\ell_i} \vec{A} \cdot d\vec{\ell} = \sum_{i=1,2} \Phi_i \int_{A_i} \vec{B} \cdot d\vec{S} = 2\Phi_1\Phi_2. \tag{26}$$

This demonstrates that K is a measure of the connectivity of the magnetic field. Note that the vector potential is only defined up to a gradient of an arbitrary function χ so that the helicity is only uniquely defined if V is bounded by a magnetic surface (so that the contribution $\int_V \nabla\chi \cdot \vec{B} dV = \oint_S \chi \vec{B} \cdot d\vec{S} = 0$ as $\nabla \cdot \vec{B} = 0$). The significance of helicity as connectivity can be extended to the general case (Field 1986)

In an *ideal plasma*, the topological structure of the field remains intact during the evolution. Consequently, *the linkage and, therefore, the helicity of every closed flux tube* inside the plasma are *invariants*. In a plasma with finite resistivity, field lines reconnect and flux tubes have no time-independent meaning. One can still define the *total helicity* of the plasma if the plasma is contained inside a magnetic surface which preserves its identity during relaxation, for example, a perfectly conducting vessel containing a laboratory plasma. We now consider what happens to the total helicity inside such a magnetic surface as the plasma evolves. For the vector potential, we choose the radiation gauge in which the electrostatic potential vanishes so that

$$\frac{\partial \vec{A}}{\partial t} = -\vec{E}. \tag{27}$$

Then, the comoving (Lagrangian) derivative of the total helicity is found to be

$$\begin{aligned}
\frac{DK}{Dt} &= \frac{dK}{dt} + \int_V dV \nabla \cdot (\vec{A} \cdot \vec{B}\vec{v}) \\
&= \int_V dV \left\{ -(\nabla \times \vec{E}) \cdot \vec{A} - \vec{E} \cdot \vec{B} \right\} + \oint_S \vec{A} \cdot \vec{B}\vec{v} \cdot d\vec{S}
\end{aligned}$$

$$= \int_V dV \left\{ -2\vec{E} \cdot \vec{B} \right\} + \oint_S \left\{ \vec{A} \times (\vec{E} + \vec{v} \times \vec{B}) + \vec{A} \cdot \vec{v}\vec{B} \right\} \cdot d\vec{S}$$

$$= \int_V dV \left\{ -2\vec{j} \cdot \vec{B}/\sigma \right\} + \oint_S \left\{ \vec{A} \times \vec{j}/\sigma \right\} \cdot d\vec{S}, \tag{28}$$

where we have used Equation (27), Faraday's equation, the definition for the vector potential, Ohm's law $\vec{j} = \sigma(\vec{E} + \vec{v} \times \vec{B})$, and, finally, $\vec{B} \cdot d\vec{S} = 0$ on a magnetic surface.

Reconnection allows a nearly ideal plasma to make a transition to a state of lower magnetic energy by finite resistivity in many (vanishingly small) regions (Figure 12). Then, Equation (28) shows that the total helicity of V is practically conserved during the reconnection processes.

It is now a straightforward mathematical operation (Freidberg 1987) to show that minimization of the total magnetic energy in V under the only constraint of conserved total helicity leads to a linear force free field $\nabla \times \vec{B} = \alpha \vec{B}$ with α constant in space and its value determined by the (invariant) total helicity. Taylor's hypothesis has been successful in understanding the relaxation of a toroidal laboratory pinch. His idea has been further developed to stellar flares (Norman and Heyvaerts 1983), to slow motions at the boundary of a stellar corona (Heyvaerts and Priest 1984; Berger and Field 1984; Berger 1984; Berger 1988), to energy injection in jets and to relaxing plasmoids (Choudhuri and Königl 1986; Dixon *et al.* 1989; de Vries and Kuijpers 1992).

In the case of a flaring magnetic structure in the corona, the enveloping flux tube is anchored in the subphotospheric layers and the observable part does not form a complete magnetic surface, as is also true for Figure 12. The concept of helicity can be extended, however, in the form of a gauge invariant *relative helicity* (Berger and Field 1984) (see also Heyvaerts and Priest 1984). This is why Taylor's hypothesis can also be applied to force free parts of anchored flux tubes. It is then possible in principle to *calculate the free energy from the difference between the actual magnetic structure and a structure with a linear force free field having the same magnetic flux distribution at the boundary and the same total helicity.*

7. Singular Current Layers

Thin current layers or *current sheets* (Strauss and Otani 1988; Bhattacharjee and Wang 1991) play an important role in a realistic corona because of at least two effects:

Firstly, the coronal magnetic field is rooted in *discrete elements* in the subphotospheric layers (Figure 14). Since neighbouring coronal field lines have a finite separation at the photospheric level, singular current layers appear in the corona at the separatrices (Low 1990).

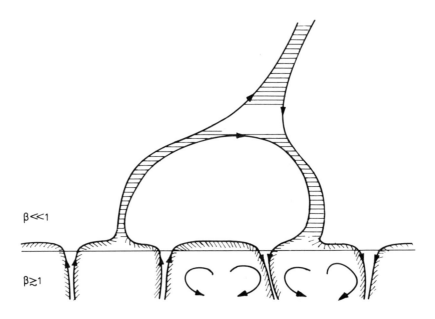

$\beta \ll 1$

$\beta \gtrsim 1$

Figure 14: Thin current layers are present at the separatrices of magnetic structures rooted at different locations in the subphotosphere.

Here, the adjective "singular" is used for infinitely thin in the mathematical sense. Of course, in practice, the thickness is at least an ion gyro radius. Since the current runs at an angle with respect to the local magnetic field, these thin layers are *not force free*. In a steady state, the Lorentz force should then be balanced either by pressure $-\nabla p$, by gravity $\rho\vec{g}$, or by inertial $\rho(\vec{v} \cdot \nabla)\vec{v}$ forces.

Secondly, if *Taylor relaxation* occurs inside a flux tube embedded in a completely force free corona, the resulting linear force free field does not match the surrounding nonlinear field. A non-force free surface current is then created surrounding the relaxed flux tube (Figure 15). This follows directly from application of Ampère's law applied to a curve around the tube: $\oint \vec{B} \cdot d\vec{\ell} = \mu_0 I_{net}$, where I_{net} is the net current through the curve. The boundary current ensures that

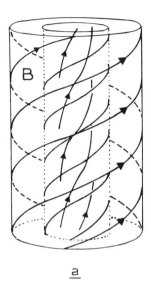 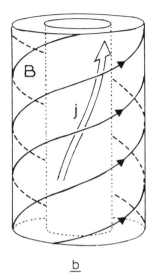

a   b

Figure 15: When the inner tube in (a) relaxes by internal reconnection a surface current $\vec{j}$ appears, matching the relaxed internal field to the practically unperturbed external field (b).

the field surrounding the tube is similar to the original external field before relaxation. What happens during internal reconnection is not dissipation of the net current but rather dissipation related to its spatial redistribution. Part of the internal magnetic energy is converted and this corresponds to *cancelling of opposite internal currents plus an expulsion of the net (!) current to the surface.* It should be noted, of course, that if a flux tube relaxes internally, it only comes to equilibrium again if the external magnetoplasma adjusts by slightly compressing the relaxed flux tube.

Since Taylor relaxation of force free structures has the tendency to expel net currents to the surface, one wonders if such currents have an observational signature. In particular are the chromospheric Hα fibrils such a system of non-force free surface currents?

Although these thin current layers have not been well studied, they may play a crucial role in the energy storage and release. From a theoretical point of view, it is to be noted that, in such layers, the drift velocity of the current carrying particles becomes large and, perhaps, even violates the stability criterion for micro instabilities (particle drift speed larger than some thermal speed) starting a sequence of enhanced resistivity and further field dissipation (see also Low and Wolfson 1988). It is interesting that observational comparison of the flare location and vector magnetograms indicates that the flare is initiated at the edges of major vertical currents (Canfield *et al.* 1991).

8. Flare Geometromagnetics in Other Stars

What can we learn from solar flares that can be applied to magnetic explosions in other objects? In the Introduction, we have summarized the conditions for flare production. Apart from the trivial extension of the flare phenomenon to single stars with surface magnetic fields and relative footpoint motion, flares are expected to occur

- in close binaries (see Figure 16) such as:

    - RS CVn stars with two interacting magnetospheres and relative footpoint motion (Bahcall, Rosenbluth, and Kulsrud 1973; DeCampli and Baliunas 1979; Simon, Linsky, and Schiffer 1980; Uchida and Sakurai 1983; Kuijpers and van der Hulst 1985; van den Oord 1988);

    - Algol stars with transient mass exchange (White *et al.* 1986; van den Oord *et al.* 1989; van den Oord and Mewe 1989);

    - contact binaries (Vilhu *et al.* 1988);

    - cataclysmic variables with interactions between infalling magnetized gas and a magnetosphere as in AM Her systems (Lamb *et al.* 1983; Kuijpers 1989);

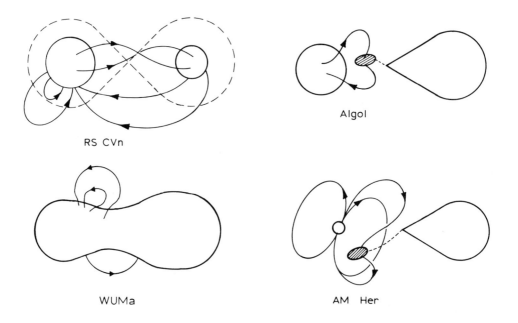

Figure 16: Possible flare generation in various close binaries: RS CVn type binaries with stellar connections; Algol type binaries with plasma penetrating a magnetic atmosphere; W UMa contact binaries; AM Her type binaries with accretion into a white dwarf magnetosphere.

- in accretion disks (see Figure 17):

    - produced by interactions between disk and magnetosphere of a white dwarf as in DQ Her systems;

- from interactions with magnetic neutron stars and associated with magnetic expulsion of plasmoids (Aly and Kuijpers 1990);
- around proto and T Tauri stars (Kuijpers 1989);
- around black holes (Volwerk, Kuijpers, and de Vries 1991). Here the magnetic field is not anchored on the hole (which would require an electrically charged hole, an unrealistic assumption in the presence of accretion) but embedded in the accretion flow;
- in Active Galactic Nuclei (de Vries and Kuijpers 1989);

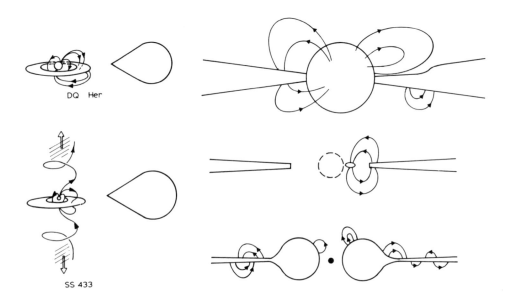

Figure 17: Flares in accretion disks around: magnetic DQ Her type white dwarfs; magnetized neutron stars; proto- or T Tauri stars; stellar mass black holes (horizon is dashed); massive black holes in active galactic nuclei.

- in jets (see Figure 18):
  - from Taylor relaxation of force free jets (Choudhuri and Königl 1986);
  - from relaxation of expanding plasmoids and coalescing flux tubes (Choudhuri 1988; Dixon *et al.* 1989; de Vries and Kuijpers 1992).

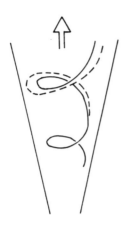
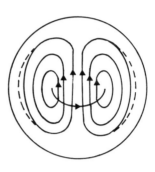

Figure 18: Flares in jets and spheromaks caused by anisotropic expansion (dashed is before relaxation).

Let us briefly consider the flare requirements from the Introduction:

1. The *driver* with differential motions at the footpoints is formed either by two stars, or by one star and an accretion disk, or by a star and infalling material, or by different parts of an accretion disk. In the case of jets and plasmoids. the "driver" is not formed by a changing external medium but rather by a changing external pressure in the comoving frame of the jet or plasmoid. The high Lundqvist number of these drivers is caused by the large dimensions of the objects, as in the solar case. This is even true on the horizon of a stellar mass black hole.

2. The condition on the *ordering of the relevant time scales* may or may not be fulfilled temporally in some of these objects. Roughly the condition implies

$$\Delta v \ll c_A, \tag{29}$$

where Δv is the velocity difference over the anchoring surface (or across the expanding plasmoid) and c_A the Alfvén speed in the corona. While this condition is easily satisfied in coronae of normal stars, it may not always be fulfilled in magnetospheres and accretion disk coronae at the inner orbits around neutron stars or black holes. Further, the coronal Alfvén speed in a magnetosphere can temporarily go down during bursts of accretion.

3. Pressure and dynamic effects are more important than magnetic effects in these drivers, as required.

4. The existence of a low beta corona depends primarily on the strength of the atmospheric magnetic field. In the case of neutron stars and white dwarfs, the field depends on the evolution of the objects. For other stars and accretion disks, the fields depend on the efficiency of a dynamo.

5. *The driver should be sufficiently powerful.* In the case of the Sun, the dynamic pressure in the photospheric layers is by far insufficient to act as a powerful driver (McClymont and Fisher 1989), basically since $\rho(\Delta v)^2 \ll p_{\text{gas}} \approx B^2(2\mu_0)^{-1}$. Indeed, the anchoring must occur in much deeper layers where the dynamic pressure of the convective eddies is relatively large; the forcing of the flux tubes is communicated to the surface layers by Maxwell stress in the form of Alfvénic motions. The case of accretion disks is treated in the next section.

We have seen that the nature of the flare products (particle acceleration, heating, bulk motion) cannot at present be predicted theoretically as a function of the physical conditions in the unstable magnetic structure. It is only the global flare physics or "geometromagnetics" that we begin to understand. This is sufficient, however, to make estimates of the typical energy release and time scale of the magnetic explosion. These estimates can then be compared to the observed explosions in particular objects to find out if magnetic flaring can or cannot be ruled out. In Table 1, we give the observed amount of energy per explosion or "flare" in classic close binaries, in cataclysmic variables, and in neutron star binaries (taken from Kuijpers 1989), together with assumed dimensions of magnetic structures in these objects (put equal to the dimension of the stellar object) and the magnetic field strengths (based on Equation (15)) required to produce the observed explosion by magnetic means.

Table 1: Possible Magnetic Origin of Observed Explosions

binary	flare (J)	R (m)	B (T)	$B^2R^3/(2\mu_0)$ (J)
RS CVn	10^{27}	10^9	10^{-2}	$4 \cdot 10^{28}$
white dwarf	10^{26}	10^7	1	$4 \cdot 10^{26}$
neutron star	10^{33}	10^4	10^8	$4 \cdot 10^{33}$

It can be seen from the Table that the required fields are realistic.

Which variations in X-ray binaries are of magnetic origin? The Quasi-Periodic Oscillations in Low Mass X-Ray Binaries (Lewin, van Paradijs, and van der Klis 1988) and some of the eruptions in dwarf novae, polars and intermediate polars may be of the magnetic flaring type. However, for the X-ray pulses of neutron star binaries with massive companions, for the type I bursts from X-ray pulsars, and for the (recurrent) novae (nuclear fusion), there is no reason to invoke magnetic activity.

Apart from such known effects of magnetic flares as X-ray bursts, radio bursts, and transient mass outflow, new physical effects also appear:

- In close detached binaries, magnetic loops can extend to the other side of the inner Lagrangian point, resulting in *inversion of gravity* and accumulation of cool plasma in the top of the loop;

- Similarly, for a fast single rotator with a strong field, plasma can accumulate at the top of a corotating loop as a result of *centrifugal forces* (Collier Cameron 1989);

- The strong radiation field in the inner parts of accretion disks determines the quality of the flare products. For instance, flare accelerated electrons radiate their energy by *inverse Compton losses*, instead of causing "evaporation" by collisional heating of the denser atmosphere as in the solar case. The radiation is, therefore, harder (de Vries and Kuijpers 1989);

- In the inner parts of disks around neutron stars or black holes, the available energy per "hydrogen atom" corresponds to a fictitious temperature of $75/r_S$ MeV, where r_S is the central distance in Schwarzschild radii. Consequently, $e^- - e^+$ *pair production* may occur in these objects.

- If the velocity field is regular instead of stochastic as in the Sun, the *flare appearance* can also be *regular in time*. A typical example is the strong and regular Keplerian velocity field of thin accretion disks;

- Angular momentum transfer between the central star and accretion disk can lead to magnetic *expulsion of bullets* in the direction of the disk rotation axis (Uchida and Hamatake 1989; Shibata 1990).

9. Flares in Accretion Disks

Let us consider a *geometrically thin accretion disk* around a magnetized star (see Figure 19). Let us assume that the radial inflow speed is much below the azimuthal speed, implying that the rotation is nearly Keplerian:

$$v_K(r) \equiv \left(\frac{GM_\star}{r}\right)^{1/2}. \tag{30}$$

Let us further assume that the vertical disk structure is determined by hydrostatic equilibrium

$$\frac{dp}{dz} = -\rho g_{\text{eff}} = -\rho \frac{GM_\star}{r^2} \frac{z}{r}, \tag{31}$$

where z is the distance from the midplane. The condition for the disk (half-thickness $h(r)$ at central distance r) to be thin can then be rewritten as a condition on the smallness of the sound speed in the disk (using $p \approx \rho c_s^2$)

$$\frac{h}{r} \approx \frac{c_s}{v_K} \ll 1. \tag{32}$$

The disk is, therefore, cool, that is, the Keplerian speed is highly supersonic.

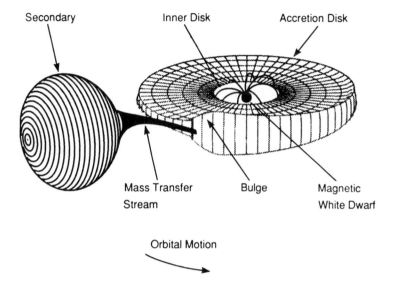

Secondary · Inner Disk · Accretion Disk

Mass Transfer Stream · Bulge · Magnetic White Dwarf

Orbital Motion

Figure 19: Sketch of an X-ray binary with a thin accretion disk and a magnetic white dwarf (Mason, Rosen, and Hellier 1988).

9.1. DISK FLARES VERSUS MAGNETOSPHERIC FLARES

In line with the foregoing, we shall only consider systems with a *sufficiently large Alfvén speed* in the corona, or roughly

$$v_K(r) \ll c_A. \tag{33}$$

Two kinds of magnetic flares are possible:

1. Flares from loops anchored in the disk (see Figure 20a). These disk flares have first been proposed by (Galeev, Rosner, and Variana 1979);

2. Flares from loops anchored in the disk at one "end" and in the star at the other "end" (see Figure 20b). These "magnetospheric" flares have been studied by (Aly and Kuijpers 1990).

The occurrence of both flares depends on the existence of a magnetic corona. The formation of magnetic loops extending on both sides of an accretion disk has first been studied by Coroniti (1985) and later by Burm and Kuperus (1988) and by Heyvaerts and Priest (1989). In many respects, the disk resembles a two-dimensional star with the energy production from fusion replaced by gravitational accretion.

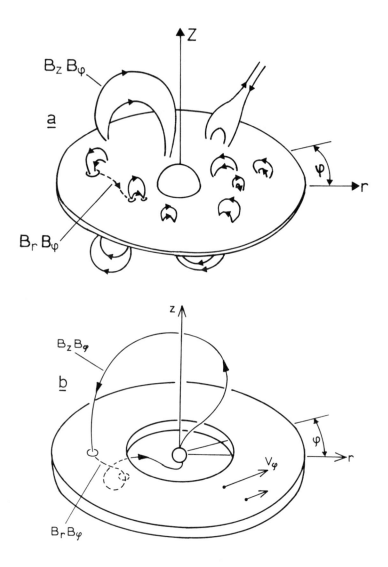

Figure 20: In (a) relative footpoint motion leads to disk flares. (b) shows how a magnetic link between central star and accretion disk leads to a much larger magnetospheric flare. Angular momentum transport occurs by the stress $B_z B_\phi/\mu_0$ along field sticking out of the disk and by the stress $B_r B_\phi/\mu_0$ along field embedded inside the disk.

The parts that have an effective temperature of several 10^3 K or more can be considered as well-conducting plasmas in view of their dimensions. Inside the disk, magnetic fields are amplified by differential Keplerian motion. As they become amplified, they become buoyant and rise on both sides out of the disk. In this picture, the disk is surrounded by a hot and dilute corona. In order of magnitude, the maximum field strength in the flux tubes at the disk photosphere is probably determined by equipartition with the disk *gas* (and not radiation) pressure (Coroniti 1985). Other, smooth magnetic disk models exist (*e.g.* Lovelace, Wang, and Sulkanen 1987) but they may not be realistic because of buoyancy and the possible presence of convection leading to an intermittent magnetic field distribution inside the disk. Moreover, the Sun and our galaxy (Figure 21) do not show a regular magnetic field pattern but rather a stochastic appearance

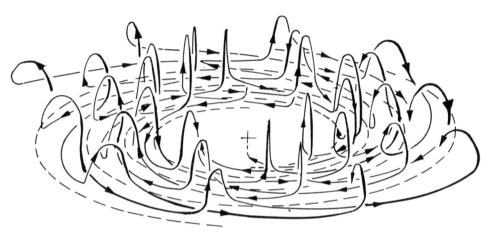

Figure 21: Sketch of the galactic magnetic field in accordance with the observations (Sofue, Fujimoto, and Wielebinski 1986).

with a fluctuating component at least as strong as the spatially averaged field (Parker 1979; Heiles 1987). We, therefore, assume that a characteristic thin accretion disk has a fragmented magnetic field distribution with discrete flux tubes and a surrounding magnetically dominated corona.

A necessary condition for a given loop to flare is that the driver is powerful enough. The difference in angular velocity of the footpoints should be so large that the two lumps of plasma cannot be brought into corotation within the build- up time of a flare. The excess angular momentum per unit surface area is $2h(r)\rho r^2 \Omega_B \delta$, where Ω_B is the difference in angular speeds of both footpoints (also called the *beat frequency*) and δ depends on the mass ratio of the lumps at both ends ($\delta = 1$ if one footpoint is on the star, and $\delta \approx 2$ if both footpoints are in the disk). The rate of magnetic transport of angular momentum out of the disk per unit surface area is $2rB_z B_\phi / \mu_0 \leq 2rB_z^2 / \mu_0$. The flare build-up time is the shearing time, of magnitude $1/\Omega_B$. Using these expressions, Equation (32) and assuming

$$B_{disk}^2 / 2\mu_0 = \rho c_s^2, \tag{34}$$

we find that a *given loop will produce a flare* if

$$\frac{B_z^2}{2\mu_0} < \frac{1}{2}\rho c_s v_K \left(\frac{\Omega_B}{\Omega_K}\right)^2, \tag{35}$$

that is, if the radial distance of its footpoints is large enough:

$$\frac{\Delta r}{r} > \left(\frac{h(r)}{r}\right)^{1/2} \frac{B_{cor}}{B_{disk}}. \tag{36}$$

Since $B_{cor} < B_{disk}$, this condition is easily satisfied in a thin disk.

If, at one end, the loop is anchored in the central object, a similar reasoning leads to the requirement that the second footpoint is not too close to the radius of corotation. The *minimum distance from the corotation radius* is again given by Equation (36). Now the *beat frequency* is

$$\Omega_B(r) \equiv |\, \Omega_K(r) - \Omega_\star\,|, \tag{37}$$

where $\Omega_K(r) = v_K(r)/r$. The *corotation radius* r_{co} is defined by

$$v_K(r_{co}) \equiv \Omega_\star r_{co}. \tag{38}$$

Further, we have used the fact that the field strength is weakest at the footpoint anchored in the disk and not on the star, so that at the former footpoint the "force free" condition $B_\phi < B_z$ is most restrictive. From Equation (35), we see that, sufficiently far from the radius of corotation (where $\Omega_B(r) \approx \Omega_K(r)$), the disk material never reaches corotation with the star since in the disk $B_z^2(2\mu_0)^{-1} \approx \rho c_s^2 \ll \rho c_s v_K$ for a thin disk.

9.2. Magnetospheric Radius

What determines the boundary between accreting material and magnetosphere? For spherically symmetric accretion, a so-called *Alfvén radius* is defined by a local balance between the ram pressure of the freely radially infalling gas and the magnetic pressure of the magnetosphere

$$\rho v_{ff}^2 = \frac{B^2(r_A)}{2\mu_0}. \tag{39}$$

For thin disk accretion (see Figure 19), a magnetospheric radius r_m is defined by a condition similar to Equation (39), now, however, with the Keplerian velocity v_K replacing the free-fall speed v_{ff} (Figure 22). Clearly, for the same density and field strength, the

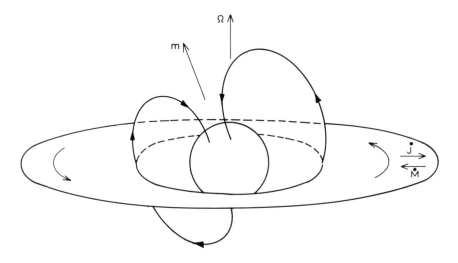

Figure 22: Disk accretion with accretion rate $\dot{M}$ and outward rate of transport of angular momentum $\vec{J}$ onto an oblique magnetic rotator with rotation $\vec{\Omega}$ and magnetic moment $\vec{m}$. If the magnetic moment is large enough, the inner part of the disk is strongly modified by the magnetosphere.

two values of the magnetospheric radius are not very different. However, for the same values of the accretion rate and field strength, the density is much higher for thin disk accretion and the magnetospheric radius is, therefore, much smaller than for spherical accretion.

It is important to realize that the magnetic field in Definition (39) is the stellar field modified by the presence of the disk. The (well-conducting) disk increases the field strength at a given radius by compression of the stellar field flux. Figure 23 shows the realistic field structure in the case of a thin disk that excludes the field from

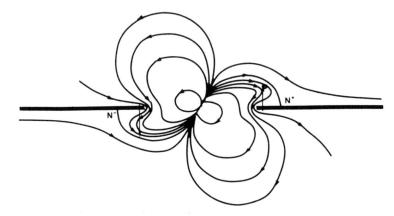

Figure 23: Modification of the magnetic field of an oblique rotator by a perfectly conducting disk with a hole at the center (Aly 1980).

an oblique rotator. The typical magnetospheric field strength at a distance $h(r)$ away from the inner disk boundary is enhanced over the dipolar "vacuum" value by a factor $(r/h(r))^{\frac{1}{2}}$ (Aly 1980; Spruit and Taam 1990). Because of the compression of the stellar flux, in principle, the Alfvén radius always lies outside the star (axial symmetry assumed).

A better definition of the radial distance within which the influence of the stellar field on the disk is important, is the *pressure balance radius*. This is the radius r_p where, coming in from infinity, *the disk gas pressure first equals the external stellar field pressure*. Within this radius r_p, the disk remnant is compressed in the

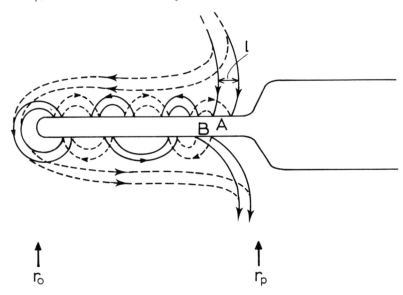

Figure 24: Within the pressure balance radius r_p, the disk is strongly compressed by the ambient stellar field in the vertical direction. Reconnection between stellar and disk field further modifies the accretion. The field lines are projected onto the meridional plane; the original field is dashed while the reconnected field is drawn. Note that only a small fraction of the stellar field (at A and B) is linked to the disk. Magnetic interactions break up the disk at an inner radius r_0 ($> r_m$).

vertical direction by the stellar magnetic rather than by the gravitational field (see Figure 24). Therefore, in the disk an inner ring exists in which the gas pressure is at any radius comparable to the stellar magnetic field pressure just outside. We first consider the magnetic transport of angular momentum and then the energy release in flares from accretion disks.

9.3. TRANSPORT OF ANGULAR MOMENTUM

Magnetic fields are efficient transmitters of disk angular momentum in the radial direction (see Figure 20a). The *magnetic torque* (rate of transport of angular momentum) on a column of unit disk surface area in a flux tube rising out of the disk compares to the *viscous torque* in a Shakura-Sunyaev alpha-disk (Shakura and Sunyaev 1973) approximately as

$$\frac{B_z B_\phi r}{\mu_0 \alpha p_{\text{gas}} 2h(r)} \approx \frac{r}{\alpha h(r)} \gg 1, \tag{40}$$

where we have put $B_z \approx B_\phi$ and $p_{\text{gas}} \approx B^2(2\mu_0)^{-1}$. This proves that the *radial motion of disk matter inside a flux tube is dominated by magnetic stress*. Of course, the transport of angular momentum by magnetic flux tubes anchored in the disk is extremely inhomogeneous. Angular momentum is transferred from the disk material at the innermost footpoint through the corona and deposited at the other footpoint. Moreover the transport between these footpoints greatly exceeds the transport of angular momentum in the ambient unmagnetized disk material.

Apart from transport of angular momentum in the disk magnetic fields can also efficiently transport angular momentum from the disk to the star and vice versa. This is sketched in Figure 20b. We make a distinction between gas inside and gas outside the corotation radius (Equation (38)). A magnetic link from the star to a disk element inside the corotation radius removes angular momentum from the disk since there the Keplerian rotation is faster than the stellar rotation (see Figure 25). Therefore, inside the corotation radius

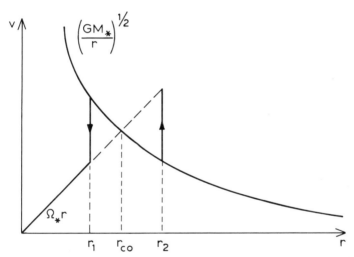

Figure 25: The figure shows the corotation radius r_{co} where disk material with the Keplerian speed corotates with the star, $v_K(r_{co}) = \Omega_* r_{co}$. Disk material at $r_1 < r_{co}$ rotates faster than the star and falls in when brought into corotation, while disk material at $r_2 > r_{co}$ is accelerated away from the star when magnetically connected to the magnetosphere.

a magnetic link strongly enhances the accretion rate. We can estimate the accretion speed from the magnetic torque on a column of unit disk area

$$2\rho h(r) \frac{d\{r^2 \Omega(r)\}}{dt} \approx -\frac{2r B_z B_\phi}{\mu_0}. \tag{41}$$

Substituting the Keplerian rotation speed for the actual rotation $\Omega(r) \approx \Omega_K(r) \equiv (GM_*/r^3)^{1/2}$ and putting $B_\phi \approx B_z$, we find from Equation (41)

$$v_r \approx -\frac{2rB_z^2}{h\rho v_K(r)\mu_0} \approx -\frac{2B_z^2}{\rho c_s \mu_0} \approx -4c_s, \tag{42}$$

where we have used $h(r)/r = c_s(r)/v_K(r)$ and $B_z^2 = 2\mu_0 p_{\text{gas}}$. First, we see that the assumption $\Omega(r) \approx \Omega_K(r)$ is *a posteriori* justified as, for a thin disk, the derived radial speed is still much less than the Keplerian speed. Secondly, we find that the inward radial speed is much larger than the viscous inward radial speed for an alpha-disk, which is of order $\alpha c_s(r)h(r)/r = \alpha c_s^2/v_K$. (Alternatively this shows that magnetic links in an alpha-disk lead to local values of the *parameter* α of order $r/h(r)$, *much larger than unity* in contrast with popular belief that α cannot exceed unity.) As a result, in the inner ring of the disk ($r < r_p$ and $r < r_{co}$), *the disk breaks up into separate blobs* which spiral inwards much quicker than the outer viscous regions (see Figure 26a).

Magnetic links can also end on the disk at a radial distance r larger than the corotation radius (as long as $r < r_p$). Such links, however, deposit stellar angular momentum into the disk (see Figure 25) and they *prohibit rather than promote accretion*. Similarly to the estimate in Equations (41) and (42), the disk material quickly acquires a supersonic radial speed, now in the outward direction, causing the formation of dissipative *shocks* in the surrounding matter. As in the picture of (Ghosh and Lamb 1978, 1979a,b), the simultaneous existence of magnetic links between star and disk inside and outside the corotation radius leads to the possibility of steady stellar rotation rates for (on average) stationary disk accretion.

Further, by assumption, the coronal Alfvén speed is large (Equation (33)) so that the link evolves through a series of force-free equilibria. *Then, the azimuthal field component can, at most, become equal to the vertical component after which a flare occurs and distortion of the newly reconnected field starts again.* We, therefore, conclude that the estimate $B_\phi \approx B_z$ is characteristic for the initial stage of a link as the matter spirals inwards. The link is periodically destroyed and reformed with different stellar field lines (see Figure 26a).

The magnetic transport of angular momentum from the disk in a boundary layer or ring near the magnetospheric radius is apparently very *inhomogeneous and time-dependent, in the form of flaring linking flux tubes.* This highly variable transport is in contrast with the smooth and stationary reconnection assumed in the standard picture (Ghosh and Lamb 1978, 1979a,b). Also, in our picture, the magnetic flux tubes connected to the star have their footpoints in a ring $r_m - r_p$, which is much larger than the "standard" boundary layer near r_m.

Whether an individual gas blob remains spiralling in the plane of the disk as in Figure 26a or starts flowing along the stellar field as in Figure 26b, depends on the spatial dependence of the stellar field. Inside the inner disk radius, the field is, of course, not limited to equipartition with the

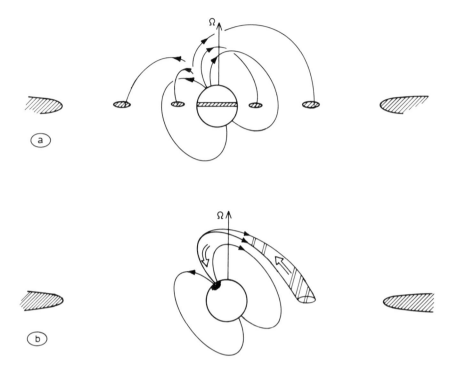

Figure 26: Accretion of gas threaded by stellar magnetic field inside the magnetosphere. In (a), the stellar field is relatively weak for the given accretion rate and gas blobs continue to spiral inwards in the disk plane, ultimately hitting the star along an equatorial belt. In (b), the stellar field is strong enough to guide the accreting material in the inner region along the field onto the magnetic poles. In (a), no sharp X-ray pulses are observed but only QPO; in (b), both X-ray pulses are observed from the guided accretion in the deep interior and QPO from the flares in the outer region.

gas pressure. A simple estimate based on a dipolar dependence for the stellar field and conservation of flux through the infalling gas blob indicates that aligned infall towards a neutron star only occurs for very large surface fields (10^8 T or more) while for smaller fields the accretion can remain in the disk plane (Aly and Kuijpers 1990). The former case of aligned infall onto the poles applies to X-ray pulsars. In the latter case, *accretion of the different blobs occurs on an equatorial ring around the star and no X-ray pulses are expected* to be observed. We suggest that this applies to *Low Mass X-Ray Binaries.* However, a self-consistent solution of the thermodynamics of the radiation dominated inner region of accretion onto the neutron star, including the role of magnetic flares, has not yet been formulated.

9.4.. FLARE ENERGY

If the electrodynamic coupling leads to flares, it is easy to estimate their energy. Using estimate (15) for the energy content of a flare, a dipolar dependence for the magnetic field of the central object, and a scaling of the linked area $S(R) \propto R^2$, one finds that the flare energy depends on link radius as $W_M \propto R^{-3}$. Therefore, we expect the largest flares to occur near compact objects with large field strength and for disk accretion with high accretion rate.

We now estimate the relative luminosity in these magnetic flares. In an axially symmetric (with respect to rotation axis) force-free magnetic link, the quantity $B_\phi \tilde{r} sin\theta$ is constant along a field line ($\tilde{r}$ is the distance from a point on the field line to the stellar center and θ is the angle of the radius vector with the rotation axis). For a dipolar field, ($sin^2(\theta)/\tilde{r}$ constant along a field line and $B_z \propto \tilde{r}^{-3}$) the main relative distortion of the field ($B_\phi/B_z \uparrow 1$), therefore, occurs near the weak field at the disk footpoint. The energy in a flare from a link to the star is then of order (compare Equation (15))

$$W_M = RS \frac{(B_z(R))^2}{2\mu_0},\tag{43}$$

where R is the distance of the footpoint on the disk to the star and S is the tube cross-section. Let us assume that the links in a particular system can, at any instant, be approximated by a spherically symmetric thin shell of equatorial radius R and radial thickness at the equator ℓ and that the flaring rate is $\Omega_B^{-1}(R)$. For a stellar dipole field of strength $B_\star$ at the stellar pole, we then find a *total flare luminosity* from one side of the disk

$$
\begin{aligned}
L_M &\approx \pi \ell R_\star^2 \left(\frac{R_\star}{R}\right)^4 \Omega_B(R) \frac{B_\star^2}{4\mu_0} \\
&= 1.3 \cdot 10^{28} \frac{\ell}{10^4 m} \left(\frac{R_\star}{10^4 m}\right)^6 \left(\frac{R}{10^5 m}\right)^{-4} \frac{\Omega_B}{200 Hz} \left(\frac{B_\star}{10^6 T}\right)^2 \text{Watt,}
\end{aligned}\tag{44}
$$

where $R_\star$ is the stellar radius. The observed luminosity in X-ray fluctuations (*Quasi Periodic Oscillations*, van der Klis 1989) is a few percent of the total source luminosity of 10^{31} Watt and can be reproduced by Equation (44) for plausible conditions.

Finally, the energy release of an individual flare from a gas blob of mass m at radius $R = r_p$ relates to the gravitational energy release at the moment of impact of the blob onto a (non-rotating) star as (use Equation (43), pressure equality, thin disk assumption, and Keplerian disk rotation)

$$\frac{W_M(R)}{W_G(R_\star)} \approx \frac{h(R)R_\star}{R^2}$$

and can easily be larger than 10^{-3}.

10. Plasma Turbulence

The dilute and hot coronae of stars and accretion disks often deviate considerably from local thermodynamic equilibrium because of two reasons: They are the sites of magnetic flares and moreover, the collisonal relaxation times are relatively long. As a result many wave modes can be present at suprathermal levels. This state is known as plasma turbulence (Tsytovich 1970). In contrast with hydrodynamical turbulence plasma turbulence is often weak. For plasma turbulence the growth rates (imaginary part of the wave frequency) are much less than the real parts of the frequencies: $|\gamma| \ll \omega$, whereas for hydrodynamical turbulence $|\gamma| \approx \omega$, where $\omega^{-1} \approx \lambda/v_\lambda$ is the eddy turnover time. This makes plasma turbulence a relatively well understood and sophisticated subject.

10.1. COLLISIONLESS PLASMA

The deflection time of a thermal electron from encounters with ions in a plasma of temperature T is of order (Spitzer 1962)

$$t_{ei} = \frac{\pi}{\ell n \Lambda} \frac{N_D}{\omega_{pe}} = 0.9 \cdot 10^3 \frac{T^{3/2}}{n} \frac{20}{\ell n \Lambda} \text{ s}, \tag{45}$$

where the Coulomb logarithm is defined by $\ell n \Lambda \equiv \ell n \left(12\pi n \lambda_D^3\right)$, n is the electron density (unit m^{-3}), n_i the ion density, T is the temperature, ω_{pe} the electron plasma frequency defined by

$$\omega_{pe} \equiv \left(\frac{n\,e^2}{m\epsilon_0}\right)^{1/2} = 56.4\, n^{1/2} \text{ rad sec}^{-1} \tag{46}$$

and the Debye length is given by

$$\lambda_D \equiv \left(\frac{\epsilon_0 K\, T}{\sum_\alpha n_\alpha (z_\alpha e)^2}\right)^{1/2} \approx 69 \left(\frac{T}{n}\right)^{1/2} \text{ m}, \tag{47}$$

where the sum is over the particle species α, e is the electron charge, m_e is the electron mass and K is the Boltzmann constant. The Debye number is the number of electrons in a sphere of radius the Debye length λ_D

$$N_D \equiv n\lambda_D^3 \approx 3.29 \cdot 10^5 \frac{T^{3/2}}{n^{1/2}}, \tag{48}$$

The plasma can be considered to be an ideal gas, *i.e.* a collection of point particles that undergo small and brief interactions or "collisions," when the characteristic interaction energy is much less than the average particle energy

$$\frac{e^2 n^{1/3}}{4\pi\epsilon_0} \ll KT, \tag{49}$$

where we have used $n^{-1/3}$ as an estimate for the typical interparticle distance. Eq. (49) can be rewritten in the form of a small number, the inverse of the Debye number (Eq. (48))

$$\alpha_0 \equiv \frac{1}{N_D} \ll 1. \tag{50}$$

It follows from Equation (45) that binary collisions become less and less important for phenomena of a given characteristic time as the plasma density decreases and the plasma temperature increases. The Debye length is the *screening* distance for the electrostatic potential of a fixed point charge placed inside a (globally) neutral plasma. The screening is caused by the presence of opposite charges: charges of equal sign are repelled and charges of opposite sign are attracted to the point charge. The Debye number then, or the number of electrons in a Debye sphere, is the instantaneous number of particles with which a given particle interacts. If this number is much larger than unity it implies that *collective* effects are much more important than binary interactions or "collisions." In order of magnitude the Debye length is the distance a "thermal" (speed $v_{te} \equiv (KT/m_e)^{1/2} \approx 3.89 \cdot 10^3$ m sec^{-1}) electron travels in a plasma period $2\pi/\omega_{pe}$.

Approximations: We shall consider completely ionized hydrogen plasmas ($n = n_i$ on average). Further we shall neglect quantum effects. This condition implies that recoil effects are unimportant and that

$$KT \gg E_F \equiv \frac{\hbar^2 n^{2/3}}{m_e}, \tag{51}$$

where E_F is the Fermi energy, and

$$KT \gg (\hbar\omega_{pe}, \hbar\omega_{ce}), \tag{52}$$

with

$$\omega_{ce} \equiv \frac{eB}{m_e} \approx 1.76 \cdot 10^{11} \text{ rad sec}^{-1}, \tag{53}$$

the electron gyrofrequency, which is the gyration frequency of a non-relativistic electron in a magnetic field of strength B (units Tesla). These approximations are well satisfied in most cosmic plasmas. Exceptions are the degenerate interiors of white dwarfs and neutron stars, and the strongly magnetized atmospheres of neutron stars (Galeev and Sudan 1983).

The *equation of motion* in a collisionless plasma is the *Vlasov equation*:

$$\frac{Df(\vec{r}, \vec{p}, t)}{Dt} = 0,$$
$$\frac{\partial f}{\partial t} + \vec{v} \cdot \frac{\partial f}{\partial \vec{r}} + q(\vec{E} + \vec{v} \times \vec{B}) \cdot \frac{\partial f}{\partial \vec{p}} = 0. \tag{54}$$

Here $\frac{D}{Dt} = \frac{\partial}{\partial t} + \vec{v} \cdot \frac{\partial}{\partial \vec{r}} + \dot{\vec{p}} \cdot \frac{\partial}{\partial \vec{p}}$ is the comoving derivative in six-dimensional phase space $\{\vec{r}, \vec{p}\}$, and $\vec{E}$ and $\vec{B}$ are the smoothed electromagnetic fields. The Vlasov equation is obtained from the exact microscopic Klimontovich equation for point-like particles by splitting the electromagnetic fields into a smooth part averaged over a distance large compared to the interparticle distance and into the residu, or fluctuation caused by the discrete nature of the charges (Ichimaru 1973). Averaging over the "mesoscopic" length scale larger than the interparticle distance then leads to the Boltzmann equation, of which the left-hand side is the Vlasov equation and the right-hand side contains the effect of fluctuations or "collisions." The right-hand side is of order $\frac{f}{t_{ei}} \approx \frac{\omega_{pe}f}{N_D}$ and can therefore be neglected if $1 \ll \omega t_{ei}$ or if $1 \ll kvt_{ei}$ (wavelength much less than the electron mean free path), where ω is the characteristic frequency and k the characteristic wave number of the phenomenon under investigation. The Vlasov equation is the equation of continuity of the particle distribution function in phase space.

10.2. SMALL AMPLITUDE WAVES

A plasma sustains various kinds of propagating electromagnetic disturbances, *eigen modes or waves*. Vacuum electromagnetic waves are modified by a plasma by the reaction of the charged particles to the wave fields, in particular at frequencies near the electron plasma frequency or lower. But also new wave modes appear such as electrostatic waves created by charge fluctuations in the plasma. To find the various wave modes of infinitesimally small amplitude one must know the *response* of the plasma to electromagnetic perturbations. In general this response consists of induced electric current and charge densities. It can be found by *linearizing* the equation of motion (in our case the Vlasov equations for the particles of various kinds) around the unperturbed equilibrium.

Wave Equation: To derive the wave equation one proceeds as follows. The current density is written as

$$\vec{j}(\vec{r}, t) = \vec{j}^{ind}(\vec{r}, t) + \vec{j}^{ext}(\vec{r}, t), \tag{55}$$

where $\vec{j}^{ind}$ is that part of the current density that is induced by the plasma as a response to the wave electromagnetic field and where $\vec{j}^{ext}$ is the rest of the current, or the "externallly" imposed current. For small electromagnetic field perturbations the response is linear and can be written as

$$\vec{j}^{ind}(\vec{r}, t) = \int_V dV \int_{-\infty}^{t} dt' \vec{\vec{\sigma}}(\vec{r}, \vec{r}'; t, t') \cdot \vec{E}(\vec{r}', t'). \tag{56}$$

Here $\vec{\vec{\sigma}}$ is called the conductivity tensor. If in equilibrium the medium is homogeneous and stationary, the conductivity tensor depends on its arguments only in the combinations $\vec{r} - \vec{r}'$ and $t - t'$. Fourier transformation in space and time, with a dependence $exp\{i(\vec{k} \cdot \vec{r} - \omega t)\}$, then leads to

$$j_i^{ind}(\vec{k}, \omega) = \sigma_{ij}(\vec{k}, \omega) E_j(\vec{k}, \omega). \tag{57}$$

The conductivity tensor can be found from the material properties of the gas. Usually one linearizes the equation of motion and then Fourier transforms to find an expression for the current density that is linear in the applied electromagnetic fields. From this expression the conductivity follows directly. For instance if the equilibrium is given by

$$\begin{aligned} \vec{E}_0 &= \vec{B}_0 = \vec{j}_0 = \tau_0 = 0, \\ n_i &= n_e = n_0, \\ f_e(\vec{r}, \vec{v}, t) &= n_0 f^0(\vec{v}) + f^1(\vec{r}, \vec{v}, t) \end{aligned} \tag{58}$$

one finds for the perturbation of the electron distribution function induced by the wave electric field

$$f^1(\vec{k}, \omega, \vec{v}) = \frac{\frac{n_0 e}{m_e} \frac{\partial f^0}{\partial \vec{v}} \cdot \left\{ \vec{\vec{I}} \left(1 - \frac{\vec{k} \cdot \vec{v}}{\omega}\right) + \frac{\vec{k}\vec{v}}{\omega} \right\} \cdot \vec{E}(\vec{k}, \omega)}{i(\vec{k} \cdot \vec{v} - \omega - i\eta)}, \tag{59}$$

where I is the identity operator. Note that poles arise in the resulting expressions which are handled by using the *Landau prescription* or imposing the *causal condition* $\eta \to 0^+$. These "artificial" prescriptions can be avoided by the use of *Laplace transformation* instead of Fourier transformation in time. The poles correspond to *resonances* between particles and waves (see below).

As a next step one Fourier transforms Maxwell's eqautions, elimates the time-dependent magnetic field in favour of the electric field with Faraday's law:

$$\vec{B}(\vec{k},\omega) = \vec{k} \times \vec{E}(\vec{k},\omega)/\omega, \tag{60}$$

and substitutes Equation (57) to find the wave equation

$$\Lambda_{ij}(\vec{k},\omega)E_j(\vec{k},\omega) = -\frac{i}{\epsilon_0\omega}j_i^{ext}(\vec{k},\omega),$$

$$\Lambda_{ij}(\vec{k},\omega) \equiv \frac{k_i k_j c^2}{\omega^2} - \left(\frac{k^2 c^2}{\omega^2} - 1\right)\delta_{ij} + \frac{i}{\epsilon_0\omega}\sigma_{ij}(\vec{k},\omega). \tag{61}$$

To find the kinds of plane waves sustained by the particular plasma as expressed by the conductivity tensor, one now puts the external current to zero and solves Eq. (61). This amounts to finding the zeros of the determinant Λ. The resulting algebraic expressions between wave vector $\vec{k}$ and wave frequency ω are the so-called *dispersion equations* for the various kinds of wave modes. In general the relations have complex and real parts. For a particular wave mode and a real $\vec{k}$ the resulting ω has a real part and a complex part. The latter describes damping (Im $\omega < 0$) or growth (Im $\omega > 0$). If the wave grows energy is transferred from the plasma into the eigenmode and one has an *instability* since the equilibrium state of the plasma is not stable to a small perturbation of the right kind.

Instabilities and Resonances: Various kinds of instabilities exist: *Configuration instabilities* arise from spatial inhomogeneities (In this case one cannot in general make a Fourier transformation in space but has to solve an ordinary differential equation to find the spatial part of the eigenmodes.) *Velocity space or micro instabilities* include instabilities in homogeneous plasmas and are caused either by an energy inversion or a velocity anisotropy of the particle distribution functions. Depending on the value of Im $\omega/$Re ω in comparison to unity one speaks of *reactive or hydrodynamic* (large growth rate) and *resistive or kinetic* (small growth rate) instabilities. Usually a particular physical instability can appear in both forms depending on the relative deviation of the particle distributions from thermodynamic equilibrium.

For a given wave mode and given wave vector there are two groups of particles which react differently to the wave. Most of the particles are set into oscillation by the wave field. On average there is no energy transfer between these particles and the wave. This is the bulk of the plasma which sustains the particular wave mode. On the other hand for some particles the projection of their velocity in the direction of the wave vector equals the phase speed of the wave. These particles feel a constant electromagnetic field and are accelerated (or retarded). They contribute to damping (or growth) of the particular wave. The general form of these wave–particle *resonances* can be found as follows: The force on a particle of speed $\vec{v}$ and orbit $\vec{r}(t)$ exerted by a wave $(\vec{k},\omega)$ is proportional to $exp\, i(\vec{k}\cdot\vec{r}(t) - \omega t)$. In a magnetic field the particle orbit is given by

$$\vec{r}(t) = \left(v_{||}t, \frac{v_\perp}{\omega_{cq}}sin(\omega_{cq}t), \frac{v_\perp}{\omega_{cq}}cos(\omega_{cq}t)\right),$$

where ω_{cq} is the cyclotron frequency of the particle of kind q. Substituting the orbit into the force and making use of Bessel identities then gives the force with dependence

$$\sum_{n=-\infty}^{+\infty} J_n \left(\frac{k_\perp v_\perp}{\omega_{cq}} \right) e^{-i(\omega - k_\parallel v_\parallel + N\omega_{cq})t}.$$

It follows that a resonance between particle and wave exist provided

$$\omega - k_\parallel v_\parallel + N\omega_{cq} = 0, \tag{62}$$

for any integer value of N and the wave has the correct polarization.

The local energy density of plasma waves of kind σ and wave vector $\vec{k}$ can be characterized by an *effective temperature* which in thermodynamic equilibrium coincides with the temperature of the plasma (Tsytovich 1973; Melrose 1986b):

$$KT^\sigma(\vec{k}) \equiv W^\sigma(\vec{k}), \tag{63}$$

where $W^\sigma(\vec{k})$, the energy density of wave mode σ per unit volume of space and per unit volume of wave vector space, is related to W^σ, the energy density of wave mode σ, by $W^\sigma \equiv \int W^\sigma(\vec{k}) d^3\vec{k}(2\pi)^{-3}$. In a collisionless plasma (with a Debye number much larger than unity $N_D \equiv n_e v_{te}^3/\omega_{pe}^3 = 3.3 \cdot 10^5 \, T_e^{1.5} n_e^{-0.5} \gg 1$, $v_{te} \equiv \sqrt{KT_e/m_e}$ is the thermal electron speed and n_e the electron density per m^3) the effective temperature of plasma waves can be much larger than the gas temperature (or the characteristic particle kinetic energy divided by Boltzmann's constant). For instance for Langmuir waves of effective temperature T^ℓ the energy density per unit volume is of order $W^\ell \approx KT^\ell(N_D 18\pi^2)^{-1} W_{gas}$ and is smaller than unity up to an effective temperature $T^\ell \approx N_D T$, where $W_{gas} = 3nKT$ is the kinetic energy density in the gas. Further the effective temperature of a given wave mode can be high only for a small range of wave vectors $\vec{k}$ so that the integrated energy density still remains small.

For such a suprathermal wave level the particle momentum distribution functions from Maxwellian distributions must deviate from Maxwellians. Specific deviations, either in the form of an *energy inversion* of one of the particle species (bump-in-tail) or an *anisotropy* (beams, loss-cones) or an *inhomogeneity* (configuration space instabilities), each have their own instabilities.

10.3. WEAK TURBULENCE

The weak turbulence approximation is based on a perturbation expansion of the combined Maxwell and Vlasov equations in the electric field (Melrose 1986b, Ch. VI). One uses the relatively small wave energy densities as given by the expansion parameter (Davidson 1972)

$$\alpha_1 \equiv \frac{W^\sigma}{nKT} \ll 1. \tag{64}$$

We shall briefly summarize the general procedure. The first part is identical to the derivation of the wave equation. Again the Maxwell and Vlasov equations are Fourier transformed in space and time and the time-dependent magnetic field is expressed in the electric field. The difference is that the equation of motion is not linearized but higher-order terms in the perturbing electric field are kept. If one uses the Vlasov equation its Fourier transform is solved by a perturbation expansion technique for the distribution function. The distribution function contains a zero-order part and a part induced by the electric field: one writes an infinite sum of terms in the form

$$f(\vec{r}, \vec{p}, t) = \sum_{n=0}^{\infty} f^{(n)}(\vec{r}, \vec{p}, t), \tag{65}$$

where $f^0(\vec{r}, \vec{p}, t) = f^0(p)$ is the stationary and homogeneous unperturbed equilibrium solution, and $f^{(n)}$ is proportional to the n-th power of the electric field. Since the acceleration term in the Vlasov equation is proportional to the electric field, a formal solution for the distribution function can readily be written down in the form of a hierarchy of contributions Equation (65) which are expressed recursively in a convolution of a lower-order contribution and the electric field. These successively higher-order (in E) terms of the distribution function lead to successively higher-order contributions to the electric current density induced in the plasma by the electric field, by straightforward multiplication with charge and velocity, integration over momentum space and summing over different species of charged particles:

$$
\begin{aligned}
j_i(\vec{k}, \omega) &= \sigma_{ij}(\vec{k}, \omega) E_j(\vec{k}, \omega) + \\
&+ \int d\lambda^{(2)} \sigma_{ij\ell}(\vec{k}, \omega, \vec{k}_1, \omega_1, \vec{k}_2, \omega_2) E_j(\vec{k}_1, \omega_1) E_\ell(\vec{k}_2, \omega_2) + \\
&+ \int d\lambda^{(3)} \sigma_{ij\ell m}(\vec{k}, \omega, \vec{k}_1, \omega_1, \vec{k}_2, \omega_2, \vec{k}_3, \omega_3) E_j(\vec{k}_1, \omega_1) E_\ell(\vec{k}_2, \omega_2) E_m(\vec{k}_3, \omega_3) \\
&+ ...,
\end{aligned}
\tag{66}
$$

where

$$d\lambda^{(n)} \equiv \frac{d^3\vec{k}_1 d\omega_1}{(2\pi)^4} ... \frac{d^3\vec{k}_n d\omega_n}{(2\pi)^4} (2\pi)^4 \delta^3(\vec{k} - \vec{k}_1 - ... - \vec{k}_n) \delta(\omega - \omega_1 - ... - \omega_n).$$

Substituting Equation (66) for the induced current density into Ampère's law, the wave equation for the electric field can now be derived in the standard manner from Maxwell's equations. Collecting the terms which are linear in the electric field on the left hand side of the equation and the rest on the other side, one arrives at the wave equation in the form

$$\Lambda_{ij}(\vec{k}, \omega) E_j(\vec{k}, \omega) = -\frac{i}{\epsilon_0 \omega} j_i^{nonl}(\vec{k}, \omega), \tag{67}$$

which differs from Equation (61) in that the non-linear induced current contributions (second and higher order in the electric field) are kept and placed on the right-hand side as an effective "external" current density. For a given nonlinear current density Equation (67) determines the corresponding electric field: The field can therefore be solved for by inversion of Equation (67). In other words the nonlinear properties of the plasma (through the action of electric fields on the particle motion) create electric perturbations at beats and harmonics (in Fourier space) of primary electric fields.

The *power per unit volume radiated* by an "exterior" current is given by

$$P = -lim_{\tau, V \to \infty} \frac{1}{\tau V} \int \frac{d^3\vec{k} d\omega}{(2\pi)^4} \vec{j}^{ext}(\vec{k}, \omega) \cdot \vec{E}^*(\vec{k}, \omega), \tag{68}$$

where we take $\vec{j}^{ext} = \vec{j}^{nonl}$ in this case.

Apart from small field amplitudes the weak turbulence approximation assumes the validity of the *Random Phase Approximation* (RPA) for the ensemble average of the product of the field Fourier components

$$< E_i(\vec{k},\omega)E'_j(\vec{k},\omega) >= \lim_{\tau,V\to\infty} \frac{(2\pi)^4}{\tau V} < E_i(\vec{k},\omega)E^*_j(\vec{k},\omega) > \delta^3(\vec{k}+\vec{k}')\delta(\omega+\omega'). \quad (69)$$

Physically this approximation is valid if the *autocorrelation time* of a wave packet τ_{ac} is much less than the interaction time τ_{int} of a characteristic particle with the wave field

$$\alpha_2 \equiv \frac{\tau_{ac}}{\tau_{int}} \ll 1. \quad (70)$$

Here τ_{ac} is the time during which a resonant particle feels a force exerted by the wave packet

$$\tau_{ac} \equiv \frac{1}{k\Delta\frac{\omega}{k}}, \quad (71)$$

and the interaction time is given by

$$\tau_{int} = min(\tau_{tr}, \tau_D, \gamma^{-1}), \quad (72)$$

with the trapping time for a particle of mass m, charge q in a wave of wave vector $\vec{k}$ and electric field amplitude E

$$\tau_{tr} = \left(\frac{m}{qkE}\right)^{1/2}, \quad (73)$$

the diffusion time of a resonant particle under the action of the waves

$$\tau_D = \frac{v^2}{D}, \quad (74)$$

where $D = D(\vec{v}, W^\sigma(\vec{k}))$ is the corresponding *diffusion coefficient* in velocity space for the process considered, and finally γ the growth rate of the waves involved. (Note that if one of the conditions for weak turbulence are not satisfied one of various forms of *"strong" plasma turbulence* exists. We refer to Similon and Sudan (1990) for a recent review on strong turbulence, to Sitenko (1982) for self-consistent nonlinear fluctuation theory and Zakharov equations, to Papadopoulos and Freund (1979) and Goldman (1984) for ponderomotive force and modulational instability, and to Freund (1982) and Melrose (1986).)

Further the electric Fourier component at arbitrary argument $(\vec{k},\omega)$ can be written as

$$E_i(\vec{k},\omega) = \sum_\sigma E_i^\sigma(\vec{k},\omega)2\pi\delta\left(\omega - \omega^\sigma(\vec{k})\right), \quad (75)$$

where the dispersion relations of the various wave modes in the plasma under consideration are indicated by the label σ. Technically for each wave mode in Equation (75) positive and negative frequencies are involved, and one must extend the summation over *"forward"* and *"backward"* waves with separate dispersion relations $\sigma^\pm$ for each mode. Emission of electromagnetic waves of kind t occurs if the integrand in Equation (68) does not vanish at the argument $(\vec{k},\omega^t(\vec{k}))$ where the label t indicates the dispersion relation $\omega = \omega^t(\vec{k})$. The radiated power per unit volume of space and per unit volume of wave-vector space is now obtained from Equation (68) with Equations (69) and (75) and integration over frequency

$$Q^t(\vec{k}) \equiv \frac{dW^t(\vec{k})}{dt} = -lim_{V \to \infty} \frac{1}{V} \, \vec{j}^{ext}(\vec{k}) \cdot \vec{E}^{t*}(\vec{k}) + C.C. \tag{76}$$

Here $W^\sigma(\vec{k})$ is the energy density of wave mode σ per unit volume of space and wave–vector space

$$W^\sigma(\vec{k}) = \frac{\epsilon_0 \, | \, \vec{E}^\sigma(\vec{k}) \, |^2}{V R_E^\sigma(\vec{k})}, \tag{77}$$

where $R_E^\sigma(\vec{k})$ is the ratio of electric to total energy density of mode σ at wave vector $\vec{k}$. In Equation (76) and later the energy density definition is chosen in such a way that the frequencies of the wave modes are always positive, in contrast with the earlier expressions.

All one has to do now is to single out the contributions to the nonlinear current for a specific process and to substitute this for the external current in Equation (68).

10.4. Observations of Waves

Direct evidence for waves in the solar corona comes from *satellite measurements* in the solar wind: In the ecliptic plane at a solar distance of 0.3 – 1 AU predominantly outward travelling Alfvén waves have been observed in the frequency range $10^{-4} - 10^{-1}$ Hz in the satellite frame (Bavassano *et al.* 1982), corresponding to *Alfvén* waves of large spatial scales (0.3 AU – $3 \cdot 10^{-4}$ AU) and long intrinsic periods (hours – mins) (Note that for outgoing Alfvén waves in the satellite's frame $\omega^A(\vec{k}^A) = \vec{k}(\vec{V}_{sw} + \vec{v}_A)$, with ω^A the Alfvén wave frequency, $\vec{k}^A$ the wave vector, $V_{sw} \approx 500 - 800$ km sec^{-1} is the solar wind speed and the Alfvén speed is $v_A \ll V_{sw}$). At a distance from the Sun larger than 1 AU the energy in both incoming and outgoing waves becomes comparable (Roberts *et al.* 1987). Further at a distance of 1 AU *Langmuir* waves and *ion-sound* waves have been observed in situ concurrent with electron beams of solar origin and radiation at the fundamental electron plasma frequency (Lin 1990). Finally Langmuir waves have been observed in situ in the vicinity of shock fronts and of shock accelerated superthermal electrons, simultaneously with interplanetary radio bursts (Type II bursts) at the second harmonic of the electron plasma frequency (Kikuchi *et al.* 1989).

At distances closer to the Sun and other stars the presence of waves can be inferred only from radiation received on earth. Since stellar coronae are hot ($T \approx 10^5 - 10^9$ K) and dilute ($n \approx 10^{12} - 10^{19}$ m^{-3}) their radiation is mainly in the X-ray and radio domain.

Unfortunately in *X-rays* wave motions of solar coronal gas are difficult to observe because of the weakness of emission lines from the dilute corona. One has to go to large wavelengths, in the optical and infrared, all coming from the dense photospheric and lower chromospheric layers, to find suitable lines to detect the presence of waves. Apart from the solar 5-min oscillations, standing pressure waves in subphotospheric cavities, (Marmolino and Severino 1990; Chapter 3) the experimental evidence for waves in the solar photosphere and lower chromosphere is rather meagre (Narain and Ulmschneider 1990). In magnetic flux tubes, which at these altitudes still cover only a small fraction (≈ 1 %) of the sphere, the highest "turbulent velocities" have been found of 2km sec^{-1} (Keller 1990; Solanki 1990). Averaged over the solar surface the wave flux inside magnetic flux tubes (probably in the form of slow–mode acoustic waves) is of the same order as the acoustic flux outside the tubes and sufficient to heat the chromosphere, of order $4 \cdot 10^3$ W m^{-2} (Solanki and Roberts 1990).

The solar corona is best observed in the *radio* domain and it is at these wavelengths that the most impressive evidence exists for the nonstationary occurrence of plasma waves at various altitudes in the corona (Dulk 1985; McLean and Labrum 1985; Krüger 1979; Kuijpers 1980). Many of the emissions cannot be explained by single–particle processes as Bremsstrahlung, cyclotron or synchrotron radiation (see Chapter 14), because of the high effective temperatures ($\geq 10^{10}$ K), the large degree of circular polarization (sometimes 100%) and the occurrence of characteristic fine structures in the emission as a a function of frequency and time. Examples of such *dynamic radio spectrograms,* in which the intensity of the radiation is presented as a function of frequency and time with a maximum resolution of 1 ms can be found in (Allaart *et al.* 1990; Slottje 1980; Güdel and Benz 1988). Apparently collective effects play a role and channel some of the free energy liberated in relaxing coronal magnetic structures (such as flares) into specific plasma modes and electromagnetic radiation. From the particular structure of the highly variable emission patterns which show up in the frequency-time plane (such as bandwidth, drift rates, harmonic structures) it is sometimes possible to infer which kinds of plasma modes are involved. From the radio data it is concluded in this way that the *following kinds of plasma waves occur in the disturbed solar corona*:

- *Langmuir* waves from electron beams (associated with so-called Type III radio bursts), and also directly observed in situ during the presence of electron beams in association with the radio bursts (Lin 1990),

- *Upper hybrid* waves from a loss-cone instability when accelerated electrons travel downwards in a converging magnetic loop anchored in the dense lower atmosphere of the Sun. The presence of these waves is inferred from radio continua emitted during and shortly after solar flares, the so-called Type IV bursts (Kuijpers 1974; Stepanov 1974) and zebra fine structures (Kuijpers 1975a; Zheleznyakov and Zlotnik 1975c),

- *Whistler* waves and *mhd* waves, again excited by a loss-cone instability in sources of Type IV bursts with so-called fiber fine structure (Kuijpers 1975b; Mann *et al.* 1989), and

- *High-frequency electromagnetic* waves in the extraordinary mode directly generated by a cyclotron maser instability of a one-sided loss-cone distribution of fast electrons (during so-called spike radio bursts) (Stepanov 1978; Wu and Lee 1979; Holman, Eichler and Kundu 1980; Melrose, Hewitt and Dulk 1984; Le Quéau *et al.* 1984; White, Melrose and Dulk 1986; McKean, Winglee and Dulk 1989).

Other wave modes have been proposed to explain certain kinds of radio bursts from the solar corona (such as *Bernstein waves* (Chiuderi, Giachetti and Rosenberg 1973; Zheleznyakov and Zlotnik 1975a,b) and *lower hybrid waves* (Spicer, Benz and Huba 1981; Thejappa 1987) but the observational evidence for their presence is not so compelling as for the wave modes listed above. From a theoretical point of view various low-frequency wave modes are certainly expected in the corona either as a result of a bump-in-tail instability, of a loss-cone instability, or of multiple reconnections in a flare. However the problem is that low-frequency waves cannot leave the corona and propagate directly towards the earth. They

582

must be *upconverted* in frequency above the relevant cut- off frequency, e.g. by wave coupling to a high-frequency wave. In this respect an interesting interactive experiment has been performed by irradiating the solar corona with strong radio waves (Benz and Fitze 1979). However as yet no signals were received at a satellite frequency displaced over a frequency interval corresponding to inelastic backscattering on coronal low-frequency waves.

10.5. PLASMA RADIATION

For incoherent emission processes the radiation intensity at most equals the sum of the contributions from the individual radiating particles. The effective temperature is then at most equal to the characteristic energy of an individual radiating particle (divided by K). However if collective interactions are important the observed brightness temperature can exceed the particle energy considerably. This is precisely what happens in the presence of a suprathermal level of waves. Two kinds of coherent radiation processes can be distinguished (Ginzburg and Zheleznyakov 1975)

1. An **antenna mechanism**, where a strong coherent plasma wave groups the particles in bunches in space on a scale smaller than the wavelength of the radiation emitted. Then the amplitudes of the radiation fields of the individual particles add constructively and the radiation temperature is enhanced by a factor equal to the number of particles in a bunch;

2. A **maser** where spontaneous radiation is amplified as a result of a deviation from thermodynamic equilibrium *(negative absorption coefficient)*. The non–equilibrium can consist of an energy inversion, an anisotropy or a spatial gradient and lead to a *linear* instability of electromagnetic waves

$$\frac{dW^t}{dt} \propto W^t. \tag{78}$$

Alternatively the deviation can consist of a suprathermal level of plasma waves W^σ which scatter on particles or couple with other plasma waves into radiation

$$\frac{dW^t}{dt} \propto W^t W^\sigma, \qquad \frac{dW^t}{dt} \propto W^\sigma W^{\sigma'}. \tag{79}$$

Strictly speaking the latter process in Equation (79) is not a maser but a nonlinear instability as $\sigma, \sigma' \neq t$. In general the observed brightness temperature can be of the order of the effective temperature of the partaking waves but depends on the details of their phase space distribution and on the optical thickness for the particular process (Melrose 1980).

The following candidates for efficient plasma radiation have been developed theoretically for Sun and stars:

1. A *linear instability for electromagnetic waves at harmonics of the electron cyclotron frequency* (Stepanov 1978; Wu and Lee 1979; Holman, Eichler and Kundu 1980; Melrose, Hewitt and Dulk 1984; Le Quéau, Pellat, and Roux 1984; White, Melrose and

Dulk 1986; McKean, Winglee and Dulk 1989). The instability depends on the occurrence of a *cyclotron resonance* between a fast electron and an electromagnetic
wave of suitable frequency, direction and polarization. In the frame moving along
the magnetic field with the projected particle speed the Lorentz transformed wave
frequency must equal an integer number of cyclotron harmonics in the same frame.
Also the wave polarization must match the sense of rotation of the particle around
the magnetic field. The resonance condition can be written as (compare Equation
(62)

$$\omega - k_\| v_\| = N\omega_{cj}, \quad \omega_{cj} \equiv \frac{q_j B}{\gamma m_j}, \quad N = \pm 1, \pm 2, ..., \tag{80}$$

where q_j and m_j are the charge and, respectively, the rest mass of the particle of
kind j, ω_{cj} is the relativistic it cyclotron frequency in the laboratory frame and γ the
Lorentz factor of the particle in the same frame. Note that the (*Landau or Čerenkov*)
resonance $N = 0$ is lacking in Equation (80) since in an ordinary plasma escaping
radiation has a phase speed above the speed of light. Further the wave polarization
determines the relevant sign on the right-hand side.

The first condition for the instability to develop is an anisotropy such as a one-sided
loss-cone distribution of fast electrons in an ambient cooler isotropic plasma. This
is characteristic for electrons accelerated in a converging magnetic field rooted in
a dense atmosphere, such as in the magnetospheres of planets, the Sun and stars.
The second condition is that either the electron Larmor frequency $\omega_{c0} \equiv eB/m_e$ is
sufficiently large with respect to the electron plasma frequency — this is the case for
the *cyclotron* maser with non- relativistic electrons $\omega_{c0}/\omega_{pe} \geq 3$ — or that the electrons
are sufficiently relativistic so that the radiation is emitted at higher harmonics of
the cyclotron frequency — this is the case for the *synchrotron* maser (Zheleznyakov
1967; Louarn, Le Quéau and Roux 1987) with mildly relativistic particles where
$\omega_{c0}/\omega_{pe} \geq 0.3$ suffices.

2. *Induced scattering* of a suprathermal level of plasma waves into electromagnetic waves
(Tsytovich 1970). The general resonance condition for scattering of a wave of kind σ,
wave vector $\vec{k}^\sigma$ and corresponding frequency $\omega^\sigma(\vec{k}^\sigma)$ into a wave of kind τ is

$$\omega^\sigma(\vec{k}^\sigma) - k_\|^\sigma v_\| = \pm(\omega^\tau(\vec{k}^\tau) - k_\|^\tau v_\| + N\omega_{cj}), \quad N = 0, \pm 1, \pm 2, ... \tag{81}$$

The + sign is for *scattering*, the − sign for *double emission* (Melrose 1982). Physically
the condition says that the Doppler shifted wave frequencies in the frame of the
scattering particle are equal (modulo its cyclotron frequency).

If the fast electrons responsible for the excitation of plasma waves have a much smaller
density than the Maxwellian background distribution, the scattering process is dominated by the thermal background. Further the (Thomson) scattering on the thermal
electrons is then practically cancelled by the (nonlinear) scattering on the oppositely
charged polarization clouds. As a result the *(nonlinear) scattering on the polarization
clouds around the ions*, which is not compensated by the inefficient Thomson scattering on the relatively heavy ions themselves, dominates. Finally the scattered wave

then undergoes exponential amplification (negative absorption, induced scattering) if it has a smaller frequency than the incoming wave.

3. *Nonlinear coupling* of two plasma waves (one or both of high frequency) into escaping radiation (*wave fusion or decay*, three wave coupling). The resonance condition is

$$\vec{k}^\sigma + \vec{k}^\tau = \vec{k}^\mu,$$
$$\omega^\sigma(\vec{k}^\sigma) + \omega^\tau(\vec{k}^\tau) = \omega^\mu(\vec{k}^\mu). \tag{82}$$

The efficiency of this process depends on the effective temperatures of both kinds of fusing plasma waves which must be sufficiently larger than the gas temperature (Melrose 1982).

4. Wave coupling processes of *higher than second order* in the wave energy densities.

5. *Emission from acceleration centers.* Recently it has been found that strong radiation is emitted if electrons are accelerated in a *locally strong electric field* (Kuijpers 1990a; Tajima *et al.* 1990). Such radiation would be very important as a diagnostic means to probe the acceleration process.

The process calculated by Kuijpers (1990) is for an *electrostatic double layer* and is physically related to the process of *linear accelerator radiation* (Wagoner 1969; Ginzburg 1970; Melrose 1980, p. 349; Ginzburg 1989). As a single–particle process the mechanism is inefficient since the electrons are accelerated parallel to their momentum. However for a strong electrostatic double layer a coherent version of the process has been found and it can produce radiation with a brightness temperature under coronal conditions of up to 10^{25} K (Kuijpers 1990a). Physically the process also shows some similarity to the *free electron laser*. In the case of the double layer however it is the low-frequency electric field "wiggle" instead of the magnetic wiggle which is upconverted by the electron beam. To a first approximation a resonance condition for scattering must be satisfied: in the frame of the scattering electron the Doppler shifted frequency of the Fourier electric field components of the double layer must equal the Doppler shifted frequency of the emitted radiation. The emission process is therefore *induced scattering*, now however starting with a very nonlinear wave in the form of the double layer. Also the electrons need not be relativistic to produce intense emission. The process depends crucially on the properties of electrostatic double layers and their occurrence in stellar coronae. Since such electrostatic double layers are sustained by the electric circuit of the flare the electric field can remain constant on the time scale of the emission and the process constitutes a true *(linear) maser*. We think that this process is responsible for some of the small band emission (spikes, Type I) observed in the solar corona.

The process in Tajima *et al.* (1990) is calculated numerically and depends on a *Čerenkov beam resonance* with a propagating electromagnetic branch in the strongly disturbed non-Maxwellian plasma (Wentzel 1991). Both radiation processes from DC fields deserve further study.

6. *Transition radiation* from fast particles encountering inhomogeneities (Ginzburg and Tsytovich 1990). (Note that the autors also include induced scattering under this name).

Apart from the last two items strictly speaking this list is based on the limiting assumption of a perturbative expansion of the electromagnetic wave–particle interactions for weak fields. So it can not be applied to situations where the orbits of the particles interacting with the electromagnetic field, are strongly disturbed. This can already occur for electro(magnetic) field energy densities much less than the gas kinetic energy density. In that case the above resonances are *"broadened"* by the plasma turbulence (Sitenko 1982).

Further when the wave energy density approaches the gas kinetic energy density, locally the relative importance of various wave-coupling processes cannot be so easily be decomposed into a hierarchy. This occurs for instance in the intense low-frequency field of radio pulsars impinging upon the interstellar medium (Tsinsadze 1989), or when a strong pulse from a cyclotron instability enters a dilute environment (Winglee, Dulk and Pritchett 1988; Karimibadi *et al.* 1990).

10.6. Conversion Problems

In the explanation of observed radiation of high-brightness temperature (relative to the gas temperature) the following problems exist:

- The cyclotron maser: It is not clear how common this mechanism is in stars as it requires *rather high coronal field strengths*. The condition for the occurrence of the cyclotron maser can be rewritten as a lower limit on the Alfvén speed of $2 \cdot 10^4$ km sec^{-1}. In the solar corona the characteristic Alfvén speed is often taken to be $500 - 1000$ km sec^{-1}s. This is in contrast with radiation from the earth and planets where the cyclotron maser is probably the most important radiation process.

  Because of this difficulty it has been suggested that the synchrotron, rather than the cyclotron, maser operates in solar and stellar coronae (Louarn, Le Quéau and Roux 1987). Indeed the mildly relativistic electrons required for this mechanism to operate, are known to be present under flare conditions. What is not clear however is whether this emission is more important than radiation from converted plasma waves, which themselves are produced by the loss-cone distribution of the much more numerous non-relativistic electrons in the flare.

- Induced scattering of plasma waves into radiation: Under characteristic conditions in the solar corona flare-accelerated electrons in a magnetic loop are expected to develop a loss-cone distribution and to generate an exponential growth of upper–hybrid waves or "Z-mode" waves (Kuijpers 1974; Stepanov 1974). Indeed it has been shown recently (Winglee and Dulk 1986) that the *cyclotron maser transforms into the upper–hybrid instability for relatively small ratios of* ω_{c0}/ω_{pe}. The difficulty in assessing the role of the electrostatic upper–hybrid instability in the observed radiation in comparison with the synchrotron maser is the *unknown conversion efficiency* of upper–hybrid waves into radiation in a realistic solar corona. The original proposal that induced scattering on the background Maxwellian distribution can be very efficient has been questioned (Melrose 1986b) on the grounds that induced emission is offset by absorption once the amplified radiation reaches less dense layers inside the source of upper–hybrid waves where the upper–hybrid frequency is smaller than the radiation frequency. In our view this criticism is not justified since absorption plays only a minor role if a strong density contrast exists between the plasma wave source and its surroundings. In particular in directions transverse to the magnetic field density scale lengths of 100 *km* or less are not exceptional in the observed *fibrous* corona (see also Bruggmann *et al.* 1990). It is even not excluded that the plasma waves transform into radiation by *linear mode coupling* on strong density gradients.

- Nonlinear coupling of plasma waves: This process competes with induced scattering for the production of radiation at the fundamental of the upper–hybrid frequency and depends on the presence of low-frequency waves and the source geometry.

Probably wave-coupling is responsible for the radio fine structures with "intermediate" drift speeds in the frequency-time plane. It seems likely that such drifts are due to wave structures propagating at a magnetoacoustic (shocks, Alfvén waves) or electron Alfvén speed (whistlers). Why these drifting fine structures are formed *quasiperiodically* is not completely understood but may be related to the formation of solitons (Kuijpers 1975c; Treumann, Güdel and Benz 1990; Mann *et al.* 1989). This quasi-periodicity is also found in other bursts (Type III bursts, spike bursts) and again there the cause of the periodicity is not agreed upon (Winglee, Dulk and Pritchett 1988; Kuijpers, van der Post and Slottje 1981; Melrose 1986a). The radiation properties of such structures need further study.

As yet there is no detailed theory satisfactorily explaining the Type II emission patterns observed in association with coronal *shocks* (Thejappa 1987). The main reason is our limited understanding of the structure of collisionless shocks.

- Other processes: The above instabilities arise when the accelerated particles have travelled away from the acceleration region. It is expected however that intense *radiation is also coming from the acceleration sites* themselves (Kuijpers 1990a; Tajima *et al.* 1990).

11. Particle Acceleration

From observations of gamma ray emission in solar flares it is now clear that in the first few seconds of the energy release both electrons and ions are accelerated to relativistic energies: electrons in 1s up to 100 MeV and ions within seconds to GeV energies (Rieger 1989; Chapter 24). At present the best candidates for acceleration in the flare involve a *multitude of nonlinear structures* either in the form of double layers or of shocks or of a high level of mhd waves (Vlahos 1989). Apart from acceleration to relativistic energies in solar and stellar flares electrons are accelerated in large number to medium relativistic energies. Below we shall concentrate on the acceleration processes believed to be operating in flares.

What is the difference between *heating and acceleration?* Heating is used for a temperature increase of gas or, more loosely, for an increase in random energy of the bulk of the particles in a specified volume. Acceleration is used either for the preferential increase of a small set of particles or for the increase of the net momentum vector of gas, resulting in a particle distribution which differs markedly from a Maxwellian.

Three groups of acceleration mechanisms can be distinguished:

- Acceleration by waves;

- Acceleration by unidirectional electric fields;

- Acceleration by shocks.

Each of these processes can lead to acceleration as opposed to heating since they operate selectively: Waves because of the resonance condition which has to be satisfied; Electric fields since they cause runaway acceleration either of a small fraction or of the entire population; Shocks since only particles above a certain injection energy gain energy on average.

A distinction is usually made between *stochastic acceleration* processes and *regular acceleration*. Stochastic acceleration is characterized by an increase of the spread in momenta of the accelerated particles, at the same time as the average energy per particle increases. Examples of stochastic acceleration are, apart from acceleration by *plasma turbulence: second-order Fermi* acceleration (Fermi 1954) and *betatron* acceleration (Melrose 1983). Physically the foundation of all these processes is the tendency for an ensemble of particles that interact with a population of energetic entities via random kicks, to strive towards energy equipartition per particle or entity.

Examples of regular acceleration are *first-order Fermi* acceleration in *quasi-parallel shocks* (shock direction parallel to the magnetic field), drift acceleration in *quasi-perpendicular shocks,* acceleration by *unidirectional fields* as in electrostatic double layers or in a current circuit.

11.1. ACCELERATION BY WAVES

Both electrons and protons can be accelerated to highly *relativistic* energies by high levels of *Alfvén* waves, *magnetosonic* waves and *whistler* waves (Miller and Ramaty 1987; Ramaty and Murphy 1987; Smith 1990; Schlickeiser 1989; Steinacker and Schlickeiser 1989; Miller, Guessoum, and Ramaty 1990; Miller 1991). The energy is transferred through a cyclotron resonance Equation (80) between particle and wave of the right frequency, direction and polarization. From this condition and the dispersion relations for these waves it follows that particles from the ambient plasma with characteristic coronal temperatures of $10^6 - 10^7$ K are accelerated only if waves of sufficiently high frequency (approaching the electron, respectively ion, cyclotron frequency) are present.

In the *subrelativistic* regime acceleration by electrostatic waves is very efficient. In particular *Langmuir* waves efficiently accelerate electrons by *Landau damping*. The waves are however not efficient accelerators for relativistic energies. The reason is that as the particle speed approaches the speed of light the wave energy available for acceleration occupies an ever smaller volume of wave vector space. Apart from Langmuir waves also *low-frequency* electrostatic waves as lower-hybrid waves are efficient accelerators. The latter can be excited by anisotropic ion distributions (Benz and Smith 1987; Thejappa 1987).

The main problem with acceleration by waves is the *origin* of the particular kind of waves invoked. Somehow magnetic energy must be converted into waves. The details of the energy liberation in the flare are however unknown. While it is true that the expected sudden changes in coronal field topology during flares must be *abundant radiators of low-frequency mhd waves* these waves primarily have periods of the order of milliseconds or larger, corresponding to the characteristic Alfvén travel time of the reconnecting regions. From the cyclotron resonance condition it then follows that these waves are of too small a frequency to accelerate low-energy particles of typical coronal energy. Therefore an *injection problem* exists. Either a cascade of mhd waves to higher frequencies (e.g. by nonlinear wave couplings) ensures efficient acceleration from coronal to subrelativistic energies or preacceleration of the coronal plasma is required by other mechanisms.

Preacceleration by high-frequency electrostatic waves does not really solve the problem if the generation of these waves is not explained. Again a process would be needed to produce such waves out of the low-frequency flare turbulence. Less than a decade ago "Turbulent Bremsstrahlung" was thought to generate Langmuir waves (and high-frequency radiation) out of low-frequency electrostatic waves via a Čerenkov resonance of particles with the low-frequency waves. It has become clear however that this mechanism does not exist (Melrose and Kuijpers 1987) although it still appears in the literature. More interesting are the *lower-hybrid* waves (Benz and Smith 1987).

Another alternative to solve the preacceleration or injection problem may be to properly take account of the *finite amplitude* of the waves: First calculations by large amplitude mhd waves (de la Beaujardière and Zweibel 1989) demonstrate that the cyclotron resonance is not a necesssary condition for particle acceleration. Instead a Čerenkov condition ($N = 0$ in Equation (80)) is sufficient by the appearance of parallel electric fields associated with charge separations created by mirroring of the particles between the large wave crests. Calculations of particle acceleration in many waves show that large amplitude perturbations cause nonlinear overlapping of particle orbits in phase space resulting in chaotic particle acceleration. The first calculations indicate maximum acceleration speeds ten times the ambient Alfvén speed. Further, strong initial mhd turbulence indeed leads to the development of solitons and nonlinear structures (Uberoi 1990) required for such acceleration. Finally as we have seen in the previous section such nonlinear mhd structures have also been inferred to exist from radio fine structures.

The second alternative way for preacceleration is through *shock* wave heating in the shocks developing in the nonsteady reconnecting flare plasma (Cargill, Goodrich and Vlahos 1988).

Finally the third alternative possibility for acceleration from coronal energies upwards consists of *electrostatic double layers* (Alfvén 1981).

Although mhd waves, shocks and double layers have different acceleration properties and effiencies, in a realistic solar flare they are all generated by reconnection, and probably in a stochastic fashion. Observational evidence for the existence of a multitude of acceleration centers comes from the spatial (Tapping *et al.* 1983; Kattenberg 1981) and spectral properties of the radio emission (Kuijpers, van der Post and Slottje 1981), from the association of spikes with Type III bursts (Benz 1985; Vlahos 1989) and from X-rays (Martens, van den Oord and Hoyng 1985; de Jager *et al.* 1987; Lin *et al.* 1984; Athay 1984; Canfield and Metcalf 1987). Note that also from a theoretical point of view the acceleration in one large double layer (as in the original Alfvén–Carlqvist model (Alfvén and Carlqvist 1967)) or in one single shock is unlikely: If in a solar flare the entire electric current were to dissipate in one strong double layer on the short timescale observed in the actual flare, primarily extremely relativistic particles would be produced (Raadu 1989). For the case of one single shock it would take too long to produce the first relativistic particles (Achterberg and Norman 1980).

The question of the nature of particle acceleration and the role of waves is connected to the problem of the *energy partitioning* in a stellar flare. *What determines the quality of the flare products?* In other words how much of the energy appears as bulk motion, how much as acceleration of a small fraction of the particle population and how much as heating (and perhaps also how much as intense radiation (Melrose, Hewitt and Dulk 1984)). What parameters determine the end products? Apparently in some cases (Mätzler and Wiehl 1980) the flare consists of pure heating only. It may however be that the primary energy release is unique, for instance in the form of pure runaway acceleration (Moghaddam-Taaheri and Goertz 1990) but that the column density of the overlying gas determines whether the accelerated particles can be observed or whether their energy is completely degraded into heating.

11.2. ACCELERATION BY UNIDIRECTIONAL ELECTRIC FIELDS

Direct acceleration by an electric field perpendicular to the magnetic field only becomes important when it approaches the value of the latter multiplied by c. This occurs for instance in the strong wave field of a pulsar impinging on a plasma (Tsintsadze 1989) but not in ordinary stellar atmospheres. Drift acceleration however resulting from particles drifting across moving magnetic fields and perpendicular to equipotential surfaces may be a rather common phenomenon.

Acceleration by an electric field parallel to the magnetic field leads to *runaway* acceleration of particles with initial speed above $(E_c/E)^{1/2} v_{te}$, where the critical field strength, at which the entire plasma "runs away" is $E_c \approx 0.16 E_D$ and $E_D \equiv e \ln \Lambda \lambda_D^{-2}$ is the Dreicer field (Dreicer 1959, 1960; Kaastra 1982). The effect is caused by the decreasing collision frequency for superthermal particles, $\nu_{coll} \propto (v/v_{te})^3$. Since particle speeds are limited by the speed of light it follows that the minimum required field strength for runaway particles to appear is

$$ E > \frac{KT}{mc^2} E_c. \tag{83} $$

Such parallel electric fields are thought to appear in regions of reconnecting fields, either because of flux annihilation (Bulanov and Cap 1988) or because of "short circuiting" effects (Haerendel 1989; Schindler, Hesse and Birn 1991, Winglee *et al.* 1991).

In particular electric circuits in plasmas are known to excite electrostatic double layers or thin structures of parallel electric fields (Alfvén 1958; Jacobsen and Carlqvist 1964; Alfvén and Carlqvist 1967; Block 1978; Alfvén 1981; Raadu 1989). Perhaps the electric current circuit dissipates in the flare by a multitude of transient electric double layers (Hénoux 1987; Kuijpers 1990; Moghaddam–Taaheri and Goertz 1990). In contrast with acceleration by mhd waves or shock waves double layers accelerate both electrons and ions, and both to the same energy. It has been objected (Holman 1985) that the rate of particle acceleration by double layers is limited to the total electric current in the flaring flux tube divided by the electron charge. In our opinion this claim is not valid and the total flux of accelerating electrons is given by the flux of electrons entering double layers which is a factor v_{te}/v_{De} larger than the current drift flux (v_{De} is the electron drift speed of the electric current). The reason is that a beam of accelerated particles is quickly current neutralized when impinging on a relatively dense plasma (van den Oord 1990) as would be the case for an electron beam coming out of a transient double layer with a potential drop that is too large for the original current circuit. A different solution for the number problem based on the geometry of a current sheet where particles drift into the current, are accelerated over a relatively short path in the current direction and drift out again can be found in Kaastra (1985) and Martens (1988).

11.3. SHOCK WAVE ACCELERATION

A first-order (in v/c, v being the speed of the relative motion) Fermi process is realized in the process of *diffuse shock acceleration* (Axford, Leer and Skadron 1977; Krimsky 1977; Bell 1978; Blandford and Ostriker 1978). Essentially one has a shock moving along the magnetic field direction. The fraction of particles that move fast enough can cross the shock front from the downstream region and back again, repeatedly. This occurs because their distribution is kept more or less isotropic by an instability for Alfvén waves or whistler waves. Effectively this then means that each fast particle is reflected by and trapped in between two approaching mirrors and is accelerated. The success of this process lies in the predicted slope of the differential spectrum of accelerated particles. For Fermi acceleration this slope is $f(E) \propto E^{-(1+\tau_{acc}/\tau_{esc})}$, where τ_{acc} is the (energy independent) acceleration time and τ_{esc} the (energy independent) escape time. Whereas the ratio of both times is undetermined in Fermi's original suggestion it is fixed for a strong (non-relativistic) shock and equal to unity. The process therefore leads to a universal spectrum of cosmic rays of slope -2 independent of source characteristics, which compares rather well with the observe slopes -2.8.

Recent reviews of diffusive shock acceleration are Blandford and Eichler (1987) and Achterberg (1990). Modern problems are: the flattening of the particle spectrum at the high energy end for relativistic shocks and the resulting time dependence of the acceleration, and the injection problem for particles (in particular electrons) to be accelerated at all.

A different shock acceleration mechanism relies on the "grad B" drift of a charged particle along the shock front for a shock moving perpendicular to the magnetic field. Particles with cyclotron radius larger than the shock front thickness gain energy by the electric field at the shock ($\vec{E} = -\vec{v} \times \vec{B}$). This process of *shock drift acceleration* (Pesses 1981; Pesses, Decker and Armstrong 1982; Webb, Axford and Terasawa 1983; Holman and Pesses 1983; Decker 1988) is not very efficient if the shock is not moving perpendicular to the field but can become appreciable for nearly perpendicular "superluminal" shocks (Begelman and Kirk 1990; Lieu and Quenby 1990).

Finally a combination of diffusive shock acceleration and shock drift acceleration has been proposed by Jokipii (1982, 1987). If it operates the maximum energy a particle gains is limited to the total potential jump along the shock front.

Both shock acceleration mechanisms are thought not to be important for solar flares, unless perhaps in the form of a collection of many small shocks (Cargill 1991).

12. Conclusion

A solar flare is probably the result of an ideal mhd instability in a force free plasma anchored in a dense driver. This forms a simple paradigm to predict the presence of magnetic flaring activity in other objects in the universe. In this respect, a prominent place is taken by magnetic accretion disks and their interaction with magnetospheres. In contrast with the global flare evolution, the microphysics determining the reconnection and particle acceleration are still very unclear. Future dynamic radiospectra of stellar flares with high temporal and frequency resolution provide a powerful diagnostic means to probe these microphysics, similar to their role in the study of solar flares.

References

Achterberg, A. 1990, in *Physical Processes in Hot Cosmic Plasmas,*, NATO ASI, eds. W. Brinkmann, A.C. Fabian and F. Giovanelli, Kluwer Acad. Publ., p. 67.

Achterberg, A. and Norman, C.A. 1980, *A&A,* **89**, 353.

Alfvén, H. 1958, *Tellus,* **10**, 104.

Alfvén, H. 1981, *Cosmic Plasma,* Reidel Publ. Cy, Dordrecht, Holland.

Alfvén, H., and Carlqvist, P. 1967, *Solar Phys.* **1**, 220.

Allaart, M.A.F., van Nieuwkoop, J., Slottje, C., and Sondaar, L.H. 1990, *Solar Phys.,* **130**, 183.

Aly, J. J. 1980, *A&A,* **86**, 192.

Aly, J.J. 1985, *A&A,* **143**, 19.

Aly, J.J. 1990, in *Flares 22 Workshop, Dynamics of Solar Flares,* eds. B. Schmieder and E.R. Priest, Observatoire de Paris, p. 29.

Aly, J.J. 1991, *ApJ,* **375**, L61.

Aly, J.J. and Kuijpers, J. 1990, *A&A,* **227**, 473.

Athay, R.G. 1984, *Solar Phys.* **93**, 123.

Axford, W.I., Leer, E., and Skadron, G. 1976, *EOS,* **57**, 780, Proc. 15th Int. Cosmic Ray Conf., Plovdiv.

Bahcall, J.N., Rosenbluth, M.N., Kulsrud, R.M. 1973, *Nature Phys. Sci.,* **243**, 27.

Barnett, A. and Olbert, S. 1986, *J. Geophys. Res.,* **91**, 10117.

Bavassano, B., Dobrowolny, M., Mariani, F. and Ness, N.F. 1982, *J. Geophys. Res.* **87**, 3617.

Begelman, M.C. and Kirk, J.G. 1990, *ApJ,* **353**, 66.

Bell, A.R. 1978, *MNRAS,* **182**, 147.

Benz, A.O.: 1985, *Solar Phys.* **96**, 357.

Benz, A.O. and Fitze, H.R. 1979, *A&A,* **76**, 354.

Benz, A.O. and Smith, D.F. 1987, *Solar Phys.* **107**, 299.

Berger, M.A. 1984, *Geophys. Astrophys. Fluid Dynamics,* **30**, 79.

Berger, M.A. 1988, *A&A,* **201**, 355.

Berger, M.A. and Field, G.B. 1984, *J. Fluid Mech.,* **147**, 133.

Bhattarchajee, A., and Wang, X. 1991 *ApJ,* **372**, 321.

Biskamp, D. 1985, in *Magnetic Reconnection and Turbulence,* Cargèse Workshop, eds. M.A. Dubois, D. Grésillon and M.N. Bussac, Les Editions de Physique, Les Ulis, France, p. 19.

592

Biskamp, D. 1989, in *Plasma Phenomena in the Solar Atmosphere,* eds. M.A. Dubois, F. Bély Dubau and D. Grésillon, Cargèse Workshop, Les Editions de Physique, Les Ulis, France, p. 125.

Biskamp, D. and Welter, H. 1989, *Solar Phys.* **120**, 49.

Blandford, R.D. and Eichler, D. 1987, *Physics Reports,* **154**, 1.

Blandford, R.D. and Ostriker, J.P. 1978, *ApJ,* **221**, L29.

Block, L.P. 1978, *Astrophys. Space Sci.,* **55**, 59.

Bruggmann, G., Benz, A.O., Magun, A. and Stehling, W. 1990, *A&A,* **240**, 506.

Bulanov, S.V. and Cap, F. 1988, *Soviet Astron.,* **32**, 436.

Burm, H., and Kuperus, M. 1988 *A&A,* **192**, 165.

Canfield, R.C. and Metcalf, T.R. 1987, *ApJ,* **321**, 586.

Canfield, R.C., Fan, Y., Leka, K.D., McClymont, A.N., Wülser, J.P., Lites, B.W., and Zirin, H. 1991, in *Solar Polarimetry,* ed. L.J. November, Nat. Solar Obs., Sacramento Peak, Sunspot, New Mexico, p. 296.

Cargill, P. 1991, *ApJ,* **376**, 771.

Cargill, R.J., Goodrich, C.C. and Vlahos, L. 1988, *A&A,* **189**, 254.

Chiuderi, C., Giachetti, R. and Rosenberg, J. 1973, *Solar Phys.* **33**, 225.

Choudhuri, A.R. 1988 *Geophys. Astrophys. Fluid Dynamics,* **40**, 261.

Choudhuri, A.R., and Königl, A. 1986 *ApJ,* **310**, 96.

Collier Cameron, A. 1989 *MNRAS,* **238**, 657.

Coroniti, F.V. 1985, in *Unstable Current Systems and Plasma Instabilities in Astrophysics,* IAU Symp. No. 107, eds. M.R. Kundu and G.D. Holman, D. Reidel Publ Cy., Dordrecht, Holland, p. 453.

Davidson, R.C. 1972, *Methods in Nonlinear Plasma Theory,* Academic Press, New York.

DeCampli, W.M. and Baliunas, S.L. 1979, *ApJ,* **230**, 815.

Decker, R.B. 1988, *Space Sci. Rev.,* **48**, 195.

Dreicer, H. 1959 *Phys. Rev.,* **115**, 238.

Dreicer, H. 1960 *Phys. Rev.,* **117**, 329.

de Jager, C., Kuijpers, J., Correia, E., and Kaufmann, P. 1987, *Solar Phys.* **110**, 317.

de la Beaujardière, J.F. and Zweibel, E.G. 1989, *ApJ,* **336**, 1059.

Dere, K.P. 1990, in *Flares 22 Workshop, Dynamics of Solar Flares,* eds. B. Schmieder and E.R. Priest, Observatoire de Paris, p. 81.

Dere, K.P., Bartoe, J.-D.F., Brueckner, G.E., Ewing, J., and Lund, P. 1991 *J.Geophys.Res.,* **96**, 9399.

de Vries, M. and Kuijpers, J. 1989, in *23rd ESLAB Symp. on Two Topics in X-ray Astronomy,,* ed. N.E. White, p. 1069.

de Vries, M. and Kuijpers, J. 1992, *Magnetic Relaxation of a Cylinder: A VLBI Jet Model,* in preparation.

Dixon, A.M., Berger, M.A., Browning, P.K., and Priest, E.R. 1989, *A&A,* **225**, 156.

Dobrowolny, M. and Veltri, P. 1986, *A&A,* **167,**, 179.

Dubois, M.A. and Samain, A. 1985, in *Magnetic Reconnection and Turbulence,* Cargèse Workshop, eds. M.A. Dubois, D. Grésillon and M.N. Bussac, Les Editions de Physique, Les Ulis, France, p. 213.

Dulk, G.A. 1985, *ARA&A,* **23**, 169.

Fermi, E. 1954, *ApJ,* **119**, 1.

Field, G. 1986, in *Magnetospheric Phenomena in Astrophysics*, eds. R.I. Epstein and W.C. Feldman, Am. Inst. of Physics, New York, p. 324.

Forbes, T.G., and Isenberg, P.A. 1991, *ApJ*, **373**, 294.

Freidberg, J.P. 1987, *Ideal Magnetohydrodynamics*, Plenum Press, New York, p. 61, p. 477.

Galeev, A.A., Rosner, R. and Vaiana, G.S. 1979, *ApJ*, **229**, 318.

Galeev, A.A., and Sudan, R.N., eds., 1983, *Handbook of Plasma Physics*, Basic Plasma Physics I, II, North Holland Publ. Cy., Amsterdam.

Ghosh, P. and Lamb, F.K. 1978, *ApJ*, **223**, L83.

Ghosh, P. and Lamb, F.K. 1979a, *ApJ*, **232**, 259.

Ghosh, P. and Lamb, F.K. 1979b, *ApJ*, **234**, 296.

Ginzburg, V.L. 1970, *Soviet Physics Usp.*, **12**, 565.

Ginzburg, V.L. 1989, *Applications of Electrodynamics in Theoretical Physics and astrophysics*, Gordon and Breach Sci. Publ., Ch. 3.

Ginzburg, V.L. and Tsytovich, V.N. 1990, *Transition Radiation and Transition Scattering*, Adam Hilger, New York.

Ginzburg, V.L. and Zheleznyakov, V.V. 1975, *ARA&A*, **13**, 511.

Goldman, M.V. 1984, *Rev. Modern Physics*, **56**, 709.

Güdel, M. and Benz, A.O. 1988, *Astron. Astrophys. Suppl. Series* **75**, 243.

Haerendel, G. 1989, in *Plasma Astrophysics*, ESA SP-285, vol. I, eds. T.D. Guyenne and J.J. Hunt, p. 37.

Haisch, B.M., and Rodonò, M., eds., 1989, *Solar and Stellar Flares*, Proc. of the 104th Coll. of the IAU, *Solar Phys.*, **121**, Nos. 1/2.

Haisch, B., Strong, K.T., and Rodonò, M. 1991, *ARA&A*, **29**, in press.

Heiles, C. 1987, in *Physical Processes in Interstellar Clouds*, eds. G.E. Morfill and M. Scholer, NATO ASI Series C, Vol. 210, p. 429.

Hénoux, J.C. 1987, in *Solar Maximum Analysis*, eds. V.E. Stepanov and V.N. Obridko, VNU Science Press, Utrecht, The Netherlands, p. 105.

Heyvaerts, J., Lasry, J.M., Schatzman, M., and Witomsky, P. 1982, *A&A*, **111**, 104.

Heyvaerts, J., and Priest, E.R. 1984, *A&A*, **137**, 63.

Heyvaerts, J., and Priest, E.R. 1989, *A&A*, **216**, 230.

Hofmann, A. 1990, in *Flares 22 Workshop, Dynamics of Solar Flares*, eds. B. Schmieder and E.R. Priest, Observatoire de Paris, p. 31.

Holman, G.D. 1985, *ApJ*, **293**, 584.

Holman, G.D. and Pesses, M.E. 1983, *ApJ*, **267**, 837.

Holman, G.D., Eichler, D. and Kundu, M.R. 1980, in *Radio Physics of the Sun*, IAU Coll. No. 86, eds. M.R. Kundu and T.E. Gergely, Reidel, Dordrecht, p. 457.

Ichimaru, S. 1973 *Basic Principles of Plasma Physics*, W.A. Benjamin Inc., London.

Jacobsen, C. and Carlqvist, P. 1964, *Icarus*, **3**, 270.

Jokipii, J.R. 1982, *ApJ*, **255**, 716.

Jokipii, J.R. 1987, *ApJ*, **313**, 842.

Kaastra, J.S. 1982, *J. Plasma Phys.*, **29**, 287.

Kaastra, J.S. 1985, *Solar Flares, An Electrodynamic Model*, Ph.D. Thesis, Utrecht University, The Netherlands, Ch.4 and 5.

Kadomtsev, B.B. 1987, *Physics Reports*, **50**, 115.

Karimabadi, H., Akimoto, K., Omidi, N., and Menyuk, C.R. 1990, *Phys. Fluids B* **2**, 606.

594

Kattenberg, A. 1981, *Solar Radio Bursts and their Relation to Coronal Magnetic Structures*, PhD Thesis, Utrecht University.

Keller, C.U. 1990, in *Solar Photosphere: Structure, Convection and Magnetic Fields*, IAU Symp. No. 138, ed. J.O. Stenflo, p. 121.

Kikuchi, H., and 9 co-authors 1989, in *Laboratory and Space Plasmas*, ed. H. Kikuchi, Springer-Verlag, Berlin, p. 415.

Klimchuk, J.A. and Sturrock, P.A. 1989, *ApJ*, **345**, 1034.

Kovitya, P. and Cram, L. 1983, *Solar Phys.* **84**, 45.

Krimsky, G.E. 1977, *Doklady Akad. Nauk. SSR* **242**, 1306.

Krüger, A. 1979, *Introduction to Solar Radio Astronomy and Radio Physics*, Dordrecht, Reidel.

Kuijpers, J. 1974, *Solar Phys.* **36**, 157.

Kuijpers, J. 1975a, *A&A*, **40**, 405.

Kuijpers, J. 1975b, *Solar Phys.* **44**, 173.

Kuijpers, J. 1975c, *Collective Wave-Particle Interactions in Solar Type IV Radio Sources*, PhD Thesis, Utrecht University, Ch. 5.

Kuijpers, J. 1980, in *Radio Physics of the Sun*, IAU Coll. No.86, eds. M.R. Kundu and T.E. Gergely, Reidel, Dordrecht, p. 341.

Kuijpers, J. 1989, *Solar Phys.* **121,**, 163.

Kuijpers, J. 1990a, in *Plasma Phenomena in the Solar Atmosphere,* 1989 Cargése Workshop, eds. M.A. Dubois, F. Bély-Dubau and D. Grésillon, Les Editions de Physique, Les Ulis, France, p.17.

Kuijpers, J. 1990b, in *Plasma Phenomena in the Solar Atmosphere,* 1989 Cargése Workshop, eds. M.A. Dubois, F. Bély-Dubau and D. Grésillon, Les Editions de Physique, Les Ulis, France, p.227.

Kuijpers, J., and van der Hulst, J.M. 1985, *A&A*, **149**, 343.

Kuijpers, J., van der Post, P., and Slottje, C. 1981, *A&A*, **103**, 331.

Kuperus, M., and van Tend, W. 1981, *Solar Phys.*, **71**, 125.

Lamb, F.K., Aly, J., Cook, M., and Lamb, D.Q. 1983, *ApJ*, **274**, L71.

Lee, L.C. 1990, *Computer Physics Comm.*, **59**, 163.

Le Quéau, D., Pellat, R., and Roux, A. 1984, *J. Geophys. Res.*, **89**, 2831.

Lewin, W.H.G., van Paradijs, J. and van der Klis, M. 1988, *Space Sci. Rev*, **46**, 273.

Lieu, R. and Quenby, J.J. 1990, *ApJ*, **350**, 692.

Lin, R.P. 1990, in *Basic Plasma Processes on the Sun*, IAU Symp. 142, eds. E.R. Priest and V. Krishan, p. 467.

Lin, R.P., Schwartz, R.A., Kane, S.R., Pelling, R.M. and Hurley, K.C. 1984, *ApJ*, **283**, 421.

Louarn, P., Le Quéau, D. and Roux, A. 1987, *Solar Phys.* **111**, 201.

Lovelace, R.V.E., Wang, J.C.L., and Sulkanen, M.E. 1987, *ApJ*, **315**, 504.

Low, B.C. 1986, *ApJ*, **307**, 205.

Low, B.C. 1990, *ARA&A*, **28**, 491.

Low, B.C. and Wolfson, R. 1988, *ApJ*, **324**, , 574.

Machado, M.E., Moore, R.L., Hernandez, A.M., Rovira, M.G., Hagyard, M.J., and Smith, J.B. 1988, *ApJ*, **326**, 425.

Mätzler, C. and Wiehl, H.J. 1980, in *Radio Physics of the Sun*, IAU Coll. No.86, eds. M.R. Kundu and T.E. Gergely, Reidel, Dordrecht, p. 177.

Mann, G., Baumgaertel, K., Chernov, G. P. and Karlicky, M. 1989, *Solar Phys.* **120**, 383.

Marmolino, C. and Severino, G. 1990, in *Solar Photosphere: Structure, Convection and Magnetic Fields*, IAU Symp. No. 138, ed. J.O. Stenflo, p. 251.

Martens, P.C.H. 1988, *ApJ*, **330**, L131.

Martens, P.C.H., and Kuin, N.P.M. 1989, *Solar Phys.*, **22**, 263.

Martens, P.C.H., van den Oord, G.H.J. and Hoyng, P. 1985, *Solar Phys.* **96**, 253.

Martin, S.F. 1990, in *Dynamics of Quiescent Prominences*, IAU Coll. No. 117, eds. V. Ruždjak and E. Tandberg-Hanssen, Springer Verlag, Berlin, p. 1.

Mason, K.O., Rosen, S.R., and Hellier, C. 1988, *Adv. Space Res.*, **8**, No. 2, p. 293.

McClymont, A.N. and Fisher, G.H. 1989, in *Solar System Plasma Physics, Geophysical Monograph 54*, eds. J.H. Waite, J.L. Burch and R.L. Moore, Am. Geophys. Union, Washington D.C., p. 219.

McKean, M.E., Winglee, R.M. and Dulk, G.A. 1989, in *Solar and Stellar Flares*, eds. B.M. Haisch and M. Rodonò, IAU coll. No. 104, Poster Volume, Publ. Catania Astrophys. Obs., p. 333.

McLean, D.J. and Labrum, N.R. (eds.) 1985, *Solar Radiophysics*, Cambridge Univ. Press, UK.

Melrose, D.B. 1980, *Plasma Astrophysics I, II*, Gordon and Breach Publ., New York.

Melrose, D.B. 1982, *Austr. J. Phys.* **35**, 67.

Melrose, D.B. 1983, *Solar Phys.*, **89**, 149.

Melrose, D.B. 1986a, *J. Geophys. Res.* **91**, 7970.

Melrose, D.B. 1986b, *Instabilities in Space and Laboratory Plasmas*, Cambridge Univ. Press, UK.

Melrose, D.B. 1990, *Proc. Astron. Soc. Aust.*, **8**, 286.

Melrose, D.B., Hewitt, R.G. and Dulk, G.A. 1984, *J. Geophys.Res.* **89**, 897.

Melrose, D.B. and Kuijpers, J. 1987, *ApJ*, **323**, 338.

Mikić, Z., Barnes, D.C., Schnack, D.D. 1988, *ApJ*, **328**, 830.

Miller, J.A. 1991, *ApJ*, **376**, 342.

Miller, J.A. and Ramaty, R. 1987, *Solar Phys.* **113**, , 195.

Miller, J.A., Guessoum, N., and Ramaty, R. 1990, *ApJ*, **361**, 701.

Moghaddam-Taaheri, E. and Goertz, C. 1990, *ApJ*, **352**, 361.

Narain, U. and Ulmschneider, P. 1990, *Space Sci.Rev.*, **54**, 377.

Neubauer, F.M. 1980, *J. Geophys. Res.*, **85**, 1171.

Norman, C.A. and Heyvaerts, J. 1983, *A&A*, **124**, L1.

Otto, A. 1990, *Computer Physics Comm.*, **59**, 185.

Parker, E.N. 1957 *J.Geophys.Res.*, **62**, 509.

Parker, E.N. 1979, *Cosmical Magnetic Fields*, Clarendon Press, Oxford, UK.

Pesses, M.E. 1981, *J. Geophys. Res.*, **86**, 150.

Pesses, M.E., Decker, R.B., and Armstrong, T.P. 1982, *Space Science Rev.*, **32**, 185.

Petschek, H.E. 1964, *AAS-NASA Symp. on the Physics of Solar Flares*, NASA ST 50, p. 425.

Priest, E.R. 1982, *Solar Magnetohydrodynamics*, D.Reidel Publ. Cy., Dordrecht, Holland.

Raadu, M.A. 1989, *Physics Reports* **178**, 25.

Ramaty, R. and Murphy, R.J. 1987 *Space Science Rev.*, **45**, 213.

Rieger, E. 1989, *Solar Phys.* **121**, 323.

Roberts, D.A., Goldstein, M.L., Klein, L.W., and Matthaeus,W.H. 1987 *J. Geophys. Res.* **92**, 12023.

Sakai, J.-I. and de Jager, C. 1991, *Solar Phys.,* **134**, 329.

Sakurai, T. 1989, *Solar Phys.* **121**, 347.

Schadee, A. 1986, *Adv. Space Res.* **6**, No. 6, 41.

Schadee, A., de Jager, C. and Svestka, Z. 1983, *Solar Phys.* **89**, 287.

Scheurwater, R. and Kuijpers, J. 1988, *A&A*, **190**, 178.

Schindler, K., Hesse, M., and Birn, J. 1991, *ApJ,* **380**, 293.

Schlickeiser, R. 1989, *ApJ,* **336**, 264.

Schnack, D.D., Mikić, Z., and Barnes, D.C. 1990, *Computer Phys. Comm.,* **59**, 21.

Shakura, N.I. and Sunyaev, R.A. 1973, *A&A,* **24**, 337.

Shibata, K. 1990 in *Galactic and Intergalactic Magnetic Fields*, IAU Symp. No. 140, eds. R. Beck, P.P. Kronberg, and R. Wielebinski , p. 419.

Shibata, K. and Uchida, Y. 1986, *Solar Phys.*, **103**, 299.

Similon, P.L. and Sudan, R.N. 1990, *Annual Review Fluid Mech.,* **22**, 317.

Simon, T., Linsky, J.L., and Schiffer III, F.H. 1980, *ApJ,* **239**, 911.

Sitenko, A.G. 1982 *Physica Scripta* **T2/1**, 67.

Slottje, C. 1978, *Nat,* **275**, 520.

Slottje, C. 1980, *Atlas of Fine Structures of Dynamic Spectra of Solar Type IV-dm and Some Type II Radio Bursts*, Ph.D. Thesis, Utrecht University.

Smith, D.F. 1990, in *Basic Plasma Processes on the Sun*, IAU Symp. 142, eds. E.R. Priest and V. Krishan, p. 375.

Sofue, Y., Fujimoto, M., and Wielebinski, R. 1986, *ARA&A*, **24**, 459.

Solanki, S.K. 1990, in *Solar Photosphere: Structure, Convectionand Magnetic Fields*, IAU Symp. No. 138, ed. J.O. Stenflo, p. 103.

Solanki, S.K. and Roberts, B. 1990, *Solar Photosphere: Structure, Convection and Magnetic Fields*, IAU Symp. No. 138, ed. J.O. Stenflo,p. 259.

Spicer, D.S., Benz, A.O. and Huba, J.D. 1981, *A&A*, **105**, 221.

Spitzer, L. 1962, *Physics of Fully Ionized Gases,* Interscience Publ., New York.

Spruit, H.C. and Taam, R.E. 1990, *A&A,* **229**, 475.

Steinacker, J. and Schlickeiser, R. 1989, *A&A,* **224**, 259.

Stepanov, A.V. 1974, *Soviet Astron.* **17**, 781.

Stepanov, A.V. 1978, *Soviet Astron. Letters* **4**, 103.

Strauss, H.R. 1988, *ApJ,* **326**, 412.

Strauss, H.R. and Otani, N.F. 1988, *ApJ,* **326**, 418.

Sweet, P.A. 1958 in *Electromagnetic Phenomena in Cosmical Physics,* IAU Symp. No. 6, p. 123.

Tajima, T., Benz, A.O., Thaker, M. and Leboeuf, J.N. 1990, *ApJ,* **353**, 666.

Tapping, K.F., Kuijpers, J., Kaastra, J.S., van Nieuwkoop,J., Graham, D., and Slottje, C. 1983, *A&A*, **122**, 177.

Taylor, J.B. 1974, *Phys. Rev. Lett.,* **33**, 1139.

Taylor, J.B. 1986, *Rev. Modern Phys.,* **58**, 741.

Thejappa, G. 1987, *Solar Phys.* **111**, 45.

Treumann, R.A., Güdel, M. and Benz, A.O. 1990, *A&A,* **236**, 242.

Tsintsadze, N.L. 1989, in *Laboratory and Space Plasmas*, ed. H. Kikuchi, Springer-Verlag, Berlin, p. 164.

Tsytovich , V.N. 1970, *Nonlinear Effects in Plasma*, Plenum Press, New York.

Tsytovich, V.N. 1973, *ARA&A*, **11**, 363.

Uberoi, C. 1990, in *Basic Plasma Processes on the Sun*, IAU Symp. 142, eds. E.R. Priest and V. Krishan, p. 245.

Uchida, Y., and Hamatake, H. 1989, in *Accretion Disks and Magnetic Fields in Astrophysics*, ed. G. Belvedere, Kluwer Acad. Publ., Dordrecht, Holland, p. 233.

Uchida, Y. and Sakurai, T. 1983, in *Activity in Red-Dwarf Stars*, IAU Coll. No. 71, eds. P.B. Byrne and M. Rodonò, D.Reidel Publ. Cy., Dordrecht, Holland, p. 629.

van den Oord, G.H.J. 1988, *A&A*, **207**, 101.

van den Oord, G.H.J. 1990, *A&A*, **234**, 496.

van den Oord, G.H.J., Kuijpers, J., White, N.E., van der Hulst, J.M., and Culhane, J.L. 1989, *A&A*, **209**, 296.

van den Oord, G.H.J. and Mewe, R. 1989, *A&A*, **213**, 245.

van der Klis, M. 1989, *ARA&A*, **27**, 517

van Tend, W. and Kuperus, M. 1978, *Solar Phys.* **59**, 115.

Vilhu, O., Caillault, J.P., Neff, J., and Heise, J. 1988, in *Activity in Cool Star Envelopes,* eds. O. Havnes, B.R. Pettersen, J.H.M.M. Schmitt and J.E. Solheim, Kluwer, Dordrecht, Holland, p. 179.

Vlahos, L. 1989, *Solar Phys.* **121**, 431.

Vlahos, L. 1990, in *Flares 22 Workshop, Dynamics of Solar Flares,* eds. B. Schmieder and E.R. Priest, Observatoire de Paris, p. 91.

Volwerk, M., Kuijpers, J., and van Oss, R. 1991 in *X-ray Binaries and the Formation of Binary and Millisecond Radio Pulsars,* eds. E.P.J. van den Heuvel and S. Rappaport, NATO ARW, Santa Barbara, USA, in press.

Wagoner, R.V. 1969, *ApJ*, **158**, 739.

Webb, G.M., Axford, W.I., and Terasawa, T. 1983, *ApJ*, **270**, 537.

Wentzel, D.G. 1991 *ApJ*, **373**, 285.

White, N.E., Culhane, J.L., Parmar, A.N., Kellett, B.J., Kahn, S., van den Oord, G.H.J. and Kuijpers, J. 1986, *ApJ*, **301**, 262.

White, S.M., Melrose, D.B. and Dulk, G.A. 1986 *Astrophys.J.,* **308**, 424.

Winglee, R.M. and Dulk, G.A. 1986, *ApJ*, **307**, 808.

Winglee, R.M., Dulk, G.A., Bornmann, P.L., and Brown, J.C. 1991, *ApJ*, **375**, 382.

Winglee, R.M., Dulk, G.A. and Pritchett, P.L. 1988, *ApJ*, **328**, 809.

Woltjer, L. 1958, *Proc. Nat. Acad. Sci.,* **44**, 489.

Wu, C.S. and Lee, L.C. 1979, *ApJ*, **230**, 621.

Zaitsev, V.V., Mityakov, N.A. and Rapoport, V.O. 1972, *Solar Phys.* **24**, 444.

Zheleznyakov, V.V. 1967, *Soviet Phys. JETP* **24**, 381.

Zheleznyakov, V.V. and Zlotnik, E.Y. 1975a, *Solar Phys.* **43**, 431.

Zheleznyakov, V.V. and Zlotnik, E.Y. 1975b, *Solar Phys.* **43**, 447.

Zheleznyakov, V.V. and Zlotnik, E.Y. 1975c, *Solar Phys.* **43**, 461.

Zirin, H. 1983, *ApJ*, **274**, 900.

Zwingmann, W. 1987, *Solar Phys.,* **111**, 309.

INDEX

The Yellow Pages of Astrophysics

A to Z

© 1991
SJLB

INDEX